VOLUME NINETY TWO

CURRENT TOPICS IN DEVELOPMENTAL BIOLOGY

Notch Signaling

VOLUME NINETY TWO

CURRENT TOPICS IN DEVELOPMENTAL BIOLOGY

Notch Signaling

Edited by

RAPHAEL KOPAN
Department of Developmental Biology and the Department of Medicine, Washington University
St. Louis, MO
USA

ELSEVIER

AMSTERDAM • BOSTON • HEIDELBERG • LONDON
NEW YORK • OXFORD • PARIS • SAN DIEGO
SAN FRANCISCO • SINGAPORE • SYDNEY • TOKYO
Academic Press is an imprint of Elsevier

AP

Academic Press is an imprint of Elsevier
525 B Street, Suite 1900, San Diego, CA 92101-4495, USA
30 Corporate Drive, Suite 400, Burlington, MA 01803, USA
32, Jamestown Road, London NW1 7BY, UK
Linacre House, Jordan Hill, Oxford OX2 8DP, UK

First edition 2010

ISBN: 978-0-12-380914-8
ISSN: 0070-2153

For information on all Academic Press publications visit our website at elsevierdirect.com

Printed and bound by CPI Group (UK) Ltd, Croydon, CR0 4YY

Transferred to Digital Print 2011

Contents

CONTRIBUTORS

Hugo J. Bellen
Program in Developmental Biology; Department of Molecular and Human Genetics; Department of Neuroscience; and Howard Hughes Medical Institute, Baylor College of Medicine, Houston, TX, USA

Fred Bernard
Department of Physiology Development and Neuroscience, University of Cambridge, Cambridge, UK

Stephen C. Blacklow
Departments of Pathology and Biological Chemistry and Molecular Pharmacology, Harvard Medical School, Dana Farber Cancer Institute and Brigham and Women's Hospital, Boston, MA, USA

Sarah Bray
Department of Physiology Development and Neuroscience, University of Cambridge, Cambridge, UK

Massimiliano Cerletti
Joslin Diabetes Center, One Joslin Place, Boston, MA, USA

Wu-Lin Charng
Program in Developmental Biology, Baylor College of Medicine, Houston, TX, USA

José Luis de la Pompa
Laboratorio de Biología Celular y del Desarrollo, Dpto. de Biología del Desarrollo Cardiovascular, Centro Nacional de Investigaciones Cardiovasculares (CNIC), Madrid, Spain

Brendan D'Souza
Department of Biological Chemistry, David Geffen School of Medicine, University of California, Los Angeles, CA, USA

Thomas Gridley
The Jackson Laboratory, Bar Harbor, ME, USA

Pascal Heitzler
Institut de Génétique et de Biologie Moléculaire et Cellulaire, Illkirch, France

Tasuku Honjo
Department of Immunology and Genomic Medicine, Kyoto University, Sakyo-ku, Kyoto, Japan

Ellen Jorissen
Center for Human Genetics, KULeuven, and Department for Molecular and Developmental Genetics, VIB, Leuven, Belgium

Ryoichiro Kageyama
Institute for Virus Research, Kyoto University; and Japan Science and Technology Agency, CREST, Kyoto, Japan

Taeko Kobayashi
Institute for Virus Research, Kyoto University; and Japan Science and Technology Agency, CREST, Kyoto, Japan

Ute Koch
Ecole Polytechnique Fédérale de Lausanne (EPFL), Swiss Institute for Experimental Cancer Research (ISREC), Lausanne, Switzerland

Rhett A. Kovall
Department of Molecular Genetics, Biochemistry, and Microbiology, University of Cincinnati, Cincinnati, OH, USA

Jianing Liu
Joslin Diabetes Center, One Joslin Place, Boston, MA, USA

Donal MacGrogan
Laboratorio de Biologı́a Celular y del Desarrollo, Dpto. de Biologı́a del Desarrollo Cardiovascular, Centro Nacional de Investigaciones Cardiovasculares (CNIC), Madrid, Spain

Laurence Meloty-Kapella
Department of Biological Chemistry, David Geffen School of Medicine, University of California, Los Angeles, CA, USA

Marc. A. T. Muskavitch
Department of Biology, Boston College, Chestnut Hill, MA, USA

Yasutaka Niwa
Institute for Virus Research, Kyoto University; and Japan Science and Technology Agency, CREST, Kyoto, Japan

Meritxell Nus
Laboratorio de Biologı́a Celular y del Desarrollo, Dpto. de Biologı́a del Desarrollo Cardiovascular, Centro Nacional de Investigaciones Cardiovasculares (CNIC), Madrid, Spain

Toshiyuki Ohtsuka
Institute for Virus Research, Kyoto University; and Japan Science and Technology Agency, CREST, Kyoto, Japan

Tetsuya Okajima
Nagoya University Graduate School of Medicine, Center for Neural Disease and Cancer, Nagoya, Japan

Freddy Radtke
Ecole Polytechnique Fédérale de Lausanne (EPFL), Swiss Institute for Experimental Cancer Research (ISREC), Lausanne, Switzerland

Chihiro Sato
Washington University, St. Louis, MO, USA

Hiromi Shimojo
Institute for Virus Research, Kyoto University; and Japan Science and Technology Agency, CREST, Kyoto, Japan

Pamela Stanley
Department of Cell Biology, Albert Einstein College Medicine, New York, USA

Bart De Strooper
Center for Human Genetics, KULeuven, and Department for Molecular and Developmental Genetics, VIB, Leuven, Belgium

Kenji Tanigaki
Research Institute, Shiga Medical Center, Moriyama, Shiga, Japan

Spyros Artavanis-Tsakonas
Department of Cell Biology, Harvard Medical School, Boston, MA, USA; and Department of Biology and Genetics of Development, College de France, Paris, France

Amy Wagers
Joslin Diabetes Center, One Joslin Place, Boston, MA, USA

Gerry Weinmaster
Department of Biological Chemistry, David Geffen School of Medicine; and Molecular Biology Institute, University of California, Los Angeles, CA, USA and Jonsson Comprehensive Cancer Center, Los Angeles, California, USA

Shinya Yamamoto
Program in Developmental Biology, Baylor College of Medicine, Houston, TX, USA

Prologue

When I was asked if I was interested in editing a book on Notch, a mixture of excitement and trepidation gripped me. I have been faithful to this one molecule for a long time—longer than I'd like to admit sometimes—working with and alongside many talented scientists. Some of them have been in the field longer than I have, others just drifted in for some time and then moved on, but all contributed to what we now know about Notch. Due to our collective efforts, much of the facts explaining how the Notch pathway works in many phyla are known. Nevertheless, a complete picture of how Notch is integrated within the larger tapestry of inputs that cells process in real time is not yet understood.

It is exciting to assemble, with a few colleagues, the story of Notch before we cross into the next frontiers, and the last two decades have been compressed into the opening paragraph of the first review.

There was also trepidation, as the majority of contributors to the field cannot participate, and those invited to contribute are being pressured by the editor to fit a timetable and a certain vision of the story. In addition, given that more than 1500 reviews on Notch can be found on PubMed by the first quarter of 2010, any book on this topic is bound to leave out hundreds of citations, ignore hundreds of wonderful papers, and omit many important observations and viewpoints. And all those omissions are the responsibility of the editor.

So why study Notch? Notch signaling provides cells with the means to probe their immediate environment and, together with less than 20 other signaling "cassettes," is responsible for the entire diversity of metazoan cell types, organs, and life forms. Think of music: only seven notes are enough to generate a near infinite number of melodies, and we are not done yet. One can think of Notch as "Sol," of Wnt as "DO," of TGFβ proteins as "Re," etc., you get the point. All of these signaling cassettes contain proteins that, when "out of tune," are linked to human disease, and every one contains pharmaceutical "targets." Notch-related therapies are just now making their way through the clinical trial process, which is another reason why bringing together many of the known facts under one roof is a timely undertaking.

A word about the book itself: on some topics, the reader will get overlapping views from several authors. On others, a single author provides an authoritative review summarizing a field. Although other experts clearly exist, each of the authors contributing to this volume is an authority on the area on which he or she writes. As a result, this volume represents a wealth of information. Indeed, despite having worked on Notch for 20 years, I have learned many important things during the preparation of this book, and so will you. I have made some effort to cross-reference chapters, but each author created their own narrative. I will use the pathway figure we published in the past (Ilagan and Kopan, 2007) to create a navigation map for the chapters, and I hope you will enjoy the fact that each chapter includes a variation on this central theme. We all see the same pathway, but from different perspectives.

I also want to thank the American taxpayer for supporting basic science for three quarters of a century before the first connection of Notch to human disease was made. It is this investment that makes progress possible, then and now. We invest so that our children will collect the dividends, and we should not lose sight of this because of our short-term needs. The Notch field resembles a tree. The trunk is made of observations in *Drosophila melanogaster*, the humble fruit fly. The limbs are made of contributions from investigations into the inner workings of a transparent worm, African clawed toads, Asian zebrafish, and descendents of field mice. As a result of this investment, the fruits of this tree include important contributions toward better management or cure of Alzheimer's disease, of cancer, and of developmental disorders in humans. We understand stem cells better because of our investment in Notch. Many unexpected discoveries have been made, and many will continue to be made, as long as basic science continues to flourish with your support.

I have asked many of my colleagues to help. Some contributed articles to this book, others reviewed, and many more provided advice. Without detailing who provided what, I would like to acknowledge contributions from Drs. Alain Israel, Stacey Huppert, Jon Epstein, Marc Vooijs, Jim Priess, Iva Greenwald, Maxene Ilagan, Matt Hass, Kory Levine, Ralf Adams, Alfonso Martinez-Arias, Tim Schedl, Doug Barrick, François Schweisguth, Jim Skeath, Robert Haltiwanger and Olivier Pourquié, who talked me into editing this book after a long, international flight when my resistance was low.

Finally, I wish to thank the many colleagues, students, postdocs, technicians, and undergraduates who had spent time collaborating with me in the study of Notch biology. I learned a great deal from all of them, and I hope that you, the reader, will enjoy learning what we have discovered.

REFERENCES

Ilagan, M. X., and Kopan, R. (2007). SnapShot: Notch signaling pathway. *Cell* **128**, 1246.

CHAPTER ONE

Notch: The Past, the Present, and the Future

Spyros Artavanis-Tsakonas* *and* Marc A. T. Muskavitch†

Contents

Abstract

Proliferating investigations of the Notch pathway have given rise to the Notch "field," which has grown exponentially over the past 30 years. This field, founded by investigations of embryology and genetics in *Drosophila*, now encompasses many metazoa, including humans. The increasingly diverse scope of the field has engendered an expanding understanding that normal Notch pathway function is central to most developmental decision-making in animals, and that pathway dysfunction is implicated in many diseases, including cancer. We provide a personal view of the foundations and rapid evolution of the Notch field; and we discuss a variety of outstanding conundrums and questions regarding Notch biology, for which answers will be found and refined during the next 30 years.

* Department of Cell Biology, Harvard Medical School, Boston, MA, USA and Department of Biology and Genetics of Development, College de France, Paris, France
† Department of Biology, Boston College, Chestnut Hill, MA, USA

Current Topics in Developmental Biology, Volume 92
ISSN 0070-2153, DOI 10.1016/S0070-2153(10)92001-2

The first review to emphasize *Notch*, published by Ted Wright (1970), started with the inspiring sentence: *If one was asked to choose the single, most important genetic variation concerned with the expression of the genome during embryogenesis in Drosophila melanogaster, the answer would have to be the Notch locus*. The second *Notch* review arrived 18 years later, after cloning and sequencing of the locus (Artavanis-Tsakonas, 1988).

Today, a casual inspection of the *Notch* gene in Flybase (http://flybase.org/) reveals some instructive statistics. There are 345 classic alleles listed, 305 alleles on transgenic constructs, and 2250 references. These statistics, which include only fly-related work, are greatly expanded if we include research in all species. Notch biology can rightfully claim "field" status today, worthy of a book, such as this one.

The goal of this review is to give the reader some perspective on the history of the Notch field, which has become a very diverse field, rather than to review comprehensively particular aspects of Notch biology, and to try to define constants of the pathway, as well as some of the current pivotal questions. Many reviews apart from the current volume, covering more comprehensively particular aspects of Notch molecular biology, have appeared in recent years (Artavanis-Tsakonas *et al.*, 1999; Baron, 2003; Bray, 2006; Fortini and Bilder, 2009; Gordon *et al.*, 2008; Kopan and Ilagan, 2009; Louvi and Artavanis-Tsakonas, 2006). Moreover, the reader will find many details in subsequent chapters. Given our goals, we wish to apologize from the outset that we fail to refer to many original and important studies, and we warn the reader that our citations from the primary literature generally serve as examples, rather than providing comprehensive coverage of any topic.

1. The Beginnings: Embryology and Genetics

Almost a century has passed since T. H. Morgan's group described a mutant in *Drosophila* that they named *Notch* because it generated serrations on the wing margin (Fig. 1.1). The *Notch* gene has thus contributed to the progress of genetics as a discipline from the very start. It also provided a fundamental link between genetics and developmental biology through the work of Donald F. Poulson.

Don Poulson, in the early 1930s, was conducting work at Caltech for his doctoral thesis under the supervision of A. H. Sturtevant and Th. Dobzhansky, studying the embryonic phenotypes associated with chromosomal deletions. The relationship between genes and embryonic development was very much in doubt at the time. The general belief, by the dominant figures in biology, who were undoubtedly the embryologists, was that the parameters followed by geneticists, i.e., phenotypes associated with mutations in genes, reflected only terminal traits—for instance, a notched wing, rather than activities that

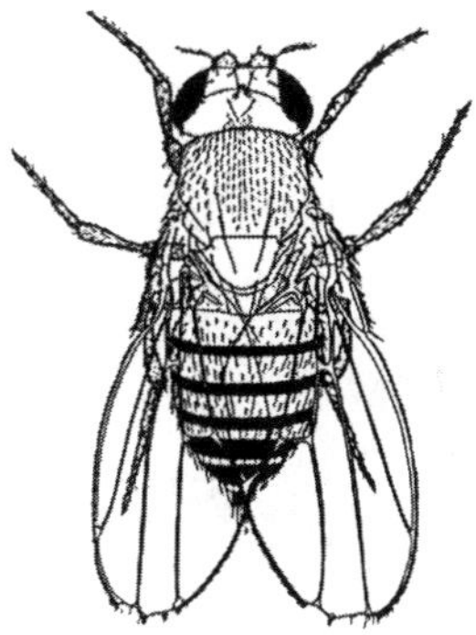

FIG. 44.
Notch-wing, a dominant sex-linked, recessive lethal of Drosophila melanogaster.

Figure 1.1 Morgan's Notch phenotype. A *Notch/+* female *Drosophila* showing the characteristic serrations of the wing, but not the bristle abnormalities, typical of *Notch* heterozygotes. Males lacking the X-linked Notch locus die as embryos. Drawing is from T. H. Morgan's book *The Theory of the Gene* originally published by Yale University Press in 1926.

governed morphogenesis of the wing. …*What is the role of the gene in development? Are there certain genes that are essential for the developmental process, or are genes only determinants of superficial characters? There are biologists to this day who believe that the latter is true, although there are few geneticists in their company*… wrote Don Poulson in the introduction to his doctoral thesis.

The evidence lacking was a clear correlation of embryonic phenotypes with specific mutations. Poulson, who in order to describe the embryonic phenotypes linked to deletions of chromosomes, the bearers of genes, single-handedly described the embryology of *Drosophila melanogaster* with an extraordinary accuracy, examined the lethal phenotypes of chromosomal deficiencies (Demerec, 1950; Poulsons, 1936). Among them was $\textit{Notch}^{8}$, a small X-linked deficiency encompassing the *Notch* locus, which had been genetically characterized in Morgan's laboratory (Dexter, 1914; Mohr, 1919) (Fig. 1.2). *Notch* behaved as a dominant, haploinsufficient, X-linked mutation; and heterozygous females had the characteristic "Notch" wings, while homozygous Notch females or hemizygous Notch males died as embryos.

Poulson's analysis of the Notch lethal phenotype revealed a specific and reproducible phenotype. In his words, *Although development in the early stages up to four hours is normal,* $\textit{Notch}^{8}$ *embryos fail to form the germ layers as evidenced by the absence of mesoderm and endoderm from the embryos at the time when the gut is normally completed. The organs and tissues, which are formed (although they may become highly abnormal) are all of ectodermal origin. There is no differentiation of ectoderm into hypoderm and the embryo is without skin. Those organs which*

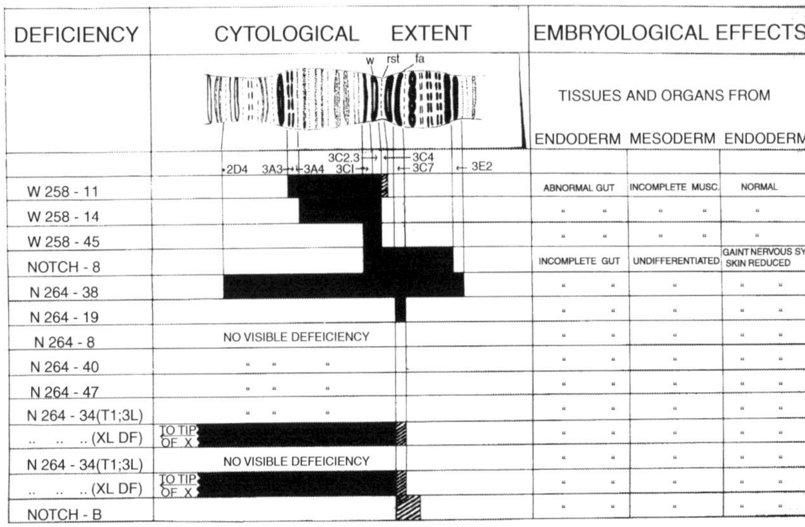

Figure 1.2 Mapping the Notch locus in *Drosophila*. Unpublished drawing by D. F. Poulson, depicting cytogenetic analyses of the X chromosome region that encompasses the *Notch* locus, including the famous *Notch*[8] deletion, which was used for much of the embryological characterization of embryos lacking Notch activity (see also Fig. 1.3).

undergo most differentiation and development are the nervous system and the hind-gut [Fig. 1.3 (Poulsons, 1936)]. One could thus argue that Poulson forged the link between the action of a genetic locus, *Notch*, and embryonic

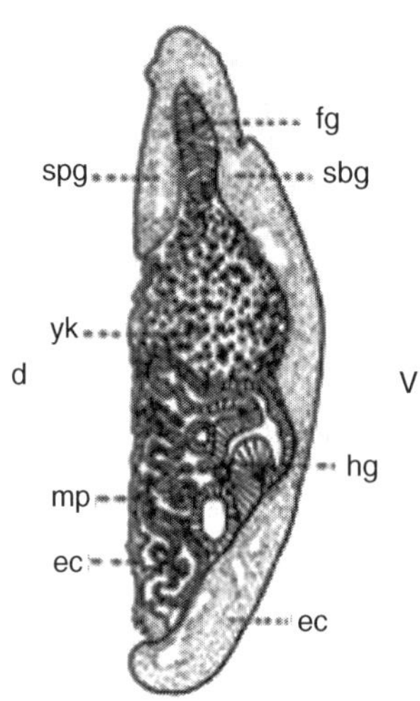

Figure 1.3 Notch embryology. A page from D. F. Poulsons Caltech doctoral thesis entitled *Chromosomal Deficiencies and Embryonic Development*. It shows the drawing of a *Notch*[8] mutant embryo in cross-section, which has no mesoderm but a hypertrophic ectoderm. Abbreviations: EC, ectoderm; YK, yolk.

morphogenesis. In our view, this seminal discovery has not been given the credit it deserves. In many ways, this link between genes and development was most famously acknowledged almost a half century later, when it was granted the weighty imprimatur of the 1995 "fly Nobel Prize." Later analyses refined and extended Poulson's observations, demonstrating conclusively that when Notch activity is lost, cells under normal circumstances would give rise to epidermal precursors, the dermoblasts, switch fate, and become neuroblasts. These excessive neuroblasts continue their normal differentiation to produce a morphologically deranged, inviable embryo that displays hypertrophy of the nervous system at the expense of epidermal structures. It was because of this neural hypertrophy that the phenotype was later baptized with the term "neurogenic" (Lehmann *et al.*, 1983).

Already in the 1920s, many *Notch* alleles had been identified by Morgan and his students, all of which yielded the typical notched-wing phenotype and bristle abnormalities in the females, as well as embryonic lethality, testifying to the pleiotropic nature of Notch activity. As *Notch* alleles started to accumulate, the spectrum of amorphic, hypomorphic, neomorphic, gain-of-function, recessive visible, and recessive lethal *Notch* alleles gave complex and often hard-to-interpret genetic complementation patterns, especially in the absence of clues as to the biochemical nature of the protein. Fanciful interpretations suggested that the *Notch* locus might be best represented by a spiral genetic map, while biochemical studies of *Notch* mutants were thought to suggest that Notch was the structural locus of several mitochondrial enzymes (Foster, 1973; Thorig *et al.*, 1981a, b).

The most sophisticated genetic analyses of the *Notch* locus were undertaken during the 1950s and 1960s, when only a single laboratory devoted serious time to *Notch*—that of Bill Welshons in Iowa. He, and his colleague and wife Jean Welshons, devoted a lifetime studying the cytology and genetics of the 3C7 region of the X chromosome, which harbors *Notch*. Tens of chromosomal rearrangements affecting *Notch* were characterized, balanced stocks that permitted intralocus genetic mapping were generated, and *Notch* mutations affecting eye, wing, and bristle traits were mapped with admirable resolution, producing a very accurate genetic map (Welshons, 1956, 1965) that was eventually completely corroborated in detail by molecular data (Grimwade *et al.*, 1985). Almost 60 years had passed since the discovery of the locus, and despite this exquisite genetic analysis the esoteric genetics defined during the 1970s did not provide further insights into the developmental or molecular biology of *Notch*. Nonetheless, the extraordinary cytogenetic analysis by the Welshons produced a rich roster of chromosomal rearrangements with breakpoints in the locus, among them $Notch^{76b8}$, an inversion with one breakpoint in the locus and the other in an adjacent region—which Welshons characterized as a single-band inversion affecting only band 3C7 within the salivary gland chromosomes! Such astonishing cytological resolution, which raised eyebrows in the uninitiated, proved crucial for the next era of Notch biology, which started with the cloning of the locus.

Chromosomal walking was made possible through the generation of genomic libraries and the pioneering work of Hogness laboratory at Stanford. When cloned chromosomal segments in the 3C10 region became available, the quest to clone *Notch* began. However, a stretch of repetitive sequences completely frustrated the march toward *Notch* at 3C7. It was the Welshons' *Notch*^{76b8} inversion that enabled jumping over the repetitive region and permitted the cloning of the breakpoint within the *Notch* locus (Artavanis-Tsakonas *et al.*, 1983; Kidd *et al.*, 1983; Fig. 1.4), initiating the molecular era of Notch and leading to its identification as the receptor of a fundamental cell interaction mechanism. Shortly thereafter, Notch proteins were cloned in the nematode *Caenorhabditis elegans* (Austin and Kimble, 1987; Greenwald *et al.*, 1987; Priess *et al.*, 1987) and in the vertebrate Xenopus (Coffman *et al.*, 1990).

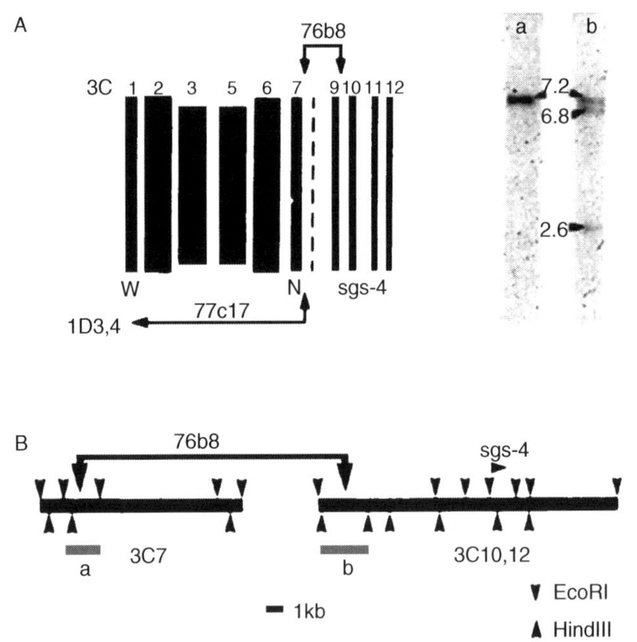

Figure 1.4 Jumping into Notch on the X chromosome. Using Welshon's *Notch*^{76b8} inversion, we were able to "jump" from 3C10 to 3C7, the location of Notch, as walking proved impossible give impassable repetitive sequences. The original drawing showing schematically the polytene chromosome region harboring *Notch*, the restriction map of the relevant genomic region, and the Southern blot showing the cloning of the inversion breakpoint are depicted.

2. The Developmental Logic of Notch: A Constant Frame of Reference

Before delving into the molecular biology of Notch, it is useful to consider its developmental functions in the fly. The most important aspects of this function can be considered even in the absence of molecular information, which in turn is best understood if one keeps the developmental biology of the locus in mind. Genetically, even the earliest analyses involving *Notch* demonstrated that the locus was on one hand pleiotropic and on the other haploinsufficient, not a common property for metazoan genes. Embryologically, the cellular phenotype of *Notch* loss-of-function mutants results from the redirection of cells into an alternative developmental program such that, in the neuroectoderm of the *Drosophila* embryo, cells destined to become dermoblasts under normal circumstances switch fates and give rise to neuroblasts. The embryological analysis of Notch action in the neuroectoderm implied an additional, crucial aspect of Notch biology, namely that the cells affected by Notch are physically adjacent (Doe and Goodman, 1985; Greenspan, 1990).

As the analysis of *Notch* phenotypes expanded into other tissues, into *C. elegans*, and later into other species, the broadly inclusive pleiotropy of Notch action was clearly evident, as was the consistent characteristic that Notch affects developmental choices of neighboring cells. In fact, today it is fair to maintain that if two adjacent cells follow different developmental paths, Notch is very likely to be involved. The existing exceptions seem to confirm, rather than challenge, this rule. Consequently, a generalization with considerable predictive power is that Notch functions in development to link the fate of a given cell to that of its next-door neighbor. This "canonical" developmental logic of Notch has fundamental consequences for morphogenesis, as it provides the means for specific cell lineages to segregate from within groups of developmentally equivalent precursors. Notch is known to contribute to the definition of boundaries between fields of cells with distinct developmental properties, such as the border that separates the ventral from the dorsal compartment of a fly wing or the somite boundaries in vertebrates (see Chapter 10). Thus, the early notion that Notch functions in development to fine-tune morphogenetic events was and still is well supported by the accumulated evidence. It is noteworthy, however, that there are instances in which Notch activity in one cell can at least indirectly affect distant cell populations, as we will briefly elaborate later (and as discussed in other chapters of this book).

Notch is not only pleiotropic in its action as judged by the near-universal array of tissues it affects throughout ontogeny but also pleiotropic in terms of the fundamental developmental processes it affects. Depending on the developmental context, differentiation, proliferation, and apoptotic cellular processes can each be profoundly affected by Notch activity. Developmental context is a crucial parameter when considering Notch, and by now it has been observed in many different species that Notch action in one tissue induces cellular proliferation, while in another it induces apoptosis. One thing that seems safe to assume is that the modulation of Notch activity, at least in cells that are not terminally differentiated, will trigger cell fate changes. The nature of the resulting cell fates is impossible to predict, *a priori*, as the fates affected will depend both on developmental context (spatial and temporal) and, as we elaborate below, on the dosage of *Notch* activity, a crucial and sometimes overlooked characteristic of the pathway.

Generally, Notch activity is associated with progressive lineage restriction of early developmental precursors and is often used reiteratively to drive decisions of precursors between two alternative fates. Two well-defined examples of repeated employment of Notch signals in successive developmental decisions are vertebrate hematopoietic lineages (see Chapter 12) and the *Drosophila* peripheral nervous system (Gering and Patient, 2010). The association of Notch with early lineages is emphasized by the growing appreciation that Notch function and stem cell biology are closely linked, as first revealed in the *C. elegans* gonad (Austin and Kimble, 1987); given the developmental logic it serves, this is not surprising. Maintenance and differentiation of stem cells depend intimately on cellular interactions between stem cells themselves, and between stem cells and their adjacent environment, or niche. Thus, as the roster of tissue-specific stem cells influenced by Notch activity is growing (see Chapter 12), the characterization of Notch as a "stem cell pathway," as some reviewers have called it, may be overstated but justified (Brack *et al.*, 2008; Casali and Batlle, 2009; Dreesen and Brivanlou, 2007; Farnie and Clarke, 2006). The involvement of Notch in the adult, i.e. in organs classically considered to be terminally differentiated, is certainly assured through its roles in stem cell biology, and through organ homeostasis and facultative injury response. However, despite reports regarding the activity of Notch in differentiated tissues, the questions of how broadly is Notch involved in physiological aspects of cellular and organ homeostasis remain to be investigated. Whether Notch activity contributes to, or is essential for, the maintenance of any differentiated state, the degree to which Notch activity is needed for organ integrity after organs are fully developed, is an actively investigated area in Notch biology.

In considering the developmental action of Notch, it is finally worth mentioning that the classic screens for embryonic phenotypes by Nusslein and Weischaus (Nusslein-Volhard *et al.*, 1984; Nusslein-Volhard and Wieschaus, 1980) identified a group of genes that displayed phenotypes similar to those seen in *Notch* embryos. Similar phenotypes could, of course, suggest that these genes affect a common developmental pathway and could, but would not necessarily, be reflective of the steps in that pathway. Remarkably, and testifying to the power of the genetic approach and the extraordinary value of model systems, even in the current era of "translational biology," all of the six zygotically acting loci identified in those embryonic screens—*Delta* (*Dl*), *Enhancer of split* [*E(spl)*], *mastermind* (*mam*), *big brain* (*bib*), *neuralized* (*neu*) and, of course, *Notch*—have been directly implicated in the Notch signaling pathway.

3. The Notch Receptor: Key Features

Cloning of an X chromosome segment that genetically behaved as a duplication of the *Notch* gene in *Drosophila* confirmed the isolation of the *Notch* locus, and the subsequent sequencing of corresponding cDNAs revealed the existence of a protein ca. 2700 amino acids in length. The extracellular domains of Notch proteins contain tandem arrays of epidermal growth factor (EGF)-like repeats (ELRs), ranging from 36 in *Drosophila* to as few as 11 in *C. elegans*, an experimental system that substantially contributed to the dissection of *Notch* function with two Notch-like receptors (Greenwald, 1985; Greenwald *et al.*, 1983; Kidd *et al.*, 1986; Wharton *et al.*, 1985). Associating the *Notch* locus with a putative transmembrane, receptor-like protein was crucial, as it implied the possible involvement of Notch in cell–cell interactions, a property compatible with the embryology of mutants and one that demanded as well the existence of ligands, downstream effectors and other components, essentially unveiling a then-novel cell interaction mechanism (Wharton *et al.*, 1985). However, it was the cloning of vertebrate Notch proteins (Coffman *et al.*, 1990) that established the pathway logic biochemically, starting with the suggestion that truncated receptors were constitutively active (Coffman *et al.*, 1993; Ellisen *et al.*, 1991), identification of Notch/RBPjk complexes in nuclear extracts (Jarriault *et al.*, 1995), and the characterization of Notch cleavage sites (see below). Twenty-five years after cloning the *Notch* locus, we know that Notch is the central element in one of the few fundamental, evolutionarily conserved, short-range cell interaction signaling pathways that govern metazoan fate cell determination. In *Drosophila* where the initial dissection of the pathway took place, two ligands—Delta and Serrate—along with the two nuclear downstream effectors—Suppressor of Hairless (RBPJK in vertebrates, Lag-1 in *C. elegans*) and mastermind—and Helix-Loop-Helix Notch target genes encoded by the *E(spl)* locus, complete the basic elements of the Notch signaling pathway (Fig. 1.5).

Vertebrates have four Notch receptors, Notch 1, 2, 3, and 4; fish have more; *C. elegans* has two. *Drosophila* has only one, making dissection of some aspects of Notch pathway easier. The paradigmatic Notch receptor is

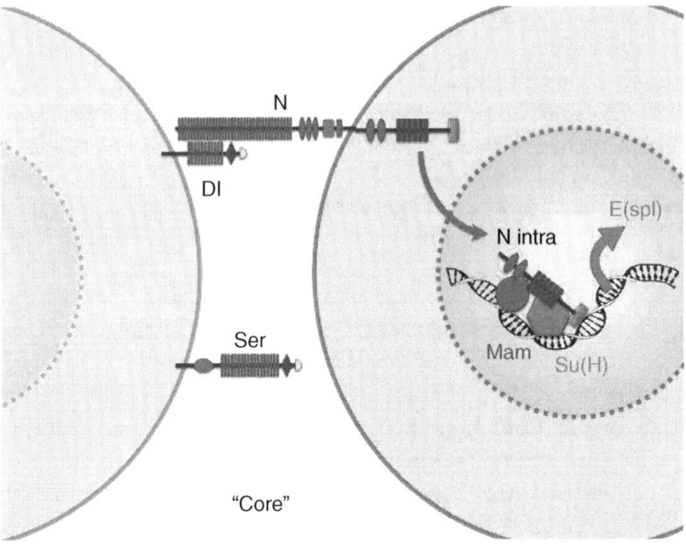

Figure 1.5 The core. A very simplified cartoon, showing the core elements of the Notch signaling pathway using *Drosophila* terminology. For a more complete description, see Kopan and Ilagan (2009). The membrane-bound ligands, Delta (Dl) or Serrate (Ser) (Jagged in mammals) link the fate of the cell expressing them to that of the neighboring cell expressing the Notch receptor. Ligand–receptor interactions are followed by a cascade of proteolytic events that release the Notch intracellular domain (NICN) from the membrane. NICN translocates into the nucleus where it forms a transcriptional complex with the coactivator Mastermind (Mam) and the DNA-binding protein. Suppressor of Hairless [Su(H)] (RBPjk, also generically called CSL) to yield the so-called CSL/N/Mam complex, which with the cooperation of other nuclear factors drives Notch-dependent transcription, for example, the classical Notch HLH targets encoded by the *Enhancer of split* locus [*E(spl)*]. (See Color Insert.)

composed of distinct domains that are essentially conserved across all species [see Chapter 2 and (Kopan and Ilagan, 2009)]. Vertebrate Notch paralogs do display differences in primary sequence, which distinguish them from each other, and they have overlapping, yet distinct, expression profiles and developmental functions. Differences in Notch primary structure translate into differential target specificity among mammalian paralogs (Ong *et al.*, 2006), especially when it comes to Notch 3, suggested at some point to be an antagonist of Notch 1 (Beatus *et al.*, 1999). It is possible, and perhaps even likely, that the functions of some vertebrate receptors may be interchangeable biochemically as they are in *C. elegans* (Fitzgerald *et al.*, 1993), but this has yet to be investigated in depth.

A crucial aspect of *Notch* that has been obvious since Morgan's first description of the gene is that the Notch receptor is haploinsufficient, as alluded to earlier. This is not a common property of genes in diploid organisms, and very few genes throughout the genome are haploinsufficient. Still, it is remarkable in addition that the *Delta* locus, which encodes one of the two Notch ligands in *Drosophila*, is also haploinsufficient, as is the ligand Jagged1 in humans (Li *et al.*, 1997). More strikingly perhaps, *Notch* is one of the two genes in the *Drosophila* genome that is also triplo-mutant. Females that carry a duplication of the locus and thus harbor three, as opposed to the normal two, copies of the gene display the so-called Confluens mutant wing-vein phenotype. Thus, the animal seems to be able to "count" *Notch* gene dosage, a fact that is presumably associated with the intensity of Notch signaling, such that too much or too little signaling will result in altered pathway function and altered development. An explanation for this counting mechanism may relay on cis-inhibition (see Chapter 3), and on the fact that the Notch signaling mechanism, which we outline below, lacks an enzymatic amplification step, relying on stoichiometric interactions among pathway components. The stoichiometric character of the Notch signaling pathway implies that many mechanisms that influence the number of ligand-engaged receptors at the cell surface—whether transcription, translation, trafficking, or turnover—can serve as Notch signaling control mechanisms. The several chapters in this book describe these details at length.

4. Cloning the Ligands: Engaging Notch

The premise that the Notch pathway mediates signaling and decision-making between by adjacent cells, elegantly described in the AC/VU decision (Seydoux and Greenwald, 1989), the vulva (Sternberg, 1988), and the fly eye (Cagan and Ready, 1989), was strongly supported when the sequences obtained for ligands that activate Notch receptors revealed that they are themselves cell surface proteins, which unexpectedly also include numerous EGF-like repeats (ELRs) within their extracellular domains. The first Notch ligands sequenced, shortly after the Notch receptor, were Delta (Kopczynski *et al.*, 1988; Vassin *et al.*, 1987) and Serrate (Fleming *et al.*, 1990) from *Drosophila* and Lag-2 from *C. elegans* (Mello *et al.*, 1994; Tax *et al.*, 1994), all of which are type I transmembrane proteins dynamically expressed by limited numbers of cells during development. Homologous organization of ligands in mammals was then discovered with the sequencing of Jagged1 (Lindsell *et al.*, 1995) and Jagged2 (Shawber *et al.*, 1996). At the time, all possessed conserved extracellular domains that include an N-terminal "DSL domain," named based on its common occurrence in Delta, Serrate, and Lag-2, followed by a tandem array of ELRs.

Adhesion assays demonstrated that a ligand-expressing cell can physically interact with a receptor-expressing cell, and deletion analyses defined specific sequences within the extracellular domains of these molecules that mediated this interaction (Fehon *et al.*, 1990). This led to a simple scheme for the mechanism by which ligands bind to and activate Notch receptors in apposing cells, which falls short of elucidating the complete mechanism of signal transduction; and while they have been misleading on occasion (Fortini and Artavanis-Tsakonas, 1994), these cell culture experiments have been very helpful in exploring molecular relationships between the main components of the pathway. Over the years, additional layers of complexity are being uncovered, many of which will be discussed in detail in other chapters.

When a ligand expressed on the surface of one cell engages Notch through two ELRs (ELR11-12) within the receptor extracellular domain, a conserved domain shown to be necessary and sufficient for physical interaction between Notch and its ligands (Rebay *et al.*, 1991), one could argue this to be the quintessential event underlying the developmental logic of the pathway: linking two adjacent cells together and mediating communication between them. But how was the signal transduced?

The "nuclear" localization of Notch took a long time to establish, in spite of early indications that proteins synthesized from mutant transgenes encoding truncated receptors encompassing only the Notch intracellular domain (NICD) clearly accumulated in the nucleus, consistent with the existence of nuclear localization signals within the NICD (Coffman *et al.*, 1993; Kopan *et al.*, 1994; Stifani *et al.*, 1992; Struhl *et al.*, 1993). The difficulty in detecting nuclear Notch antigens in wild-type cells called into question the biological relevance of these exciting initial findings. Definitive acceptance of this key feature of Notch signaling required careful quantitative studies in conjunction with biological readouts examined mainly in mammalian cultured cells to persuade most in the field that the undetectability by conventional means of the miniscule but potent quantities of NICD in the nuclei of wild-type cells did not mean that the nuclear translocation of the NICD encoded by transgenes was an artifact of its overexpression (Schroeter *et al.*, 1998). A genetic requirement for liberation of NICD by intramembrane proteolysis, was established by site-directed mutagenesis producing a single amino acid substitution within the transmembrane domain of mouse Notch 1, which resulted in a null phenotype despite normal protein expression (Huppert *et al.*, 2000). Genetic screens in *C. elegans* identified the components of the Presenilin complex that release NICD (Francis *et al.*, 2002; Goutte *et al.*, 2000, 2002; Levitan and Greenwald, 1995) and a flurry of papers confirmed the role of Presenilin as an enzyme, and its activity as the Notch intramembrane protease (De Strooper *et al.*, 1999; Struhl and Greenwald, 1999; Wolfe *et al.*, 1999).

The Notch receptor usually presents itself on the cell surface in a cleaved, heterodimeric form (Blaumueller *et al.*, 1997). There is evidence that Notch is cleaved in the trans-Golgi apparatus, possibly by furin (Logeat *et al.*, 1998) at "Site 1"(or "S1") cleavage (Kopan *et al.*, 1996) and is presented on the surface as a heterodimer that seems to be held together by noncovalent interactions (Rand *et al.*, 1997). The discovery of receptor ectodomain shedding at Site 2 (Brou *et al.*, 2000; Mumm *et al.*, 2000), and its trans-endocytosis in flies (Parks *et al.*, 2000), have suggested a set of possible mechanisms by which ligand-receptor interactions would promote receptor dissociation following Site 2 proteolysis by metalloproteases known to be required for receptor activation. Receptor cleavage at Site 2 *in vitro* by tumor necrosis factor-α converting enzyme (TACE) (Brou *et al.*, 2000) and *in vivo* by the metalloprotease Kuzbanian (Kuz)/ADAM10 (Pan and Rubin, 1997; van Tetering *et al.*, 2009; Wen *et al.*, 1997) is permissive for subsequent proteolysis of the resulting Notch extracellular truncation (NEXT) derivative of the mature Notch receptor (Mumm *et al.*, 2000). It should be noted, however, that the relationship between Notch signaling and metalloproteases may be more complex than this, especially in light of the finding that the Delta extracellular domain can be cleaved by Kuzbanian in a manner that down-regulates ligand levels (Mishra-Gorur *et al.*, 2002; Qi *et al.*, 1999), as discussed in detail in Chapter 3. Despite substantial effort over the past decade, the exact mechanism by which ligand endocytosis promotes receptor dissociation and Site 2 proteolysis remains unclear. The recent description of monoclonal antibodies that appear capable of detecting the Notch receptor in either an inactive ("off" or "closed") state or an active ("on" or "open") state (Aste-Amezaga *et al.*, 2010; Li *et al.*, 2008; Wu *et al.*, 2010) offers the prospect of using these antibodies to probe "transition state" receptors generated by site-directed mutagenesis or by interaction with ligand-expressing cells in which endocytosis or ubiquitination is impeded. Further work along these and other lines will be required to determine whether mechanical force, allostery or unmasking is centrally permissive for shifting receptors from the "off" state to the "on" state and for promoting activating proteolysis of Notch receptors.

Even though essentially all Notch pathway models depict the Site 3 cleavage as occurring on the cell surface, the subcellular compartment in which this critical cleavage takes place and the factors that dictate where such cleavage can or must occur are not well understood and are discussed in Chapter 5. In general, the generation and trafficking into the nucleus of the active NICD seems to involve subcellular steps that can regulate signaling at various levels, explaining, refining, and dissecting the long-ago observation that *Drosophila* embryos with reductions in endocytosis due to reduced dynamin function exhibit neurogenic phenotypes indistinguishable from those of *Notch* loss-of-function mutants (Poodry, 1990). Which of these steps are ligand-dependent is still being determined, but internalized Delta and Notch molecules can traffic to specialized endosomes, which can cleave and thus activate Notch away from the surface (Coumailleau *et al.*, 2009). The biology of NICD in the nucleus is discussed in three chapters within this book.

As a rule in developmental genetics, noncanonical behavior is interesting and often instructive, making claims for the existence of Notch activity that is independent of SuH an intriguing problem. Claims for the existence of such events have been reported in the literature for a long time, but they still remain enigmatic. Two independent instances are most noteworthy. The first claim is that the NICD reaches the nucleus and interacts with transcription factors other than SuH (e.g., NF-κB, p53), thus directly influencing gene expression. The second is that Notch, in either its cleaved (NICD) or full-length form, interacts with proteins that are participating in various cellular functions (e.g., ABL kinase (Giniger, 1998), the ubiquitin ligase Deltex (Hori *et al.*, 2004), armadillo (Hayward *et al.*, 2008; Sanders *et al.*, 2009) etc.), thus influencing their specific activities and producing phenotypes that do not depend on SuH. In this regard, it is worth pointing out that over the years, many proteins have been suggested to interact directly with Notch—apart from SuH, Mastermind, and Deltex, for which interactions have been corroborated by biochemical, genetic, and, indeed, structural analyses. We will refrain in this context from comprehensively listing all of the various proteins that have been reported to interact physically with Notch, but see Chapter 14 and Arias *et al.*, 2002.

5. Ligand–Receptor Interactions: Not a One-Way Street

Various lines of evidence suggest that activating and inhibiting the Notch pathway may be of comparable importance for Notch signaling in many contexts, and some aspects of ligand–receptor engagement are worth considering in this regard. The notion of trans-interactions between Notch and Delta is well established. At a cellular level, however, often, if not usually, a pair of adjacent cells engaged in signal-sending and signal-receiving may each express both the ligand and the receptor. Yet, eventually one must become the signal-receiving cell and one the signal-sending cell, a decision of crucial importance from a developmental point of view. A key mechanism underlying the genesis of this asymmetry, inferred from genetic arguments, is the apparently critical ratio between functional ligand and receptor levels on each cell (Gibert and Simpson, 2003; Heitzler and Simpson, 1991; Wilkinson *et al.*, 1994). The cell expressing more ligand becomes the signal-sending cell, while the one expressing more receptor becomes the signal-receiving cell. The resulting asymmetry is thought to be mediated and stabilized by feedback loops that amplify small, perhaps even stochastic, differentials between the levels of ligand and receptor in each cell. The nature of such feedback loops remains obscure, notwithstanding the attractive nature of the concept and some evidence in favor of transcriptional feedback. Surprisingly, after all these years of Notch studies, the elements controlling transcription of the ligands and the receptors have yet to be examined in any great detail. As the importance of cis-inhibitory interactions between the receptor and the ligand expressed in the same cell is beginning to emerge (Chapter 3 and Sprinzak *et al.*, 2010) and may well be involved in creating this developmentally crucial asymmetry, the relative roles of these mechanisms remain unclear and their context-dependence remains to be examined.

Genetic data, especially the negative complementation associated with the *Abruptex* mutations in *Drosophila*, a group of gain-of-function, ligand-dependent mutations affecting the extracellular domain of the receptor, argue in favor of a multimeric receptor quaternary structure (Foster, 1975). Given the difficulties associated with protein purification and structural analysis of the receptor, single-molecule electron microscopy (EM) does offer novel possibilities for analysis. EM studies determined unambiguously the quaternary structure of the entire Notch extracellular domain (NECD). Single-particle EM reconstructions of both human Notch 1 and *Drosophila* NECDs yield a dimer, which intriguingly adopts three defined yet distinct conformations. The significance of this finding remains to be functionally explored, especially in view of experiments from cultured cells using transgene reporters suggesting that dimers and monomers exist on the cell surface, and neither is active without a ligand (Vooijs *et al.*, 2004). Further studies will be required to establish what these different NECD conformations represent and whether mutations such as the *Abruptex* mutations or mutations underlying the catastrophic neurodegenerative disease CADASIL in humans affect the conformations or quaternary structures of the NECD (Kelly, 2010).

6. Targets, Signal Integration, and the Genetic Circuitry of Notch: On Being Old

Molecular biology may tell us how things work, but only evolutionary considerations can touch the why. There is an important and most interesting conundrum that has been offered to us by genomic analyses of canonical signaling pathways in many organisms. The fact that a very similar, almost identical genetic framework sustains the development and homeostasis of all metazoans raises the question of how such a "rigid" genetic scaffold can support development of such a diverse spectrum of animal complexity. For evolution to occur, there must clearly be flexibility within this genetic framework.

The small roster of basic, core signaling pathways controlling metazoan morphogenesis and cell interactions, in general, includes Notch, RTK, Wnt, Hh, TGFbeta, JAK–STAT, and a handful of others; and these ancient mechanisms have retained a remarkable degree of conservation, reinforcing the notion of a "rigid signaling scaffold" and indeed the problem of how the same roster of pathways can have such diverse developmental outcomes (Gerhart, 1999; Kirschner and Gerhart, 2005). The answer to this question may define the quintessence of developmental genetics and must rely on how signals are integrated so as to produce differential outcomes. Moreover, even though the core of a pathway may be defined, the control of its activity may be diverse and dependent on the cellular context. It is intuitively reasonable to suggest that for a pathway marching up the evolutionary tree, chance and necessity will generate diverse means of controlling activity. Shutting down or activating Notch signals can be achieved via many different mechanisms, including the transcription of core pathway elements, their trafficking through the cell, their posttranslational modification, their response to microRNA activity, etc. The older a process is, the more diverse, we would argue, is the array of controls' evolution has devised to govern it. Thus, we propose that the complexity of signaling pathway control, such as that we increasingly understand for Notch, is proportional to the "age" of the pathway.

In this light, it is perhaps not surprising to find that *in vivo*, the complexity of the genetic circuitry capable of modulating Notch activity is very extensive, a fact that is often ignored because it substantially complicates interpretations of individual observations and also, in a sense, reduces the relative importance of discovering and characterizing specific Notch-modulating pathways. Thus, if for instance, we inhibit Notch signaling by expressing a dominant-negative form of Mastermind (Weng *et al.*, 2003), a potent Notch inhibitor in invertebrates and vertebrates and ask the question of how many genes are capable of modulating the phenotype associated with the resulting loss of Notch function, the answer in *Drosophila* lies in the hundreds (Kankel *et al.*, 2007). Not only does one identify hundreds of genes for which modulation clearly affects Notch signaling *in vivo*, but the gamut of functional categories we identify among these genes is also very broad. For example, gene ontogeny analysis has associated Notch modifiers with metabolism or RNA processing, processes that have not been associated with Notch signaling previously in the fly (Kankel *et al.*, 2007). Consistent with these functional studies, transcriptional profiling experiments in mammals and vertebrates alike point to a complex roster of genes that are affected by Notch modulation (Hurlbut, 2008; Krejci *et al.*, 2009; Palomero *et al.*, 2006; Weng *et al.*, 2006).

The rather restricted repertoire of cell signaling mechanisms available to guide metazoan cell signaling may well be the basis of the repeated usage of these mechanisms in development, as well as their pleiotropic actions. Notch pleiotropy in development does not manifest itself in the same way, in all contexts. Activating the pathway in one context may stimulate proliferation, while in another, it may induce apoptosis. The signal can be both instructive and permissive; Notch activity may either allow a cell to proceed or prevent a cell from proceeding, to the next developmental stage or restrict the developmental potential of a cell. All depends on the developmental context, spatially and temporally. Noteworthy in this respect is the observation that a given genetic interaction, say between Notch and Wnt, may be conserved across species, but it may function synergistically in flies and antagonistically in mammalian cells (Hayward *et al.*, 2008; Kankel *et al.*, 2007). The underlying mechanism(s) may not be known, but perhaps "chance" in evolution has "tinkered" with an existing interaction to suit the "necessity" of a new developmental process, rather than inventing a mechanism *ab initio*.

The pleiotropy of Notch signaling is ultimately manifested in the target genes that Notch activates or suppresses. As alluded to earlier, the architecture of Notch-sensitive promoters is slowly being dissected, and while the SuH/RBPjk binding site seems to be a constant component, much about the number and arrangement of these sites within promoter regions and cooperating enhancers remains to be elucidated. For many years, the classic Notch targets were HLH transcription factors in both vertebrates and invertebrates [HESR, Hey, and E(spl)-related family], but the inventory of Notch targets has begun, not unexpectedly, to expand (Hurlbut, 2008; Hurlbut *et al.*, 2009; Krejci *et al.*, 2009; Mummery-Widmer *et al.*, 2009). Target selection for a pleiotropic pathway such as Notch rests upon how it integrates its action with other signaling mechanisms; and if we did understand how signals synergize to influence downstream developmental events, we would have made long strides into the quintessence of Developmental Genetics. That Notch integrates its action with essentially all major signaling pathways is clear. What, however, is quite unknown and most interesting is whether this integration follows some logic that reflects evolutionary history.

As we have argued before, it is conceivable and testable that the Notch and, say, RTK pathways engage in "crosstalk" using specific nodes (Hurlbut *et al.*, 2009; 2007; Yoo *et al.*, 2004). This hypothesis holds that the activities of specific targets can be controlled by both pathways simultaneously, so that a target gene can, in turn, integrate both signals at the same time. Several such nodes have been identified, but it remains to be seen how general these nodes are. If there is an underlying "code" of crosstalk, then such a node should be valid in diverse developmental contexts. Indeed, studies thus far have identified such nodes in *Drosophila* in more than one developmental context, and it is tempting to suggest that a network of signaling pathway nodes defining interactions among the signaling pathways is waiting to be discovered (Krejci *et al.*, 2009). The technology to address this question directly now exists, for instance, with the help of microarrays; and genetic analysis, at least in model organisms, can in a straightforward, albeit nontrivial, fashion probe the functional significance of a given signaling node.

The possible existence of rules that underlie Notch crosstalk is of fundamental interest and brings us back to the question of how a rigid genetic framework sustains evolutionary novelty. If there is a logic that governs crosstalk of Notch with the other signaling pathways, then we expect that critical genes that integrate signaling action would be conserved not only in ontogeny but also in phylogeny, defining a network of genes that define the logic of signal integration in development and evolution.

7. Disease and Notch: The Pathobiology of Gain and Loss of Function

In an introductory chapter such as this, predicting the "future" is always subjective as it inevitably relies on interpretations and extrapolations that may not be shared by everyone. Nevertheless, we find it useful and hopefully thought-provoking to offer the reader some of our thoughts on the problematic and hypotheses that are guiding our work on aspects of the involvement of Notch in disease.

That a pathway of such fundamental importance in development may be associated with disease is certainly not surprising. Notch signaling has been linked thus far to three inherited syndromes involving mutations in both ligands and receptors (Gridley, 2003) and we can predict fairly safely that more diseases associated with Notch malfunctions will be further unveiled. It is fair to say that the precise cellular pathogenic focus of some of these diseases is still elusive, we do not always know with certainty if they reflect loss- or gain-of-function mutations. For example, CADASIL was associated with specific mutations in the extracellular domain of Notch 3 more than a decade ago (Joutel *et al.* 1996) but we still do not know the nature of the mutations, a reflection of not having clear assays to measure receptor functionality, a difficulty compounded by the dosage dependence of the signaling process. Of course knowing whether gain-or loss-of-function is the cause of pathogenesis is essential for contemplating therapeutic avenues. Needless to say that both loss-of-function and, even more challengingly, neomorphic mutations define therapeutically challenging problems, notwithstanding the generic fact that attacking any pleiotropic pathway systemically is quite difficult. The lack of an enzymatic step in Notch signaling does not offer a classical target for drug intervention and thus antibodies as well as more exotic biologicals (Moellering *et al.*, 2009; Wu *et al.*, 2010) are being examined to target Notch. Given the density of the genetic circuitry capable of modulating Notch activity, it may be possible that classical "druggable" targets capable of modulating Notch activity may indeed exist. Thus, it is conceivable that a therapeutic modulation of Notch may be attained via targeting a modifier of the pathway.

Gain-of-function Notch pathology has been clearly associated with cancer. Cancer is a developmental disease par excellence and pathways such as Notch that can affect so profoundly cell fates—and thus the balance among differentiation, apoptosis, and proliferation—are almost bound to be involved in oncogenesis. Sklar and his colleagues were the first to associate altered Notch function with cancer, as they defined chromosomal rearrangements in T-cell acute lymphoblastic leukemias that caused the truncation of the Notch 1 receptor and resulted in what we now know to be a constitutively active form of the receptor (Ellisen *et al.*, 1991). The significance of this original finding was highlighted by the determination that more than 50% of human T-ALLs harbor activating mutations in Notch 1 (Weng *et al.*, 2004). This emphasizes the fact that Notch can act as an oncogene, but the involvement of Notch signaling in cancer beyond leukemias has been increasingly appreciated defining Notch as an actively pursued drug target; Koch and Radke review in this tome the remarkable expansion of Notch studies in tumorigenesis we have witnessed over the past years and thus we will refrain from discussing them here. Nevertheless a few generalities we consider important are worth pointing out here.

True to its context-dependent nature and testifying to how difficult it is to generalize with such a pleiotropic pathway, Notch can function as a tumor suppressor rather than an oncogene in skin tumor mouse models (Nicolas *et al.*, 2003) (but see Demehri *et al.*, 2009). In many tumors with Notch involvement, and in spite of extensive searches, mutations in Notch are lacking, in contrast to the case in T-ALLs where the oncogenic nature of Notch is clear. The presumed links between these oncogenic events and Notch function are based on correlations of the clinical outcome and Notch pathway activity, usually as measured by the expression of Notch pathway elements. Thus, notwithstanding the fact that the Notch receptor can behave as an oncogene, the nature of its involvement in solid tumors remains enigmatic (see Chapter 13) as the links between a broad spectrum of solid tumors and Notch are based on correlations of the clinical outcome and Notch pathway activity, usually as revealed by the expression levels of Notch pathway elements. Considering the involvement of Notch in proliferation together with results from mouse tumor models, an indirect yet important involvement of Notch in tumorigenesis ought to be contemplated.

8. Notch and Cancer: Affecting Proliferation Where it Matters?

Cancer is a developmental disease par excellence and pathways such as Notch that can affect so profoundly cell fates—and thus the balance between differentiation, apoptosis, and proliferation—are almost bound to be involved in oncogenesis. We know that Notch activation can dramatically affect proliferation. It has been shown that receptor activation can induce cell proliferation in a variety of cellular contexts in both vertebrates and invertebrates. It is important to keep in mind that while much of this evidence comes from gain-of-function studies involving ligand-independent, activated Notch receptors, there is good evidence showing that ligand-dependent activation can stimulate cell proliferation. What has received little attention, even though it has been recognized for years, is the capacity of Notch to affect proliferative (and other?) events in a cell nonautonomous fashion. Indeed, many studies have documented the fact that the activation of Notch in a cell can affect the proliferative capacity in nonadjacent cells (de Celis *et al.*, 1998; Go *et al.*, 1998). There are likely many ways this can be achieved, and such effects are presumably mediated by the induction of diffusible gene products for which expression is influenced by Notch. Indeed, in *Drosophila*, Notch activation in one cell, at least in the context of the developing eye, was demonstrated to induce the expression of the JAK–STAT pathway diffusible ligand unpaired (Upd), which can in turn affect proliferative properties in distant cells (Moberg *et al.*, 2005; Thompson *et al.*, 2005; Vaccari *et al.*, 2005). Such nonautonomous behavior can have profound consequences for oncogenesis, including perhaps an involvement in stroma–tumor interactions.

Notch activation on its own can induce cell proliferation, but it is becoming clear that it is the synergy between Notch and other gene activities that can make a very dramatic difference in proliferation. Just as an example we can refer to cooperating effects that have been established in imaginal discs in which activated Notch acts in synergy with up-regulation of vestigial, a pleiotropic nuclear protein (Rabinow *et al.*, 1990; Couso *et al.*, 1995; Neumann *et al.*, 1996; Kim *et al.*, 1997; Go *et al.*, 1998) or where the coactivation of Notch and RTK (Fig. 1.6) pathways results in enormous discs ("teradiscs") grossly distorted by overgrowth. We do know, through genetic analyses, that such dramatic overgrowth can result from the synergy of activated Notch with many different genes. This synergistic proliferation circuitry is of interest as such synergies could actually define novel oncogenic processes.

Notch activation in the intestine and the mammary gland on its own may not be necessarily associated with malignant transformations, but it certainly affects proliferation. In one mouse tumor model for instance (Fre *et al.*, 2005, 2009), the consequence of activating Notch in the crypts of the adult mouse intestine is a dramatic increase in the proliferation within this compartment,

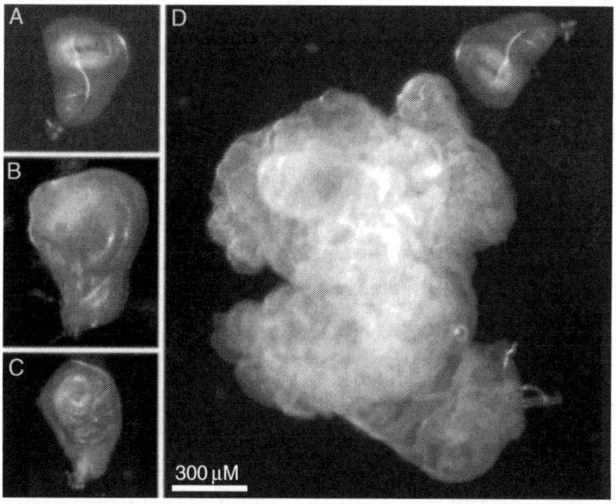

Figure 1.6 Notch synergies and proliferation. Dissected *Drosophila* wing imaginal discs shown at identical magnification, indicated by scale bar in panel D. (A) Wild-type disc (genotype: w^{1118}; Vg-GAL4). (B) Disc expressing a constitutively activated Notch receptor shows increased disc size and aberrant morphology (genotype: UAS-$Notch^{ICD}$ X Vg-GAL4). (C) Disc expressing an activated form of Ras (genotype: UAS-$Ras1^{V12}$ X Vg-GAL4) shows wild-type disc size, but altered disc morphology. (D) Disc expressing both the activated form of Notch and the activated form of Ras shows dramatically increased disc size, altered disc morphology (genotype: a dual transgenic carrying UAS-$Notch^{ICD}$, UAS-$Ras1^{V12}$, and Vg-GAL4). In D, a wild-type disc (upper right) is included for comparison. Although not shown, all transgenic line discs are wild-type in the absence of the GAL4 driver (Hurlbut, 2007).

which also harbors the stem cells responsible for the regular regeneration of the intestine. This effect, which by itself is not tumorigenic, can be completely blocked when Wnt signaling is inhibited through the deletion of TCF, the major effector of Wnt. Thus, the crosstalk of the two pathways is essential for the Notch-dependent proliferation in the crypt, and it is a relationship that is valid across species, as it is conserved in *Drosophila.* If we modulate Wnt signaling by generating a mouse that is heterozygous for the Wnt inhibitor APC, the intestine in the mouse, as does the intestine of humans heterozygous for APC, develops benign polyps, which upon loss of heterozygosity develop into *bona fide* adenocarcinomas. If in such a heterozygous mouse we also activate the Notch receptor in the crypts, the number of polyps seen jumps by a factor of more than 20. These mice develop tumors much more frequently than simple APC heterozygotes (Fre *et al.*, 2009).

Therefore, the activation of Notch in the crypts of the intestine is not tumourigenic *per se*, but is likely responsible for the expansion of a cell population (possibly stem cells?) that can, in this case, give rise to premalignant growths under the correct circumstances. Such premalignant cells can then accumulate more mutations and eventually give rise to malignant carcinomas. Activation of Notch in the mammary gland can also in certain circumstances give rise to premalignant growths that eventually progress to adenocarcinomas (Kiaris *et al.*, 2004) (see also Chapter 13).

In our view these observations may have broad implications. Namely, while the abnormal activation of the Notch receptor *per se* may in certain circumstances triggers malignant transformations, the activation of Notch signaling, presumably in synergy with other factors, may result in dramatically expanding cell populations that are prone to accumulate oncogenic mutations. Given the association of Notch activity with early precursors in general, and indeed with stem cells, this is compatible with known Notch biology. If true, the implication is that Notch signaling fluctuations in the "right" environment may result in amplifying "dangerous" cell populations. Such an involvement of Notch in cancer may be quite widespread, but difficult to identify and prevent. Moreover, given the complexity of the circuitry that is capable of modulating Notch activity, it would mean that many parameters including metabolic processes may result in what can be a potentially pathogenic up-regulation of the pathway.

9. Notch: What's Next

In spite of the extraordinary expansion of Notch-related studies over the past two decades, there are many questions at the mechanistic and cellular levels for which we currently envisage answers only through a glass, darkly. The rules of receptor–ligand engagement are poorly defined and undoubtedly will be the subject of many studies in the near future. Amalgamation of structural approaches, including EM and live imaging, with more classical biochemistry and genetics will provide us with a better functional picture of how these molecules interact on the surface of the cell. This should, in turn, shed more light on the fundamental question of how signaling and receiving cells are delineated within an adjacent pair of cells that express both ligand and receptor on their surfaces. The search for the identity of the cellular and membrane compartments critical for the processing and posttranslational modification of receptors and their ligands begins to define a field of its own, especially as this effort links signaling to classical cell biology and the many emerging live-imaging technologies that are becoming more broadly available. The path to the nucleus of the activated receptor remains a black box in Notch biology. And, while the extraordinary complexity of the genetic circuitry capable of modulating Notch signaling will be better understood through many detailed molecular genetic studies by many investigators devoted to the dissection of the signaling pathway, the intelligent pursuit of genomic and proteomic approaches to Notch pathway function will clearly be required, as well.

It seems certain that many successors of this present tome will appear in the years to come, as the encompassing nature of Notch signaling ensures an expansion of the field and many more years of Notch research. Notwithstanding the increasingly understood medical relevance of Notch, which points toward a path that will be trodden by ever more travelers over time, worms, flies, and mice will continue to provide the fuel for expansion of basic and translational investigations of the pathway. In fact, the evolution of the Notch field continues to elaborate an interesting and instructive paradigm, illustrating not only how and why model systems are invaluable for pathway dissection—a fact sometimes forgotten in this era of translational biology—but illustrating, as well, how model systems can and should be used to address clinically relevant problems.

ACKNOWLEDGMENTS

We thank our colleagues A. Louvi, D. Ho, A. Sen, Harsha Guruharsha, and J. Arboleda for their comments. The work in the S. Artavanis-Tsakonas laboratory is funded by the National Institute of Health (NIH) and the Spinal Muscular Atrophy Foundation. The work in the Muskavitch laboratory has been funded by NIH, the American Cancer Society, and the DeLuca Professorship from Boston College.

REFERENCES

Arias, A., Zecchini, V., and Brennan, K. (2002). CSL-independent Notch signalling: a checkpoint in cell fate decisions during development? *Curr. Opin. Genet. Dev.* **12**, 524–533.

Artavanis-Tsakonas, S. (1988). The molecular biology of the Notch locus and the fine tuning of differentiation in *Drosophila. Trends Genet.* **4**, 95–100.

Artavanis-Tsakonas, S., Muskavitch, M. A., and Yedvobnick, B. (1983). Molecular cloning of Notch, a locus affecting neurogenesis in *Drosophila melanogaster. Proc. Natl. Acad. Sci. U.S.A.* **80**, 1977–1981.

Artavanis-Tsakonas, S., Rand, M. D., and Lake, R. J. (1999). Notch signaling: cell fate control and signal integration in development. *Science* **284**, 770–776.

Aste-Amezaga, M., Zhang, N., Lineberger, J. E., Arnold, B. A., Toner, T. J., Gu, M., Huang, L., Vitelli, S., Vo, K. T., Haytko, P., *et al.* (2010). Characterization of Notch1 antibodies that inhibit signaling of both normal and mutated Notch1 receptors. *PLoS One* **5**, e9094.

Austin, J., and Kimble, J. (1987). glp-1 is required in the germ line for regulation of the decision between mitosis and meiosis in *C. elegans. Cell* **51**, 589–599.

Baron, M. (2003). An overview of the Notch signalling pathway. *Semin. Cell. Dev. Biol.* **14**, 113–119.

Beatus, P., Lundkvist, J., Oberg, C., and Lendahl, U. (1999). The Notch 3 intracellular domain represses Notch 1-mediated activation through Hairy/Enhancer of split (HES) promoters. *Development* **126**, 3925–3935.

Blaumueller, C. M., Qi, H., Zagouras, P., and Artavanis-Tsakonas, S. (1997). Intracellular cleavage of Notch leads to a heterodimeric receptor on the plasma membrane. *Cell* **90**, 281–291.

Brack, A. S., Conboy, I. M., Conboy, M. J., Shen, J., and Rando, T. A. (2008). A temporal switch from notch to Wnt signaling in muscle stem cells is necessary for normal adult myogenesis. *Cell Stem Cell* **2**, 50–59.

Bray, S. J. (2006). Notch signalling: a simple pathway becomes complex. *Nat. Rev. Mol. Cell Biol.* **7**, 678–689.

Brou, C., Logeat, F., Gupta, N., Bessia, C., LeBail, O., Doedens, J. R., Cumano, A., Roux, P., Black, R. A., and Israel, A. (2000). A novel proteolytic cleavage involved in Notch signaling: the role of the disintegrin-metalloprotease TACE. *Mol. Cell* **5**, 207–216.

Cagan, R. L., and Ready, D. F. (1989). Notch is required for successive cell decisions in the developing Drosophila retina. *Genes Dev.* **3**, 1099–1112.

Casali, A., and Batlle, E. (2009). Intestinal stem cells in mammals and Drosophila. *Cell Stem Cell* **4**, 124–127.

Coffman, C., Harris, W., and Kintner, C. (1990). Xotch, the Xenopus homolog of Drosophila notch. *Science* **249**, 1438–1441.

Coffman, C. R., Skoglund, P., Harris, W. A., and Kintner, C. R. (1993). Expression of an extracellular deletion of Xotch diverts cell fate in Xenopus embryos. *Cell* **73**, 659–671.

Coumailleau, F., Furthauer, M., Knoblich, J. A., and Gonzalez-Gaitan, M. (2009). Directional Delta and Notch trafficking in Sara endosomes during asymmetric cell division. *Nature* **458**(7241), 1051–1055.

de Celis, J. F., Tyler, D. M., de Celis, J., and Bray, S. J. (1998). Notch signalling mediates segmentation of the Drosophila leg. *Development* **125**, 4617–4626.

Demehri, S., Turkoz, A., and Kopan, R. (2009). Epidermal Notch1 loss promotes skin tumorigenesis by impacting the stromal microenvironment. *Cancer Cell* **16**, 55–66.

Demerec, M. (1950). *The Biology of Drosophila*. John Wiley, New York.

De Strooper, B., Annaert, W., Cupers, P., Saftig, P., Craessaerts, K., Mumm, J. S., Schroeter, E. H., Schrijvers, V., Wolfe, M. S., Ray, W. J., *et al.* (1999). A presenilin-1-dependent gamma-secretase-like protease mediates release of Notch intracellular domain. *Nature* **398**, 518–522.

Dexter, J. S. (1914). The analysis of a case of continuous variation in *Drosophila* by a linkage relation. *Am. Nat.* **48**, 712–758.

Doe, C. Q., and Goodman, C. S. (1985). Early events in insect neurogenesis. II. The role of cell interactions and cell lineage in the determination of neuronal precursor cells. *Dev. Biol.* **111**, 206–219.

Dreesen, O., and Brivanlou, A. H. (2007). Signaling pathways in cancer and embryonic stem cells. *Stem Cell Rev.* **3**, 7–17.

Ellisen, L. W., Bird, J., West, D. C., Soreng, A. L., Reynolds, T. C., Smith, S. D., and Sklar, J. (1991). TAN-1, the human homolog of the *Drosophila* notch gene, is broken by chromosomal translocations in T lymphoblastic neoplasms. *Cell* **66**, 649–661.

Farnie, G., and Clarke, R. B. (2006). Breast stem cells and cancer. *Ernst Schering Found Symp Proc* **5**, 141–153.

Fehon, R. G., Kooh, P. J., Rebay, I., Regan, C. L., Xu, T., Muskavitch, M. A., and Artavanis-Tsakonas, S. (1990). Molecular interactions between the protein products of the neurogenic loci Notch and Delta, two EGF-homologous genes in *Drosophila*. *Cell* **61**, 523–534.

Fitzgerald, K., Wilkinson, H. A., and Greenwald, I. (1993). glp-1 can substitute for lin-12 in specifying cell fate decisions in Caenorhabditis elegans. *Development* **119**, 1019–1027.

Fleming, R. J., Scottgale, T. N., Diederich, R. J., and Artavanis-Tsakonas, S. (1990). The gene Serrate encodes a putative EGF-like transmembrane protein essential for proper ectodermal development in *Drosophila melanogaster*. *Genes Dev.* **4**, 2188–2201.

Fortini, M. E., and Artavanis-Tsakonas, S. (1994). The suppressor of hairless protein participates in Notch receptor signaling. *Cell* **79**, 273–282.

Fortini, M. E., and Bilder, D. (2009). Endocytic regulation of Notch signaling. *Curr. Opin. Genet. Dev.* **19**(4), 323–328.

Foster, G. G. (1973). Is the notch locus of *Drosophila melanogaster* a tandem repeat? Correlation of the genetic map and complementation pattern of selected mutations. *Genetics* **73**, 435–438.

Foster, G. G. (1975). Negative complementation at the notch locus of *Drosophila melanogaster*. *Genetics* **81**, 99–120.

Francis, R., McGrath, G., Zhang, J., Ruddy, D. A., Sym, M., Apfeld, J., Nicoll, M., Maxwell, M., Hai, B., and Ellis, M. C., *et al.* (2002). aph-1 and pen-2 are required for Notch pathway signaling, gamma-Secretase cleavage of betaAPP, and presenilin protein accumulation. *Dev. Cell* **3**, 85–97.

Fre, S., Huyghe, M., Mourikis, P., Robine, S., Louvard, D., and Artavanis-Tsakonas, S. (2005). Notch signals control the fate of immature progenitor cells in the intestine. *Nature* **435**, 964–968.

Fre, S., Pallavi, S. K., Huyghe, M., Lae, M., Janssen, K. P., Robine, S., Artavanis-Tsakonas, S., and Louvard, D. (2009). Notch and Wnt signals cooperatively control cell proliferation and tumorigenesis in the intestine. *Proc. Natl. Acad. Sci. U.S.A.* **106**(15), 6309–6314.

Gerhart, J. (1999). Warkany lecture: signaling pathways in development. *Teratology* **60**, 226–239.

Gering, M., and Patient, R. (2010). Notch signalling and haematopoietic stem cell formation during embryogenesis. *J. Cell Physiol.* **222**(1), 11–16.

Gibert, J. M., and Simpson, P. (2003). Evolution of cis-regulation of the proneural genes. *Int. J. Dev. Biol.* **47**, 643–651.

Giniger, E. (1998). A role for Abl in Notch signaling. *Neuron* **20**, 667–681.

Go, M. J., Eastman, D. S., and Artavanis-Tsakonas, S. (1998). Cell proliferation control by Notch signaling in *Drosophila* development. *Development* **125**, 2031–2040.

Gordon, W. R., Arnett, K. L., and Blacklow, S. C. (2008). The molecular logic of Notch signaling—a structural and biochemical perspective. *J. Cell Sci.* **121**, 3109–3119.

Goutte, C., Hepler, W., Mickey, K. M., and Priess, J. R. (2000). aph-2 encodes a novel extracellular protein required for GLP-1-mediated signaling. *Development* **127**, 2481–2492.

Goutte, C., Tsunozaki, M., Hale, V. A., and Priess, J. R. (2002). APH-1 is a multipass membrane protein essential for the Notch signaling pathway in Caenorhabditis elegans embryos. *Proc. Natl. Acad. Sci. U.S.A.* **99**, 775–779.

Greenspan, R. J. (1990). The Notch gene, adhesion, and developmental fate in the *Drosophila* embryo. [Review]. *New Biol.* **2**, 595–600.

Greenwald, I. (1985). lin-12, a nematode homeotic gene, is homologous to a set of mammalian proteins that includes epidermal growth factor. *Cell* **43**, 583–590.

Greenwald, I., Coulson, A., Sulston, J., and Priess, J. (1987). Correlation of the physical and genetic maps in the lin-12 region of Caenorhabditis elegans. *Nucleic Acids Res.* **15**, 2295–2307.

Greenwald, I. S., Sternberg, P. W., and Horvitz, H. R. (1983). The lin-12 locus specifies cell fates in Caenorhabditis elegans. *Cell* **34**, 435–444.

Gridley, T. (2003). Notch signaling and inherited disease syndromes. *Hum. Mol. Genet.* **12**(*Spec No 1*), R9–R13.

Grimwade, B. G., Muskavitch, M. A., Welshons, W. J., Yedvobnick, B., and Artavanis-Tsakonas, S. (1985). The molecular genetics of the Notch locus in *Drosophila melanogaster*. *Dev. Biol.* **107**, 503–519.

Hayward, P., Kalmar, T., and Martinez Arias, A. (2008). Wnt/Notch signalling and information processing during development. *Development* **135**, 411–424.

Heitzler, P., and Simpson, P. (1991). The choice of cell fate in the epidermis of *Drosophila*. *Cell* **64**, 1083–1092.

Hori, K., Fostier, M., Ito, M., Fuwa, T. J., Go, M. J., Okano, H., Baron, M., and Matsuno, K. (2004). *Drosophila* Deltex mediates Suppressor of Hairless-independent and late-endosomal activation of Notch signaling. *Development* **131**, 5527–5537.

Huppert, S., Le, A., Schroeter, E. H., Mumm, J. S., Saxena, M. T., Milner, L. A., and Kopan, R. (2000). Embryonic lethality in mice homozygous for a processing deficient allele of Notch1. *Nature* **405**, 966–970.

Hurlbut, G.D. (2007). Notch and RTK signaling: transcriptional output and signal integration. PhD Thesis, Universite de Paris XI Orsay.

Hurlbut, G., Kankel, M., Artavanis Tsakonas, S., (2008). Notch-Ras signal integration in *Drosophila* development: nodal points and complexity, PNAS. (in press)

Hurlbut, G., Kankel, M., and Artavanis Tsakonas, S. (2009a). Notch-Ras signal integration in *Drosophila* development: Nodal points and complexity. *Proc. Natl. Acad. Sci. U.S.A.* **106**(7), 2218–2223.

Hurlbut, G. D., Kankel, M. W., and Artavanis-Tsakonas, S. (2009b). Nodal points and complexity of Notch-Ras signal integration. *Proc. Natl. Acad. Sci. U.S.A.* **106**, 2218–2223.

Hurlbut, G. D., Kankel, M. W., Lake, R. J., and Artavanis-Tsakonas, S. (2007). Crossing paths with Notch in the hyper-network. *Curr. Opin. Cell Biol.* **19**, 166–175.

Jarriault, S., Brou, C., Logeat, F., Schroeter, E. H., Kopan, R., and Israel, A. (1995). Signalling downstream of activated mammalian Notch. *Nature* **377**, 355–358.

Kankel, M. W., Hurlbut, G. D., Upadhyay, G., Yajnik, V., Yedvobnick, B., and Artavanis-Tsakonas, S. (2007). Investigating the genetic circuitry of mastermind in *Drosophila*, a notch signal effector. *Genetics* **177**, 2493–2505.

Kelly, D. F., Lake, R. J., Middelkoop, T. C., Fan, H.-Y., Artavanis-Tsakonas, S., and Walz, T. (2010). Molecular structure and dimeric organization of the Notch extracellular domain as revealed by electron microscopy. *PLoS One* **5**(5), e10532.

Kiaris, H., Politi, K., Grimm, L. M., Szabolcs, M., Fisher, P., Efstratiadis, A., and Artavanis-Tsakonas, S. (2004). Modulation of notch signaling elicits signature tumors and inhibits hras1-induced oncogenesis in the mouse mammary epithelium. *Am. J. Pathol.* **165**, 695–705.

Kidd, S., Kelley, M. R., and Young, M. W. (1986). Sequence of the notch locus of *Drosophila melanogaster*: Relationship of the encoded protein to mammalian clotting and growth factors. *Mol. Cell. Biol.* **6**, 3094–3108.

Kidd, S., Lockett, T. J., and Young, M. W. (1983). The Notch locus of *Drosophila melanogaster*. *Cell* **34**, 421–433.

Kirschner, M., and Gerhart, J. (2005). *The Plausibility of Life: Great Leaps of Evolution*. Yale University Press, New Haven, CT.

Kopan, R., and Ilagan, M. X. (2009). The canonical Notch signaling pathway: unfolding the activation mechanism. *Cell* **137**, 216–233.

Kopan, R., Nye, J. S., and Weintraub, H. (1994). The intracellular domain of mouse Notch: a constitutively activated repressor of myogenesis directed at the basic helix-loop-helix region of MyoD. *Development* **120**, 2385–2396.

Kopan, R., Schroeter, E. H., Weintraub, H., and Nye, J. S. (1996). Signal transduction by activated mNotch: importance of proteolytic processing and its regulation by the extracellular domain. *Proc. Natl. Acad. Sci. U.S.A.* **93**, 1683–1688.

Kopczynski, C. C., Alton, A. K., Fechtel, K., Kooh, P. J., and Muskavitch, M. A. (1988). Delta, a *Drosophila* neurogenic gene, is transcriptionally complex and encodes a protein related to blood coagulation factors and epidermal growth factor of vertebrates. *Genes Dev.* **2**, 1723–1735.

Krejci, A., Bernard, F., Housden, B. E., Collins, S., and Bray, S. J. (2009). Direct response to Notch activation: Signaling crosstalk and incoherent logic. *Sci. Signal.* **2**, ra1.

Lehmann, R., Jimenez, F., Dietrich, U., and Campos-Ortega, J. A. (1983). On the phenotype and development of mutants of early neurogenesis in *Drosophila melanogaster*. *Roux. Arch. Dev. Biol.* **192**, 62–74.

Levitan, D., and Greenwald, I. (1995). Facilitation of lin-12-mediated signalling by sel-12, a Caenorhabditis elegans S182 Alzheimer's disease gene. *Nature* **377**, 351–354.

Li, L. H., Krantz, I. D., Deng, Y., Genin, A., Banta, A. B., Collins, C. C., Qi, M., Trask, B. J., Kuo, W. L., Cochran, J., *et al.* (1997). Alagille-syndrome is caused by mutations in human jagged1, which encodes a ligand for Notch1. *Nat. Genet.* **16**, 243–251.

Li, K., Li, Y., Wu, W., Gordon, W. R., Chang, D. W., Lu, M., Scoggin, S., Fu, T., Vien, L., Histen, G., *et al.* (2008). Modulation of Notch signaling by antibodies specific for the extracellular negative regulatory region of NOTCH3. *J. Biol. Chem.* **283**, 8046–8054.

Lindsell, C. E., Shawber, C. J., Boulter, J., and Weinmaster, G. (1995). Jagged: a mammalian ligand that activates Notch1. *Cell* **80**, 909–917.

Logeat, F., Bessia, C., Brou, C., LeBail, O., Jarriault, S., Seidah, N. G., and Israel, A. (1998). The Notch1 receptor is cleaved constitutively by a furin-like convertase. *Proc. Natl. Acad. Sci. U.S.A.* **95**, 8108–8112.

Louvi, A., and Artavanis-Tsakonas, S. (2006). Notch signalling in vertebrate neural development. *Nat. Rev. Neurosci.* **7**, 93–102.

Mello, C. C., Draper, B. W., and Priess, J. R. (1994). The maternal genes apx-1 and glp-1 and establishment of dorsal-ventral polarity in the early C. elegans embryo. *Cell* **77**, 95–106.

Mishra-Gorur, K., Rand, M. D., Perez-Villamil, B., and Artavanis-Tsakonas, S. (2002). Down-regulation of Delta by proteolytic processing. *J. Cell Biol.* **159**, 313–324.

Moberg, K. H., Schelble, S., Burdick, S. K., and Hariharan, I. K. (2005). Mutations in erupted, the *Drosophila* ortholog of mammalian tumor susceptibility gene 101, elicit non-cell-autonomous overgrowth. *Dev. Cell* **9**, 699–710.

Moellering, R. E., Cornejo, M., Davis, T. N., Del Bianco, C., Aster, J. C., Blacklow, S. C., Kung, A. L., Gilliland, D. G., Verdine, G. L., and Bradner, J. E. (2009). Direct inhibition of the NOTCH transcription factor complex. *Nature* **462**, 182–188.

Mohr, O. L. (1919). Character changes caused by mutation of an entire region of a chromosome in *Drosophila. Genetics* **4**, 275–232.

Mumm, J. S., Schroeter, E. H., Saxena, M. T., Griesemer, A., Tian, X., Pan, D. J., Ray, W. J., and Kopan, R. (2000). A ligand-induced extracellular cleavage regulates γ-secretase-like proteolytic activation of Notch1. *Mol. Cell* **5**, 197–206.

Mummery-Widmer, J. L., Yamazaki, M., Stoeger, T., Novatchkova, M., Bhalerao, S., Chen, D., Dietzl, G., Dickson, B. J., and Knoblich, J. A. (2009). Genome-wide analysis of Notch signalling in *Drosophila* by transgenic RNAi. *Nature* **458**(7241), 987–992.

Nicolas, M., Wolfer, A., Raj, K., Kummer, J. A., Mill, P., Van Noort, M., Hui, C. C., Clevers, H., Dotto, G. P., and Radtke, F. (2003). Notch1 functions as a tumor suppressor in mouse skin. *Nat. Genet.* **33**, 416–421.

Nusslein-Volhard, C., and Wieschaus, E. (1980). Mutations affecting segment number and polarity in *Drosophila. Nature* **287**, 795–801.

Nusslein-Volhard, C., Wieschaus, E., and Kluding, H. (1984). Mutations affecting the pattern of the larval cuticle in *Drosophila melanogaster*. I. Zygotic loci on the second chromosome. Wilhelm *Roux's. Arch. Dev. Biol.* **193**, 267–282.

Okochi, M., Fukumori, A., Jiang, J., Itoh, N., Kimura, R., Steiner, H., Haass, C., Tagami, S., and Takeda, M. (2006). Secretion of the Notch-1 Abeta-like peptide during Notch signaling. *J. Biol. Chem.* **281**, 7890–7898.

Ong, C., Cheng, H., Chang, L. W., Ohtsuka, T., Kageyama, R., Stormo, D. G., and Kopan, R. (2006). Target selectivity of vertebrate Notch proteins: collaboration between discrete domains and CSL binding site architecture determine activation probability. *J. Biol. Chem.* **281**, 5106–5119.

Palomero, T., Lim, W. K., Odom, D. T., Sulis, M. L., Real, P. J., Margolin, A., Barnes, K. C., O'Neil, J., Neuberg, D., Weng, A. P., Aster, J. C., Sigaux, F., et al. (2006). NOTCH1 directly regulates c-MYC and activates a feed-forward-loop transcriptional network promoting leukemic cell growth. *Proc. Natl. Acad. Sci. USA.* 103, 18261–18266.

Pan, D. J., and Rubin, G. M. (1997). Kuzbanian controls proteolytic processing of Notch and mediates lateral inhibition during *Drosophila* and vertebrate neurogenesis. *Cell* **90**, 271–280.

Parks, A. L., Klueg, K. M., Stout, J. R., and Muskavitch, M. A. (2000). Ligand endocytosis drives receptor dissociation and activation in the Notch pathway. *Development* **127**, 1373–1385.

Poodry, C. A. (1990). Shibire, a neurogenic mutant of *Drosophila*. *Dev. Biol.* **138**, 464–472.

Poulsons, D.F. (1936). Chromosomal deficiencies and embryonic development. PhD Thesis Caltech.

Priess, J. R., Schnabel, H., and Schnabel, R. (1987). The glp-1 locus and cellular interactions in early C. elegans embryos. *Cell* **51**, 601–611.

Qi, H., Rand, M. D., Wu, X., Sestan, N., Wang, W., Rakic, P., Xu, T., and Artavanis-Tsakonas, S. (1999). Processing of the notch ligand delta by the metalloprotease Kuzbanian. *Science* **283**, 91–94.

Rand, M. D., Lindblom, A., Carlson, J., Villoutreix, B. O., and Stenflo, J. (1997). Calcium binding to tandem repeats of EGF-like modules. Expression and characterization of the EGF-like Modules of human Notch-1 implicated in receptor-ligand interactions. *Protein Sci.* **6**, 2059–2071.

Rebay, I., Fleming, R. J., Fehon, R. G., Cherbas, L., Cherbas, P., and Artavanis-Tsakonas, S. (1991). Specific EGF repeats of Notch mediate interactions with Delta and Serrate: implications for Notch as a multifunctional receptor. *Cell* **67**, 687–699.

Sanders, P. G., Munoz-Descalzo, S., Balayo, T., Wirtz-Peitz, F., Hayward, P., and Arias, A. M. (2009). Ligand-independent traffic of Notch buffers activated Armadillo in *Drosophila*. *PLoS Biol.* **7**, e1000169.

Schroeter, E. H., Kisslinger, J. A., and Kopan, R. (1998). Notch-1 signalling requires ligand-induced proteolytic release of intracellular domain [see comments]. *Nature* **393**, 382–386.

Seydoux, G., and Greenwald, I. (1989). Cell autonomy of lin-12 function in a cell fate decision in C. elegans. *Cell* **57**, 1237–1245.

Shawber, C., Boulter, J., Lindsell, C. E., and Weinmaster, G. (1996). Jagged2: a serrate-like gene expressed during rat embryogenesis. *Dev. Biol.* **180**, 370–376.

Sprinzak, D., Lakhanpal, A., LeBon, L., Santat, L. A., Fontes, M. E., Anderson, G. A., Garcia-Ojalvo, J., and Elowitz, M. E. (2010). Cis-interactions beween Notch and Delta generate mutally exclusive signaling states. *Nature* **465**(7294), 86–90.

Sternberg, P. W. (1988). Lateral inhibition during vulval induction in Caenorhabditis elegans. *Nature* **335**, 551–554.

Stifani, S., Blaumueller, C. M., Redhead, N. J., Hill, R. E., and Artavanis, T. S. (1992). Human homologs of a *Drosophila* Enhancer of split gene product define a novel family of nuclear proteins. *Nat. Genet.* **2**, 119–127.

Struhl, G., Fitzgerald, K., and Greenwald, I. (1993). Intrinsic activity of the Lin-12 and Notch intracellular domains in vivo. *Cell* **74**, 331–345.

Struhl, G., and Greenwald, I. (1999). Presenilin is required for activity and nuclear access of Notch in *Drosophila*. *Nature* **398**, 522–525.

Tax, F. E., Yeargers, J. J., and Thomas, J. H. (1994). Sequence of C. elegans lag-2 reveals a cell-signalling domain shared with Delta and Serrate of *Drosophila*. *Nature* **368**, 150–154.

Thorig, G. E., Heinstra, P. W., and Scharloo, W. (1981a). The action of the notch locus in *Drosophila melanogaster*. I. Effects of the notch8 deficiency on mitochondrial enzymes. *Mol. Gen. Genet.* **182**, 31–38.

Thorig, G. E., Heinstra, P. W., and Scharloo, W. (1981b). The action of the notchlocus in *Drosophila melanogaster*. II. Biochemical effects of recessive lethals on mitochondrial enzymes. *Genetics* **99**, 65–74.

van Tetering, G., van Diest, P., Verlaan, I., van der Wall, E., Kopan, R., and Vooijs, M. (2009). The metalloprotease ADAM10 is required for notch1 S2 cleavage. *J. Biol. Chem.* **284**(45), 31018–31027.

Vassin, H., Bremer, K. A., Knust, E., and Campos-Ortega, J. A. (1987). The neurogenic gene Delta of *Drosophila melanogaster* is expressed in neurogenic territories and encodes a putative transmembrane protein with EGF-like repeats. *EMBO J.* **6**, 3431–3440.

Vooijs, M., Schroeter, E. H., Pan, Y., Blandford, M., and Kopan, R. (2004). Ectodomain shedding and intramembrane cleavage of mammalian notch proteins is not regulated through oligomerization. *J. Biol. Chem.* **279**, 50864–50873.

Welshons, W. J. (1956). Dosage experiments with *split* mutants in the presence of an *enhancer of split. Drosoph. Inf. Serv.* **30**, 157–158.

Welshons, W. J. (1965). Analysis of a gene in *drosophila. Science* **150**(*700*), 1122–1129.

Wen, C., Metzstein, M. M., and Greenwald, I. (1997). SUP-17, a Caenorhabditis elegans ADAM protein related to *Drosophila* KUZBANIAN, and its role in LIN-12/NOTCH signalling. *Development* **124**, 4759–4767.

Weng, A. P., Millholland, J. M., Yashiro-Ohtani, Y., Arcangeli, M. L., Lau, A., Wai, C., Del Bianco, C., Rodriguez, C. G., Sai, H., Tobias, J., Li, Y., Wolfe, M. S., Shachaf, C., Felsher, D., Blacklow, S. C., Pear, W. S., and Aster, J. C. (2006). c-Myc is an important direct target of Notch1 in T-cell acute lymphoblastic leukemia/lymphoma. *Genes Dev.* **20**, 2096-2109

Weng, A. P., Ferrando, A. A., Lee, W., Morris, J. P.,4th, Silverman, L. B., Sanchez-Irizarry, C., Blacklow, S. C., Look, A. T., and Aster, J. C. (2004). Activating mutations of NOTCH1 in human T cell acute lymphoblastic leukemia. *Science* **306**, 269–271.

Weng, A. P., Nam, Y., Wolfe, M. S., Pear, W. S., Griffin, J. D., Blacklow, S. C., and Aster, J. C. (2003). Growth suppression of pre-T acute lymphoblastic leukemia cells by inhibition of notch signaling. *Mol. Cell. Biol.* **23**, 655–664.

Wharton, K. A., Johansen, K. M., Xu, T., and Artavanis-Tsakonas, S. (1985). Nucleotide sequence from the neurogenic locus notch implies a gene product that shares homology with proteins containing EGF-like repeats. *Cell* **43**, 567–581.

Wilkinson, H. A., Fitzgerald, K., and Greenwald, I. (1994). Reciprocal changes in expression of the receptor lin-12 and its ligand lag-2 prior to commitment in a C. elegans cell fate decision. *Cell* **79**, 1187–1198.

Wolfe, M. S., Xia, W. M., Ostaszewski, B. L., Diehl, T. S., Kimberly, W. T., and Selkoe, D. J. (1999). Two transmembrane aspartates in presenilin-1 required for presenilin endoproteolysis and gamma-secretase activity. *Nature* **398**, 513–517.

Wright, T. R. (1970). The genetics of embryogenesis in *Drosophila. Adv. Genet.* **15**, 261–395.

Wu, Y., Cain-Hom, C., Choy, L., Hagenbeek, T. J., de Leon, G. P., Chen, Y., Finkle, D., Venook, R., Wu, X., Ridgway, J., *et al.* (2010). Therapeutic antibody targeting of individual Notch receptors. *Nature* **464**, 1052–1057.

Yoo, A. S., Bais, C., and Greenwald, I. (2004). Crosstalk between the EGFR and LIN-12/Notch pathways in C. elegans vulval development. *Science* **303**, 663–666.

CHAPTER TWO

Mechanistic Insights into Notch Receptor Signaling from Structural and Biochemical Studies

Rhett A. Kovall* *and* Stephen C. Blacklow†

Contents

* Department of Molecular Genetics, Biochemistry, and Microbiology, University of Cincinnati, Cincinnati, OH, USA
† Departments of Pathology and Biological Chemistry and Molecular Pharmacology, Harvard Medical School, Dana Farber Cancer Institute and Brigham and Women's Hospital, Boston, MA, USA

Current Topics in Developmental Biology, Volume 92
ISSN 0070-2153, DOI 10.1016/S0070-2153(10)92002-4

Abstract

Notch proteins are the receptors in a highly conserved signal transduction system used to communicate signals between cells that contact each other. Studies investigating structure–function relationships in Notch signaling have gained substantial momentum in recent years. Here, we summarize the current understanding of the molecular logic of Notch signal transduction, emphasizing structural and biochemical studies of Notch receptors, their ligands, and complexes of intracellular Notch proteins with their target transcription factors. Recent advances in the structure-based modulation of Notch-signaling activity are also discussed.

1. Introduction

Notch receptors are large transmembrane proteins that normally communicate signals upon binding to transmembrane ligands expressed on adjacent cells. Because both the receptors and their ligands are transmembrane proteins, Notch signals rely on cell–cell contact. Evolutionary divergence of invertebrates and vertebrates has been accompanied by at least two rounds of gene duplication: flies possess a single *Notch* gene, worms two (*GLP-1* and *LIN-12*), and mammals four (*NOTCH1-4*).

Notch signals influence a wide spectrum of cell fate decisions, both during development and in the adult organism. However, dysregulated signaling has also been implicated in a number of different human diseases ranging from neurodegeneration to cancer, most notably in the case of T-cell acute lymphoblastic leukemia/lymphoma (T-ALL) (Aster *et al.*, 2008; Weng *et al.*, 2004).

Canonical Notch signals are transduced by a process called regulated intramembrane proteolysis (Brown *et al.*, 2000). Notch receptors are normally maintained in a resting, proteolytically resistant conformation on the cell surface, but ligand binding initiates a proteolytic cascade that releases the intracellular portion of the receptor (ICN) from the membrane. The critical, regulated cleavage step is effected using ADAM metalloproteases and occurs at a site called S2 immediately external to the plasma membrane (Brou *et al.*, 2000; Mumm *et al.*, 2000). This truncated receptor, dubbed NEXT (for Notch extracellular truncation), remains membrane tethered until it is processed at site S3 and additional sites by gamma secretase, a multiprotein enzyme complex (De Strooper *et al.*, 1999; Struhl and Greenwald, 1999; Wolfe *et al.*, 1999; Ye *et al.*, 1999).

After gamma secretase cleavage, ICN ultimately enters the nucleus, where it assembles a transcriptional activation complex that contains a DNA-binding transcription factor called CSL, and a transcriptional coactivator of the Mastermind family (Doyle *et al.*, 2000; Petcherski and Kimble,

2000; Wu *et al.*, 2000). This complex then engages additional coactivator proteins such as p300 to recruit the basal transcription machinery and turn on the expression of downstream target genes (Fryer *et al.*, 2002; Wallberg *et al.*, 2002).

Drosophila and mammalian Notch receptors are first synthesized in precursor form as 300–350 kD type I single-pass transmembrane glycoproteins. During maturation, mammalian Notch precursor polypeptides are proteolytically processed (Kopan *et al.*, 1996) by a furin-like convertase (Logeat *et al.*, 1998) at a site called S1 (Fig. 2.1), yielding two non-covalently associated subunits. The resulting two associated subunits, here termed extracellular Notch (NEC) and transmembrane Notch (NTM), constitute the mature heterodimeric form of the protein present at the cell surface (Blaumueller *et al.*, 1997).

Notch receptors have a modular domain organization. The ectodomains of Notch receptors consist of a series of N-terminal epidermal growth factor (EGF)-like repeats that are responsible for ligand binding. The number of EGF-like repeats varies by species and receptor subtype, as the EGF repeat regions of fly and mammalian proteins are much longer than those found in the *Caenorhabditis elegans* proteins LIN-12 and GLP-1. O-linked glycosylation of these EGF repeats, including modification by O-fucose, Fringe, and Rumi glycosyltransferases (Fig. 2.1), also modulates the activity of Notch receptors in response to different ligand subtypes in flies and mammals (see Chapter 4 for a comprehensive review).

The EGF repeats are followed by three LIN-12/Notch repeat (LNR) modules, which are unique to Notch receptors and participate in preventing premature receptor activation (Greenwald and Seydoux, 1990; Rand *et al.*, 2000; Sanchez-Irizarry *et al.*, 2004). The heterodimerization (HD) domain of Notch1 is divided by furin cleavage, so that its N-terminal part (HD-N; Fig. 2.1) terminates the NEC subunit, and its C-terminal half (HD-C) constitutes the beginning of the NTM subunit. Following the extracellular HD-C region, NEC has a transmembrane segment and an intracellular region (ICN), which consists of a RAM domain (originally denoted the "RAM23" domain; (Tamura *et al.*, 1995)), seven ankyrin (ANK) repeats flanked by two nuclear localization signals (NLS), a transactivation domain (TAD), and a PEST region that participates in protein degradation (Fryer *et al.*, 2004; Rechsteiner and Rogers, 1996).

Canonical Notch ligands in flies and higher eukaryotes fall into two general classes, depending on whether they are homologous to the *Drosophila* prototypes Delta and Serrate. Mammals have three Delta-like proteins, called Delta-like 1 (DLL1), Delta-like 3 (DLL3), and Delta-like 4 (DLL4), and two homologues of Serrate, called Jagged-1 and Jagged-2 (JAG1, and JAG2, respectively).

Both Delta and Serrate ligands also exhibit a modular domain arrangement, with an N-terminal MNNL (Module at the N-terminus of Notch

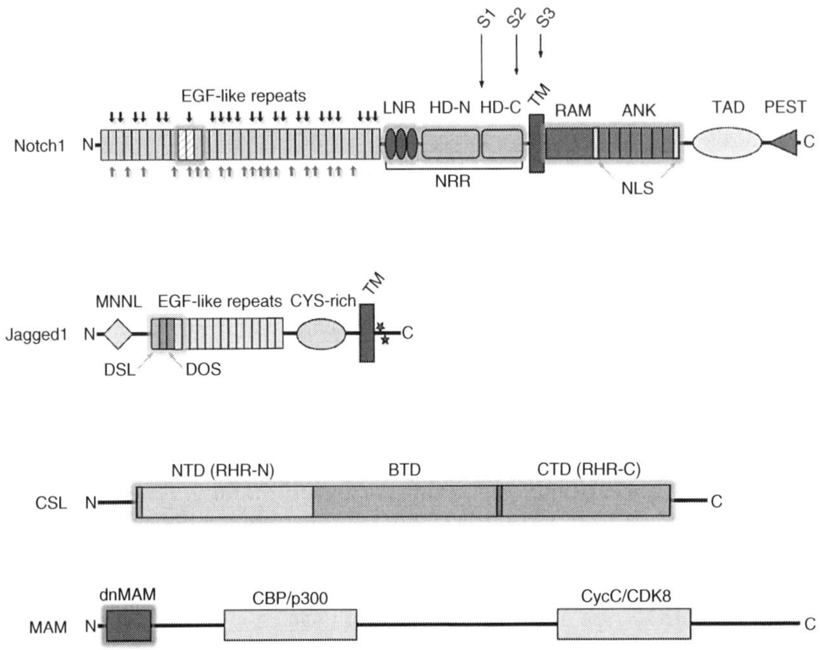

Figure 2.1 Domain organization of Notch pathway components. Figure shows domain schematics of Notch pathway components for which structures have been determined, including Notch1, Jagged1, CSL, and Mastermind. Notch1 is a typical Notch pathway receptor and is composed of multiple modular domains. Extracellularly, Notch1 consists of multiple EGF-like repeats followed by the NRR, which is composed of the LNR (Lin12-Notch Repeats) and the HD (Heterodimerization Domain). There is a single transmembrane-spanning segment (TM). EGF-like repeats that are modified by the glycotransferases Fringe and Rumi are denoted with black and red arrows, respectively. Intracellularly, Notch1 is composed of RAM (RBP-J Associated Molecule) and ANK (ankyrin repeats) domains that are required to interact with CSL, as well as TAD and PEST (proline, glutamate, serine, and threonine) domains that are important for transcriptional activation and degradation, respectively. NLS and S1, S2, and S3 cleavage sites are indicated. Jagged1 is a mammalian Notch pathway ligand related to Serrate in flies. Jagged is also a modular multidomain protein, containing a single transmembrane spanning (TM) region. Jagged1 contains MNNL (Module at N-terminus of Notch Ligands), DSL (Delta/Serrate/LAG-2), DOS (Delta and OSM-11-like), and EGF-like repeats, along with a membrane-proximal cysteine-rich domain. Red stars denote sites of potential ubiquitination of Jagged1, based on the identification of these sites in Serrate (Glittenberg *et al.*, 2006). CSL (CBF1/RBP-J, Su(H), Lag1) is the nuclear effector of the Notch pathway and a DNA binding transcription factor that consists of three domains: NTD, BTD, and CTD. The NTD and CTD are structurally similar to RHR-N and RHR-C (Rel-homology region) domains, respectively. Mastermind proteins are transcriptional coactivator proteins that contain a short NTD (dnMAM) that is required to form a complex with CSL and Notch, and CTD that are required for interaction with the general transcription factor CBP/p300 and the cyclin-dependent kinase CycC/CDK8. Highlighted regions of Notch1, Jagged1, CSL, and Mastermind correspond to regions for which high-resolution structural data is available. (See Color Insert.)

ligands) domain, followed by a Delta–Serrate–LAG2 (DSL) domain. In some ligands, including all Serrate proteins and DLL1 in mammals, the DSL domain is followed by two variant EGF-like repeats also referred to as the "DOS" domain. All ligands then include a variable number of additional EGF-like repeats before the transmembrane segment and a C-terminal cytoplasmic tail (Fig. 2.1). Serrate family ligands are distinguished from Delta ligands by a larger number of EGF repeat domains and by the presence of an additional cysteine-rich domain homologous to the von Willebrand Factor C domain immediately preceding the transmembrane region.

As with the Notch receptors, the ligand proteins in the worm *C. elegans* also exhibit additional divergence from their counterparts in higher eukaryotes. The MNNL domain is not found at the N-terminus of the ligand molecules in *C. elegans*. In addition, the ligands in *C. elegans* typically contain either a DSL or a DOS domain, but not both. It has thus been proposed that the DSL ligands utilize soluble DOS co-ligands (or vice versa) to stimulate signals in receptor-bearing cells (Komatsu *et al.*, 2008).

Transduction of Notch signals relies on three key events: (i) ligand recognition, (ii) conformational exposure of the ligand-dependent cleavage site, and (iii) assembly of nuclear transcriptional activation complexes. Here, we will focus on the progress made in understanding the structural, biochemical, and mechanistic underpinnings of these three steps in canonical Notch signaling.

2. Notch–Ligand Interactions

Notch signal transduction normally requires that a canonical ligand on one cell binds to a Notch receptor on another cell in *trans* (Fehon *et al.*, 1990; Rebay *et al.*, 1991). On the other hand, expression of ligands and Notch receptors in the same cell in *cis* leads to the inhibition of signaling (de Celis and Bray, 1997; Klein *et al.*, 1997; Ladi *et al.*, 2005; Micchelli *et al.*, 1997). Whether direct association of receptor and ligand is required for *cis*-inhibition, and if so, whether the same binding interface is used for both *trans*-activation and *cis*-inhibition is not clear. It is known that the integrity of the *Abruptex* region in fly Notch (characterized by both gain and loss of function phenotypes and by complex complementation patterns), which includes EGF repeats 24–29, is required for *cis*-inhibition of Notch by ligands (de Celis and Bray, 2000). Subsequent genetic studies have also shown that *cis*-inhibition of ligand on Notch is not inhibited by Fringe modification of Notch (Glittenberg *et al.*, 2006). On the other hand, a recent report also identified a potential role for NEC as a *cis*-inhibitor of ligand activity as a signal inducer in the ligand-expressing cells, and argues that EGF repeats 10–12, but not the Abruptex region, are required for such *cis*-inhibition (Becam *et al.*, 2010).

It thus remains an open question whether *trans*-activation and the two identified *cis*-inhibtion binding modes differ or largely overlap.

Direct interaction between ligands and receptors in *trans* was first inferred from cell aggregation assays. This approach was also employed to show that EGF repeats 11–12 of Notch are necessary and sufficient to promote aggregation of Notch cells with both Delta- and Serrate-bearing cells (Fehon *et al.*, 1990; Rebay *et al.*, 1991). The interaction with Serrate-expressing cells was perceived as qualitatively weaker; though the explanation for this observation is unclear, it may have resulted in part from the influence of O-linked glycosylation on Notch–ligand interactions (see below).

A number of lines of evidence support the conclusion that the same region of Notch1 is also necessary for binding of its cognate Delta-like and Jagged ligands. First, knockin mice expressing mutated Notch1 receptors lacking EGF repeats 8–12 in place of wild-type receptors phenocopy Notch1 null mice (Ge *et al.*, 2008) even though surface receptor levels are not detectably altered, indicating that EGF repeats 8–12 are required for function. Second, when biotinylated minreceptors consisting of EGF repeats 11–13 of human Notch1 are captured with fluorescently labeled avidin, binding to DLL1-expressing cells is readily detectable using flow cytometry (Hambleton *et al.*, 2004). Finally, biochemical studies using purified proteins have shown that EGF repeats 11–14 of Notch1 bind to a fragment of DLL1 containing the DSL domain and the first three EGF repeats with an estimated K_d of 130 μM, based on equilibrium surface plasmon resonance measurements (Cordle *et al.*, 2008b).

It is widely assumed that the same region of mammalian Notches 2–4 is also responsible for ligand binding. Though ligand-binding studies of these receptors have been limited, a solid phase-binding assay has estimated an affinity of 0.7 nM for the binding of EGF repeats 1–15 from Notch2 to soluble Jagged1-Fc (Shimizu *et al.*, 1999). This stronger apparent affinity compared with the Notch1–DLL1 interaction described above may result from the use of a dimeric ligand (as an F_c domain fusion), differences in the identities of the interacting proteins, differences in the size of the fragments used for the binding studies, or some combination thereof. Studies in transfected cells have shown that the deletion of EGF repeats 10–11 (which aligns with EGF repeats 11–12 of *Drosophila* Notch and mammalian Notch1) prevents ligand binding by Notch3 (Peters *et al.*, 2004). However, studies investigating the ligand-binding activity of Notch4 are lacking.

2.1. Structural studies of ligand and receptor binding-active fragments

Structures have now been reported for a fragment of human Notch1-containing EGF repeats 11–13, and for a region of human Jagged1 that includes its DSL domain, its two DOS-subtype EGF repeats, and its third EGF repeat. These fragments bind to each other in a calcium-dependent manner, but it has not been possible to quantify the binding affinity for this interaction, which is weak *in vitro* (Cordle *et al.*, 2008a).

The structure of the Notch1 fragment was solved both by solution NMR methods (Hambleton *et al.*, 2004) and by X-ray crystallography (Cordle *et al.*, 2008a). In the X-ray structure, the three repeats adopt an elongated conformation, with an interdomain orientation defined by the coordination of a calcium ion between adjacent repeats and by the packing of a tyrosine residue from one repeat against an isoleucine residue from the next repeat (Fig. 2.2A). In the NMR structure, the position of EGF repeat 13 was less constrained with respect to the preceding repeat, but the same interdomain interactions were observed. Models for the entire ectodomain of Notch receptors have been proposed invoking rigidity at inter-repeat linkers that contain the consensus sequence for calcium coordination, and different degrees of intrinsic flexibility at linkers lacking a predicted calcium-binding site.

The X-ray structure of the Jagged1 receptor-binding region showed that this four-domain fragment is also found in a rod-like conformation (Cordle *et al.*, 2008a). The structure, when combined with a multiple sequence alignment to identify conserved residues, pointed the authors to a surface patch on the DSL domain as a potential site for binding to Notch (Fig. 2.2B). Mutation of residues at this interface interferes with the formation of receptor–ligand complexes, and causes loss-of-function to varying degrees in an *in vivo* assay in transgenic flies. In a comprehensive review, Kopan has also noted (Kopan and Ilagan, 2009) that missense mutations associated with inner ear malformations in mice (*headturner, slalom*, and *nodder*), as well as certain human mutations of Jagged1 associated with Alagille's syndrome or Tetralogy of Fallot, map onto the same face of the protein within the DOS repeats (Fig. 2.2B). Docking studies, combined with this observation, suggest that the contact interface between Jagged1 and Notch1 lies along an extended surface, but the nature of this interface remains unknown.

2.2. Influence of post-translational sugar modification on Notch-ligand interactions

O-linked glycosylation of Notch receptors is essential for Notch activity in both flies and mammals (Okajima and Irvine, 2002; Shi and Stanley, 2003) though a role for O-linked glycosylation of Notch has not yet been confirmed in *C. elegans*. The most well-characterized sugar modifications are initiated by an O-fucosyltransferase called O-Fut1 (POFUT1 in mammals), which transfers an O-fucose moiety to a serine or threonine

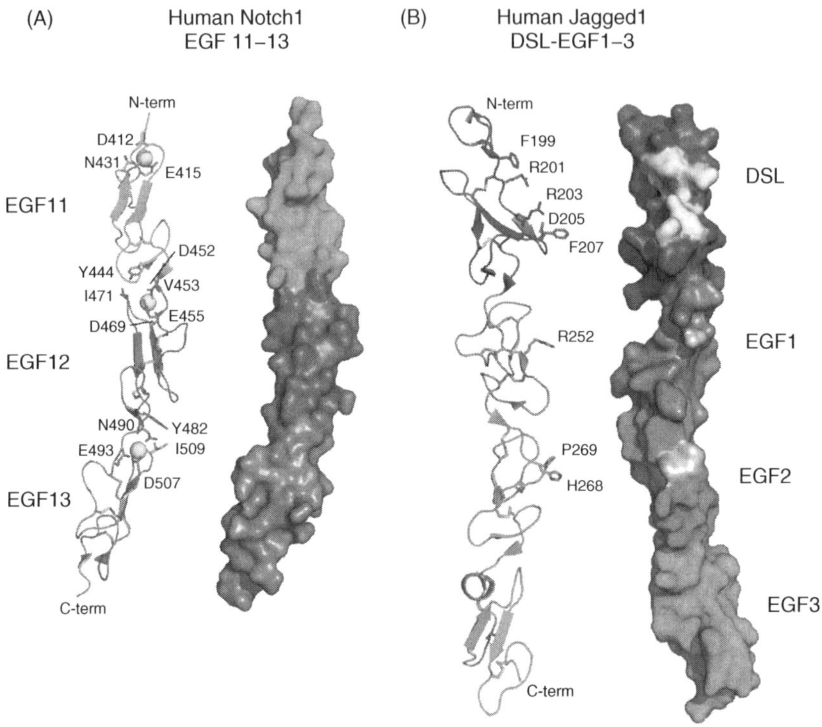

Figure 2.2 Structures of human Notch1 and Jagged1 ectodomain fragments (Cordle *et al.*, 2008a). (A). Ribbon (left panel) and surface (right panel) representations of EGF repeats 11–13 from human Notch1 (pdb ID code 2VJ3). EGF11 is blue, EGF12 is green, and EGF13 is orange. In the ribbon trace, calcium coordinating residues and interdomain contact residues are labeled and rendered as sticks. V453, a residue suggested to be in contact with Jagged1 based on NMR line broadening data, is also labeled and shown in stick representation. The calcium ions are shown as yellow spheres. In the surface representation, both V453 and G472, which was also implicated as a contact site from the NMR studies, are shown in red. T466, which undergoes O-linked glycosylation, is on the back face of the protein and is not visible in this view. (B). Ribbon (left panel) and surface (right panel) representations of the DSL-EGF3 region of human Jagged1 (pdb ID code 2VJ2). The surface view is rotated 90 degrees counterclockwise with respect to the ribbon diagram. The DSL domain is red, EGF1 (DOS repeat 1) is orange, EGF2 (DOS repeat 2) is green, and EGF3 is blue. DSL domain residues implicated in Notch binding (Cordle *et al.*, 2008a) are rendered as sticks (left panel), or in light yellow (right panel). A mutation analogous to F207A generates a null allele in flies; mutations analogous to F199A and R203A retain *cis*-inhibitory activity under some conditions, and mutations analogous to R201A and D205A exhibit only weak loss of function effects. Residues R252, H268, and P269, which lie on the same face of the protein and which correspond to sites that are mutated in human and murine developmental anomalies, are also rendered as sticks (left panel), or in white (right panel). The R252K mutation is found in Alagille's syndrome. The H268Q, and P269S mutations are found in the *nodder* and *slalom* mice, respectively. Loss-of-function mutations of cysteine and glycine residues (e.g. in Alagille's syndrome and other developmental syndromes such as Tetralogy of Fallot), most likely associated with substantial structural disruption of the protein, are also frequently found in this region of Jagged1. (See Color Insert.)

residue situated right before the third cysteine of an EGF repeat (Harris and Spellman, 1993; Okajima and Irvine, 2002; Shi and Stanley, 2003; Wang and Spellman, 1998; Wang *et al.*, 1996; Wang *et al.*, 2001). Loss-of-function mutations in the O-fucosyltransferase enzyme result in phenotypes resembling Notch loss of function in both flies and mammals (Okajima and Irvine, 2002; Shi and Stanley, 2003). Fringe glycosyltransferases (including Lunatic, Radical, and Manic Fringe in mammals) catalyze the β1,3 addition of N-acetylglucosamine to the primary O-fucose (Bruckner *et al.*, 2000; Moloney *et al.*, 2000), and the extension of this disaccharide into a tetrasaccharide in mammals is catalyzed by the sequential action of a β1,4 galactosyl transferase and either an α-2,3 or -2,6 sialyltransferase (see Chapter 4 for a review).

Of the 23 putative consensus sites for O-fucosylation on human Notch1 (Luther and Haltiwanger, 2009), 13 are evolutionarily conserved (Shao *et al.*, 2003), including threonine 466 within EGF repeat 12 from the central ligand-binding region. A T466A mutation, which eliminates this conserved O-fucosylation site from EGF repeat 12, results in a hypomorphic allele that is embryonic lethal when paired with a null allele, indicating that O-fucose modification within the critical ligand-binding region of the mammalian receptor is needed for optimal receptor activity (Ge and Stanley, 2008).

Drosophila Fringe potentiates the ability of Notch to respond to Delta, but inhibits the responsiveness of Notch to Serrate (Panin *et al.*, 1997). The functional effects of Fringe modification in the *Drosophila* system correlate with the influence of Fringe on the binding of Serrate and Delta to fly Notch : Fringe modification of affinity-purified Notch receptors increases its ability to recover Delta, but interferes with its ability to recover Serrate (Xu *et al.*, 2007).

It is generally believed that Fringe modification of mammalian Notch receptors similarly modulates their ability to bind Jagged and Delta-like ligands, but the increased number of receptors, ligands, and fringe enzymes results in the potential for vastly greater complexity (Kopan and Ilagan, 2009). In the case of Notch1, the majority of data support the idea that fringe modification potentiates signaling from Delta family ligands, but limits the signaling activity of Jagged ligands or leaves them unchanged. On the other hand, studies with Notch2 have yielded conflicting results (Hicks *et al.*, 2000; Shimizu *et al.*, 2001). More generally ligand and receptor fragments that have been purified without post-translational glycosylation are capable of binding to one another *in vitro* (Cordle *et al.*, 2008a; Cordle *et al.*, 2008b), suggesting that the influence of O-linked glycosylation on both Notch conformation and ligand-binding activity are likely to be subtle, rather than black and white. Clearly, a more complete understanding of the mechanism(s) by which fringe modulates ligand responsiveness awaits further and more complete structural and biochemical analysis of receptor–ligand complexes.

More recently, an O-glucosyltransferase called Rumi was identified in *Drosophila* (Acar *et al.*, 2008). Rumi loss of function is associated with Notch-like phenotypes in the fly. The enzyme is capable of transferring glucose to serine residues in the EGF repeats of Notch carrying a consensus sequence of C (I) -X-S-X-P-C (II) and the loss of Rumi activity under restrictive conditions in temperature-sensitive mutants results in the intracellular accumulation of Notch receptors. Thus, the O-glucosylation of Notch by Rumi seems to enable receptor folding and/or trafficking to permit export and subsequent signaling at the cell membrane.

2.3. Summary and outstanding questions

Though remarkable progress has been made in the past several years, a number of key questions revolving around the structure and biochemistry of Notch receptor–ligand complexes remain. What are the overall architectures of the full-length ligands and receptor proteins? Are they extended rods like the structures of the smaller binding-active fragments might suggest or is there interdomain flexibility at some of the linkers connecting the tandem EGF-like repeats? Why are the affinities of the small fragments for one another so weak, and difficult to measure, yet ligand- and receptor-bearing cells adhere tightly to one another, as judged by cell aggregation assays (Cordle *et al.*, 2008a; Fehon *et al.*, 1990; Rebay *et al.*, 1991) and atomic force microscopy studies of cell–cell contact (Ahimou *et al.*, 2004)? The smaller, recombinant Notch fragments lack the O-linked sugar modifications required for productive Delta signaling *in vivo*, but these fragments exhibit low affinity in their interactions with Jagged as well. Perhaps larger regions of the receptors and the ligands are in contact with one another when the full-length proteins interact, leading to higher affinity. Another possibility, not mutually exclusive, is that clustering of receptors and ligands results in a multivalency or avidity effect that enhances complex stability *in vivo.*

3. The Activation Event

3.1. The LNR domain prevents metalloprotease access to the S2 site

The key regulated step in the activation of Notch receptors is proteolysis at S2, which creates the substrate for intramembrane proteolysis at S3 by gamma secretase. Early genetic and molecular studies pointed to the importance of the juxtamembrane region of Notch receptors as an important negative regulatory region (NRR) controlling signaling (Kopan *et al.*, 1996; Lieber *et al.*, 1993; Rebay *et al.*, 1993). This NRR, which consists of the three LNR modules and the juxtamembrane "heterodimerization (HD) domain" that is divided by S1 cleavage during maturation, constitutes the activation switch of the receptor.

More recent work has since established that the LNR repeats prevent metalloprotease cleavage of mammalian Notch receptors in the absence of ligand, and that the HD domain is sufficient to maintain the non-covalent association of the two Notch subunits after division by furin cleavage at S1 (Sanchez-Irizarry *et al.*, 2004). Moreover, the vast majority of the activating mutations in human T-ALLs alter residues in the HD domain, further highlighting the importance of this region of Notch receptors as a critical regulatory switch (Weng *et al.*, 2004).

Each LNR module is 40 residues long and contains three disulfide bonds that pair in a characteristic pattern: cys I–cys V, cys II–cys IV, and cys III–cys VI (Aster *et al.*, 1999), whereas the HD is approximately 150 residues long and contains a single disulfide bond between two cysteines that lie between the S1 and the S2 sites, within ten residues of each other in the primary sequence. The NMR structure of the first LNR module from human Notch1 (Vardar *et al.*, 2003) shows that a prototypical repeat has an irregular fold with little secondary structure, constrained by the three disulfide bonds and by ligation of a calcium ion by acidic residues that are highly conserved among the various LNR modules among different species.

The X-ray structure of the Notch2 NRR in its autoinhibited conformation first revealed how metalloprotease cleavage of Notch receptors is prevented prior to ligand stimulation (Gordon *et al.*, 2007). More recently, the X-ray structure of the Notch1 NRR was also solved in its autoinhibited conformation (Gordon *et al.*, 2009a). The overall architecture of the Notch1 and Notch2 NRR structures exhibits remarkable overall similarity, providing strong support for the conclusion that the structural basis for autoinhibition applies generally to all Notch receptors.

Overall, the NRR adopts a compact conformation, with the three LNR modules wrapped around the HD domain resembling a mushroom cap covering its stem (Fig. 2.3A). Extensive interactions between the HD domain and the LNR A-B linker, the LNR-B domain, and the LNR-C domain combine to bury a total of approximately 3000 $Å^2$ in the interdomain interface for both Notch1 and Notch2 and provide stability to the autoinhibited conformation.

The HD domain in the core of the structure adopts an alpha–beta fold that bears structural similarity to the SEA domains found in mucins (Macao *et al.*, 2006; Maeda *et al.*, 2004). In human Notch1, the S2 cleavage site lies on the terminal beta strand of the HD domain, at the amide bond connecting A1720–V1721 (A1710–V1711 in murine Notch1; Brou *et al.*, 2000; Mumm *et al.*, 2000). A three residue hydrophobic plug from the linker connecting the first and the second LNR modules, anchored by the highly conserved Leu 1482, directly prevents metalloprotease access by sterically occluding the S2 site (Fig. 2.3B).

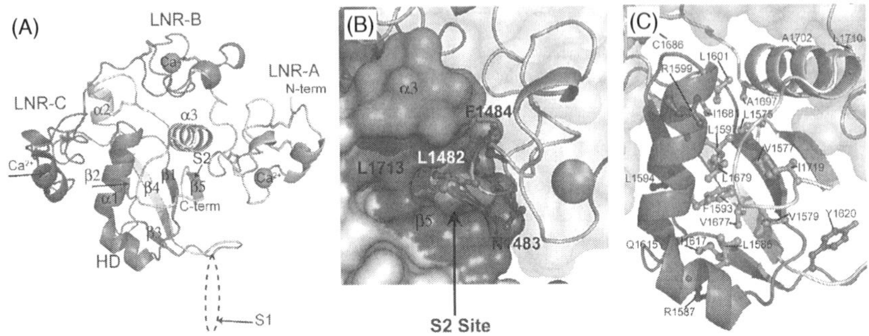

Figure 2.3 Structure of the Notch1 NRR in its autoinhibited conformation (pdb ID code 3ETO). (A). Ribbon representation. The three LIN12/Notch repeats are colored different shades of pink and purple. The HD domain is colored in light blue and turquoise N- and C-terminal, respectively, to the furin-cleavage loop (S1). The three bound Ca^{2+} ions are green, and the ten disulfide bonds are yellow. The positions of S1 and S2 cleavage are indicated with arrows. (B). Close-up around the metalloprotease cleavage site. The HD domain is rendered as a molecular surface, with helix three in turquoise and the terminal beta strand in dark blue. L1713, which lies in the connecting loop between these two secondary structural elements is colored green. V1723, which is the residue C-terminal to the scissile bond, is colored bright orange. The LNR domain is rendered in ribbon representation, with the key residues of the autoinhibitory plug (L1482, N1483, and F1484) rendered as red sticks. The calcium ion bound to LNR-A is shown as a green sphere. (C). Sites of tumor associated mutations. Residues in the hydrophobic core of the HD domain are colored green, residues at the interface between the HD and LNR domains are orange, and residues that are partially exposed in the structure are colored purple. Panels A and C are adapted from Gordon *et al.* (2009a) and are used with permission. (See Color Insert.)

The structure of the Notch1 NRR also made it possible to directly identify the residues harboring T-ALL-associated mutations. Tumor-associated mutations in the HD domain of Notch1 map predominantly to the hydrophobic interior of the domain, suggesting that tumor-associated mutations increase metalloprotease susceptibility by decreasing the thermodynamic stability of the NRR and/or increasing kinetic exposure of the S2 site (Fig. 2.3C). Concurrent biochemical studies of isolated NRRs took advantage of the maturation cleavage step at site S1 to probe the stability of Notch1 NRRs with tumor-associated mutations. These studies found that the mutated forms of the NRR are less stable to denaturant-induced subunit dissociation, arguing that strategies designed to increase the stability of mutated forms of the NRR may have therapeutic potential in T-ALL (Malecki *et al.*, 2006).

To facilitate crystallization, the original NRR structures were determined using proteins that were modified by deletion of the S1-cleavage loop. A consensus site for cleavage by furin-like proteases is present in *Drosophila* Notch and mammalian Notches 1–3, though the overall length and sequence of this loop is otherwise poorly conserved. The X-ray structure of the Notch1 NRR, determined after furin cleavage *in vitro*, exhibits few detectable changes when compared with the X-ray structure of the Notch1 NRR determined after deletion of the cleavage loop. NMR studies of the isolated HD domain of Notch2 suggest that its S1-cleavage loop is poorly ordered in the intact domain and that S1 cleavage exerts little effect on the overall conformation of the isolated HD domain (Gordon *et al.*, 2009b).

Even though S1 cleavage appears to have little effect on the structure of the NRR, mutations that remove the S1 site, either by point mutation or by loop deletion, have different effects on receptor transport and function, depending on the identity of the receptor. Deletion of the S1-cleavage loop from Notch2 prevents processing by furin *in vitro* and eliminates detectable S1 processing in cells, but these receptors are fully competent for delivery to the cell surface and retain wild-type levels of signaling in reporter gene assays. In contrast, point mutations and internal deletions (of up to 47 residues) eliminating the S1 site of Notch1 substantially decrease the fraction of receptors that reach the cell surface. Nevertheless, it is clear that some receptors can be detected at the cell surface and that they are competent to convey signals in reporter gene assays (Gordon *et al.*, 2009b). Mutations preventing S1 cleavage of *Drosophila* Notch produce a receptor that fails to reach the cell surface, but it is not clear whether these receptors are folding competent because the residues mutated align with positions in the core of the HD domain of the mammalian proteins, and structural studies of the *Drosophila* Notch NRR have not yet been performed (Lake *et al.*, 2009).

3.2. How does ligand engagement overcome autoinhibition?

The ligand-binding region of the receptor lies more than 1000 residues away from the metalloprotease cleavage site. Thus, a key question that has drawn much speculation over many years is: how does ligand binding at such a distant site overcome autoinhibition? Early studies suggested that ligand binding might promote the dissociation of S2-resistant Notch oligomers into metalloprotease-sensitive monomers, but more recent studies have challenged this view (Vooijs *et al.*, 2004). In addition, the NRR structures solved to date show that the autoinhibited conformation is an intrinsic property of the protein monomer, and purified NRR monomers appear to be intrinsically resistant to metalloprotease cleavage (Cheryl Sanchez-Irizarry and SCB, unpublished data). These findings indicate that the changes in receptor oligomerization are unlikely to play a direct role in exposing the metalloprotease cleavage site.

The structure of the "off state" shows that the physical displacement of the LNR modules must occur in order to reveal the metalloprotease cleavage site. Experiments testing nested NOTCH1 and NOTCH2

NRR deletions confirmed that LNR-A, the LNR-A-B linker, and LNR-B must all be removed before significant activation occurs, consistent with the prediction from the structure that ligand-mediated S2 cleavage is preceded by large-scale displacement of the LNR region protecting the S2 site (Gordon *et al.*, 2007). In addition, the structure of the catalytic domain of metalloproteases such as TNF-α converting enzyme (TACE) shows that the active site of the protease lies in a deep cleft, which cannot gain access to the Notch S2 site after mere stripping of the LNR modules away from the HD domain (Maskos *et al.*, 1998).

The leading model for ligand-induced activation proposes that endocytosis of bound ligand exerts a mechanical force on the receptor, pulling the protective LNR cap away from the HD domain to promote exposure of the S2 site (for example, see (Kopan and Ilagan, 2009)). Consistent with this idea, genetic studies have shown that endocytosis in the ligand-expressing cell is required to convey Notch signals into the signal-receiving cell (Seugnet *et al.*, 1997). A number of more recent studies have elucidated a critical role for ligand ubiquitylation in this process, implicating specific lysine-containing motifs of Serrate family ligands as ubiquitylation targets (Glittenberg *et al.*, 2006) and E3 ligases of the Neuralized and Mindbomb families as the enzymes responsible for ubiquitin transfer in different cellular contexts (see, e.g. Pitsouli and Delidakis, (2005)).

Cellular assays have also shown that *trans*-endocytosis of Notch receptor ectodomains into the ligand-bearing cells correlates with the transduction of signals in signal-receiving cells (Nichols *et al.*, 2007; Parks *et al.*, 2000). Soluble ligands generally fail to activate Notch receptors (Sun and Artavanis-Tsakonas, 1997), but when attached to beads (Varnum-Finney *et al.*, 1998) or to a solid surface (Varnum-Finney *et al.*, 2000) they can bypass the ligand endocytosis requirement to activate signaling. The large interface area between the LNR modules and the HD domain also suggests that a substantial amount of energy is required to overcome autoinhibition.

Although all of these observations are consistent with the force-based model for activation, several perplexing questions remain unresolved. Must ligand drive dissociation of the extracellular and transmembrane subunits of Notch receptors to initiate activating proteolysis, as some studies have proposed (Nichols *et al.*, 2007)? If monovalent receptor–ligand interactions are weak, as affinity measurements for mammalian minireceptors binding to DLL1 and Jagged-1 ligands suggest, are "catch bonds" created to enable the application of force by ligand onto the receptor (Thomas, 2009; Thomas *et al.*, 2008)? Alternatively, is clustering of ligand–receptor complexes at sites of cell–cell contact required to drive Notch proteolysis? If so, what are the biochemical and cellular events that promote clustering, and how are they related to the requirement for endocytosis in signal-sending cells? Does the association of ligand with receptor promote formation of oligomeric receptor–ligand assemblies, and if so, are there specific regions of the receptor and/or ligand molecules that lead to ordered self-assembly of complexes, analogous to the self-assembly of fibrillin, which contains many EGF-like repeats in its extracellular domain? These and other questions about Notch activation should be fertile ground for future study using biochemical and biophysical probes of conformation and signal transduction.

4. Effector Function

All canonical Notch signals are ultimately transduced into changes in gene expression via the nuclear effector of the pathway, CSL. CSL is a DNA-binding protein that regulates the transcription of genes that are responsive to Notch signaling. The capacity of CSL to act as either an activator or a repressor is dependent upon whether it is bound by transcriptional coactivator or corepressor proteins, respectively.

Genetic studies in the model organisms *Drosophila melanogaster* (Fortini and Artavanis-Tsakonas, 1994) and *C. elegans* (Christensen *et al.*, 1996) first established that CSL, in conjunction with Notch, functions as a transcriptional activator; however, these initial findings did not suggest that CSL also plays a role in transcriptional repression. Experiments in cultured mammalian cells first demonstrated that RBP-J—the mammalian CSL ortholog—functions as a repressor (Dou *et al.*, 1994; Hsieh and Hayward, 1995) and that the addition of activated forms of Notch could convert CSL from a repressor to an activator of transcription (Hsieh *et al.*, 1996; Waltzer *et al.*, 1995).

Additional studies identified multiple transcriptional corepressor proteins that interact with RBP-J (Hsieh *et al.*, 1999; Kao *et al.*, 1998; Zhou *et al.*, 2000a), thereby linking RBP-J to the repression machinery in the nucleus. Subsequent experiments in flies with null alleles of Su(H)—the CSL ortholog in flies—demonstrated that at some target genes loss of Su(H) resulted in ectopic gene expression (Furriols and Bray, 2001; Morel and Schweisguth, 2000; Muller and Littlewood-Evans, 2001), suggesting that Su(H) acts as a repressor at those sites. While there is not an abundance of genetic data in worms that suggest LAG-1—the CSL ortholog in worms—also functions as a transcriptional repressor, there is at least one study that demonstrates loss of LAG-1 results in ectopic expression from the gene hlh-6 (Ghai and Gaudet, 2008). Taken together, these studies have led to current working models in the field, reviewed in Kovall (2008) and Gordon *et al.* (2008), that suggest prior to pathway activation, CSL functions as a repressor by directly interacting with transcriptional corepressor proteins. Activation of the pathway leads to translocation of the intracellular domain of Notch (NICD) from the plasma membrane to the nucleus, where it directly binds CSL. The binding of NICD to CSL is thought to displace corepressors from CSL and allows for recruitment of the transcriptional coactivator Mastermind (MAM) to the

CSL–NICD binary complex. Formation of the CSL–NICD–MAM ternary complex results in the activation of transcription from Notch target genes. Structural studies have illuminated the molecular details of these Notch transcription complexes, and in conjunction with more recent biophysical and biochemical studies have, in some cases, validated this paradigm, and in other cases, challenged it.

4.1. The structure of CSL

All CSL proteins are composed of a highly conserved core region—approximately 420 amino acids, flanked by additional N- and C-terminal extensions of varying lengths (Fig. 2.1). These extensions can vary within a species due to alternative splicing, but are not conserved among different species orthologs, nor have any functions yet been ascribed to these regions. Primary sequence analysis of these N- and C-terminal extensions suggests that these regions are largely devoid of any secondary structure and contain multiple segments of low compositional complexity. While most metazoans encode a single CSL protein within their genomes, interestingly, vertebrates encode an additional CSL paralog termed RBP-L (RBPJ-like) (Minoguchi *et al.*, 1997), which contains the conserved CSL core, but appears to function as a constitutive transcriptional activator independent of Notch (Beres *et al.*, 2006). Also of note, putative CSL orthologs have been identified in several fungal species (Prevorovský *et al.*, 2007), e.g., *Schizosaccharomyces pombe*; however, in light of the fact that fungi do not encode other Notch pathway components, the structure and function of these proteins remains to be determined.

Not surprisingly, all X-ray structures of CSL complexes solved (Friedmann and Kovall, 2009; Friedmann *et al.*, 2008; Kovall and Hendrickson, 2004; Nam *et al.*, 2006; Wilson and Kovall, 2006) to date include only the conserved core of CSL, and it seems unlikely, due to the lack of secondary structure, that structural studies of full-length CSL would provide any additional structural or functional insights. At the primary sequence level, CSL proteins from nematodes (*C. elegans and Caenorhabditis briggsae*) are the most divergent members of the group with approximately 54% sequence identity between worm and mammalian CSL orthologs; whereas for comparative purposes, there is approximately 75% sequence identity between fly and mammalian orthologs, and >90% identity between Zebrafish or Xenopus with mammalian CSL. Despite these differences, the structures of core CSL from worms (Friedmann *et al.*, 2008; Kovall and Hendrickson, 2004; Wilson and Kovall, 2006) and mammals (Friedmann *et al.*, 2008; Nam *et al.*, 2006) are remarkably similar, suggesting that the structures of all CSL orthologs have a similarly conserved fold.

As shown in Figs. 2.1 and 2.4A, the conserved core of CSL consists of three structural domains termed the NTD (N-terminal domain), the BTD

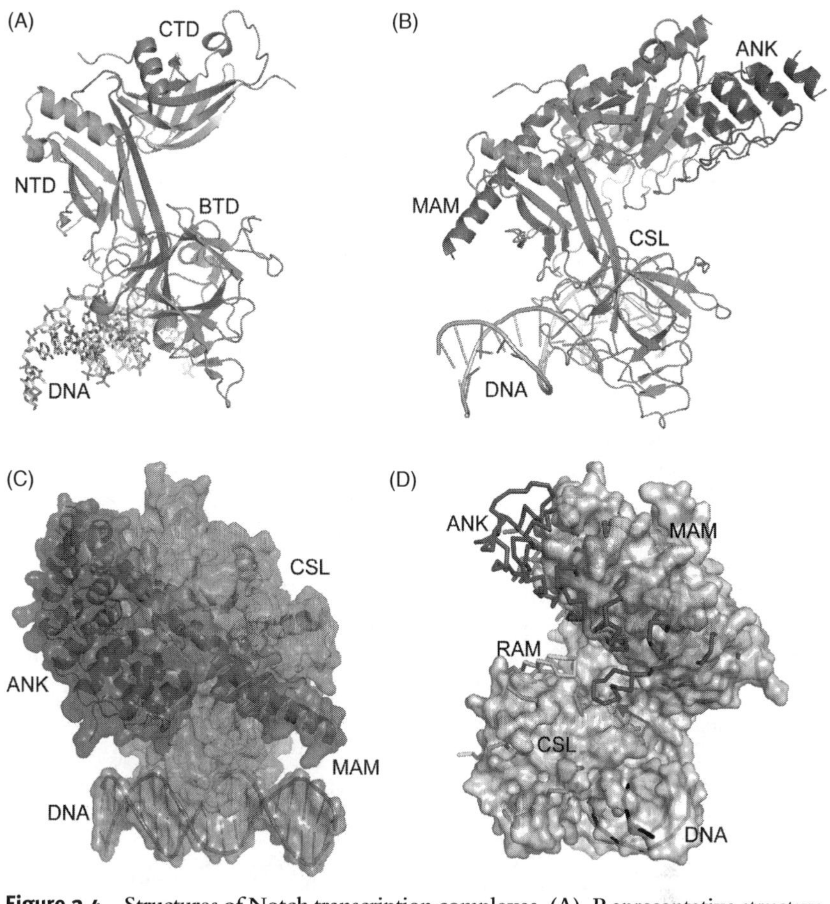

Figure 2.4 Structures of Notch transcription complexes. (A). Representative structure of a CSL-DNA complex. Panel shows ribbon diagram of mouse CSL bound to a cognate DNA site from the hes-1 promoter element (PDB ID: 3IAG) (Friedmann and Kovall, 2009). The NTD and BTD, which interact with the DNA, are colored cyan and green, respectively; the CTD is colored orange. The DNA is in a stick representation with carbon, oxygen, nitrogen, phosphorous atoms colored yellow, red, blue, and orange, respectively. (B). CSL-NICD-Mastermind active transcription complex. Panel shows ribbon diagram of the human ternary complex structure (PDB ID: 2F8X) composed of CSL (green), the ANK domain of NICD (blue), and Mastermind (red), bound to DNA (grey ribbon) (Nam *et al.*, 2006). (C). Panel shows a surface representation of a CSL–NICD–MAM complex, highlighting the groove formed by the CTD–ANK interface and the NTD that binds the N- and C-terminal regions of Mastermind, respectively. (D). Structural overlay of worm (PDB ID: 2FO1) and human (PDB ID: 2F8X) ternary complex structures. The human complex is depicted as a transparent grey surface and the worm proteins CSL, RAM, ANK, and MAM are colored green, yellow, blue, and red, respectively. Panel shows the more compact nature of the worm complex compared to the human. Magenta arrow points to an insertion in the fifth ankyrin repeat of LIN-12 that is conserved in nematodes, but not other metazoans, and makes additional contacts with the BTD of CSL. (See Color Insert.)

(β-trefoil domain), and the CTD (C-terminal domain) (Kovall and Hendrickson, 2004); due to their structural similarity with the Rel Homology Region (RHR) family of transcription factors (Kovall and Hendrickson, 2004; Nam *et al.*, 2003), the NTD and CTD are also commonly referred to as RHR-N and RHR-C domains, respectively. Overall, the structure of CSL is primarily composed of β-strands with very little α-helical content. The three domains of CSL are integrated into one overall fold, primarily through a long β-strand structural element that spans all three domains (Fig. 2.4A, magenta colored strand). While the NTD and CTD are structurally similar to the corresponding RHR domains in other Rel proteins, e.g. NF-κB and NFAT, the overall topology of CSL is quite unique amongst Rel family members—the CSL fold has an insertion of the BTD between the NTD and the CTD, which does not occur in other Rel proteins, and the relative three-dimensional arrangement of these domains is particular to CSL (Kovall and Hendrickson, 2004). The BTD derives its name from the 12-stranded capped β-barrel structure it forms that was first observed in cytokine structures such as fibroblast growth factor and interleukin-1 (Murzin *et al.*, 1992).

CSL binds DNA as a monomer, in contrast to other Rel proteins, which bind DNA as homo- or heterodimers (Chung *et al.*, 1994; Nam *et al.*, 2003). Early work identified a consensus DNA sequence –C/tGTGGGAA– for CSL binding (Tun *et al.*, 1994); however, more recent quantitative binding measurements have shown that the relative selectivity for consensus over nonconsensus bases at certain sites is modest (Friedmann and Kovall, 2009). A more comprehensive study using protein-binding microarrays or other methods to rank site preferences for CSL binding would be a welcome addition to current knowledge.

In the structures of all complexes containing both CSL and DNA, the NTD and BTD of CSL form an extensive electropositive surface that provides both specific and nonspecific contacts with the DNA (Kovall and Hendrickson, 2004). In a manner very similar to other Rel proteins, the NTD inserts a β-hairpin loop into the major groove of DNA to specifically recognize the –GGGA– base pairs in the second half of the consensus-binding site. Remarkably, unlike other Rel proteins, the CTD of CSL does not function in DNA binding and the BTD provides additional DNA specificity via a loop that inserts into the minor groove of DNA, recognizing the –C/tG– base pairs that comprise the first half of the consensus-binding site.

4.2. Structure of the Notch transcriptionally active complex

Upon pathway activation, the NICD translocates to the nucleus where it directly binds CSL. NICD is composed of multiple modular domains (Fig. 2.1), these include from N- to C-terminus—RAM (RBP-J associated molecule), ANK (ankyrin repeats), TAD, and a PEST sequence (*for rich in* proline, glutamate, serine, and threonine); however, only the RAM and ANK domains are necessary and sufficient for interacting with CSL (Nam *et al.*, 2003; Tamura *et al.*, 1995). Once NICD binds CSL this complex is competent for binding a transcriptional coactivator of the Mastermind family (MAM) (Doyle *et al.*, 2000; Petcherski and Kimble, 2000; Wu *et al.*, 2000). MAM proteins are generally about 1000 residues in length, in which only a short NTD of ~60 amino acids is required to bind CSL–NICD (Petcherski and Kimble, 2000; Wu *et al.*, 2000) (Fig. 2.1). Its long C-terminal tail is polyglutamine rich, binds the coactivator CBP (CREB binding protein), and is critical for activating transcription, as truncation of this region results in dominant-negative forms of Mastermind (Fryer *et al.*, 2002; Wallberg *et al.*, 2002; Weng *et al.*, 2003).

Formation of the CSL–NICD–MAM ternary complex results in a sequence of events that ultimately activate transcription from Notch target genes. As described in the ternary complex structures from both worm and human Notch pathway components (Nam *et al.*, 2006; Wilson and Kovall, 2006), the RAM and ANK domains of NICD interact with the BTD and CTD of CSL, respectively (Fig. 2.4B). The ANK domains of all Notch receptors contain seven iterative ankyrin repeat motifs (Zweifel and Barrick, 2001; Zweifel *et al.*, 2003) (Fig. 2.1). An irregular N-terminal capping repeat was observed in the worm ternary complex structure (Wilson and Kovall, 2006), and although the sequence identity corresponding to this region between Notch orthologs is low, presumably the N-terminal capping repeat is a structural motif found in all Notch proteins. Several structures of the isolated ANK domains from mammals and flies have also been determined (Ehebauer *et al.*, 2005; Lubman *et al.*, 2005; Nam *et al.*, 2006; Zweifel *et al.*, 2003); however, in all of these structures the N-terminal cap is missing from the crystallized protein and the first repeat is structurally disordered. It is thus unclear whether the folding of the N-terminal repeats is coupled to formation of the CSL–NICD–MAM ternary complex, and if so, what influence the induced folding of this region exerts on effector function.

The most striking feature of the ternary complex structures is the elongated α-helix formed by the Mastermind proteins (Fig. 2.4B, C). In both the human and the worm complexes, the Mastermind helix exhibits a distinctive bend, in which its N- and C-terminal helical regions bind a continuous groove formed by the CTD–ANK interface and a β-sheet from the NTD of CSL, respectively (Nam *et al.*, 2006; Wilson and Kovall, 2006) (Fig. 2.4C).

As with the core CSL structures described above, the worm and human CSL–NICD–MAM X-ray structures only contain the regions of NICD and Mastermind that are required for ternary complex formation, i.e., these complex structures are missing the long C-terminal tail regions

(>500 residues) of NICD and Mastermind (Fig. 2.1). Primary sequence analysis of the NICD and MAM C-terminal tails indicates that these regions contain large blocks of low compositional complexity, suggesting that they are largely disordered. While one study has biophysically characterized full-length NICD constructs (Kelly *et al.*, 2007), gleaning some functional data, it is likely that in the absence of a binding partner the low-complexity regions of NICD and Mastermind are random coil in solution, which precludes any meaningful analysis by X-ray crystallography.

While the overall architecture of the worm and human ternary complexes has been remarkably conserved through evolution, there are a number of notable structural differences between the two complexes (Gordon *et al.*, 2008; Kovall, 2007). First, the worm ternary complex structure contains the RAM and ANK domains of NICD, but the human complex only contains ANK. Previous *in vitro* and *in vivo* studies demonstrated that the ANK domain of NICD could form a ternary complex with CSL and Mastermind (Aster *et al.*, 2000; Jeffries and Capobianco, 2000; Nam *et al.*, 2003; Roehl *et al.*, 1996), and activate transcription, though less efficiently than NICD constructs that contain both RAM and ANK in some studies [(Jarriault *et al.*, 1995) and see chapter 7]. Interestingly, a β-hairpin motif in the BTD that binds RAM, becoming structurally ordered, forms a very similar structural element in the human ternary complex in the absence of RAM (Nam *et al.*, 2006; Wilson and Kovall, 2006). Second, the worm ternary complex forms a more compact structure overall, in which the BTD and CTD squeeze toward one another, as much as 15 Å, when compared with structures of isolated CSL–DNA complexes. In contrast, the human ternary complex does not assume this compact form and the three domains of CSL are in a more extended arrangement, resembling the conformation of CSL in the worm CSL–DNA complex (Fig. 2.4D). Third, there are striking differences in the protein side chain interactions at the CTD–ANK interface for the two ternary complexes, even between highly conserved protein side chain pairs (Kovall, 2007). Initially, it was speculated that the RAM domain may be the source of the structural differences observed between the human and worm complexes (Barrick and Kopan, 2006); however, subsequent structural studies of CSL–RAM complexes (described below) and of human CSL–NICD–MAM ternary complex crystals, which were grown with an 18-residue peptide corresponding to RAM (Yunsun Nam and SCB unpublished data), argue that RAM binding does not account for the observed differences (Friedmann *et al.*, 2008). Whether these structural differences are functionally relevant, organism specific, or even a consequence of crystallization remain open questions that have yet to be addressed. Certainly, additional structures of CSL from different model organisms, e.g., *Drosophila, Xenopus, or Zebrafish*, would provide some insights into these questions.

4.3. Assembly of the Notch transcription complex

The structures of the worm and human CSL–NICD–MAM ternary complexes enabled a thorough exploration of the biochemistry of assembly for the transcriptionally active nuclear complexes (Bertagna *et al.*, 2008; Del Bianco *et al.*, 2008; Friedmann *et al.*, 2008; Lubman *et al.*, 2007). A variety of quantitative binding assays by several research groups confirmed that the RAM domain of NICD forms a high-affinity interaction with the BTD of CSL, but the isolated ANK domain interacts very weakly with CSL—almost beyond the detection limits of the techniques utilized (Del Bianco *et al.*, 2008; Friedmann *et al.*, 2008). However, an elegant theoretical study that modeled RAM as a *worm-like* chain suggested that the tethering of ANK to CSL, via RAM, would increase the local concentration of NICD to millimolar levels—counteracting the low affinity between ANK and CSL—and would position NICD to ideally interact with the CTD of CSL (Bertagna *et al.*, 2008).

Structurally, several additional complexes were determined, including two worm CSL-RAM structures and a coregulator-free murine CSL structure (Friedmann *et al.*, 2008) (Fig. 2.5). Comparison of these structures with previous ones suggested that RAM binding to the BTD is associated with a conformational change at a distant site within the NTD of CSL; this movement eliminates steric restraints expected to interfere with binding of Mastermind to the CSL–NICD binary complex. Biochemical gel-shift assays using worm components showed that a peptide corresponding to

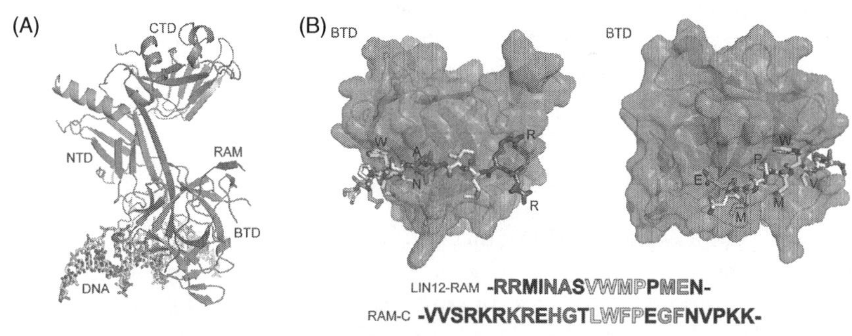

Figure 2.5 CSL-RAM structure. (A). Ribbon diagram of a worm CSL–RAM (LAG1–LIN-12) complex bound to DNA (PDB ID: 3BRD) (Friedmann *et al.*, 2008). RAM, colored red, exclusively interacts with the BTD. The three domains of CSL are colored as in Fig. 2.4A. (B). Zoomed-in views of the BTD–RAM interface, highlighting the sequence differences between the LIN-12 RAM peptide used in the structural studies (Friedmann *et al.*, 2008) and the RAM consensus peptide (RAM-C) used in the binding study by Johnson *et al.* (2009). Basic, HG, hydrophobic tetrapeptide (ΦWΦP), and GF motifs that were shown to be imporant for binding to BTD are colored blue, magenta, yellow, and cyan, respectively. While the basic region and ΦWΦP are conserved in worm RAM the HG and GF motifs are not and correspond to residues NA and ME, respectively. (See Color Insert.)

RAM could act *in trans,* promoting the assembly of complexes containing CSL, ANK, and MAM, as compared with ternary complexes formed without RAM (Friedmann *et al.*, 2008). Interestingly, though the formation of worm ternary complexes was absolutely dependent on RAM in these *in vitro* binding studies, the effect of including RAM on the formation of mouse ternary complexes was more subtle, highlighting yet another molecular difference between organisms for the assembly of the CSL–NICD–MAM ternary complex. Taken together, a remarkably lucid stepwise model for the assembly of the active transcription complex arises from these studies—first, the RAM domain mediates the high-affinity interaction between CSL and NICD, which likely targets NICD to CSL in the nucleus. In some species and/or contexts RAM binding is associated with an allosteric change in the NTD of CSL, removing steric restraints to permit Mastermind binding; second, the increased local concentration of ANK mediates formation of its weak interaction with the CTD of CSL; and third, these two molecular events between CSL and NICD create the binding site for Mastermind to interact with the complex, as Mastermind does not interact pairwise with either CSL or NICD.

4.4. The CSL–RAM Interaction

Largely due to its tractability in biophysical binding assays, the complex formed between the RAM domain of NICD and the BTD of CSL has been the most scrutinized interaction of the CSL–NICD–MAM ternary complex (Bertagna *et al.*, 2008; Del Bianco *et al.*, 2008; Friedmann *et al.*, 2008; Lubman *et al.*, 2007). Prior to its interaction with CSL, multiple studies have shown that RAM is an unstructured random coil (Bertagna *et al.*, 2008; Nam *et al.*, 2003). Binding of RAM to CSL is enthalpically driven with K_d values ranging from 30 nM to about 1 μM, depending on the specific complex studied, the conditions of binding, and the method of detection (Del Bianco *et al.*, 2008; Friedmann *et al.*, 2008; Lubman *et al.*, 2007).

The affinities of the RAM domains of mammalian Notch1-4 for BTD are similar (Lubman *et al.*, 2007). Differences in the affinity of RAM for CSL, therefore, cannot account for the differences in transcriptional potencies observed for the mammalian Notch paralogs (Ong *et al.*, 2006). There are also orthologous differences in CSL-RAM binding, as an approximately 50-fold higher affinity was observed for RBPJ–RAM complexes when compared with LAG1–RAM complexes (Friedmann *et al.*, 2008). Interestingly, the differences in affinity mapped exclusively to the CSL ortholog—RBP-J bound LIN-12 and GLP-1 RAM with similar affinity as mouse Notch1 RAM.

A comprehensive binding study of the BTD–RAM complex analyzed the energetic contribution of four conserved stretches of residues within RAM, and addressed the question of whether RAM and an Epstein–Barr virus nuclear antigen 2 (EBNA2)-derived peptide bind to independent or overlapping sites on RBP-J (Johnson *et al.* (2009) and reviewed in Chapter 7). While previous studies had demonstrated the importance of the hydrophobic tetrapeptide motif (ΦWΦP, Φ = hydrophobic residue) and the N-terminal basic residues of RAM for binding to RBP-J (Ling and Hayward, 1995; Lubman *et al.*, 2007; Tamura *et al.*, 1995), this study also uncovered two other regions in RAM that made significant contributions to binding. These regions correspond to two dipeptide motifs –HG– and –GF– that are located directly upstream and downstream, respectively, of the ΦWΦP motif (Fig. 2.5B); –HG– and –GF– contribute approximately 1.6 and 0.6 kcal/mol of binding energy, respectively, to complex formation. Interestingly, the –HG– and –GF– dipeptide motifs are conserved in most metazoans, but are not conserved in worms; however, the only high-resolution CSL-RAM structural information we have comes from the complex crystal structures of the worm proteins LAG-1 and LIN-12 (Friedmann *et al.*, 2008; Wilson and Kovall, 2006) (Fig. 2.5B). Another unexpected finding from this study showed that EBNA2 and RAM have largely overlapping binding sites on RBP-J, but the affinity of EBNA2 is at least 60-fold weaker. Binding studies of mutant BTD proteins revealed distinctive RAM- and EBNA2-binding modes as well (Johnson *et al.*, 2009). Taken together, these data challenge the simple assumption that all RAM orthologs and viral peptides such as EBNA2 interact with RBP-J in a structurally identical manner. Differences in binding modes might account for the observed differences in binding affinities between worm and mammalian CSL–RAM complexes noted above. Finally, these data also illustrate how the details among the interactions of orthologous Notch components can vary substantially, providing impetus for pursuing structural studies of mammalian, or other metazoan, CSL–RAM complexes and CSL–EBNA2 complexes.

4.5. How do corepressors interact with CSL in order to repress transcription from Notch target genes?

In stark contrast to active Notch transcription complexes, in which numerous structural and biophysical studies have provided important functional insights, our knowledge at the structural level for how corepressors interact with CSL and compete with NICD for binding surfaces on CSL is very limited. A number of corepressor proteins have been identified and shown biochemically to interact with CSL, including Hairless in flies (Brou *et al.*, 1994), and SMRT (Kao *et al.*, 1998), SKIP (Zhou *et al.*, 2000b), CIR (Hsieh *et al.*, 1999), KyoT2 (Taniguchi *et al.*, 1998), ETO(MTG8) (Salat *et al.*, 2008), MTG16 (Engel *et al.*, 2010), and MINT/SHARP in mammals (Kuroda *et al.*, 2003; Oswald *et al.*, 2002). At the primary sequence level these corepressors are unrelated, but functionally are thought, in general, to link CSL to the HDAC machinery within the nucleus. More recently, KDM5/LID histone demethylases and histone chaperones, e.g., ASF1 and NAP1, have also been shown to interact with CSL and function in transcriptional repression of Notch target genes (Goodfellow *et al.*, 2007; Liefke *et al.*, 2010; Moshkin *et al.*, 2009).

Using pulldown assays from cells, multiple groups have shown that corepressor and NICD binding to RBP-J is mutually exclusive and likely competitive in nature (Kao *et al.*, 1998; Kuroda *et al.*, 2003; Zhou *et al.*, 2000b). These findings have led to a generalized model in the field that suggest NICD binding to CSL displaces, or outcompetes, corepressors from CSL; however, at the molecular level how this displacement/competition occurs and whether this model holds true for all corepressors is poorly defined. Moreover, a number of these corepressors have been reported to interact with each other, e.g., SMRT-SKIP (Zhou *et al.*, 2000b) and MINT-SMRT (Shi *et al.*, 2001), raising the question as to which corepressors directly contact CSL and which corepressors are merely a component of a multiprotein CSL-mediated repression complex.

In terms of supporting genetic and biochemical data, Hairless and MINT are the best-characterized CSL-interacting corepressors (Borggrefe and Oswald, 2009; Maier, 2006). Hairless is an approximately 1000 residue protein that has been shown genetically to interact with Su(H) and Notch (Bang *et al.*, 1991; Schweisguth and Posakony, 1994), as well as other components of the Notch pathway. It has been shown biochemically to interact with Su(H) (Brou *et al.*, 1994; Maier *et al.*, 1997) and the repressor proteins Groucho and CtBP (Barolo *et al.*, 2002). Secondary structure prediction analysis suggests that Hairless is largely devoid of α/β structure, and consistent with this analysis, only relatively short peptide-like sequences in Hairless are required to interact with Su(H), Groucho, and CtBP. While Hairless is conserved in insects, there are no clear orthologs in mammals or worms. However, it has been suggested that the corepressor MINT/SHARP may be the functional analog of Hairless in mammals (Kuroda *et al.*, 2003; Oswald *et al.*, 2005).

MINT is an approximately 6600 residue multidomain nuclear protein and has been assigned to the SPEN family of proteins based on its domain organization (Ariyoshi and Schwabe, 2003). While genetic knockouts of MINT are embryonic lethal, MINT has been shown *in vivo* to function as a Notch antagonist in B and T cells (Kuroda *et al.*, 2003; Tsuji *et al.*, 2007), and in the kidney (Surendran *et al.*, 2010). Biochemically, multiple groups have shown that MINT interacts with RBP-J (Kuroda *et al.*, 2003; Oswald *et al.*, 2002), as well as with members of the nuclear hormone receptor family and the transcriptional repressor homeodomain MSX2 (Newberry *et al.*, 1999; Shi *et al.*, 2001). Similar to Hairless, only an approximately 50 residue sequence of MINT is necessary and sufficient to interact with RBP-J (Kuroda *et al.*, 2003; Oswald *et al.*, 2002). MINT has also been shown to interact with CtBP (Oswald *et al.*, 2005); however, this interaction is mediated by CtIP. In addition to CtBP and CtIP, MINT has been shown to interact with a number of other corepressors, including ETO (Salat *et al.*, 2008), SMRT, and NCor (Shi *et al.*, 2001), suggesting that MINT functions as a central organizing element, mediating contacts with several distinct multiprotein transcriptional repression complexes. It should also be mentioned that while SPEN proteins are found in flies and worms (Ariyoshi and Schwabe, 2003), as pointed out by Oswald *et al.*, the region of MINT that interacts with CSL is not conserved in fly and worm SPEN proteins (Oswald *et al.*, 2005), suggesting that the involvement of MINT in Notch transcription complexes may be a more recent evolutionary event.

4.6. Is CSL constitutively bound to DNA?

Most models of transcriptional regulation in the Notch pathway posit that CSL is predominantly localized to the nucleus and constitutively bound to DNA. The exchange of transcriptional coregulators (corepressors and coactivators) associated with DNA-bound CSL is thus proposed to mediate whether or not the Notch target gene is turned on. This static view of DNA binding and nuclear localization of CSL stems in part from very early studies in the field that found CSL primarily resides in the nucleus (Fortini and Artavanis-Tsakonas, 1994; Jarriault *et al.*, 1995; Roehl *et al.*, 1996) and binds DNA with a low nanomolar affinity (Matsunami *et al.*, 1989). However, several recent studies challenge this model, suggesting that subcellular localization and DNA binding by CSL is more dynamic in nature, and likely to be influenced by cooperative binding mechanisms. In one study, Krejci *et al.* showed that the occupancy of Su(H) at binding sites within the Enhancer of Split complex transiently increases following the activation of Notch signaling (Krejci and Bray, 2007). Another key finding from this study demonstrated that there is a cytoplasmic pool of Su(H) protein, which translocates to the nucleus following stimulation of the Notch pathway.

Other studies have pointed to the assembly of higher order complexes as potential contributors to the stable loading of CSL complexes onto DNA. Posakony's group first noted that genes in the enhancer of split locus of *Drosophila* contained conserved Su(H)-binding sites that were oriented head-to-head and separated by 15–22 nucleotides (Bailey and Posakony, 1995). This "paired site" architecture is also found in the mammalian Hes-1 promoter, which contains two RBP-J binding sites in a head-to-head arrangement separated by 16 nucleotides (Jarriault *et al.*, 1995). Paired site CSL-binding elements are conserved in mammals, zebrafish, and frogs, but interestingly do not appear to be prevalent in worms (Nam *et al.*, 2007; Yoo *et al.*, 2004).

Crystal contacts between the Notch ankyrin-repeat domains in the structure of the human transcriptional complex suggested that these interactions might mediate cooperative assembly of higher order complexes on the Hes-1 paired site (Nam *et al.*, 2007; Nam *et al.*, 2006) (Fig. 2.6A).

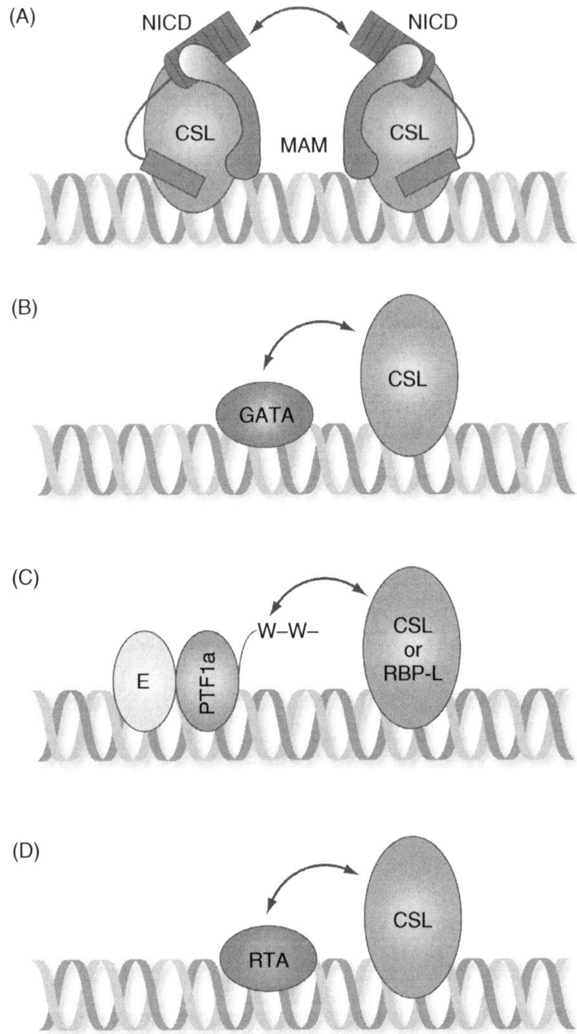

Figure 2.6 Models of higher-order transcription complexes. (A). Model of the cooperative assembly of two CSL–NICD–MAM ternary complexes onto the paired CSL binding sites of the hes-1 promoter (Nam *et al.*, 2007). Cooperative binding is mediated by contacts between the ANK domains of NICD. (B). Model of the putative cooperative assembly of GATA-CSL transcription complexes at the ref-1 enhancer element in worms (Neves *et al.*, 2007). Cooperative binding of LAG-1/CSL and ELT-2/GATA is likely mediated by direct interactions between these two transcription factors; however, other potential interactions with the CSL–NICD–MAM ternary complex cannot be formally excluded. (C). Model of the Notch-independent cooperative assembly of the trimeric PTF1 complex (Beres *et al.*, 2006). Cooperative binding of the E-protein/PTF1a heterodimer with either RBP-J/CSL or RBP-L is likely mediated by interactions between the C-terminal tail of PTF1a, which consists of two tryptophan (W) containing motifs, and RBP-J or RBP-L. (D). Model of the putative cooperative assembly of the viral transactivator protein (RTA) with CSL (Carroll *et al.*, 2006). Cooperative binding is likely mediated by direct interactions between RTA and CSL, and is also likely Notch independent.

Biochemical studies provided evidence to support the idea that ANK–ANK contacts, together with Mastermind binding, promote cooperative dimerization of these complexes on the Hes-1 paired-site element, and that the integrity of the ANK–ANK interface is required for dimerization *in vitro* and for transcriptional activation in cells (Nam *et al.*, 2007).

Other studies have identified transcriptional regulatory regions that are defined by CSL binding sites adjacent to other transcription factor sites, leading to the suggestion that the loading of complexes at these loci are also cooperative or synergistic in nature. In worms, there is strong evidence for the coordinated action of LAG-1 and the GATA transcription factor Elt-2 in regulating the transcription of ref-1 (Neves *et al.*, 2007). This study showed that an enhancer element within the ref-1 promoter conferred tissue-specific expression of ref-1 that was dependent upon adjacent LAG-1 and GATA DNA-binding sites (Fig. 2.6B). In addition, the authors demonstrated that LAG-1 and ELT-2 directly interact *in vitro*, which they speculated might be the molecular basis for the observed synergy. An analogous RBP-J–GATA synergy has been observed for the expression of IL-4 in mice (Amsen *et al.*, 2007; Fang *et al.*, 2007). In yet another study, cooperative binding mechanisms underlie the formation of the trimeric PTF1 complex, which is composed of the basic helix–loop–helix heterodimer PTF1a and E-protein, and either RBP-J or its paralog RBP-L (Beres *et al.*, 2006) (Fig. 2.6C). Interestingly, cooperative assembly of the trimeric complex on tandem E- and TC-boxes that typify PTF1 DNA-binding sites requires short conserved peptide sequences in the C-terminus of PTF1a that bind either RBP-J or RBP-L, and are reminiscent of the ΦWΦP motif found in NICD RAM domains. In addition, other work suggests that a cooperative binding mechanism, involving the viral transactivator RTA, may underlie the recruitment of RBP-J to viral and cellular genes, in order to reactivate the herpesvirus KSHV from latency (Carroll *et al.*, 2006) (Fig. 2.6D).

A more recent reexamination of the affinity of CSL for DNA, using highly purified recombinant protein, revealed that CSL has at least a 100-fold weaker affinity for DNA than previously reported (Friedmann and Kovall, 2009). Moreover, this property is shared amongst all CSL orthologs examined—mouse, worm, and fly. While the affinity of CSL–NICD–MAM and CSL–corepressor complexes for DNA has not been directly measured, previous binding studies observed that the affinity of NICD for RBP-J or RBP-J prebound to DNA were similar. Due to the properties of linked equilibria, this implies that the affinity of CSL or CSL-NICD for DNA is the same and, therefore, suggests that NICD does not change the affinity of CSL for DNA. Taken together, these aforementioned studies strongly suggest that CSL binding to DNA is more dynamic than previously thought, and given its only moderate affinity for DNA, cooperative binding mechanisms are important for increasing the occupancy of CSL-mediated transcription complexes at some sites. These cooperative binding mechanisms are likely critical for providing a robust burst of transcription, following pathway activation, such as has been observed at the Hes-1 locus (Ong *et al.*, 2006). Furthermore, it appears that DNA binding by CSL is another step in the Notch pathway that can be regulated, and it would not be at all surprising if future studies identify many more examples of this form of regulation. Accompanying structural studies of these higher order protein–DNA complexes will provide critical molecular insights into the basis for cooperative assembly of such complexes onto DNA.

4.7. Post-translational modifications

Post-translational modifications of the intracellular components of the Notch pathway have been described in a number of studies, and a comprehensive cataloguing of the various modifications, their sites, and their functional effects extends beyond the scope of this review. Phosphorylation of the C-terminal region of Notch itself occurs at several different sites, with the specific modification pattern and usage likely to depend on cell type and context. There are well-documented examples of C-terminal phosphorylation sites associated with negative regulation that are used to promote Notch degradation (Chiang *et al.*, 2006; Fryer *et al.*, 2004).

The ANK domains of Notch1–3, but not Notch4, are hydroxylated on conserved asparagine residues within repeats 2 and 4 by factor-inhibiting HIF (Coleman *et al.*, 2007). The level of hydroxylated ANK varies as a function of oxygen tension (elevated in normoxia and reduced under hypoxia), but the physiologic significance of hydroxylation remains unclear. Structural and biochemical studies have revealed that asparaginyl hydroxylation does not change the overall fold of ANK (Coleman *et al.*, 2007), but hydroxylation of individual ankyrin repeats does increase the stability of the ANK domain (Kelly *et al.*, 2009). Mapping the hydroxylation sites on the CSL–NICD–MAM transcription complex reveals that the modified asparagine residues lie on a face ANK that is neither involved in interactions with CSL nor MAM and, therefore, unlikely to have any effect on the assembly of this transcription complex.

Mastermind has also been shown to undergo a number of post-translational modifications. MAM is acetylated by p300 at a set of conserved lysine doublet motifs (Saint Just Ribeiro *et al.*, 2007); MAM is phosphorylated by GSK3β (Saint Just Ribeiro *et al.*, 2009) and MAM is sumoylated at two conserved lysine residues by the E2 conjugating enzyme and E3 ligase UBC9 and PIAS1, respectively, which results in downregulation of signaling and increased association of MAM with HDAC7 (Lindberg *et al.*, 2010).

4.8. Summary and outstanding questions

While great strides have been made in resolving the structures of Notch transcription complexes and biophysically characterizing the interactions that compose these complexes, many key structural and biochemical questions remain. Are corepressor interactions with CSL regulated in a stepwise manner similar to the assembly of Notch pathway active transcription complexes? Do all identified corepressors interact with CSL in a common or distinct manner? Are corepressor complexes target-gene or tissue specific? Are corepressor complexes important for the regulation of transcription from Notch target genes in worms? How do post-translational modifications, e.g., phosphorylation, sumoylation, ubiquitination, and acetylation, affect the function and stability of CSL *in vivo*?

5. Therapeutic Implications Of Structural Insights

5.1. Targeting Notch–ligand interactions

The Notch–DLL4-signaling axis lies downstream of vascular endothelial growth factor signaling in normal and tumor angiogenesis. Thus, the inhibition of Notch–DLL4 signaling has emerged as a potential target for next-generation anti-angiogenic therapeutics in cancer. Toward this end, monoclonal antibodies (Ridgway *et al.*, 2006), decoy ligands (Noguera-Troise *et al.*, 2006), and decoy receptor molecules (Funahashi *et al.*, 2008) have been developed to block signaling by competitively inhibiting Notch1–DLL4 receptor–ligand interactions. The appeal of this approach, however, has been considerably tempered by the recent report that chronic blockade of DLL4-Notch signaling by anti-DLL4 antibodies is associated with the development of vascular neoplasms and liver toxicity (Yan *et al.*, 2010), suggesting that more refined strategies or dosing regimens with small molecule inhibitors might be needed to target Notch signaling in tumor angiogenesis without prohibitively toxic side effects.

5.2. The activation switch as a potential therapeutic target

T-ALL-associated mutations in the Notch1 NRR lead to increased ligand-independent signaling in these tumors. More recently, mutations of the Notch1 gene have also been reported in some non-small-cell lung cancers (Westhoff *et al.*, 2009), suggesting that Notch1 might also be a therapeutic target in these cancers. Notch3 gene amplification has been reported in approximately 20% of ovarian cancers (Park *et al.*, 2006), and there is also evidence pointing to a role for Notch3 signaling in promoting the development of pulmonary arterial hypertension (Li *et al.*, 2009). Controlled activation of Notch receptors may also have clinical utility in the *ex vivo* expansion of certain stem cell populations, such as in the hematopoetic system (Delaney *et al.*, 2005).

Zhou and colleagues were the first to identify allosteric antibodies modulating Notch signaling by binding to the NRR (Li *et al.*, 2008). They searched for monoclonal antibodies directed against the entire extracellular region of Notch3, and found both inhibitory and activating antibodies directed against the NRR. Importantly, the inhibitory antibodies showed potent activity that was independent of the identity of the stimulating ligand. The epitope for the inhibitory antibodies was discontinuous, including residues from both the LNR-A module and the HD domain, whereas the site bound by the activating antibody only mapped to residues in the LNR-A domain. When the epitope of the inhibitory antibodies was mapped onto a homology model for the structure of the Notch3 NRR, the location of the mapped contact site suggested that the inhibitory antibody was acting as a clamp holding the NRR in the closed conformation to prevent metalloprotease access.

More recently, inhibitory antibodies directed against the Notch1 (Aste-Amezaga *et al.*, 2010; Wu *et al.*, 2010) and Notch2 NRRs (Wu *et al.*, 2010) have also been reported. The structure of a complex between the Notch1 NRR and one of its inhibitory antibodies has now been solved, and it too covers a discontinuous epitope encompassing residues from both the LNR and HD domains (Wu *et al.*, 2010) (Fig. 2.7). Anti-Notch1 antibodies directed against the NRR inhibit signaling from both normal Notch1 and

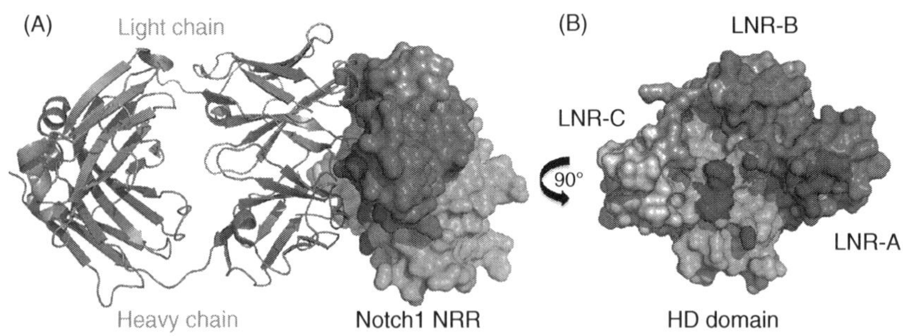

Figure 2.7 Structure of a complex between an inhibitory F_{ab} and the Notch1 NRR. (A). Side view of the complex. The F_{ab} is rendered as a ribbon diagram, whereas the Notch1 NRR is shown in surface representation. The light chain is orange and the heavy chain is purple. The LNR modules are colored different shades of purple and pink, and the HD domain is aquamarine. Residues of the NRR in contact with the heavy chain are colored blue, and those in contact with the light-chain are colored red. (B). 90 degree rotation from the view in A, showing a surface view of the Notch1 NRR to highlight the antibody-binding epitope. The F_{ab} has been removed for clarity. (See Color Insert.)

Notch1 variants possessing tumor-associated mutations in the HD domain (Aste-Amezaga *et al.*, 2010; Wu *et al.*, 2010), whereas antibodies directed against the ligand-binding region of the receptor have no effect on the aberrant signaling observed with Notch1 receptors carrying T-ALL-associated mutations (Aste-Amezaga *et al.*, 2010). The antibodies generated by different groups appear to exhibit variation with respect to their growth inhibitory effect on T-ALL cell lines in culture. The origin of these observed differences remains unclear, and it remains to be seen whether the antibodies consistently exert as strong a growth-inhibitory effect on T-ALL tumor lines as do conventional GSIs. Nevertheless, selective anti-Notch inhibitory antibodies remain among the strongest therapeutic candidates directed against the Notch pathway because of their receptor specificity and their accessible extracellular target site.

5.3. Targeting the MAML-1 binding groove of nuclear ternary complexes

Studies in T-ALL cell lines first established that an N-terminal peptide spanning residues 13–74 of human MAML-1 acts as a potent dominant negative inhibitor of Notch signaling. The extended helical conformation of this part of MAML-1 in the human ternary complex on DNA suggested that a hydrocarbon-stapled α-helical peptide or other stabilized alpha-helix mimetic might be capable of inhibiting Notch signaling by competing for binding to the native MAML-1 binding pocket. Bradner, Verdine, and colleagues exploited this idea to design a series of stapled α-helical peptides to span the MAML-1-binding groove and examined a stapled peptide called stapled alpha-helix from MAML (SAHM-1) spanning residues 21–36 in more detail. The initial characterization of this molecule suggests that it antagonizes Notch activity in cells and in a murine T-ALL model by targeting Notch nuclear complexes, but more work is needed to confirm the presumed mechanism of action of this compound (Moellering *et al.*, 2009). It also remains to be determined whether the peptide exhibits any selectivity for binding to one mammalian Notch nuclear complex over any of the others. Because the four mammalian Notch receptors exhibit 70% sequence identity in their ANK domains, it may also be challenging to develop selectivity for one complex over the other three by additional tailoring of the SAHM-1 scaffold.

6. Summary

Recent pioneering structural and biochemical studies focusing on three key steps in canonical Notch signaling have substantially enhanced current understanding of the molecular logic controlling signal activation.

Prototype structures of ligand binding-active and receptor binding-active fragments from Notch1 and Jagged1 have shown that the recognition sites lie on concatenated domains with an elongated overall conformation, and that Notch–ligand interactions do not require post-translational modifications, even though such modifications influence ligand responsiveness *in vivo*. The structures of the Notch1 and Notch2 NRRs in their autoinhibited conformations revealed the basis for the intrinsic proteolytic resistance of these receptors prior to ligand binding and established the need for a substantial conformational opening of this domain to permit activating proteolysis. Finally, structural and biochemical studies of worm and mammalian nuclear complexes have clarified how Mastermind-family proteins are only captured by preformed Notch–CSL complexes, revealing new insights into how Notch functions as a transcriptional activation switch.

On the other hand, clear differences between receptors and complexes from different species also highlight the limitations of models that generalize from a single structural example. Moreover, the emergence of different components of the Notch pathway as potential therapeutic targets in cancer and other diseases underscores the future importance of acquiring high-resolution structural data to gain additional insights into the differences among receptors and among the various functionally relevant receptor–ligand complexes that have eluded such analysis to date.

REFERENCES

Acar, M., Jafar-Nejad, H., Takeuchi, H., Rajan, A., Ibrani, D., Rana, N. A., Pan, H., Haltiwanger, R. S., and Bellen, H. J. (2008). Rumi is a CAP10 domain glycosyltransferase that modifies Notch and is required for Notch signaling. *Cell* **132**, 247–258.

Ahimou, F., Mok, L. P., Bardot, B., and Wesley, C. (2004). The adhesion force of Notch with Delta and the rate of Notch signaling. *J. Cell Biol.* **167**, 1217–1229.

Amsen, D., Antov, A., Jankovic, D., Sher, A., Radtke, F., Souabni, A., Busslinger, M., McCright, B., Gridley, T., and Flavell, R. A. (2007). Direct regulation of Gata3 expression determines the T helper differentiation potential of Notch. *Immunity* **27**, 89–99.

Ariyoshi, M., and Schwabe, J.W.R. (2003). A conserved structural motif reveals the essential transcriptional repression function of Spen proteins and their role in developmental signaling. *Genes Dev.* **17**, 1909–1920.

Aste-Amezaga, M., Zhang, N., Lineberger, J. E., Arnold, B. A., Toner, T. J., Gu, M., Huang, L., Vitelli, S., Vo, K. T., Haytko, P., *et al.* (2010). Characterization of Notch1 antibodies that inhibit signaling of both normal and mutated Notch1 receptors. *PLoS One* **5**, e9094.

Aster, J. C., Pear, W. S., and Blacklow, S. C. (2008). Notch signaling in leukemia. *Annu. Rev. Pathol.* **3**, 587–613.

Aster, J. C., Simms, W. B., Zavala-Ruiz, Z., Patriub, V., North, C. L., and Blacklow, S. C. (1999). The folding and structural integrity of the first LIN-12 module of human Notch1 are calcium-dependent. *Biochemistry* **38**, 4736–4742.

Aster, J. C., Xu, L., Karnell, F. G., Patriub, V., Pui, J. C., and Pear, W. S. (2000). Essential roles for ankyrin repeat and transactivation domains in induction of T-cell leukemia by notch1. *Mol. Cell. Biol.* **20**, 7505–7515.

Bailey, A. M., and Posakony, J. W. (1995). Suppressor of hairless directly activates transcription of enhancer of split complex genes in response to Notch receptor activity. *Genes Dev.* **9**, 2609–2622.
Bang, A. G., Hartenstein, V., and Posakony, J. W. (1991). Hairless is required for the development of adult sensory organ precursor cells in Drosophila. *Development* **111**, 89–104.
Barolo, S., Stone, T., Bang, A. G., and Posakony, J. W. (2002). Default repression and Notch signaling: Hairless acts as an adaptor to recruit the corepressors Groucho and dCtBP to Suppressor of Hairless. *Genes Dev.* **16**, 1964–1976.
Barrick, D., and Kopan, R. (2006). The Notch transcription activation complex makes its move. *Cell* **124**, 883–885.
Becam, I., Fiuza, U. M., Arias, A. M., and Milan, M. (2010). A Role of Receptor Notch in Ligand cis-Inhibition in Drosophila. *Curr. Biol.* **20**, 554–560.
Beres, T. M., Masui, T., Swift, G. H., Shi, L., Henke, R. M., and MacDonald, R. J. (2006). PTF1 is an organ-specific and Notch-independent basic helix-loop-helix complex containing the mammalian Suppressor of Hairless (RBP-J) or its paralogue, RBP-L. *Mol. Cell. Biol.* **26**, 117–130.
Bertagna, A., Toptygin, D., Brand, L., and Barrick, D. (2008). The effects of conformational heterogeneity on the binding of the Notch intracellular domain to effector proteins: a case of biologically tuned disorder. *Biochem. Soc. Trans.* **36**, 157–166.
Blaumueller, C. M., Qi, H., Zagouras, P., and Artavanis-Tsakonas, S. (1997). Intracellular cleavage of Notch leads to a heterodimeric receptor on the plasma membrane. *Cell* **90**, 281–291.
Borggrefe, T., and Oswald, F. (2009). The Notch signaling pathway: Transcriptional regulation at Notch target genes. *Cell Mol. Life Sci.*
Brou, C., Logeat, F., Gupta, N., Bessia, C., LeBail, O., Doedens, J. R., Cumano, A., Roux, P., Black, R. A., and Israel, A. (2000). A novel proteolytic cleavage involved in Notch signaling: the role of the disintegrin-metalloprotease TACE. *Mol. Cell.* **5**, 207–216.
Brou, C., Logeat, F., Lecourtois, M., Vandekerckhove, J., Kourilsky, P., Schweisguth, F., and Israel, A. (1994). Inhibition of the DNA-binding activity of Drosophila suppressor of hairless and of its human homolog, KBF2/RBP-J kappa, by direct protein-protein interaction with Drosophila hairless. *Genes Dev.* **8**, 2491–2503.
Brown, M. S., Ye, J., Rawson, R. B., and Goldstein, J. L. (2000). Regulated intramembrane proteolysis: a control mechanism conserved from bacteria to humans. *Cell* **100**, 391–398.
Bruckner, K., Perez, L., Clausen, H., and Cohen, S. (2000). Glycosyltransferase activity of Fringe modulates Notch-Delta interactions. *Nature* **406**, 411–415.
Carroll, K. D., Bu, W., Palmeri, D., Spadavecchia, S., Lynch, S. J., Marras, S.A.E., Tyagi, S., and Lukac, D. M. (2006). Kaposi's Sarcoma-associated herpesvirus lytic switch protein stimulates DNA binding of RBP-Jk/CSL to activate the Notch pathway. *J Virol.* **80**, 9697–9709.
Chiang, M. Y., Xu, M. L., Histen, G., Shestova, O., Roy, M., Nam, Y., Blacklow, S. C., Sacks, D. B., Pear, W. S., and Aster, J. C. (2006). Identification of a conserved negative regulatory sequence that influences the leukemogenic activity of NOTCH1. *Mol. Cell. Biol.* **26**, 6261–6271.
Christensen, S., Kodoyianni, V., Bosenberg, M., Friedman, L., and Kimble, J. (1996). lag-1, a gene required for lin-12 and glp-1 signaling in Caenorhabditis elegans, is homologous to human CBF1 and Drosophila Su(H). *Development* **122**, 1373–1383.
Chung, C. N., Hamaguchi, Y., Honjo, T., and Kawaichi, M. (1994). Site-directed mutagenesis study on DNA binding regions of the mouse homologue of Suppressor of Hairless, RBP-J kappa. *Nucleic Acids Res.* **22**, 2938–2944.
Coleman, M. L., McDonough, M. A., Hewitson, K. S., Coles, C., Mecinovic, J., Edelmann, M., Cook, K. M., Cockman, M. E., Lancaster, D. E., Kessler, B. M., *et al.* (2007).

Asparaginyl hydroxylation of the Notch ankyrin repeat domain by factor inhibiting hypoxia-inducible factor. *J. Biol. Chem.* **282**, 24027–24038.

Cordle, J., Johnson, S., Zi Yan Tay, J., Roversi, P., Wilkin, M. B., de Madrid, B. H., Shimizu, H., Jensen, S., Whiteman, P., Jin, B., *et al.* (2008a). A conserved face of the Jagged/Serrate DSL domain is involved in Notch trans-activation and cis-inhibition. *Nat. Struct. Mol. Biol.* **15**, 849–857.

Cordle, J., Redfieldz, C., Stacey, M., van der Merwe, P. A., Willis, A. C., Champion, B. R., Hambleton, S., and Handford, P. A. (2008b). Localization of the delta-like-1-binding site in human Notch-1 and its modulation by calcium affinity. *J. Biol. Chem.* **283**, 11785–11793.

de Celis, J. F., and Bray, S. (1997). Feed-back mechanisms affecting Notch activation at the dorsoventral boundary in the Drosophila wing. *Development* **124**, 3241–3251.

de Celis, J. F., and Bray, S. J. (2000). The Abruptex domain of Notch regulates negative interactions between Notch, its ligands and Fringe. *Development* **127**, 1291–1302.

De Strooper, B., Annaert, W., Cupers, P., Saftig, P., Craessaerts, K., Mumm, J. S., Schroeter, E. H., Schrijvers, V., Wolfe, M. S., Ray, W. J., *et al.* (1999). A presenilin-1-dependent gamma-secretase-like protease mediates release of Notch intracellular domain. *Nature* **398**, 518–522.

Del Bianco, C., Aster, J. C., and Blacklow, S. C. (2008). Mutational and energetic studies of Notch 1 transcription complexes. *J. Mol. Biol.* **376**, 131–140.

Delaney, C., Varnum-Finney, B., Aoyama, K., Brashem-Stein, C., and Bernstein, I. D. (2005). Dose-dependent effects of the Notch ligand Delta1 on ex vivo differentiation and in vivo marrow repopulating ability of cord blood cells. *Blood* **106**, 2693–2699.

Dou, S., Zeng, X., Cortes, P., Erdjument-Bromage, H., Tempst, P., Honjo, T., and Vales, L. D. (1994). The recombination signal sequence-binding protein RBP-2N functions as a transcriptional repressor. *Mol. Cell. Biol.* **14**, 3310–3319.

Doyle, T. G., Wen, C., and Greenwald, I. (2000). SEL-8, a nuclear protein required for LIN-12 and GLP-1 signaling in Caenorhabditis elegans. *Proc. Natl. Acad. Sci. U.S.A* **97**, 7877–7881.

Ehebauer, M. T., Chirgadze, D. Y., Hayward, P., Martinez Arias, A., and Blundell, T. L. (2005). High-resolution crystal structure of the human Notch 1 ankyrin domain. *Biochem. J.* **392**, 13–20.

Engel, M. E., Nguyen, H. N., Mariotti, J., Hunt, A., and Hiebert, S. W. (2010). Myeloid translocation gene (MTG)-16 interacts with Notch transcription complex components to integrate Notch signaling in hematopoietic cell fate specification. *Mol. Cell. Biol.*

Fang, T. C., Yashiro-Ohtani, Y., Del Bianco, C., Knoblock, D. M., Blacklow, S. C., and Pear, W. S. (2007). Notch directly regulates Gata3 expression during T helper 2 cell differentiation. *Immunity* **27**, 100–110.

Fehon, R. G., Kooh, P. J., Rebay, I., Regan, C. L., Xu, T., Muskavitch, M. A., and Artavanis-Tsakonas, S. (1990). Molecular interactions between the protein products of the neurogenic loci Notch and Delta, two EGF-homologous genes in Drosophila. *Cell* **61**, 523–534.

Fortini, M. E., and Artavanis-Tsakonas, S. (1994). The suppressor of hairless protein participates in notch receptor signaling. *Cell* **79**, 273–282.

Friedmann, D. R., and Kovall, R. A. (2010). Thermodynamic and structural insights into CSL-DNA complexes. *Protein Sci.* **19**, 34–46.

Friedmann, D. R., Wilson, J. J., and Kovall, R. A. (2008). RAM-induced allostery facilitates assembly of a notch pathway active transcription complex. *J. Biol. Chem.* **283**, 14781–14791.

Fryer, C. J., Lamar, E., Turbachova, I., Kintner, C., and Jones, K. A. (2002). Mastermind mediates chromatin-specific transcription and turnover of the Notch enhancer complex. *Genes Dev.* **16**, 1397–1411.

Fryer, C. J., White, J. B., and Jones, K. A. (2004). Mastermind recruits CycC:CDK8 to phosphorylate the Notch ICD and coordinate activation with turnover. *Mol. Cell.* **16**, 509–520.

Funahashi, Y., Hernandez, S. L., Das, I., Ahn, A., Huang, J., Vorontchikhina, M., Sharma, A., Kanamaru, E., Borisenko, V., Desilva, D. M., *et al.* (2008). A notch1 ectodomain construct inhibits endothelial notch signaling, tumor growth, and angiogenesis. *Cancer Res.* **68**, 4727–4735.

Furriols, M., and Bray, S. (2001). A model Notch response element detects Suppressor of Hairless-dependent molecular switch. *Curr. Biol.* **11**, 60–64.

Ge, C., Liu, T., Hou, X., and Stanley, P. (2008). In vivo consequences of deleting EGF repeats 8-12 including the ligand binding domain of mouse Notch1. *BMC Dev. Biol.* **8**, 48.

Ge, C., and Stanley, P. (2008). The O-fucose glycan in the ligand-binding domain of Notch1 regulates embryogenesis and T cell development. *Proc. Natl. Acad. Sci. U.S.A* **105**, 1539–1544.

Ghai, V., and Gaudet, J. (2008). The CSL transcription factor LAG-1 directly represses hlh-6 expression in C. *elegans*. *Dev. Biol.* **322**, 334–344.

Glittenberg, M., Pitsouli, C., Garvey, C., Delidakis, C., and Bray, S. (2006). Role of conserved intracellular motifs in Serrate signalling, cis-inhibition and endocytosis. *EMBO J.* **25**, 4697–4706.

Goodfellow, H., Krejcí, A., Moshkin, Y., Verrijzer, C. P., Karch, F., and Bray, S. J. (2007). Gene-specific targeting of the histone chaperone asf1 to mediate silencing. *Dev. Cell.* **13**, 593–600.

Gordon, W. R., Arnett, K. L., and Blacklow, S. C. (2008). The molecular logic of Notch signaling – a structural and biochemical perspective. *J. Cell. Sci.* **121**, 3109–3119.

Gordon, W. R., Roy, M., Vardar-Ulu, D., Garfinkel, M., Mansour, M. R., Aster, J. C., and Blacklow, S. C. (2009a). Structure of the Notch1-negative regulatory region: implications for normal activation and pathogenic signaling in T-ALL. *Blood* **113**, 4381–4390.

Gordon, W. R., Vardar-Ulu, D., Histen, G., Sanchez-Irizarry, C., Aster, J. C., and Blacklow, S. C. (2007). Structural basis for autoinhibition of Notch. *Nat. Struct. Mol. Biol.* **14**, 295–300.

Gordon, W. R., Vardar-Ulu, D., L'Heureux, S., Ashworth, T., Malecki, M. J., Sanchez-Irizarry, C., McArthur, D. G., Histen, G., Mitchell, J. L., Aster, J. C., *et al.* (2009b). Effects of S1 cleavage on the structure, surface export, and signaling activity of human Notch1 and Notch2. *PLoS One* **4**, e6613.

Greenwald, I., and Seydoux, G. (1990). Analysis of gain-of-function mutations of the lin-12 gene of Caenorhabditis elegans. *Nature* **346**, 197–199.

Hambleton, S., Valeyev, N. V., Muranyi, A., Knott, V., Werner, J. M., McMichael, A. J., Handford, P. A., and Downing, A. K. (2004). Structural and functional properties of the human notch-1 ligand binding region. *Structure* **12**, 2173–2183.

Harris, R. J., and Spellman, M. W. (1993). O-linked fucose and other post-translational modifications unique to EGF modules. *Glycobiology* **3**, 219–224.

Hicks, C., Johnston, S. H., diSibio, G., Collazo, A., Vogt, T. F., and Weinmaster, G. (2000). Fringe differentially modulates Jagged1 and Delta1 signalling through Notch1 and Notch2. *Nat Cell Biol* **2**, 515–520.

Hsieh, J. J., and Hayward, S. D. (1995). Masking of the CBF1/RBPJ kappa transcriptional repression domain by Epstein-Barr virus EBNA2. *Science* **268**, 560–563.

Hsieh, J. J., Henkel, T., Salmon, P., Robey, E., Peterson, M. G., and Hayward, S. D. (1996). Truncated mammalian Notch1 activates CBF1/RBPJk-repressed genes by a mechanism resembling that of Epstein-Barr virus EBNA2. *Mol. Cell. Biol.* **16**, 952–959.

Hsieh, J. J., Zhou, S., Chen, L., Young, D. B., and Hayward, S. D. (1999). CIR, a corepressor linking the DNA binding factor CBF1 to the histone deacetylase complex. *Proc. Natl. Acad. Sci. U.S.A* **96**, 23–28.

Jarriault, S., Brou, C., Logeat, F., Schroeter, E. H., Kopan, R., and Israel, A. (1995). Signalling downstream of activated mammalian Notch. *Nature* **377**, 355–358.

Jeffries, S., and Capobianco, A. J. (2000). Neoplastic transformation by Notch requires nuclear localization. *Mol. Cell. Biol.* **20**, 3928–3941.

Johnson, S. E., Ilagan, M.X.G., Kopan, R., and Barrick, D. (2010). Thermodynamic analysis of the CSL:Notch binding interaction: Distribution of free energy of binding and implications for competition between EBNA2 and RAM on the beta-trefoil domain. *J. Biol. Chem.* **285**, 6681–6692.

Kao, H. Y., Ordentlich, P., Koyano-Nakagawa, N., Tang, Z., Downes, M., Kintner, C. R., Evans, R. M., and Kadesch, T. (1998). A histone deacetylase corepressor complex regulates the Notch signal transduction pathway. *Genes Dev.* **12**, 2269–2277.

Kelly, D. F., Lake, R. J., Walz, T., and Artavanis-Tsakonas, S. (2007). Conformational variability of the intracellular domain of Drosophila Notch and its interaction with Suppressor of Hairless. *Proc. Natl. Acad Sci. USA* **104**, 9591–9596.

Kelly, L., McDonough, M. A., Coleman, M. L., Ratcliffe, P. J., and Schofield, C. J. (2009). Asparagine beta-hydroxylation stabilizes the ankyrin repeat domain fold. *Mol. Biosyst.* **5**, 52–58.

Klein, T., Brennan, K., and Arias, A. M. (1997). An intrinsic dominant negative activity of serrate that is modulated during wing development in Drosophila. *Dev. Biol.* **189**, 123–134.

Komatsu, H., Chao, M. Y., Larkins-Ford, J., Corkins, M. E., Somers, G. A., Tucey, T., Dionne, H. M., White, J. Q., Wani, K., Boxem, M., *et al.* (2008). OSM-11 facilitates LIN-12 Notch signaling during Caenorhabditis elegans vulval development. *PLoS Biol.* **6**, e196.

Kopan, R., and Ilagan, M. X. (2009). The canonical Notch signaling pathway: unfolding the activation mechanism. *Cell* **137**, 216–233.

Kopan, R., Schroeter, E. H., Weintraub, H., and Nye, J. S. (1996). Signal transduction by activated mNotch: importance of proteolytic processing and its regulation by the extracellular domain. *Proc. Natl. Acad. Sci. U.S.A* **93**, 1683–1688.

Kovall, R. A. (2007). Structures of CSL, Notch and Mastermind proteins: piecing together an active transcription complex. *Curr. Opin. Struct. Biol.* **17**, 117–127.

Kovall, R. A. (2008). More complicated than it looks: assembly of Notch pathway transcription complexes. *Oncogene* **27**, 5099–5109.

Kovall, R. A., and Hendrickson, W. A. (2004). Crystal structure of the nuclear effector of Notch signaling, CSL, bound to DNA. *EMBO J.* **23**, 3441–3451.

Krejci, A., and Bray, S. (2007). Notch activation stimulates transient and selective binding of Su(H)/CSL to target enhancers. *Genes Dev.* **21**, 1322–1327.

Kuroda, K., Han, H., Tani, S., Tanigaki, K., Tun, T., Furukawa, T., Taniguchi, Y., Kurooka, H., Hamada, Y., Toyokuni, S., *et al.* (2003). Regulation of marginal zone B cell development by MINT, a suppressor of Notch/RBP-J signaling pathway. *Immunity* **18**, 301–312.

Ladi, E., Nichols, J. T., Ge, W., Miyamoto, A., Yao, C., Yang, L. T., Boulter, J., Sun, Y. E., Kintner, C., and Weinmaster, G. (2005). The divergent DSL ligand Dll3 does not activate Notch signaling but cell autonomously attenuates signaling induced by other DSL ligands. *J. Cell Biol.* **170**, 983–992.

Lake, R. J., Grimm, L. M., Veraksa, A., Banos, A., and Artavanis-Tsakonas, S. (2009). In vivo analysis of the Notch receptor S1 cleavage. *PLoS One* **4**, e6728.

Li, K., Li, Y., Wu, W., Gordon, W. R., Chang, D. W., Lu, M., Scoggin, S., Fu, T., Vien, L., Histen, G., *et al.* (2008). Modulation of Notch signaling by antibodies specific for the extracellular negative regulatory region of NOTCH3. *J. Biol. Chem.* **283**, 8046–8054.

Li, X., Zhang, X., Leathers, R., Makino, A., Huang, C., Parsa, P., Macias, J., Yuan, J. X., Jamieson, S. W., and Thistlethwaite, P. A. (2009). Notch3 signaling promotes the development of pulmonary arterial hypertension. *Nat. Med.* **15**, 1289–1297.

Lieber, T., Kidd, S., Alcamo, E., Corbin, V., and Young, M. W. (1993). Antineurogenic phenotypes induced by truncated Notch proteins indicate a role in signal transduction and may point to a novel function for Notch in nuclei. *Genes Dev.* **7**, 1949–1965.

Liefke, R., Oswald, F., Alvarado, C., Ferres-Marco, D., Mittler, G., Rodriguez, P., Dominguez, M., and Borggrefe, T. (2010). Histone demethylase KDM5A is an integral part of the core Notch-RBP-J repressor complex. *Genes Dev.* **24**, 590–601.

Lindberg, M. J., Popko-Scibor, A. E., Hansson, M. L., and Wallberg, A. E. (2010). SUMO modification regulates the transcriptional activity of MAML1. *FASEB J.* **24**, 2396–2404.

Ling, P. D., and Hayward, S. D. (1995). Contribution of conserved amino acids in mediating the interaction between EBNA2 and CBF1/RBPJk. *J Virol.* **69**, 1944–1950.

Logeat, F., Bessia, C., Brou, C., LeBail, O., Jarriault, S., Seidah, N. G., and Israel, A. (1998). The Notch1 receptor is cleaved constitutively by a furin-like convertase. *Proc. Natl. Acad. Sci. U.S.A.* **95**, 8108–8112.

Lubman, O. Y., Ilagan, M. X., Kopan, R., and Barrick, D. (2007). Quantitative dissection of the Notch:CSL interaction: insights into the Notch-mediated transcriptional switch. *J. Mol. Biol.* **365**, 577–589.

Lubman, O. Y., Kopan, R., Waksman, G., and Korolev, S. (2005). The crystal structure of a partial mouse Notch-1 ankyrin domain: repeats 4 through 7 preserve an ankyrin fold. *Protein Sci.* **14**, 1274–1281.

Luther, K. B., and Haltiwanger, R. S. (2009). Role of unusual O-glycans in intercellular signaling. *Int. J. Biochem. Cell Biol.* **41**, 1011–1024.

Macao, B., Johansson, D.G.A., Hansson, G. C., and Hard, T. (2006). Autoproteolysis coupled to protein folding in the SEA domain of the membrane-bound MUC1 mucin. *Nat. Struct. Mol. Biol.* **13**, 71–76.

Maeda, T., Inoue, M., Koshiba, S., Yabuki, T., Aoki, M., Nunokawa, E., Seki, E., Matsuda, T., Motoda, Y., Kobayashi, A., *et al.* (2004). Solution structure of the SEA Domain from the murine homologue of ovarian cancer antigen CA125 (MUC16). *J. Biol. Chem.* **279**, 13174–13182.

Maier, D. (2006). Hairless: the ignored antagonist of the Notch signalling pathway. *Hereditas* **143**, 212–221.

Maier, D., Marquart, J., Thompson-Fontaine, A., Beck, I., Wurmbach, E., and Preiss, A. (1997). In vivo structure-function analysis of Drosophila Hairless. *Mech. Dev.* **67**, 97–106.

Malecki, M. J., Sanchez-Irizarry, C., Mitchell, J. L., Histen, G., Xu, M. L., Aster, J. C., and Blacklow, S. C. (2006). Leukemia-associated mutations within the NOTCH1 heterodimerization domain fall into at least two distinct mechanistic classes. *Mol. Cell. Biol.* **26**, 4642–4651.

Maskos, K., Fernandez-Catalan, C., Huber, R., Bourenkov, G. P., Bartunik, H., Ellestad, G. A., Reddy, P., Wolfson, M. F., Rauch, C. T., Castner, B. J., *et al.* (1998). Crystal structure of the catalytic domain of human tumor necrosis factor-alpha-converting enzyme. *Proc. Natl. Acad. Sci. U.S.A* **95**, 3408–3412.

Matsunami, N., Hamaguchi, Y., Yamamoto, Y., Kuze, K., Kangawa, K., Matsuo, H., Kawaichi, M., and Honjo, T. (1989). A protein binding to the J kappa recombination sequence of immunoglobulin genes contains a sequence related to the integrase motif. *Nature* **342**, 934–937.

Micchelli, C. A., Rulifson, E. J., and Blair, S. S. (1997). The function and regulation of cut expression on the wing margin of Drosophila: Notch, Wingless and a dominant negative role for Delta and Serrate. *Development* **124**, 1485–1495.

Minoguchi, S., Taniguchi, Y., Kato, H., Okazaki, T., Strobl, L. J., Zimber-Strobl, U., Bornkamm, G. W., and Honjo, T. (1997). RBP-L, a transcription factor related to RBP-Jkappa. *Mol. Cell. Biol.* **17**, 2679–2687.

Moellering, R. E., Cornejo, M., Davis, T. N., Del Bianco, C., Aster, J. C., Blacklow, S. C., Kung, A. L., Gilliland, D. G., Verdine, G. L., and Bradner, J. E. (2009). Direct inhibition of the NOTCH transcription factor complex. *Nature* **462**, 182–188.

Moloney, D. J., Panin, V. M., Johnston, S. H., Chen, J., Shao, L., Wilson, R., Wang, Y., Stanley, P., Irvine, K. D., Haltiwanger, R. S., *et al.* (2000). Fringe is a glycosyltransferase that modifies Notch. *Nature* **406**, 369–375.

Morel, V., and Schweisguth, F. (2000). Repression by suppressor of hairless and activation by Notch are required to define a single row of single-minded expressing cells in the Drosophila embryo. *Genes Dev.* **14**, 377–388.

Moshkin, Y. M., Kan, T. W., Goodfellow, H., Bezstarosti, K., Maeda, R. K., Pilyugin, M., Karch, F., Bray, S. J., Demmers, J. A. A., and Verrijzer, C. P. (2009). Histone chaperones ASF1 and NAP1 differentially modulate removal of active histone marks by LID-RPD3 complexes during NOTCH silencing. *Mol. Cell.* **35**, 782–793.

Muller, U., and Littlewood-Evans, A. (2001). Mechanisms that regulate mechanosensory hair cell differentiation. *Trends Cell Biol.* **11**, 334–342.

Mumm, J. S., Schroeter, E. H., Saxena, M. T., Griesemer, A., Tian, X., Pan, D. J., Ray, W. J., and Kopan, R. (2000). A ligand-induced extracellular cleavage regulates gamma-secretase-like proteolytic activation of Notch1. *Mol. Cell.* **5**, 197–206.

Murzin, A. G., Lesk, A. M., and Chothia, C. (1992). beta-Trefoil fold. Patterns of structure and sequence in the Kunitz inhibitors interleukins-1 beta and 1 alpha and fibroblast growth factors. *J. Mol. Biol.* **223**, 531–543.

Nam, Y., Sliz, P., Pear, W. S., Aster, J. C., and Blacklow, S. C. (2007). Cooperative assembly of higher-order Notch complexes functions as a switch to induce transcription. *Proc. Natl. Acad. Sci. U.S.A* **104**, 2103–2108.

Nam, Y., Sliz, P., Song, L., Aster, J. C., and Blacklow, S. C. (2006). Structural basis for cooperativity in recruitment of MAML coactivators to Notch transcription complexes. *Cell* **124**, 973–983.

Nam, Y., Weng, A. P., Aster, J. C., and Blacklow, S. C. (2003). Structural requirements for assembly of the CSL.intracellular Notch1.Mastermind-like 1 transcriptional activation complex. *J. Biol. Chem.* **278**, 21232–21239.

Neves, A., English, K., and Priess, J. R. (2007). Notch-GATA synergy promotes endoderm-specific expression of ref-1 in C. elegans. *Development* **134**, 4459–4468.

Newberry, E. P., Latifi, T., and Towler, D. A. (1999). The RRM domain of MINT, a novel Msx2 binding protein, recognizes and regulates the rat osteocalcin promoter. *Biochemistry* **38**, 10678–10690.

Nichols, J. T., Miyamoto, A., Olsen, S. L., D'Souza, B., Yao, C., and Weinmaster, G. (2007). DSL ligand endocytosis physically dissociates Notch1 heterodimers before activating proteolysis can occur. *J. Cell Biol.* **176**, 445–458.

Noguera-Troise, I., Daly, C., Papadopoulos, N. J., Coetzee, S., Boland, P., Gale, N. W., Lin, H. C., Yancopoulos, G. D., and Thurston, G. (2006). Blockade of Dll4 inhibits tumour growth by promoting non-productive angiogenesis. *Nature* **444**, 1032–1037.

Okajima, T., and Irvine, K. D. (2002). Regulation of notch signaling by o-linked fucose. *Cell* **111**, 893–904.

Ong, C. T., Cheng, H. T., Chang, L. W., Ohtsuka, T., Kageyama, R., Stormo, G. D., and Kopan, R. (2006). Target selectivity of vertebrate notch proteins. *J. Biol. Chem.* **281**, 5106–5119.

Oswald, F., Kostezka, U., Astrahantseff, K., Bourteele, S., Dillinger, K., Zechner, U., Ludwig, L., Wilda, M., Hameister, H., Knochel, W., *et al.* (2002). SHARP is a novel component of the Notch/RBP-Jkappa signalling pathway. *EMBO J.* **21**, 5417–5426.

Oswald, F., Winkler, M., Cao, Y., Astrahantseff, K., Bourteele, S., Knöchel, W., and Borggrefe, T. (2005). RBP-Jkappa/SHARP recruits CtIP/CtBP corepressors to silence Notch target genes. *Mol. Cell. Biol.* **25**, 10379–10390.

Panin, V. M., Papayannopoulos, V., Wilson, R., and Irvine, K. D. (1997). Fringe modulates Notch-ligand interactions. *Nature* **387**, 908–912.

Park, J. T., Li, M., Nakayama, K., Mao, T. L., Davidson, B., Zhang, Z., Kurman, R. J., Eberhart, C. G., Shih Ie, M., and Wang, T. L. (2006). Notch3 gene amplification in ovarian cancer. *Cancer Res.* **66**, 6312–6318.

Parks, A. L., Klueg, K. M., Stout, J. R., and Muskavitch, M. A. (2000). Ligand endocytosis drives receptor dissociation and activation in the Notch pathway. *Development* **127**, 1373–1385.

Petcherski, A. G., and Kimble, J. (2000). LAG-3 is a putative transcriptional activator in the C. elegans Notch pathway. *Nature* **405**, 364–368.

Peters, N., Opherk, C., Zacherle, S., Capell, A., Gempel, P., and Dichgans, M. (2004). CADASIL-associated Notch3 mutations have differential effects both on ligand binding and ligand-induced Notch3 receptor signaling through RBP-Jk. *Exp. Cell Res.* **299**, 454–464.

Pitsouli, C., and Delidakis, C. (2005). The interplay between DSL proteins and ubiquitin ligases in Notch signaling. *Development* **132**, 4041–4050.

Prevorovský, M., Půta, F., and Folk, P. (2007). Fungal CSL transcription factors. *BMC Genomics* **8**, 233.

Rand, M. D., Grimm, L. M., Artavanis-Tsakonas, S., Patriub, V., Blacklow, S. C., Sklar, J., and Aster, J. C. (2000). Calcium depletion dissociates and activates heterodimeric notch receptors. *Mol. Cell. Biol.* **20**, 1825–1835.

Rebay, I., Fehon, R. G., and Artavanis-Tsakonas, S. (1993). Specific truncations of Drosophila Notch define dominant activated and dominant negative forms of the receptor. *Cell* **74**, 319–329.

Rebay, I., Fleming, R. J., Fehon, R. G., Cherbas, L., Cherbas, P., and Artavanis-Tsakonas, S. (1991). Specific EGF repeats of Notch mediate interactions with Delta and Serrate: implications for Notch as a multifunctional receptor. *Cell* **67**, 687–699.

Rechsteiner, M., and Rogers, S. W. (1996). PEST sequences and regulation by proteolysis. *Trends Biochem. Sci.* **21**, 267–271.

Ridgway, J., Zhang, G., Wu, Y., Stawicki, S., Liang, W. C., Chanthery, Y., Kowalski, J., Watts, R. J., Callahan, C., Kasman, I., *et al.* (2006). Inhibition of Dll4 signalling inhibits tumour growth by deregulating angiogenesis. *Nature* **444**, 1083–1087.

Roehl, H., Bosenberg, M., Blelloch, R., and Kimble, J. (1996). Roles of the RAM and ANK domains in signaling by the C. elegans GLP-1 receptor. *EMBO J.* **15**, 7002–7012.

Saint just ribeiro, M., Hansson, M. L., Lindberg, M. J., Popko-Scibor, A. E., and Wallberg, A. E. (2009). GSK3beta is a negative regulator of the transcriptional coactivator MAML1. *Nucleic Acids Res.* **37**, 6691–6700.

Saint Just Ribeiro, M., Hansson, M. L. ,, and Wallberg, A. E. (2007). A proline repeat domain in the Notch co-activator MAML1 is important for the p300-mediated acetylation of MAML1. *Biochem. J.* **404**, 289–298.

Salat, D., Liefke, R., Wiedenmann, J., Borggrefe, T., and Oswald, F. (2008). ETO, but not leukemogenic fusion protein AML1/ETO, augments RBP-Jkappa/SHARP-mediated repression of notch target genes. *Mol. Cell. Biol.* **28**, 3502–3512.

Sanchez-Irizarry, C., Carpenter, A. C., Weng, A. P., Pear, W. S., Aster, J. C., and Blacklow, S. C. (2004). Notch subunit heterodimerization and prevention of ligand-independent proteolytic activation depend, respectively, on a novel domain and the LNR repeats. *Mol. Cell. Biol.* **24**, 9265–9273.

Schweisguth, F., and Posakony, J. W. (1994). Antagonistic activities of Suppressor of Hairless and Hairless control alternative cell fates in the Drosophila adult epidermis. *Development* **120**, 1433–1441.

Seugnet, L., Simpson, P., and Haenlin, M. (1997). Requirement for dynamin during Notch signaling in Drosophila neurogenesis. *Dev. Biol.* **192**, 585–598.

Shao, L., Moloney, D. J., and Haltiwanger, R. (2003). Fringe modifies O-fucose on mouse Notch1 at epidermal growth factor-like repeats within the ligand-binding site and the Abruptex region. *J. Biol. Chem.* **278**, 7775–7782.

Shi, S., and Stanley, P. (2003). Protein O-fucosyltransferase 1 is an essential component of Notch signaling pathways. *Proc. Natl. Acad. Sci. U.S.A* **100**, 5234–5239.

Shi, Y., Downes, M., Xie, W., Kao, H. Y., Ordentlich, P., Tsai, C. C., Hon, M., and Evans, R. M. (2001). Sharp, an inducible cofactor that integrates nuclear receptor repression and activation. *Genes Dev.* **15**, 1140–1151.

Shimizu, K., Chiba, S., Kumano, K., Hosoya, N., Takahashi, T., Kanda, Y., Hamada, Y., Yazaki, Y., and Hirai, H. (1999). Mouse jagged1 physically interacts with notch2 and other notch receptors. *J. Biol. Chem.* **274**, 32961–32969.

Shimizu, K., Chiba, S., Saito, T., Kumano, K., Takahashi, T., and Hirai, H. (2001). Manic fringe and lunatic fringe modify different sites of the Notch2 extracellular region, resulting in different signaling modulation. *J. Biol. Chem.* **276**, 25753–25758.

Struhl, G., and Greenwald, I. (1999). Presenilin is required for activity and nuclear access of Notch in Drosophila. *Nature* **398**, 522–525.

Sun, X., and Artavanis-Tsakonas, S. (1997). Secreted forms of DELTA and SERRATE define antagonists of Notch signaling in Drosophila. *Development* **124**, 3439–3448.

Surendran, K., Boyle, S., Barak, H., Kim, M., Stomberski, C., McCright, B., and Kopan, R. (2010). The contribution of Notch1 to nephron segmentation in the developing kidney is revealed in a sensitized Notch2 background and can be augmented by reducing Mint dosage. *Dev. Biol.* **337**, 386–395.

Tamura, K., Taniguchi, Y., Minoguchi, S., Sakai, T., Tun, T., Furukawa, T., and Honjo, T. (1995). Physical interaction between a novel domain of the receptor Notch and the transcription factor RBP-J kappa/Su(H). *Curr. Biol.* **5**, 1416–1423.

Taniguchi, Y., Furukawa, T., Tun, T., Han, H., and Honjo, T. (1998). LIM protein KyoT2 negatively regulates transcription by association with the RBP-J DNA-binding protein. *Mol. Cell. Biol.* **18**, 644–654.

Thomas, W. E. (2009). Mechanochemistry of receptor-ligand bonds. *Curr. Opin. Struct. Biol.* **19**, 50–55.

Thomas, W. E., Vogel, V., and Sokurenko, E. (2008). Biophysics of catch bonds. *Annu. Rev. Biophys.* **37**, 399–416.

Tsuji, M., Shinkura, R., Kuroda, K., Yabe, D., and Honjo, T. (2007). Msx2-interacting nuclear target protein (Mint) deficiency reveals negative regulation of early thymocyte differentiation by Notch/RBP-J signaling. *Proc. Natl. Acad Sci. USA* **104**, 1610–1615.

Tun, T., Hamaguchi, Y., Matsunami, N., Furukawa, T., Honjo, T., and Kawaichi, M. (1994). Recognition sequence of a highly conserved DNA binding protein RBP-J kappa. *Nucleic Acids Res.* **22**, 965–971.

Vardar, D., North, C. L., Sanchez-Irizarry, C., Aster, J. C., and Blacklow, S. C. (2003). Nuclear magnetic resonance structure of a prototype Lin12-Notch repeat module from human Notch1. *Biochemistry* **42**, 7061–7067.

Varnum-Finney, B., Purton, L. E., Yu, M., Brashem-Stein, C., Flowers, D., Staats, S., Moore, K. A., Le Roux, I., Mann, R., Gray, G., *et al.* (1998). The Notch ligand, Jagged-1, influences the development of primitive hematopoietic precursor cells. *Blood* **91**, 4084–4091.

Varnum-Finney, B., Wu, L., Yu, M., Brashem-Stein, C., Staats, S., Flowers, D., Griffin, J. D., and Bernstein, I. D. (2000). Immobilization of Notch ligand, Delta-1, is required for induction of notch signaling. *J. Cell. Sci.* **113**(*Pt 23*), 4313–4318.

Vooijs, M., Schroeter, E. H., Pan, Y., Blandford, M., and Kopan, R. (2004). Ectodomain shedding and intramembrane cleavage of mammalian Notch proteins is not regulated through oligomerization. *J. Biol. Chem.* **279**, 50864–50873.

Wallberg, A. E., Pedersen, K., Lendahl, U., and Roeder, R. G. (2002). p300 and PCAF act cooperatively to mediate transcriptional activation from chromatin templates by notch intracellular domains in vitro. *Mol. Cell. Biol.* **22**, 7812–7819.

Waltzer, L., Bourillot, P. Y., Sergeant, A., and Manet, E. (1995). RBP-J kappa repression activity is mediated by a co-repressor and antagonized by the Epstein-Barr virus transcription factor EBNA2. *Nucleic Acids Res.* **23**, 4939–4945.

Wang, Y., Lee, G. F., Kelley, R. F., and Spellman, M. W. (1996). Identification of a GDP-L-fucose:polypeptide fucosyltransferase and enzymatic addition of O-linked fucose to EGF domains. *Glycobiology* **6**, 837–842.

Wang, Y., Shao, L., Shi, S., Harris, R. J., Spellman, M. W., Stanley, P., and Haltiwanger, R. S. (2001). Modification of epidermal growth factor-like repeats with O-fucose. *J. Biol. Chem.* **276**, 40338–40345.

Wang, Y., and Spellman, M. W. (1998). Purification and characterization of a GDP-fucose: polypeptide fucosyltransferase from Chinese hamster ovary cells. *J. Biol. Chem.* **273**, 8112–8118.

Weng, A. P., Ferrando, A. A., Lee, W., Morris, J. Pt., Silverman, L. B., Sanchez-Irizarry, C., Blacklow, S. C., Look, A. T., and Aster, J. C. (2004). Activating mutations of NOTCH1 in human T cell acute lymphoblastic leukemia. *Science* **306**, 269–271.

Weng, A. P., Nam, Y., Wolfe, M. S., Pear, W. S., Griffin, J. D., Blacklow, S. C., and Aster, J. C. (2003). Growth suppression of pre-T acute lymphoblastic leukemia cells by inhibition of notch signaling. *Mol. Cell. Biol.* **23**, 655–664.

Westhoff, B., Colaluca, I. N., D'Ario, G., Donzelli, M., Tosoni, D., Volorio, S., Pelosi, G., Spaggiari, L., Mazzarol, G., Viale, G., *et al.* (2009). Alterations of the Notch pathway in lung cancer. *Proc. Natl. Acad. Sci. U.S.A* **106**, 22293–22298.

Wilson, J. J., and Kovall, R. A. (2006). Crystal structure of the CSL-Notch-Mastermind ternary complex bound to DNA. *Cell* **124**, 985–996.

Wolfe, M. S., Xia, W., Ostaszewski, B. L., Diehl, T. S., Kimberly, W. T., and Selkoe, D. J. (1999). Two transmembrane aspartates in presenilin-1 required for presenilin endoproteolysis and gamma-secretase activity. *Nature* **398**, 513–517.

Wu, L., Aster, J. C., Blacklow, S. C., Lake, R., Artavanis-Tsakonas, S., and Griffin, J. D. (2000). MAML1, a human homologue of Drosophila mastermind, is a transcriptional co-activator for NOTCH receptors. *Nat. Genet.* **26**, 484–489.

Wu, Y., Cain-Hom, C., Choy, L., Hagenbeek, T. J., de Leon, G. P., Chen, Y., Finkle, D., Venook, R., Wu, X., Ridgway, J., *et al.* (2010). Therapeutic antibody targeting of individual Notch receptors. *Nature* **464**, 1052–1057.

Xu, A., Haines, N., Dlugosz, M., Rana, N. A., Takeuchi, H., Haltiwanger, R. S., and Irvine, K. D. (2007). In vitro reconstitution of the modulation of Drosophila Notch-ligand binding by Fringe. *J. Biol. Chem.* **282**, 35153–35162.

Yan, M., Callahan, C. A., Beyer, J. C., Allamneni, K. P., Zhang, G., Ridgway, J. B., Niessen, K., and Plowman, G. D. (2010). Chronic DLL4 blockade induces vascular neoplasms. *Nature* **463**, E

Ye, Y., Lukinova, N., and Fortini, M. E. (1999). Neurogenic phenotypes and altered Notch processing in Drosophila Presenilin mutants. *Nature* **398**, 525–529.

Yoo, A. S., Bais, C., and Greenwald, I. (2004). Crosstalk between the EGFR and LIN-12/ Notch pathways in C. elegans vulval development elegans Notch pathway. *Nat Sci.* **303**, 663–666.

Zhou, S., Fujimuro, M., Hsieh, J. J., Chen, L., and Hayward, S. D. (2000a). A role for SKIP in EBNA2 activation of CBF1-repressed promoters. *J Virol.* **74**, 1939–1947.

Zhou, S., Fujimuro, M., Hsieh, J. J., Chen, L., Miyamoto, A., Weinmaster, G., and Hayward, S. D. (2000b). SKIP, a CBF1-associated protein, interacts with the ankyrin repeat domain of NotchIC To facilitate NotchIC function. *Mol. Cell. Biol.* **20**, 2400–2410.

Zweifel, M. E., and Barrick, D. (2001). Studies of the ankyrin repeats of the Drosophila melanogaster Notch receptor. 1. Solution conformational and hydrodynamic properties. *Biochemistry* **40**, 14344–14356.

Zweifel, M. E., Leahy, D. J., Hughson, F. M., and Barrick, D. (2003). Structure and stability of the ankyrin domain of the Drosophila Notch receptor. *Protein Sci.* **12**, 2622–2632.

CHAPTER THREE

Canonical and Non-Canonical Notch Ligands

Brendan D'Souza,[*] Laurence Meloty-Kapella,[*] *and* Gerry Weinmaster[*,†,‡]

Contents

[*] Department of Biological Chemistry, David Geffen School of Medicine, University of California, Los Angeles, CA, USA
[†] Molecular Biology Institute, University of California, Los Angeles, CA, USA
[‡] Jonsson Comprehensive Cancer Center, Los Angeles, California, USA

Current Topics in Developmental Biology, Volume 92
ISSN 0070-2153, DOI 10.1016/S0070-2153(10)92003-6

Abstract

Notch signaling induced by canonical Notch ligands is critical for normal embryonic development and tissue homeostasis through the regulation of a variety of cell fate decisions and cellular processes. Activation of Notch signaling is normally tightly controlled by direct interactions with ligand-expressing cells, and dysregulated Notch signaling is associated with developmental abnormalities and cancer. While canonical Notch ligands are responsible for the majority of Notch signaling, a diverse group of structurally unrelated noncanonical ligands has also been identified that activate Notch and likely contribute to the pleiotropic effects of Notch signaling. Soluble forms of both canonical and noncanonical ligands have been isolated, some of which block Notch signaling and could serve as natural inhibitors of this pathway. Ligand activity can also be indirectly regulated by other signaling pathways at the level of ligand expression, serving to spatiotemporally compartmentalize Notch signaling activity and integrate Notch signaling into a molecular network that orchestrates developmental events. Here, we review the molecular mechanisms underlying the dual role of Notch ligands as activators and inhibitors of Notch signaling. Additionally, evidence that Notch ligands function independent of Notch is presented. We also discuss how ligand posttranslational modification, endocytosis, proteolysis, and spatiotemporal expression regulate their signaling activity.

1. Introduction

The Notch pathway functions as a core signaling system during embryonic development and is also required for the regulation of tissue homeostasis and stem cell maintenance in the adult (Artavanis-Tsakonas *et al.*, 1999; Gridley, 1997, 2003). Ligand-induced Notch signaling directs the specification of a variety of cell types and contributes to tissue patterning and morphogenesis through effects on cellular differentiation, proliferation, survival, and apoptosis (Bray, 2006; Fiuza and Arias, 2007). Given the widespread usage of the Notch pathway in different cell types and cellular processes, it is not surprising that defects in Notch ligands are associated with hereditary diseases such as Alagille syndrome and spondylocostal dysostosis and that aberrant ligand expression is detected in several cancers (Koch and Radtke, 2007; Leong and Karsan, 2006; Piccoli and Spinner, 2001; Turnpenny *et al.*, 2007).

The canonical ligands that bind and activate Notch receptors are integral cell surface proteins, and thus activation of Notch signaling is dependent on direct cell-to-cell interactions. The transmembrane nature of Notch ligands serves to limit signaling to local cell interactions and additionally provides a signaling system for cells to communicate directly with their neighbors. Interestingly, during certain developmental processes, ligands have been found to activate Notch expressed on the surface of distantly located cells. Such long range signaling may utilize actin-based cellular projections to deliver activating signals to Notch at distant sites (de Joussineau *et al.*, 2003). In support of such a model, the ligand Delta appears to concentrate in filopodia-like projections, possibly inducing and stabilizing these structures to facilitate long-range signaling (de Joussineau *et al.*, 2003; Renaud and Simpson, 2001). Similarly, the *Caenorhabditis elegans*, distal tip cell has long cellular processes that contain the ligand Lag2 and appear to extend all the way to the mitotic/meiotic border where they regulate proliferation of the germ line through activation of the Notch homolog Glp1 (Fitzgerald and Greenwald, 1995).

Signaling induced by Notch cells following engagement with ligand cells involves a series of proteolytic cleavages in Notch to release the intracellular domain (ICD) that functions directly as the biologically active signal transducer (Kopan and Ilagan, 2009). During maturation and trafficking to the cell surface, the Notch receptor is processed by a furin-like protease to produce an intramolecular heterodimer that predisposes Notch to proteolytic activation by ligand. Interactions with ligand cells result in an extracellular juxtamembrane cleavage in Notch catalyzed by an A-Disintegrin-And-Metalloprotease (ADAM), which is followed by an intramembrane cleavage by γ-secretase to release the Notch intracellular domain (NICD) from the membrane (Fig. 3.1). NICD translocates to the nucleus where it functions directly in signal transduction through complexing with the CSL (*C*BF1, *S*u(H), *L*AG1) DNA binding protein and transcriptional coactivators to switch on expression of Notch target genes such as hairy and enhancer of split (HES) family. The mechanism and details of Notch transcriptional activation are covered extensively in Chapter 8. In addition to the well-characterized role for the activation of Notch signaling through cell–cell interactions (*trans*-interactions), ligands can also interact with Notch cell autonomously (*cis*-interactions) leading to inhibition of Notch signaling. The nature and mechanisms underlying the inhibitory role of

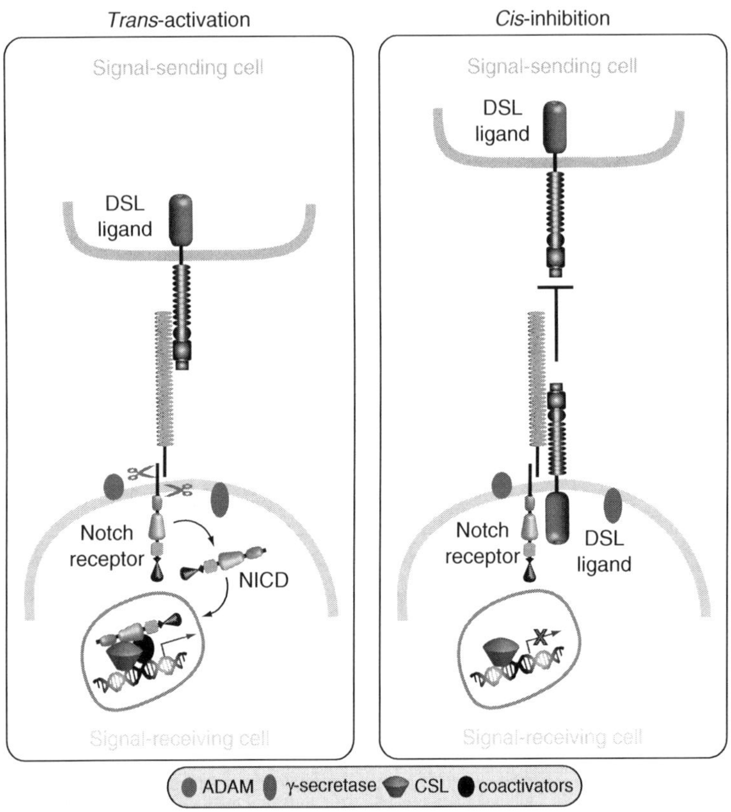

Fig. 3.1 Models for DSL ligand *trans*-activation and *cis*-inhibition in Notch signaling. Ligand expressed on the surface of the signal-sending cell binds to Notch expressed on the surface of the signal-receiving cell (*trans*-interactions) and induces sequential cleavages by *A*-Disintegrin-*A*nd-*M*etalloprotease (ADAM) and γ-secretase in Notch releasing the Notch intracellular domain (NICD) from the membrane. NICD translocates to the nucleus where it directly interacts with the CSL (*C*BF1, *S*u(H), *L*AG1) transcription factor and recruits coactivators to induce Notch target gene expression. Ligand binding to Notch expressed in the same cell (*cis*-interactions) prevents Notch activation by *trans*-ligand by competing with *trans*-ligand for Notch binding. (See Color Insert.)

Notch ligands will be discussed in Section 3 of this review. Additional characteristics of canonical and noncanonical Notch ligands required to activate signaling are discussed below.

2. Canonical Notch Ligand Structure

The majority of Notch signaling is induced by a family of DSL ligands that are characterized by the presence of a DSL (*D*elta, *S*errate, and *L*ag2) domain (Henderson *et al.*, 1994; Tax *et al.*, 1994). The mammalian DSL ligands are classified as either Delta-like (Dll1, Dll3, and Dll4) or Serrate (Jagged)-like (Jagged1 and Jagged2) based on homology to their *Drosophila* prototypes Delta and Serrate (Kopan and Ilagan, 2009). DSL ligands are type 1 transmembrane proteins that share a common modular arrangement in their extracellular domains (ECDs)comprising an N-terminal (NT) domain followed by the DSL domain and multiple tandemly arrayed epidermal growth factor (EGF)-like repeats (both calcium binding and noncalcium binding, see Fig. 3.2).

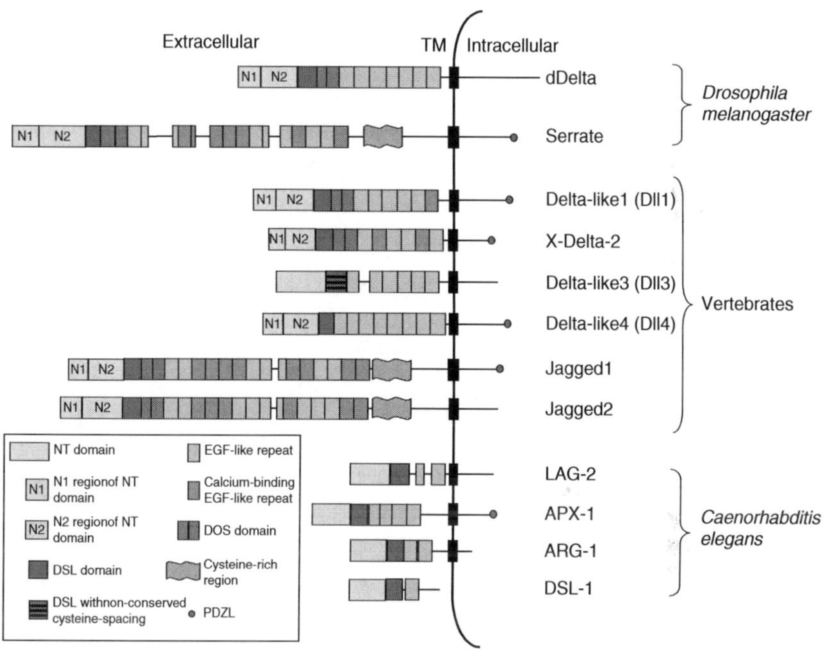

Fig. 3.2 Structural domains of canonical ligands. The extracellular domains of canonical ligands are characterized by the presence of an N-terminal (NT) domain followed by a Delta/Serrate/LAG-2 (DSL) domain and multiple tandemly arranged epidermal growth factor (EGF)-like repeats (see text for details). The DSL domain together with the flanking NT domain and the first two EGF repeats containing the Delta and OSM-11-like proteins (DOS) motif are required for canonical ligands to bind Notch. The NT domain of vertebrate and *Drosophila* ligands is subdivided into a region containing six conserved cysteine residues, N1 and a cysteine-free region, N2. Serrate/Jagged ligands contain an additional cysteine-rich region not present in Delta-like ligands. The intracellular domains of some canonical ligands contain a carboxy-terminal PSD-95/Dlg/ZO-1-ligand (PDZL) motif that plays a role independent of Notch signaling. *C. elegans* DSL ligands lack a DOS motif but have been proposed to cooperate with DOS-only containing ligands (not depicted) to activate Notch signaling. Dll3 is the most structurally divergent vertebrate DSL ligand and lacks structural features required by other DSL ligands to bind and activate Notch. (See Color Insert.)

The DSL is a degenerate EGF-like repeat that is necessary but not sufficient for interactions with Notch (Shimizu *et al.*, 1999). Mutations in conserved residues within the DSL domain are associated with losses in Notch signaling in both vertebrate and invertebrates (Henderson *et al.*, 1994, 1997; Morrissette *et al.*, 2001; Parks *et al.*, 2006; Tax *et al.*, 1994; Warthen *et al.*, 2006). In particular, DSL mutations in Jagged1 are linked to Alagille syndrome. In addition, a conserved motif called DOS (*D*elta and *OS*M-11-like proteins) has been identified within the first two EGF-like repeats that are proposed to cooperate with the DSL domain (Komatsu *et al.*, 2008). Mutational and structural studies indicate a contributory role for the DOS domain in Notch binding and signaling distinguishing them from the remaining EGF-like repeats (Cordle *et al.*, 2008; Komatsu *et al.*, 2008; Parks *et al.*, 2006; Shimizu *et al.*, 1999). In particular, the sequence and spacing within the DOS are important for signaling (Geffers *et al.*, 2007). Furthermore, mutations associated with Alagille syndrome and the congenital disorder tetralogy of Fallot map to the DOS motif of Jagged1, highlighting the importance of this region in Notch signaling (Eldadah *et al.*, 2001; Guarnaccia *et al.*, 2009; Warthen *et al.*, 2006). Surprisingly, Dll4 and Dll3 and all *C. elegans* DSL ligands lack a DOS motif and it has been proposed that optimal activation of Notch signaling by DSL domain-only containing ligands requires cooperative Notch binding by DOS domain-containing noncanonical ligands (Komatsu *et al.*, 2008).

In addition to the DSL and DOS domains, sequences NT to the DSL are also conserved among the canonical ligands that appear important for function (Fleming, 1998; Henderson *et al.*, 1997; Parks *et al.*, 2006). The NT domain can be subdivided into two distinct regions based on differential cysteine content: N1 is cysteine-rich while N2 is cysteine-free (Parks *et al.*, 2006), and Alagille mutations map to the N1 and N2 regions of Jagged1 (Morrissette *et al.*, 2001; Warthen *et al.*, 2006). More recently, a conserved glycosphingolipid (GSL)-binding motif (GBM) has been identified within the N2 region that may regulate ligand membrane association and endocytosis (Hamel *et al.*, 2010).

Despite the similarity in the overall modular organization of the ECDs (Fig. 3.2), some structural differences exist among the DSL ligands. For example, the number of EGF-like repeats varies as does the spacing between this motif. Moreover, the Serrate-like Jagged ligands have a cysteine-rich region sharing partial homology with the von Willebrand factor type C domain that is absent from Delta ligands (Vitt *et al.*, 2001). Although the non-DOS-containing EGF-like repeats have not been reported to regulate signaling activity (Henderson *et al.*, 1997; Parks *et al.*, 2006), mutations in some of these repeats in Jagged1 are associated with Alagille syndrome (Morrissette *et al.*, 2001; Warthen *et al.*, 2006).

The ICDs of DSL ligands exhibit the lowest level of overall sequence homology (Pintar *et al.*, 2007). With the exception of Dll3, they contain multiple lysine residues that are potential sites for modification by distinct E3 ubiquitin ligases (outlined in Section 4). Although ubiquitination is critical for ligands to activate Notch signaling, the role that this modification plays is poorly defined. In addition, most, but not all DSL, ligands have a C-terminal PDZ (PSD-95/Dlg/ZO-1)-ligand motif that promotes interactions with the actin cytoskeleton and appears to play a role independent of Notch signaling (Pintar *et al.*, 2007). Although PDZ-ligand motifs are predicted for some invertebrate DSL ligands, including *Drosophila* Serrate and *C. elegans* APX-1 (Sheng and Sala, 2001), the functional relevance remains to be determined.

Finally, Dll3 is the most structurally divergent DSL ligand having a degenerate DSL domain (Dunwoodie *et al.*, 1997) and lacking both a DOS motif (Komatsu *et al.*, 2008) and an ICD lysine residues (Pintar *et al.*, 2007). Although these structural features are critical for ligand signaling activity, losses in Dll3 are associated with vertebral segmentation and rib malformations similar to those caused by defects in Notch signaling (Dunwoodie, 2009). Dll3, however, does not bind Notch in *trans* or activate Notch signaling (Ladi *et al.*, 2005), and the majority of Dll3 is detected in the Golgi, with relatively little, if any, cell surface expression (Geffers *et al.*, 2007). Gene replacement studies in mice clearly show that Dll3 cannot substitute for the loss of Dll1 (Geffers *et al.*, 2007), indicating that these DSL ligands are not functionally equivalent. In contrast to Dll1 that both activates and inhibits Notch signaling, Dll3 functions exclusively as a Notch antagonist (Ladi *et al.*, 2005). Despite these findings, it is still unclear how Dll3 functions in Notch signaling, and while the Dll3 structural differences are predicted to perturb ligand signaling activity, it is difficult to reconcile how Dll3 in the Golgi would participate in Notch signaling.

3. Canonical Ligands as Inhibitors of Notch Signaling

The Notch receptors and DSL ligands are widely expressed during development, and in many cases, interacting cells express both ligands and receptors. Cells take on distinct fates because Notch signaling is consistently activated in only one of the two interacting cells, indicating that the signaling polarity must be highly regulated. The relative levels of Notch and its ligands present on interacting cells are thought to establish the signaling polarity necessary to ensure that the correct cell fates are generated at the right time in development. In fact, developmental processes are sensitive to Notch ligand and receptor gene dosage, underscoring the importance of Notch ligand and receptor expression levels for normal signaling. In humans, haploinsufficiency of either Jagged1 or Notch2 is associated with Alagille syndrome

(McDaniell *et al.*, 2006), while Notch1 haploinsufficiency is implicated in a subtype of inherited aortic disease (Garg *et al.*, 2005).

Studies in flies and worms have identified positive and negative transcriptional feedback mechanisms that amplify small differences in Notch and DSL ligand expression that could introduce a bias for which of the interacting cells sends or receives the Notch signal (Greenwald and Rubin, 1992; Seugnet *et al.*, 1997b). If this were the case, cells competent to send a signal would be expected to display higher DSL ligand levels than cells receiving the Notch signal; however, Delta expression appears uniform among cells undergoing lateral inhibition during selection of the neural fate (Kooh *et al.*, 1993; Kopczynski and Muskavitch, 1989). Therefore, mechanisms in addition to transcription must exist to ensure fidelity in cell fate decisions regulated by Notch signaling. In this regard, interactions between Notch and its ligand in the same cell may provide additional mechanisms to regulate the cell's potential to send or receive a Notch signal.

3.1. *Cis*-interactions between ligand and Notch inhibit signaling by *trans*-ligand

In contrast to the *trans*-interactions between Notch ligand and receptor cells that activate signaling (Fig. 3.1), interactions between Notch ligands and receptors in the same cell result in inhibition of signaling through a poorly defined process of *cis*-inhibition (Glittenberg *et al.*, 2006; Jacobsen *et al.*, 1998; Klein and Arias, 1998; Klein *et al.*, 1997; Ladi *et al.*, 2005; Micchelli *et al.*, 1997; Sakamoto *et al.*, 2002a). Nonetheless, *cis*-inhibition appears to be a particularly important mechanism to establish and maintain the signaling polarity required for specific Notch-dependent cell fate determinations (Becam *et al.*, 2010; de Celis and Bray, 1997; Jacobsen *et al.*, 1998; Klein and Arias, 1998; Klein *et al.*, 1997; Matsuda and Chitnis, 2009; Miller *et al.*, 2009; Sprinzak *et al.*, 2010). Ectopic expression of truncated ligands lacking most of the ICD function cell autonomously to block Notch signaling and promote retinal neurogenesis and neurite outgrowth as well as inhibit keratinocyte differentiation within the epidermal stem cell niche (Dorsky *et al.*, 1997; Franklin *et al.*, 1999; Henrique *et al.*, 1997; Lowell *et al.*, 2000; Lowell and Watt, 2001). Although these studies have relied on overexpression of DSL ligands, loss-of-function studies have also demonstrated that endogenous ligands can function in a *cis*-inhibitory manner (Micchelli *et al.*, 1997; Miller *et al.*, 2009).

3.2. *Cis*-interactions between ligand and Notch determine signal polarity

A recent study using mammalian cell culture to manipulate ligand expression levels in Notch expressing cells has provided insight into understanding how *cis-* versus *trans*-ligand expression could influence the Notch signaling response (Sprinzak *et al.*, 2010). Specifically, cells adopt mutually exclusive signaling states so that depending on their relative levels of Notch ligand and receptor they either "send" or "receive," but not both. According to this model, an "ultrasensitive switch" between these states is capable of amplifying small differences between interacting cells even in the absence of transcriptional feedback. One way to set up this signaling asymmetry is to control levels of both ligand and receptor on the cell surface such that the signal-sending cell maintains high ligand surface expression while the signal-receiving cell has high surface Notch. Previously this asymmetry has been explained solely by a feedback mechanism through which activation of Notch downregulates ligand expression at the level of transcription (Greenwald and Rubin, 1992; Seugnet *et al.*, 1997b). Changes in signal-sending and signal-receiving potential, however, have been observed that do not involve overall changes in ligand or receptor transcription (Becam *et al.*, 2010; Sprinzak *et al.*, 2010). These studies suggest that *cis*-interactions between ligands and receptors would mutually inhibit the potential of ligands to signal as well as restrict Notch activation to the receiving cell. Studies in the developing fly eye have provided additional support for ligand *cis*-inhibition in establishing unidirectional signaling and have also suggested a role for maintaining signaling polarity once cell fates have been determined (Miller *et al.*, 2009).

3.3. Molecular mechanisms for ligand *cis*-inhibition of Notch signaling

The molecular mechanism underlying *cis*-inhibition is poorly understood and has remained highly controversial. Competition between *trans*- and *cis*-ligand binding to Notch is likely to underlie the ability of ligands to activate or inhibit Notch signaling. This hypothesis assumes that the ligand–Notch binding interfaces overlap. Consistent with this idea, the Jagged1 DSL domain has been proposed to contain a highly conserved binding site for both *trans*- and *cis*-interactions with Notch (Cordle *et al.*, 2008). At odds with the competition model, the binding sites in Notch for *cis*- and *trans*-interactions might not overlap. Extensive data indicate that the 11th and 12th EGF repeats in Notch are critical for *trans*-ligand binding and signaling activity (see Chapter 2 for details); however, early studies in flies implicated EGF-like repeats 24–29 in *cis*-inhibition (de Celis and Bray, 2000). More recent studies report a requirement for the 11th and 12th EGF repeats in *cis*-inhibition (Becam *et al.*, 2010; Cordle *et al.*, 2008; Fiuza *et al.*, 2010), suggesting that the *cis*- and *trans*-ligand-binding sites in Notch do overlap. Together these findings support a competitive mechanism for ligand–Notch interactions that ultimately results in either *trans*-activation or *cis*-inhibition of Notch signaling.

Interestingly, the switch from an active to inhibited signaling state requires a high threshold of *cis*-ligand, while signaling responses over a range of *trans*-ligand are linear (Sprinzak *et al.*, 2010). Since low levels of activated Notch are sufficient to induce Notch target gene expression (Schroeter *et al.*, 1998), it seems likely that most, if not all, Notch receptors would need to interact with ligand in *cis* for signaling by *trans*-ligand to be suppressed.

Even though both *cis*- and *trans*-interactions with Notch may involve overlapping binding sites, only *trans*-ligand interactions activate Notch. Based on structural studies discussed below, *trans*-ligand is thought to induce conformational changes in Notch that facilitate proteolytic activation required for downstream signaling. Since *cis*-interactions do not lead to Notch activation, ligand–Notch interactions formed within the plane of the same membrane must not be able to induce the conformational changes required to activate Notch. In support of this idea, a recent study has suggested that ligand *cis*-interactions with Notch prevent proteolytic activation (Fiuza *et al.*, 2010).

Although the majority of findings are consistent with *cis*-inhibition involving ligand–receptor interactions at the cell surface, inhibitory *cis*-interactions formed in the secretory pathway have been proposed to prevent Notch receptors from reaching the cell surface to account for losses in signaling (Sakamoto *et al.*, 2002a). At odds with this notion, ligands retained within the biosynthetic pathway are defective in *cis*-inhibition, providing indirect support that ligand–Notch *cis*-interactions occur at the cell surface (Glittenberg *et al.*, 2006; Ladi *et al.*, 2005). Consistent with this, defects in ligand endocytosis that promote accumulation of ligand on the cell surface diminish *trans*-activation yet potentiate *cis*-inhibition (Glittenberg *et al.*, 2006). Together these findings suggest that mechanisms must exist to coordinate the *trans*- and *cis*-activities mediated by ligands.

In addition to ligand–receptor *cis*-interactions inhibiting the ability of Notch cells to receive a signal, similar *cis*-interactions also inhibit the ability of ligand cells to send signals (Becam *et al.*, 2010; Matsuda and Chitnis, 2009; Miller *et al.*, 2009; Sprinzak *et al.*, 2010), indicating that ligand–receptor interactions in the same cell can be mutually inactivating for sending or receiving signals. Although these studies did not detect losses in protein expression, Notch-stimulated endocytosis has been reported to result in a decrease of cell surface ligand available for activation of signaling in adjacent cells (Becam *et al.*, 2010; Matsuda and Chitnis, 2009). Specifically, studies in both zebra fish and flies report that under Notch knock-down conditions the ligands DeltaD and Serrate accumulate on the cell surface, suggesting Notch *cis*-interactions result in removal of cell surface ligand through endocytosis (Becam *et al.*, 2010; Matsuda and Chitnis, 2009). Further, DeltaD–Notch *cis*-interactions have been proposed to inhibit Notch signaling through removing both the ligand and the receptor from the cell surface (Matsuda and Chitnis, 2009). Studies in flies have found that a truncated form of Notch lacking ICD sequences accumulates at the cell surface without increasing levels of Serrate (Becam *et al.*, 2010). This suggests that interaction with the Notch extracellular domain (NECD) is sufficient to promote clearance of cell surface Serrate alone without simultaneous receptor internalization. Importantly, the 11th and 12th EGF-like repeats that function in *trans*-ligand binding are also required for this clearance and inhibitory affect on Serrate signaling. Interestingly, not all DSL ligands are susceptible to downregulation by Notch *cis*-interactions, and the molecular basis and biological relevance of these findings are unclear.

Even though *cis*-inhibition has been proposed to involve intercellular ligand–ligand interactions leading to a decrease in ligand available for *trans*-activation of Notch, a recent report has challenged this view by demonstrating that cells coexpressing both Notch and Delta form cell aggregates with Delta cells even though these same cell–cell interactions do not activate Notch signaling (Fiuza *et al.*, 2010). Together these findings support the idea that *cis*-inhibition involves ligand–receptor interactions at the surface of the same cell to restrict signaling to one of the two interacting cells.

4. Regulation of Ligand-Induced Notch Signaling by Posttranslational Modification

4.1. Glycosylation

The Notch ligands and receptors undergo O- and N-linked glycan modifications at conserved sequences within specific EGF repeats; however, only O-fucose and O-glucose additions to Notch have so far been reported to affect signaling. N-glycan modifications of Notch, on the other hand, do not appear to alter Notch-dependent development in mice (Haltiwanger and Lowe, 2004). Glycosylation of Notch both positively and negatively regulates signaling induced by ligands, presumably through modulating the strength of the ligand–receptor interactions. Although DSL ligands are glycosylated as reported for Notch (Panin *et al.*, 2002), affects on ligand signaling activity have so far not been detected. Roles for glycosylation in Notch signaling are the subject of the Chapter 4 and the reader is encouraged to consult the indicated chapter for further details.

4.2. Ubiquitination

Posttranslational modification of Notch ligands by ubiquitination regulates cell surface levels and is an absolute requirement for ligand signaling activity

(Chitnis, 2006; Le Borgne, 2006; Le Borgne and Schweisguth, 2003a; Nichols *et al.*, 2007b). As found for *Drosophila* Delta and Serrate, the ICDs of Dll1, Dll4, Jagged1, and Jagged2 contain multiple lysine residues that can serve as potential sites for the addition of ubiquitin by E3 ligases. Two structurally distinct RING-containing E3 ligases, Neuralized (Neur) and Mind bomb (Mib), influence Notch signaling through interacting with and ubiquitinating DSL ligands to enhance their endocytosis. Initial studies in *Drosophila* and *Xenopus* reported that Neur had intrinsic ubiquitin ligase activity and interacted with Delta to promote its internalization and degradation through ubiquitination (Deblandre *et al.*, 2001; Lai *et al.*, 2001; Pavlopoulos *et al.*, 2001; Yeh *et al.*, 2001). Given that Neur is required for Notch signaling these findings are difficult to reconcile; however, based on the cell autonomous activity identified for Neur (Lai and Rubin, 2001a, b; Yeh *et al.*, 2000) a model was suggested in which the loss of cell surface Delta induced by Neur might indirectly enhance Notch signaling through relieving *cis*-inhibition imposed by Delta (Deblandre *et al.*, 2001). More recent studies, however, have clearly shown that *cis*-inhibition does not require ligand ubiquitination (Glittenberg *et al.*, 2006). Moreover, Neur expression is enhanced in signal-sending cells where it is asymmetrically localized and functions to direct cell fate decisions regulated by Notch signaling (Bardin and Schweisguth, 2006; Le Borgne and Schweisguth, 2003b; Morel *et al.*, 2003; Pavlopoulos *et al.*, 2001), providing support for the idea that Neur-induced endocytosis functions to stimulate ligand signaling activity.

Although studies in flies and frogs support a role for Neur in regulating cell surface levels and generating a productive signal, mice lacking the mammalian Neur homolog do not display any obvious Notch developmental phenotypes (Ruan *et al.*, 2001; Vollrath *et al.*, 2001). This surprising finding suggested that mammalian Neur might not be an essential component of the Notch signaling pathway. Alternatively, additional E3 ubiquitin ligases could exist to modify DSL ligands and facilitate Notch activation. Supporting the latter idea, a structurally distinct E3 ligase was subsequently identified as the target of the Mib neurogenic mutant in zebra fish (Chen and Casey Corliss, 2004; Itoh *et al.*, 2003). Mib binds and ubiquitinates Delta and upregulates Delta endocytosis as reported for Neur, but in contrast to Neur, Mib functions exclusively in the ligand cell to activate Notch signaling and is unable to reverse the *cis*-inhibitory effects of Delta on Notch reception (Itoh *et al.*, 2003; Koo *et al.*, 2005a).

Neur and Mib homologs have been isolated from a number of different species, and despite being conserved and having similar molecular activities, Neur and Mib genes may have evolved to serve different roles in vertebrate Notch signaling. *Drosophila* has a single Neur gene (*dNeur*) and two related Mib genes (*dMib1* and *dMib2*) that regulate distinct Notch-dependent developmental events (Lai *et al.*, 2005; Le Borgne *et al.*, 2005; Pitsouli and

Delidakis, 2005; Wang and Struhl, 2005), apparently through differential expression. Both Neur and Mib ubiquitinate the *Drosophila* ligands, Delta and Serrate, and stimulate ligand endocytosis and signaling activities. Importantly, gene rescue experiments indicate that for the most part these structurally distinct E3 ligases are functionally redundant. In contrast to these findings in flies, studies in mice indicate the surprising findings that the mammalian *Neur1* and *Neur2* genes are dispensable for normal development. Additionally, animals defective in *Neur1*, *Neur2*, and *Mib2* gene expression do not display any Notch-dependent phenotypes, while additional removal of *Mib1* produces embryonic lethal pheonotypes associated with losses in Notch signaling (Koo *et al.*, 2007). Importantly, disruption of only the *Mib1* gene produces the known constellation of Notch mutant phenotypes in developing mouse embryos (Barsi *et al.*, 2005; Koo *et al.*, 2005a). Although Mib1 and Mib2 appear functionally redundant (Zhang *et al.*, 2007a, b), Mib2 is not strongly expressed during embryonic development accounting for the absolute requirement for Mib1 in Notch-dependent developmental processes (Koo *et al.*, 2007). In contrast to findings reported for the functionally redundant E3 ligases in flies, Mib2 but not Neur1 or Neur2 can rescue the Mib1 mutant neurogenic phenotype in zebra fish (Koo *et al.*, 2005b). Further, while both Neur1 and Neur2 are dispensable for normal neurogenesis in mice, Mib1 mutant embryos display strong neurogenic phenotypes in the developing brain and neural tube (Koo *et al.*, 2005b, 2007). Therefore, while Neur and Mib appear to perform similar roles in Notch signaling in flies, the vertebrate Neur and Mib proteins do not appear to be functionally equivalent.

Findings from mammalian cells have suggested that Mib, not Neur, is the E3 ligase responsible for DSL ligand endocytosis that activates Notch signaling, while Neur functions downstream of Mib to direct lysosomal degradation of internalized ligands and thereby regulate the level of ligand available for Notch activation (Song *et al.*, 2006). Consistent with this idea, overexpression of Neur1 monoubiqutinates Jagged1 leading to degradation and attenuation of Jagged1-induced Notch signaling (Koutelou *et al.*, 2008); however, Mib2 (skeletrophin) ubiquitination of Jagged2 is associated with activation of Notch signaling (Takeuchi *et al.*, 2005). The different functional roles for Neur and Mib ligases in Notch signaling might reflect different ubiquitin states of DSL ligands mediated by these structurally distinct E3 ligases. Notch ligands have been reported to be mono- and/or polyubiquitinated; however, the functional consequences of these types of ubiquitination to Notch signaling are not well documented. Polyubiquitination is associated with proteasome degradation, while both mono and multi-mono ubiqutination can signal endocytosis of membrane proteins from the cell surface and further influence intracellular trafficking (Staub and Rotin, 2006). Trafficking events that degrade internalized DSL ligands could function to downregulate Notch signaling, while recognition of ubiquitinated ligands by specific adaptor/sorting molecules might promote signaling.

In addition to inducing different types of ubiquitination, Mib and Neur could potentially regulate ligand activity by modifying distinct lysine residues. Ligand ICDs contain multiple lysine residues that could potentially be modified by the addition of ubiquitin. Mutation of two intracellular lysine residues in Serrate produces signaling defects that are associated with losses in endocytosis and accumulation of ligand at the cell surface (Glittenberg *et al.*, 2006). In contrast, mutation of all 17 intracellular lysine residues in Dll1 did not prevent internalization or promote accumulation of this lysine-less mutant at the cell surface (Heuss *et al.*, 2008). Nonetheless, the internalized lysine-less mutant was unable to recycle or activate Notch signaling, and these defects were associated with decreased fractionation to detergent-resistant lipid microdomains compared to wild-type Dll1. In addition, the inability of the lysine-less mutant to recycle also correlated with defects in binding a soluble form of Notch. In contrast, a Dll1/Dll3 chimeric ligand containing the ECD of Dll1 and the transmembrane and ICDs of Dll3, internalized, recycled, and displayed high-affinity binding to Notch despite lacking lysines required for ubiquitination (Heuss *et al.*, 2008). Underscoring the importance of ligand ubiquitination in signaling activity, the Dll1/Dll3 chimera did not activate Notch signaling, and this also correlated with a loss in fractionation to lipid microdomains. Based on these findings, the authors concluded that ubiquitination is not required for ligand endocytosis but rather functions to direct ligand to a specific recycling pathway where it acquires high-affinity binding to Notch. As exciting as these findings are, the authors failed to unravel the connections between ubiquitination, recycling, lipid microdomain fractionation, and high-affinity binding in the generation of an active ligand. Importantly, this study did not determine the signaling activity of wild-type Dll1 when either protein recycling or lipid raft formation is disrupted.

Studies in flies indicate that Neur may play additional roles in Notch ligand endocytosis to enhance signaling activity beyond ubiquitination (Pitsouli and Delidakis, 2005; Skwarek *et al.*, 2007). Specifically, a phosphoinositide-binding domain was identified in Neur that is necessary for its interactions with the plasma membrane. Although Neur membrane localization is not required for Neur to interact with or ubiquitinate Delta, membrane association of Neur is required for Delta endocytosis (Skwarek *et al.*, 2007). In this regard, a recent study identified a link between the GSL content of the plasma membrane and Mib-dependent endocytosis of Delta that is required to activate Notch signaling (Hamel *et al.*, 2010). A conserved GBM was identified in the NT region of Delta and Serrate that conferred binding to specific GSLs, which is proposed to modulate ligand membrane association and in turn ligand endocytosis. Together these studies underscore the importance of membrane lipids in modulating ligand endocytosis and signaling activity.

5. Ligand Endocytosis in Activation of Notch Signaling

A requirement for direct cell-to-cell interactions is a hallmark of Notch signaling; however, the transmembrane property of the ligands may underlie the basic mechanism of Notch activation that is dependent on ligand endocytosis. Specifically, in the absence of endocytosis, ligands accumulate at the cell surface but fail to activate signaling (Itoh *et al.*, 2003; Nichols *et al.*, 2007a; Parks *et al.*, 2000). That ligands need to be internalized by the signal-sending cell to activate Notch on the signal-receiving cell represents a fundamentally new paradigm for endocytic activation of a signaling pathway. Nonetheless, the exact mechanism by which ligand endocytosis triggers Notch signaling has remained a mystery.

5.1. Identifying the endocytic machinery required for ligand cells to activate Notch

The majority of cell surface proteins are internalized via clathrin-mediated endocytosis (CME); however, additional portals of entry exist that do not involve clathrin (Conner and Schmid, 2003; Doherty and McMahon, 2009). The specific endocytic pathways used by Notch ligands are poorly characterized, but what is certain is that only ubiquitinated ligands internalized in an epsin-dependent manner are competent to signal (Chen and Casey Corliss, 2004; Deblandre *et al.*, 2001; Glittenberg *et al.*, 2006; Haltiwanger and Lowe, 2004; Itoh *et al.*, 2003; Koo *et al.*, 2005a; Lai *et al.*, 2001; Overstreet *et al.*, 2004; Pavlopoulos *et al.*, 2001; Wang and Struhl, 2005; Yeh *et al.*, 2001). Genetic and cellular studies indicate that ligand cells require the key endocytic factor dynamin to activate Notch (Nichols *et al.*, 2007a; Parks *et al.*, 2000; Seugnet *et al.*, 1997a), however, dynamin functions to release endocytic vesicles from the plasma membrane during both clathrin-dependent and clathrin-independent endocytosis (Conner and Schmid, 2003), so either or both pathways could function in ligand activity. In addition, the clathrin adaptor epsin that is critical for ligand activity has also been implicated in endocytosis independent of clathrin (Chen and De Camilli, 2005; Sigismund *et al.*, 2005).

Indirect support for CME in ligand signaling activity has come from genetic studies indicating that Notch-dependent developmental events require auxilin and the ubiquitious cyclin G-associated kinase that functions at multiple steps in clathrin-coated pit formation and un-coating of clathrin-coated vesicles (Eisenberg and Greene, 2007; Yim *et al.*, 2010). Moreover, the Notch signaling defects identified with auxilin mutants can be partially rescued by ectopic clathrin expression (Eun *et al.*, 2008), suggesting that losses in auxilin produce un-coating defects that limit clathrin availability for ligand endocytosis. Together with findings from mammalian cell culture indicating that blockade of CME in ligand cells inhibits Notch signaling (Nichols *et al.*, 2007a), it seems likely that ligand endocytosis required for activation of Notch is clathrin dependent. Nonetheless, a role for ligand endocytosis independent of clathrin for signaling activity in specific cellular contexts cannot be ruled out.

Although it is clear that endocytosis by the ligand cell is critical for activation of signaling in the Notch cell, the exact role that ligand endocytosis serves in signaling has remained poorly defined. Studies in flies and mammalian cells have suggested that ligands undergo two distinct endocytic events to activate Notch (Fig. 3.3). The first ligand endocytic event occurs prior to engagement of Notch and is proposed to facilitate recycling to generate an active ligand. Following interactions with Notch on adjacent cells, a second ligand endocytic event is proposed to generate a pulling force to allow activating Notch proteolysis. It is important to note that whether the first, second, or both endocytic events is necessary for ligand activation of Notch is controversial.

5.2. Recycling to generate an active ligand

The recycling model assumes that newly synthesized ligand delivered to the cell surface cannot activate Notch and requires endocytosis, trafficking, and recycling back to the cell surface to gain signaling activity (Heuss *et al.*, 2008; Rajan *et al.*, 2009; Wang and Struhl, 2004). To account for the absolute requirements for epsin and ligand ubiquitination in signaling activity, this model further proposes that epsin selectively promotes endocytosis and/or trafficking of a sub-population of ubiquitinated ligand for conversion in the recycling endosome into an active ligand. The changes conferred by recycling to obtain signaling activity are completely unknown; however concentration, clustering, and proteolytic processing of ligand, as well as localization of ligand to a specific microdomain or recruitment of cofactors, have all been suggested as possible modifications (Chitnis, 2006; Le Borgne, 2006; Nichols *et al.*, 2007b).

Even though Notch ligands are known to recycle (Heuss *et al.*, 2008; Rajan *et al.*, 2009), the role that ligand recycling plays, if any, in activating Notch is poorly defined. In addition to returning internalized proteins and membrane to the cell surface, recycling is used to establish distinct apical and basolateral membranes in polarized cells (Grant and Donaldson, 2009; Maxfield and McGraw, 2004). Therefore, it may not be surprising that the strongest support for ligand recycling in activation of Notch signaling comes from studies on cell fates derived from sensory organ precursors (SOP) that involve polarized cells (see Chapter 5). Specifically, SOP

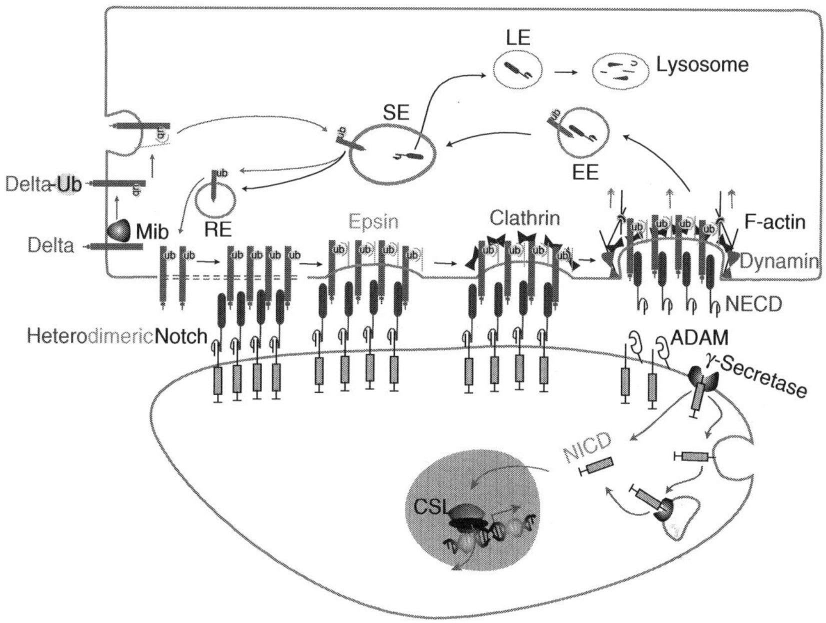

Fig. 3.3 Models for distinct endocytic events by the ligand cell to activate signaling in the Notch cell. Prior to Notch engagement, endocytosis allows ligand to enter the sorting endosome (SE) or recycling endosome (RE) where it is processed into an active ligand and returned to the cell surface to activate Notch. Ligand ubiquitination by Mib may facilitate interactions with epsin that direct the required endocytosis and/or trafficking. Alternatively, ligand binding to Notch may induce ligand ubiquitination for recruitment of epsin to orchestrate the formation of a clathrin-coated endocytic structure specialized in force generation to pull the noncovalent heterodimeric Notch apart. Heterodimer dissociation would account for the observed uptake of the Notch extracellular domain (NECD) by ligand cells. In the early endosome (EE), internalized NECD dissociates from the ligand and trafficks to the late endsome (LE) where it is targeted for lysosomal degradation. Ligand dissociated in the EE traffics to the SE or RE for return to the cell surface where it is available to activate Notch on adjacent cells. Removal of the NECD exposes the ADAM site in the membrane-bound heterodimer subunit to facilitate γ-secretase cleavage and release of the Notch intracellular domain (NICD) from the membrane. Released NICD translocates to nucleus where it interacts with CSL to activate Notch target gene transcription. As discussed in the text, these models that account for the critical requirement for endocytosis by the ligand cell to activate signaling in the Notch cell may not be mutually exclusive. (See Color Insert.)

progeny that activates Notch signaling in neighboring cells is enriched in Rab11 recycling endosomes that concentrate Delta and apically internalized Delta must traffic from the basolateral membrane to an apical actin-rich structure for SOP progeny to acquire signaling activity (Emery *et al.*, 2005; Jafar-Nejad *et al.*, 2005; Rajan *et al.*, 2009). However, Sec15 that functions with Rab11 in the recycling endosome to regulate SOP-derived cell fates is not required in every developmental event regulated by Notch (Jafar-Nejad *et al.*, 2005; Windler and Bilder, 2010). Additionally, loss of Rab11 activity does not perturb Delta signaling in the germ line (Windler and Bilder, 2010), and Rab11 mutants do not display Notch eye phenotypes (Li *et al.*, 2007a) as expected if ligand recycling is an absolute requirement for Notch signaling. Moreover, Rab11 does not overlap with Delta in the morphogenetic furrow (Hagedorn *et al.*, 2006) where Notch signaling directs normal eye development. If recycling is absolutely required to generate an active ligand, then Rab5 that is a prerequisite for entry into the Rab11 recycling pathway should also be required; however, defects in Rab5 do not perturb Delta signaling activity (Windler and Bilder, 2010). Together these findings suggest that ligand recycling, at least that dependent on Rab11 and Sec15, is not a general requirement of Notch signaling.

5.3. Ligand endocytosis in force generation to activate Notch

A general requirement for ligand endocytosis has been proposed to reflect the need for Notch to undergo conformational changes to effect activating proteolysis that ligand binding alone would not induce (Gordon *et al.*, 2008a, b). Proteolytic activation of Notch signaling involves the specific uptake of the NECD by the ligand cell (Nichols *et al.*, 2007a; Parks *et al.*, 2000), and although ligand cells defective in endocytosis bind and cluster Notch they do not internalize NECD or activate signaling (Nichols *et al.*, 2007a). These findings first suggested a role for ligand endocytosis in activation of signaling that involved a mechanical force to dissociate the NECD from intact Notch. The force produced by ligand endocytosis is thought to induce conformational changes that destabilize the noncovalent interactions that keep the Notch heterodimer intact and inactive in the absence of ligand. The identification and characterization of a negative regulatory region (NRR) in the Notch ectodomain that stabilizes the Notch heterodimer and prevents activating proteolysis provide additional support for the ligand endocytosis pulling-force model (see Chapter 2). Specifically, structural analyses of the NRR confirm that multiple noncovalent interactions stabilize the structure and serve to occlude the ADAM cleavage site that is required to initiate activating Notch proteolysis (Gordon *et al.*, 2007). Moreover, these findings have suggested that ligand binding alone would not be sufficient to induce the required global conformational changes, but rather, endocytosis of ligand-bound Notch would be necessary to produce a force to pull on Notch and expose the ADAM cleavage site for activating proteolysis. Although endocytosis is a good force-generating candidate, it is not known if force is produced during the process of ligand endocytosis, or if such a force could destabilize the NRR structure and expose the ADAM cleavage site.

Endocytosis of ligand-bound Notch by the ligand cell may be mechanistically different than constitutive ligand endocytosis. Specifically, cells may experience a resistance to ligand internalization of Notch attached to the surface of an adjacent cell. To overcome such a resistance, ligand cells may need to recruit specific cellular factors to form an endocytic structure specialized in force generation to effect ligand endocytosis of cell surface Notch. Specifically, ligand endocytic force-induced conformational changes in Notch could physically release NECD by dissociating the heterodimer subunits or unmasking the ADAM cleavage site and both mechanisms have been proposed (discussed in Chapter 2). In any event, removal of NECD from the intact Notch heterodimer would be necessary for activating proteolysis of the remaining membrane-bound Notch for downstream signaling (Fig. 3.3).

The requirement for epsin in ligand signaling activity (Overstreet *et al.*, 2003, 2004; Tian *et al.*, 2004) has been proposed to reflect a role for epsin in ligand endocytosis and/or trafficking to allow access to a specific recycling pathway for conversion into an active ligand (Wang and Struhl, 2004). Nonetheless, epsin is not known to regulate protein recycling (Vanden Broeck and De Wolf, 2006), and data are lacking to show that losses in epsin actually perturb ligand recycling. Although it is clear that epsin is required for ligand signaling activity, it is possible that this does not involve ligand recycling prior to engagement with Notch. Rather, we propose that epsin may function downstream of ligand binding to Notch to induce the formation of a force-producing endocytic structure. Notch binding may induce ligand ubiquitination and/or clustering to amass multiple ubiquitin-binding sites for epsin. By assembling multiple low-affinity mono-ubiquitin interactions, strong epsin-UIM/ubiquitinated DSL interactions could be generated (Barriere *et al.*, 2006; Hawryluk *et al.*, 2006), and this may be necessary for ligand to overcome resistance to internalization when bound to cell surface Notch. In fact, replacement of the Delta ICD with a single ubiquitin motif that can undergo polyubiquitination promotes internalization and signaling activity in zebra fish (Itoh *et al.*, 2003). However, a nonextendable ubiquitin only weakly signals even though it promotes endocytosis (Wang and Struhl, 2004), supporting the idea that multiple ubiquitin interaction sites are required for ligands to activate Notch, possibly through providing stable associations with epsin-containing endocytic vesicles.

Consistent with these ideas, ligand cells require epsin, dynamin, and the actin cytoskeleton to activate signaling in Notch cells, and all of these cellular factors have been implicated in inducing membrane constriction and tension that could contribute to force generation during the process of endocytosis (Itoh *et al.*, 2005; Roux *et al.*, 2006). Therefore, it is tempting to speculate that ligand cells require epsin to orchestrate the formation of a molecularly distinct clathrin-coated endocytic structure specialized in force generation. In addition to membrane bending, epsin has also been reported to regulate the actin cytoskeleton during endocytosis (Horvath *et al.*, 2007; Maldonado-Baez and Wendland, 2006), which together could endow cells with sufficient endocytic force to induce conformational changes in ligand-bound Notch required to initiate activating proteolysis. Although epsin participates in endocytosis through simultaneous binding to the plasma membrane, clathrin endocytic vesicles, and ubiquitinated cargo (Horvath *et al.*, 2007), interactions between ligands and epsin have yet to be reported, and it is still unclear how epsin and ubiquitinated ligands contribute to Notch activation.

Implicit in the pulling-force model is the need for ligand–Notch interactions to survive the endocytic force that induces conformational changes required for NECD transendocytosis and activating Notch proteolysis. That NECD transendocytosis by ligand cells is required for activation of Notch (Heuss *et al.*, 2008; Nichols *et al.*, 2007a; Parks *et al.*, 2000) indicates that ligand–Notch interactions do indeed survive the putative endocytic force required for global conformational changes in Notch to expose the ADAM cleavage site. In this regard, reported atomic force microscopy (AFM) measurements for Delta cells binding to uncleaved Notch are stronger than those detected for furin-cleaved Notch (Ahimou *et al.*, 2004), suggesting that Delta–Notch interactions are indeed stronger than the noncovalent interactions that hold the heterodimer subunits together (see Chapter 2 for further discussion). Therefore, ligand endocytosis could function first to allow recycling to produce a high-affinity ligand for avid binding to Notch, and this in turn would enable ligand–Notch interactions to survive the pulling force produced by ligand endocytosis of Notch bound to adjacent cells (Fig. 3.3).

Recycling has been suggested to generate a high-affinity ligand by directing ligand to a specific membrane microdomain (Heuss *et al.*, 2008), and this could provide a mechanism to produce strong ligand–Notch interactions. While it is attractive to propose that ligand endocytosis regulates recycling to generate a high-affinity ligand, the fact that soluble ligands that have never recycled can signal when attached to surfaces (Varnum-Finney *et al.*, 2000) argues against a requirement for endosomal processing to generate an active ligand. Additionally, the dependence of soluble ligands on surface attachment to activate signaling is consistent with the proposed role for force in exposing Notch to activating proteolysis; however, in this case the Notch cell would provide the force to disrupt the NRR structure through cell migration. Finally, in contrast to the absolute requirement for ligand endocytosis in signaling, studies have failed to establish a firm correlation between ligand recycling and signaling activity (Glittenberg *et al.*, 2006; Heuss *et al.*, 2008), implying that endocytosis rather than recycling is a general requirement for ligands to activate Notch signaling.

Confirmation of endocytic force in ligand signaling activity awaits biophysical analyses to establish that ligand cells can indeed produce a mechanical force following interactions with Notch. AFM studies have provided support that Delta binds Notch with high avidity (Ahimou *et al.*, 2004), but whether ligands need to recycle to acquire strong binding potential is unknown. To confirm a requirement for ligand recycling in nonpolarized cells it will be necessary to show that losses in ligand recycling produce signaling defects. Moreover, elucidating how epsin functions to regulate signaling activity of ubiquitinated ligands is critical to understanding the molecular, cellular, and physical basis of ligand endocytosis in Notch activation. Biophysical studies will also be required to determine if activating Notch proteolysis and downstream signaling are regulated by mechanical force; however, the ultimate challenge will be to obtain evidence for endocytic force in regulating Notch signaling in whole animals.

6. Regulation of DSL Ligand Activity by Proteolysis

DSL ligands undergo proteolytic cleavage by ADAMs and γ-secretase as described for Notch; however, in contrast to signaling induced by Notch proteolysis, proteolytic removal of cell surface ligand can either inhibit or enhance Notch signaling. Although Notch proteolysis generates an intracellular fragment that acts as the signal transducer, it is less clear if the cleavage products generated by ligand proteolysis have intrinsic activity (Fig. 3.4). A detailed review describing the proteases that cleave DSL ligands and the biological significance has been previously published (Zolkiewska, 2008); here we discuss possible mechanisms by which ligand proteolysis could affect Notch signaling. While mammalian DSL ligands are cleaved by several ADAMs (ADAM9, ADAM10, ADAM12, ADAM17), *Drosophila* ligands have been reported to be cleaved by only the homologs of ADAM10 (Kuzbanian/Kuz and Kuzbanian-like/Kul) and ADAM17 (DTACE).

6.1. ADAM ectodomain shedding of DSL ligands as regulators of Notch signaling

One of the consequences of ADAM cleavage of DSL ligands is shedding of the ectodomain that contains the Notch binding site. Accordingly, ADAM shedding of ligands would decrease ligand–Notch interactions both in *trans* and in *cis*; however, these scenarios would produce opposing outcomes on Notch signaling. Specifically, losses in *trans*-interactions would lead to losses in Notch signaling while losses in *cis*-interactions would relieve *cis*-inhibition and thereby enhance signaling. Therefore, in addition to transcriptional feedback loops and endocytosis discussed in Section 3, ligand shedding provides an additional mechanism to regulate the cell's potential to send or receive a Notch signal. Furthermore, in addition to regulating signal polarity, ligand shedding could also determine the intensity and duration of Notch signaling.

Several lines of evidence suggest a role for ADAM-mediated ligand ectodomain shedding in establishing and/or maintaining an asymmetric distribution of cell surface ligand between signal-sending and signal-receiving cells. In flies, this has been best demonstrated for the ADAM Kul, where both losses and gains in Kul activity produce wing vein defects characteristic of aberrant Notch signaling (Lieber *et al.*, 2002; Sapir *et al.*, 2005). These studies suggest that Kul, which exclusively cleaves ligands and not Notch, is required to maintain asymmetric distribution of Delta in the developing wing to facilitate unidirectional signaling. In the signal-receiving cell, Kul acts as a positive regulator of Notch signaling by maintaining low levels of ligand at the cell surface to prevent *cis*-inhibition and ensure efficient signal reception necessary for normal wing margin formation (Sapir *et al.*, 2005). Similar to the requirement for Kul in the signal-receiving cell, ectopic expression of ADAM12 (an ADAM that cleaves Dll1 but not Notch) results in Dll1 shedding and enhanced Notch signaling in mammalian cells again, presumably by relieving *cis*-inhibition (Dyczynska *et al.*, 2007; Sun *et al.*, 2008a). Dll1 shedding is also thought to deplete ligand available for activation of Notch signaling that would result in decreases in signaling. Such asymmetry in Notch signaling among initially equivalent myogenic progenitors, created through Dll1 shedding, is proposed to maintain the balance between self-renewal and differentiation (Sun *et al.*, 2008a).

ADAM expression and activity could regulate ligand ectodomain shedding and thus Notch signaling. In this regard, transforming growth factor (TGF)-β3 downregulates ADAM10 expression and correlates with activation of Notch signaling in cultured chick leg bud mesenchymal cells (Jin *et al.*, 2007). In this scenario, TGF-β3-induced downregulation of ADAM10 prevents Dll1 ectodomain shedding, and this correlates with an inhibition in cell proliferation and subsequent precartilage condensation through increases in Notch signaling. The glycosylphosphatidyl-anchored cell surface protein, RECK (*re*version-inducing *c*ysteine-rich protein with *k*azal motifs), specifically inhibits ADAM10 activity leading to inhibition of ectodomain shedding of DSL ligands and activation of Notch signaling (Muraguchi *et al.*, 2007). Consistent with this role, RECK-deficient mouse embryos exhibit a loss in Notch target gene expression and display some Notch-dependent developmental defects, presumably due to loss of cell surface ligand available for interaction with Notch in *trans* (Muraguchi *et al.*, 2007).

6.2. Activity of the ADAM-shed ectodomain of DSL ligands in Notch signaling

ADAM proteolysis of DSL ligands generates several cleavage products that could potentially affect Notch signaling (Fig. 3.4A). The putative activity of soluble ligand ECD (Fig. 3.4B) has best been examined through the use of recombinant ligands containing ECD sequences. While some studies have suggested that the ECDs are inactive others have suggested that they can either activate or inhibit Notch signaling depending on the cellular context. Nonetheless, soluble forms of Delta have been detected in *Drosophila* embryos (Klueg *et al.*, 1998; Qi *et al.*, 1999) and ectopic expression of Delta or Serrate ECDs antagonize Notch signaling (Hukriede *et al.*, 1997; Sun and Artavanis-Tsakonas, 1997). A requirement for ADAM10/Kuzbanian in Notch signaling was initially interpreted to reflect shedding of Delta to produce an active ligand (Qi *et al.*, 1999); however, subsequent findings from this same group have questioned this idea (Mishra-Gorur *et al.*, 2002). The agonistic activity of soluble ligands is not easy to reconcile given the strict requirement for ligand endocytosis in Notch activation. Providing insight into this paradox, pre-fixed Delta cells that are presumably endocytosis defective can activate Notch target genes (Delwig and Rand, 2008; Mishra-Gorur *et al.*, 2002), suggesting that a physical force required to dissociate the Notch heterodimer may be provided by other mechanisms. Perhaps movement of Notch cells away from soluble ligand attached to the extracellular matrix or cell surface could produce the required force for heterodimer dissociation. In support of this idea, several studies have demonstrated that recombinant soluble ligands need to be pre-clustered or immobilized to activate Notch signaling and induce biological responses (Hicks *et al.*, 2002; Karanu *et al.*, 2000; Morrison *et al.*, 2000; Shimizu *et al.*, 2002; Varnum-Finney *et al.*, 2000; Vas *et al.*, 2004) (Fig. 3.4C). Additionally, while unclustered soluble ligands can bind Notch, they are unable to activate signaling but rather appear to antagonize signaling induced by *trans*-ligands (Hicks *et al.*, 2002; Shimizu *et al.*, 2002; Varnum-Finney *et al.*, 2000; Vas *et al.*, 2004) (Fig. 3.4D). In these cases, soluble ligands may compete with membrane-bound ligands for Notch binding, providing a mechanistic basis for the antagonistic activities identified for putative soluble forms of *Drosophila* (Hukriede *et al.*, 1997; Sun and Artavanis-Tsakonas, 1997) and mammalian DSL ligands (Li *et al.*, 2007b; Lobov *et al.*, 2007; Noguera-Troise *et al.*, 2006; Small *et al.*, 2001; Trifonova *et al.*, 2004).

Naturally occurring soluble DSL ligands that function as Notch agonists have been identified in *C. elegans* and mammalian cells (Aho, 2004; Chen and Greenwald, 2004; Komatsu *et al.*, 2008). In fact 5 of the 10 *C. elegans* DSL ligands are soluble which represent the highest proportion of soluble DSL ligands identified for any phylum. Interestingly, neither the soluble nor the membrane-bound *C. elegans* DSL ligands have a DOS motif, which

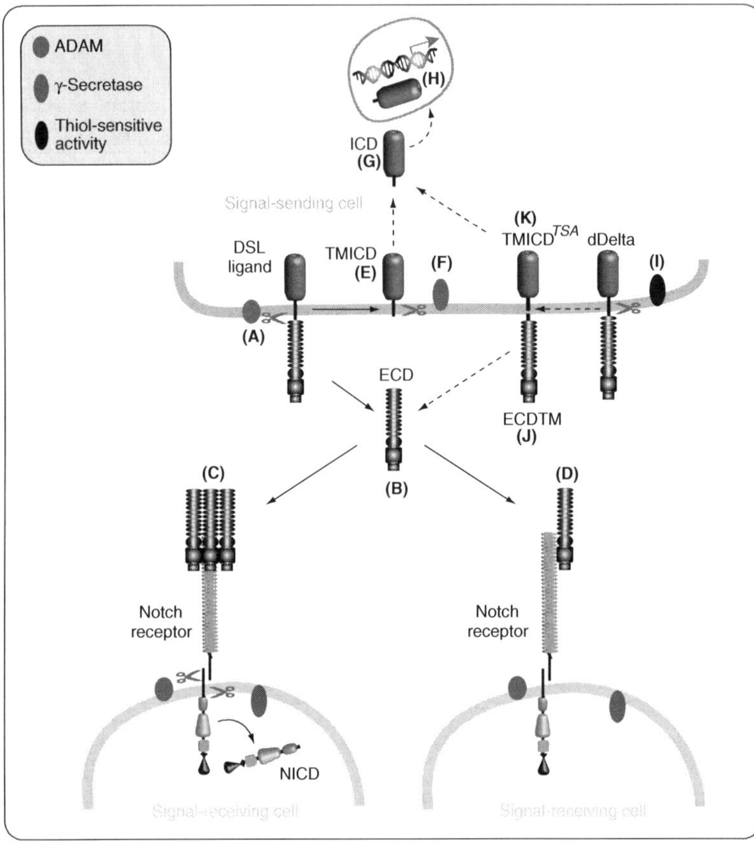

Fig. 3.4 Regulation of DSL ligand signaling activity by proteolysis. Mammalian and *Drosophila* DSL ligands undergo proteolytic cleavages within the juxtamembrane and intramembrane regions. A-Disintegrin-And-Metalloprotease (ADAM) mediated cleavage (A) of mammalian and *Drosophila* DSL (*D*elta/*S*errate/*L*AG-2) ligands within the juxtamembrane region results in shedding of the extracellular domain (B, ECD). The shed ECD requires clustering to activate Notch signaling (C). Although unclustered soluble ECD can bind Notch, it may antagonize Notch signaling (D). In mammalian cells, the remaining membrane-tethered ADAM cleavage product, that contains the intracellular domain (TMICD, E) may undergo further cleavage by γ-secretase (F) to release the intracellular domain (ICD) from the membrane (G) allowing it translocate to the nucleus and activate gene transcription (H) (see text for details). However, the *Drosophila* Delta (dDelta) TMICD (5) is not further processed and could antagonize Notch signaling (see text for details). Like mammalian DSL ligands, dDelta also undergoes intramembrane cleavage, however, this event does not require prior ADAM cleavage and is catalyzed by a thiol-sensitive activity (TSA, I). It is unclear if the resulting cleavage products remain membrane-tethered. If the ECD containing fragment (ECDTM) remains membrane-tethered (J), it could antagonize Notch signaling, but if released from the membrane, ECDTM could function as proposed for soluble ECD (B, C, D) (see text for details). If the ICD-containing intramembrane cleavage product $TMICD^{TSA}$ remains membrane-bound (K), it could antagonize Notch signaling, but if released from the membrane (G), $TMICD^{TSA}$ could translocate to the nucleus and activate gene transcription (H) (see text for details). (See Color Insert.)

is present in most, but not all, DSL ligands from other phyla (Komatsu *et al.*, 2008). In flies and mammalian systems, both the DSL and the DOS domains are required for high-affinity binding to Notch receptors and activation of signaling (Parks *et al.*, 2006; Shimizu *et al.*, 1999). Genetic studies in *C. elegans* have identified the existence of soluble proteins that contain a DOS domain that are required in some developmental contexts for DSL ligands to activate the Notch-related LIN-12 receptor (Komatsu *et al.*, 2008). To account for the biological activity observed for the DOS-containing proteins in DSL ligand activation of Notch signaling, the authors propose that optimal signaling requires the formation of a bipartite ligand system comprising distinct DSL and DOS domain-containing ligands. These findings emphasize the cooperative action of DSL and DOS domains for optimal Notch signaling, irrespective of whether these domains are present collinearly (as in the case of *Drosophila* Delta and Serrate and vertebrate ligands Dll1, Jagged1 and Jagged2) or within distinct proteins (as in the case of *C. elegans* ligands). Of the vertebrate DSL ligands, only Dll4 and Dll3 lack DOS domains (Komatsu *et al.*, 2008) and similar to *C. elegans* DSL ligands their signaling activity may be dependent on collaboration with DOS domain-containing ligands (Kopan and Ilagan, 2009). At odds with this idea, Dll4 has been reported to be the most avid DSL ligand (Funahashi *et al.*, 2008; Karanu *et al.*, 2001; Sun *et al.*, 2008b) and Dll3 is unable to bind or activate Notch (Ladi *et al.*, 2005). While DSL and DOS domains may cooperate to activate Notch signaling, it is possible that on their own they function to antagonize Notch signaling as discussed in Section 9.1.

6.3. Activity of the ADAM-cleaved membrane-tethered fragment in signaling

ADAM cleavage of DSL ligands also produces a membrane-tethered fragment containing the intracellular domain (TMICD; Fig. 3.4E), which in mammalian cells undergoes further cleavage by γ-secretase (Ikeuchi and Sisodia, 2003; LaVoie and Selkoe, 2003; Six *et al.*, 2003) (Fig. 3.4F). Several studies have indicated that the released ligand ICD translocates to the nucleus (Hiratochi *et al.*, 2007; Ikeuchi and Sisodia, 2003; Kolev *et al.*, 2005; LaVoie and Selkoe, 2003; Six *et al.*, 2003) and activates gene transcription (Hiratochi *et al.*, 2007; Kolev *et al.*, 2005; LaVoie and Selkoe, 2003 6) (Fig. 3.4G, H), similar to that identified for cleaved Notch. In support of this idea, the ICDs contain positively charged amino acids that when mutated prevent nuclear translocation and transcriptional activation (Kolev *et al.*, 2005; LaVoie and Selkoe, 2003). Although these studies provide some support for the idea that DSL ligands undergo reverse signaling it is important to note that this has mostly relied on the use of engineered fragments, rather than physiological proteolytic cleavage of full-length ligands. Nonetheless, the possibility that DSL ligand–Notch signaling is bidirectional is exciting and awaits a clear demonstration of signaling events triggered in both DSL ligand and Notch cells following ligand–Notch interactions as established for the prototypic EphB/ephrinB bidirectional signaling system (Aoto and Chen, 2007; Dravis *et al.*, 2004; Holland *et al.*, 1996), that also involves transmembrane ligands and receptors.

Unlike mammalian DSL ligands, the TMICD fragment produced by ADAM cleavage of *Drosophila* Delta does not appear to undergo further processing and likely remains membrane bound (Bland *et al.*, 2003; Delwig *et al.*, 2006) (Fig. 3.4E). Although this fragment lacks a Notch binding domain, it could potentially compete with full-length ligands for the ubiquitination and/or endocytic machinery and thus antagonize ligand signaling activity. Another distinguishing feature of the proteolytic cleavage of *Drosophila* Delta is that although intramembrane cleavage occurs, this event does not require prior ADAM cleavage and does not involve γ-secretase (Delwig *et al.*, 2006). Rather, this cleavage is induced by a thiol-sensitive activity and occurs close to the extracellular face of the membrane (Fig. 3.4I). Hence, it is uncertain whether the ICD would be readily released as proposed for ligand ICDs generated by γ-secretase (Delwig *et al.*, 2006). If the ECD containing fragment (ECDTM) remains membrane tethered (Fig. 3.4J), it could function like ICD-truncated ligands, which are endocytosis-defective and unable to activate signaling but are efficient *cis*-inhibitors (Chitnis *et al.*, 1995; Henrique *et al.*, 1997; Nichols *et al.*, 2007a; Shimizu *et al.*, 2002), but, if released, the ECDTM could function as proposed for soluble DSL ligands (Fig. 3.4B–D). The corresponding ICD-containing intramembrane cleavage product ($TMICD^{TSA}$, Fig. 3.4K) would be expected to function similarly to the *Drosophila* Delta TMICD (Fig. 3.4E) if it remained membrane bound; however, if released (Fig. 3.4G), it could translocate to the nucleus and activate gene transcription (Fig. 3.4H). In this regard, nuclear staining of Delta has only been detected using engineered ICD forms (Bland *et al.*, 2003; Sun and Artavanis-Tsakonas, 1996), and hence, it is unclear whether the ICD is in fact released from full-length Delta and moves to the nucleus. Like Delta, Serrate also undergoes ADAM cleavage (Sapir *et al.*, 2005); however, intramembrane cleavage of Serrate has not been reported to date.

6.4. Regulation of ligand proteolysis

Compared to the proteolytic activation of Notch that is tightly regulated by ligand, it is less clear if or how ligand proteolysis is induced or regulated. DSL ligands are actively cleaved in cell culture (Bland *et al.*, 2003; Delwig *et al.*, 2006; Dyczynska *et al.*, 2007; LaVoie and Selkoe, 2003; Six *et al.*, 2003; Yang *et al.*, 2005); however, this proteolysis could be induced by signaling pathways trigged by serum components (Seals and Courtneidge,

2003). In fact, phorbol esters are known to activate intracellular signaling as well as ADAMs, both of which can induce DSL ligand proteolysis (Seals and Courtneidge, 2003). The extracellular matrix protein Mircrofibril-associated glycoprotein (MAGP)-2 has also been reported to regulate DSL ligand proteolysis (Nehring *et al.*, 2005). Interestingly, MAGP2 interacts with several DSL ligands, yet only the Jagged1 ectodomain is shed in a metalloprotease-dependent manner following interactions with MAGP2. Direct cell–cell interactions also contribute to ADAM cleavage of DSL ligands and both homotypic ligand–ligand and ligand–Notch interactions have been implicated (Bland *et al.*, 2003; Delwig *et al.*, 2006; Dyczynska *et al.*, 2007; Hiratochi *et al.*, 2007; LaVoie and Selkoe, 2003). Finally, gains and losses in Neur activity have been shown to be associated with Delta proteolytic processing in flies (Delwig *et al.*, 2006; Haltiwanger and Lowe, 2004; Pavlopoulos *et al.*, 2001), raising the possibility that ligand cleavage may occur intracellularly and involve endocytosis.

7. DSL Ligand Interactions with PDZ-Domain Containing Proteins

The vertebrate DSL ligands Dll1, Dll4, and Jagged1 have PDZ-binding motifs at their carboxy-termini (Pintar *et al.*, 2007), which mediate interactions with PDZ-containing scaffold/adaptor proteins (Ascano *et al.*, 2003; Estrach *et al.*, 2007; Mizuhara *et al.*, 2005; Pfister *et al.*, 2003; Six *et al.*, 2004; Wright *et al.*, 2004). While being dispensable for both ligand activation (Ascano *et al.*, 2003; Mizuhara *et al.*, 2005; Six *et al.*, 2004; Wright *et al.*, 2004) and inhibition of Notch signaling (Glittenberg *et al.*, 2006), the PDZ-binding sequences are required to mediate the effects of ligands on cell adhesion (Estrach *et al.*, 2007; Mizuhara *et al.*, 2005), migration (Six *et al.*, 2004; Wright *et al.*, 2004), and oncogenic transformation (Ascano *et al.*, 2003). DSL ligands exhibit some preference for binding specific PDZ-containing proteins, most likely a reflection of the sequence differences in their PDZ-binding motifs (Pintar *et al.*, 2007). For example, Jagged1 is unable to bind the PDZ domain proteins, MAGI-1 (membrane-associated guanylate kinase with inverted domain arrangement-1) and Dlg1 (human homolog of *Drosophila* discs large 1) (Mizuhara *et al.*, 2005; Six *et al.*, 2004), while the closely related Dll1 and Dll4 proteins both bind Dlg1 (Six *et al.*, 2004). Although PDZ interactions do not mediate activation of Notch signaling, loss of the PDZ motif enhances the signaling activity of Delta (Estrach *et al.*, 2007). These findings raise the intriguing possibility that PDZ-based interactions may restrict access of ligands to specific endocytic pathways necessary for their signaling activity.

PDZ-containing proteins play an important role in organizing specialized sites of cell–cell contact at adherens junctions as well as facilitating the cytoskeletal attachment of membrane proteins (Brone and Eggermont, 2005; Harris and Lim, 2001; Jelen *et al.*, 2003). In fact, DSL ligands colocalize with actin (Lowell and Watt, 2001) and their specific PDZ domain partners at regions of cell–cell contact (Estrach *et al.*, 2007; Mizuhara *et al.*, 2005; Six *et al.*, 2004; Wright *et al.*, 2004), consistent with the proposed role for DSL ligands in promoting cell adhesion and inhibiting cell motility. Additionally, Jagged1-PDZ interactions may produce changes in gene expression that promote oncogenic transformation (Ascano *et al.*, 2003). How such interactions at the cell surface lead to transcriptional events in the nucleus is unknown, but PDZ domain proteins such as calcium/calmodulin-dependent serine protein kinase (CASK), Bridge-1, or glutamate receptor interacting protein (GRIP)-tau are known to directly act as transcriptional activators (Hsueh *et al.*, 2000; Lee *et al.*, 2005; Nakata *et al.*, 2004) whereas others such as the Dll1 interacting PDZ domain protein Acvrinp1 and the Jagged1 PDZ domain partner afadin/AF6 could indirectly effect gene transcription by binding the signal transducers Smad3 (Pfister *et al.*, 2003; Shoji *et al.*, 2000) or Ras (Ascano *et al.*, 2003; Quilliam *et al.*, 1999). Finally, that the cellular responses associated with DSL–PDZ interactions require both the extracellular and the ICDs of DSL ligands suggests that homotypic ligand–ligand interactions could activate ligand signaling (Lowell *et al.*, 2000; Lowell and Watt, 2001), while ligand–Notch interactions could induce bidirectional signaling (Ascano *et al.*, 2003). Interestingly, a model in which fringe could block Jagged1-induced Notch1 signaling yet allows Jagged1 to mediate PDZ-dependent intracellular signaling has been proposed (Ascano *et al.*, 2003).

8. Regulation of DSL Ligand Expression Patterns

Notch signaling can both positively and negatively regulate DSL ligand expression, such that defects in Notch signaling are associated with increased expression of Dll1 (Barrantes *et al.*, 1999; de la Pompa *et al.*, 1997) or Dll4 (Suchting *et al.*, 2007). On the other hand, Notch inductive signals upregulate DSL ligand expression, which is necessary for proper wing margin formation in flies (Doherty *et al.*, 1996) as well as somite formation and patterning in vertebrates (Barrantes *et al.*, 1999; Cheng *et al.*, 2007, 2003; de la Pompa *et al.*, 1997; Doherty *et al.*, 1996; Takahashi *et al.*, 2003).

8.1. Cellular factors that regulate Notch ligand expression

In addition to Notch, other signaling systems are thought to intersect with the Notch pathway at the level of ligand expression (Hurlbut *et al.*, 2007). In particular, the signaling pathways outlined in Table 3.1 are known to regulate ligand expression and produce specific cellular responses. These include vascular EGF (VEGF) (Benedito *et al.*, 2009; Hellstrom *et al.*, 2007; Limbourg *et al.*, 2007; Liu *et al.*, 2003; Lobov *et al.*, 2007; Patel *et al.*, 2005; Seo *et al.*, 2006; Williams *et al.*, 2006), tumor necrosis factor alpha (TNFα) (Benedito *et al.*, 2009), fibroblast growth factor (Akai *et al.*, 2005; Faux *et al.*, 2001; Limbourg *et al.*, 2007), platelet-derived growth factor (PDGF) (Campos *et al.*, 2002), TGFβ (Zavadil *et al.*, 2004), lipopolysaccharide (LPS) (Amsen *et al.*, 2004; Liotta *et al.*, 2008), interleukin-6 (IL6) (Sansone *et al.*, 2007; Studebaker *et al.*, 2008), Hedgehog (McGlinn *et al.*, 2005), *Drosophila* epidermal growth factor receptor (Carmena *et al.*, 2002; Tsuda *et al.*, 2002), and Wnt (Estrach *et al.*, 2006; Hofmann *et al.*, 2004; Pannequin *et al.*, 2009; Rodilla *et al.*, 2009).

The majority of these signaling pathways enhance ligand expression, such as canonical Wnt signaling that activates Jagged1 transcription during hair follicle differentiation (Hofmann *et al.*, 2004). In the angiogenic vasculature, VEGF induces Dll4 expression in endothelial cells to prevent sprouting angiogenesis (Roca and Adams, 2007; Sainson and Harris, 2008; Thurston *et al.*, 2007; Yan and Plowman, 2007) while TNFα-induced Jagged1 expression has the opposite effect (Benedito *et al.*, 2009; Sainson *et al.*, 2008). The differential regulation of expression of Dll4 and Jagged1 with opposing roles in angiogenesis has been proposed to guide the specification of tip cells and stalk cells to regulate the number of sprouting vessels (see Chapter 9). In the immune system, specific inflammatory responses upregulate expression of either Delta-like or Jagged1 ligands in dendritic cells to guide activated CD4+ T cells toward either a T-helper (Th)-1 or Th-2 response, respectively (Amsen *et al.*, 2004; Maekawa *et al.*, 2003). However, more recent findings have questioned the role of Notch signaling in T-cell fate acquisition (Ong *et al.*, 2008). Nevertheless, ligand-specific effects of Notch signaling have also been reported in nonsmall cell lung cancer cells and hematopoietic progenitors (Choi *et al.*, 2009; de La Coste and Freitas, 2006). Fringe-mediated modulation of the sensitivity of Notch for different ligands as well as interaction of different ligands with distinct Notch receptors have been proposed to regulate some of these ligand-dependent effects (Amsen *et al.*, 2004; Cheng and Gabrilovich, 2007; de La Coste and Freitas, 2006; Maekawa *et al.*, 2003; Raymond *et al.*, 2007).

Table 1 Cellular factors that regulate DSL ligand expression

Effector of DSL ligand expression	DSL ligand	Effect on ligand expression: Upregulation (+) Downregulation (−)	Cell type	Biological effect	References
VEGF[a]	Dll4[b]	+	Endothelial	Inhibition of angiogenic sprouting; arterial specification	Hellstrom *et al.* (2007), Liu *et al.* (2003), Lobov *et al.* (2007), Patel *et al.* (2005), Seo *et al.* (2006), Williams *et al.* (2006)
TNFα[c]	Jagged1	+	Endothelial	Promotion of angiogenic sprouting	Benedito *et al.* (2009), Sainson *et al.*, (2008)
FGF[d]	Dll1[b]	+	Neural stem cells	Maintenance of spinal cord stem cells	Akai *et al.*, (2005)
LPS[e]	Dll4[b]	+	Dendritic cells	CD4$^+$ Th1[k] polarization	Amsen *et al.* (2004)
LPS[e]/PGE2[f]	Jagged1	+	Dendritic cells	CD4$^+$ Th2[k] polarization	Amsen *et al.*,(2004)
IL6[g]	Jagged1	+	Mammary epithelial cells	Proliferation and invasion	Sansone *et al.* (2007), Studebaker *et al.* (2008)
Hedgehog	Jagged1	+	Mesenchymalcells	Limb development	McGlinn *et al.*, (2005)
VEGF[a] + FGF2[d]	Dll1[b]	+	Endothelial cells	Postnatal Arteriogenesis	Limbourg *et al.*, (2007)
Wnt	Jagged1	+	Hair follicle precortex	Hair follicle differentiation	Estrach *et al.*, (2006)
Wnt	Jagged1	+			

			Intestinal epithelial cells	Proliferation (tumorigenesis)	Rodilla *et al.* (2009), Pannequin *et al.* (2009)
Wnt	Dll1[b]	+	Presomitic mesoderm	Somitogenesis	Hofmann *et al.* (2004)
DER[h] and/or Heartless	*Drosophila* Delta	+	Embryonic mesoderm	Specification of muscle and heart progenitors, photo-receptor and nonneuronal cone cells	Carmena *et al.* (2002), Tsuda *et al.*, (2002)
TGFβ[i]	Jagged1	+	Epithelial cells	Epithelial -mesenchymal transformation	Zavadil *et al.* (2004)
FGF1[d]/ FGF2[d]	Dll1[b]	-	Neuroepithelium	Maintenance of neuroepithelial precursors	Faux *et al.* (2001)
PDGF[j]/ angiotensin II	Jagged1	-	Vascular smooth muscle cells	Growth retardation	Campos *et al.* (2002)
LPS[e]	Jagged1	-	Bone-marrow mesenchymal stem cells	Proliferation of $CD4^+$ T cells	Liotta *et al.* (2008)

[a] Vascular Endothelial Growth Factor.
[b] Dll: Delta-like.
[c] Tumor necrosis factor α.
[d] FGF: Fibroblast growth factor.
[e] Lipopolysaccharide.
[f] Prostaglandin E2.
[g] Interleukin 6.
[h] *Drosophila* epidermal growth factor receptor.
[i] Transforming growth factor β,
[j] Platelet-derived growth factor,
[k] Th: T helper cell.

Contrasting with the positive regulatory effects of signaling pathways on DSL ligand expression, downregulation of Jagged1 expression by PDGF and angiotensin II restricts vascular smooth muscle cell growth *in vitro* (Campos *et al.*, 2002), while LPS-mediated downregulation of Jagged1 expression in bone marrow mesenchymal stem cells inhibits proliferation of CD4+ T cells (Liotta *et al.*, 2008). The regulation of ligand expression not only plays a role in coordinating normal cellular responses but also in promoting cancer. In fact, upregulation of Jagged1 expression has recently emerged as a "pathological link" between Wnt and IL6 signaling pathways and Notch activation in colon (Pannequin *et al.*, 2009; Rodilla *et al.*, 2009) and breast (Sansone *et al.*, 2007; Studebaker *et al.*, 2008) cancer.

8.2. Spatio-temporal regulation of Notch ligand expression

The existence of mechanisms to regulate ligand expression provides a means to temporally and/or spatially compartmentalize Notch signaling activity and coordinate specific Notch-dependent responses. In fact, the establishment of developmental boundaries and the segmentation of limbs and appendages is dependent on Notch signaling and coordination of these processes can be regulated by the spatio-temporal distribution of ligand expression (Bishop *et al.*, 1999; de Celis *et al.*, 1998; Klein and Arias, 1998; Panin *et al.*, 1997; Rauskolb and Irvine, 1999). For instance, in the developing wing disc of flies, Serrate is expressed dorsally while higher Delta expression occurs ventrally and Notch signaling directs this ligand expression pattern (Blair, 2000; Doherty *et al.*, 1996). The coexpression of Fringe in the dorsal compartment ensures that Serrate can only signal to adjacent ventral cells that lack Fringe, while ventral Delta signals preferentially to adjacent dorsal cells. In this manner, reciprocal Notch signaling between dorsal and ventral cells restricts Notch activation to cells along the dorso-ventral boundary required for proper wing margin formation. Further, establishment of leg segments in *Drosophila* which is dependent on Notch signaling is regulated by the "leg gap genes" *homothorax, dachshund*, and *Distal-less* that temporally control the segmental pattern of Notch ligand expression as well as the glycosyltransferase Fringe (Bishop *et al.*, 1999; Rauskolb, 2001). Together these findings indicate that the regulation of DSL ligand expression by other signaling pathways serves to spatio-temporally compartmentalize Notch signaling activity. This allows Notch signaling to be integrated into a highly ordered and complex molecular network (Hurlbut *et al.*, 2007), which could regulate embryonic development, the induction of immune and vascular responses, and contribute to disease states such as cancer in the adult.

9. Noncanonical Ligands

In contrast to other signaling systems that employ a large number of activating ligands, there are only four mammalian ligands known to activate Notch receptors. It is difficult to account for the pleiotropic affects of Notch given this limited number of DSL ligands; however, the identification of noncanonical ligands expands the repertoire of ligands reported to activate signaling. Unlike the activating canonical ligands that contain a DSL domain required to interact with Notch (Fig. 3.2), noncanonical ligands lack this essential motif and comprise a group of structurally diverse proteins that include integral and glycosylphosphatidylinositol (GPI)-linked membrane as well as secreted proteins outlined in Fig. 3.5.

9.1. Membrane-tethered noncanonical ligands

Delta-like 1 (Dlk-1), also known as Pref-1, or FA-1 is one of the first reported noncanonical ligands for Notch (Bachmann *et al.*, 1996; Laborda *et al.*, 1993; Smas and Sul, 1993) and is best known for its role in preventing adipogenesis (Wang *et al.*, 2006). While lacking a DSL domain, Dlk-1 is otherwise structurally similar to Delta-like proteins. Dlk1 is also cleaved by ADAMs and is negatively regulated by Notch signaling (Ross *et al.*, 2004; Wang and Sul, 2006). Most evidence support the idea that Dlk-1 and Notch only interact in *cis* and the affects of Dlk-1 overexpression on Notch target gene expression and phenotype are consistent with Dlk-1 functioning as a *cis*-inhibitor of Notch signaling (Baladron *et al.*, 2005; Bray *et al.*, 2008; Nueda *et al.*, 2007). Interestingly, an ADAM-resistant, membrane-bound form of Dlk-1 is more potent than wild-type or soluble forms in functioning in *cis*-inhibition, suggesting that Dlk-1-mediated antagonism of Notch signaling may require low cellular ADAM activity to maintain membrane-bound Dlk-1 (Bray *et al.*, 2008). The molecular basis for Dlk-1-mediated Notch antagonism is unclear, but given the overlap in the binding sites for Dlk-1 and DSL ligands on Notch (Baladron *et al.*, 2005), it seems plausible that Dlk-1 antagonizes Notch signaling by competing with DSL ligands for Notch binding. Although Dlk-1 and Notch have been shown to interact by yeast two-hybrid analysis (Baladron *et al.*, 2005; Komatsu *et al.*, 2008), interactions between these proteins has not been demonstrated for endogenous or ectopic proteins. Neither is there a consensus on whether Dlk-1-induced loss of Hes-1 expression directly involves Notch, since Hes-1 is regulated by other signaling pathways (Hatakeyama *et al.*, 2004; Kluppel and Wrana, 2005; Ross *et al.*, 2004).

More recently, the identification of a DOS domain in Dlk-1 and Dlk-2/EGFL9 has led to the proposal that these proteins may also function as coactivating noncanonical Notch ligands (Komatsu *et al.*, 2008). In fact, genetic studies have shown that Dlk-1 can functionally substitute for

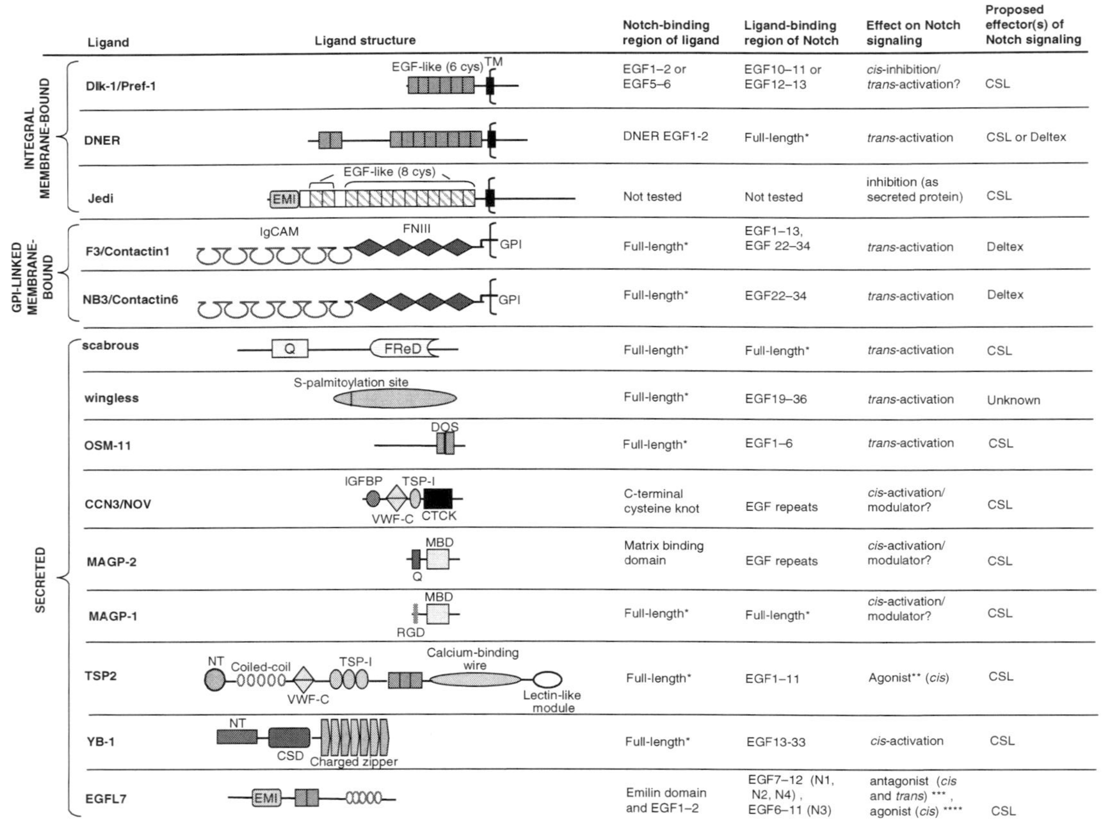

Fig. 3.5 *(Continued)*

the *C. elegans* DOS-only containing ligand, OSM-11, in cooperating with the DSL-only ligand, DSL-1, to activate Notch signaling, suggesting that the role of DOS-motif in Notch signaling may be conserved across species (Komatsu *et al.*, 2008). However, these findings are difficult to reconcile given that Dlk-1 has been suggested to antagonize Jagged1-induced Notch signaling (Baladron *et al.*, 2005). In light of the fact that Jagged1 contains both a DSL domain and a DOS motif, it has been proposed that the DOS-only containing ligand Dlk-1 competes with Jagged1 for Notch binding and thereby antagonizes Jagged1 signaling (Komatsu *et al.*, 2008). Thus, DOS domain proteins may function as Notch agonists in cooperation with DSL domain-only containing ligands but antagonize signaling by ligands containing both DSL and DOS domains.

Another integral membrane-bound Notch ligand lacking a DSL domain is Delta/Notch-like EGF-related receptor (DNER) that like Dlk-1 also contains extracellular tandem EGF repeats (Eiraku *et al.*, 2002). In contrast to Dlk-1, DNER binds Notch when presented in *trans* and DNER-expressing cells activate a CSL reporter in cocultured cells (Eiraku *et al.*, 2005). Both *in vitro* and *in vivo* studies support DNER's function as a *trans*-ligand to effect glial morphological changes through activation of Notch. DNER, however, does not affect glial cell number *in vivo*, suggesting that it functions at later stages of differentiation. Consistent with the expression of DNER in Purkinje cells and Notch in the adjacent Bergmann glia, DNER mutant mice exhibit morphological defects in Bergmann glia (Eiraku *et al.*, 2005). A soluble recombinant form of DNER can also affect Bergmann glia morphology *in vitro* in a γ-secretase-dependent but CSL-independent manner. Instead of CSL, the E3 ubiquitin ligase Deltex has been implicated as an alternative downstream effector of Notch in DNER-induced glial morphological changes. Deltex can bind directly to

Fig. 3.5 Noncanonical ligand structure and proposed effects on Notch signaling. Noncanonical ligands lack a DSL domain (*D*elta/*S*errate/*L*AG-2), are structurally diverse and include integral- and GPI-linked membrane proteins as well as secreted proteins (see text for details). EGF-like (6 cys), 6-cysteine epidermal growth factor-like repeat as found in canonical ligands; cys, cysteine; TM, transmembrane domain, CSL (CBF1, Su(H), LAG1); EMI, emilin-like domain; EGF-like (8 cys), EGF-like motif with 8 cysteines that is not laminin-like; Ig-CAM, immunoglobulin-containing cell adhesion molecule domain; FNIII, fibronectin type III domain; GPI, glycosylphosphatidylinositol; Q, glutamine-rich region; FReD, fibrinogen-related domain; DOS, Delta and OSM-11-like proteins; IGFBP, insulin-like growth factor-binding protein-like domain; VWF-C, von Willebrand factor type C-like domain; TSP-1, thrombospondin type 1-like domain; CTCK, C-terminal cysteine knot domain; MBD, matrix binding domain; RGD, integrin-binding motif; NT, N-terminal domain; CSD, cold shock domain, N1, Notch1; N2, Notch2; N3, Notch3; N4, Notch4. *Only full-length constructs were tested for binding **Agonist of Jagged1 signaling ***Antagonist of Jagged1 signaling **** Agonist of Dll4 (Delta-like 4) signaling (See Color Insert.)

the NICD and mediate a trimeric complex between itself, full-length Notch, and β-arrestin (Mukherjee *et al.*, 2005), making it possible that Notch could activate signaling through β-arrestin that would require Deltex but not CSL. Whether the effects of DNER are dependent on Notch receptor expression in Bergmann glia have yet to be determined.

A putative DSL ligand-like protein called Jagged and Delta protein (Jedi) has been identified based on sequence data (Krivtsov *et al.*, 2007). However, the putative DSL and EGF repeats of Jedi lack the conserved cysteine spacing common to either the signature motif of canonical ligands or EGF repeats present in DNER and Dlk-1. Instead, the Jedi ECD contains an NT EMILIN domain followed by multiple tandem repeats of an eight-cysteine variation of the EGF domain interspersed with two single six-cysteine EGF repeats (Krivtsov *et al.*, 2007; Nanda *et al.*, 2005). In fact, Jedi has not been reported to interact with any of the Notch receptors and lacks *trans*-activating or *cis*-inhibitory activity. Although soluble Jedi added to Notch-expressing cells weakly inhibits a Notch reporter, there is currently no strong evidence linking Jedi to Notch signaling. The closely related Jedi family member, multiple EGF-like domains 10 protein (MEGF10) (Krivtsov *et al.*, 2007), has also been proposed to interact with the Notch signaling pathway (Holterman *et al.*, 2007); however, like Jedi, there has been no formal demonstration that MEGF10 can directly interact with Notch receptors.

9.2. GPI-linked noncanonical ligands

Structurally distinct from the integral membrane noncanonical ligands are the GPI-linked neural cell adhesion molecules, F3/contactin1 and NB3/contactin6 that activate Notch signaling to induce oligodendrocyte (OL) differentiation (Cui *et al.*, 2004; Hu *et al.*, 2003). Although binding and fractionation studies have indicated that these contactins interact with Notch in *trans, cis* interactions cannot be ruled out since both endogenous F3 and NB3 coimmunoprecipitate with Notch. Both contactins interact with Notch EGF repeats distal to the DSL binding site; however, F3 can also interact with Notch EGF repeats 1–13 that includes the DSL ligand-binding site at EGF 11–12. While this interaction would initially suggest that F3 competes for the DSL ligand-binding site, further studies are required to determine whether the F3 and DSL ligand-binding sites actually overlap.

As found for DSL ligand treatment, soluble forms of either contactin induce γ-secretase-dependent NICD production in OL cells. However, F3-Notch signaling does not activate Hes-1 transcription, and there is no evidence of NB3 activating canonical CSL-induced Notch signaling (Hu *et al.*, 2003; Lu *et al.*, 2008). Instead of CSL, both contactins utilize Deltex as an effector of Notch signaling to induce glial maturation. An interesting conundrum is raised in these *in vitro* assays since the same cells (that presumably utilize the same Notch receptors) differentiate in response to contactins but remain progenitors in response to DSL ligand or NICD expression. It is thought that differences in the temporal expression of DSL ligands and contactins could dictate which effect takes precedence *in vivo* since DSL ligands and contactins are expressed at distinct developmental time points. Therefore, like DNER, the contactins appear to utilize Notch to effect changes late in differentiation as opposed to DSL ligands that can impact early cell fate decisions (Hu *et al.*, 2003).

The important role for contactin-induced Notch signaling in OL differentiation has led to speculation that in multiple sclerosis lesions, contactin expression may be lost in demyelinated axons that would normally activate Notch in neighboring OL precursor cells. Recent findings, however, have demonstrated that not only is contactin expression maintained in demyelinated axons but also that NICD is generated in the OL precursor cells within these lesions (Nakahara *et al.*, 2009). Instead, translocation of NICD to the nucleus was inhibited in these cells and the proapototic factor TAT-interacting protein 30 that prevents nuclear transport has been implicated in this process. These findings identify a novel mechanism for regulation of Notch activity downstream of NICD generation.

9.3. Secreted noncanonical ligands

Two secreted non-DSL ligands have been identified in *Drosophila.* The first, Scabrous (Sca), plays a role in Notch-dependent patterning of eye ommatidia and sensory bristles (Baker *et al.*, 1990; Mlodzik *et al.*, 1990). Sca binds to Notch in *trans* and activates transcription of the Notch target gene *E(spl)C m3* (Mok *et al.*, 2005; Powell *et al.*, 2001). It is not known, however, if the effects of Sca require γ-secretase cleavage of Notch, the Notch downstream effector Su(H), or indeed activation of some other signaling pathway. Another reported *Drosophila* secreted non-DSL ligand for Notch is Wingless (Wg), the fly ortholog of mammalian Wnt proteins. Wg was identified as a Notch-binding protein in a screen of a phage display library expressing *Drosophila* embryo transcripts and immunoprecipitation of endogenous Notch and Wg in fly embryos supports such an interaction *in vivo* (Wesley, 1999). Although the gene *shaggy* can be transcriptionally activated in a Wg- and Notch-dependent manner, it is not clear if binding of Wg to Notch is required for its transcription or which Notch downstream effectors are required. While many vertebrate Wnt proteins exist, none have been reported to bind Notch as demonstrated for *Drosophila* Wg.

In *C. elegans*, five secreted putative Notch ligands lacking DSL domains have been identified: OSM11, OSM7, DOS1, DOS2, and DOS3 (Komatsu *et al.*, 2008). Interestingly, all five proteins contain conserved amino acids within a common motif called DOS and although this motif is lacking in all *C. elegans* DSL ligands, it is present in most DSL ligands from other phyla. The best characterized of these *C. elegans* DOS-containing ligands, OSM11, interacts with the ECD of LIN-12 in yeast two-hybrid assays and genetic analyses suggest that OSM11 enhances LIN-12 signaling during vulval development by acting upstream of or during LIN-12 receptor activation. While losses of *osm11* produce defects in vulval precursor cell specification associated with losses in Notch signaling, loss of the DSL domain-containing ligand *dsl-1* potentiates *osm11* loss-of-function defects, suggesting that DOS domain-only and DSL domain-only containing ligands cooperate to activate signaling in some developmental contexts in *C. elegans*. Although these results suggest the existence of a bipartite ligand system for activating Notch in *C. elegans*, biochemical studies are necessary to confirm that OSM11 directly interacts with endogenous LIN-12 to activate signaling. Furthermore, it is not clear from these studies whether the effects of OSM11 on Notch signaling are indeed mediated by the DOS motif. In this regard, a role for the DOS motif in Notch binding and signaling has been extrapolated (Komatsu *et al.*, 2008; Kopan and Ilagan, 2009) from mutational and structural studies of *Drosophila* and mammalian DSL ligands (Cordle *et al.*, 2008; Parks *et al.*, 2006; Shimizu *et al.*, 1999). Significantly, mutations that map to the DOS motif of Jagged1 are associated with human syndromes (Eldadah *et al.*, 2001; Guarnaccia *et al.*, 2009; Warthen *et al.*, 2006) and genetic malformations in mice (Kiernan *et al.*, 2001; Tsai *et al.*, 2001). Although X-ray crystallography and NMR-based binding studies have suggested that the binding interface between Jagged1 and Notch1 includes amino acid residues from not only the DSL domain but also the DOS domain, the topography of this interface is not known (Cordle *et al.*, 2008). Understanding how the DOS and DSL domains cooperatively bind Notch will require crystal structure studies of the ligand–Notch complex.

Secreted Notch ligands lacking a DSL domain have also been identified in vertebrates. One of these is the Connective Tissue Growth Factor/ *c*ysteine-rich 61/*N*ephroblastoma Overexpressed Gene family member, CCN3. When coexpressed, CCN3 interacts with Notch via the CCN3 C-terminal cysteine knot that appears to be a general tandem EGF repeat-binding domain (Sakamoto *et al.*, 2002b; Thibout *et al.*, 2003). Coexpression of CCN3 potentiates endogenous CSL-dependent Notch signaling in reporter assays and losses in CCN3 reduce *trans*-DSL ligand-induced activation of a CSL reporter (Gupta *et al.*, 2007; Minamizato *et al.*, 2007; Sakamoto *et al.*, 2002b). Further supporting a role for CCN3 as an activating cofactor for canonical ligand-induced signaling is the observation that soluble CCN3 can enhance hematopoietic precursor cell colony formation induced by Jagged-1 (Gupta *et al.*, 2007). Additionally, gains and losses in CCN3 expression produce corresponding changes in Hes-1 expression, suggesting that CCN3 may activate Notch in an autocrine fashion (Gupta *et al.*, 2007; Minamizato *et al.*, 2007; Sakamoto *et al.*, 2002b). Autocrine Notch signaling by CCN3 may be relevant to cell types such as chondrocytes and vascular smooth muscle cells that secrete extracellular matrix and consequently are isolated and unable to undergo juxtacrine signaling by canonical ligands. Consistent with this notion, both chondrocytes and vascular smooth muscle cells express CCN3 (Ellis *et al.*, 2000; Perbal, 2004).

A second secreted, non-DSL vertebrate protein that can activate Notch signaling is the microfibril-associated glycoprotein family, MAGP-1 and MAGP-2 (Gibson *et al.*, 1991, 1996). Both MAGP proteins can interact with Notch leading to γ-secretase-dependent NICD generation and activation of CSL-dependent reporter constructs (Miyamoto *et al.*, 2006). Like CCN3, MAGP-2 only activates Notch when coexpressed in the same cell, and vascular smooth muscle cells that express MAGP2 may use this noncanonical ligand to activate Notch signaling in an autocrine manner (Albig *et al.*, 2008; Miyamoto *et al.*, 2006). Interestingly, similar to DSL ligands, MAGP-2 can activate Notch by inducing ADAM-independent dissociation of the Notch heterodimer and in fact, MAGP-2 is the only noncanonical ligand that has so far been demonstrated to cause nonenzymatic dissociation of Notch (Miyamoto *et al.*, 2006). The biological significance of MAGP-2-induced Notch signaling is as yet unclear and it appears that MAGP-2 can also inhibit Notch signaling in certain cell types; however, the molecular basis for these cell type differences are not understood (Albig *et al.*, 2008).

In addition to CCN3/NOV and MAGP proteins, a third vertebrate matrix protein, thrombospondin2 (TSP2), has been implicated as a noncanonical Notch ligand (Meng *et al.*, 2009). As found for CCN3/NOV, TSP2 enhances signaling induced by *trans*-DSL ligands either when coexpressed or when exposed to Notch cells as a soluble recombinant protein (Meng *et al.*, 2009). The effect of TSP2 on Notch signaling is γ-secretase-dependent and requires the Notch extracellular sequences. Consistent with these findings, coimmunoprecipitation studies suggest that TSP2 interacts with the Notch3 ECD at the cell surface. *In vitro* binding assays using recombinant proteins further suggest that TSP2 can directly interact with the first 11 EGF-like repeats of Notch3. It is surprising that TSP2 enhances rather than inhibits ligand-induced Notch signaling given that the region of Notch3 includes the DSL ligand-binding domain; however, it is not known if the TSP2 and DSL ligand binding sites actually overlap. Interestingly, TSP2 can also interact with Jagged1 and enhance binding to Notch3 EGF repeats suggesting a molecular basis for increased Notch signaling by TSP2. Further supporting interactions between TSP2 and Notch is the observation that arterial tissue from TSP2 knockout mice exhibits significant reduction in Notch target expression. Although the closely related thrombospondin, TSP1, also interacts with Notch3 and Jagged1 (Meng *et al.*, 2009), it is neither able to enhance binding between the ligand–receptor pair nor enhance Notch signaling—subtle structural differences between these TSP family members may account for their differential effects on Notch signaling.

A fourth secreted vertebrate non-DSL ligand that activates Notch signaling, Y-*box* (YB) protein-1, belongs to the cold shock protein family (Frye *et al.*, 2009; Rauen *et al.*, 2009). Yeast two-hybrid and coimmunoprecipitation studies have demonstrated that YB1 interacts with EGF repeats 13–33 of Notch3 (Rauen *et al.*, 2009), a region distinct from that required for DSL ligand binding. Furthermore, confocal microscopy and fluorescence-activated cell sorter analyses suggest that YB1–Notch3 interactions occur at the cell surface and soluble YB1–Notch3 interactions activate CSL-dependent reporter constructs in a γ-secretase-dependent manner. Interestingly, YB1 does not bind Notch1 and although interactions with Notch2 and Notch4 have not been examined, it is tempting to speculate that YB1 interactions may be specific for Notch3. YB1–Notch3 interactions may be relevant to kidney disease since in a mouse model of mesanglioproliferative disease in which YB1 and Notch3 expression are coordinately upregulated during the course of the disease, both the cleaved ECD of Notch3 and YB-1 are detected in urine samples of diseased animals. The molecular basis of these findings is unclear; however, they could reflect YB-1-induced dissociation of the Notch3 heterodimer.

More recently, a fifth vertebrate secreted non-DSL Notch ligand, EGF-like domain 7 (EGFL7) was identified in yeast two-hybrid screens (Schmidt *et al.*, 2009). EGFL7 interacts with a set of EGF-like repeats that includes the DSL binding sites of all four human Notch receptors and antagonizes Notch signaling induced by Jagged1-type ligands. The inhibitory effects of EGFL7 on Jagged1-induced Notch signaling were demonstrated both *in cis* and *in trans* using biochemical studies as well as a neurosphere model and appear to result from competition with Jagged1 for Notch binding. Expression of EGFL7 prevents self-renewal of neural stem cells cultured as neurospheres, a process dependent on Jagged1–Notch1 interactions. Further supporting a role for EGFL7 as a Notch antagonist, EGFL7 expression promotes differentiation of neural stem cells into neurons and oligodendrocytes at the expense of astrocytes. Surprisingly, the inhibitory effects of EGFL7 seem specific to Jagged-type ligands. Ablation of EGFL7 reduces Dll4-induced activation of a CSL-dependent reporter suggesting that EGFL7 enhances Dll4-induced Notch signaling. Although it is difficult to reconcile the differential DSL ligand-dependent effects of EGFL7 on Notch signaling, yeast two-hybrid studies have shown that EGFL7 can also interact with Dll4 but not with Jagged1 or Jagged2. It is possible that as proposed for TSP2, EGFL7 enhances ligand–Notch interactions accounting for its ability to potentiate Dll4-induced Notch signaling. It is important to note that the opposing DSL ligand-specific effects of EGFL7 were demonstrated in different cell contexts and thus could reflect cell type-specific differences for EGFL7 on Notch signaling as found for MAGP2.

In summary, noncanonical ligands represent a subset of Notch ligands that despite lacking a DSL domain can activate Notch signaling. Compared to the canonical ligands that all require binding to Notch EGF repeats 11 and 12 to activate signaling, noncanonical ligands do not appear to have a consensus Notch binding site, yet some activate Notch γ-secretase cleavage and CSL-dependent transcription. Given that nonenzymatic dissociation of Notch leads to signaling, it would appear that any protein that can bind Notch and dissociate the heterodimeric structure activates Notch signaling. Indeed, binding of MAGP2 can cause nonenzymatic dissociation of Notch and activation of Notch signaling, and it remains to be demonstrated if other noncanonical ligands also follow a similar mechanism for Notch activation. Interestingly, like the membrane-bound DSL ligands, all type 1 transmembrane noncanonical ligands contain lysines in their ICDs that could serve as ubiquitination sites to facilitate transendocytosis; however, it is not known if endocytosis is required for activity of these noncanonical ligands. Even less clear is how Notch binding to secreted noncanonical ligands like MAGP2 could produce force for heterodimer dissociation, but perhaps cooperative binding with membrane-bound ligands or tethering to the extracellular matrix would induce a pulling force on Notch.

While noncanonical ligands may contribute to the pleiotropic nature of Notch signaling, the effects of many have only been demonstrated using *in vitro* assays and need to be confirmed *in vivo*. In this regard, it is noteworthy that while DSL ligands are crucial for embryonic development and viability in the mouse, none of the reported noncanonical ligands are similarly required. It thus appears that if noncanonical ligands function *in vivo*, they may do so as modulators of Notch signaling in the adult animal.

10. Conclusions and Future Directions

Although unique ligand–receptor combinations have been identified that induce specific cellular responses, the molecular mechanisms underlying ligand-specific signaling remains an outstanding question in the field. Moreover, given the direct and somewhat simple signaling mechanism ascribed to Notch, it is unclear how different Notch ligands could induce distinct signaling responses. It will be important to determine if different ligand–Notch complexes recruit unique signaling effectors and whether the distinct responses involve activation of cytoplasmic and/or nuclear signaling pathways. In this regard, identification of the endocytic machinery used by ligand cells to activate Notch signaling and the potential role for endocytosis in force generation are critical avenues that remain to be tested. That ligands have intrinsic signaling activity independent of Notch as well as their potential to participate in bidirectional signaling is exciting but relatively unexplored areas of ligand biology that warrant further investigation. The importance of Notch ligands in cancer and other pathological states involving aberrant angiogenesis have identified Notch ligands as potential and promising therapeutic targets (Roca and Adams, 2007; Sainson and Harris, 2008; Thurston *et al.*, 2007; Yan and Plowman, 2007). Finally, the use of Notch ligands in the expansion and maintenance of stem cells for tissue regeneration and replacement underscores their fundamental biological importance (Dallas *et al.*, 2005; Delaney *et al.*, 2005).

ACKNOWLEDGMENTS

We thank Abdiwahab Musse and Jason Tchieu for help with the illustrations and Alison Miyamoto for contributions to the material previously published in Oncogene (2008) 27, 5148–5167. The authors acknowledge the National Institute of Health NIH (GW), Jonsson Comprehensive Cancer Center (JCCF) (LMK), and Association of International Cancer Research (AICR) (BD) for financial support.

REFERENCES

Ahimou, F., Mok, L. P., Bardot, B., and Wesley, C. (2004). The adhesion force of Notch with Delta and the rate of Notch signaling. *J. Cell Biol.* **167**, 1217–1229.

Aho, S. (2004). Soluble form of Jagged1: unique product of epithelial keratinocytes and a regulator of keratinocyte differentiation. *J. Cell. Biochem.* **92**, 1271–1281.

Akai, J., Halley, P. A., and Storey, K. G. (2005). FGF-dependent Notch signaling maintains the spinal cord stem zone. *Genes Dev.* **19**, 2877–2887.

Albig, A. R., Becenti, D. J., Roy, T. G., and Schiemann, W. P. (2008). Microfibril-associate glycoprotein-2 (MAGP-2) promotes angiogenic cell sprouting by blocking notch signaling in endothelial cells. *Microvasc. Res.* **76**(1), 7–14.

Amsen, D., Blander, J. M., Lee, G. R., Tanigaki, K., Honjo, T., and Flavell, R. A. (2004). Instruction of distinct CD4 T helper cell fates by different notch ligands on antigen-presenting cells. *Cell* **117**, 515–526.

Aoto, J., and Chen, L. (2007). Bidirectional ephrin/Eph signaling in synaptic functions. *Brain Res.* **1184**, 72–80.

Artavanis-Tsakonas, S., Rand, M. D., and Lake, R. J. (1999). Notch signaling: cell fate control and signal integration in development. *Science* **284**, 770–776.

Ascano, J. M., Beverly, L. J., and Capobianco, A. J. (2003). The C-terminal PDZ-ligand of JAGGED1 is essential for cellular transformation. *J. Biol. Chem.* **278**, 8771–8779.

Bachmann, E., Krogh, T. N., Hojrup, P., Skjodt, K., and Teisner, B. (1996). Mouse fetal antigen 1 (mFA1), the circulating gene product of mdlk, pref-1 and SCP-1: Isolation, characterization and biology. *J. Reprod. Fertil.* **107**, 279–285.

Baker, N. E., Mlodzik, M., and Rubin, G. M. (1990). Spacing differentiation in the developing *Drosophila* eye: a fibrinogen-related lateral inhibitor encoded by scabrous. *Science* **250**, 1370–1377.

Baladron, V., Ruiz-Hidalgo, M. J., Nueda, M. L., Diaz-Guerra, M. J., Garcia-Ramirez, J. J., Bonvini, E., Gubina, E., and Laborda, J. (2005). Dlk acts as a negative regulator of Notch1 activation through interactions with specific EGF-like repeats. *Exp. Cell Res.* **303**, 343–359.

Bardin, A. J., and Schweisguth, F. (2006). Bearded family members inhibit Neuralized-mediated endocytosis and signaling activity of Delta in *Drosophila. Dev. Cell* **10**, 245–255.

Barrantes, I. B., Elia, A. J., Wunsch, K., Hrabe de Angelis, M. H., Mak, T. W., Rossant, J., Conlon, R. A., Gossler, A., and de la Pompa, J. L. (1999). Interaction between Notch signalling and Lunatic fringe during somite boundary formation in the mouse. *Curr. Biol.* **9**, 470–480.

Barriere, H., Nemes, C., Lechardeur, D., Khan-Mohammad, M., Fruh, K., and Lukacs, G. L. (2006). Molecular basis of oligoubiquitin-dependent internalization of membrane proteins in Mammalian cells. *Traffic* **7**, 282–297.

Barsi, J. C., Rajendra, R., Wu, J. I., and Artzt, K. (2005). Mind bomb1 is a ubiquitin ligase essential for mouse embryonic development and Notch signaling. *Mech. Dev.* **122**, 1106–1117.

Becam, I., Fiuza, U. M., Arias, A. M., and Milan, M. (2010). A role of receptor Notch in ligand cis-inhibition in *Drosophila. Curr. Biol.* **20**, 554–560.

Benedito, R., Roca, C., Sorensen, I., Adams, S., Gossler, A., Fruttiger, M., and Adams, R. H. (2009). The notch ligands Dll4 and Jagged1 have opposing effects on angiogenesis. *Cell* **137**, 1124–1135.

Bishop, S. A., Klein, T., Arias, A. M., and Couso, J. P. (1999). Composite signalling from Serrate and Delta establishes leg segments in *Drosophila* through Notch. *Development* **126**, 2993–3003.

Blair, S. S. (2000). Notch signaling: Fringe really is a glycosyltransferase. *Curr. Biol.* **10**, R608–R612.

Bland, C. E., Kimberly, P., and Rand, M. D. (2003). Notch induced proteolysis and nuclear localization of the Delta ligand. *J. Biol. Chem.* **278**, 13607–13610.

Bray, S. J. (2006). Notch signalling: A simple pathway becomes complex. *Nat. Rev. Mol. Cell Biol.* **7**, 678–689.

Bray, S. J., Takada, S., Harrison, E., Shen, S. C., and Ferguson-Smith, A. C. (2008). The atypical mammalian ligand Delta-like homologue 1 (Dlk1) can regulate Notch signalling in *Drosophila. BMC Dev. Biol.* **8**, 11.

Brone, B., and Eggermont, J. (2005). PDZ proteins retain and regulate membrane transporters in polarized epithelial cell membranes. *Am. J. Physiol. Cell Physiol.* **288**, C20–C29.

Campos, A. H., Wang, W., Pollman, M. J., and Gibbons, G. H. (2002). Determinants of Notch-3 receptor expression and signaling in vascular smooth muscle cells: implications in cell-cycle regulation. *Circ. Res.* **91**, 999–1006.

Carmena, A., Buff, E., Halfon, M. S., Gisselbrecht, S., Jimenez, F., Baylies, M. K., and Michelson, A. M. (2002). Reciprocal regulatory interactions between the Notch and Ras signaling pathways in the *Drosophila* embryonic mesoderm. *Dev. Biol.* **244**, 226–242.

Chen, W., and Casey Corliss, D. (2004). Three modules of zebrafish Mind bomb work cooperatively to promote Delta ubiquitination and endocytosis. *Dev. Biol.* **267**, 361–373.

Chen, H., and De Camilli, P. (2005). The association of epsin with ubiquitinated cargo along the endocytic pathway is negatively regulated by its interaction with clathrin. *Proc. Natl. Acad. Sci. U.S.A.* **102**, 2766–2771.

Chen, N., and Greenwald, I. (2004). The lateral signal for LIN-12/Notch in *C. elegans* vulval development comprises redundant secreted and transmembrane DSL proteins. *Dev. Cell* **6**, 183–192.

Cheng, P., and Gabrilovich, D. (2007). Notch signaling in differentiation and function of dendritic cells. *Immunol. Res.* **41**(1), 1–14.
Cheng, H. T., Kim, M., Valerius, M. T., Surendran, K., Schuster-Gossler, K., Gossler, A., McMahon, A. P., and Kopan, R. (2007). Notch2, but not Notch1, is required for proximal fate acquisition in the mammalian nephron. *Development* **134**, 801–811.
Cheng, H. T., Miner, J. H., Lin, M., Tansey, M. G., Roth, K., and Kopan, R. (2003). Gamma-secretase activity is dispensable for mesenchyme-to-epithelium transition but required for podocyte and proximal tubule formation in developing mouse kidney. *Development* **130**, 5031–5042.
Chitnis, A. (2006). Why is delta endocytosis required for effective activation of notch?. *Dev. Dyn.* **235**, 886–894.
Chitnis, A., Henrique, D., Lewis, J., Ish-Horowicz, D., and Kintner, C. (1995). Primary neurogenesis in Xenopus embryos regulated by a homologue of the *Drosophila* neurogenic gene Delta [see comments]. *Nature* **375**, 761–766.
Choi, K., Ahn, Y. H., Gibbons, D. L., Tran, H. T., Creighton, C. J., Girard, L., Minna, J. D., Qin, F. X., and Kurie, J. M. (2009). Distinct biological roles for the notch ligands Jagged-1 and Jagged-2. *J. Biol. Chem.* **284**, 17766–17774.
Conner, S. D., and Schmid, S. L. (2003). Regulated portals of entry into the cell. *Nature* **422**, 37–44.
Cordle, J., Johnson, S., Tay, J. Z., Roversi, P., Wilkin, M. B., de Madrid, B. H., Shimizu, H., Jensen, S., Whiteman, P., Jin, B., Redfield, C., Baron, M., *et al.* (2008). A conserved face of the Jagged/Serrate DSL domain is involved in Notch trans-activation and cis-inhibition. *Nat. Struct. Mol. Biol.* **15**, 849–857.
Cui, X. Y., Hu, Q. D., Tekaya, M., Shimoda, Y., Ang, B. T., Nie, D. Y., Sun, L., Hu, W. P., Karsak, M., Duka, T., Tukeda, Y., Ou, L. Y., *et al.* (2004). NB-3/Notch1 pathway via Deltex1 promotes neural progenitor cell differentiation into oligodendrocytes. *J. Biol. Chem.* **279**, 25858–25865.
Dallas, M. H., Varnum-Finney, B., Delaney, C., Kato, K., and Bernstein, I. D. (2005). Density of the Notch ligand Delta1 determines generation of B and T cell precursors from hematopoietic stem cells. *J. Exp. Med.* **201**, 1361–1366.
Deblandre, G. A., Lai, E. C., and Kintner, C. (2001). Xenopus neuralized is a ubiquitin ligase that interacts with XDelta1 and regulates Notch signaling. *Dev. Cell* **1**, 795–806.
de Celis, J. F., and Bray, S. (1997). Feed-back mechanisms affecting Notch activation at the dorsoventral boundary in the *Drosophila* wing. *Development* **124**, 3241–3251.
de Celis, C. F., and Bray, S. J. (2000). The Abruptex domain of Notch regulates negative interactions between Notch, its ligands and Fringe. *Development* **127**, 1291–1302.
de Celis, J. F., Tyler, D. M., de Celis, J., and Bray, S. J. (1998). Notch signalling mediates segmentation of the *Drosophila* leg. *Development* **125**, 4617–4626.
de Joussineau, C., Soule, J., Martin, M., Anguille, C., Montcourrier, P., and Alexandre, D. (2003). Delta-promoted filopodia mediate long-range lateral inhibition in *Drosophila*. *Nature* **426**, 555–559.
de La Coste, A., and Freitas, A. A. (2006). Notch signaling: distinct ligands induce specific signals during lymphocyte development and maturation. *Immunol. Lett.* **102**, 1–9.
Delaney, C., Varnum-Finney, B., Aoyama, K., Brashem-Stein, C., and Bernstein, I. D. (2005). Dose-dependent effects of the Notch ligand Delta1 on ex vivo differentiation and in vivo marrow repopulating ability of cord blood cells. *Blood* **106**, 2693–2699.
de la Pompa, J. L., Wakeham, A., Correia, K. M., Samper, E., Brown, S., Aguilera, R. J., Nakano, T., Honjo, T., Mak, T. W., Rossant, J., and Conlon, R. A., (1997). Conservation of the Notch signalling pathway in mammalian neurogenesis. *Development* **124**, 1139–1148.
Delwig, A., Bland, C., Beem-Miller, M., Kimberly, P., and Rand, M. D. (2006). Endocytosis-independent mechanisms of Delta ligand proteolysis. *Exp. Cell Res.* **312**, 1345–1360.

Delwig, A., and Rand, M. D. (2008). Kuz and TACE can activate Notch independent of ligand. *Cell Mol. Life Sci.* **65**, 2232–2243.

Doherty, D., Feger, G., Younger-Shepherd, S., Jan, L. Y., and Jan, Y. N. (1996). Delta is a ventral to dorsal signal complementary to Serrate, another Notch ligand, in *Drosophila* wing formation. *Genes Dev.* **10**, 421–434.

Doherty, G. J., and McMahon, H. T. (2009). Mechanisms of endocytosis. *Annu. Rev. Biochem.* **78**, 857–902.

Dorsky, R. I., Chang, W. S., Rapaport, D. H., and Harris, W. A. (1997). Regulation of neuronal diversity in the Xenopus retina by Delta signalling. *Nature* **385**, 67–70.

Dravis, C., Yokoyama, N., Chumley, M. J., Cowan, C. A., Silvany, R. E., Shay, J., Baker, L. A., and Henkemeyer, M. (2004). Bidirectional signaling mediated by ephrin-B2 and EphB2 controls urorectal development. *Dev. Biol.* **271**, 272–290.

Dunwoodie, S. L. (2009). The role of Notch in patterning the human vertebral column. *Curr. Opin. Genet. Dev.* **19**, 329–337.

Dunwoodie, S. L., Henrique, D., Harrison, S. M., and Beddington, R. S. (1997). Mouse Dll3: a novel divergent Delta gene which may complement the function of other Delta homologues during early pattern formation in the mouse embryo. *Development* **124**, 3065–3076.

Dyczynska, E., Sun, D., Yi, H., Sehara-Fujisawa, A., Blobel, C. P., and Zolkiewska, A. (2007). Proteolytic processing of delta-like 1 by ADAM proteases. *J. Biol. Chem.* **282**, 436–444.

Eiraku, M., Hirata, Y., Takeshima, H., Hirano, T., and Kengaku, M. (2002). Delta/notch-like epidermal growth factor (EGF)-related receptor, a novel EGF-like repeat-containing protein targeted to dendrites of developing and adult central nervous system neurons. *J. Biol. Chem.* **277**, 25400–25407.

Eiraku, M., Tohgo, A., Ono, K., Kaneko, M., Fujishima, K., Hirano, T., and Kengaku, M. (2005). DNER acts as a neuron-specific Notch ligand during Bergmann glial development. *Nat. Neurosci.* **8**, 873–880.

Eisenberg, E., and Greene, L. E. (2007). Multiple roles of auxilin and hsc70 in clathrin-mediated endocytosis. *Traffic* **8**, 640–646.

Eldadah, Z. A., Hamosh, A., Biery, N. J., Montgomery, R. A., Duke, M., Elkins, R., and Dietz, H. C. (2001). Familial Tetralogy of Fallot caused by mutation in the jagged1 gene. *Hum. Mol. Genet.* **10**, 163–169.

Ellis, P. D., Chen, Q., Barker, P. J., Metcalfe, J. C., and Kemp, P. R. (2000). Nov gene encodes adhesion factor for vascular smooth muscle cells and is dynamically regulated in response to vascular injury. *Arterioscler. Thromb. Vasc. Biol.* **20**, 1912–1919.

Emery, G., Hutterer, A., Berdnik, D., Mayer, B., Wirtz-Peitz, F., Gaitan, M. G., and Knoblich, J. A. (2005). Asymmetric Rab 11 endosomes regulate delta recycling and specify cell fate in the *Drosophila* nervous system. *Cell* **122**, 763–773.

Estrach, S., Ambler, C. A., Lo Celso, C., Hozumi, K., and Watt, F. M. (2006). Jagged 1 is a beta-catenin target gene required for ectopic hair follicle formation in adult epidermis. *Development* **133**, 4427–4438.

Estrach, S., Legg, J., and Watt, F. M. (2007). Syntenin mediates Delta1-induced cohesiveness of epidermal stem cells in culture. *J. Cell. Sci.* **120**, 2944–2952.

Eun, S. H., Banks, S. M., and Fischer, J. A. (2008). Auxilin is essential for Delta signaling. *Development* **135**, 1089–1095.

Faux, C. H., Turnley, A. M., Epa, R., Cappai, R., and Bartlett, P. F. (2001). Interactions between fibroblast growth factors and Notch regulate neuronal differentiation. *J. Neurosci.* **21**, 5587–5596.

Fitzgerald, K., and Greenwald, I. (1995). Interchangeability of *Caenorhabditis elegans* DSL proteins and intrinsic signalling activity of their extracellular domains in vivo. *Development* **121**, 4275–4282.

Fiuza, U. M., and Arias, A. M. (2007). Cell and molecular biology of Notch. *J. Endocrinol.* **194**, 459–474.

Fiuza, U. M., Klein, T., Martinez Arias, A., and Hayward, P. (2010). Mechanisms of ligand-mediated inhibition in Notch signaling activity in *Drosophila*. *Dev. Dyn.* **239**, 798–805.

Fleming, R. J. (1998). Structural conservation of Notch receptors and ligands. *Semin. Cell Dev. Biol.* **9**, 599–607.

Franklin, J. L., Berechid, B. E., Cutting, F. B., Presente, A., Chambers, C. B., Foltz, D. R., Ferreira, A., and Nye, J. S. (1999). Autonomous and non-autonomous regulation of mammalian neurite development by Notch1 and Delta1. *Curr. Biol.* **9**, 1448–1457.

Frye, B. C., Halfter, S., Djudjaj, S., Muehlenberg, P., Weber, S., Raffetseder, U., En-Nia, A., Knott, H., Baron, J. M., Dooley, S., Bernhagen, J. and Mertens, P. R. (2009). Y-box protein-1 is actively secreted through a non-classical pathway and acts as an extracellular mitogen. *EMBO Rep.* **10**, 783–789.

Funahashi, Y., Hernandez, S. L., Das, I., Ahn, A., Huang, J., Vorontchikhina, M., Sharma, A., Kanamaru, E., Borisenko, V., Desilva, D. M., Suzuki, A., Wang, X., *et al.* (2008). A notch1 ectodomain construct inhibits endothelial notch signaling, tumor growth, and angiogenesis. *Cancer Res.* **68**, 4727–4735.

Garg, V., Muth, A. N., Ransom, J. F., Schluterman, M. K., Barnes, R., King, I. N., Grossfeld, P. D., and Srivastava, D. (2005). Mutations in NOTCH1 cause aortic valve disease. *Nature* **437**, 270–274.

Geffers, I., Serth, K., Chapman, G., Jaekel, R., Schuster-Gossler, K., Cordes, R., Sparrow, D. B., Kremmer, E., Dunwoodie, S. L., Klein, T., and Gossler, A. (2007). Divergent functions and distinct localization of the Notch ligands DLL1 and DLL3 in vivo. *J. Cell Biol.* **178**, 465–476.

Gibson, M. A., Hatzinikolas, G., Kumaratilake, J. S., Sandberg, L. B., Nicholl, J. K., Sutherland, G. R., and Cleary, E. G. (1996). Further characterization of proteins associated with elastic fiber microfibrils including the molecular cloning of MAGP-2 (MP25). *J. Biol. Chem.* **271**, 1096–1103.

Gibson, M. A., Sandberg, L. B., Grosso, L. E., and Cleary, E. G. (1991). Complementary DNA cloning establishes microfibril-associated glycoprotein (MAGP) to be a discrete component of the elastin-associated microfibrils. *J. Biol. Chem.* **266**, 7596–7601.

Glittenberg, M., Pitsouli, C., Garvey, C., Delidakis, C., and Bray, S. (2006). Role of conserved intracellular motifs in Serrate signalling, cis-inhibition and endocytosis. *EMBO J.* **25**, 4697–4706.

Gordon, W. R., Arnett, K. L., and Blacklow, S. C. (2008a). The molecular logic of Notch signaling—a structural and biochemical perspective. *J. Cell. Sci.* **121**, 3109–3119.

Gordon, W. R., Roy, M., Vardar-Ulu, D., Garfinkel, M., Mansour, M. R., Aster, J. C., and Blacklow, S. C. (2008b). Structure of the Notch1 negative regulatory region: implications for normal activation and pathogenic signaling in T-ALL. *Blood* **113**(18), 4381–4390.

Gordon, W. R., Vardar-Ulu, D., Histen, G., Sanchez-Irizarry, C., Aster, J. C., and Blacklow, S. C. (2007). Structural basis for autoinhibition of Notch. *Nat. Struct. Mol. Biol.* **14**, 295–300.

Grant, B. D., and Donaldson, J. G. (2009). Pathways and mechanisms of endocytic recycling. *Nat. Rev. Mol. Cell Biol.* **10**, 597–608.

Greenwald, I., and Rubin, G. M. (1992). Making a difference: the role of cell-cell interactions in establishing separate identities for equivalent cells. *Cell* **68**, 271–281.

Gridley, T. (1997). Notch signaling in vertebrate development and disease. *Mol. Cell. Neurosci.* **9**, 103–108.

Gridley, T. (2003). Notch signaling and inherited disease syndromes. *Hum. Mol. Genet.* **12**(*Spec No 1*), R9–R13.

Guarnaccia, C., Dhir, S., Pintar, A., and Pongor, S. (2009). The tetralogy of Fallot-associated G274D mutation impairs folding of the second epidermal growth factor repeat in Jagged-1. *FEBS J.* **276**, 6247–6257.

Gupta, R., Hong, D., Iborra, F., Sarno, S., and Enver, T. (2007). NOV (CCN3) functions as a regulator of human hematopoietic stem or progenitor cells. *Science* **316**, 590–593.

Hagedorn, E. J., Bayraktar, J. L., Kandachar, V. R., Bai, T., Englert, D. M., and Chang, H. C. (2006). *Drosophila* melanogaster auxilin regulates the internalization of Delta to control activity of the Notch signaling pathway. *J. Cell Biol.* **173**, 443–452.

Haltiwanger, R. S., and Lowe, J. B. (2004). Role of glycosylation in development. *Annu. Rev. Biochem.* **73**, 491–537.

Hamel, S., Fantini, J., and Schweisguth, F. (2010). Notch ligand activity is modulated by glycosphingolipid membrane composition in *Drosophila* melanogaster. *J. Cell Biol.* **188**, 581–594.

Harris, B. Z., and Lim, W. A. (2001). Mechanism and role of PDZ domains in signaling complex assembly. *J. Cell. Sci.* **114**, 3219–3231.

Hatakeyama, J., Bessho, Y., Katoh, K., Ookawara, S., Fujioka, M., Guillemot, F., and Kageyama, R. (2004). Hes genes regulate size, shape and histogenesis of the nervous system by control of the timing of neural stem cell differentiation. *Development* **131**, 5539–5550.

Hawryluk, M. J., Keyel, P. A., Mishra, S. K., Watkins, S. C., Heuser, J. E., and Traub, L. M. (2006). Epsin 1 is a polyubiquitin-selective clathrin-associated sorting protein. *Traffic* **7**, 262–281.

Hellstrom, M., Phng, L. K., Hofmann, J. J., Wallgard, E., Coultas, L., Lindblom, P., Alva, J., Nilsson, A. K., Karlsson, L., Gaiano, N., Yoon, K., Rossant, J., *et al.* (2007). Dll4 signalling through Notch1 regulates formation of tip cells during angiogenesis. *Nature* **445**, 776–780.

Henderson, S. T., Gao, D., Christensen, S., and Kimble, J. (1997). Functional domains of LAG-2, a putative signaling ligand for LIN-12 and GLP-1 receptors in *Caenorhabditis elegans*. *Mol. Biol. Cell* **8**, 1751–1762.

Henderson, S. T., Gao, D., Lambie, E. J., and Kimble, J. (1994). lag-2 may encode a signaling ligand for the GLP-1 and LIN-12 receptors of *C. elegans*. *Development* **120**, 2913–2924.

Henrique, D., Hirsinger, E., Adam, J., Le Roux, I., Pourquie, O., Ish-Horowicz, D., and Lewis, J. (1997). Maintenance of neuroepithelial progenitor cells by Delta-Notch signalling in the embryonic chick retina. *Curr. Biol.* **7**, 661–670.

Heuss, S. F., Ndiaye-Lobry, D., Six, E. M., Israel, A., and Logeat, F. (2008). The intracellular region of Notch ligands Dll1 and Dll3 regulates their trafficking and signaling activity. *Proc. Natl. Acad. Sci. U.S.A.* **105**, 11212–11217.

Hicks, C., Ladi, E., Lindsell, C., Hsieh, J., Hayward, S., Collazo, A., and Weinmaster, G. (2002). A secreted Delta1-Fc fusion protein functions both as an activator and inhibitor of Notch1 signaling. *J. Neurosci. Res.* **69**, 60–71.

Hiratochi, M., Nagase, H., Kuramochi, Y., Koh, C. S., Ohkawara, T., and Nakayama, K. (2007). The Delta intracellular domain mediates TGF-beta/Activin signaling through binding to Smads and has an important bi-directional function in the Notch-Delta signaling pathway. *Nucleic Acids Res.* **35**, 912–922.

Hofmann, M., Schuster-Gossler, K., Watabe-Rudolph, M., Aulehla, A., Herrmann, B. G., and Gossler, A. (2004). WNT signaling, in synergy with T/TBX6, controls Notch signaling by regulating Dll1 expression in the presomitic mesoderm of mouse embryos. *Genes Dev.* **18**, 2712–2717.

Holland, S. J., GAle, N. W., Mbamalu, G., Yancopoulos, G., Henkemeyer, M., and Pawson, T. (1996). Bidirectional signaling through the EPH family receptor Nuk and its transmembrane ligands. *Nature* **383**, 722–725.

Holterman, C. E., Le Grand, F., Kuang, S., Seale, P., and Rudnicki, M. A. (2007). Megf10 regulates the progression of the satellite cell myogenic program. *J. Cell Biol.* **179**, 911–922.

Horvath, C. A., Vanden Broeck, D., Boulet, G. A., Bogers, J., and De Wolf, M. J. (2007). Epsin: inducing membrane curvature. *Int. J. Biochem. Cell Biol.* **39**, 1765–1770.

Hsueh, Y. P., Wang, T. F., Yang, F. C., and Sheng, M. (2000). Nuclear translocation and transcription regulation by the membrane-associated guanylate kinase CASK/LIN-2. *Nature* **404**, 298–302.

Hu, Q. D., Ang, B. T., Karsak, M., Hu, W. P., Cui, X. Y., Duka, T., Takeda, Y., Chia, W., Sankar, N., Ng, Y. K., Ling, E. A., Maciag, T., *et al.* (2003). F3/contactin acts as a functional ligand for Notch during oligodendrocyte maturation. *Cell* **115**, 163–175.

Hukriede, N. A., Gu, Y., and Fleming, R. J. (1997). A dominant-negative form of Serrate acts as a general antagonist of Notch activation. *Development* **124**, 3427–3437.

Hurlbut, G. D., Kankel, M. W., Lake, R. J., and Artavanis-Tsakonas, S. (2007). Crossing paths with Notch in the hyper-network. *Curr. Opin. Cell Biol.* **19**, 166–175.

Ikeuchi, T., and Sisodia, S. S. (2003). The Notch ligands, Delta1 and Jagged2, are substrates for presenilin-dependent "gamma-secretase" cleavage. *J. Biol. Chem.* **278**, 7751–7754.

Itoh, T., Erdmann, K. S., Roux, A., Habermann, B., Werner, H., and De Camilli, P. (2005). Dynamin and the actin cytoskeleton cooperatively regulate plasma membrane invagination by BAR and F-BAR proteins. *Dev. Cell* **9**, 791–804.

Itoh, M., Kim, C. H., Palardy, G., Oda, T., Jiang, Y. J., Maust, D., Yeo, S. Y., Lorick, K., Wright, G. J., Ariza-McNaughton, L., Weissman, A. M., Lewis, J., *et al.* (2003). Mind bomb is a ubiquitin ligase that is essential for efficient activation of Notch signaling by Delta. *Dev. Cell* **4**, 67–82.

Jacobsen, T. L., Brennan, K., Arias, A. M., and Muskavitch, M. A. (1998). Cis-interactions between Delta and Notch modulate neurogenic signalling in *Drosophila*. *Development* **125**, 4531–4540.

Jafar-Nejad, H., Andrews, H. K., Acar, M., Bayat, V., Wirtz-Peitz, F., Mehta, S. Q., Knoblich, J. A., and Bellen, H. J. (2005). Sec15, a component of the exocyst, promotes notch signaling during the asymmetric division of *Drosophila* sensory organ precursors. *Dev. Cell* **9**, 351–363.

Jelen, F., Oleksy, A., Smietana, K., and Otlewski, J. (2003). PDZ domains—common players in the cell signaling. *Acta. Biochim. Pol.* **50**, 985–1017.

Jin, E. J., Choi, Y. A., Sonn, J. K., and Kang, S. S. (2007). Suppression of ADAM 10-induced Delta-1 shedding inhibits cell proliferation during the chondro-inhibitory action of TGF-beta3. *Mol. Cells* **24**, 139–147.

Karanu, F. N., Murdoch, B., Gallacher, L., Wu, D. M., Koremoto, M., Sakano, S., and Bhatia, M. (2000). The notch ligand jagged-1 represents a novel growth factor of human hematopoietic stem cells. *J. Exp. Med.* **192**, 1365–1372.

Karanu, F. N., Murdoch, B., Miyabayashi, T., Ohno, M., Koremoto, M., Gallacher, L., Wu, D., Itoh, A., Sakano, S., and Bhatia, M. (2001). Human homologues of Delta-1 and Delta-4 function as mitogenic regulators of primitive human hematopoietic cells. *Blood* **97**, 1960–1967.

Kiernan, A. E., Ahituv, N., Fuchs, H., Balling, R., Avraham, K. B., Steel, K. P., and Hrabe de Angelis, M. (2001). The Notch ligand Jagged1 is required for inner ear sensory development. *Proc. Natl. Acad. Sci. U.S.A.* **98**, 3873–3878.

Klein, T., and Arias, A. M. (1998). Interactions among Delta, Serrate and Fringe modulate Notch activity during *Drosophila* wing development. *Development* **125**, 2951–2962.

Klein, T., Brennan, K., and Arias, A. M. (1997). An intrinsic dominant negative activity of serrate that is modulated during wing development in *Drosophila*. *Dev. Biol.* **189**, 123–134.

Klueg, K. M., Parody, T. R., and Muskavitch, M. A. (1998). Complex proteolytic processing acts on Delta, a transmembrane ligand for Notch, during *Drosophila* development. *Mol. Biol. Cell* **9**, 1709–1723.

Kluppel, M., and Wrana, J. L. (2005). Turning it up a Notch: cross-talk between TGF beta and Notch signaling. *Bioessays* **27**, 115–118.

Koch, U., and Radtke, F. (2007). Notch and cancer: a double-edged sword. *Cell Mol. Life Sci.* **64**, 2746–2762.

Kolev, V., Kacer, D., Trifonova, R., Small, D., Duarte, M., Soldi, R., Graziani, I., Sideleva, O., Larman, B., Maciag, T., and Prudovsky, I., (2005). The intracellular domain of Notch ligand Delta1 induces cell growth arrest. *FEBS Lett.* **579**, 5798–5802.

Komatsu, H., Chao, M. Y., Larkins-Ford, J., Corkins, M. E., Somers, G. A., Tucey, T., Dionne, H. M., White, J. Q., Wani, K., Boxem, M., and Hart, A. C., (2008). OSM-11 facilitates LIN-12 Notch signaling during *Caenorhabditis elegans* vulval development. *PLoS Biol.* **6**, e196.

Koo, B. K., Lim, H. S., Song, R., Yoon, M. J., Yoon, K. J., Moon, J. S., Kim, Y. W., Kwon, M. C., Yoo, K. W., Kong, M. P., Lee, J., Chitnis, A. B. (2005a). Mind bomb 1 is essential for generating functional Notch ligands to activate Notch. *Development* **132**, 3459–3470.

Koo, B. K., Yoon, K. J., Yoo, K. W., Lim, H. S., Song, R., So, J. H., Kim, C. H., and Kong, Y. Y. (2005b). Mind bomb-2 is an E3 ligase for Notch ligand. *J. Biol. Chem.* **280**, 22335–22342.

Koo, B. K., Yoon, M. J., Yoon, K. J., Im, S. K., Kim, Y. Y., Kim, C. H., Suh, P. G., Jan, Y. N., and Kong, Y. Y. (2007). An obligatory role of mind bomb-1 in notch signaling of Mammalian development. *PLoS One* **2**, e1221.

Kooh, P. J., Fehon, R. G., and Muskavitch, M. A. (1993). Implications of dynamic patterns of Delta and Notch expression for cellular interactions during *Drosophila* development. *Development* **117**, 493–507.

Kopan, R., and Ilagan, M. X. (2009). The canonical Notch signaling pathway: unfolding the activation mechanism. *Cell* **137**, 216–233.

Kopczynski, C. C., and Muskavitch, M. A. (1989). Complex spatio-temporal accumulation of alternative transcripts from the neurogenic gene Delta during *Drosophila* embryogenesis. *Development* **107**, 623–636.

Koutelou, E., Sato, S., Tomomori-Sato, C., Florens, L., Swanson, S. K., Washburn, M. P., Kokkinaki, M., Conaway, R. C., Conaway, J. W., and Moschonas, N. K. (2008). Neuralized-like 1 (Neurl1) targeted to the plasma membrane by N-myristoylation regulates the Notch ligand Jagged1. *J. Biol. Chem.* **283**, 3846–3853.

Krivtsov, A. V., Rozov, F. N., Zinovyeva, M. V., Hendrikx, P. J., Jiang, Y., Visser, J. W., and Belyavsky, A. V. (2007). Jedi–a novel transmembrane protein expressed in early hematopoietic cells. *J. Cell. Biochem.* **101**, 767–784.

LaVoie, M. J., and Selkoe, D. J. (2003). The Notch ligands, Jagged and Delta, are sequentially processed by alpha-secretase and presenilin/gamma-secretase and release signaling fragments. *J. Biol. Chem.* **278**, 34427–34437.

Laborda, J., Sausville, E. A., Hoffman, T., and Notario, V. (1993). dlk, a putative mammalian homeotic gene differentially expressed in small cell lung carcinoma and neuroendocrine tumor cell line. *J. Biol. Chem.* **268**, 3817–3820.

Ladi, E., Nichols, J. T., Ge, W., Miyamoto, A., Yao, C., Yang, L. T., Boulter, J., Sun, Y. E., Kintner, C., and Weinmaster, G. (2005). The divergent DSL ligand Dll3 does not activate Notch signaling but cell autonomously attenuates signaling induced by other DSL ligands. *J. Cell Biol.* **170**, 983–992.

Lai, E. C., Deblandre, G. A., Kintner, C., and Rubin, G. M. (2001). *Drosophila* neuralized is a ubiquitin ligase that promotes the internalization and degradation of delta. *Dev. Cell* **1**, 783–794.

Lai, E. C., Roegiers, F., Qin, X., Jan, Y. N., and Rubin, G. M. (2005). The ubiquitin ligase *Drosophila* Mind bomb promotes Notch signaling by regulating the localization and activity of Serrate and Delta. *Development* **132**, 2319–2332.

Lai, E. C., and Rubin, G. M. (2001a). neuralized functions cell-autonomously to regulate a subset of notch- dependent processes during adult *Drosophila* development. *Dev. Biol.* **231**, 217–233.

Lai, E. C., and Rubin, G. M. (2001b). Neuralized is essential for a subset of Notch pathway-dependent cell fate decisions during *Drosophila* eye development. *Proc. Natl. Acad. Sci. U.S.A.* **98**, 5637–5642.

Le Borgne, R. (2006). Regulation of Notch signalling by endocytosis and endosomal sorting. *Curr. Opin. Cell Biol.* **18**, 213–222.

Le Borgne, R., Remaud, S., Hamel, S., and Schweisguth, F. (2005). Two distinct E3 ubiquitin ligases have complementary functions in the regulation of delta and serrate signaling in *Drosophila*. *PLoS Biol.* **3**, e96.

Le Borgne, R., and Schweisguth, F. (2003a). Notch signaling: endocytosis makes delta signal better. *Curr. Biol.* **13**, R273–R275.

Le Borgne, R., and Schweisguth, F. (2003b). Unequal segregation of Neuralized biases Notch activation during asymmetric cell division. *Dev. Cell* **5**, 139–148.

Lee, J. H., Volinic, J. L., Banz, C., Yao, K. M., and Thomas, M. K. (2005). Interactions with p300 enhance transcriptional activation by the PDZ-domain coactivator Bridge-1. *J. Endocrinol.* **187**, 283–292.

Leong, K. G., and Karsan, A. (2006). Recent insights into the role of Notch signaling in tumorigenesis. *Blood* **107**, 2223–2233.

Li, J. L., Sainson, R. C., Shi, W., Leek, R., Harrington, L. S., Preusser, M., Biswas, S., Turley, H., Heikamp, E., Hainfellner, J. A. and Harris, A. L. (2007b). Delta-like 4 Notch ligand regulates tumor angiogenesis, improves tumor vascular function, and promotes tumor growth in vivo. *Cancer Res.* **67**, 11244–11253.

Li, B. X., Satoh, A. K., and Ready, D. F. (2007a). Myosin V, Rab11, and dRip11 direct apical secretion and cellular morphogenesis in developing *Drosophila* photoreceptors. *J. Cell Biol.* **177**, 659–669.

Lieber, T., Kidd, S., and Young, M. W. (2002). kuzbanian-mediated cleavage of *Drosophila* Notch. *Genes Dev.* **16**, 209–221.

Limbourg, A., Ploom, M., Elligsen, D., Sorensen, I., Ziegelhoeffer, T., Gossler, A., Drexler, H., and Limbourg, F. P. (2007). Notch ligand Delta-like 1 is essential for postnatal arteriogenesis. *Circ. Res.* **100**, 363–371.

Liotta, F., Angeli, R., Cosmi, L., Fili, L., Manuelli, C., Frosali, F., Mazzinghi, B., Maggi, L., Pasini, A., Lisi, V., Santarlasci, V., Consoloni, L., *et al.* (2008). Toll-like receptors 3 and 4 are expressed by human bone marrow-derived mesenchymal stem cells and can inhibit their T-cell modulatory activity by impairing Notch signaling. *Stem Cells* **26**, 279–289.

Liu, Z. J., Shirakawa, T., Li, Y., Soma, A., Oka, M., Dotto, G. P., Fairman, R. M., Velazquez, O. C., and Herlyn, M. (2003). Regulation of Notch1 and Dll4 by vascular endothelial growth factor in arterial endothelial cells: Implications for modulating arteriogenesis and angiogenesis. *Mol. Cell. Biol.* **23**, 14–25.

Lobov, I. B., Renard, R. A., Papadopoulos, N., Gale, N. W., Thurston, G., Yancopoulos, G. D., and Wiegand, S. J. (2007). Delta-like ligand 4 (Dll4) is induced by VEGF as a negative regulator of angiogenic sprouting. *Proc. Natl. Acad. Sci. U.S.A.* **104**, 3219–3224.

Lowell, S., Jones, P., Le Roux, I., Dunne, J., and Watt, F. M. (2000). Stimulation of human epidermal differentiation by delta-notch signalling at the boundaries of stem-cell clusters. *Curr. Biol.* **10**, 491–500.

Lowell, S., and Watt, F. M. (2001). Delta regulates keratinocyte spreading and motility independently of differentiation. *Mech. Dev.* **107**, 133–140.

Lu, L., Chen, X., Zhang, C. W., Yang, W. L., Wu, Y. J., Sun, L., Bai, L. M., Gu, X. S., Ahmed, S., Dawe, G. S. and Xiao, Z. C. (2008). Morphological and functional characterization of predifferentiation of myelinating glia-like cells from human bone marrow stromal cells through activation of F3/Notch signaling in mouse retina. *Stem Cells* **26**, 580–590.

Maekawa, Y., Tsukumo, S., Chiba, S., Hirai, H., Hayashi, Y., Okada, H., Kishihara, K., and Yasutomo, K. (2003). Delta1-Notch3 interactions bias the functional differentiation of activated CD4+ T cells. *Immunity* **19**, 549–559.

Maldonado-Baez, L., and Wendland, B. (2006). Endocytic adaptors: recruiters, coordinators and regulators. *Trends Cell Biol.* **16**, 505–513.

Matsuda, M., and Chitnis, A. B. (2009). Interaction with Notch determines endocytosis of specific Delta ligands in zebrafish neural tissue. *Development* **136**, 197–206.
Maxfield, F. R., and McGraw, T. E. (2004). Endocytic recycling. *Nat. Rev. Mol. Cell Biol.* **5**, 121–132.
McDaniell, R., Warthen, D. M., Sanchez-Lara, P. A., Pai, A., Krantz, I. D., Piccoli, D. A., and Spinner, N. B. (2006). NOTCH2 mutations cause Alagille syndrome, a heterogeneous disorder of the notch signaling pathway. *Am. J. Hum. Genet.* **79**, 169–173.
McGlinn, E., van Bueren, K. L., Fiorenza, S., Mo, R., Poh, A. M., Forrest, A., Soares, M. B., Bonaldo Mde, F., Grimmond, S., Hui, C. C., Wainwright, B. and Wicking, C. (2005). Pax9 and Jagged1 act downstream of Gli3 in vertebrate limb development. *Mech. Dev.* **122**, 1218–1233.
Meng, H., Zhang, X., Hankenson, K. D., and Wang, M. M. (2009). Thrombospondin 2 potentiates notch3/jagged1 signaling. *J. Biol. Chem.* **284**, 7866–7874.
Micchelli, C. A., Rulifson, E. J., and Blair, S. S. (1997). The function and regulation of cut expression on the wing margin of *Drosophila*: Notch, Wingless and a dominant negative role for Delta and Serrate. *Development* **124**, 1485–1495.
Miller, A. C., Lyons, E. L., and Herman, T. G. (2009). cis-Inhibition of Notch by endogenous Delta biases the outcome of lateral inhibition. *Curr. Biol.* **19**, 1378–1383.
Minamizato, T., Sakamoto, K., Liu, T., Kokubo, H., Katsube, K., Perbal, B., Nakamura, S., and Yamaguchi, A. (2007). CCN3/NOV inhibits BMP-2-induced osteoblast differentiation by interacting with BMP and Notch signaling pathways. *Biochem. Biophys. Res. Commun.* **354**, 567–573.
Mishra-Gorur, K., Rand, M. D., Perez-Villamil, B., and Artavanis-Tsakonas, S. (2002). Down-regulation of Delta by proteolytic processing. *J. Cell Biol.* **159**, 313–324.
Miyamoto, A., Lau, R., Hein, P. W., Shipley, J. M., and Weinmaster, G. (2006). Microfibrillar proteins MAGP-1 and MAGP-2 induce Notch1 extracellular domain dissociation and receptor activation. *J. Biol. Chem.* **281**, 10089–10097.
Mizuhara, E., Nakatani, T., Minaki, Y., Sakamoto, Y., Ono, Y., and Takai, Y. (2005). MAGI1 recruits Dll1 to cadherin-based adherens junctions and stabilizes it on the cell surface. *J. Biol. Chem.* **280**, 26499–26507.
Mlodzik, M., Baker, N. E., and Rubin, G. M. (1990). Isolation and expression of scabrous, a gene regulating neurogenesis in *Drosophila*. *Genes Dev.* **4**, 1848–1861.
Mok, L. P., Qin, T., Bardot, B., LeComte, M., Homayouni, A., Ahimou, F., and Wesley, C. (2005). Delta activity independent of its activity as a ligand of Notch. *BMC Dev. Biol.* **5**, 6.
Morel, V., Le Borgne, R., and Schweisguth, F. (2003). Snail is required for Delta endocytosis and Notch-dependent activation of single-minded expression. *Dev. Genes Evol.* **213**, 65–72.
Morrison, S. J., Perez, S. E., Qiao, Z., Verdi, J. M., Hicks, C., Weinmaster, G., and Anderson, D. J. (2000). Transient Notch activation initiates an irreversible switch from neurogenesis to gliogenesis by neural crest stem cells. *Cell* **101**, 499–510.
Morrissette, J. D., Colliton, R. P., and Spinner, N. B. (2001). Defective intracellular transport and processing of JAG1 missense mutations in Alagille syndrome. *Hum. Mol. Genet.* **10**, 405–413.
Mukherjee, A., Veraksa, A., Bauer, A., Rosse, C., Camonis, J., and Artavanis-Tsakonas, S. (2005). Regulation of Notch signalling by non-visual beta-arrestin. *Nat. Cell Biol.* **7**, 1191–1201.
Muraguchi, T., Takegami, Y., Ohtsuka, T., Kitajima, S., Chandana, E. P., Omura, A., Miki, T., Takahashi, R., Matsumoto, N., Ludwig, A., Noda, M. and Takahashi, C. (2007). RECK modulates Notch signaling during cortical neurogenesis by regulating ADAM10 activity. *Nat. Neurosci.* **10**, 838–845.
Nakahara, J., Kanekura, K., Nawa, M., Aiso, S., and Suzuki, N. (2009). Abnormal expression of TIP30 and arrested nucleocytoplasmic transport within oligodendrocyte precursor cells in multiple sclerosis. *J. Clin. Invest.* **119**, 169–181.

Nakata, A., Ito, T., Nagata, M., Hori, S., and Sekimizu, K. (2004). GRIP1tau, a novel PDZ domain-containing transcriptional activator, cooperates with the testis-specific transcription elongation factor SII-T1. *Genes Cells* **9**, 1125–1135.

Nanda, N., Bao, M., Lin, H., Clauser, K., Komuves, L., Quertermous, T., Conley, P. B., Phillips, D. R., and Hart, M. J. (2005). Platelet endothelial aggregation receptor 1 (PEAR1), a novel epidermal growth factor repeat-containing transmembrane receptor, participates in platelet contact-induced activation. *J. Biol. Chem.* **280**, 24680–24689.

Nehring, L. C., Miyamoto, A., Hein, P. W., Weinmaster, G., and Shipley, J. M. (2005). The extracellular matrix protein MAGP-2 interacts with Jagged1 and induces its shedding from the cell surface. *J. Biol. Chem.* **280**, 20349–20355.

Nichols, J. T., Miyamoto, A., Olsen, S. L., D'Souza, B., Yao, C., and Weinmaster, G. (2007a). DSL ligand endocytosis physically dissociates Notch1 heterodimers before activating proteolysis can occur. *J. Cell Biol.* **176**, 445–458.

Nichols, J. T., Miyamoto, A., and Weinmaster, G. (2007b). Notch signaling–constantly on the move. *Traffic* **8**, 959–969.

Noguera-Troise, I., Daly, C., Papadopoulos, N. J., Coetzee, S., Boland, P., Gale, N. W., Lin, H. C., Yancopoulos, G. D., and Thurston, G. (2006). Blockade of Dll4 inhibits tumour growth by promoting non-productive angiogenesis. *Nature* **444**, 1032–1037.

Nueda, M. L., Baladron, V., Sanchez-Solana, B., Ballesteros, M. A., and Laborda, J. (2007). The EGF-like protein dlk1 inhibits notch signaling and potentiates adipogenesis of mesenchymal cells. *J. Mol. Biol.* **367**, 1281–1293.

Ong, C. T., Sedy, J. R., Murphy, K. M., and Kopan, R. (2008). Notch and presenilin regulate cellular expansion and cytokine secretion but cannot instruct Th1/Th2 fate acquisition. *PLoS ONE* **3**, e2823.

Overstreet, E., Chen, X., Wendland, B., and Fischer, J. A. (2003). Either part of a *Drosophila* epsin protein, divided after the ENTH domain, functions in endocytosis of delta in the developing eye. *Curr. Biol.* **13**, 854–860.

Overstreet, E., Fitch, E., and Fischer, J. A. (2004). Fat facets and Liquid facets promote Delta endocytosis and Delta signaling in the signaling cells. *Development* **131**, 5355–5366.

Panin, V. M., Papayannopoulos, V., Wilson, R., and Irvine, K. D. (1997). Fringe modulates Notch-ligand interactions. *Nature* **387**, 908–912.

Panin, V. M., Shao, L., Lei, L., Moloney, D. J., Irvine, K. D., and Haltiwanger, R. S. (2002). Notch ligands are substrates for EGF protein O-fucosyltransferase and Fringe. *J. Biol. Chem.* **29**, 29.

Pannequin, J., Bonnans, C., Delaunay, N., Ryan, J., Bourgaux, J. F., Joubert, D., and Hollande, F. (2009). The wnt target jagged-1 mediates the activation of notch signaling by progastrin in human colorectal cancer cells. *Cancer Res.* **69**, 6065–6073.

Parks, A. L., Klueg, K. M., Stout, J. R., and Muskavitch, M. A. (2000). Ligand endocytosis drives receptor dissociation and activation in the Notch pathway. *Development* **127**, 1373–1385.

Parks, A. L., Stout, J. R., Shepard, S. B., Klueg, K. M., Dos Santos, A. A., Parody, T. R., Vaskova, M., and Muskavitch, M. A. (2006). Structure-function analysis of delta trafficking, receptor binding and signaling in *Drosophila*. *Genetics* **174**, 1947–1961.

Patel, N. S., Li, J. L., Generali, D., Poulsom, R., Cranston, D. W., and Harris, A. L. (2005). Up-regulation of delta-like 4 ligand in human tumor vasculature and the role of basal expression in endothelial cell function. *Cancer Res.* **65**, 8690–8697.

Pavlopoulos, E., Pitsouli, C., Klueg, K. M., Muskavitch, M. A., Moschonas, N. K., and Delidakis, C. (2001). neuralized Encodes a peripheral membrane protein involved in delta signaling and endocytosis. *Dev. Cell* **1**, 807–816.

Perbal, B. (2004). CCN proteins: multifunctional signalling regulators. *Lancet* **363**, 62–64.

Pfister, S., Przemeck, G. K., Gerber, J. K., Beckers, J., Adamski, J., and Hrabe de Angelis, M. (2003). Interaction of the MAGUK family member Acvrinp1 and the cytoplasmic domain of the Notch ligand Delta1. *J. Mol. Biol.* **333**, 229–235.

Piccoli, D. A., and Spinner, N. B. (2001). Alagille syndrome and the Jagged1 gene. *Semin. Liver Dis.* **21**, 525–534.

Pintar, A., De Biasio, A., Popovic, M., Ivanova, N., and Pongor, S. (2007). The intracellular region of Notch ligands: does the tail make the difference?. *Biol. Direct* **2**, 19.

Pitsouli, C., and Delidakis, C. (2005). The interplay between DSL proteins and ubiquitin ligases in Notch signaling. *Development* **132**, 4041–4050.

Powell, P. A., Wesley, C., Spencer, S., and Cagan, R. L. (2001). Scabrous complexes with Notch to mediate boundary formation. *Nature* **409**, 626–630.

Qi, H., Rand, M. D., Wu, X., Sestan, N., Wang, W., Rakic, P., Xu, T., and Artavanis-Tsakonas, S. (1999). Processing of the notch ligand delta by the metalloprotease Kuzbanian. *Science* **283**, 91–94.

Quilliam, L. A., Castro, A. F., Rogers-Graham, K. S., Martin, C. B., Der, C. J., and Bi, C. (1999). M-Ras/R-Ras3, a transforming ras protein regulated by Sos1, GRF1, and p120 Ras GTPase-activating protein, interacts with the putative Ras effector AF6. *J. Biol. Chem.* **274**, 23850–23857.

Rajan, A., Tien, A. C., Haueter, C. M., Schulze, K. L., and Bellen, H. J. (2009). The Arp2/3 complex and WASp are required for apical trafficking of Delta into microvilli during cell fate specification of sensory organ precursors. *Nat. Cell Biol.* **11**, 815–824.

Rauen, T., Raffetseder, U., Frye, B. C., Djudjaj, S., Muhlenberg, P. J., Eitner, F., Lendahl, U., Bernhagen, J., Dooley, S., and Mertens, P. R. (2009). YB-1 acts as a ligand for Notch-3 receptors and modulates receptor activation. *J. Biol. Chem.* **284**, 26928–26940.

Rauskolb, C. (2001). The establishment of segmentation in the *Drosophila* leg. *Development* **128**, 4511–4521.

Rauskolb, C., and Irvine, K. D. (1999). Notch-mediated segmentation and growth control of the *Drosophila* leg. *Dev. Biol.* **210**, 339–350.

Raymond, T., Schaller, M., Hogaboam, C. M., Lukacs, N. W., Rochford, R., and Kunkel, S. L. (2007). Toll-like receptors, Notch ligands, and cytokines drive the chronicity of lung inflammation. *Proc. Am. Thorac. Soc.* **4**, 635–641.

Renaud, O., and Simpson, P. (2001). scabrous modifies epithelial cell adhesion and extends the range of lateral signalling during development of the spaced bristle pattern in *Drosophila*. *Dev. Biol.* **240**, 361–376.

Roca, C., and Adams, R. H. (2007). Regulation of vascular morphogenesis by Notch signaling. *Genes Dev.* **21**, 2511–2524.

Rodilla, V., Villanueva, A., Obrador-Hevia, A., Robert-Moreno, A., Fernandez-Majada, V., Grilli, A., Lopez-Bigas, N., Bellora, N., Alba, M. M., Torres, F., Dunach, M., Sanjuan, X., *et al.* (2009). Jagged1 is the pathological link between Wnt and Notch pathways in colorectal cancer. *Proc. Natl. Acad. Sci. U.S.A.* **106**, 6315–6320.

Ross, D. A., Rao, P. K., and Kadesch, T. (2004). Dual roles for the Notch target gene Hes-1 in the differentiation of 3T3-L1 preadipocytes. *Mol. Cell. Biol.* **24**, 3505–3513.

Roux, A., Uyhazi, K., Frost, A., and De Camilli, P. (2006). GTP-dependent twisting of dynamin implicates constriction and tension in membrane fission. *Nature* **441**, 528–531.

Ruan, Y., Tecott, L., Jiang, M. M., Jan, L. Y., and Jan, Y. N. (2001). Ethanol hypersensitivity and olfactory discrimination defect in mice lacking a homolog of *Drosophila* neuralized. *Proc. Natl. Acad. Sci. U.S.A.* **98**, 9907–9912.

Sainson, R. C., and Harris, A. L. (2008). Regulation of angiogenesis by homotypic and heterotypic notch signalling in endothelial cells and pericytes: from basic research to potential therapies. *Angiogenesis* **11**, 41–51.

Sainson, R. C., Johnston, D. A., Chu, H. C., Holderfield, M. T., Nakatsu, M. N., Crampton, S. P., Davis, J., Conn, E., and Hughes, C. C. (2008). TNF primes endothelial cells for angiogenic sprouting by inducing a tip cell phenotype. *Blood* **111**, 4997–5007.

Sakamoto, K., Ohara, O., Takagi, M., Takeda, S., and Katsube, K. (2002a). Intracellular cell-autonomous association of Notch and its ligands: a novel mechanism of Notch signal modification. *Dev. Biol.* **241**, 313–326.
Sakamoto, K., Yamaguchi, S., Ando, R., Miyawaki, A., Kabasawa, Y., Takagi, M., Li, C. L., Perbal, B., and Katsube, K. (2002b). The nephroblastoma overexpressed gene (NOV/ccn3) protein associates with Notch1 extracellular domain and inhibits myoblast differentiation via Notch signaling pathway. *J. Biol. Chem.* **277**, 29399–29405.
Sansone, P., Storci, G., Tavolari, S., Guarnieri, T., Giovannini, C., Taffurelli, M., Ceccarelli, C., Santini, D., Paterini, P., Marcu, K. B., Chieco, P. and Bonafe, M. (2007). IL-6 triggers malignant features in mammospheres from human ductal breast carcinoma and normal mammary gland. *J. Clin. Invest.* **117**, 3988–4002.
Sapir, A., Assa-Kunik, E., Tsruya, R., Schejter, E., and Shilo, B. Z. (2005). Unidirectional Notch signaling depends on continuous cleavage of Delta. *Development* **132**, 123–132.
Schmidt, M. H., Bicker, F., Nikolic, I., Meister, J., Babuke, T., Picuric, S., Muller-Esterl, W., Plate, K. H., and Dikic, I. (2009). Epidermal growth factor-like domain 7 (EGFL7) modulates Notch signalling and affects neural stem cell renewal. *Nat. Cell Biol.* **11**, 873–880.
Schroeter, E., Kisslinger, J., and Kopan, R. (1998). Notch1 signalling requires ligand-induced proteolytic release of the intracellular domain. *Nature* **393**, 382–386.
Seals, D. F., and Courtneidge, S. A. (2003). The ADAMs family of metalloproteases: multidomain proteins with multiple functions. *Genes Dev.* **17**, 7–30.
Seo, S., Fujita, H., Nakano, A., Kang, M., Duarte, A., and Kume, T. (2006). The forkhead transcription factors, Foxc1 and Foxc2, are required for arterial specification and lymphatic sprouting during vascular development. *Dev. Biol.* **294**, 458–470.
Seugnet, L., Simpson, P., and Haenlin, M. (1997a). Requirement for dynamin during Notch signaling in *Drosophila* neurogenesis. *Dev. Biol.* **192**, 585–598.
Seugnet, L., Simpson, P., and Haenlin, M. (1997b). Transcriptional regulation of Notch and Delta: requirement for neuroblast segregation in *Drosophila*. *Development* **124**, 2015–2025.
Sheng, M., and Sala, C. (2001). PDZ domains and the organization of supramolecular complexes. *Annu. Rev. Neurosci.* **24**, 1–29.
Shimizu, K., Chiba, S., Kumano, K., Hosoya, N., Takahashi, T., Kanda, Y., Hamada, Y., Yazaki, Y., and Hirai, H. (1999). Mouse Jagged1 Physically Interacts with Notch2 and Other Notch Receptors. Assessment by quantitative methods. *J. Biol. Chem.* **274**, 32961–32969.
Shimizu, K., Chiba, S., Saito, T., Takahashi, T., Kumano, K., Hamada, Y., and Hirai, H. (2002). Integrity of intracellular domain of Notch ligand is indispensable for cleavage required for release of the Notch2 intracellular domain. *EMBO J.* **21**, 294–302.
Shoji, H., Tsuchida, K., Kishi, H., Yamakawa, N., Matsuzaki, T., Liu, Z., Nakamura, T., and Sugino, H. (2000). Identification and characterization of a PDZ protein that interacts with activin type II receptors. *J. Biol. Chem.* **275**, 5485–5492.
Sigismund, S., Woelk, T., Puri, C., Maspero, E., Tacchetti, C., Transidico, P., Di Fiore, P. P., and Polo, S. (2005). Clathrin-independent endocytosis of ubiquitinated cargos. *Proc. Natl. Acad. Sci. U.S.A.* **102**, 2760–2765.
Six, E., Ndiaye, D., Laabi, Y., Brou, C., Gupta-Rossi, N., Israel, A., and Logeat, F. (2003). The Notch ligand Delta1 is sequentially cleaved by an ADAM protease and gamma-secretase. *Proc. Natl. Acad. Sci. U.S.A.* **100**, 7638–7643.
Six, E. M., Ndiaye, D., Sauer, G., Laabi, Y., Athman, R., Cumano, A., Brou, C., Israel, A., and Logeat, F. (2004). The notch ligand Delta1 recruits Dlg1 at cell-cell contacts and regulates cell migration. *J. Biol. Chem.* **279**, 55818–55826.
Skwarek, L. C., Garroni, M. K., Commisso, C., and Boulianne, G. L. (2007). Neuralized contains a phosphoinositide-binding motif required downstream of ubiquitination for delta endocytosis and notch signaling. *Dev. Cell* **13**, 783–795.

Small, D., Kovalenko, D., Kacer, D., Liaw, L., Landriscina, M., Di Serio, C., Prudovsky, I., and Maciag, T. (2001). Soluble Jagged 1 represses the function of its transmembrane form to induce the formation of the Src-dependent chord-like phenotype. *J. Biol. Chem.* **276**, 32022–32030.

Smas, C. M., and Sul, H. S. (1993). Pref-1, a protein containing EGF-like repeats, inhibits adipocyte differentiation. *Cell* **73**, 725–734.

Song, R., Koo, B. K., Yoon, K. J., Yoon, M. J., Yoo, K. W., Kim, H. T., Oh, H. J., Kim, Y. Y., Han, J. K., Kim, C. H. and Kang, Y. Y. (2006). Neuralized-2 regulates a Notch ligand in cooperation with Mind bomb-1. *J. Biol. Chem.* **281**, 36391–36400.

Sprinzak, D., Lakhanpal, A., LeBon, L., Santat, L. A., Fontes, M. E., Anderson, G. A., Garcia-Ojalvo, J., and Elowitz, M. E. (2010). Cis Interactions between Notch and Delta generate mutually exclusive signaling states. *Nature* **465**(7294), 86–90.

Staub, O., and Rotin, D. (2006). Role of ubiquitylation in cellular membrane transport. *Physiol. Rev.* **86**, 669–707.

Studebaker, A. W., Storci, G., Werbeck, J. L., Sansone, P., Sasser, A. K., Tavolari, S., Huang, T., Chan, M. W., Marini, F. C., Rosol, T. J., Bonafe, M. and Hall, B. M. (2008). Fibroblasts isolated from common sites of breast cancer metastasis enhance cancer cell growth rates and invasiveness in an interleukin-6-dependent manner. *Cancer Res.* **68**, 9087–9095.

Suchting, S., Freitas, C., le Noble, F., Benedito, R., Breant, C., Duarte, A., and Eichmann, A. (2007). The Notch ligand Delta-like 4 negatively regulates endothelial tip cell formation and vessel branching. *Proc. Natl. Acad. Sci. U.S.A.* **104**, 3225–3230.

Sun, X., and Artavanis-Tsakonas, S. (1996). The intracellular deletions of Delta and Serrate define dominant negative forms of the *Drosophila* Notch ligands. *Development* **122**, 2465–2474.

Sun, X., and Artavanis-Tsakonas, S. (1997). Secreted forms of DELTA and SERRATE define antagonists of Notch signaling in *Drosophila. Development* **124**, 3439–3448.

Sun, J., Krawczyk, C. J., and Pearce, E. J. (2008b). Suppression of Th2 cell development by Notch ligands Delta1 and Delta4. *J. Immunol.* **180**, 1655–1661.

Sun, D., Li, H., and Zolkiewska, A. (2008a). The role of Delta-like 1 shedding in muscle cell self-renewal and differentiation. *J. Cell. Sci.* **121**, 3815–3823.

Takahashi, Y., Inoue, T., Gossler, A., and Saga, Y. (2003). Feedback loops comprising Dll1, Dll3 and Mesp2, and differential involvement of Psen1 are essential for rostrocaudal patterning of somites. *Development* **130**, 4259–4268.

Takeuchi, T., Adachi, Y., and Ohtsuki, Y. (2005). Skeletrophin, a novel ubiquitin ligase to the intracellular region of Jagged-2, is aberrantly expressed in multiple myeloma. *Am. J. Pathol.* **166**, 1817–1826.

Tax, F. E., Yeargers, J. J., and Thomas, J. H. (1994). Sequence of *C. elegans* lag-2 reveals a cell-signalling domain shared with Delta and Serrate of *Drosophila. Nature* **368**, 150–154.

Thibout, H., Martinerie, C., Creminon, C., Godeau, F., Boudou, P., Le Bouc, Y., and Laurent, M. (2003). Characterization of human NOV in biological fluids: an enzyme immunoassay for the quantification of human NOV in sera from patients with diseases of the adrenal gland and of the nervous system. *J. Clin. Endocrinol. Metab.* **88**, 327–336.

Thurston, G., Noguera-Troise, I., and Yancopoulos, G. D. (2007). The Delta paradox: DLL4 blockade leads to more tumour vessels but less tumour growth. *Nat. Rev. Cancer* **7**, 327–331.

Tian, X., Hansen, D., Schedl, T., and Skeath, J. B. (2004). Epsin potentiates Notch pathway activity in *Drosophila* and *C. elegans. Development* **131**, 5807–5815.

Trifonova, R., Small, D., Kacer, D., Kovalenko, D., Kolev, V., Mandinova, A., Soldi, R., Liaw, L., Prudovsky, I., and Maciag, T. (2004). The non-transmembrane form of Delta1, but not of Jagged1, induces normal migratory behavior accompanied by fibroblast growth factor receptor 1-dependent transformation. *J. Biol. Chem.* **279**, 13285–13288.

Tsai, H., Hardisty, R. E., Rhodes, C., Kiernan, A. E., Roby, P., Tymowska-Lalanne, Z., Mburu, P., Rastan, S., Hunter, A. J., Brown, S. D. and Steel, K. P. (2001). The mouse slalom mutant demonstrates a role for Jagged1 in neuroepithelial patterning in the organ of Corti. *Hum. Mol. Genet.* **10**, 507–512.

Tsuda, L., Nagaraj, R., Zipursky, S. L., and Banerjee, U. (2002). An EGFR/Ebi/Sno pathway promotes delta expression by inactivating Su(H)/SMRTER repression during inductive notch signaling. *Cell* **110**, 625–637.

Turnpenny, P. D., Alman, B., Cornier, A. S., Giampietro, P. F., Offiah, A., Tassy, O., Pourquie, O., Kusumi, K., and Dunwoodie, S. (2007). Abnormal vertebral segmentation and the notch signaling pathway in man. *Dev. Dyn.* **236**, 1456–1474.

Vanden Broeck, D., and De Wolf, M. J. (2006). Selective blocking of clathrin-mediated endocytosis by RNA interference: epsin as target protein. *Biotechniques* **41**, 475–484.

Varnum-Finney, B., Wu, L., Yu, M., Brashem-Stein, C., Staats, S., Flowers, D., Griffin, J. D., and Bernstein, I. D. (2000). Immobilization of Notch ligand, Delta-1, is required for induction of notch signaling. *J. Cell. Sci.* **113**(*Pt 23*), 4313–4318.

Vas, V., Szilagyi, L., Paloczi, K., and Uher, F. (2004). Soluble Jagged-1 is able to inhibit the function of its multivalent form to induce hematopoietic stem cell self-renewal in a surrogate in vitro assay. *J. Leukoc. Biol.* **75**, 714–720.

Vitt, U. A., Hsu, S. Y., and Hsueh, A. J. (2001). Evolution and classification of cystine knot-containing hormones and related extracellular signaling molecules. *Mol. Endocrinol.* **15**, 681–694.

Vollrath, B., Pudney, J., Asa, S., Leder, P., and Fitzgerald, K. (2001). Isolation of a murine homologue of the *Drosophila* neuralized gene, a gene required for axonemal integrity in spermatozoa and terminal maturation of the mammary gland. *Mol. Cell. Biol.* **21**, 7481–7494.

Wang, Y., Kim, K. A., Kim, J. H., and Sul, H. S. (2006). Pref-1, a preadipocyte secreted factor that inhibits adipogenesis. *J. Nutr.* **136**, 2953–2956.

Wang, W., and Struhl, G. (2004). *Drosophila* Epsin mediates a select endocytic pathway that DSL ligands must enter to activate Notch. *Development* **131**, 5367–5380.

Wang, W., and Struhl, G. (2005). Distinct roles for Mind bomb, Neuralized and Epsin in mediating DSL endocytosis and signaling in *Drosophila. Development* **132**, 2883–2894.

Wang, Y., and Sul, H. S. (2006). Ectodomain shedding of preadipocyte factor 1 (Pref-1) by tumor necrosis factor alpha converting enzyme (TACE) and inhibition of adipocyte differentiation. *Mol. Cell. Biol.* **26**, 5421–5435.

Warthen, D. M., Moore, E. C., Kamath, B. M., Morrissette, J. J., Sanchez, P., Piccoli, D. A., Krantz, I. D., and Spinner, N. B. (2006). Jagged1 (JAG1) mutations in Alagille syndrome: increasing the mutation detection rate. *Hum. Mutat.* **27**, 436–443.

Wesley, C. S. (1999). Notch and wingless regulate expression of cuticle patterning genes. *Mol. Cell. Biol.* **19**, 5743–5758.

Williams, C. K., Li, J. L., Murga, M., Harris, A. L., and Tosato, G. (2006). Up-regulation of the Notch ligand Delta-like 4 inhibits VEGF-induced endothelial cell function. *Blood* **107**, 931–939.

Windler, S. L., and Bilder, D. (2010). Endocytic Internalization Routes Required for Delta/Notch Signaling. *Curr. Biol.* **20**(6), 538–543.

Wright, G. J., Leslie, J. D., Ariza-McNaughton, L., and Lewis, J. (2004). Delta proteins and MAGI proteins: n interaction of Notch ligands with intracellular scaffolding molecules and its significance for zebrafish development. *Development* **131**, 5659–5669.

Yan, M., and Plowman, G. D. (2007). Delta-like 4/Notch signaling and its therapeutic implications. *Clin. Cancer Res.* **13**, 7243–7246.

Yang, L. T., Nichols, J. T., Yao, C., Manilay, J. O., Robey, E. A., and Weinmaster, G. (2005). Fringe glycosyltransferases differentially modulate Notch1 proteolysis induced by Delta1 and Jagged1. *Mol. Biol. Cell* **16**, 927–942.

Yeh, E., Dermer, M., Commisso, C., Zhou, L., McGlade, C. J., and Boulianne, G. L. (2001). Neuralized functions as an E3 ubiquitin ligase during *Drosophila* development. *Curr. Biol.* **11**, 1675–1679.

Yeh, E., Zhou, L., Rudzik, N., and Boulianne, G. L. (2000). Neuralized functions cell autonomously to regulate *Drosophila* sense organ development. *EMBO J.* **19**, 4827–4837.

Yim, Y. I., Sun, T., Wu, L. G., Raimondi, A., De Camilli, P., Eisenberg, E., and Greene, L. E. (2010). Endocytosis and clathrin-uncoating defects at synapses of auxilin knockout mice. *Proc. Natl. Acad. Sci. U.S.A.* **107**, 4412–4417.

Zavadil, J., Cermak, L., Soto-Nieves, N., and Bottinger, E. P. (2004). Integration of TGF-beta/Smad and Jagged1/Notch signalling in epithelial-to-mesenchymal transition. *EMBO J.* **23**, 1155–1165.

Zhang, C., Li, Q., and Jiang, Y. J. (2007a). Zebrafish Mib and Mib2 are mutual E3 ubiquitin ligases with common and specific delta substrates. *J. Mol. Biol.* **366**, 1115–1128.

Zhang, C., Li, Q., Lim, C. H., Qiu, X., and Jiang, Y. J. (2007b). The characterization of zebrafish antimorphic mib alleles reveals that Mib and Mind bomb-2 (Mib2) function redundantly. *Dev. Biol.* **305**, 14–27.

Zolkiewska, A. (2008). ADAM proteases: Ligand processing and modulation of the Notch pathway. *Cell Mol. Life Sci.* **65**(13), 2056–2068.

CHAPTER FOUR

Roles of Glycosylation in Notch Signaling

Pamela Stanley* *and* Tetsuya Okajima†

Contents

Abstract

Notch and the DSL Notch ligands Delta and Serrate/Jagged are glycoproteins with a single transmembrane domain. The extracellular domain (ECD) of both Notch receptors and Notch ligands contains numerous epidermal growth factor

* Department of Cell Biology, Albert Einstein College Medicine, New York, USA
† Nagoya University Graduate School of Medicine, Center for Neural Disease and Cancer, Nagoya, Japan

Current Topics in Developmental Biology, Volume 92
ISSN 0070-2153, DOI 10.1016/S0070-2153(10)92004-8

(EGF)-like repeats which are post-translationally modified by a variety of glycans. Inactivation of a subset of genes that encode glycosyltransferases which initiate and elongate these glycans inhibits Notch signaling. In the formation of developmental boundaries in *Drosophila* and mammals, in mouse T-cell and marginal zone B-cell development, and in co-culture Notch signaling assays, the regulation of Notch signaling by glycans is to date a cell-autonomous effect of the Notch-expressing cell. The regulation of Notch signaling by glycans represents a new paradigm of signal transduction. *O*-fucose glycans modulate the strength of Notch binding to DSL Notch ligands, while *O*-glucose glycans facilitate juxtamembrane cleavage of Notch, generating the substrate for intramembrane cleavage and Notch activation. Identifying precisely how the addition of particular sugars at specific locations on Notch modifies Notch signaling is a challenge for the future.

1. Introduction

Notch receptors are covered with a variety of glycans (Fig. 4.1). Mutations that prevent their synthesis cause Notch signaling defects of varying severity (Figs. 4.2 and 4.3). While general populations of glycans might be important in promoting the biologically active conformation, trafficking, and membrane stability of Notch, mutations that affect a single glycan site cause Notch signaling phenotypes. *In vitro* and cell-based assays show that *O*-fucose glycans modulate the degree of binding between Notch and Delta or Serrate/Jagged, but it is not known if Notch ligands bind sugars directly. *O*-glucose on Notch promotes Notch cleavage and activation. The first indication that glycans on Notch may be important for Notch signaling came from hydrophobic cluster analyses of Fringe proteins which led to the proposal that Fringe may encode a glycosyltransferase (Yuan *et al.*, 1997). Fringe was discovered in *Drosophila* in a screen for novel genes that modulate Notch signaling (Irvine and Wieschaus, 1994). It was soon shown to be necessary for Notch signaling at the wing margin (Fleming *et al.*, 1997; Panin *et al.*, 1997) and at other tissue boundaries in *Drosophila* (Irvine, 1999). Mammalian homologues of Fringe (termed Lunatic, Manic, and Radical Fringe) were shown to have conserved functions in *Drosophila* (Johnston *et al.*, 1997). Consistent with a requirement in Notch signaling, inactivation of *Lfng* in the mouse was found to cause defective somitogenesis leading to profound skeletal aberrations (Evrard *et al.*, 1998; Zhang and Gridley, 1998). Meanwhile, several groups were investigating whether Fringe had the sugar transfer ability of a glycosyltransferase. A hint came with the finding that Notch1 in mammals carries two unusual glycans (Moloney *et al.*, 2000b). One began with fucose linked to Ser or Thr located between the second and the third cysteine of an EGF repeat in the consensus $C^2X_{4-5}S/TC^3$, and the other began with glucose linked to

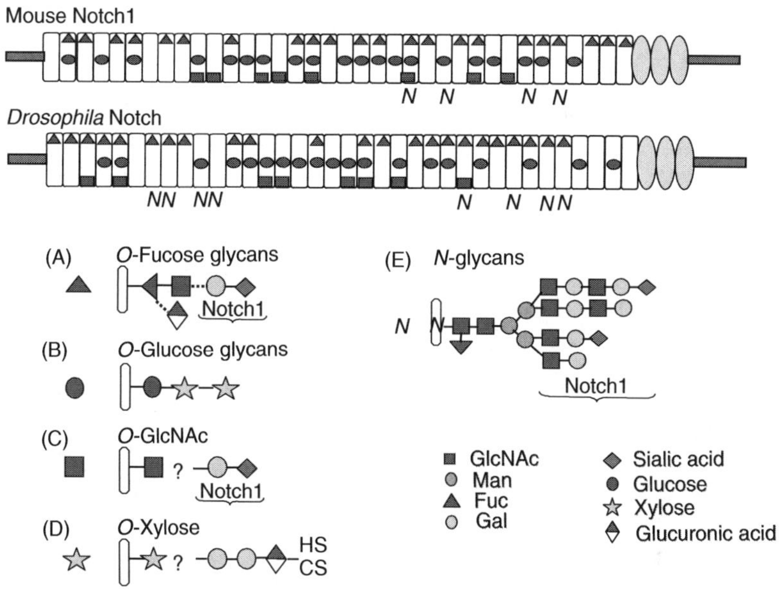

Figure 4.1 Glycans on Notch. A diagram representing the ECDs of mouse Notch1 and *Drosophila* Notch which contain 36 EGF repeats (white ovals) and 3 Lin repeats (blue ovals). Symbols in the EGF repeats identify consensus motifs for *O*-fucose (A), *O*-glucose (B), *O*-GlcNAc (C), *O*-xylose (D), and *N*-glycans (E) that have the potential to contain the sugars shown in the structures below the diagram. *O*-fucose glycans of *Drosophila* Notch may contain a glucuronic acid (Aoki *et al.*, 2008) and Notch1 *O*-fucose glycans may contain Gal and SA (Moloney *et al.*, 2000b) as noted. *N*-glycans in *Drosophila* are mainly oligomannosyl and rarely contain Gal and SA (Aoki *et al.*, 2007; Koles *et al.*, 2007), whereas Notch1 probably has complex *N*-glycans (Moloney *et al.*, 2000b) as noted. Several of the glycosylation sites in *Drosophila* Notch and mammalian Notch1 are conserved, for example, in EGF12 in the DSL Notch ligand-binding domain. Each sugar of the *O*-fucose (A), *O*-glucose (B), and *O*-GlcNAc (C) glycans is transferred by a specific glycosyltransferase described in the text. *N*-glycans (E) and GAGs (D) are synthesized by the concerted action of many glycosyltransferases and other glycosylation activities (Stanley *et al.*, 2009; Esko *et al.*, 2009). (See Color Insert.)

Ser or Thr between the first and the second cysteine of an EGF repeat with the consensus C^1XSXPC^2 (Moloney *et al.*, 2000b; Panin *et al.*, 2002). Consensus sites for *O*-fucose glycans (Fig. 4.1A) and *O*-glucose glycans (Fig. 4.1B) present in mouse Notch1 and *Drosophila* Notch are shown in Fig. 4.1.

The structural observations suggested substrates for *in vitro* assays which led to the discovery that Fringe is a glycosyltransferase which transfers *N*-acetylglucosamine (GlcNAc) to fucose (Fuc) on Notch EGF repeats to generate GlcNAcβ1,3Fuc-*O*-EGF (Bruckner *et al.*, 2000; Moloney *et al.*, 2000a). EGF repeats with an *O*-fucose consensus site occur in a number of proteins, including the DSL Notch ligands (Rampal *et al.*, 2007). Nevertheless, the cell-autonomous and phenotypic consequences of blocking these glycosylation pathways as described below indicate that the modification of Notch receptors by sugars is a key factor in regulating Notch signaling *in vivo*. Since publication of the glycosyltransferase activity of Fringe in 2000, there have been a host of investigations into the roles of *O*-fucose, *O*-glucose, and other glycans in Notch signaling. Most studies to date have been performed in *Drosophila* or mammals, although the zebra fish, Xenopus, and *Caenorhabditis elegans* genomes encode protein *O*-fucosyltransferase (Ofut1/Pofut1) homologues. Protein *O*-fucosyltransferases

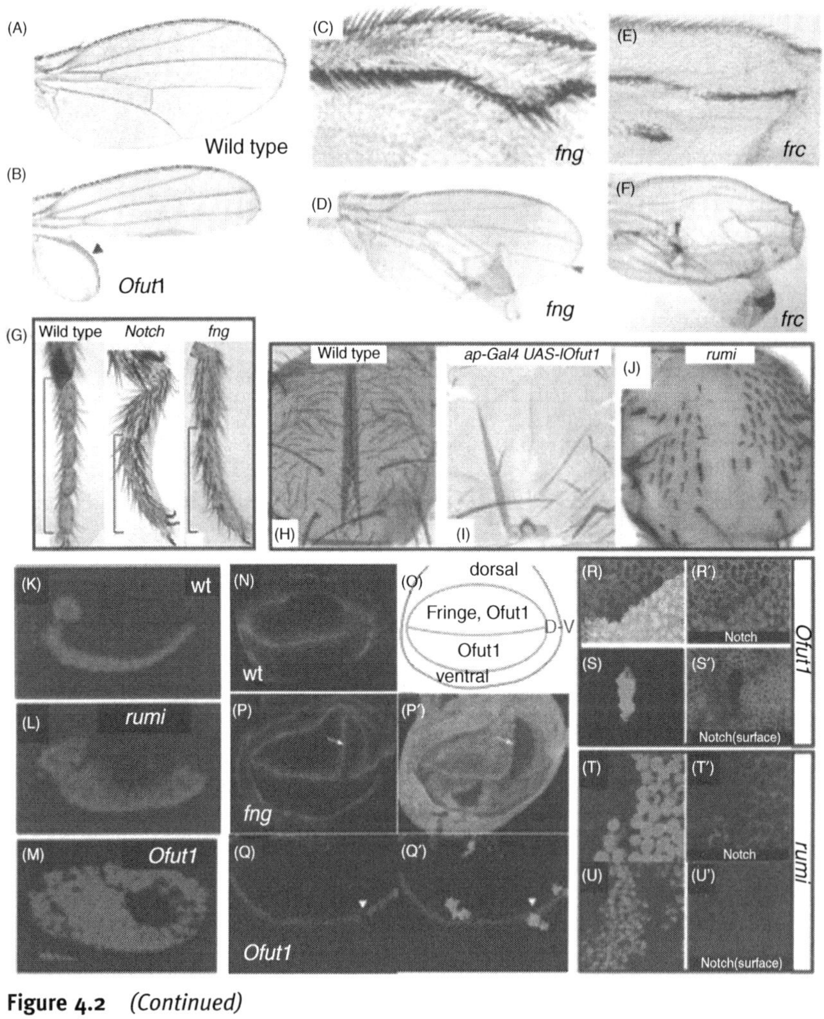

Figure 4.2 *(Continued)*

transfer fucose directly to Ser or Thr in an EGF-like domain with the appropriate consensus (Wang *et al.*, 2001). The chick (Sakamoto *et al.*, 1997), zebra fish (Qiu *et al.*, 2004), and Xenopus (Wu *et al.*, 1996) express up to three Fringe genes, but Fringe does not appear to be present in *C. elegans* based on phylogenetic comparisons (Haines and Irvine, 2003).

2. Glycans of Notch Receptors and DSL Notch Ligands

A variety of glycans can be added to the portion of Notch that transits the secretory pathway—the Notch ECD—and the intracellular domain of Notch is potentially modified by O-GlcNAc which is found on many cytoplasmic and nuclear proteins (Butkinaree *et al.*, 2010). The ECD of Notch and the DSL Notch ligands in *Drosophila* and mammals have

Figure 4.2 Notch signaling phenotypes of *Drosophila ofut1, fng, frc*, and *rumi* mutants. (A) A wild-type adult wing. (B) A wing-bearing clones of *Ofut1* mutant cells shows wing nicking and vein thickening (arrowhead), indicating defective Notch signaling. (C and D) Wings bearing *Fringe* mutant clones show duplications of wing margins (C) and an additional wing outgrowth from the wing blade (D). (E and F) *frc* mutant clones in the wings exhibit similar phenotypes to *Fringe* mutant clones. (G) Adult legs with *Notch* or *Fringe* clones show shortened legs and fused joints. A wild-type leg is shown *left*. Tarsal segments 2–5 are indicated by brackets. (H) A wild-type notum. (I) RNAi-mediated suppression of *Ofut1* in notum (*ap-Gal4 UAS-iOfut1*) results in loss of bristles, indicating defective Notch signaling. (J) Loss of bristles was also observed in a notum bearing *rumi* clones. (K) A wild-type embryo stained with a neuronal marker, ELAV. (L) *rumi* embryos lacking zygotic expression show a neurogenic phenotype at 28°C, in which ectodermal cells are replaced by excess neural cells. (M) *Ofut1* embryos lacking maternal and zygotic expression also exhibit the neurogenic phenotype. (N) A wild-type third instar wing disc stained for Wingless (WG) expression (red). (O) Schematic drawing. WG expression is indicated in red. D–V indicates dorsal–ventral boundary. *Ofut1* is expressed in both compartments whereas *Fringe* is expressed only dorsally. WG expression at D–V depends on Notch signaling. (P) *Fringe* mutant clones, marked by absence of GFP (green). Ectopic WG is indicated (arrow). (Q) *Ofut1* mutant clones, marked by presence of GFP (green). Loss of WG is indicated (arrowhead). (R) A wing disc with *ofut1* mutant clones (green), stained with antibodies against the Notch ECD (NECD; red) after detergent treatment. Increased and mislocalized Notch protein is observed within *ofut1* mutant cells. (S) A wing disc stained without detergent treatment. An *ofut1* mutant clone (green) is devoid of cell surface Notch (red). (T) *rumi* clones marked by GFP (green) also showed an accumulation of Notch (red). (U) A wing disc with *rumi* clones (green) stained with Notch (red) antibody in the absence of detergent. Notch expression is elevated at the apical cell surface. Panels A and B are adapted from (Sasamura *et al.*, 2003); C and D are from (Irvine and Wieschaus, 1994); E and F are from (Selva *et al.*, 2001); G is from (Rauskolb and Irvine, 1999); H and I are from (Okajima and Irvine, 2002); J, K, L, T, and U are from (Acar *et al.*, 2008); M, N, O, P, Q, R, and S are from (Okajima *et al.*, 2008) with permission of the publishers. (See Color Insert.)

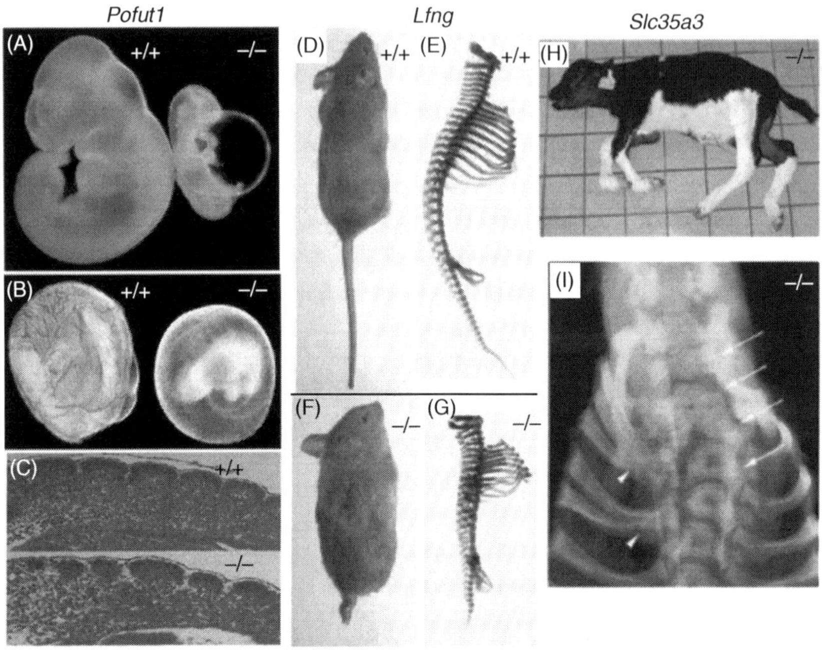

Figure 4.3 Notch signaling phenotypes of *Pofut1*, *Lfng*, and *Slc35a3* mutants. Inactivation of the *Pofut1* gene is embryonic lethal in mice with embryos showing defective development of the heart (A), defective vasculogenesis (B), and defective somitogenesis (A and C). Deletion of the *Lfng* gene causes marked skeletal defects apparent in *Lfng* mutant mice that lack a tail (D, wild-type $Lfng^{+/+}$; F, affected $Lfng^{-/-}$) and in skeletal preparations (E, wild-type $Lfng^{+/+}$; G, affected $Lfng^{-/-}$). A calf homozygous for a mutant *Slc35a3* allele is moribund due to skeletal defects (H), highlighted by arrows in an X-ray of the skeleton (I). Panels A, B, C are adapted from (Shi and Stanley, 2003); panels D, E, F, G are modified from (Serth *et al.*, 2003); and panels H and I are adapted from (Thomsen *et al.*, 2006) with permission of the publishers. (See Color Insert.)

consensus sites for the addition of *N*-glycans (at N-X-S/T or N-X-C where X is not proline), as well as *O*-Fuc, *O*-Glc, and *O*-GlcNAc glycans at Ser or Thr residues (Fig. 4.1). *O*-glycosylations of *Drosophila* Notch have been extensively studied using one of the common *Drosophila* cell lines, Schneider-2 (S2), while Chinese hamster ovary (CHO), NIH-3T3, and COS-7 cells have been used to investigate *O*-glycosylation in mammals. Cell lines are noted because glycosyltransferase gene expression and other factors affecting glycosylation may vary between cell lines. Proof of occupancy of individual EGF repeats by *O*-glycans is available for certain *O*-fucose (Fig. 4.1A), *O*-glucose (Fig. 4.1B), and *O*-GlcNAc (Fig. 4.1C) sites based on either radioactive labeling or western analysis of Notch EGF fragments or mass spectroscopy of Notch EGF fragments produced in cultured cells (Table 4.1), as discussed below. While this is important

Table 4.1 Sites of O-glycosylation on *Drosophila* Notch, Delta, and Serrate

	Fuc-O-	GlcNAcβ1, 3Fuc-O-	Peptide	References
Serrate	Yes	Yes	ND	(Panin *et al.*, 2002)
Serrate 4M	Yes	Yes	ND	(Panin *et al.*, 2002)
Serrate 8M	No	No		(Panin *et al.*, 2002)
Delta	Yes	Yes	ND	(Panin *et al.*, 2002)
Notch				
EGF3	Yes	Yes	CLNGGTC	(Xu *et al.*, 2007)
EGF5	Yes	Yes	CKYGGTC	(Xu *et al.*, 2007)
EGF7	Yes	Yes	CQNGGTC	(Xu *et al.*, 2007)
EGF12	Yes	Yes	CQNEGSC	(Lei *et al.*, 2003)
EGF17	Yes	Yes	CNNGATC	(Xu *et al.*, 2007)
EGF20	Yes	Yes	CQHGGTC	(Matsuura *et al.*, 2008)(Xu *et al.*, 2007)
EGF23	Yes	Yes	CRNGASC	(Xu *et al.*, 2007)
EGF25	Yes	Yes	CQNGATC	(Panin *et al.*, 2002)
	Glc-O-	**Xylα1, 3Glc-O-**		
EGF14	Yes	ND	CQSQPC	(Acar *et al.*, 2008)
EGF16	Yes	ND	CESNPC	(Acar *et al.*, 2008)
EGF17	Yes	ND	CHSNPC	(Acar *et al.*, 2008)
EGF19	Yes	ND	CASNPC	(Acar *et al.*, 2008)
EGF20	Yes	YES	CSSNPC	(Matsuura *et al.*, 2008)
EGF35	Yes	ND	CDSNPC	(Acar *et al.*, 2008)
	GlcNAc-O			
EGF20	Yes		CMPGYTG	(Matsuura *et al.*, 2008)
EGF1-10	Yes		ND	(Matsuura *et al.*, 2008)
EGF22-32	Yes		ND	(Matsuura *et al.*, 2008)
Delta	Yes		ND	(Matsuura *et al.*, 2008)

and represents the current state-of-the-art, identifying sites in Notch that are occupied by glycans *in vivo,* under conditions of endogenous and regulated expression of enzymes and their substrates, in a particluar cell type at a specific time in development, is the goal. Clearly the latter is needed to eventually understand how and why glycans modulate Notch signaling.

2.1. *N*-glycans and *O*-GalNAc glycans

Drosophila Notch was inferred to be a glycoprotein based on the ability of Notch ECD to bind to a lentil lectin affinity column and be eluted with α-methylmannoside (Johansen *et al.*, 1989). These properties are consistent with modification by oligomannosyl or simple complex *N*-glycans (Fig. 4.1E) found in *Drosophila* (Aoki *et al.*, 2007). Mammalian Notch1 was shown to carry *N*-glycans based on its sensitivity to peptide *N*-glycosidase F (*N*-glycanase) (Shao *et al.*, 2003) which cleaves *N*-glycans from Asn, thereby generating Asp. It is not known if Notch carries *O*-GalNAc or mucin *O*-glycans (Brockhausen *et al.*, 2009), although predictions of the NetOGlyc 3.1 database (Julenius *et al.*, 2005) suggest that neither *Drosophila* Notch nor mammalian Notch1 ECDs have potential sites of *O*-GalNAc glycosylation.

2.2. *O*-fucose glycans

Modification of Notch EGF repeats with ^{3}H-fucose was discovered in Lec1 CHO cells that incorporate very little fucose into *N*-glycans (Moloney *et al.*, 2000b). Previous studies had shown that EGF repeats of tissue plasminogen activator, blood clotting factor VII, and factor IX contain *O*-fucose at a Ser or Thr residue just before the third Cys of the EGF repeat (Harris and Spellman, 1993). Notch1 EGF repeat sequences were examined for Ser or Thr at this position, and a consensus motif for *O*-fucosylation was proposed (Moloney *et al.*, 2000b). This was later modified based on experimental evidence and theoretical considerations to $C^{2}XXX(A/G/S)\mathbf{S/T}C^{3}$ based on the fact that EGF15 in mouse Notch1 ($C^{1}HYG\mathbf{S}C^{2}$) is not modified (Li *et al.*, 2003; Rampal *et al.*, 2005a; Shao *et al.*, 2003). The *O*-fucose in coagulation factors is elongated to a tetrasaccharide by the addition of GlcNAc, Gal, and sialic acid (SA) (Fig. 4.1A). This fact, and the suspicion that Fringe might be a glycosyltransferase, led to *in vitro* assays of sugar transfer using pNP-*O*-fucose as substrate and Fringe on beads. These experiments revealed that Fringe in *Drosophila* and mammals is a GlcNAc transferase which generates GlcNAcβ1,3Fuc-*O*-EGF on Notch (Moloney *et al.*, 2000a). While *Drosophila* has only one Fringe (*Fringe*) and one Notch gene (*N*), mammals have four Notch genes (*Notch1–4*) and three Fringe genes as noted above. *In vitro* comparisons of the mammalian Fringes identify mouse Lfng as the most active followed by Mfng and then Rfng (Rampal *et al.*, 2005b). Although all three mammalian Fringe proteins have a single transmembrane domain and are thought to reside and function in the Golgi, both Lfng and Mfng are secreted from cells (Johnston *et al.*,

1997), perhaps as a way to regulate their activity (Shifley and Cole, 2008), whereas Rfng remains predominantly intracellular. *Drosophila* Fringe was also shown to be secreted (Irvine and Wieschaus, 1994).

All Fringe enzymes transfer GlcNAc to *O*-Fuc on a folded EGF repeat with a ~1000-fold improved efficiency over a denatured EGF repeat (Moloney *et al.*, 2000a) and ~10-fold better than to a simple fucose acceptor (Luther *et al.*, 2009). Fringe glycosyltransferases are glycoproteins (Rampal *et al.*, 2005b) and probably need their *N*-glycans to be active in the cell. Two predominant transcripts of Lfng exist, and a large number of differentially spliced forms of Mfng, as well as numerous splice forms of Rfng, have been isolated as cDNAs from different sources (see AceView at http://www.ncbi.nlm.nih.gov/IEB/Research/Acembly/ (Thierry-Mieg and Thierry-Mieg, 2006). Attempts to determine whether the different mammalian Fringes are regulated primarily by transcription or have different substrate specificities with respect to the amino acid sequence of the EGF repeat to which *O*-Fuc is attached have provided interesting insights (Rampal *et al.*, 2005b). The general conclusion from *in vitro* assays is that there is no simple motif for Fringe recognition of a Fuc-*O*-EGF repeat and that differences observed between the Fringes may primarily reflect differences in their catalytic efficiency (Rampal *et al.*, 2005b). However, only a limited number of EGF repeat substrates have been explored in this context, and evidence discussed below for additive effects amongst Fringe genes argues for some degree of variation in the sites modified by different Fringe enzymes.

Once Fringe has acted in a mammalian cell there is a possibility of further elongation of the disaccharide with Gal followed by SA to generate the tetrasaccharide SAα2,3Galβ1,4GlcNAcβ1,3Fuc-*O*-EGF (Fig. 4.1A). This elongation is variable, however, so that any Fuc-*O*-EGF may not be modified further, or may be a disaccharide, a trisaccharide or a tetrasaccharide. The functional significance of this diversity is an important question for the future. In S2 cells, Fringe expression is negligible but *O*-fucose is present on Notch (Okajima and Irvine, 2002). Upon expression of Fringe in S2 cells, GlcNAcβ1,3Fuc disaccharide is synthesized, but unlike mammalian *O*-fucose glycans, no further elongation has been observed. This could be because the *O*-fucose glycan *in vivo* is not faithfully replicated in cultured S2 cells. Interestingly, a novel glucuronyl trisaccharide *O*-fucose glycan, GlcNAcβ1,3(GlcAβ1,4)-Fucitol, was amongst the *O*-glycans released from glycoproteins of *Drosophila* embryos (Aoki *et al.*, 2008). This trisaccharide is enriched in the dorsal compartment of the wing imaginal disc, which is consistent with the dorsal expression of Fringe. However, it is not known whether this unique *O*-fucose glycan is actually attached to Notch *in vivo*. Intriguingly, while it was reduced in amount in embryos lacking Fringe, the trisaccharide was still detected (Aoki *et al.*, 2008), though this may be due to maternal Fringe.

After Fringe was found to elongate Fuc-*O*-EGF on Notch, the search was on for the *O*-fucosyltransferase that transfers the fucose to Notch. This is encoded by the *Ofut1* gene in *Drosophila* and the *Pofut1* gene in mammals (Wang *et al.*, 2001). A second distantly related gene termed *Pofut2* transfers fucose to Ser or Thr on thrombospondin repeats, but not to EGF repeats (Luo *et al.*, 2006; Shi *et al.*, 2007). Ofut1 and Pofut1 have a KDEL-like sequence at their C-terminus and are luminal proteins of the endoplasmic reticulum (ER)/*cis*-Golgi network (Luo and Haltiwanger, 2005). When Ofut1 and Notch are overexpressed in S2 cells, they physically associate (Okajima *et al.*, 2005; Sasamura *et al.*, 2007). This finding, and the fact that Ofut1 aids in the folding of Notch as discussed below, supports the proposal that Ofut1 is a chaperone for Notch in *Drosophila* (Okajima *et al.*, 2005; Sasamura *et al.*, 2007).

While most studies of *O*-fucose glycans have focused on their presence on the Notch ECD, it was shown early on that the DSL Notch ligands Delta and Serrate are also modified by both *O*-fucose and Fringe (Panin *et al.*, 2002). Sequence comparisons indicate that a cohort of about 50 proteins are potential carriers of *O*-fucose glycans (Rampal *et al.*, 2007). However, mechanistic studies described below indicate that it is necessary to determine for each site of modification, whether the presence of an *O*-fucose glycan affects biological activity, and if so, how.

2.3. *O*-glucose glycans

The presence of *O*-glucose glycans on Notch1 was discovered along with *O*-fucose glycans in Lec1 CHO cells (Moloney *et al.*, 2000b). The Glc-*O*-EGF modification of Notch1 is found at a Ser or Thr adjacent to the second Cys in the consenus C^1XSXPC^2. The Glc-*O*-EGF is elongated by the addition of xylose (Moloney *et al.*, 2000b) and was proposed to form Xylα1,3Xylα1,3Glc-*O*-EGF (Fig. 4.1B), as detected on bovine coagulation factors VII and IX (Hase *et al.*, 1988). This structure and the glycosyltransferases that generate it have now been confirmed. The *O*-glucosyltransferase is encoded by the *rumi* gene in *Drosophila* (Acar *et al.*, 2008), and two genes in mammals encode a xylosyltransferase that transfers Xyl to Glc-*O*-EGF (Sethi *et al.*, 2010). The gene encoding the Xyl-to-Xyl xylosyltransferase remains to be identified. While the nature and distribution of *O*-glucose glycans on endogenous Notch is not known, *Drosophila* Notch fragments expressed in S2 cells transfected with Rumi carry Glc- or a Xly-Glc-disaccharide (Acar *et al.*, 2008; Matsuura *et al.*, 2008; Table 4.1). As for *O*-fucose glycans, *O*-glucose glycans are potentially present on any EGF repeat that contains the acceptor motif for the *O*-glucosyltransferase. Most surprisingly, Rumi has also been found to transfer xylose to the same EGF consensus motif and thereby to generate Xyl-*O*-EGF (Fig. 4.1D; R. S. Haltiwanger and H. Takeuchi, personal communication). It is not known if xylose occurs on EGF repeats *in vivo*, whether a second Xyl or other sugars are subsequently added, or if this Xyl serves as an initiator of proteoglycan synthesis, as discussed below.

2.4. Glycosaminoglycans

The transfer of xylose to Ser or Thr residues that usually occur in a cluster but may be isolated residues initiates glycosaminoglycan (GAG) synthesis (Esko *et al.*, 2009). The subsequent addition of two Gal residues and a glucuronic acid provides the core structure on which long GAG chains are synthesized to generate heparan sulfate or chondroitin sulfate. To date there has been no structural evidence that Notch or its ligands are modified by GAGs. However, this certainly is a possibility since elimination of the GAG-specific sulfotransferase Hst-3b in *Drosophila* affects Notch signaling and trafficking (Kamimura *et al.*, 2004).

2.5. A novel *O*-GlcNAc modification

The presence of O-GlcNAc at Ser or Thr in cytoplasmic and nuclear proteins is now well established (Butkinaree *et al.*, 2010), and Notch ICD may carry O-GlcNAc, which could regulate the expression of Notch target genes. However, it was a big surprise to find O-GlcNAc as a modification of the ECD of *Drosophila* Notch (Matsuura *et al.*, 2008). Based on galactosyltransferase labeling, β-*N*-acetylhexosaminidase digestion and immunoblotting with O-GlcNAc-specific antibody (CTD110.6), the modification was determined to be GlcNAc-β-O-EGF (Fig. 4.1C). Intracellular O-GlcNAc transfer is catalyzed by a single O-GlcNAc transferase (OGT) (Kreppel *et al.*, 1997). However, this OGT is not responsible for the O-GlcNAc glycosylation of Notch ECD, since O-glycosylation of Notch EGF repeats occurs in the secretory pathway. Consistent with this, RNAi-mediated reduction of OGT did not decrease O-GlcNAc levels on *Drosophila* Notch. Furthermore, OGT activity was detected in a membrane fraction prepared from S2 cells (Matsuura *et al.*, 2008). Thus, it appears that the O-GlcNAc modification on EGF domains occurs independently of the action of OGT. The O-GlcNAc on Notch is found at Ser or Thr located between the fifth and sixth cysteines of a Notch EGF domain, C-terminal to the site of Ofut modification (Matsuura *et al.*, 2008). For the structure of an EGF domain see Chapter 2. Notch EGF domains like those of Factors VII, IX, and XII as well as plasminogen activators and Protein Z are O-glycosylated by fucose and/or glucose as discussed above (Rampal *et al.*, 2007). However, with the exception of Factor XII, these plasma glycoproteins do not contain Ser or Thr at the corresponding site that might receive O-GlcNAc. By contrast, potential O-GlcNAc sites are present in many EGF repeats of Notch receptors (Fig. 4.1) and Notch ligands, Delta, and Serrate. In fact, it was shown that the O-GlcNAc modification occurs at multiple sites in Notch EGF repeats and the ECD of Delta (Matsuura *et al.*, 2008; Table 4.1). This extracellular O-GlcNAc might be employed to modulate specific biological processes during animal development.

The O-GlcNAc modification is present on mammalian Notch EGF repeats secreted from CHO cells and preliminary data have detected an EGF repeat OGT activity in a membrane fraction from mammalian cells (C. Saito, Y. Tashima, P. Stanley, and T. Okajima; unpublished observations). This is consistent with the presence of Thr/Ser residues at conserved consensus sites (C^5XXXXS/T^6) in mammalian Notch receptors and DSL ligands (Matsuura *et al.*, 2008). In addition, it was previously reported that O-GlcNAc is present at the luminal face of the ER (Abeijon and Hirschberg, 1988). It should be noted that in mammals O-GlcNAc glycans on secreted or membrane proteins are likely to be elongated since O-GlcNAc is readily modified by β1,4galactosyltransferase in the Golgi (Whelan and Hart, 2006).

2.6. General overview

It is now clear that the ECDs of Notch and the DSL Notch ligands are coated with sugars (Fig. 4.1). For the most part, these sugars are transferred to specific motifs recognized by an initiating glycosyltransferase. Subsequently, glycosyltransferases like Fringe may recognize the initial sugar in the context of the EGF motif. Based on knowledge of the specificity of the glycosyltransferases for EGF repeats, biochemical properties of Notch, ligands, and recombinant mutants lacking individual glycosylation sites, and structural analyses of Notch fragments, a general picture of mature, glycosylated, *Drosophila* Notch, and mammalian Notch1, as they would be expected to be expressed *in vivo,* has emerged (Fig. 4.1). In the case of *Drosophila* Notch, concrete structural information has been obtained in several instances by mass spectrometry of tryptic peptides (Table 4.1). The other mammalian Notch receptors and the DSL *Drosophila* and mammalian Notch ligands should be similarly glycosylated on their EGF repeats. For example, mouse Notch1 EGF4 (C^1**A S**NPC^2) has been shown to be O-glucosylated (Bakker *et al.*, 2009; Sethi *et al.*, 2010).

The glycans associated with Notch may confer direct or indirect effects on Notch activity. For example, glycan-binding proteins may bind to Notch glycans and thereby link Notch with other glycoproteins on the same or an adjacent cell surface. Other sources of indirect effects of glycans on Notch signaling are the glycolipids formed by Brainiac and Egghead in *Drosophila* (Muller *et al.*, 2002; Schwientek *et al.*, 2002; Wandall *et al.*, 2003, 2005). Mutants in these genes cannot make the complete glycan part of the glycolipid, and one consequence is that Notch signaling is defective.

However, inactivation of genes encoding GalNAcTs, which transfer GalNAc to the substrates generated by Egghead and Brainiac, does not result in severe Notch signaling defects (Stolz *et al.*, 2008). Glycolipids are presumably important in regulating Notch conformation or stability in the membrane (Pizette *et al.*, 2009; Hamel *et al.*, 2010).

3. Consequences of Glycan Removal for Notch Signaling

3.1. *N*-glycans or *O*-GalNAc glycans

The removal of all *N*-glycans leads to embryonic death in the mouse at the peri-implantation stage (Marek *et al.*, 1999). However, when only two major classes of *N*-glycans are eliminated, the complex (Fig. 4.1E) and hybrid type embryos survive until mid-gestation (Ioffe and Stanley, 1994; Metzler *et al.*, 1994). The phenotype is not identical to a Notch-null phenotype (Bolos *et al.*, 2007) but has some features which suggest that Notch signaling may be partly affected. Thus the heart is underdeveloped and remains as a loop, and some embryos exhibit *situs inversus* which is consistent with inhibition of Notch signaling (Raya *et al.*, 2003). However, this may be an indirect effect. The loss of complex and hybrid *N*-glycans is expected to reduce the time that cell surface glycoproteins interact with the extracellular galectin lattice (Dennis *et al.*, 2009), thereby enhancing the endocytosis of growth factor receptors, and potentially Notch receptors, leading to reduced Notch signaling.

O-GalNAc glycans are initiated by polypeptide GalNAc transferases (Ten Hagen *et al.*, 2003). There are ~20 ppGalNAcTs in mammals and to date the inactivation of a subset of these enzymes has not led to Notch phenotypes. However, removal of the single core 1 GalT termed T-synthase, which transfers Gal to GalNAc-*O*-Ser/Thr, is embryonic lethal in mouse at ~E14 (Xia *et al.*, 2004). The embryos die of brain hemorrhage and exhibit defective angiogenesis. Conditional deletion of T-synthase in endothelial cells revealed that core 1 (and/or core 2) *O*-GalNAc glycans control the separation between blood and lymphatic vessels, in part by affecting the function of podoplanin (Fu *et al.*, 2008). Interestingly, *Drosophila* has a number of genes potentially encoding a core 1 GalT. Deletion of C1Galt1A that is expressed in the amnioserosa and the central nervous system is lethal (Lin *et al.*, 2008). Larval brain hemisheres are misshapen and the ventral nerve cord is elongated. Thus, the elongation of *O*-GalNAc on glycoproteins in the developing central nervous system is essential for morphogenesis of the larval brain in *Drosophila*. Notch signaling has not been investigated in the mouse or fly *O*-glycan mutants, though it is potentially affected.

3.2. *O*-fucose glycans

3.2.1. Inactivation of Fringe genes

Fringe functions in the Golgi compartment, where it transfers GlcNAc onto Fucose in EGF repeats with the appropriate consensus as discussed above (Fig. 4.1). Catalytic activity is required for Fringe function. In third instar wing discs, Fringe is exclusively expressed in the dorsal compartment and acts in signal-receiving cells. Fringe inhibits Notch activation by Serrate in dorsal cells, which limits Serrate-Notch signaling from dorsal cells to ventral cells. In contrast, Fringe potentiates Notch activation by Delta, which allows Delta-Notch signaling from ventral to dorsal cells. Thus, Fringe is key to the positioning of strong Notch activation at the D-V boundary. Such positioning of Notch activation is also required for boundary formation of the leg and eye imaginal discs. In *Drosophila*, Fringe is required for a subset of Notch-dependent processes including inductive signaling, but it is not required in lateral inhibition or asymmetric cell division processes regulated by Notch signaling (Haines and Irvine, 2003; Irvine, 1999).

In mammals the mouse *Lfng* gene was the first to be inactivated by gene disruption (Evrard *et al.*, 1998; Zhang and Gridley, 1998). Consistent with roles in boundary formation in *Drosophila*, mice lacking Lfng have defective somitogenesis and severe skeletal defects. A missense mutation in human LFNG also gives rise to skeletal defects (Sparrow *et al.*, 2006). The expression of Lfng must be tightly controlled for somitogenesis to proceed correctly. Overexpression or underexpression of Lfng causes similar skeletal defects (Barrantes *et al.*, 1999; Serth *et al.*, 2003). A regulatory element upstream of the Lfng coding sequence termed FCE is required to maintain transcriptional oscillation of Lfng during somitogenesis (Cole *et al.*, 2002; Morales *et al.*, 2002). In mice lacking the FCE in which Lfng is expressed but transcriptional oscillation is lost, it was revealed that Lfng oscillation is critical for the segmentation of the anterior but not the posterior skeleton (Shifley *et al.*, 2008). By rescuing $Lfng^{-/-}$ mice with a chicken Lfng cDNA controlled by up to 5 kb of the mouse Lfng promoter, oscillation of Lfng was found to be necessary for cervical, thoracic and lumbar somite, and vertebrae development, but not for sacral and tail somite or vertebrae development (Stauber *et al.*, 2009). Lfng expression is also regulated at the protein level by processing via a specific proprotein convertase (Shifley and Cole, 2008). Thus, precise timing of Lfng modification of Notch is essential for the proper formation of somites and the skeleton (Cinquin, 2007). Deletion of Mfng (Moran *et al.*, 2009) or Rfng (Zhang *et al.*, 2002) has no discernable effects on somitogenesis or skeletal development. Most importantly, it was found that mice lacking all three *Fringe* genes may be viable, and two females were fertile (Moran *et al.*, 2009). Therefore, unless there is another gene that can substitute for Fringe, it must be concluded that Notch signaling proceeds through embryogenesis, with the exception of somitogenesis, and postnatal development in the mouse with Notch receptors modified solely by *O*-fucose. Mfng and Rfng do not play obvious roles during these developmental stages.

The consequences of inhibiting Fringe expression have also been investigated during development in chick, fish, and frog. Somitogenesis requires oscillating expression of lunatic fringe in the chick as it does in mammals (Dale *et al.*, 2003). Lunatic fringe is important in zebra fish for induction of mesoderm (Peterson and McClay, 2005), the generation of segmental boundaries (Prince *et al.*, 2001), and development of the notochord (Appel *et al.*, 2003), though its expression does not oscillate. In Xenopus, Notch, Delta, and Lunatic fringes are important in regulating the outgrowth of the tail bud (Beck and Slack, 2002).

Because *Lfng* null mice survive poorly, conditional mutants and bone marrow or fetal liver transfer experiments were used to identify requirements for Lfng in T-cell and marginal zone B (MZB) cell development (Stanley and Guidos, 2009; Visan *et al.*, 2006b). Lfng is expressed in double-negative (DN) T cells but not in double-positive (DP) T cells (Visan *et al.*, 2006a). Misexpression of Lfng in DP T cells blocks T-cell development by preventing DN T cells from interacting with thymic stroma and allows B cells to develop in the thymus (Koch *et al.*, 2001). In the spleen, not only Lfng but also Mfng is required for the maximal generation of MZB cells (Tan *et al.*, 2009). This interesting result shows that Lfng and Mfng are not redundant but play complementary roles in generating MZB cells. In a disease-related model of Alagille syndrome, removal of one copy of Mfng (or Rfng or Lfng) along with one copy of Jagged1 causes proliferation of bile ducts in mouse liver (Ryan *et al.*, 2008). This was the first indication of a role for Rfng *in vivo*. However, based on expression levels, Rfng may also play a role in angiogenic sprouting of tip cells during vascularization of the retina (Benedito *et al.*, 2009). The three Fringe genes are expressed in tip cells and loss of Lfng leads to an increase in sprouting.

3.2.2. Overexpression or misexpression of fringe

Both temporal and spatial regulation of Fringe expression is necessary for appropriate control of Notch signaling and cell fate determination. Misexpression of Fringe in the *Drosophila* ventral wing disc inhibits Notch signaling and results in wing loss (Irvine and Wieschaus, 1994). Ectopic expression of Fringe in the fly rescues neurogenic defects induced by overexpression of Serrate but gives reduced viability when ubiquitously overexpressed under a heat-shock promoter (Fleming *et al.*, 1997). Expression of Fringe throughout the wing from early development results in wing loss (Klein and Arias, 1998), and an overexpression screen for modulators of Notch signaling identified Fringe (Hall *et al.*, 2004). In the mouse, misexpression of Lfng in the thymus causes T cell precursors to become B cells (Koch *et al.*, 2001). The mechanism is non cell-autonomous. Competition experiments showed that DP T cells expressing Lfng take up the stromal niche of DN cells and prevent their interaction with the stroma, thereby preventing their development into DP cells (Visan *et al.*, 2006a). During somitogenesis, Fringe expression must oscillate in a tight cycle or skeletal formation is disrupted (Serth *et al.*, 2003). As noted above, Fringe proteins are secreted (Irvine and Wieschaus, 1994). Although it was shown that tethering Fringe in the Golgi so that it cannot be secreted preserves its functions in wing development (Bruckner *et al.*, 2000) and that most probably Fringe is secreted in order to reduce its intracellular concentration (Shifley and Cole, 2008), it is also possible that Fringe has extracellular function(s).

3.2.3. Inactivation of protein *O*-fucosyltransferase

The functional significance of *O*-fucosylation was investigated by mutation or RNAi-mediated suppression of Ofut1 in *Drosophila* (Okajima and Irvine, 2002; Sasamura *et al.*, 2003) and by targeted mutation in the mouse (Shi and Stanley, 2003). In both mouse and fly the loss of Ofut1/Pofut1 leads to phenotypes characterized by the absence of all Notch signaling. Not only fringe-dependent inductive signaling but also fringe-independent lateral inhibition and lineage decision processes were impaired in *Drosophila*, suggesting that Ofut1 is universally required for Notch signaling. Similarly in mouse, the phenotype of *Pofut1*$^{-/-}$ embryos is like that of embryos defective in global Notch signaling (Lu and Stanley, 2006). A spontaneous mutation in the mouse *Pofut1* gene that gives a milder phenotype has also been described (Schuster-Gossler *et al.*, 2009). This mouse revealed that Pofut1 expression is most important in the paraxial mesoderm during skeletal development.

3.2.4. Inhibition of GDP-fucose synthesis

The donor substrate for Ofut1 and Pofut1 is GDP-fucose which is synthesized by two enzymes termed GMD (GDP-mannose-4-6-dehydratase) and FX (3-5-epimerase/4-reductase). A GMD mutant cell line Lec13 with markedly reduced Notch signaling first indicated that the addition of fucose to Notch is necessary for optimal Notch signaling (Moloney *et al.*, 2000a). Like Fringe mutants, *Gmd* mutants in *Drosophila* show impairment of Notch activation at the D-V boundary of wing discs (Okajima *et al.*, 2005; Sasamura *et al.*, 2007). FX mouse embryos are partially rescued by GDP-fucose from maternal sources (Becker *et al.*, 2003; Smith *et al.*, 2002). However, when FX$^{-/-}$ bone marrow cells were used to form chimeras, myelopoiesis (Zhou *et al.*, 2008), and intestinal development (Waterhouse *et al.*, 2010) were impaired due to defective Notch signaling. Therefore, the transfer of fucose to Notch, not just the presence of Pofut1, is required for optimal Notch signaling (Stahl *et al.*, 2008).

3.2.5. Inactivation of nucleotide sugar transporters

GDP-fucose and UDP-GlcNAc must be imported from the cytosol into the secretory pathway to be utilized by Ofut1/Pofut1 or Fringe, respectively. In *Drosophila*, two GDP-fucose transporters, GFR (Golgi GDP-fucose transporter) and EFR (ER GDP-fucose transporter), are required (Ishikawa *et al.*, 2005, 2010). *EfrGfr* double mutants exhibit loss of Notch activation at the D-V boundary, whereas single mutants have only temperature-sensitive Notch signaling defects. EFR is a multifunctional nucleotide sugar transporter that also contributes to heparan sulfate biosynthesis. In the mouse, mutants that lack the Golgi GDP-Fuc transporter GFR homologue Slc35c1 do not have markedly defective Notch signaling, but mimic the symptoms of a human leukocyte adhesion deficiency termed LADII (Hellbusch *et al.*, 2007; Yakubenia *et al.*, 2008). Consistent with this, the synthesis of *O*-fucose glycans on Notch1 EGF fragments was shown not to be impaired in fibroblasts from LADII patients (Sturla *et al.*, 2003). However, knockdown of the mouse Golgi GDP-fucose transporter in C2C12 muscle cells caused a slight decrease in Notch signaling in a co-culture assay (Ishikawa *et al.*, 2005). Nevertheless, it is clear that one or more additional transporters are required to *O*-fucosylate mammalian Notch receptors. The human homologue of *Drosophila* EFR, SLC35B4, would seem not to be a candidate for the ER GDP-fucose transporter in mammals, since it transports only UDP-sugars and specifically did not transport GDP-fucose in a cell-free assay (Ashikov *et al.*, 2005).

Another transporter which is directly involved in the synthesis of *O*-fucose glycans on Notch is the UDP-sugar transporter fringe-connection (FRC). This transporter, discovered in *Drosophila*, is multifunctional and transports UDP-glucuronic acid, UDP-GlcNAc and possibly UDP-xylose and UDP-glucose (Selva *et al.*, 2001). The homologue is SLC35D2 in humans and it transports UDP-GlcNAc (Ishida *et al.*, 2005; Suda *et al.*, 2004). *Drosophila frc* mutants exhibit defective Notch signaling as well as heparan sulfate-defective phenotypes (Selva *et al.*, 2001). They display a neurogenic phenotype as well as Notch processing defects (Goto *et al.*, 2001). Thus, a subset of Notch phenotypes observed in the *frc* mutant may in part be attributable to other glycosylation defects, including O-glucose glycosylation.

A transporter termed Slc35a3 that may be more specific for UDP-GlcNAc (Ishida *et al.*, 1999) is mutated in cattle with congenital skeletal malformations (Thomsen *et al.*, 2006; Fig. 4.3), a phenotype typical of mice lacking Lfng (Evrard *et al.*, 1998; Zhang and Gridley, 1998) and the human spondylocostal disease due to mutated *LFNG* (Sparrow *et al.*, 2006). Slc35a3 must have a predominant role in delivering UDP-GlcNAc to Fringe.

3.2.6. Inactivation *of* β1,4galactosyltransferase 1

Investigations of Notch signaling in a co-culture assay using CHO glycosylation mutants identified a requirement for Gal on *O*-fucose glycans for the inhibition of Jagged1-induced Notch signaling by Lfng or Mfng (Chen *et al.*, 2001). Consistent with a biological function for Gal *in vivo*, reduced expression of a subset of Notch target genes involved in somitogenesis in the mouse was observed in mice lacking β4galt1 (Chen *et al.*, 2006). The effects were quite subtle, perhaps because there are several other β4galts that may modify *O*-fucose glycans in mammals (Lo *et al.*, 1998).

Based on the fact that *O*-fucose glycans from mammalian cells may carry a tetrasaccharide (Fig. 4.1), it was expected that *Drosophila* β1,4galactosyltransferase and sialyltransferase genes might affect Notch signaling. Although there is no evidence that the GlcNAcβ1,3Fuc disaccharide is elongated by galactose or SA in *Drosophila*, genes that could encode the relevant glycosyltransferases are present in the genome. Two *Drosophila* homologoues of mammalian β1,4galactosyltransferases (β4GalNAcTA and β4GalNAcTB) turn out to be β1,4*N*-acetylgalactosaminyltransferases and do not transfer Gal *in vitro* (Chen *et al.*, 2007; Haines and Irvine, 2005; Stolz *et al.* 2008). They are involved in the biosynthesis of insect-specific glycosphingolipids but not glycoproteins and transfer GalNAc to GlcNAcβ1-3Manβ1-4Glcβ1-Ceramide (Chen *et al.*, 2007; Stolz *et al.*, 2008). Mutations in these genes affect ventralization of ovarian follicle cells due to defective EGFR signaling between the oocyte and dorsal follicle cells. Mutations affecting other steps in this biothynthetic pathway block the generation of Manβ1-4Glcβ1-Cer from Glcβ1-Cer (*egghead*), or the subsequent generation of GlcNAcβ1-3Manβ1-4Glcβ1-Cer from Manβ1-4Glcβ1-Cer (*brainiac*) (Wandall *et al.*, 2003, 2005). Interestingly, *egghead* and *brainiac* mutations cause abnormal neurogenesis during embryogenesis and compound egg chambers and dorsal appendage fusion during oogenesis. These phenotypes have been explained by defects in Notch and EGFR signaling and suggest that the extended form of glycosphingolipids may play a role in the modulation of receptor activities or the distribution of signaling molecules (Pizette *et al.*, 2009). Consistent with this proposal, recent evidence shows that DSL Notch ligand signaling is modulated by the composition of glycosphingolipids in a membrane (Hamel *et al.*, 2010). Like *egghead* and *brainiac*, β4GalNAcTB mutant animals display ventralization of ovarian follicle cells due to defective EGFR signaling (Chen *et al.*, 2007), whereas the β4GalNAcTA mutant exhibits abnormal neuromuscular system and behavioral defects (Chen *et al.*, 2007; Haines and Irvine, 2005). Both mutations do not give a neurogenic phenotype, although the possibility remains that these two enzymes are functionally redundant during embryogenesis (Chen *et al.*, 2007).

The *Drosophila* genome encodes a sole α2,6-sialyltransferase (SiaT) (Koles *et al.*, 2004). This enzyme acts on oligosaccharides and glycoproteins *in vitro* and *in vivo* (Koles *et al.*, 2007). *Drosophila SiaT* is expressed in a limited number of cells in the late stages of the developing embryonic central nervous system. Thus, it appears that modification with SA in *Drosophila* does not affect general functions of glycoproteins, but rather it affects specific glycoprotein functions in the nervous system (Repnikova *et al.*, 2010).

3.2.7. Elimination or addition of an *O*-fucose site

An *O*-fucose site highly conserved across the metazoa resides in EGF12 of all Notch receptors (Haines and Irvine, 2003). Deletion experiments in *Drosophila* Notch identified EGF11 and EGF12 as the DSL Notch ligand-binding site (Rebay *et al.*, 1993; Xu *et al.*, 2005), suggesting that the *O*-fucose in EGF12 may be important in Notch ligand binding. Deletion of this ligand-binding region in mouse Notch1 reduced binding of Delta1 and Jagged1 and the ability of both ligands to induce Notch1 signaling (Ge *et al.*, 2008). Elimination of solely the *O*-fucose site in EGF12 of *Drosophila* Notch prevented inhibition by Fringe of Serrate-induced Notch signaling (Lei *et al.*, 2003). This was reflected in the inability of Fringe to inhibit Notch binding to S2 cells expressing Serrate, leading to the conclusion that Fringe action at Notch EGF12 is important for downregulation of Notch signaling by Serrate at the dorsal/ventral wing boundary. Mutation of three other *O*-fucose sites in the *Abruptex* region of Notch (EGF24, EGF26, EGF24, and EGF26, or EGF31) did not affect Notch activation (Lei *et al.*, 2003).

Interestingly, similar experiments in mouse Notch1 gave rise to rather different results. First, the EGF12 mutation to remove *O*-fucose gave Notch1 that was inactive in co-culture signaling assays (Rampal *et al.*, 2005a; Shi *et al.*, 2007). Removal of *O*-fucose in EGF26 gave a hyperactive Notch1 for both Delta1 and Jagged1 ligands and removal of *O*-fucose from EGF27 reduced cell surface expression of Notch1 (Rampal *et al.*, 2005a). *In vivo*, the consequences of mutating EGF12 was also different. Mice homozygous for Notch1 lacking *O*-fucose in EGF12 are viable and fertile (Ge and Stanley, 2008). However, the *Notch1*^{12f} allele is hypomorphic, as shown by its inability to rescue a Notch1 ligand-binding domain mutant allele. T-cell development is markedly compromised in *Notch1*^{12f} homozygotes due to reduced Notch1 signaling and ligand binding to T cells. This hypomorphic allele is of interest because it affects only Notch1 signaling in the context of a viable mouse.

There is also the *split* mutation in *Drosophila* Notch which results in the introduction of an *O*-fucose site in EGF14 (Li *et al.*, 2003). This mutation causes activation of Notch in proneural cells of the ommatidium, thereby preventing their differentiation which normally follows after Delta inhibits Notch signaling by lateral inhibition. This phenotype is not dependent on Fringe, indicating that the addition of *O*-fucose to EGF14 is enough to activate Notch inappropriately during eye development.

3.3. *O*-glucose glycans

The first insight into biological functions of *O*-glucose glycans (Fig. 4.1B) in Notch signaling was obtained by the identification of the *Drosophila* mutant *rumi* (Acar *et al.*, 2008) which encodes a protein *O*-glucosyltransferase. Rumi mutations cause a global Notch pathway defective phenotype. However, the requirement for Rumi is temperature dependent; severe Notch signaling defects are observed when flies are raised at elevated temperatures (28°C), but not the lower temperature of 18°C. Like Ofut1, Rumi is a soluble protein of the ER. Although both Notch and Notch ligands can be modified with *O*-glucose (Moloney *et al.*, 2000b), Rumi acts cell-autonomously in Notch signal-receiving cells and not in signal-sending cells that present Notch ligands, suggesting that O-glucosylation is required for Notch functions. To date there are no mutants in Rumi in mammals, nor in the xylosyltransferases that subsequently add xylose to Glc-*O*-EGF (Sethi *et al.*, 2010).

3.4. Glycosaminoglycans

There are a variety of mutants in GAG synthesis in *Drosophila*, mice, and *C. elegans* (Bulow and Hobert, 2006). While many of these mutations affect developmental processes, none give rise to strong Notch signaling mutant phenotypes. However, targeted knockdown of a specific heparan sulfate sulfotransferase in *Drosophila* (Hst3b) causes neurogenic phenotypes indicative of a role for a specific form of heparan sulfate in Notch signaling (Kamimura *et al.*, 2004).

3.5. General overview

Mutant organisms lacking the ability to initiate or elongate *O*-fucose glycans display Notch signaling defects that reflect cell-autonomous effects of Notch receptor functions (Figs. 4.2 and 4.3). Notch receptors lacking *O*-fucose glycans (Fig. 4.1A) altogether are functionally inactive, but do not behave in a dominant-negative manner in heterozygotes. Overexpression of Ofut1 gives a similar phenotype to loss of Ofut1 in *Drosophila*. The loss of *Drosophila* Fringe, or Lfng in other organisms, gives rise to defects in the formation of segmental boundaries, a subset of the developmental fate decisions under the control of Notch signaling. Again, overexpression or misexpression of Fringe may also give Notch mutant phenotypes. By contrast, the loss of *O*-glucose glycans (Fig. 4.1B) generates milder, temperature-sensitive Notch signaling phenotypes in *Drosophila*. Nevertheless, all these phenotypes appear to arise from the altered glycosylation of Notch because they mimic Notch signaling phenotypes generated by mutations in Notch receptors themselves, or in downstream members of Notch signaling pathways. The same cannot be said for Notch phenotypes related to removal of other glycans such as *N*-glycans (Fig. 4.1E), *O*-GalNAc (mucin) glycans (Xia *et al.*, 2004) or GAGs (Fig. 4.1D), which are known to play key roles in embryogenesis, but may not directly affect Notch signaling. One way to address roles for known glycans is to generate Notch mutants that lack a particular site of glycosylation. To date this has been done with *O*-fucose sites only and care must be taken in interpreting results. For example, mutation of the *O*-fucose site in Cripto inactivates its ability to stimulate Nodal signaling (Schiffer *et al.*, 2001). However, this was found to be due to the amino acid change rather than to the loss of the *O*-fucose glycan (Shi *et al.*, 2007). Also, roles for an *O*-fucose glycan may be inhibitory, as in *Drosophila* Notch EGF12 (Lei *et al.*, 2003) or muscle agrin (Kim *et al.*, 2008), or stimulatory, as in mammalian Notch1 EGF12 (Ge and Stanley, 2008).

4. Mechanisms of Glycan Regulation of Notch Signaling

Identifying biological roles for glycans by targeted knockdown of glycosyltransferase genes is effective and a necessary first step, but rarely identifies the key substrate of the missing glycosyltransferase responsible for a given phenotype. All activities involved in glycosylation act on multiple substrates. Thus, Notch phenotypes arising from defective *N*-glycan, *O*-GalNAc or glycosaminoglycan synthesis are difficult to investigate at a mechanistic level because these glycans are ubiquitously expressed on many cell surface receptors. Determining whether their removal causes an effect on Notch signaling is a challenge. Nevertheless, it is possible to identify specific substrates that give rise to a particular phenotype. For example, loss of the *N*-glycan branching transferase GlcNAcT-V causes reduced signaling from certain growth factor receptors (Partridge *et al.*, 2004); loss of a different branching GlcNAcT causes reduced glucose transport by Glut2 (Ohtsubo *et al.*, 2005); and removal of SA by Klotho from the ion channel TRPV5 increases its activity (Cha *et al.*, 2008). All of these effects reflect changes in cell surface retention time due to interactions with cell surface galectins. Alterations in Gal or SA residues of the *N*-glycans of Notch might likewise alter cell surface residence time.

The number of glycoproteins modified by *O*-fucose, *O*-glucose or *O*-GlcNAc glycans are far fewer because these *O*-glycans are found only at specific sites in certain EGF repeats (Fig. 4.1). Nevertheless, there are numerous glycoproteins that possess such EGF repeats (Matsuura *et al.*, 2008; Rampal *et al.*, 2007). Identification of the glycoprotein whose altered activity leads to a mutant phenotype involves determining whether an effect is cell-autonomous or non-cell-autonomous and carefully characterizing the phenotype of various mutant alleles. It is by these approaches that *O*-fucose and *O*-glucose glycans have been associated directly with Notch receptor signaling activity. The important question is—how exactly do the individual sugars of *O*-fucose and *O*-glucose glycans regulate signaling by Notch receptors?

4.1. *O*-fucose glycans

Further analysis of *Drosophila Gmd* mutants revealed that, unlike *Ofut1* mutants, Notch-dependent lateral inhibition and cell lineage decision processes are not affected during embryogenesis, as evidenced by the lack of a neurogenic phenotype in maternal and zygotic *Gmd* mutants (Okajima *et al.*, 2008). The different phenotypes of *Gmd* and *Ofut1* mutant embryos suggested that Ofut1 might possess additional functions besides acting as a fucosyltransferase. To test this possibility, rescue experiments were performed using the *Ofut1*^{R245A} allele that lacks fucosyltransferase activity but is expressed at normal levels. Expression of *Ofut1*^{R245A} in *Ofut1*$^{-/-}$ embryos results in robust neurogenesis, suggesting that O-fucosylation may be dispensable for Notch receptor function. Moreover, as in the case of *fringe* mutant clones, clones of cells expressing only *Ofut1*^{R245A} show ectopic *wg* expression in the dorsal wing disc. Thus, O-fucosylation of Notch is not absolutely required for Notch to signal in *Drosophila.*

In mammalian ES cells lacking Pofut1, partial rescue of Notch signaling and Notch ligand binding was observed with a cDNA encoding *Pofut1*^{R245A} (Stahl *et al.*, 2008). However, similar levels of rescue were obtained following transfection with an unrelated ER glucosidase, suggesting a non-specific effect of upregulating the unfolded protein response. Nevertheless, these results show that Notch1 lacking *O*-fucose can signal. However, cells having a normal amount of Pofut1 but reduced GDP-fucose levels, and therefore reduced O-fucosylation of Notch receptors, exhibit markedly reduced Notch signaling (Moloney *et al.*, 2000a; Chen *et al.*, 2001). Thus mammalian Notch receptors may signal poorly when they do not carry *O*-fucose.

Therefore the mechanisms by which Ofut1/Pofut1 affects Notch signaling are multifaceted. In both flies (Ahimou *et al.*, 2004; Okajima *et al.*, 2005; Sasaki *et al.*, 2007; Sasamura *et al.*, 2007) and mouse somites (Okamura and Saga, 2008), loss of Ofut1/Pofut1 causes Notch to be expressed at reduced levels at the cell surface. *Ofut1*^{R245A} partially restores the localization of Notch to the apical cell surface (Okajima *et al.*, 2005), whereas extracellular Ofut1 is proposed to stabilize Notch at the cell surface (Sasamura *et al.*, 2007). Moreover, Ofut1 expression rescues defective secretion and ligand binding of *Drosophila* Notch EGF point mutations (Okajima *et al.*, 2005). Accumulation of Notch in the ER of *Drosophila Ofut1* mutant cells has been identified as one mechanism preventing cell surface localization (Okajima *et al.*, 2005), whereas accumulation in novel endocytic vesicles following normal trafficking to the cell surface has been identified as another (Sasamura *et al.*, 2007). While these observations are hard to reconcile, it is possible that the endocytic compartment is closely apposed to the ER. It is difficult to understand why Ofut1 is not also a required chaperone of other glycoproteins such as Crumbs which, like Notch, has many EGF repeats (Okajima and Irvine, 2002). In contrast to *Drosophila* wing disc and mouse somites, surface expression levels of Notch are unaffected by the removal of Pofut1 in ES or CHO cells (Stahl *et al.*, 2008).

Ofut1/Pofut1 appears to be required for Notch to acquire the correct conformation for recognition by ligands (Okajima *et al.*, 2005; Stahl *et al.*, 2008). Interestingly, when Ofut1 is overexpressed, Notch signaling is inhibited both inside and outside of the regions where it is expressed. This non-autonomous effect of Ofut1 does not depend on its enzyme activity (Sasamura *et al.*, 2007). It is not known whether, under physiological conditions, Ofut1 is secreted and acts outside the cell. Nonetheless, this possibility is of potential interest from a pharmacological point of view, since secreted Ofut1 might serve as a soluble inhibitor of Notch signaling. Thus, it may be that Ofut1 possesses a third activity, which depends on neither its enzyme nor its chaperone activities. As an example, Ofut1 promotes transcytosis of Notch from the apical plasma membrane to the adherens junctions (Sasaki *et al.*, 2007).

In summary, both enzymatic and non-enzymatic activities of Ofut1 contribute to the absolute requirement of Ofut1 for Notch signaling in *Drosophila*. Non-enzymatic activities of Ofut1 are involved in folding and endocytosis of Notch receptors, and these activities are sufficient for a subset of Notch receptor functions. It is conceivable that *O*-glycans such as *O*-glucose and *O*-GlcNAc rescue requirements for *O*-fucose monosaccharide in Notch signaling. Removal of these *O*-glycans might reveal roles of O-fucosylation of Notch receptors. By contrast, it is clear that mammalian Notch receptors in cells unable to transfer fucose but containing normal levels of Pofut1 function poorly (Stahl *et al.*, 2008). The glycosyltransferase activity of Ofut1/Pofut1 is essential to provide the substrate for Fringe and it has been suggested that this is the major function of *O*-fucose on Notch in *Drosophila* (Okajima *et al.*, 2008). On the other hand, *O*-fucose may be required for Fringe-independent Notch signaling. *In vitro* binding assays show that Drosophila Notch fragments lacking fucose bind to Delta and Serrate expressed by S2 cells, albeit at low levels (Okajima *et al.*, 2003, 2005). In addition, a human Notch1 EGF fragment EGF11-13 lacking post-translational modifications can bind to Notch ligand-expressing cells (Hambleton *et al.*, 2004). However, tetramerization of the Notch fragment was necessary to observe binding. Subsequent studies identified calcium as a key requisite and EGF12 to be the major Delta1-binding site (Cordle *et al.*, 2008b). The X-ray structure of a Jagged1 N-terminal fragment DSL-EGF3 that binds to Notch1 revealed a conserved face, and mutations designed to alter this face caused *cis*-inhibition and *trans*-regulation Notch phenotypes in *Drosophila* (Cordle *et al.*, 2008a). Based on an NMR structure of the Notch1 ligand-binding domain fragment that also revealed a conserved face, the nature of the complex was proposed. Interestingly, the model places the *O*-fucose glycan in EGF12 on the opposite side to the Jagged1-binding face. This makes it difficult to understand how Fringe could alter Jagged1-induced

Notch signaling even indirectly, because previous NMR studies indicate that the presence of *O*-fucose would not be expected to change the conformation of an EGF repeat (Kao *et al.*, 1999).

The addition of GlcNAc to *O*-fucose on Notch EGF repeats by Fringe markedly enhances Delta binding to *Drosophila* Notch and inhibits Serrate binding (Lei *et al.*, 2003; Okajima *et al.*, 2003; Xu *et al.*, 2005, 2007). At least in *Drosophila*, the simple addition of GlcNAc is sufficient to produce the effects of Fringe on Notch-ligand binding (Xu *et al.*, 2007). However, the trisaccharide GlcNAc[GlcA]Fuc is found in flies and is reduced in *Drosophila* Fringe mutants (Aoki *et al.*, 2008). If the trisaccharide occurs on Notch, the function of GlcA in Notch ligand binding is of interest to determine.

In mammals, there are also effects on the binding of Jagged1 and Delta1 by Fringe modification of Notch. However, the effects vary for different Notch receptors and ligands such that the binding of Delta ligands is not always increased by the action of Fringe, nor is the binding of Jagged ligands always reduced (Hicks *et al.*, 2000; Ladi *et al.*, 2005; Yang *et al.*, 2005). In addition, co-culture signaling assays in which cells expressing a Notch reporter activated by the released intracellular domain of Notch are stimulated by cells expressing DSL Notch ligands are not always affected by Fringe in a manner directly reflected by changes in soluble ligand binding (Yang *et al.*, 2005). This may be because soluble Notch ligands do not bind with the same properties as membrane-bound ligands. For example, initial assays could only detect soluble Jagged1 binding after clustering (Hicks *et al.*, 2000). A decrease in Jagged1 binding could not be observed under conditions in which Fringe inhibited Jagged1-induced Notch1 cleavage (Yang *et al.*, 2005). In addition, Lfng, Mfng, and Rfng have been reported to have different effects on signaling through the same exogenous Notch receptor (Shimizu *et al.*, 2001). There is also evidence of a requirement for the Gal residue on *O*-fucose glycans to observe the effects of Lfng or Mfng on Jagged1-induced Notch signaling (Chen *et al.*, 2001). In this case, Fringe action was necessary but not sufficient to modulate ligand-induced Notch signaling. Ligand-binding assays support a role for Gal in Jagged1 binding to endogenous Notch receptors acted on by Lfng or Mfng (Y. Tashima and P. Stanley; unpublished observations).

In summary therefore, it is clear that Notch and Delta/Jagged ECDs physically interact (Shimizu *et al.*, 1999, 2000; Xu *et al.*, 2007) and that in *Drosophila* Fringe increases binding of Notch to Delta and reduces binding to Serrate (Okajima *et al.*, 2003; Xu *et al.*, 2005, 2007). This suggests that the mechanism by which *O*-fucose glycans regulate Notch signaling is by directly altering the ability of DSL Notch ligands to bind to Notch. However, the situation may be more complicated. For example, it is proposed that in order to bind to the ligand-binding domain of *Drosophila* Notch, Delta must displace the Abruptex region of Notch EGF repeats, and it is not known if Fringe affects this intramolecular interaction (Pei and Baker,

2008). Structural studies of bacterially produced Notch1 and Jagged1 ECD fragments suggest that sugars are not essential to their interaction, but this is hard to reconcile with *in vitro* effects of Fringe and the effects of *O*-fucose glycans on Notch signaling in co-culture assays. Only when structures of complexes between Notch and ligand ECD fragments have been compared before and after Fringe modification will it be possible to begin to understand how *O*-fucose glycans regulate Notch/ligand interactions. In this regard it is encouraging that mutations that eliminate or add a single site of O-fucosylation affect Notch signaling and, in the case of EGF12, cause altered Notch ligand binding (Ge and Stanley, 2008; Lei *et al.*, 2003; Xu *et al.*, 2005). Understanding how the loss of one *O*-fucose glycan affects Notch signaling when 22 other *O*-fucose glycans are presumably present, is a challenge for the future.

4.2. *O*-glucose glycans

Although the loss of protein *O*-glucosyltransferase in *Drosophila rumi* mutants results in a slight accumulation of Notch intracellularly, cell surface expression of Notch is maintained and rather elevated compared to wild-type cells (Acar *et al.*, 2008). Thus, unlike *Ofut1, rumi* is not required for the folding of Notch receptors. Furthermore, RNAi-mediated reduction of *rumi* in a cell-based assay suggests that *O*-glucose is not required for Notch binding to the Delta ligand. Based on comparisons of cleaved Notch forms in *rumi* mutants, it appears that *O*-glucosylation may be required for conformational changes in Notch that occur subsequent to ligand binding, which make Notch a substrate for S2 cleavage by an ADAM protease (Acar *et al.*, 2008). This is a cell-autonomous effect of the signal-receiving cell. Notch ligands lacking *O*-glucose appear to function normally. There is currently no mouse *rumi* mutant, nor are there *Drosophila* or mouse mutants lacking the xylose residues added to *O*-glucose on Notch (Sethi *et al.*, 2010). Finally, no *in vitro* assays of ligand binding to Notch ECD lacking *O*-glucose have been performed.

4.3. General overview

In terms of mechanistic studies, roles for the *O*-fucose glycans on Notch have been those most investigated to date. Removal of single sites of O-fucosylation alters those Notch signaling and the action of Fringe alters DSL Notch ligand binding in *in vitro* assays. However, it is not clear if Notch ligands bind directly to the *O*-fucose glycans of Notch and thereby regulate Notch activation. The structures of complexes between modified and unmodified Notch and its ligands will be necessary to know if *O*-fucose glycans modulate Notch signaling directly or indirectly.

4.4. Conclusions

It is now clear that O-fucose and O-glucose glycans modulate Notch signaling events critical to cell fate determination and tissue development. However, much work remains to understand exactly how this occurs and also to identify roles for xylose and O-GlcNAc on Notch. Meanwhile, it is clear that the glycans of Notch are not just the icing on the cake!

ACKNOWLEDGMENTS

The authors wish to thank their collaborators and lab members for their many contributions over the years. This work was supported by National Caner Institute grant RO1 95022 to P.S. and grants from the Japanese Ministry of Education, Science, Sports and Culture and Human Frontier Science Program to T.O.

REFERENCES

Abeijon, C., and Hirschberg, C. B. (1988). Intrinsic membrane glycoproteins with cytosol-oriented sugars in the endoplasmic reticulum. *Proc. Natl. Acad. Sci. U.S.A.* **85**, 1010–1014.

Acar, M., Jafar-Nejad, H., Takeuchi, H., Rajan, A., Ibrani, D., Rana, N. A., Pan, H., Haltiwanger, R. S., and Bellen, H. J. (2008). Rumi is a CAP10 domain glycosyltransferase that modifies Notch and is required for Notch signaling. *Cell* **132**, 247–258.

Ahimou, F., Mok, L. P., Bardot, B., and Wesley, C. (2004). The adhesion force of Notch with Delta and the rate of Notch signaling. *J. Cell Biol.* **167**, 1217–1229.

Aoki, K., Perlman, M., Lim, J. M., Cantu, R., Wells, L., and Tiemeyer, M. (2007). Dynamic developmental elaboration of N-linked glycan complexity in the Drosophila melanogaster embryo. *J. Biol. Chem.* **282**, 9127–9142.

Aoki, K., Porterfield, M., Lee, S. S., Dong, B., Nguyen, K., McGlamry, K. H., and Tiemeyer, M. (2008). The diversity of O-linked glycans expressed during Drosophila melanogaster development reflects stage- and tissue-specific requirements for cell signaling. *J. Biol. Chem.* **283**, 30385–30400.

Appel, B., Marasco, P., McClung, L. E., and Latimer, A. J. (2003). lunatic fringe regulates Delta-Notch induction of hypochord in zebrafish. *Dev. Dyn.* **228**, 281–286.

Ashikov, A., Routier, F., Fuhlrott, J., Helmus, Y., Wild, M., Gerardy-Schahn, R., and Bakker, H. (2005). The human solute carrier gene SLC35B4 encodes a bifunctional nucleotide sugar transporter with specificity for UDP-xylose and UDP-N-acetylglucosamine. *J. Biol. Chem.* **280**, 27230–27235.

Bakker, H., Oka, T., Ashikov, A., Yadav, A., Berger, M., Rana, N. A., Bai, X., Jigami, Y., Haltiwanger, R. S., Esko, J. D., and Gerardy-Schahn, R. (2009). Functional UDP-xylose transport across the endoplasmic reticulum/Golgi membrane in a Chinese hamster ovary cell mutant defective in UDP-xylose Synthase. *J. Biol. Chem.* **284**, 2576–2583.

Barrantes, I. B., Elia, A. J., Wunsch, K., De Angelis, M. H., Mak, T. W., Rossant, J., Conlon, R. A., Gossler, A., and de la Pompa, J. L. (1999). Interaction between Notch signalling and Lunatic fringe during somite boundary formation in the mouse. *Curr. Biol.* **9**, 470–480.

Beck, C. W., and Slack, J. M. (2002). Notch is required for outgrowth of the Xenopus tail bud. *Int. J. Dev. Biol.* **46**, 255–258.

Becker, D. J., Myers, J. T., Ruff, M. M., Smith, P. L., Gillespie, B. W., Ginsburg, D. W., and Lowe, J. B. (2003). Strain-specific modification of lethality in fucose-deficient mice. *Mamm. Genome* **14**, 130–139.

Benedito, R., Roca, C., Sorensen, I., Adams, S., Gossler, A., Fruttiger, M., and Adams, R. H. (2009). The notch ligands Dll4 and Jagged1 have opposing effects on angiogenesis. *Cell* **137**, 1124–1135.

Bolos, V., Grego-Bessa, J., and de la Pompa, J. L. (2007). Notch signaling in development and cancer. *Endocr. Rev.* **28**, 339–363.

Brockhausen, I., Schachter, S., and Stanley, P. (2009). O-GalNAc glycans. Chapter 9. In "Essentials of Glycobiology" (A. Varki, R. D. Cummings, J. D. Esko, H. H. Freeze, P. Stanley, C. R. Bertozzi, G. W. Hart and M. E. Etlzler, eds.). *Cold Spring Harbor laboratory Press, New York.*

Bruckner, K., Perez, L., Clausen, H., and Cohen, S. (2000). Glycosyltransferase activity of Fringe modulates Notch-Delta interactions. *Nature* **406**, 411–415.

Bulow, H. E., and Hobert, O. (2006). The molecular diversity of glycosaminoglycans shapes animal development. *Annu. Rev. Cell Dev. Biol.* **22**, 375–407.

Butkinaree, C., Park, K., and Hart, G. W. (2010). O-linked beta-N-acetylglucosamine (O-GlcNAc): Extensive crosstalk with phosphorylation to regulate signaling and transcription in response to nutrients and stress. *Biochim. Biophys. Acta.* **1800**, 96–106.

Cha, S. K., Ortega, B., Kurosu, H., Rosenblatt, K. P., Kuro, O. M., and Huang, C. L. (2008). Removal of sialic acid involving Klotho causes cell-surface retention of TRPV5 channel via binding to galectin-1. *Proc. Natl. Acad. Sci. U.S.A.* **105**, 9805–9810.

Chen, J., Lu, L., Shi, S., and Stanley, P. (2006). Expression of Notch signaling pathway genes in mouse embryos lacking b4galactosyltransferase-1. *Gene Expr. Patterns* **6**, 376–382.

Chen, J., Moloney, D. J., and Stanley, P. (2001). Fringe modulation of Jagged1-induced Notch signaling requires the action of beta 4galactosyltransferase-1. *Proc. Natl. Acad. Sci. U.S.A.* **98**, 13716–13721.

Chen, Y. W., Pedersen, J. W., Wandall, H. H., Levery, S. B., Pizette, S., Clausen, H., and Cohen, S. M. (2007). Glycosphingolipids with extended sugar chain have specialized functions in development and behavior of Drosophila. *Dev. Biol.* **306**, 736–749.

Cinquin, O. (2007). Understanding the somitogenesis clock: what's missing?. *Mech. Dev.* **124**, 501–517.

Cole, S. E., Levorse, J. M., Tilghman, S. M., and Vogt, T. F. (2002). Clock regulatory elements control cyclic expression of Lunatic fringe during somitogenesis. *Dev. Cell* **3**, 75–84.

Cordle, J., Johnson, S., Tay, J. Z., Roversi, P., Wilkin, M. B., de Madrid, B. H., Shimizu, H., Jensen, S., Whiteman, P., Jin, B., Redfield, C., Baron, M., Lea, S. M., and Handford, P. A., (2008a). A conserved face of the Jagged/Serrate DSL domain is involved in Notch trans-activation and cis-inhibition. *Nat. Struct. Mol. Biol.* **15**, 849–857.

Cordle, J., Redfieldz, C., Stacey, M., van der Merwe, P. A., Willis, A. C., Champion, B. R., Hambleton, S., and Handford, P. A. (2008b). Localization of the delta-like-1-binding site in human Notch-1 and its modulation by calcium affinity. *J. Biol. Chem.* **283**, 11785–11793.

Dale, J. K., Maroto, M., Dequeant, M. L., Malapert, P., McGrew, M., and Pourquie, O. (2003). Periodic notch inhibition by lunatic fringe underlies the chick segmentation clock. *Nature* **421**, 275–278.

Dennis, J. W., Lau, K. S., Demetriou, M., and Nabi, I. R. (2009). Adaptive regulation at the cell surface by N-glycosylation. *Traffic* **10**, 1569–1578.

Esko, J. D., Kimata, K., Lindahl, U. (2009). Proteoglycans and sulfated glycosaminoglycans. Chapter 16. In "Essentials of Glycobiology" (A. Varki, R. D. Cummings, J. D. Esko, H. H. Freeze, P. Stanley, C. R. Bertozzi, G. W. Hart and M. E. Etlzler, eds.). *Cold Spring Harbor laboratory Press, New York.*

Evrard, Y. A., Lun, Y., Aulehla, A., Gan, L., and Johnson, R. L. (1998). lunatic fringe is an essential mediator of somite segmentation and patterning. *Nature* **394**, 377–381.

Fleming, R. J., Gu, Y., and Hukriede, N. A. (1997). Serrate-mediated activation of Notch is specifically blocked by the product of the gene fringe in the dorsal compartment of the Drosophila wing imaginal disc. *Development* **124**, 2973–2981.

Fu, J., Gerhardt, H., McDaniel, J. M., Xia, B., Liu, X., Ivanciu, L., Ny, A., Hermans, K., Silasi-Mansat, R., McGee, S., Nye, E., Ju, T., Ramirez, M. I., Carmeliet, P., Cummings, R. D., Lupu, F., and Xia, L., (2008). Endothelial cell O-glycan deficiency causes blood/lymphatic misconnections and consequent fatty liver disease in mice. *J. Clin. Invest.* **118**, 3725–3737.

Ge, C., Liu, T., Hou, X., and Stanley, P. (2008). In vivo consequences of deleting EGF repeats 8-12 including the ligand binding domain of mouse Notch1. *BMC Dev Biol* **8**, 48.

Ge, C., and Stanley, P. (2008). The O-fucose glycan in the ligand-binding domain of Notch1 regulates embryogenesis and T cell development. *Proc. Natl. Acad. Sci. U.S.A.* **105**, 1539–1544.

Goto, S., Taniguchi, M., Muraoka, M., Toyoda, H., Sado, Y., Kawakita, M., and Hayashi, S. (2001). UDP-sugar transporter implicated in glycosylation and processing of Notch. *Nat. Cell Biol.* **3**, 816–822.

Haines, N., and Irvine, K. D. (2003). Glycosylation regulates Notch signalling. *Nat. Rev. Mol. Cell Biol.* **4**, 786–797.

Haines, N., and Irvine, K. D. (2005). Functional analysis of Drosophila beta1,4-N-acetlygalactosaminyltransferases. *Glycobiology* **15**, 335–346.

Hall, L. E., Alexander, S. J., Chang, M., Woodling, N. S., and Yedvobnick, B. (2004). An EP overexpression screen for genetic modifiers of Notch pathway function in Drosophila melanogaster. *Genet. Res.* **83**, 71–82.

Hambleton, S., Valeyev, N. V., Muranyi, A., Knott, V., Werner, J. M., McMichael, A. J., Handford, P. A., and Downing, A. K. (2004). Structural and functional properties of the human Notch-1 ligand binding region. *Structure* **12**, 2173–2183.

Hamel, S., Fantini, J., and Schweisguth, F. (2010). Notch ligand activity is modulated by glycosphingolipid membrane composition in Drosophila melanogaster. *J. Cell Biol.* **188**, 581–594.

Harris, R. J., and Spellman, M. W. (1993). O-linked fucose and other post-translational modifications unique to EGF modules. *Glycobiology* **3**, 219–224.

Hase, S., Ikenaka, K., Mikoshiba, K., and Ikenaka, T. (1988). Analysis of tissue glycoprotein sugar chains by two-dimensional high-performance liquid chromatographic mapping. *J. Chromatogr* **434**, 51–60.

Hellbusch, C. C., Sperandio, M., Frommhold, D., Yakubenia, S., Wild, M. K., Popovici, D., Vestweber, D., Grone, H. J., von Figura, K., Lubke, T., and Korner, C. (2007). Golgi GDP-fucose transporter-deficient mice mimic congenital disorder of glycosylation IIc/leukocyte adhesion deficiency II. *J. Biol. Chem.* **282**, 10762–10772.

Hicks, C., Johnston, S. H., diSibio, G., Collazo, A., Vogt, T. F., and Weinmaster, G. (2000). Fringe differentially modulates Jagged1 and Delta1 signalling through Notch1 and Notch2. *Nat. Cell Biol.* **2**, 515–520.

Ioffe, E., and Stanley, P. (1994). Mice lacking N-acetylglucosaminyltransferase I activity die at mid-gestation, revealing an essential role for complex or hybrid N-linked carbohydrates. *Proc. Natl. Acad. Sci. U.S.A.* **91**, 728–732.

Irvine, K. D. (1999). Fringe, Notch, and making developmental boundaries. *Curr Opin Genet Dev* **9**, 434–441.

Irvine, K. D., and Wieschaus, E. (1994). fringe, a Boundary-specific signaling molecule, mediates interactions between dorsal and ventral cells during Drosophila wing development. *Cell* **79**, 595–606.

Ishida, N., Kuba, T., Aoki, K., Miyatake, S., Kawakita, M., and Sanai, Y. (2005). Identification and characterization of human Golgi nucleotide sugar transporter SLC35D2, a novel member of the SLC35 nucleotide sugar transporter family. *Genomics* **85**, 106–116.

Ishida, N., Yoshioka, S., Chiba, Y., Takeuchi, M., and Kawakita, M. (1999). Molecular cloning and functional expression of the human Golgi UDP-N-acetylglucosamine transporter. *J. Biochem.* **126**, 68–77.

Ishikawa, H. O., Ayukawa, T., Nakayama, M., Higashi, S., Kamiyama, S., Nishihara, S., Aoki, K., Ishida, N., Sanai, Y., and Matsuno, K. (2010). Two pathways for importing GDP-fucose into the endoplasmic reticulum lumen function redundantly in the O-fucosylation of Notch in Drosophila. *J. Biol. Chem.* **285**, 4122–4129.

Ishikawa, H. O., Higashi, S., Ayukawa, T., Sasamura, T., Kitagawa, M., Harigaya, K., Aoki, K., Ishida, N., Sanai, Y., and Matsuno, K. (2005). Notch deficiency implicated in the pathogenesis of congenital disorder of glycosylation IIc. *Proc. Natl. Acad. Sci. U.S.A.* **102**, 18532–18537.

Johansen, K. M., Fehon, R. G., and Artavanis-Tsakonas, S. (1989). The notch gene product is a glycoprotein expressed on the cell surface of both epidermal and neuronal precursor cells during Drosophila development. *J. Cell Biol.* **109**, 2427–2440.

Johnston, S. H., Rauskolb, C., Wilson, R., Prabhakaran, B., Irvine, K. D., and Vogt, T. F. (1997). A family of mammalian Fringe genes implicated in boundary determination and the Notch pathway. *Development* **124**, 2245–2254.

Julenius, K., Molgaard, A., Gupta, R., and Brunak, S. (2005). Prediction, conservation analysis, and structural characterization of mammalian mucin-type O-glycosylation sites. *Glycobiology* **15**, 153–164.

Kamimura, K., Rhodes, J. M., Ueda, R., McNeely, M., Shukla, D., Kimata, K., Spear, P. G., Shworak, N. W., and Nakato, H. (2004). Regulation of Notch signaling by Drosophila heparan sulfate 3-O sulfotransferase. *J. Cell Biol.* **166**, 1069–1079.

Kao, Y. H., Lee, G. F., Wang, Y., Starovasnik, M. A., Kelley, R. F., Spellman, M. W., and Lerner, L. (1999). The effect of O-fucosylation on the first EGF-like domain from human blood coagulation factor VII. *Biochemistry* **38**, 7097–7110.

Kim, M. L., Chandrasekharan, K., Glass, M., Shi, S., Stahl, M. C., Kaspar, B., Stanley, P., and Martin, P. T. (2008). O-fucosylation of muscle agrin determines its ability to cluster acetylcholine receptors. *Mol. Cell. Neurosci.* **39**, 452–464.

Klein, T., and Arias, A. M. (1998). Interactions among Delta, Serrate and Fringe modulate Notch activity during Drosophila wing development. *Development* **125**, 2951–2962.

Koch, U., Lacombe, T. A., Holland, D., Bowman, J. L., Cohen, B. L., Egan, S. E., and Guidos, C. J. (2001). Subversion of the T/B lineage decision in the thymus by lunatic fringe-mediated inhibition of Notch-1. *Immunity* **15**, 225–236.

Koles, K., Irvine, K. D., and Panin, V. M. (2004). Functional characterization of Drosophila sialyltransferase. *J. Biol. Chem.* **279**, 4346–4357.

Koles, K., Lim, J. M., Aoki, K., Porterfield, M., Tiemeyer, M., Wells, L., and Panin, V. (2007). Identification of N-glycosylated proteins from the central nervous system of Drosophila melanogaster. *Glycobiology* **17**, 1388–1403.

Kreppel, L. K., Blomberg, M. A., and Hart, G. W. (1997). Dynamic glycosylation of nuclear and cytosolic proteins. Cloning and characterization of a unique O-GlcNAc transferase with multiple tetratricopeptide repeats. *J. Biol. Chem.* **272**, 9308–9315.

Ladi, E., Nichols, J. T., Ge, W., Miyamoto, A., Yao, C., Yang, L. T., Boulter, J., Sun, Y. E., Kintner, C., and Weinmaster, G. (2005). The divergent DSL ligand Dll3 does not activate Notch signaling but cell autonomously attenuates signaling induced by other DSL ligands. *J. Cell Biol.* **170**, 983–992.

Lei, L., Xu, A., Panin, V. M., and Irvine, K. D. (2003). An O-fucose site in the ligand binding domain inhibits Notch activation. *Development* **130**, 6411–6421.

Li, Y., Lei, L., Irvine, K. D., and Baker, N. E. (2003). Notch activity in neural cells triggered by a mutant allele with altered glycosylation. *Development* **130**, 2829–2840.

Lin, Y. R., Reddy, B. V., and Irvine, K. D. (2008). Requirement for a core 1 galactosyltransferase in the Drosophila nervous system. *Dev. Dyn.* **237**, 3703–3714.

Lo, N. W., Shaper, J. H., Pevsner, J., and Shaper, N. L. (1998). The expanding beta 4-galactosyltransferase gene family: messages from the databanks. *Glycobiology* **8**, 517–526.

Lu, L., and Stanley, P. (2006). Roles of O-fucose glycans in notch signaling revealed by mutant mice. *Methods Enzymol* **417**, 127–136.

Luo, Y., and Haltiwanger, R. S. (2005). O-fucosylation of notch occurs in the endoplasmic reticulum. *J. Biol. Chem.* **280**, 289–294.

Luo, Y., Nita-Lazar, A., and Haltiwanger, R. S. (2006). Two distinct pathways for O-fucosylation of epidermal growth factor-like or thrombospondin type 1 repeats. *J. Biol. Chem.* **281**, 9385–9392.

Luther, K. B., Schindelin, H., and Haltiwanger, R. S. (2009). Structural and mechanistic insights into lunatic fringe from a kinetic analysis of enzyme mutants. *J. Biol. Chem.* **284**, 3294–3305.

Marek, K. W., Vijay, I. K., and Marth, J. D. (1999). A recessive deletion in the GlcNAc-1-phosphotransferase gene results in peri-implantation embryonic lethality. *Glycobiology* **9**, 1263–1271.

Matsuura, A., Ito, M., Sakaidani, Y., Kondo, T., Murakami, K., Furukawa, K., Nadano, D., Matsuda, T., and Okajima, T. (2008). O-linked N-acetylglucosamine is present on the extracellular domain of notch receptors. *J. Biol. Chem.* **283**, 35486–35495.

Metzler, M., Gertz, A., Sarkar, M., Schachter, H., Schrader, J. W., and Marth, J. D. (1994). Complex asparagine-linked oligosaccharides are required for morphogenic events during post-implantation development. *EMBO Journal* **13**, 2056–2065.

Moloney, D. J., Panin, V. M., Johnston, S. H., Chen, J., Shao, L., Wilson, R., Wang, Y., Stanley, P., Irvine, K. D., Haltiwanger, R. S., and Vogt, T. F. (2000a). Fringe is a glycosyltransferase that modifies Notch. *Nature* **406**, 369–375.

Moloney, D. J., Shair, L. H., Lu, F. M., Xia, J., Locke, R., Matta, K. L., and Haltiwanger, R. S. (2000b). Mammalian Notch1 is modified with two unusual forms of O-linked glycosylation found on epidermal growth factor-like modules. *J. Biol. Chem.* **275**, 9604–9611.

Morales, A. V., Yasuda, Y., and Ish-Horowicz, D. (2002). Periodic Lunatic fringe expression is controlled during segmentation by a cyclic transcriptional enhancer responsive to notch signaling. *Dev. Cell* **3**, 63–74.

Moran, J. L., Shifley, E. T., Levorse, J. M., Mani, S., Ostmann, K., Perez-Balaguer, A., Walker, D. M., Vogt, T. F., and Cole, S. E. (2009). Manic fringe is not required for embryonic development, and fringe family members do not exhibit redundant functions in the axial skeleton, limb, or hindbrain. *Dev. Dyn.* **238**, 1803–1812.

Muller, R., Altmann, F., Zhou, D., and Hennet, T. (2002). The Drosophila melanogaster brainiac protein is a glycolipid-specific beta 1,3N-acetylglucosaminyltransferase. *J. Biol. Chem.* **277**, 32417–32420.

Ohtsubo, K., Takamatsu, S., Minowa, M. T., Yoshida, A., Takeuchi, M., and Marth, J. D. (2005). Dietary and genetic control of glucose transporter 2 glycosylation promotes insulin secretion in suppressing diabetes. *Cell* **123**, 1307–1321.

Okajima, T., and Irvine, K. D. (2002). Regulation of Notch Signaling by O-Linked Fucose. *Cell* **111**, 893–904.

Okajima, T., Reddy, B., Matsuda, T., and Irvine, K. D. (2008). Contributions of chaperone and glycosyltransferase activities of O-fucosyltransferase 1 to Notch signaling. *BMC Biol.* **6**, 1.

Okajima, T., Xu, A., and Irvine, K. D. (2003). Modulation of notch-ligand binding by protein O-fucosyltransferase 1 and fringe. *J. Biol. Chem.* **278**, 42340–42345.

Okajima, T., Xu, A., Lei, L., and Irvine, K. D. (2005). Chaperone Activity of Protein O-Fucosyltransferase 1 Promotes Notch Receptor Folding. *Science* **307**, 1599–1603.

Okamura, Y., and Saga, Y. (2008). Pofut1 is required for the proper localization of the Notch receptor during mouse development. *Mech. Dev.* **125**, 663–673.

Panin, V. M., Papayannopoulos, V., Wilson and R., Irvine, K. D. (2002). Fringe modulates Notch-ligand interactions. *Nature* **387**, 908–912.

Panin, V. M., Shao, L., Lei, L., Moloney, D. J., Irvine, K. D., and Haltiwanger, R. S. (2002). Notch ligands are substrates for protein O-fucosyltransferase-1 and Fringe. *J. Biol. Chem.* **277**, 29945–29952.

Partridge, E. A., Le Roy, C., Di Guglielmo, G. M., Pawling, J., Cheung, P., Granovsky, M., Nabi, I. R., Wrana, J. L., and Dennis, J. W. (2004). Regulation of cytokine receptors by Golgi N-glycan processing and endocytosis. *Science* **306**, 120–124.

Pei, Z., and Baker, N. E. (2008). Competition between Delta and the Abruptex domain of Notch. *BMC Dev. Biol.* **8**, 4.

Peterson, R. E., and McClay, D. R. (2005). A Fringe-modified Notch signal affects specification of mesoderm and endoderm in the sea urchin embryo. *Dev. Biol.* **282**, 126–137.

Pizette, S., Rabouille, C., Cohen, S. M., and Therond, P. (2009). Glycosphingolipids control the extracellular gradient of the Drosophila EGFR ligand Gurken. *Development* **136**, 551–561.

Prince, V. E., Holley, S. A., Bally-Cuif, L., Prabhakaran, B., Oates, A. C., Ho, R. K., and Vogt, T. F. (2001). Zebrafish lunatic fringe demarcates segmental boundaries. *Mech. Dev.* **105**, 175–180.

Qiu, X., Xu, H., Haddon, C., Lewis, J., and Jiang, Y. J. (2004). Sequence and embryonic expression of three zebrafish fringe genes: lunatic fringe, radical fringe, and manic fringe. *Dev. Dyn.* **231**, 621–630.

Rampal, R., Arboleda-Velasquez, J., Nita-Lazar, A., Kosik, K. S., and Haltiwanger, R. S. (2005a). Highly conserved O-fucose sites have distinct effects on notch1 function. *J. Biol. Chem.* **280**, 32133–32140.

Rampal, R., Li, A. S., Moloney, D. J., Georgiou, S. A., Luther, K. B., Nita-Lazar, A., and Haltiwanger, R. S. (2005b). Lunatic fringe, manic fringe, and radical fringe recognize similar specificity determinants in O-fucosylated epidermal growth factor-like repeats. *J. Biol. Chem.* **280**, 42454–42463.

Rampal, R., Luther, K. B., and Haltiwanger, R. S. (2007). Notch signaling in normal and disease States: possible therapies related to glycosylation. *Curr. Mol. Med.* **7**, 427–445.

Rauskolb, C., and Irvine, K. D. (1999). Notch-mediated segmentation and growth control of the Drosophila leg. *Dev. Biol.* **210**, 339–350.

Raya, A., Kawakami, Y., Rodriguez-Esteban, C., Buscher, D., Koth, C. M., Itoh, T., Morita, M., Raya, R. M., Dubova, I., Bessa, J. G., de la Pompa, J. L., and Izpisua Belmonte, J. C. (2003). Notch activity induces Nodal expression and mediates the establishment of left-right asymmetry in vertebrate embryos. *Genes Dev.* **17**, 1213–1218.

Rebay, I., Fehon, R. G., and Artavanis-Tsakonas, S. (1993). Specific truncations of Drosophila Notch define dominant activated and dominant negative forms of the receptor. *Cell* **74**, 319–329.

Repnikova, E., Koles, K., Nakamura, M., Pitts, J., Li, H., Ambavane, A., Zoran, M. J., and Panin, V. M. (2010). Sialyltransferase regulates nervous system function in Drosophila. *J Neurosci* **30**, 6466–6476.

Ryan, M. J., Bales, C., Nelson, A., Gonzalez, D. M., Underkoffler, L., Segalov, M., Wilson-Rawls, J., Cole, S. E., Moran, J. L., Russo, P., Spinner, N. B., Kusumi, K., and Loomes, K. M. (2008). Bile duct proliferation in Jag1/fringe heterozygous mice identifies candidate modifiers of the Alagille syndrome hepatic phenotype. *Hepatology* **48**, 1989–1997.

Sakamoto, K., Yan, L., Imai, H., Takagi, M., Nabeshima, Y., Takeda, S., and Katsube, K. (1997). Identification of a chick homologue of Fringe and C-Fringe 1: involvement in the neurogenesis and the somitogenesis. *Biochem. Biophys. Res. Commun.* **234**, 754–759.

Sasaki, N., Sasamura, T., Ishikawa, H. O., Kanai, M., Ueda, R., Saigo, K., and Matsuno, K. (2007). Polarized exocytosis and transcytosis of Notch during its apical localization in Drosophila epithelial cells. *Genes Cells* **12**, 89–103.

Sasamura, T., Ishikawa, H. O., Sasaki, N., Higashi, S., Kanai, M., Nakao, S., Ayukawa, T., Aigaki, T., Noda, K., Miyoshi, E., Taniguchi, N., and Matsuno, K. (2007). The O-fucosyltransferase O-fut1 is an extracellular component that is essential for the constitutive endocytic trafficking of Notch in Drosophila. *Development* **134**, 1347–1356.

Sasamura, T., Sasaki, N., Miyashita, F., Nakao, S., Ishikawa, H. O., Ito, M., Kitagawa, M., Harigaya, K., Spana, E., Bilder, D., Perrimon, N., and Matsuno, K. (2003). neurotic, a novel maternal neurogenic gene, encodes an O-fucosyltransferase that is essential for Notch-Delta interactions. *Development* **130**, 4785–4795.

Schiffer, S. G., Foley, S., Kaffashan, A., Hronowski, X., Zichittella, A. E., Yeo, C. Y., Miatkowski, K., Adkins, H. B., Damon, B., Whitman, M., Salomon, D., Sanicola, M., and Williams, K. P. (2001). Fucosylation of Cripto is required for its ability to facilitate nodal signaling. *J. Biol. Chem.* **276**, 37769–37778.

Schuster-Gossler, K., Harris, B., Johnson, K. R., Serth, J., and Gossler, A. (2009). Notch signalling in the paraxial mesoderm is most sensitive to reduced Pofut1 levels during early mouse development. *BMC Dev. Biol.* **9**, 6.

Schwientek, T., Keck, B., Levery, S. B., Jensen, M. A., Pedersen, J. W., Wandall, H. H., Stroud, M., Cohen, S. M., Amado, M., and Clausen, H. (2002). The Drosophila gene brainiac encodes a glycosyltransferase putatively involved in glycosphingolipid synthesis. *J. Biol. Chem.* **277**, 32421–32429.

Selva, E. M., Hong, K., Baeg, G. H., Beverley, S. M., Turco, S. J., Perrimon, N., and Hacker, U. (2001). Dual role of the fringe connection gene in both heparan sulphate and fringe-dependent signalling events. *Nat. Cell Biol.* **3**, 809–815.

Serth, K., Schuster-Gossler, K., Cordes, R., and Gossler, A. (2003). Transcriptional oscillation of lunatic fringe is essential for somitogenesis. *Genes Dev.* **17**, 912–925.

Sethi, M. K., Buettner, F. F., Krylov, V. B., Takeuchi, H., Nifantiev, N. E., Haltiwanger, R. S., Gerardy-Schahn, R., and Bakker, H. (2010). Identification of glycosyltransferase 8 family members as xylosyltransferases acting on o-glucosylated notch epidermal growth factor repeats. *J. Biol. Chem.* **285**, 1582–1586.

Shao, L., Moloney, D. J., and Haltiwanger, R. (2003). Fringe modifies O-fucose on mouse Notch1 at epidermal growth factor-like repeats within the ligand-binding site and the Abruptex region. *J. Biol. Chem.* **278**, 7775–7782.

Shi, S., Ge, C., Luo, Y., Hou, X., Haltiwanger, R. S., and Stanley, P. (2007). The threonine that carries fucose, but not fucose, is required for Cripto to facilitate Nodal signaling. *J. Biol. Chem.* **282**, 20133–20141.

Shi, S., and Stanley, P. (2003). Protein O-fucosyltransferase 1 is an essential component of Notch signaling pathways. *Proc. Natl. Acad. Sci. U.S.A.* **100**, 5234–5239.

Shifley, E. T., and Cole, S. E. (2008). Lunatic fringe protein processing by proprotein convertases may contribute to the short protein half-life in the segmentation clock. *Biochim. Biophys. Acta* **1783**, 2384–2390.

Shifley, E. T., Vanhorn, K. M., Perez-Balaguer, A., Franklin, J. D., Weinstein, M., and Cole, S. E. (2008). Oscillatory lunatic fringe activity is crucial for segmentation of the anterior but not posterior skeleton. *Development* **135**, 899–908.

Shimizu, K., Chiba, S., Kumano, K., Hosoya, N., Takahashi, T., Kanda, Y., Hamada, Y., Yazaki, Y., and Hirai, H. (1999). Mouse jagged1 physically interacts with notch2 and other notch receptors. Assessment by quantitative methods. *J. Biol. Chem.* **274**, 32961–32969.

Shimizu, K., Chiba, S., Saito, T., Kumano, K., and Hirai, H. (2000). Physical interaction of Delta1, Jagged1, and Jagged2 with Notch1 and Notch3 receptors. *Biochem. Biophys. Res. Commun.* **276**, 385–389.

Shimizu, K., Chiba, S., Saito, T., Kumano, K., Takahashi, T., and Hirai, H. (2001). Manic fringe and lunatic fringe modify different sites of the Notch2 extracellular region, resulting in different signaling modulation. *J. Biol. Chem.* **276**, 25753–25758.

Smith, P. L., Myers, J. T., Rogers, C. E., Zhou, L., Petryniak, B., Becker, D. J., Homeister, J. W., and Lowe, J. B. (2002). Conditional control of selectin ligand expression and global fucosylation events in mice with a targeted mutation at the FX locus. *J. Cell Biol.* **158**, 801–815.

Sparrow, D. B., Chapman, G., Wouters, M. A., Whittock, N. V., Ellard, S., Fatkin, D., Turnpenny, P. D., Kusumi, K., Sillence, D., and Dunwoodie, S. L. (2006). Mutation of the LUNATIC FRINGE Gene in Humans Causes Spondylocostal Dysostosis with a Severe Vertebral Phenotype. *Am. J. Hum. Genet.* **78**, 28–37.

Stahl, M., Uemura, K., Ge, C., Shi, S., Tashima, Y., and Stanley, P. (2008). Roles of Pofut1 and O-fucose in mammalian Notch signaling. *J. Biol. Chem.* **283**, 13638–13651.

Stanley, P., and Guidos, C. J. (2009). Regulation of Notch signaling during T- and B-cell development by O-fucose glycans. *Immunol. Rev.* **230**, 201–215.

Stanley, P., Schachter, H., and Taniguchi, N. (2009). N-glycans. Chapter 8. *In* "Essentials of Glycobiology" (A. Varki, R. D. Cummings, J. D. Esko, H. H. Freeze, P. Stanley, C. R. Bertozzi, G. W. Hart and M. E. Etlzler, eds.). *Cold Spring Harbor laboratory Press, New York.*

Stauber, M., Sachidanandan, C., Morgenstern, C., and Ish-Horowicz, D. (2009). Differential axial requirements for lunatic fringe and Hes7 transcription during mouse somitogenesis. *PLoS ONE* **4**, e7996.

Stolz, A., Haines, N., Pich, A., Irvine, K. D., Hokke, C. H., Deelder, A. M., Gerardy-Schahn, R., Wuhrer, M., and Bakker, H. (2008). Distinct contributions of beta 4GalNAcTA and beta 4GalNAcTB to Drosophila glycosphingolipid biosynthesis. *Glycoconj. J.* **25**, 167–175.

Sturla, L., Rampal, R., Haltiwanger, R. S., Fruscione, F., Etzioni, A., and Tonetti, M. (2003). Differential terminal fucosylation of N-linked glycans versus protein O-fucosylation in leukocyte adhesion deficiency type II (CDG IIc). *J. Biol. Chem.* **278**, 26727–26733.

Suda, T., Kamiyama, S., Suzuki, M., Kikuchi, N., Nakayama, K., Narimatsu, H., Jigami, Y., Aoki, T., and Nishihara, S. (2004). Molecular cloning and characterization of a human multisubstrate specific nucleotide-sugar transporter homologous to Drosophila fringe connection. *J. Biol. Chem.* **279**, 26469–26474.

Tan, J. B., Xu, K., Cretegny, K., Visan, I., Yuan, J. S., Egan, S. E., and Guidos, C. J. (2009). Lunatic and Manic Fringe Cooperatively Enhance Marginal Zone B Cell Precursor Competition for Delta-like 1 in Splenic Endothelial Niches. *Immunity.* **30**, 254–263.

Ten Hagen, K. G., Fritz, T. A., and Tabak, L. A. (2003). All in the family: the UDP-GalNAc:polypeptide N-acetylgalactosaminyltransferases. *Glycobiology* **13**(1), R–16R.

Thierry-Mieg, D., and Thierry-Mieg, J. (2006). AceView: a comprehensive cDNA-supported gene and transcripts annotation. *Genome Biol.* **7 Suppl. 1**, S12.1–14.

Thomsen, B., Horn, P., Panitz, F., Bendixen, E., Petersen, A. H., Holm, L. E., Nielsen, V. H., Agerholm, J. S., Arnbjerg, J., and Bendixen, C. (2006). A missense mutation in the bovine SLC35A3 gene, encoding a UDP-N-acetylglucosamine transporter, causes complex vertebral malformation. *Genome Res.* **16**, 97–105.

Visan, I., Tan, J. B., Yuan, J. S., Harper, J. A., Koch, U., and Guidos, C. J. (2006a). Regulation of T lymphopoiesis by Notch1 and Lunatic fringe-mediated competition for intrathymic niches. *Nat. Immunol.* **7**, 634–643.

Visan, I., Yuan, J. S., Tan, J. B., Cretegny, K., and Guidos, C. J. (2006b). Regulation of intrathymic T-cell development by Lunatic Fringe- Notch1 interactions. *Immunol. Rev.* **209**, 76–94.

Wandall, H. H., Pedersen, J. W., Park, C., Levery, S. B., Pizette, S., Cohen, S. M., Schwientek, T., and Clausen, H. (2003). Drosophila egghead encodes a beta 1,4-mannosyltransferase predicted to form the immediate precursor glycosphingolipid substrate for brainiac. *J. Biol. Chem.* **278**, 1411–1414.

Wandall, H. H., Pizette, S., Pedersen, J. W., Eichert, H., Levery, S. B., Mandel, U., Cohen, S. M., and Clausen, H. (2005). Egghead and brainiac are essential for glycosphingolipid biosynthesis in vivo. *J. Biol. Chem.* **280**, 4858–4863.

Wang, Y., Shao, L., Shi, S., Harris, R. J., Spellman, M. W., Stanley, P., and Haltiwanger, R. S. (2001). Modification of epidermal growth factor-like repeats with O-fucose.

Molecular cloning and expression of a novel GDP-fucose protein O- fucosyltransferase. *J. Biol. Chem.* **276**, 40338–40345.
Waterhouse, C., Johnson, S., Phillipson, M., Zbytnuik, L., Petri, B., Kelly, M., Lowe, J., and Kubes, P. (2010). Secretory Cell Hyperplasia and Defects in Notch Activity in a Mouse Model of Leukocyte Adhesion Deficiency Type II. *Gastroenterology*. **138**, 1079–1090.
Whelan, S. A., and Hart, G. W. (2006). Identification of O-GlcNAc sites on proteins. *Meth. Enzymol.* **415**, 113–133.
Wu, J. Y., Wen, L., Zhang, W. J., and Rao, Y. (1996). The secreted product of Xenopus gene lunatic Fringe, a vertebrate signaling molecule. *Science* **273**, 355–358.
Xia, L., Ju, T., Westmuckett, A., An, G., Ivanciu, L., McDaniel, J. M., Lupu, F., Cummings, R. D., and McEver, R. P. (2004). Defective angiogenesis and fatal embryonic hemorrhage in mice lacking core 1-derived O-glycans. *J. Cell Biol.* **164**, 451–459.
Xu, A., Haines, N., Dlugosz, M., Rana, N. A., Takeuchi, H., Haltiwanger, R. S., and Irvine, K. D. (2007). In vitro reconstitution of the modulation of Drosophila Notch-ligand binding by Fringe. *J. Biol. Chem.* **282**, 35153–35162.
Xu, A., Lei, L., and Irvine, K. D. (2005). Regions of Drosophila notch that contribute to ligand binding and the modulatory influence of fringe. *J. Biol. Chem.* **280**, 30158–30165.
Yakubenia, S., Frommhold, D., Scholch, D., Hellbusch, C. C., Korner, C., Petri, B., Jones, C., Ipe, U., Bixel, M. G., Krempien, R., Sperandio, M., and Wild, M. K. (2008). Leukocyte trafficking in a mouse model for leukocyte adhesion deficiency II/congenital disorder of glycosylation IIc. *Blood* **112**, 1472–1481.
Yang, L. T., Nichols, J. T., Yao, C., Manilay, J. O., Robey, E. A., and Weinmaster, G. (2005). Fringe glycosyltransferases differentially modulate Notch1 proteolysis induced by Delta1 and Jagged1. *Mol. Biol. Cell* **16**, 927–942.
Yuan, Y. P., Schultz, J., Mlodzik, M., and Bork, P. (1997). Secreted fringe-like signaling molecules may be glycosyltransferases [letter]. *Cell* **88**, 9–11.
Zhang, N., and Gridley, T. (1998). Defects in somite formation in lunatic fringe-deficient mice. *Nature* **394**, 374–377.
Zhang, N., Norton, C. R., and Gridley, T. (2002). Segmentation defects of Notch pathway mutants and absence of a synergistic phenotype in lunatic fringe/radical fringe double mutant mice. *Genesis* **33**, 21–28.
Zhou, L., Li, L. W., Yan, Q., Petryniak, B., Man, Y., Su, C., Shim, J., Chervin, S., and Lowe, J. B. (2008). Notch-dependent control of myelopoiesis is regulated by fucosylation. *Blood* **112**, 308–319.

CHAPTER FIVE

Endocytosis and Intracellular Trafficking of Notch and Its Ligands

Shinya Yamamoto,[*,1] Wu-Lin Charng,[*,1] *and* Hugo J. Bellen[*,†,‡,§]

Contents

* Program in Developmental Biology, Baylor College of Medicine, Houston, TX, USA
† Department of Molecular and Human Genetics, Baylor College of Medicine, Houston, TX, USA
‡ Department of Neuroscience, Baylor College of Medicine, Houston, TX, USA
§ Howard Hughes Medical Institute, Baylor College of Medicine, Houston, TX, USA
1 Shinya Yamamoto and Wu-Lin Charng have contributed equally to the manuscript.

Current Topics in Developmental Biology, Volume 92
ISSN 0070-2153, DOI 10.1016/S0070-2153(10)92005-X

Abstract

Notch signaling occurs through direct interaction between Notch, the receptor, and its ligands, presented on the surface of neighboring cells. Endocytosis has been shown to be essential for Notch signal activation in both signal-sending and signal-receiving cells, and numerous genes involved in vesicle trafficking have recently been shown to act as key regulators of the pathway. Defects in vesicle trafficking can lead to gain- or loss-of-function defects in a context-dependent manner. Here, we discuss how endocytosis and vesicle trafficking regulate Notch signaling in both signal-sending and signal-receiving cells. We will introduce the key players in different trafficking steps, and further illustrate how they impact the signal outcome. Some of these players act as general factors and modulate Notch signaling in all contexts, whereas others modulate signaling in a context-specific fashion. We also discuss Notch signaling during mechanosensory organ development in the fly to exemplify how endocytosis and vesicle trafficking are effectively used to determine correct cell fates. In summary, endocytosis plays an essential role in Notch signaling, whereas intracellular vesicle trafficking often plays a context-dependent or regulatory role, leading to divergent outcomes in different developmental contexts.

1. Notch Signaling and its Regulation by Endocytosis and Vesicle Trafficking

1.1. Introduction

Notch signaling is an evolutionally conserved signaling pathway which takes place between neighboring cells. When Notch receptors are activated by DSL (Delta/Serrate/LAG-2) ligands, Notch undergoes a set of serial proteolytic cleavages resulting in the release of the Notch intracellular domain (NICD). NICD translocates into the nucleus to form a positive transcriptional complex with a key transcription factor CSL for CBF-1/Su(H)/LAG-1 (C-promoter binding factor-1/Suppressor of Hairless/Lin-12-and-GLP-1) and a coactivator, Mastermind (Kopan and Ilagan, 2009). This CSL-dependent process is referred to as canonical Notch signaling. It has also been shown that in certain contexts, Notch signaling activity can be mediated through a CSL-independent pathway, which is usually referred to as noncanonical Notch signaling (Ligoxygakis *et al.*, 1998; Ordentlich *et al.*, 1998; Ramain *et al.*, 2001; Zecchini *et al.*, 1999). Since both ligands and receptors are transmembrane proteins, endocytosis and vesicle trafficking play a critical role in the regulation of this signaling pathway.

1.2. Intracellular trafficking of Notch and DSL ligands

Notch receptors and DSL ligands are produced in the endoplasmic reticulum (ER) and traffic through the Golgi apparatus to reach the plasma membrane (Fig. 5.1). From the cell surface, they re-enter the cell via endocytosis, a process by which vesicles invaginate from the plasma membrane into the cytoplasm. These endocytic vesicles typically fuse with an early endosome, a sorting center of the endocytic pathway, often referred to as the "sorting endosome." From this early/sorting endosomes, proteins can be recycled back to the plasma membrane, transported to the Golgi apparatus, or transported to the late endosome, which eventually fuses with the lysosome for protein degradation (Doherty and McMahon, 2009). In the past, endocytosis was considered to only play a negative role in signaling pathways by removing receptors from the membrane. However, more and more evidence suggests that endocytosis also plays a positive role. Signaling may occur not only at the cell membrane but also in endocytosed vesicles or endosomes. Indeed, numerous signaling pathways, including Notch signaling, have been shown to depend on endocytosis for their full activation (Sorkin and von Zastrow, 2009).

1.3. Endocytosis is essential for Notch signaling

Endocytosis and endosomal trafficking have been shown to play an important role in the activation and regulation of Notch signaling. The first hint came from the phenotype associated with the *Drosophila shibire (shi)* mutant. *shi* was initially identified as a temperature-sensitive mutation that leads to embryonic lethality at restrictive temperatures (Poodry *et al.*, 1973). The gene was later shown to encode dynamin, a GTPase essential for most, if not all, forms of endocytosis (Chen *et al.*, 1991; van der Bliek and Meyerowitz, 1991). Interestingly, shi^{ts1} embryos, raised at the restrictive temperature during neuroblasts segregation, contain excessive neuroblasts and neurons (Poodry, 1990), a neurogenic phenotype that resembles the loss of Notch phenotype (Poulson, 1937). Further studies based on clonal analysis and genetic interaction assays provided the first evidence that endocytosis is required for ligand-dependent Notch activation in both signal-sending and signal-receiving cells (Seugnet *et al.*, 1997).

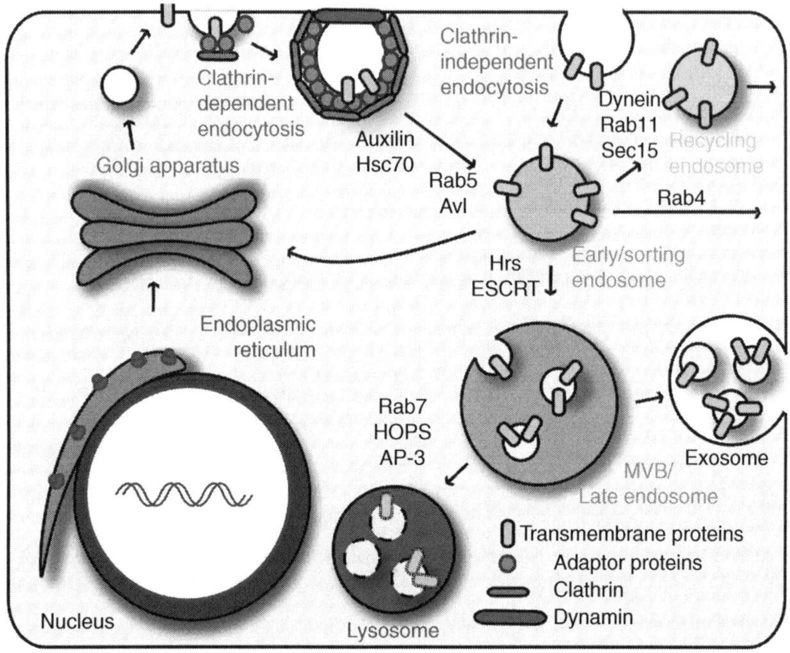

Figure 5.1 Overview of endocytosis and vesicle trafficking. Transmembrane proteins are made in the ER and traffic through the Golgi apparatus to reach the plasma membrane. From the cell surface, these proteins can re-enter the cell via various endocytosis pathways. Clathrin-dependent endocytosis is usually referred to as "canonical endocytosis." Clathrin adaptor proteins, such as the AP-2 complex, recruit clathrin and cargo transmembrane proteins to the site of endocytosis. The clathrin-coated endocytic vesicle is pinched off by the action of dynamin GTPase, and the clathrin coat is then removed by molecular chaperone Hsc70 via the assistance of auxilin. On the other hand, endocytosis can also occur without clathrin and is referred to as "noncanonical endocytosis" or "clathrin-independent endocytosis." After endocytosis, small GTPase Rab5 and SNARE protein Avalanche (Avl) mediate the fusion of endocytic vesicles with the early/sorting endosome. From the early endosome, endocytosed proteins can recycle back to the plasma membrane directly in a Rab4-dependent manner or indirectly through the recycling endosome in a Rab11-dependent manner. Alternatively, they can return back to the Golgi or travel to the late endosome and lysosome for degradation. Proteins destined for degradation are sorted into Rab7-positive late endosome or multivesicular bodies (MVB). Packaging of transmembrane proteins into intraluminal vesicles is mediated by the ESCRT complexes. In certain cell contexts, MVB can secrete their contents to extracellular regions. These secreted MVBs are referred to as exosomes. Finally, through HOPS and AP3 complexes, MVB/late endosomes fuse with the lysosome and transmembrane proteins are degraded by proteases and acid hydrolases. (See Color Insert.)

Based on these pioneering studies, various labs have focused on understanding how endocytosis regulates Notch signaling through forward and reverse genetic approaches. First, we will briefly review key steps in endocytosis and the molecular players that have been shown to affect

Notch signaling. Specific players that seemingly only affect endocytosis of Notch signaling components in a cell context-dependent manner will be discussed later.

1.4. Proteins and molecules involved in endocytosis

Canonical endocytosis requires the assembly of a clathrin lattice to form a clathrin-coated pit, which is then pinched off by the action of a GTPase, dynamin (Seugnet *et al.*, 1997; Traub, 2009). Clathrin is composed of heavy and light chains which form a triskelion upon multimerization. Clathrin is recruited to the site of endocytosis in the membrane through adaptor proteins, including the assembly protein-2 (AP-2) complex (Berdnik *et al.*, 2002). These and other adaptor proteins bind to transmembrane proteins that are targeted for endocytosis and recruited into clathrin-coated pits. The lipid composition of the plasma membrane also plays an important role in endocytosis. For example, phosphatidylinositol (4,5) diphosphate (PI(4,5) P2) is enriched in the plasma membrane at sites where endocytosis occurs, and the recruitment of many adaptor proteins depends on their binding to this lipid (Di Paolo and De Camilli, 2006; Poccia and Larijani, 2009).

A key signal to promote endocytosis of transmembrane proteins relies on the monoubiquitination of intracellular lysine residues by E3 ubiquitin ligases. The ubiquitin tag can promote the interaction with adaptor proteins and lead to recruitment and enrichment into clathrin-coated pits (d'Azzo *et al.*, 2005). Ubiquitinated proteins can be recognized by proteins that contain ubiquitin interaction motifs. Upon invagination and pinching off, vesicles are stripped of their clathrin coat by molecular chaperones such as Hsc70 with the assistance of auxilin (Eisenberg and Greene, 2007; Eun *et al.*, 2008; Hagedorn *et al.*, 2006).

Alternatively, endocytosis can also occur without the assembly of clathrin-coated pits, a process often referred to as noncanonical endocytosis or clathrin-independent endocytosis (Doherty and McMahon, 2009; Hansen and Nichols, 2009). However, compared to the well-established role of clathrin-dependent endocytosis in signaling pathways, its involvement in signal regulation is poorly understood.

1.5. Proteins involved in endocytic trafficking, sorting, recycling, and degradation

Upon the uncoating of internalized vesicles, the small GTPase Rab5 and the SNARE (Soluble N-Ethylmaleimide-Sensitive Factor Adaptor Protein Receptor) protein syntaxin 7 mediate fusion of the endocytosed vesicles with the early endosome (Lu and Bilder, 2005; Vaccari *et al.*, 2008). From the early endosome, endocytosed proteins can either be recycled to the plasma membrane, return to the Golgi and ER, or travel to the late endosome and lysosome for degradation. Some proteins can recycle to the plasma membrane directly from the early endosome in a GTPase Rab4-dependent manner, whereas most proteins enter the recycling endosomes prior to returning to the cell surface (Grant and Donaldson, 2009). The latter slower recycling process depends on the function of Rab11 and the exocyst complex, a multiprotein complex including Sec15 (Emery *et al.*, 2005; Jafar-Nejad *et al.*, 2005). In addition, some internalized proteins can travel to the Golgi apparatus and further to the ER with the assistance of the retromer complex, but the role of this trafficking route in Notch signaling has not been investigated. Proteins destined for degradation are sorted into Rab7-positive late endosomal compartments. During the transition between the early and the late endosome, transmembrane proteins are packaged into intraluminal vesicles also called multivesicular bodies (MVB) (Doherty and McMahon, 2009).

Sorting of cargos into intraluminal vesicles is mediated by the endosomal sorting complex required for transport (ESCRT) complexes. ESCRT-0 recognizes ubiquitinated receptors and recruits ESCRT-I, resulting in the activation of ESCRT-II, which assists the assembly of ESCRT-III (Herz and Bergmann, 2009; Raiborg and Stenmark, 2009). In certain cell contexts, MVB can be recycled to the plasma membrane and secrete their contents, the intraluminal vesicles (Simons and Raposo, 2009). These secreted MVBs, called exosomes, have been proposed to play a role in Notch signaling through secretion of active Delta (Chitnis, 2006; Le Borgne and Schweisguth, 2003a) but their *in vivo* role in Notch signaling awaits testing.

Finally, MVB/late endosomes fuse with the lysosome where the internalized cargos are degraded by proteases and acid hydrolases. The AP-3 complex is involved in endosomal trafficking to the lysosome, and the homotypic fusion and vacuole protein sorting (HOPS) complex is involved in the late endosome maturation/lysosomal fusion step (Dell'Angelica, 2009; Wilkin *et al.*, 2008).

Along the endocytic trafficking process, the luminal pH of endosomal compartments becomes gradually more acidic (Marshansky and Futai, 2008). The low pH assists in the dissociation of certain protein–protein interactions, as well as provides the optimal environment for enzymatic activity of certain proteases. Therefore, proteins involved in the acidification of endosomes can influence the strength of protein interactions and efficiency of protein cleavage/degradation. For example, vacuolar (H^+)-ATPase (V-ATPase), a proton transporter involved in the acidification of endosomal compartments, has been reported to influence the processing and activation of Notch receptors (Yan *et al.*, 2009).

Here, we will discuss how DSL ligands and Notch receptors functions are regulated/affected by endocytosis and intracellular vesicle trafficking.

It is important to note that endocytosis and vesicle trafficking play distinct functions in signal-sending and in signal-receiving cells, respectively. Since most of these studies used *Drosophila* as a model organism, we will mainly focus on the results from *Drosophila* and cover findings in other organisms where appropriate.

2. Ligand Endocytosis and Trafficking

2.1. The role of endocytosis of DSL ligands in the signal-sending cells

The DSL ligands are type-1 transmembrane proteins that contain a characteristic DSL domain at their N terminus followed by multiple epithelial growth factor-like repeats (EGF-r), a single transmembrane domain (TMD), and an intracellular domain (Kopan and Ilagan, 2009). These ligands can be subdivided into the Serrate/Jagged group ligands, which contain a cysteine-rich domain between the EGF-r and TMD, and Delta group ligands that lack this motif. The N terminus of DSL ligands including the DSL domain is required for ligand–receptor interaction and signaling activity (Glittenberg *et al.*, 2006; Henderson *et al.*, 1997; Parks *et al.*, 2006; Shimizu *et al.*, 1999). Early studies on the subcellular localization of Delta in *Drosophila* embryos and imaginal discs documented the presence of Delta in intracellular vesicles (Kooh *et al.*, 1993). Analysis of endocytic mutants such as *shi* (Parks *et al.*, 1995; Seugnet *et al.*, 1997) and *hook* (Kramer and Phistry, 1996, 1999) revealed that these vesicles are endocytic in nature, and a block in endocytosis of DSL ligands attenuated Notch signaling (Parks *et al.*, 2000; Seugnet *et al.*, 1997). Although many agree that endocytosis is essential for the activity of DSL ligands for canonical Notch signaling, the precise function of endocytosis is still debated. Here, we will first introduce the players in DSL ligand endocytosis and trafficking, and then discuss two nonmutually exclusive theories that have been proposed (Fig. 5.2).

2.2. The role of ubiquitin, E3 ligases, and ubiquitin interacting proteins in DSL ligand trafficking

In vivo structure function analysis of Delta using specific point mutant alleles showed that certain EGF-r as well as certain intracellular lysine residues are necessary for endocytosis and proper signaling (Parks *et al.*, 2006). Similar results have been obtained from structure function analysis of Serrate using an ectopic overexpression assay system *in vivo* (Glittenberg *et al.*, 2006). Lysine residues can be posttranslationally modified by ubiquitin, which serves as a signal for endocytosis, sorting, and/or degradation (Acconcia

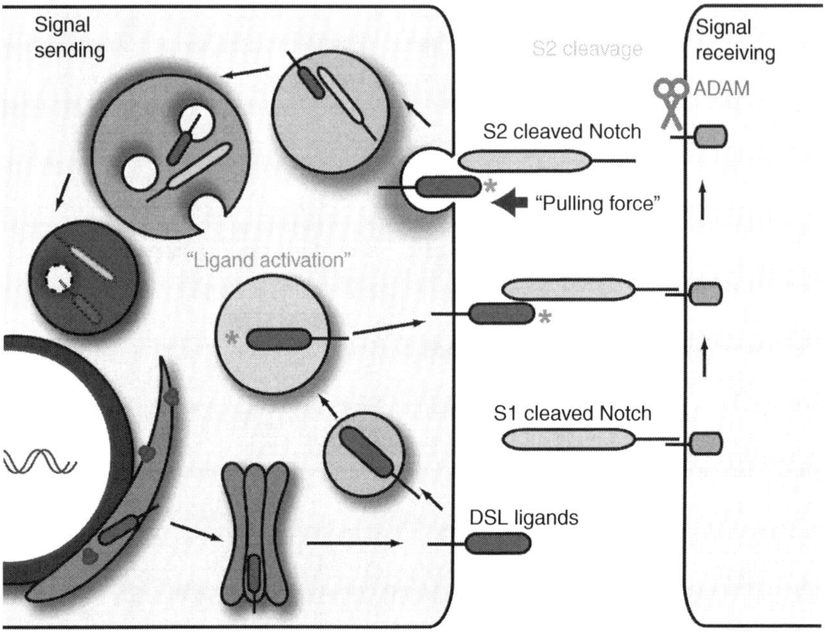

Figure 5.2 Endocytosis and trafficking of DSL ligands in the signal-sending cell. DLS ligands are synthesized in the ER and traffic to the cell surface through the secretory pathway. Endocytosis is required in the signal-sending cell for activation of the canonical Notch signaling pathway, but there are two nonmutually exclusive hypotheses to explain how endocytosis promotes DSL ligand activity. In the "Ligand Activation" theory, DSL proteins that have just been synthesized and reached the cell surface are still inactive and do not have the capacity to activate the Notch receptor on the signal-receiving cell. DSL are endocytosed and sorted into a unique endocytic compartment where they become "activated." The activated ligands return to the cell surface via the recycling pathway where they interact with and activate the Notch receptor. In contrast, the "pulling force" model insists that when DSL ligands and Notch receptor interact, endocytosis in the signal-sending cell generates a mechanical force that leads to a conformational change in the Notch receptor. This force mediates the separation of the Notch heterodimer and allows the S2 cleavage mediated by ADAM proteases. The extracellular portion of Notch is trans-endocytosed into the signal-sending cell and assumed to be degraded through the lysosomal pathway along with the DSL ligands. (See Color Insert.)

et al., 2009; Hicke and Dunn, 2003). Ubiquitination is mediated by E3 ligases which recognize their specific target proteins and recruit E2 ligases for transfer of ubiquitin on to the lysine residues. The *neuralized* (*neur*) gene, whose loss causes a neurogenic phenotype similar to *Notch* and *Delta* mutants (Lehmann *et al.*, 1981), encodes an E3 ligase with a C-terminal RING domain that is necessary for its E3 activity and two neuralized homology repeats (NHR1 and NHR2) (Deblandre *et al.*, 2001; Lai *et al.*, 2001; Pavlopoulos *et al.*, 2001; Yeh *et al.*, 2001). NRH1 has been shown to be necessary and sufficient for its interaction with Delta (Commisso and Boulianne, 2007). *Mind bomb* (*Mib*), first isolated in zebrafish to cause a neurogenic phenotype, encodes an E3 ligase with a C-terminal RING domain that ubiquitinates DSL ligands (Itoh *et al.*, 2003). Both *Neur* and *Mib* are conserved in *Drosophila* and vertebrates, and loss of function of either gene in *Drosophila* shows a loss of Notch signaling phenotype in a tissue-specific manner (Boulianne *et al.*, 1991; Lai *et al.*, 2005; Le Borgne *et al.*, 2005; Pitsouli and Delidakis, 2005; Wang and Struhl, 2005). These two E3 ligases are necessary for Notch signaling in the signal-sending cells, and loss of function leads to defects in endocytosis/sorting of DSL ligands. It has been suggested that Neur is likely to be involved in Delta-dependent signaling events, whereas Mib functions in Serrate/Jagged-dependent signaling events. However, the two proteins overlap in their functions, since they can rescue the loss-of-function phenotype of each other upon ectopic expression in a mutant background (Lai *et al.*, 2005; Le Borgne *et al.*, 2005; Pitsouli and Delidakis, 2005; Wang and Struhl, 2005). Both Neur and Mib are localized to the plasma membrane, where they can interact with DSL proteins. Neuralized can be recruited to the membrane via interaction with DSL ligands through its NHR1 domain (Commisso and Boulianne, 2007), via an interaction with phosphoinositides through its N-terminal polybasic domain (Skwarek *et al.*, 2007), and/or via N-myristoylation of the N-terminal glycine residue (Koutelou *et al.*, 2008). Notch signaling can be fine-tuned by regulating the activity of Neur by Bearded family proteins. Bearded family proteins, such as Bearded (Brd) and Twin of m4 (Tom), are negative regulators of Neur function, encoded by multiple genes clustered in the *Bearded* complex locus and *Enhancer of Split* complex locus (Bardin and Schweisguth, 2006; De Renzis *et al.*, 2006; Fontana and Posakony, 2009; Leviten *et al.*, 1997). Initially, gain-of-function mutations of Bearded were identified to cause loss of Notch signaling during lateral inhibition of mechanosensory organ precursors in *Drosophila* (Leviten and Posakony, 1996). Bearded family proteins posses Neur interaction motifs that allow them to bind to Neur and inhibit its function. However, in contrast to their gain-of-function phenotype, loss of function of all eight Bearded family genes show only partial Notch signaling defects during mechanosensory organ precursor specification and mesectoderm specification during embryogenesis, suggesting a context-specific role *in vivo* (Chanet *et al.*, 2009).

Ubiquitinated DSL ligands are potentially recognized by Epsin (encoded by the *liquid facet* gene in *Drosophila*). Epsin is a ubiquitin binding protein that interacts with PI(4,5)P2 as well as several endocytic proteins such as clathrin and AP-2 (Chen *et al.*, 1998; De Camilli *et al.*, 2002; Polo *et al.*, 2002). Loss of function of epsin shows a loss of Notch signaling phenotype in flies (Overstreet *et al.*, 2003, 2004; Tian *et al.*, 2004; Wang and Struhl, 2004) and in mice (Chen *et al.*, 2009). Epsin is required in the signal-sending cell, supporting the idea that epsin mediates the trafficking of ubiquitinated

DSL ligands. The activity of epsin is positively regulated by Fat facets (Faf), a de-ubiquitinating enzyme that stabilizes epsin. Loss of function of Faf causes defects similar to those observed on epsin mutants. However, some epsin-dependent Notch signaling events are Faf-independent, suggesting that epsin activity can be regulated by other factors than Faf (Overstreet *et al.*, 2004). How does epsin regulate Notch signaling? Several groups propose that epsin affects Notch signaling by promoting endocytosis of Delta. However, one group has observed that the bulk endocytosis of Delta is not affected in epsin mutants (Wang and Struhl, 2005), leading to an alternative hypothesis that epsin is required for sorting of internalized DSL ligands into a recycling pathway that leads to activation of Delta. Ubiquitinated DSL proteins have been proposed to undergo clathrin-dependent endocytosis, since mutations in auxilin, an adaptor molecule that recruits Hsc70 to clathrin-coated vesicles for uncoating, exhibit defects in DSL endocytosis (Eun *et al.*, 2008; Kandachar *et al.*, 2008). However, there is evidence indicating that DSL ligands are endocytosed through multiple distinct endocytic routes (Wang and Struhl, 2005), and epsin has been implicated in nonclathrin-mediated endocytosis (Sigismund *et al.*, 2005). A recent study supports this model based on the observation that loss of clathrin heavy chain in the signal-sending cell is capable of signaling during oogenesis. However, epsin is essential, suggesting that endocytosis of Delta by epsin is clathrin independent (Windler and Bilder, 2010). Since the DSL ligands are present in lipid raft compartments and cofractionate with caveolin (Heuss *et al.*, 2008), DSL ligand endocytosis may depend on a clathrin-independent lipid raft-mediated endocytic pathway for ligand activation. Further studies on endocytosis and trafficking of DSL ligands are needed to resolve these controversies.

2.3. Two theories on the function of DSL ligand endocytosis

DSL ligand endocytosis is necessary for canonical Notch signaling activation, and two models have been proposed to explain how DSL endocytosis leads to successful signal activation. One is the "ligand activation" model; the other is the "pulling force" hypothesis. It is important to note that these two models are not mutually exclusive.

2.3.1. The "ligand activation" theory

Bulk endocytosis of Delta is not affected in *epsin* mutants, yet Delta is unable to signal (Wang and Struhl, 2004). Based on this observation, it was proposed that newly synthesized Delta does not have the capacity to signal and that it needs to be endocytosed and sorted into a specialized endocytic compartment that depends on epsin and ubiquitination of Delta. Delta then traffics back to the cell surface through a recycling pathway. During this process, Delta is thought to become "activated" and acquires its signaling capability. Following this study, several mutants that exhibit Delta recycling defects and lead to Notch loss-of-function phenotypes were identified from forward genetic screens. Mutations in *sec15*, a component of the exocyst complex, cause a Notch signaling defect in the mechanosensory organ lineage in *Drosophila* (Jafar-Nejad *et al.*, 2005). In these mutants, Delta can be detected at the cell surface and can be internalized from the cell surface, suggesting that there is no defect in exo- or endocytosis. In wild-type cells, Delta enters recycling endosomes and returns to the plasma membrane, but this recycling fails to take place in *sec15* mutant cells. In addition, mutations in *Arp3*, a subunit of the Arp2/3 complex which regulates actin polymerization (Goley and Welch, 2006), cause very similar Delta recycling defects (Rajan *et al.*, 2009). Similarly, other proteins in Arp2/3 complex and WASp, an activator of the Arp2/3 complex, are required for Notch signaling (Ben-Yaacov *et al.*, 2001; Tal *et al.*, 2002). The activity of actin polymerization via the Arp2/3–WASp complex is therefore proposed to be critical for Delta recycling upon its internalization for successful canonical Notch signaling (see Section 4). Studies in mammalian cultured cells have shown that overexpression of dominant-negative forms of Rab11 in signal-sending cells leads to recycling defects and a less active signaling ability of Dll1 (Emery *et al.*, 2005), suggesting that the role of recycling pathway in DSL ligand activation may be evolutionally conserved. Further studies in a cell culture model support this ligand activation theory, and these authors also proposed the involvement of lipid rafts in the activation of Dll1 (Heuss *et al.*, 2008). However, the molecular mechanism of this mysterious "activation" remains to be identified. The activation has been proposed to consist of clustering of ligands, trafficking into lipid microdomains, proteolytic cleavage, or other posttranscriptional modification (Chitnis, 2006; Le Borgne and Schweisguth, 2003a; Wang and Struhl, 2004). In addition, although the role of the recycling pathway and actin polymerization in Delta activation have been well established in the cell fate determination during *Drosophila* mechanosensory lineage, in development of the wing margin, and in oogenesis (Ben-Yaacov *et al.*, 2001; Jafar-Nejad *et al.*, 2005; Rajan *et al.*, 2009; Tal *et al.*, 2002), this requirement may be context specific. *Sec15* mutant cells undergo normal photoreceptor development (Mehta *et al.*, 2005), a process in which Notch signaling is utilized reiteratively. Moreover, WASp mutant cells have been reported to not exhibit defects in lateral inhibition during mechanosensory organ development, another context where Notch signaling is required (Ben-Yaacov *et al.*, 2001). Furthermore, a recent study reports that during oogenesis, ligand internalization through dynamin in the signal-sending germ line cell is necessary but clathrin heavy chain, Rab5, Rab11, and Sec15 are dispensable in the germ line cell during Notch signaling in this context (Windler and Bilder, 2010). Therefore, Delta activation through endocytosis and trafficking through the recycling pathway may be required to further potentiate the activity of Delta in contexts where robust Notch signaling is required.

2.3.2. The "Pulling force" theory

Studies using *Drosophila* cell lines have indicated that Notch can be "trans-endocytosed" into the signal-sending cell. This trans-endocytosis can be inhibited upon blockage of dynamin in the signal-sending cell (Klueg and Muskavitch, 1999; Klueg *et al.*, 1998). Trans-endocytosis of Notch into Delta expressing cell has been proposed to take place *in vivo* as well, based on immunohistochemistry studies of *Drosophila* eye and wing vein tissue (Parks *et al.*, 2000). Initially, Delta had also been suggested to be trans-endocytosed into Notch expressing cells based on cell culture studies (Fehon *et al.*, 1990; Klueg *et al.*, 1998), but this has not been confirmed *in vivo*. In addition, early studies have suggested that full-length Notch is trans-endocytosed into Delta expressing cells (Klueg and Muskavitch, 1999; Klueg *et al.*, 1998), but later studies in *Drosophila* and in mammalian cultured cells have identified that it is only the extracellular portion of Notch that is trans-endocytosed and that the NICD remains in the signal-receiving cell (Nichols *et al.*, 2007; Parks *et al.*, 2000). Based on these observations, and together with the fact that endocytosis is necessary for activation of the Notch receptor, Parks *et al.* (2000) initially proposed that the interaction of DSL ligands with Notch and subsequent endocytosis of DSL ligands mediate some kind of a conformational change in Notch that leads to successful S2 cleavage via the ADAM (A Disintegrin And Metalloprotease) family proteases. This in turn would allow the extracellular portion of Notch to be trans-endocytosed. In mammalian cells, it has been shown that blocking S2 cleavage does not affect trans-endocytosis of Notch, suggesting that endocytosis of DSL ligands may generate a physical force to separate the Notch heterodimer that is linked together by non-covalent interactions within the extracellular heterodimerization domain (Nichols *et al.*, 2007). Indeed, this heterodimerization of Notch is mediated by furin-dependent S1 cleavage in the Golgi, after which Notch traffics to the cell surface for ligand-mediated activation (Logeat *et al.*, 1998). Together, endocytosis is proposed to generate a pulling force that separates the S1 cleaved heterodimer and as a consequence, the extracellular domain of Notch becomes trans-endocytosed into the signal-sending cell. The stretched Notch receptor becomes a substrate of ADAM-mediated S2 cleavage which leads to generation of Notch extracellular truncation (NEXT). Membrane attached NEXT then is cleaved by γ-secretase, termed S3 cleavage, to generate NICD. Furin-mediated S1 cleavage of Notch is still somewhat controversial in *Drosophila* as Kidd and Lieber (2002) have argued that furin cleavage is not required for Notch function and the majority of *Drosophila* Notch proteins do not undergo S1 cleavage. However, recent studies suggest that S1 cleavage plays a positive regulatory role, and they also indicate that the S1 cleavage that is mediated by a furin-like protease does take place in *Drosophila* (Lake *et al.*, 2009). Hence, the model proposed by Nichols *et al.* (2007) may be evolutionally conserved. In support of this model, structural studies using X-ray crystallography have determined that the S2 cleavage site of Notch is buried deep within the heterodimerization domain and protected by three LNR domains, suggesting that a physical pulling force is required to expose this site for interaction with the ADAM protease (Gordon *et al.*, 2007, 2008, 2009). In addition, using atomic force microscopy, the binding force between Notch and DSL ligands has been shown to be relatively strong (Ahimou *et al.*, 2004). Furthermore, work from various groups has shown that most secreted form of DSL ligands can interact with Notch but cannot activate Notch signaling. Rather they act in a dominant-negative fashion (Hukriede *et al.*, 1997; Sun and Artavanis-Tsakonas, 1997). However, when secreted DSL ligands are cross-linked, clustered or immobilized, they can activate Notch signaling in cultured cells (Morrison *et al.*, 2000; Varnum-Finney *et al.*, 2000), in support of the idea that tension and force generated between the Notch and DSL ligand complex is necessary and sufficient for Notch activation.

In summary, there are data that support both the "ligand activation" and the "pulling force" models. It is important to keep in mind that these two models are not mutually exclusive. In addition, there may be some context specificity for the requirement for the recycling pathway to modify Delta to make it a more potent ligand, together with the pulling force generated by ligand endocytosis to promote the conformational change in Notch in order to expose the S2 cleavage site.

3. Notch Receptor Endocytosis and Endosomal Trafficking

3.1. The role of endocytosis of the Notch receptor in signal-receiving cells

Endocytosis of the Notch receptor in signal-receiving cells plays both positive and negative roles in Notch signaling. As mentioned earlier, endocytosis is required for ligand-dependent Notch activation (canonical Notch signaling) in both signal-sending and receiving cells. However, the exact step in which endocytosis is required in the signal-receiving cells remains unclear and has been a topic of debate (Fig. 5.3). Moreover, Notch receptors that do not bind to DSL ligands, and remain inactive,

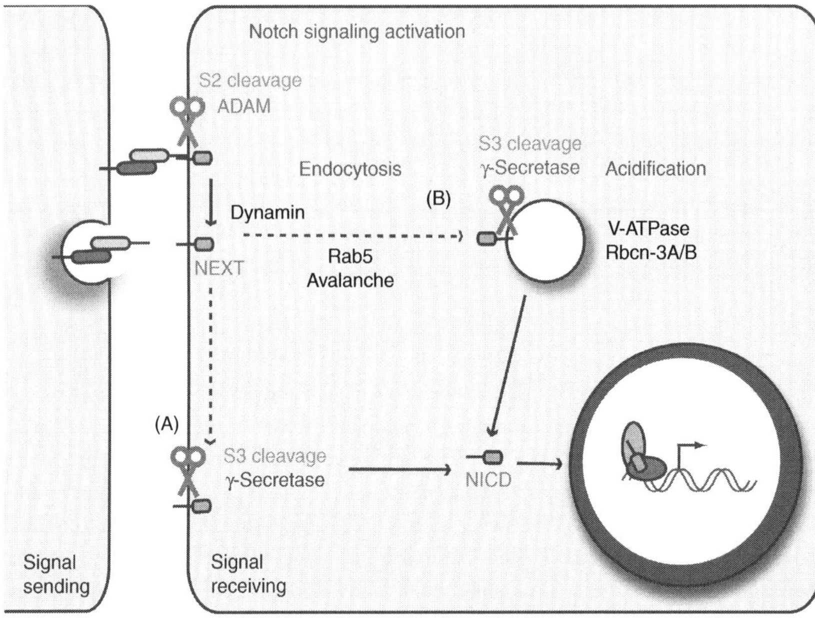

Figure 5.3 Endocytosis and trafficking of Notch during canonical signal activation. Canonical Notch signaling occurs in a cell–cell contact dependent manner, in which membrane-bounded DSL (Delta/Serrate/LAG-2) ligands activate Notch receptor on the neighboring cell. This interaction with ligands leads to a conformational change in Notch receptors and exposes the S2 cleavage site, which is cleaved by ADAM metalloprotease to produce Notch extracellular truncation (NEXT). NEXT then undergoes S3/S4 cleavage via γ-secretase to generate Notch intracellular domain (NICD). During this process, whether endocytosis is required for NEXT cleavage by γ-secretase is controversial. Two possible models are shown as dashed lines: (A) S3 cleavage takes place on the cell surface and does not require endocytosis. Instead, endocytosis might play a negative tuning role since γ-secretase prefers the generation of unstable form of NICD in endosomal compartments. (B) Endocytosis is required for S3 cleavage. Dynamin and two early endosomal proteins, Rab5 and Avl, are important in internalization and endocytic trafficking of NEXT fragment. Rbcn-3A, Rbcn-3B, and V-ATPase function in acidification of endosomal compartments where γ-secretase is more active and S3 cleavage is thus more efficient. (See Color Insert.)

are constitutively endocytosed and recycled to the cell surface (McGill *et al.*, 2009) or degraded in the lysosome (Jehn *et al.*, 2002) of cultured cells (Fig. 5.4). Recent data indicate that when endocytic trafficking of the Notch receptor destined for lysosomal degradation is disrupted, the Notch receptor can undergo proteolytic cleavage in a ligand-independent manner (Fortini, 2009; Fortini and Bilder, 2009; Furthauer and Gonzalez-Gaitan, 2009). This in turn can lead to ectopic activation of Notch signaling. As several components in this pathway have been associated with tumor progression, ligand-independent constitutive activation

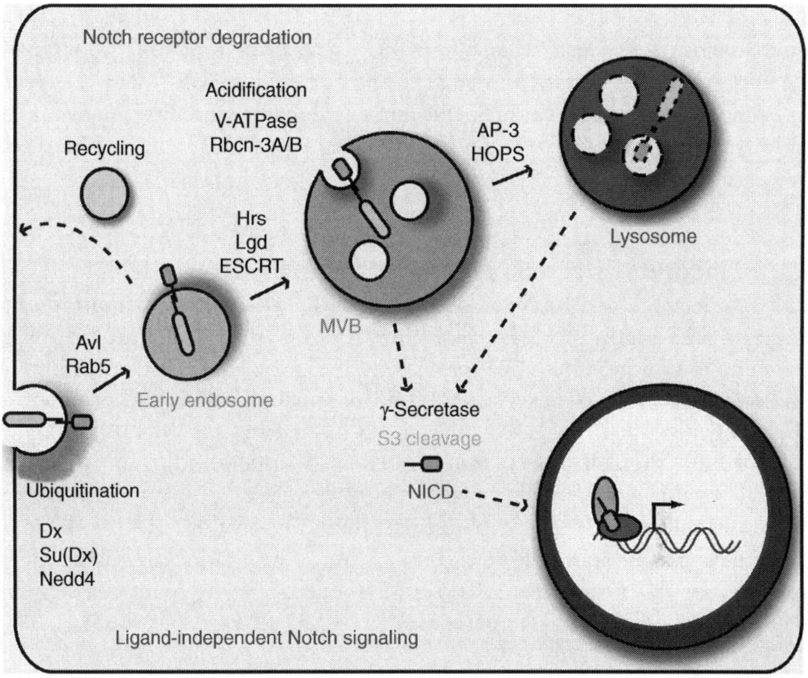

Figure 5.4 Endocytosis and trafficking of Notch receptor during lysosomal degradation and noncanonical activation. Inactive Notch receptors undergo constitutive endocytosis, endocytic trafficking, and are finally degraded in the lysosome. Full-length Notch receptors are firstly monoubiquitinated by E3 ligase Deltex, Su(dx), or DNedd4 for internalization. With the help of Rab5 and Avl, Notch receptors enter into the early endosome, from which they can recycle back to the cell surface or further progress into MVB/late endosomes. Sorting of Notch receptors into intraluminal vesicles is mediated by Hrs and ESCRT complexes (I, II, and III). Lgd, a C2-containing phospholipid binding protein, is placed between Hrs and ESCRT complexes based on epistatic analysis results. HOPS and AP-3 complexes are involved in endocytic trafficking/fusion between late endosome and lysosome. They are required for Dx-dependent Notch signaling activation. When endocytic trafficking of full-length Notch receptor for lysosomal degradation is disrupted, Notch receptor can undergo proteolytic cleavages in a ligand-independent manner (dashed line), which leads to ectopic activation of Notch signaling. (See Color Insert.)

of Notch signaling is proposed to be critical in certain cancers (Tanaka *et al.*, 2008).

3.2. The controversy on the requirement of endocytosis for S3 cleavage

Proteins involved in early steps of endocytosis, including dynamin, Rab5, and Avl, are required in the signal-receiving cell for Notch activation

(Lu and Bilder, 2005; Vaccari *et al.*, 2008). However, it is still not clear which step in canonical Notch signaling activation requires endocytosis. Since the ligand–receptor interaction and S2 cleavage of Notch receptors takes place on the cell surface, the current debate is focused on whether endocytosis is essential for effective S3 cleavage and precisely where the cleavage occurs.

3.2.1. Endocytosis is required for S3 cleavage

The S3 cleavage of Notch is mediated by the γ-secretase, an intramembrane protease complex (De Strooper *et al.*, 1999). Since mutations in γ-secretase components, like *presenilin,* are associated with Alzheimer's disease via aberrant cleavage of amyloid precursor protein and production of pathogenic Aβ (Levy-Lahad *et al.*, 1995; Rogaev *et al.*, 1995; Sherrington *et al.*, 1995), there has been much interest in understanding how, when, and where this cleavage takes place. The γ-secretase complex is present on plasma membranes, endocytic compartments, lysosomes, ER, and Golgi apparatus (Small and Gandy, 2006). Since the endocytic pathway is involved in Aβ production (Koo and Squazzo, 1994), endocytosis may play a positive role in γ-secretase-mediated S3 cleavage of Notch.

Several observations support a requirement of endocytosis for S3 cleavage. The activity of γ-secretase has been suggested to be higher in acidic environments, implicating that the S3 cleavage is more efficient in the endocytic compartments where the pH is lower (Pasternak *et al.*, 2003). In support of this idea, defects in proteins involved in acidification of endosome affect Notch signaling. Mutations in Rabconnectin (Rbcn)-3A, and Rbcn-3B, which assist the assembly of V-ATPase, as well as mutations in a subunit of V-ATPase, impair the acidification of endosomal compartments and lead to accumulation of the Notch receptor in enlarged late endosomes in *Drosophila* follicle cells and imaginal disc cells. This disruption of Notch signaling occurs after S2 cleavage in the receiving cells, supporting the idea that γ-secretase cleavage may be defective in these mutants (Yan *et al.*, 2009). A mutation in *big brain* (*bib*), a gene encoding a monovalent cation (including H+) transporter (Yanochko and Yool, 2002), was initially reported to cause a Notch endocytic trafficking and signaling defect (Kanwar and Fortini, 2008). Note that the endocytic trafficking defects in *bib* mutants were later found to be the result of a second-site mutation while the cleaned *bib* mutant chromosome still exhibits Notch signaling defects (Fortini and Bilder, 2009). The Notch signaling defects in *bib* mutants may be due to improper pH environment in endosomes and lysosomes as acidification of these organelles in *bib* mutant cells is strongly attenuated (Fortini and Bilder, 2009; Kanwar and Fortini, 2008).

In *Drosophila Rab5* and *Avl* mutants, Notch accumulates at or near the cell surface and NICD production by γ-secretase is largely reduced, suggesting that S3 cleavage occurs less efficiently at the plasma membrane (Vaccari *et al.*, 2008). In addition, with a point mutation at the S3 cleavage site in the NEXT-like fragment, this modified fragment can be detected in the endosomes, suggesting that S3 cleavage might occur in the endosomal compartments (Gupta-Rossi *et al.*, 2004). These data suggest that upon S2 cleavage, Notch is endocytosed into an acidified endosomal compartment where S3 cleavage takes place by the g-secretase complex.

3.2.2. S3 cleavage can occur WITHOUT endocytosis of Notch

Some data argue that there is no requirement of dynamin for γ-secretase-mediated S3 cleavage of Notch following removal of the Notch ectodomain and generation of NEXT (Struhl and Adachi, 2000). For example, NEXT remains associated with the apical membrane in γ-secretase mutant cells. Importantly, overexpressed NEXT-like fragments can be cleaved in *shi* mutant pupal nota and embryos (Lopez-Schier and St Johnston, 2002; Seugnet *et al.*, 1997). Furthermore, in a mammalian cell culture system, active γ-secretase complex can be purified from the plasma membrane that still contains Notch fragment cleavage activity (Chyung *et al.*, 2005). Several reports document that the optimal pH environment for γ-secretase is 6.8–7.4, suggesting that acidic endosomes are not required for S3 cleavage (Lee *et al.*, 2002; McLendon *et al.*, 2000; Zhang *et al.*, 2001). Moreover, a NEXT-like fragment with a point mutation at the ubiquitination site cannot be monoubiquitinated and endocytosed. Though the mutated NEXT fragment was thought to be unable to be cleaved into NICD (Gupta-Rossi *et al.*, 2004), it was found later that this fragment can still be cleaved at the plasma membrane producing a less stable form of NICD with a shift in the cleavage position (Tagami *et al.*, 2008). Tagami *et al.* (2008) further argued that during S3 cleavage, γ-secretase can process NEXT into various forms of NICD depending on the exact position of the cleavage, including NICD-S(+3), NICD-L(+1), NICD-L(+2), and NICD-V (the most stable one; refer to the original NICD), with similar transactivation activity. Cleavage at the plasma membrane preferred the production of the more stable NICD-V, while cleavage in the endosomes leads to the production of the less stable NICD-S(+3), arguing against the idea that endocytosis promotes γ-secretase cleavage. All together, these data suggest that endocytosis is not essential for S3 cleavage of Notch and that γ-secretase is able to mediate the cleavage at the plasma membrane.

In summary, whether endocytosis promotes γ-secretase cleavage or not is still a matter of debate. The controversy may be due to the fact that conclusions are based on different model systems (*Drosophila* vs. mammalian cell culture) and that many of the data are based on overexpression strategies to generate Notch fragment intermediates. Considering the stringent dosage dependence of Notch signaling in different contexts, endocytosis may play either a positive or a negative role in γ-secretase cleavage in different contexts. Additional studies are required to reveal a more detailed picture of the relationship between endocytosis and canonical Notch signal activation.

3.3. Degradation of Notch receptors through the lysosomal pathway

Inactive Notch receptors undergo constitutive endocytosis, endocytic trafficking, and are eventually degraded in the lysosome (Jehn *et al.*, 2002). However, the acidic environment during trafficking might promote the dissociation of the Notch heterodimer and produce membrane-tethered NEXT, which can be further processed by γ-secretase in the endosomes. Production of NICD may bypass the requirement for ligand binding as well as S2 cleavage and even occur on lysosomal membranes (Wilkin *et al.*, 2008).

Mutations in genes involved in the degradation pathway can lead to an accumulation of Notch in endosomes and ectopic activation of Notch signaling in a ligand-independent manner. In other words, some proteins must prevent the ectopic activation of Notch and act as negative regulators of Notch signaling. These include proteins that regulate the ubiquitination of Notch, proteins involved in the maturation of early endosomes into MVB, and proteins that mediate the endocytic trafficking and fusion between late endosomes and lysosomes. The regulation of Notch signaling by these proteins is discussed below.

3.5.1. E3 ligases for ubiquitination of Notch receptors

Multiple E3 ligases ubiquitinate Notch and promote its internalization to regulate its signaling activity. These include Deltex (*dx*), Su(dx) (Suppressor of *deltex*), and DNedd4 (Brennan and Gardner, 2002; Kanwar and Fortini, 2004).

Dx, a RING finger E3 ubiquitin ligase, was originally identified as a positive regulator of Notch signaling based on genetic studies (Diederich *et al.*, 1994; Matsuno *et al.*, 1995; Xu and Artavanis-Tsakonas, 1990). Loss of *dx* leads to a Notch signaling impairment in certain cell contexts (*Drosophila* eye and wing imaginal discs). In addition, overexpression of *dx* results in Notch signaling activation in the dorsal–ventral boundary of the wing, independent of DSL ligands and CSL. This ectopic signaling activation requires the internalization of Notch into Rab7-positive late endosomes

(Fuwa *et al.*, 2006; Hori *et al.*, 2004; Wilkin *et al.*, 2004, 2008). However, in contrast to its positive role in Notch signaling activation, Dx was also found to promote Notch receptor degradation when it forms a complex with Kurtz (Krz), the *Drosophila* homolog of a nonvisual β-arrestin (Mukherjee *et al.*, 2005). Therefore, Dx can act both in a positive and in a negative manner in Notch signaling depending on the context and interacting partners.

The *Drosophila* HECT (Homologous to the E6-AP Carboxyl Terminus) as full form of HECT domain-containing family of E3 ligases includes three members: Su(dx), DNedd4, and D-smurf. Much of data related to these studies were obtained from studies in the *Drosophila* wing margin, ovary development, and cultured S2 cells. Su(dx) and dNedd4 can ubiquitinate full-length Notch in a ligand-independent manner and promote its entry in the lysosomal degradation pathway (Sakata *et al.*, 2004; Wilkin *et al.*, 2004). Loss of *Su(dx)* and *dNedd4* causes Notch gain-of-function phenotypes, while overexpression causes Notch loss-of-function phenotypes. *Su(dx)* and *dNedd4* mutations can also suppress Notch partial loss-of-function phenotypes, supporting the idea that these proteins play negative roles in Notch signaling (Cornell *et al.*, 1999; Fostier *et al.*, 1998; Mazaleyrat *et al.*, 2003; Qiu *et al.*, 2000; Sakata *et al.*, 2004; Wilkin *et al.*, 2004). The third member of Nedd4 family, D-Smurf, has been suspected to have some functional redundancy with the other two members, but a direct role in Notch signaling has yet to be demonstrated (Wilkin *et al.*, 2004). Finally, the protein levels of Dx are negatively correlated to the expression level of Nedd4 family proteins, implicating their role in regulation of the Dx protein level (Wilkin *et al.*, 2004). Thus, Nedd4 family proteins might regulate Notch signaling by directly promoting lysosomal Notch degradation and by regulating the protein level of Dx.

In mammals, the homolog of Su(dx) (named AIP4/Itch) has also been reported to ubiquitinate full-length Notch-1 in a ligand-independent manner and to promote its lysosomal degradation (Chastagner *et al.*, 2008; Qiu *et al.*, 2000). The adaptor protein Numb can interact with AIP4/Itch to promote this degradation process (McGill *et al.*, 2009; McGill and McGlade, 2003). In addition, AIP4/Itch was also shown to mediate lysosomal degradation of Dx through ubiquitination (Chastagner *et al.*, 2006). Another mammalian RING type E3 ubiquitin ligase c-Cbl can also promote the degradation of the Notch-1 NEXT fragment (Jehn *et al.*, 2002). However, whether c-Cbl can target full-length Notch-1 for degradation is still unclear. Although c-Cbl is conserved in *Drosophila,* it awaits to be tested whether it can function in regulating Notch signaling.

3.3.2. Lgd and ESCRT complex

Proteins involved in the maturation of early endosomes into MVB, such as ESCRT complexes and Lethal (2) giant discs (Lgd), have been shown to function in Notch degradation. ESCRT 0 or STAM/Hrs is localized to early endosomes and can bind and transfer ubiquitinated cargo to the ESCRT I/II/III complexes (Raiborg and Stenmark, 2009). Lgd is a C2 domain-containing protein that binds to phospholipids. *lgd* mutants have been shown to exhibit general protein sorting defects. Loss of *lgd* leads to accumulation of Notch in Hrs-positive endosomes as well as ectopic Notch signaling activation. This ectopic Notch activation is ligand-independent since the activation is also observed in *lgd Dl Ser* triple mutant clones (Childress *et al.*, 2006; Jaekel and Klein, 2006). In *hrs lgd* double mutant clones, this ligand-independent Notch activation is blocked while ligand-dependent activation remains unaffected. Thus, Lgd functions to prevent ligand-independent Notch activation in an Hrs-dependent manner (Childress *et al.*, 2006; Gallagher and Knoblich, 2006; Jaekel and Klein, 2006; Klein, 2003). Interestingly, in *hrs* single mutant, Notch receptors accumulate in the Avl-positive early endosomes but remain inactive (Jekely and Rorth, 2003; Lloyd *et al.*, 2002; Lu and Bilder, 2005; Thompson *et al.*, 2005). Therefore, Hrs is only required for ectopic activation of Notch signaling in the *lgd* mutant background but not in a wild-type fly.

Mutations in *Drosophila* ESCRT I (*tsg101/erupted* and *vps28*), ESCRT II (*vps22, vps25,* and *vps36*), and ESCRT III (*vps2, vps20,* and *vps32*) complexes all result in accumulation of Notch receptors in early endosomes and most of them cause ectopic Notch signaling activation in developing imaginal discs (Herz *et al.*, 2006, 2009; Moberg *et al.*, 2005; Thompson *et al.*, 2005; Vaccari and Bilder, 2005; Vaccari *et al.*, 2009). The ectopic activation of Notch signaling up-regulates the expression of the ligand of JAK/STAT pathway, Unpaired, which in turn promotes the overgrowth of surrounding wild-type cells in a nonautonomous manner. Although *ESCRT* and *lgd* mutants both exhibit accumulation of Notch receptors and ectopic Notch signaling activity, they have also distinct phenotypes. *ESCRT* mutant cells lose epithelial organization and eventually die while inducing non-cell-autonomous tissue growth (Herz *et al.*, 2006, 2009; Moberg *et al.*, 2005; Thompson *et al.*, 2005; Vaccari and Bilder, 2005; Vaccari *et al.*, 2009). Conversely, *lgd* mutant cells display cell-autonomous overgrowth and apoptosis while still maintaining their epithelial organization (Childress *et al.*, 2006; Gallagher and Knoblich, 2006; Jaekel and Klein, 2006; Klein, 2003).

3.3.3. Other genes involved in Notch receptor trafficking

Several other genes/proteins have been shown to affect Notch signaling by affecting endocytosis and trafficking of Notch receptors. Mutations in genes involved in trafficking/fusion between late endosomes and lysosomes, such as the HOPS and AP-3 complexes, play a regulatory role in Notch degradation (Wilkin *et al.*, 2008). Loss-of-function mutations in these genes act as genetic modifiers of ectopic Notch signaling caused by overexpression of *dx*. However, these mutants only show minor, if any, Notch signaling defects on their own. Tumor suppressors *Merlin* (*Mer*) and *expanded* (*ex*) encodes proteins that belong to the FERM (four-point one, ezrin, radixin, moesin) domain superfamily and promote endocytosis and clearance of Notch receptors from the cell surface (Maitra *et al.*, 2006). In *Mer ex* double mutant flies, Notch accumulates at the plasma membrane, but these mutants do not exhibit any obvious defects in Notch signaling.

Proteins involved in membrane lipid biosynthesis can also modulate Notch signaling activity, possibly through their effect on the endocytic processes by altering the phospholipid composition in the plasma membrane and endosomal membranes. For example, the *Caenorhabditis elegans* BRE-5 (BT-toxin resistance) catalyzes the biosynthesis of glycosphingolipids (GSL), which are enriched in the lipid raft. Knockdown of BRE-5 can suppress hypermorphic LIN-12 (*C. elegans* homolog of Notch) egg-laying phenotypes (Katic *et al.*, 2005). In *Drosophila*, mutations in cytidylyltransferase-1, a rate-limiting enzyme in phosphatidylcholine (PC) biosynthesis, show reduction in Notch signaling and an increased late endosomal localization of Notch receptor (Weber *et al.*, 2003). These data suggest a positive role for GSL and PC in Notch signaling.

In brief, degradation of inactive Notch receptors through the endocytic pathway provides a mechanism to prevent ectopic Notch activation. Mutations in endocytosis-related genes, including E3 ubiquitin ligases, endocytic trafficking proteins, and enzymes involved in phospholipid biosynthesis, can cause abnormal Notch signaling activity. These similar but distinct mutant phenotypes, combined with their context dependence, reveal their different roles in Notch degradation and their partial redundancy in protein sorting and vesicle trafficking.

4. Regulation of Notch Signaling by Endocytosis and Vesicle Trafficking During Mechanosensory Organ Development in Drosophila

4.1. Introduction to mechanosensory organ development

The development of the mechanosensory organs of the *Drosophila* peripheral nervous system has served as a model system to understand many aspects of Notch signaling including endocytic trafficking. The body of an adult fly is covered by hundreds of mechanosensory bristles that act as sensors. Each bristle is composed of four different cell types: socket, hair, sheath, and neuron (Hartenstein and Posakony, 1989). These cells arise from series of asymmetric cell division of a sensory organ precursor (SOP) cell and subsequent unidirectional Notch signaling between the daughter cells during pupariation (Jan and Jan, 2001). In the dorsal thorax or notum, the division of the SOP occurs in parallel to the anterior–posterior body axis. The anterior cell becomes the signal-sending cell (pIIb), whereas the posterior cell becomes the signal-receiving cell (pIIa), leading to asymmetric activation of Notch signaling. The pIIb gives rise to the internal cells (sheath and neuron), whereas the pIIa becomes the progenitor of the external cells (socket and hair). When Notch signaling is lost, the pIIa transforms into a pIIb, leading to loss of external cells and gain of internal cells (de Celis *et al.*, 1991; Hartenstein and Posakony, 1990; Zeng *et al.*, 1998). Conversely, when Notch signaling is ectopically activated in both cells, there is a pIIb-to-pIIa cell fate change, leading to the gain of external cells and loss of internal cells (Frise *et al.*, 1996; Guo *et al.*, 1996). Notch signaling needs to be tightly controlled and the cells of the bristle lineage achieve this via an asymmetric segregation of endocytic factors, which are often referred to as "cell fate determinants." Here, we will especially focus on the specification of the pIIa and pIIb cells, and discuss how the endocytic and trafficking machinery is employed to bias Notch signaling.

4.2. Setting up the asymmetry in the SOP cell

During division of the SOP, the cell fate determinants Neur and Numb become enriched at the anterior pole, forming a crescent (Le Borgne and Schweisguth, 2003b; Rhyu *et al.*, 1994) (Fig. 5.5A). The division then allows the anterior pIIb to inherit these two factors, whereas the posterior pIIa does not. The formation of the anterior crescent is determined by cell polarity factors (Bardin *et al.*, 2004; Betschinger and Knoblich, 2004) and planar cell polarity cues to assure that the division of the SOP occurs along the anterior–posterior axis so that the cell fate determinants are properly segregated into the pIIb cell. In parallel to the asymmetric segregation of Neur and Numb, the inheritance of endocytic compartments are also biased between pIIa and pIIb cells. Rab11-positive recycling endosomes become enriched in pIIb cells, due to asymmetric enrichment of nuclear fallout (Nuf) around the pIIb cell centrosome after mitosis. Nuf is the *Drosophila* homolog of arfophilins, an effector of Rab11 that is required for recycling endosome formation and function (Emery *et al.*, 2005). On the other hand, endosomes that are positive for Sara becomes asymmetrically segregated into the pIIa cell (Coumailleau *et al.*, 2009). The asymmetric inheritance of cell fate determinants together with asymmetric redistribution of endocytic compartments works in concert to assure that Notch signaling occurs unidirectionally. Here, we will next discuss how each of these factors contributes to establishment of a signal-sending cell and a signal-receiving cell during pIIa vs. pIIb cell fate determination.

4.3. Role of asymmetrically segregated Neuralized and Delta recycling in the pIIb cell

In the pIIb cell, Neur ubiquitinates Delta and promotes its endocytosis (Le Borgne and Schweisguth, 2003b) (Fig. 5.5B). Endocytosed Delta is then sorted into a basal Rab11-positive compartment (Emery *et al.*, 2005; Jafar-Nejad *et al.*, 2005). In cells mutant for Sec15, a component of the exocyst complex and an effector of Rab11 (Wu *et al.*, 2005), Delta is stuck in this basal compartment and it cannot recycle back to the cell surface (Jafar-Nejad *et al.*, 2005). Similar phenotypes are observed in cells mutant for the Arp2/3 complex and WASp, mediators of branched actin polymerization (Rajan *et al.*, 2009). Based on detailed phenotypic analysis of these mutants, a model has been proposed where Delta recycles back to the apical plasma membrane from the basal recycling endosomes with the help of the exocyst complex and via actin polymerization through the Arp2/3–WASp complex in the pIIb cell. One interesting possibility is that basal Delta-positive vesicles may recruit Arp2/3 complex to propel them to the apical region via a force generated by actin polymerization, analogous to *Listeria monocytogenes* recruiting and activating Arp2/3 for intracellular motility (Lambrechts *et al.*, 2008). This Neur-mediated recycling of Delta is essential for Notch signaling in the mechanosensory lineage since mutations in Sec15 and Arp2/3–WASp complex lead to loss of Notch signaling. Since the distribution of Neur is restricted to the pIIb cell, Delta in the pIIa cell cannot be endocytosed/recycled, and hence cannot activate the Notch receptor on the pIIb cell.

Recently, it was reported that the apical membranes of the pIIa and pIIb cells are enriched in polymerized actin (Rajan *et al.*, 2009). This structure, which was referred to as an apical actin-rich structure (ARS), is rich in microvilli and recycled Delta. The microvilli of the ARS may be the site of ligand–receptor interaction, and the microvilli may promote Notch signaling by increasing the surface area between the signaling cells. In Arp2/3–WASp complex mutants, the ARS becomes smaller, which may contribute to the Notch loss-of-function defects in these cells.

4.4. Role of asymmetrically segregated Numb in the pIIb cell

The pIIb cell not only inherits Neuralized but also inherits Numb, which acts as a negative regulator of signal reception (Rhyu *et al.*, 1994) (Fig. 5.5B). Numb is an endocytic protein that interacts with the AP-2 complex (Santolini *et al.*, 2000). In the pIIb cell, Numb binds α-adaptin of the AP-2 complex (Berdnik *et al.*, 2002) and promotes the endocytosis of

Sanpodo (Spdo) (Hutterer and Knoblich, 2005; Langevin *et al.*, 2005; Roegiers *et al.*, 2005). Spdo is a four transmembrane protein that acts as a positive regulator of Notch signaling when present at the plasma membrane through an unknown mechanism (Babaoglan *et al.*, 2009; O'Connor-Giles

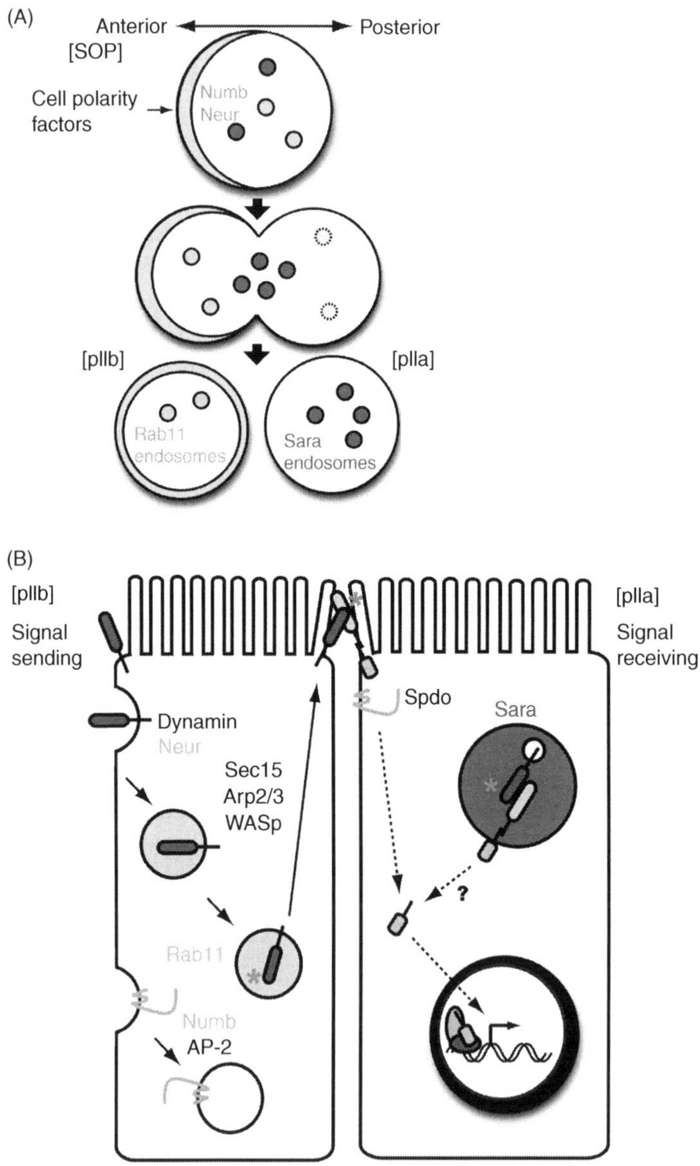

Figure 5.5 *(Continued)*

and Skeath, 2003). In the pIIb cell, Spdo is sequestered away from the plasma membrane in an endocytic compartment, where it is thought to be nonfunctional. In contrast, Spdo is localized at the plasma membrane in the pIIa cell, where Numb is absent, and Notch signaling reception is promoted. Thus, by segregating Neur and Numb into the pIIb cell, this cell becomes the signal-sending cell, whereas the pIIa becomes the signal-receiving cell.

4.5. Role of asymmetrically segregated Sara-endosomes in the pIIa cell

Recently, endosomes that are marked by Sara (Sara-endosomes) have been identified as a third cell fate determinant during pIIa–pIIb cell fate specification (Coumailleau *et al.*, 2009). Sara is a FYVE domain-containing adaptor protein that localizes to a subpopulation of PI3P-containing endosomes (Bokel *et al.*, 2006; Tsukazaki *et al.*, 1998). In the SOP, a population of Notch and Delta are endocytosed into Sara-endosomes that become segregated into the pIIa cell. Although loss of function of Sara does not exhibit any defect in the bristle lineage, overexpression of Sara or inheritance of a

Figure 5.5 Regulation of Notch signaling via endocytosis and vesicle trafficking during mechanosensory organ development in *Drosophila.* (A) During mitosis of the sensory organ precursor (SOP) cell, cell fate determinants Neuralized (Neur) and Numb are asymmetrically segregated into the anterior crescent which is determined through interactions between cell polarity factors. Upon cytokinesis, Neur and Numb are both inherited by the anterior pIIb cell, whereas the posterior cell fails lacks these factors. In parallel to the asymmetric segregation of the two cell fate determinants, Rab11-positive recycling endosomes become enriched in the pIIb cell, whereas Sara-positive endosomes are sorted into the pIIa cell. This asymmetric segregation of cell fate determinants and specific endocytic compartment biases the following Notch signaling between the pIIa and the pIIb cell. (B) In the pIIb cell, Neur promotes the endocytosis and sorting of Delta for activation. Activated Delta traffics through Rab11-positive endosomes and recycle back to the apical cell surface where there is an enrichment of actin filaments and microvilli, referred to as the ARS (apical actin-rich structure). Sec15, a Rab11 effector and component of the exocyst complex, is required for this apical recycling of Delta. In addition, Arp2/3 and WASp, positive regulators of actin polymerization, are also required for recycling of Delta, possibly through mobilization of Delta-positive vesicles and/or facilitation of ARS formation. Activated Delta that returned to the cell surface interacts with and activates Notch in the pIIa cell. Sanpodo (Spdo), a four transmembrane domain protein, is present at the cell surface of the pIIa cell to promote the reception of this signal. pIIa cell cannot signal to the pIIb since they are not able to activate Delta due to lack of Neur. In addition, pIIb cannot receive Notch signaling since Spdo is endocytosed by Numb and kept in an inactive form. As a third mechanism, Sara-positive endosome has been recently been proposed to bias Notch signaling by actively recruiting Notch and Delta into the pIIa cell. γ-secretase cleavage of Notch has been proposed to be happening at or before Notch entering the Sara-positive endosome, but the exact mechanism and role of Sara is not fully understood. (See Color Insert.)

single giant Sara-endosome generated by constitutively active Rab5 expression into the pIIb cell, can mediate pIIb-to-pIIa cell fate transformation, caused by a Notch gain of function. Together with the observation that γ-secretase-dependent cleavage of Notch may be taking place at or before entry into Sara-endosomes in the pIIa, segregation of Delta and Notch through Sara-endosomes into pIIa may create a small asymmetry in the signaling activity between the pIIa and pIIb, which is further amplified by actions of asymmetric segregated Neur and Numb. However, it is important to note that loss of function of *Sara* as well as artificial segregation of Sara-endosome away from pIIa cell does not lead to loss of Notch signaling defects, suggesting that the Sara-endosome is not an essential component of Notch activation and plays a regulatory role during bristle development.

In summary, cells of the mechanosensory lineage utilize the endocytic pathway in order to restrict and regulate the activity of Delta and Notch to achieve the proper fate via unidirectional Notch signaling. Since loss of function in genes such as Numb and Sec15 show defects in bristle development and other binary cell fate determination events but not in all Notch signaling-dependent events (Jafar-Nejad *et al.*, 2005; Mehta *et al.*, 2005; O'Connor-Giles and Skeath, 2003), there seems to be context specificity in the utilization of the endocytic pathway to achieve successful Notch signaling mediated decisions *in vivo*. Since defects in mechanosensory organ development can be subjected to high-throughput forward genetic screens using clonal analysis (Berdnik *et al.*, 2002; Jafar-Nejad *et al.*, 2005) identification of novel genes and endocytic pathways that regulate Notch signaling are likely to continue to be discovered using this model system.

5. Conclusion and Future Directions

In both signal-sending and receiving cells, the vesicle trafficking routes not only activate Notch signaling but also fine tune the signal output. Endocytosis and vesicle trafficking mediate the activation of DSL ligands through the recycling pathway, generate a pulling force to promote the S2 cleavage of Notch upon ligand–receptor interaction, may regulate the S3 cleavage to release the active NICD fragment, promote degradation of inactive Notch, and control ligand-independent activation of the pathway. It is important to keep in mind that different cell types have distinct trafficking properties and that cell context is important.

There are numerous questions that remain unanswered in this area. What is the activated state of DSL ligands upon entry into the specialized recycling pathway? What is the exact function of endocytosis in the signal-receiving cells? Which endocytic factors are true universal regulators and which factors act in context/species-specific manner? Are ARS and apically enriched microvilli the sites where Notch signaling takes place and are these structures seen in other cell types that signal though Notch? What other genes and trafficking pathways regulate Notch? What fraction of developmental disorders and human diseases are caused by defects in endocytosis and trafficking of Notch signal components? We believe that the answers to these and many other questions will come from integration of various studies from different fields. We have emphasized genetic approaches in *Drosophila* in this chapter because much of the key observations related to endocytosis and vesicle trafficking in Notch signaling were first made in *Drosophila*. We hope that further insights into the importance and various roles of endocytosis and vesicle trafficking in Notch signaling will be forthcoming, not only from the fly field but also from experiments in vertebrates.

ACKNOWLEDGMENTS

We would like to thank Mark Fortini, Nikolaos Giagtzoglou, and An-Chi Tien for useful suggestions. We apologize to all our colleagues for not being able to cite their work given the length restrictions. SY is supported by the Nakajima Foundation and HJB is Investigator with the Howard Hughes Medical Institute.

REFERENCES

Acconcia, F., Sigismund, S., and Polo, S. (2009). Ubiquitin in trafficking: the network at work. *Exp. Cell Res.* **315**, 1610–1618.

Ahimou, F., Mok, L. P., Bardot, B., and Wesley, C. (2004). The adhesion force of Notch with Delta and the rate of Notch signaling. *J. Cell Biol.* **167**, 1217–1229.

Babaoglan, A. B., O'Connor-Giles, K. M., Mistry, H., Schickedanz, A., Wilson, B. A., and Skeath, J. B. (2009). Sanpodo: a context-dependent activator and inhibitor of Notch signaling during asymmetric divisions. *Development* **136**, 4089–4098.

Bardin, A. J., Le Borgne, R., and Schweisguth, F. (2004). Asymmetric localization and function of cell-fate determinants: a fly's view. *Curr. Opin. Neurobiol.* **14**, 6–14.

Bardin, A. J., and Schweisguth, F. (2006). Bearded family members inhibit Neuralized-mediated endocytosis and signaling activity of Delta in Drosophila. *Dev. Cell* **10**, 245–255.

Ben-Yaacov, S., Le Borgne, R., Abramson, I., Schweisguth, F., and Schejter, E. D. (2001). Wasp, the Drosophila Wiskott-Aldrich syndrome gene homologue, is required for cell fate decisions mediated by Notch signaling. *J. Cell Biol.* **152**, 1–13.

Berdnik, D., Torok, T., Gonzalez-Gaitan, M., and Knoblich, J. A. (2002). The endocytic protein alpha-Adaptin is required for numb-mediated asymmetric cell division in Drosophila. *Dev. Cell* **3**, 221–231.

Betschinger, J., and Knoblich, J. A. (2004). Dare to be different: asymmetric cell division in Drosophila, C. elegans and vertebrates. *Curr. Biol.* **14**, R674–R685.

Bokel, C., Schwabedissen, A., Entchev, E., Renaud, O., and Gonzalez-Gaitan, M. (2006). Sara endosomes and the maintenance of Dpp signaling levels across mitosis. *Science* **314**, 1135–1139.

Boulianne, G. L., de la Concha, A., Campos-Ortega, J. A., Jan, L. Y., and Jan, Y. N. (1991). The Drosophila neurogenic gene neuralized encodes a novel protein and is expressed in precursors of larval and adult neurons. *EMBO J.* **10**, 2975–2983.

Brennan, K., and Gardner, P. (2002). Notching up another pathway. *Bioessays* **24**, 405–410.

Chanet, S., Vodovar, N., Mayau, V., and Schweisguth, F. (2009). Genome engineering-based analysis of Bearded family genes reveals both functional redundancy and a non-essential function in lateral inhibition in Drosophila. *Genetics* **182**, 1101–1108.
Chastagner, P., Israel, A., and Brou, C. (2006). Itch/AIP4 mediates Deltex degradation through the formation of K29-linked polyubiquitin chains. *EMBO Rep.* **7**, 1147–1153.
Chastagner, P., Israel, A., and Brou, C. (2008). AIP4/Itch regulates Notch receptor degradation in the absence of ligand. *PLoS ONE* **3**, e2735.
Chen, H., Fre, S., Slepnev, V. I., Capua, M. R., Takei, K., Butler, M. H., Di Fiore, P. P., and De Camilli, P. (1998). Epsin is an EH-domain-binding protein implicated in clathrin-mediated endocytosis. *Nature* **394**, 793–797.
Chen, H., Ko, G., Zatti, A., Di Giacomo, G., Liu, L., Raiteri, E., Perucco, E., Collesi, C., Min, W., Zeiss, C., De Camilli, P., and Cremona, O. (2009). Embryonic arrest at midgestation and disruption of Notch signaling produced by the absence of both epsin 1 and epsin 2 in mice. *Proc. Natl. Acad. Sci. U.S.A.* **106**, 13838–13843.
Chen, M. S., Obar, R. A., Schroeder, C. C., Austin, T. W., Poodry, C. A., Wadsworth, S. C., and Vallee, R. B. (1991). Multiple forms of dynamin are encoded by shibire, a Drosophila gene involved in endocytosis. *Nature* **351**, 583–586.
Childress, J. L., Acar, M., Tao, C., and Halder, G. (2006). Lethal giant discs, a novel C2-domain protein, restricts notch activation during endocytosis. *Curr. Biol.* **16**, 2228–2233.
Chitnis, A. (2006). Why is delta endocytosis required for effective activation of notch? *Dev. Dyn.* **235**, 886–894.
Chyung, J. H., Raper, D. M., and Selkoe, D. J. (2005). Gamma-secretase exists on the plasma membrane as an intact complex that accepts substrates and effects intramembrane cleavage. *J. Biol. Chem.* **280**, 4383–4392.
Commisso, C., and Boulianne, G. L. (2007). The NHR1 domain of Neuralized binds Delta and mediates Delta trafficking and Notch signaling. *Mol. Biol. Cell* **18**, 1–13.
Cornell, M., Evans, D. A., Mann, R., Fostier, M., Flasza, M., Monthatong, M., Artavanis-Tsakonas, S., and Baron, M. (1999). The Drosophila melanogaster Suppressor of deltex gene, a regulator of the Notch receptor signaling pathway, is an E3 class ubiquitin ligase. *Genetics* **152**, 567–576.
Coumailleau, F., Furthauer, M., Knoblich, J. A., and Gonzalez-Gaitan, M. (2009). Directional Delta and Notch trafficking in Sara endosomes during asymmetric cell division. *Nature* **458**, 1051–1055.
d'Azzo, A., Bongiovanni, A., and Nastasi, T. (2005). E3 ubiquitin ligases as regulators of membrane protein trafficking and degradation. *Traffic* **6**, 429–441.
De Camilli, P., Chen, H., Hyman, J., Panepucci, E., Bateman, A., and Brunger, A. T. (2002). The ENTH domain. *FEBS Lett.* **513**, 11–18.
de Celis, J. F., Mari-Beffa, M., and Garcia-Bellido, A. (1991). Cell-autonomous role of Notch, an epidermal growth factor homologue, in sensory organ differentiation in Drosophila. *Proc. Natl. Acad. Sci. U.S.A.* **88**, 632–636.
De Renzis, S., Yu, J., Zinzen, R., and Wieschaus, E. (2006). Dorsal-ventral pattern of Delta trafficking is established by a Snail-Tom-Neuralized pathway. *Dev. Cell* **10**, 257–264.
De Strooper, B., Annaert, W., Cupers, P., Saftig, P., Craessaerts, K., Mumm, J. S., Schroeter, E. H., Schrijvers, V., Wolfe, M. S., Ray, W. J., Goate, A., and Kopan, R., (1999). A presenilin-1-dependent gamma-secretase-like protease mediates release of Notch intracellular domain. *Nature* **398**, 518–522.
Deblandre, G. A., Lai, E. C., and Kintner, C. (2001). Xenopus neuralized is a ubiquitin ligase that interacts with XDelta1 and regulates Notch signaling. *Dev. Cell* **1**, 795–806.
Dell'Angelica, E. C. (2009). AP-3-dependent trafficking and disease: the first decade. *Curr. Opin. Cell Biol.* **21**, 552–559.
Di Paolo, G., and De Camilli, P. (2006). Phosphoinositides in cell regulation and membrane dynamics. *Nature* **443**, 651–657.

Diederich, R. J., Matsuno, K., Hing, H., and Artavanis-Tsakonas, S. (1994). Cytosolic interaction between deltex and Notch ankyrin repeats implicates deltex in the Notch signaling pathway. *Development* **120**, 473–481.

Doherty, G. J., and McMahon, H. T. (2009). Mechanisms of endocytosis. *Annu. Rev. Biochem.* **78**, 857–902.

Eisenberg, E., and Greene, L. E. (2007). Multiple roles of auxilin and Hsc70 in clathrin-mediated endocytosis. *Traffic* **8**, 640–646.

Emery, G., Hutterer, A., Berdnik, D., Mayer, B., Wirtz-Peitz, F., Gaitan, M. G., and Knoblich, J. A. (2005). Asymmetric Rab 11 endosomes regulate delta recycling and specify cell fate in the Drosophila nervous system. *Cell* **122**, 763–773.

Eun, S. H., Banks, S. M., and Fischer, J. A. (2008). Auxilin is essential for Delta signaling. *Development* **135**, 1089–1095.

Fehon, R. G., Kooh, P. J., Rebay, I., Regan, C. L., Xu, T., Muskavitch, M. A., and Artavanis-Tsakonas, S. (1990). Molecular interactions between the protein products of the neurogenic loci Notch and Delta, two EGF-homologous genes in Drosophila. *Cell* **61**, 523–534.

Fontana, J. R., and Posakony, J. W. (2009). Both inhibition and activation of Notch signaling rely on a conserved Neuralized-binding motif in Bearded proteins and the Notch ligand Delta. *Dev. Biol.* **333**, 373–385.

Fortini, M. E. (2009). Notch signaling: the core pathway and its posttranslational regulation. *Dev. Cell* **16**, 633–647.

Fortini, M. E., and Bilder, D. (2009). Endocytic regulation of Notch signaling. *Curr. Opin. Genet. Dev.* **19**, 323–328.

Fostier, M., Evans, D. A., Artavanis-Tsakonas, S., and Baron, M. (1998). Genetic characterization of the Drosophila melanogaster Suppressor of deltex gene: a regulator of notch signaling. *Genetics* **150**, 1477–1485.

Frise, E., Knoblich, J. A., Younger-Shepherd, S., Jan, L. Y., and Jan, Y. N. (1996). The Drosophila Numb protein inhibits signaling of the Notch receptor during cell-cell interaction in sensory organ lineage. *Proc. Natl. Acad. Sci. U.S.A.* **93**, 11925–11932.

Furthauer, M., and Gonzalez-Gaitan, M. (2009). Endocytic regulation of notch signalling during development. *Traffic* **10**, 792–802.

Fuwa, T. J., Hori, K., Sasamura, T., Higgs, J., Baron, M., and Matsuno, K. (2006). The first deltex null mutant indicates tissue-specific deltex-dependent Notch signaling in Drosophila. *Mol. Genet. Genomics* **275**, 251–263.

Gallagher, C. M., and Knoblich, J. A. (2006). The conserved c2 domain protein lethal (2) giant discs regulates protein trafficking in Drosophila. *Dev. Cell* **11**, 641–653.

Glittenberg, M., Pitsouli, C., Garvey, C., Delidakis, C., and Bray, S. (2006). Role of conserved intracellular motifs in Serrate signalling, cis-inhibition and endocytosis. *EMBO J.* **25**, 4697–4706.

Goley, E. D., and Welch, M. D. (2006). The ARP2/3 complex: an actin nucleator comes of age. *Nat. Rev.* **7**, 713–726.

Gordon, W. R., Arnett, K. L., and Blacklow, S. C. (2008). The molecular logic of Notch signaling–a structural and biochemical perspective. *J. Cell Sci.* **121**, 3109–3119.

Gordon, W. R., Roy, M., Vardar-Ulu, D., Garfinkel, M., Mansour, M. R., Aster, J. C., and Blacklow, S. C. (2009). Structure of the Notch1-negative regulatory region: implications for normal activation and pathogenic signaling in T-ALL. *Blood* **113**, 4381–4390.

Gordon, W. R., Vardar-Ulu, D., Histen, G., Sanchez-Irizarry, C., Aster, J. C., and Blacklow, S. C. (2007). Structural basis for autoinhibition of Notch. *Nat. Struct. Mol. Biol.* **14**, 295–300.

Grant, B. D., and Donaldson, J. G. (2009). Pathways and mechanisms of endocytic recycling. *Nat. Rev.* **10**, 597–608.

Guo, M., Jan, L. Y., and Jan, Y. N. (1996). Control of daughter cell fates during asymmetric division: interaction of Numb and Notch. *Neuron* **17**, 27–41.

Gupta-Rossi, N., Six, E., LeBail, O., Logeat, F., Chastagner, P., Olry, A., Israel, A., and Brou, C. (2004). Monoubiquitination and endocytosis direct gamma-secretase cleavage of activated Notch receptor. *J. Cell Biol.* **166**, 73–83.

Hagedorn, E. J., Bayraktar, J. L., Kandachar, V. R., Bai, T., Englert, D. M., and Chang, H. C. (2006). Drosophila melanogaster auxilin regulates the internalization of Delta to control activity of the Notch signaling pathway. *J. Cell Biol.* **173**, 443–452.

Hansen, C. G., and Nichols, B. J. (2009). Molecular mechanisms of clathrin-independent endocytosis. *J. Cell Sci.* **122**, 1713–1721.

Hartenstein, V., and Posakony, J. W. (1989). Development of adult sensilla on the wing and notum of Drosophila melanogaster. *Development* **107**, 389–405.

Hartenstein, V., and Posakony, J. W. (1990). Sensillum development in the absence of cell division: the sensillum phenotype of the Drosophila mutant string. *Dev. Biol.* **138**, 147–158.

Henderson, S. T., Gao, D., Christensen, S., and Kimble, J. (1997). Functional domains of LAG-2, a putative signaling ligand for LIN-12 and GLP-1 receptors in Caenorhabditis elegans. *Mol. Biol. Cell* **8**, 1751–1762.

Herz, H. M., and Bergmann, A. (2009). Genetic analysis of ESCRT function in Drosophila: a tumour model for human Tsg101. *Biochem. Soc. Trans.* **37**, 204–207.

Herz, H. M., Chen, Z., Scherr, H., Lackey, M., Bolduc, C., and Bergmann, A. (2006). vps25 mosaics display non-autonomous cell survival and overgrowth, and autonomous apoptosis. *Development* **133**, 1871–1880.

Herz, H. M., Woodfield, S. E., Chen, Z., Bolduc, C., and Bergmann, A. (2009). Common and distinct genetic properties of ESCRT-II components in Drosophila. *PLoS ONE* **4**, e4165.

Heuss, S. F., Ndiaye-Lobry, D., Six, E. M., Israel, A., and Logeat, F. (2008). The intracellular region of Notch ligands Dll1 and Dll3 regulates their trafficking and signaling activity. *Proc. Natl. Acad. Sci. U.S.A.* **105**, 11212–11217.

Hicke, L., and Dunn, R. (2003). Regulation of membrane protein transport by ubiquitin and ubiquitin-binding proteins. *Annu. Rev. Cell Dev. Biol.* **19**, 141–172.

Hori, K., Fostier, M., Ito, M., Fuwa, T. J., Go, M. J., Okano, H., Baron, M., and Matsuno, K. (2004). Drosophila deltex mediates suppressor of Hairless-independent and late-endosomal activation of Notch signaling. *Development* **131**, 5527–5537.

Hukriede, N. A., Gu, Y., and Fleming, R. J. (1997). A dominant-negative form of Serrate acts as a general antagonist of Notch activation. *Development* **124**, 3427–3437.

Hutterer, A., and Knoblich, J. A. (2005). Numb and alpha-Adaptin regulate Sanpodo endocytosis to specify cell fate in Drosophila external sensory organs. *EMBO Rep.* **6**, 836–842.

Itoh, M., Kim, C. H., Palardy, G., Oda, T., Jiang, Y. J., Maust, D., Yeo, S. Y., Lorick, K., Wright, G. J., Ariza-McNaughton, L., *et al.* (2003). Mind bomb is a ubiquitin ligase that is essential for efficient activation of Notch signaling by Delta. *Dev. Cell* **4**, 67–82.

Jaekel, R., and Klein, T. (2006). The Drosophila Notch inhibitor and tumor suppressor gene lethal (2) giant discs encodes a conserved regulator of endosomal trafficking. *Dev. Cell* **11**, 655–669.

Jafar-Nejad, H., Andrews, H. K., Acar, M., Bayat, V., Wirtz-Peitz, F., Mehta, S. Q., Knoblich, J. A., and Bellen, H. J. (2005). Sec15, a component of the exocyst, promotes notch signaling during the asymmetric division of Drosophila sensory organ precursors. *Dev. Cell* **9**, 351–363.

Jan, Y. N., and Jan, L. Y. (2001). Asymmetric cell division in the Drosophila nervous system. *Nat. Rev.* **2**, 772–779.

Jehn, B. M., Dittert, I., Beyer, S., von der Mark, K., and Bielke, W. (2002). c-Cbl binding and ubiquitin-dependent lysosomal degradation of membrane-associated Notch1. *J. Biol. Chem.* **277**, 8033–8040.

Jekely, G., and Rorth, P. (2003). Hrs mediates downregulation of multiple signalling receptors in Drosophila. *EMBO Rep.* **4**, 1163–1168.

Kandachar, V., Bai, T., and Chang, H. C. (2008). The clathrin-binding motif and the J-domain of Drosophila Auxilin are essential for facilitating Notch ligand endocytosis. *BMC Dev. Biol.* **8**, 50.

Kanwar, R., and Fortini, M. E. (2004). Notch signaling: a different sort makes the cut. *Curr. Biol.* **14**, R1043–R1045.

Kanwar, R., and Fortini, M. E. (2008). The big brain aquaporin is required for endosome maturation and notch receptor trafficking. *Cell* **133**, 852–863.

Katic, I., Vallier, L. G., and Greenwald, I. (2005). New positive regulators of lin-12 activity in Caenorhabditis elegans include the BRE-5/Brainiac glycosphingolipid biosynthesis enzyme. *Genetics* **171**, 1605–1615.

Kidd, S., and Lieber, T. (2002). Furin cleavage is not a requirement for Drosophila Notch function. *Mech. Dev.* **115**, 41–51.

Klein, T. (2003). The tumour suppressor gene l(2)giant discs is required to restrict the activity of Notch to the dorsoventral boundary during Drosophila wing development. *Dev. Biol.* **255**, 313–333.

Klueg, K. M., and Muskavitch, M. A. (1999). Ligand-receptor interactions and trans-endocytosis of Delta, Serrate and Notch: members of the Notch signalling pathway in Drosophila. *J. Cell Sci.* **112**(Pt 19), 3289–3297.

Klueg, K. M., Parody, T. R., and Muskavitch, M. A. (1998). Complex proteolytic processing acts on Delta, a transmembrane ligand for Notch, during Drosophila development. *Mol. Biol. Cell* **9**, 1709–1723.

Koo, E. H., and Squazzo, S. L. (1994). Evidence that production and release of amyloid beta-protein involves the endocytic pathway. *J. Biol. Chem.* **269**, 17386–17389.

Kooh, P. J., Fehon, R. G., and Muskavitch, M. A. (1993). Implications of dynamic patterns of Delta and Notch expression for cellular interactions during Drosophila development. *Development* **117**, 493–507.

Kopan, R., and Ilagan, M. X. (2009). The canonical Notch signaling pathway: unfolding the activation mechanism. *Cell* **137**, 216–233.

Koutelou, E., Sato, S., Tomomori-Sato, C., Florens, L., Swanson, S. K., Washburn, M. P., Kokkinaki, M., Conaway, R. C., Conaway, J. W., and Moschonas, N. K. (2008). Neuralized-like 1 (Neurl1) targeted to the plasma membrane by N-myristoylation regulates the Notch ligand Jagged1. *J. Biol. Chem.* **283**, 3846–3853.

Kramer, H., and Phistry, M. (1996). Mutations in the Drosophila hook gene inhibit endocytosis of the boss transmembrane ligand into multivesicular bodies. *J. Cell Biol.* **133**, 1205–1215.

Kramer, H., and Phistry, M. (1999). Genetic analysis of hook, a gene required for endocytic trafficking in drosophila. *Genetics* **151**, 675–684.

Lai, E. C., Deblandre, G. A., Kintner, C., and Rubin, G. M. (2001). Drosophila neuralized is a ubiquitin ligase that promotes the internalization and degradation of delta. *Dev. Cell* **1**, 783–794.

Lai, E. C., Roegiers, F., Qin, X., Jan, Y. N., and Rubin, G. M. (2005). The ubiquitin ligase Drosophila Mind bomb promotes Notch signaling by regulating the localization and activity of Serrate and Delta. *Development* **132**, 2319–2332.

Lake, R. J., Grimm, L. M., Veraksa, A., Banos, A., and Artavanis-Tsakonas, S. (2009). In vivo analysis of the Notch receptor S1 cleavage. *PLoS ONE* **4**, e6728.

Lambrechts, A., Gevaert, K., Cossart, P., Vandekerckhove, J., and Van Troys, M. (2008). Listeria comet tails: the actin-based motility machinery at work. *Trends Cell Biol.* **18**, 220–227.

Langevin, J., Le Borgne, R., Rosenfeld, F., Gho, M., Schweisguth, F., and Bellaiche, Y. (2005). Lethal giant larvae controls the localization of notch-signaling regulators numb, neuralized, and Sanpodo in Drosophila sensory-organ precursor cells. *Curr. Biol.* **15**, 955–962.

Le Borgne, R., Remaud, S., Hamel, S., and Schweisguth, F. (2005). Two distinct E3 ubiquitin ligases have complementary functions in the regulation of delta and serrate signaling in Drosophila. *PLoS Biol.* **3**, e96.

Le Borgne, R., and Schweisguth, F. (2003). Notch signaling: endocytosis makes delta signal better. *Curr. Biol.* **13**, R273–R275.

Le Borgne, R., and Schweisguth, F. (2003). Unequal segregation of Neuralized biases Notch activation during asymmetric cell division. *Dev. Cell* **5**, 139–148.

Lee, J. M., Petrucelli, L., Fisher, G., Ramdath, S., Castillo, J., Di Fiore, M. M., and D'Aniello, A. (2002). Evidence for D-aspartyl-beta-amyloid secretase activity in human brain. *J. Neuropathol. Exp Neurol.* **61**, 125–131.

Lehmann, R., Dietrich, U., Jiménez, F., and Campos-Ortega, J. A. (1981). Mutations of early neurogenesis in Drosophila. *Dev. Genes Evol.* **190**, 226–229.

Leviten, M. W., Lai, E. C., and Posakony, J. W. (1997). The Drosophila gene Bearded encodes a novel small protein and shares 3′ UTR sequence motifs with multiple Enhancer of split complex genes. *Development* **124**, 4039–4051.

Leviten, M. W., and Posakony, J. W. (1996). Gain-of-function alleles of Bearded interfere with alternative cell fate decisions in Drosophila adult sensory organ development. *Dev. Biol.* **176**, 264–283.

Levy-Lahad, E., Wasco, W., Poorkaj, P., Romano, D. M., Oshima, J., Pettingell, W. H., Yu, C. E., Jondro, P. D., Schmidt, S. D., Wang, K., *et al.* (1995). Candidate gene for the chromosome 1 familial Alzheimer's disease locus. *Science* **269**, 973–977.

Ligoxygakis, P., Yu, S. Y., Delidakis, C., and Baker, N. E. (1998). A subset of notch functions during Drosophila eye development require Su(H) and the E(spl) gene complex. *Development* **125**, 2893–2900.

Lloyd, T. E., Atkinson, R., Wu, M. N., Zhou, Y., Pennetta, G., and Bellen, H. J. (2002). Hrs regulates endosome membrane invagination and tyrosine kinase receptor signaling in Drosophila. *Cell* **108**, 261–269.

Logeat, F., Bessia, C., Brou, C., LeBail, O., Jarriault, S., Seidah, N. G., and Israel, A. (1998). The Notch1 receptor is cleaved constitutively by a furin-like convertase. *Proc. Natl. Acad. Sci. U.S.A.* **95**, 8108–8112.

Lopez-Schier, H., and St Johnston, D. (2002). Drosophila nicastrin is essential for the intramembranous cleavage of notch. *Dev. Cell* **2**, 79–89.

Lu, H., and Bilder, D. (2005). Endocytic control of epithelial polarity and proliferation in Drosophila. *Nat. Cell Biol.* **7**, 1232–1239.

Maitra, S., Kulikauskas, R. M., Gavilan, H., and Fehon, R. G. (2006). The tumor suppressors Merlin and Expanded function cooperatively to modulate receptor endocytosis and signaling. *Curr. Biol.* **16**, 702–709.

Marshansky, V., and Futai, M. (2008). The V-type H+-ATPase in vesicular trafficking: targeting, regulation and function. *Curr. Opin. Cell Biol.* **20**, 415–426.

Matsuno, K., Diederich, R. J., Go, M. J., Blaumueller, C. M., and Artavanis-Tsakonas, S. (1995). Deltex acts as a positive regulator of Notch signaling through interactions with the Notch ankyrin repeats. *Development* **121**, 2633–2644.

Mazaleyrat, S. L., Fostier, M., Wilkin, M. B., Aslam, H., Evans, D. A., Cornell, M., and Baron, M. (2003). Down-regulation of Notch target gene expression by Suppressor of deltex. *Dev. Biol.* **255**, 363–372.

McGill, M. A., Dho, S. E., Weinmaster, G., and McGlade, C. J. (2009). Numb regulates post-endocytic trafficking and degradation of Notch1. *J. Biol. Chem.* **284**, 26427–26438.

McGill, M. A., and McGlade, C. J. (2003). Mammalian numb proteins promote Notch1 receptor ubiquitination and degradation of the Notch1 intracellular domain. *J. Biol. Chem.* **278**, 23196–23203.

McLendon, C., Xin, T., Ziani-Cherif, C., Murphy, M. P., Findlay, K. A., Lewis, P. A., Pinnix, I., Sambamurti, K., Wang, R., Fauq, A., *et al.* (2000). Cell-free assays for gamma-secretase activity. *FASEB J.* **14**, 2383–2386.

Mehta, S. Q., Hiesinger, P. R., Beronja, S., Zhai, R. G., Schulze, K. L., Verstreken, P., Cao, Y., Zhou, Y., Tepass, U., Crair, M. C., *et al.* (2005). Mutations in Drosophila sec15 reveal a function in neuronal targeting for a subset of exocyst components. *Neuron* **46**, 219–232.

Moberg, K. H., Schelble, S., Burdick, S. K., and Hariharan, I. K. (2005). Mutations in erupted, the Drosophila ortholog of mammalian tumor susceptibility gene 101, elicit non-cell-autonomous overgrowth. *Dev. Cell* **9**, 699–710.

Morrison, S. J., Perez, S. E., Qiao, Z., Verdi, J. M., Hicks, C., Weinmaster, G., and Anderson, D. J. (2000). Transient Notch activation initiates an irreversible switch from neurogenesis to gliogenesis by neural crest stem cells. *Cell* **101**, 499–510.

Mukherjee, A., Veraksa, A., Bauer, A., Rosse, C., Camonis, J., and Artavanis-Tsakonas, S. (2005). Regulation of Notch signalling by non-visual beta-arrestin. *Nat. Cell Biol.* **7**, 1191–1201.

Nichols, J. T., Miyamoto, A., Olsen, S. L., D'Souza, B., Yao, C., and Weinmaster, G. (2007). DSL ligand endocytosis physically dissociates Notch1 heterodimers before activating proteolysis can occur. *J. Cell Biol.* **176**, 445–458.

Ordentlich, P., Lin, A., Shen, C. P., Blaumueller, C., Matsuno, K., Artavanis-Tsakonas, S., and Kadesch, T. (1998). Notch inhibition of E47 supports the existence of a novel signaling pathway. *Mol. Cell Biol.* **18**, 2230–2239.

Overstreet, E., Chen, X., Wendland, B., and Fischer, J. A. (2003). Either part of a Drosophila epsin protein, divided after the ENTH domain, functions in endocytosis of delta in the developing eye. *Curr. Biol.* **13**, 854–860.

Overstreet, E., Fitch, E., and Fischer, J. A. (2004). Fat facets and Liquid facets promote Delta endocytosis and Delta signaling in the signaling cells. *Development* **131**, 5355–5366.

O'Connor-Giles, K. M., and Skeath, J. B. (2003). Numb inhibits membrane localization of Sanpodo, a four-pass transmembrane protein, to promote asymmetric divisions in Drosophila. *Dev. Cell* **5**, 231–243.

Parks, A. L., Klueg, K. M., Stout, J. R., and Muskavitch, M. A. (2000). Ligand endocytosis drives receptor dissociation and activation in the Notch pathway. *Development* **127**, 1373–1385.

Parks, A. L., Stout, J. R., Shepard, S. B., Klueg, K. M., Dos Santos, A. A., Parody, T. R., Vaskova, M., and Muskavitch, M. A. (2006). Structure-function analysis of delta trafficking, receptor binding and signaling in Drosophila. *Genetics* **174**, 1947–1961.

Parks, A. L., Turner, F. R., and Muskavitch, M. A. (1995). Relationships between complex Delta expression and the specification of retinal cell fates during Drosophila eye development. *Mech. Dev.* **50**, 201–216.

Pasternak, S. H., Bagshaw, R. D., Guiral, M., Zhang, S., Ackerley, C. A., Pak, B. J., Callahan, J. W., and Mahuran, D. J. (2003). Presenilin-1, nicastrin, amyloid precursor protein, and gamma-secretase activity are co-localized in the lysosomal membrane. *J. Biol. Chem.* **278**, 26687–26694.

Pavlopoulos, E., Pitsouli, C., Klueg, K. M., Muskavitch, M. A., Moschonas, N. K., and Delidakis, C. (2001). Neuralized Encodes a peripheral membrane protein involved in delta signaling and endocytosis. *Dev. Cell* **1**, 807–816.

Pitsouli, C., and Delidakis, C. (2005). The interplay between DSL proteins and ubiquitin ligases in Notch signaling. *Development* **132**, 4041–4050.

Poccia, D., and Larijani, B. (2009). Phosphatidylinositol metabolism and membrane fusion. *Biochem. J.* **418**, 233–246.

Polo, S., Sigismund, S., Faretta, M., Guidi, M., Capua, M. R., Bossi, G., Chen, H., De Camilli, P., and Di Fiore, P. P. (2002). A single motif responsible for ubiquitin recognition and monoubiquitination in endocytic proteins. *Nature* **416**, 451–455.

Poodry, C. A. (1990). Shibire, a neurogenic mutant of Drosophila. *Dev. Biol.* **138**, 464–472.
Poodry, C. A., Hall, L., and Suzuki, D. T. (1973). Developmental properties of Shibire: a pleiotropic mutation affecting larval and adult locomotion and development. *Dev. Biol.* **32**, 373–386.
Poulson, D. F. (1937). Chromosomal Deficiencies and the Embryonic Development of Drosophila Melanogaster. *Proc. Natl. Acad. Sci. U.S.A.* **23**, 133–137.
Qiu, L., Joazeiro, C., Fang, N., Wang, H. Y., Elly, C., Altman, Y., Fang, D., Hunter, T., and Liu, Y. C. (2000). Recognition and ubiquitination of Notch by Itch, a hect-type E3 ubiquitin ligase. *J. Biol. Chem.* **275**, 35734–35737.
Raiborg, C., and Stenmark, H. (2009). The ESCRT machinery in endosomal sorting of ubiquitylated membrane proteins. *Nature* **458**, 445–452.
Rajan, A., Tien, A. C., Haueter, C. M., Schulze, K. L., and Bellen, H. J. (2009). The Arp2/3 complex and WASp are required for apical trafficking of Delta into microvilli during cell fate specification of sensory organ precursors. *Nat. Cell Biol.* **11**, 815–824.
Ramain, P., Khechumian, K., Seugnet, L., Arbogast, N., Ackermann, C., and Heitzler, P. (2001). Novel Notch alleles reveal a Deltex-dependent pathway repressing neural fate. *Curr. Biol.* **11**, 1729–1738.
Rhyu, M. S., Jan, L. Y., and Jan, Y. N. (1994). Asymmetric distribution of numb protein during division of the sensory organ precursor cell confers distinct fates to daughter cells. *Cell* **76**, 477–491.
Roegiers, F., Jan, L. Y., and Jan, Y. N. (2005). Regulation of membrane localization of Sanpodo by lethal giant larvae and neuralized in asymmetrically dividing cells of Drosophila sensory organs. *Mol. Biol. Cell* **16**, 3480–3487.
Rogaev, E. I., Sherrington, R., Rogaeva, E. A., Levesque, G., Ikeda, M., Liang, Y., Chi, H., Lin, C., Holman, K., Tsuda, T., *et al.* (1995). Familial Alzheimer's disease in kindreds with missense mutations in a gene on chromosome 1 related to the Alzheimer's disease type 3 gene. *Nature* **376**, 775–778.
Sakata, T., Sakaguchi, H., Tsuda, L., Higashitani, A., Aigaki, T., Matsuno, K., and Hayashi, S. (2004). Drosophila Nedd4 regulates endocytosis of notch and suppresses its ligand-independent activation. *Curr. Biol.* **14**, 2228–2236.
Santolini, E., Puri, C., Salcini, A. E., Gagliani, M. C., Pelicci, P. G., Tacchetti, C., and Di Fiore, P. P. (2000). Numb is an endocytic protein. *J. Cell Biol.* **151**, 1345–1352.
Seugnet, L., Simpson, P., and Haenlin, M. (1997). Requirement for dynamin during Notch signaling in Drosophila neurogenesis. *Dev. Biol.* **192**, 585–598.
Sherrington, R., Rogaev, E. I., Liang, Y., Rogaeva, E. A., Levesque, G., Ikeda, M., Chi, H., Lin, C., Li, G., Holman, K., *et al.* (1995). Cloning of a gene bearing missense mutations in early-onset familial Alzheimer's disease. *Nature* **375**, 754–760.
Shimizu, K., Chiba, S., Kumano, K., Hosoya, N., Takahashi, T., Kanda, Y., Hamada, Y., Yazaki, Y., and Hirai, H. (1999). Mouse jagged1 physically interacts with notch2 and other notch receptors. Assessment by quantitative methods. *J. Biol. Chem.* **274**, 32961–32969.
Sigismund, S., Woelk, T., Puri, C., Maspero, E., Tacchetti, C., Transidico, P., Di Fiore, P. P., and Polo, S. (2005). Clathrin-independent endocytosis of ubiquitinated cargos. *Proc. Natl. Acad. Sci. U.S.A.* **102**, 2760–2765.
Simons, M., and Raposo, G. (2009). Exosomes–vesicular carriers for intercellular communication. *Curr. Opin. Cell Biol.* **21**, 575–581.
Skwarek, L. C., Garroni, M. K., Commisso, C., and Boulianne, G. L. (2007). Neuralized contains a phosphoinositide-binding motif required downstream of ubiquitination for delta endocytosis and notch signaling. *Dev. Cell* **13**, 783–795.
Small, S. A., and Gandy, S. (2006). Sorting through the cell biology of Alzheimer's disease: intracellular pathways to pathogenesis. *Neuron* **52**, 15–31.

Sorkin, A., and von Zastrow, M. (2009). Endocytosis and signalling: intertwining molecular networks. *Nat. Rev. Mol. Cell Biol.* **10**, 609–622.

Struhl, G., and Adachi, A. (2000). Requirements for presenilin-dependent cleavage of notch and other transmembrane proteins. *Mol. Cell* **6**, 625–636.

Sun, X., and Artavanis-Tsakonas, S. (1997). Secreted forms of DELTA and SERRATE define antagonists of Notch signaling in Drosophila. *Development* **124**, 3439–3448.

Tagami, S., Okochi, M., Yanagida, K., Ikuta, A., Fukumori, A., Matsumoto, N., Ishizuka-Katsura, Y., Nakayama, T., Itoh, N., Jiang, J., *et al.* (2008). Regulation of Notch signaling by dynamic changes in the precision of S3 cleavage of Notch-1. *Mol. Cell Biol.* **28**, 165–176.

Tal, T., Vaizel-Ohayon, D., and Schejter, E. D. (2002). Conserved interactions with cytoskeletal but not signaling elements are an essential aspect of Drosophila WASp function. *Dev. Biol.* **243**, 260–271.

Tanaka, N., Kyuuma, M., and Sugamura, K. (2008). Endosomal sorting complex required for transport proteins in cancer pathogenesis, vesicular transport, and non-endosomal functions. *Cancer Sci.* **99**, 1293–1303.

Thompson, B. J., Mathieu, J., Sung, H. H., Loeser, E., Rorth, P., and Cohen, S. M. (2005). Tumor suppressor properties of the ESCRT-II complex component Vps25 in Drosophila. *Dev. Cell* **9**, 711–720.

Tian, X., Hansen, D., Schedl, T., and Skeath, J. B. (2004). Epsin potentiates Notch pathway activity in Drosophila and C. elegans. *Development* **131**, 5807–5815.

Traub, L. M. (2009). Clathrin couture: fashioning distinctive membrane coats at the cell surface. *PLoS Biol.* **7**, e1000192.

Tsukazaki, T., Chiang, T. A., Davison, A. F., Attisano, L., and Wrana, J. L. (1998). SARA, a FYVE domain protein that recruits Smad2 to the TGFbeta receptor. *Cell* **95**, 779–791.

Vaccari, T., and Bilder, D. (2005). The Drosophila tumor suppressor vps25 prevents non-autonomous overproliferation by regulating notch trafficking. *Dev. Cell* **9**, 687–698.

Vaccari, T., Lu, H., Kanwar, R., Fortini, M. E., and Bilder, D. (2008). Endosomal entry regulates Notch receptor activation in Drosophila melanogaster. *J. Cell Biol.* **180**, 755–762.

Vaccari, T., Rusten, T. E., Menut, L., Nezis, I. P., Brech, A., Stenmark, H., and Bilder, D. (2009). Comparative analysis of ESCRT-I, ESCRT-II and ESCRT-III function in Drosophila by efficient isolation of ESCRT mutants. *J. Cell Sci.* **122**, 2413–2423.

van der Bliek, A. M., and Meyerowitz, E. M. (1991). Dynamin-like protein encoded by the Drosophila shibire gene associated with vesicular traffic. *Nature* **351**, 411–414.

Varnum-Finney, B., Wu, L., Yu, M., Brashem-Stein, C., Staats, S., Flowers, D., Griffin, J. D., and Bernstein, I. D. (2000). Immobilization of Notch ligand, Delta-1, is required for induction of notch signaling. *J. Cell Sci.* **113**(*Pt 23*), 4313–4318.

Wang, W., and Struhl, G. (2004). Drosophila Epsin mediates a select endocytic pathway that DSL ligands must enter to activate Notch. *Development* **131**, 5367–5380.

Wang, W., and Struhl, G. (2005). Distinct roles for Mind bomb, Neuralized and Epsin in mediating DSL endocytosis and signaling in Drosophila. *Development* **132**, 2883–2894.

Weber, U., Eroglu, C., and Mlodzik, M. (2003). Phospholipid membrane composition affects EGF receptor and Notch signaling through effects on endocytosis during Drosophila development. *Dev. Cell* **5**, 559–570.

Wilkin, M. B., Carbery, A. M., Fostier, M., Aslam, H., Mazaleyrat, S. L., Higgs, J., Myat, A., Evans, D. A., Cornell, M., and Baron, M. (2004). Regulation of notch endosomal sorting and signaling by Drosophila Nedd4 family proteins. *Curr. Biol.* **14**, 2237–2244.

Wilkin, M., Tongngok, P., Gensch, N., Clemence, S., Motoki, M., Yamada, K., Hori, K., Taniguchi-Kanai, M., Franklin, E., Matsuno, K., *et al.* (2008). Drosophila HOPS and AP-3 complex genes are required for a Deltex-regulated activation of notch in the endosomal trafficking pathway. *Dev. Cell* **15**, 762–772.

Windler, S. L., and Bilder, D. (2010). Endocytic internalization routes required for Delta/Notch signaling. *Curr. Biol.* **20**, 538–543.

Wu, S., Mehta, S. Q., Pichaud, F., Bellen, H. J., and Quiocho, F. A. (2005). Sec15 interacts with Rab11 via a novel domain and affects Rab11 localization in vivo. *Nat. Struct. Mol. Biol.* **12**, 879–885.

Xu, T., and Artavanis-Tsakonas, S. (1990). Deltex, a locus interacting with the neurogenic genes, Notch, Delta and mastermind in Drosophila melanogaster. *Genetics* **126**, 665–677.

Yan, Y., Denef, N., and Schupbach, T. (2009). The vacuolar proton pump, V-ATPase, is required for notch signaling and endosomal trafficking in Drosophila. *Dev. Cell* **17**, 387–402.

Yanochko, G. M., and Yool, A. J. (2002). Regulated cationic channel function in Xenopus oocytes expressing Drosophila big brain. *J. Neurosci.* **22**, 2530–2540.

Yeh, E., Dermer, M., Commisso, C., Zhou, L., McGlade, C. J., and Boulianne, G. L. (2001). Neuralized functions as an E3 ubiquitin ligase during Drosophila development. *Curr. Biol.* **11**, 1675–1679.

Zecchini, V., Brennan, K., and Martinez-Arias, A. (1999). An activity of Notch regulates JNK signalling and affects dorsal closure in Drosophila. *Curr. Biol.* **9**, 460–469.

Zeng, C., Younger-Shepherd, S., Jan, L. Y., and Jan, Y. N. (1998). Delta and Serrate are redundant Notch ligands required for asymmetric cell divisions within the Drosophila sensory organ lineage. *Genes Dev.* **12**, 1086–1091.

Zhang, L., Song, L., Terracina, G., Liu, Y., Pramanik, B., and Parker, E. (2001). Biochemical characterization of the gamma-secretase activity that produces beta-amyloid peptides. *Biochemistry* **40**, 5049–5055.

CHAPTER SIX

γ-Secretase and the Intramembrane Proteolysis of Notch

Ellen Jorissen *and* Bart De Strooper

Contents

Abstract

γ-secretase is the crucial proteolytic activity that releases the Notch intracellular domain and is therefore a central player in the canonical Notch-signaling transduction pathway. We discuss here briefly the discovery of γ-secretase and what is known on its structure and function. Recent work also indicates that the assembly and activity of γ-secretase might be regulated by novel cell biological mechanisms. Finally we explore the recent insight that there are several γ-secretase complexes in mammalian and discuss possibilities to use γ-secretase as a drug target in Alzheimer's disease and cancer.

Center for Human Genetics, KULeuven, and Department for Molecular and Developmental Genetics, VIB, Leuven, Belgium

Current Topics in Developmental Biology, Volume 92
ISSN 0070-2153, DOI 10.1016/S0070-2153(10)92006-1

1. Introduction

The study of the Notch-signaling pathway has contributed fundamentally to the understanding of a novel type of cellular signaling process, called regulated intramembrane proteolysis (Brown *et al.*, 2000). This signaling process releases protein fragments at two sides of the cellular membrane, providing cell autonomous signals to the cell interior and sending instructions to neighboring cells via ligand–receptor interactions. Regulated intramembrane proteolysis is now touching a broad field of cell biological research from unicellular organisms to man, and is involved in a myriad of cellular signaling processes (Freeman, 2008; Rawson, 2003; Urban, 2009). The Notch-signaling pathway is of particular interest in this regard as receptors (Notch1-4) and ligands (Delta, Jagged in mammalian) undergo regulated intramembrane proteolysis (De Strooper *et al.*, 1999; Ikeuchi and Sisodia, 2003; LaVoie and Selkoe, 2003; Saxena *et al.*, 2001; Struhl and Greenwald, 1999). The study of the processing of Notch has been instrumental for our understanding of γ-secretase, the main subject of the current review. γ-Secretase is in fact a generic name, coming from the Alzheimer's research field (Haass and Selkoe, 1993). Indeed, the amyloid precursor protein (APP), which is the precursor of the Aβ or amyloid peptide causally related to the pathogenesis of Alzheimer's disease (AD), undergoes a very similar consecutive proteolysis as Notch (Annaert and De Strooper, 1999). γ-Secretase occurs after either α- or β-secretase cleavage of APP (Haass and Selkoe, 1993) and results in the release of the notorious Aβ peptide.

2. Regulated Intramembrane Proteolysis of Notch

Signaling of the Notch receptor is dependent on three types of proteolytic events. After the first cleavage, known as the S1 cleavage, by furin-like convertase in the secretory pathway (Blaumueller *et al.*, 1997; Logeat *et al.*, 1998), the heterodimeric receptor (Blaumueller *et al.*, 1997; Logeat *et al.*, 1998; Rand *et al.*, 2000) proceeds to the cell surface where it is able to interact with Notch ligands presented on neighboring cells. Binding of Delta, Serrate, or Lag-2 ligands like Delta or Jagged to the Notch-1 receptor triggers an additional extracellular proteolysis (S2 cleavage) by a membrane-tethered metalloprotease, within the extracellular juxtamembrane region (Brou *et al.*, 2000; Mumm and Kopan, 2000). Both "a disintegrin and metalloprotease" (ADAM) 10 and ADAM17 have been implicated in this S2 cleavage. Recent *in vitro* studies suggest that ADAM10 is required for ligand-induced Notch-1 signaling (van Tetering *et al.*, 2009), while ligand-independent Notch-1 signaling requires ADAM17 (Cagavi Bozkulak and Weinmaster, 2009). However *in vivo* only *Adam10* knockout (KO) mice display Notch-1 loss-of-function phenotypes, whereas Adam17 KO do not (Hartmann *et al.*, 2002). Also in adult brain ADAM10 loss of function causes significant Notch-dependent alterations (Jorissen *et al.*, 2010) confirming the central role of ADAM10 in the Notch-signaling pathway. The membrane-associated remnant is then cleaved within its transmembrane domain by γ-secretase, releasing the Notch intracellular domain (NICD) (De Strooper *et al.*, 1999; Struhl and Greenwald, 1999). NICD translocates to the nucleus and activates transcription after associating to nuclear proteins of the CLS (CBP/RBPjk, Su (H), Lag-1) family. Notch signaling elevates Hes and Hey expression which are two families of basic helix–loop–helix transcription factors. These in turn reduce the expression of downstream proneural effectors such as Neurogenin, Mash, and MyoD (Kopan and Ilagan, 2009).

The regulation of the intramembrane proteolysis of Notch occurs in the first place at the level of the ligand-induced cleavage by ADAM10. One mechanism proposes that ligand binding causes the opening of a "negative regulatory domain" in Notch that blocks the ADAM10 cleavage site (Gordon *et al.*, 2007). In the most simplistic scheme, the γ-secretase cleavage is then executed by default, but as we will see below, recent work suggests that γ-secretase might be regulated as well. An additional interesting novel insight is that several γ-secretases might differentially be involved in Notch signaling, adding sophistication to the system, and opening novel ways to think about targeting this proteolytic activity for therapeutic purposes.

3. Discovery of γ-Secretase

In the mid-1990s genetic linkage analysis of families with autosomal dominant forms of familial AD gave the first crucial clues to the molecular identification of γ-secretase (Alzheimer's_Disease_Collaborative_Group, 1995; Levy-Lahad *et al.*, 1995; Rogaev *et al.*, 1995; Sherrington *et al.*, 1995). Missense mutations in two previous unknown genes, *Presenilin* 1&2 (Psen-1 and Psen-2), were sufficient to cause an aggressive, dominant inherited form of AD. It should be said that it was initially unclear that the novel genes were involved in proteolysis and even the relation with APP processing remained for several years unclear, until it became clear that genetic KO of Psen-1 basically wiped out Aβ generation (De Strooper *et al.*, 1998).

The Psens were discovered independently and from a completely different perspective in elegant genetic work focusing on the basic understanding of the Notch/Lin-12 signaling pathway in the model organism

Caenorhabditis elegans. The Psen ortholog sel-12 was found to be a suppressor of a lin-12 gain-of-function mutation (Levitan and Greenwald, 1995).

The underlying molecular biological mechanism linking the two lines of genetic investigation was established some years later: Psen turned out to be directly responsible for the intramembrane proteolysis of APP and Notch (De Strooper *et al.*, 1999; Struhl and Greenwald, 1999), with Psen processing of APP yielding the Aβ peptide causing AD and Psen processing of Notch leading to the release of the NICD responsible for signaling (Annaert and De Strooper, 1999). This observation has largely set the research agenda for γ-secretase research in the next 10 years. A large number of studies have confirmed that the Psens are required for intramembranous Notch proteolysis in *Drosophila*, mice, and human cells (Berechid *et al.*, 1999; Berezovska *et al.*, 2000; De Strooper *et al.*, 1999; Herreman *et al.*, 2000; Song *et al.*, 1999; Zhang *et al.*, 2000). In addition, the phenotype of total loss of the Psen genes in *C. elegans* (Li and Greenwald, 1997; Westlund *et al.*, 1999), mice (Donoviel *et al.*, 1999; Hartmann *et al.*, 1999; Herreman *et al.*, 1999), and *Drosophila* (Struhl and Greenwald, 1999; Ye and Fortini, 1999) bears a striking resemblance to the phenotype of complete loss of Notch signaling (de la Pompa *et al.*, 1997; Oka *et al.*, 1995).

4. γ-Secretase Cleaves Many Substrates

More than 60 different substrates are known to be processed by the combination sheddase/γ-secretase (reviewed in (McCarthy *et al.*, 2009; Wakabayashi and De Strooper, 2008)) in a similar way as APP and Notch. The sheddases usually belong to the family of the ADAMs, but other sheddases such as BACE1 may cleave a more restricted panel of substrates (reviewed in (Cole and Vassar, 2008)). The released ectodomain can become a soluble ligand (e.g., secreted APP or APPs) or, in the case of Notch, bind to Delta or Jagged on the opposing cells and become internalized by the signal emitting cells (D'Souza *et al.*, 2008). The remaining membrane-embedded carboxyterminal fragments (CTF) have a shortened extracellular domain of less than 30 amino acids, which is a prerequisite to become a substrate for proteolysis within the transmembrane domain by γ-secretase (Struhl and Adachi, 2000). This results in the generation of an intracellular domain (ICD) (Haass, 2004; Wolfe and Kopan, 2004) and the release of smaller amino-terminal fragments into the extracellular space. These fragments have been sequenced only for a few substrates and have been called "Xβ" in analogy to Aβ, e.g., Nβ, APLP1β, and CD44β for Notch, APLP1, and CD44, respectively (Lammich *et al.*, 2002; Okochi *et al.*, 2002; Yanagida *et al.*, 2009).

The function, if any, of the Xβ fragments is unclear. The ICD, in analogy with NICD (Schroeter *et al.*, 1998), are usually thought to have a function in the regulation of nuclear transcription. Indeed, when overexpressed, such fragments, like the APP ICD (AICD), are transported to the nucleus and can activate reporter gene constructs. Some caution however is indicated (Wakabayashi and De Strooper, 2008) as is illustrated by the example of AICD. Several laboratories have indeed succeeded in identifying candidate genes regulated by AICD, but the proposed target genes vary considerably and no consistent hypothesis has linked these observations with any of the phenotypes observed in APP/APLP1&2 KO mice (for a more elaborate discussion see (Wakabayashi and De Strooper, 2008)). This is in contrast with the Notch-signaling cascade where the link between NICD, the "CSL" (CBF1, Su(H), LAG1) transcription factors and coactivators and the regulation of Notch target genes such as the hairy and enhancer of split family is fully supported by a whole battery of genetic experimental evidence (Kopan and Ilagan, 2009). These considerations suggest also that the Psen/γ-secretase complex may in principle be as well considered a mechanism to clear the membrane-embedded stubs of many different type I proteins, operating as the "proteasome of the membrane" (Kopan and Ilagan, 2004). Probably the situation is more complicated, and the cleavage of the transmembrane domain might have biological relevance in more than one way. γ-Secretase cleavage of E-cadherin for instance dissociates E-cadherins from its bound protein partners, α- and β-catenin, thereby promoting disassembly of the adherens junctions (Marambaud *et al.*, 2002). The E-cadherin ICD on the other hand is not known to be involved in nuclear signaling. It is an interesting question whether the concomitant release of the membrane bound pool of β-catenin could affect or bypass Wnt signaling (De Strooper and Annaert, 2001).

5. Unraveling the γ-Secretase Complex

5.1. Presenilin: the catalytic subunit and substrate docking

Psens are multipass transmembrane proteins consisting of nine transmembrane domains (TM), with the N-terminus facing the cytosol and the short C-terminus oriented to the extracellular space (Fig. 6.1A) (Henricson *et al.*, 2005; Laudon *et al.*, 2005; Oh and Turner, 2005; Spasic *et al.*, 2006). The catalytic site of Psen consists of two conserved aspartyl residues, located within TM6 en TM7 (Wolfe *et al.*, 1999). Mutations in either of these residues results in loss of γ-secretase activity (Kimberly *et al.*, 2000; Steiner *et al.*, 1999a; Wolfe *et al.*, 1999), as well as loss of binding to transition state inhibitors (Esler *et al.*, 2000; Li *et al.*, 2000; Seiffert *et al.*, 2000), without affecting the formation of the high molecular weight

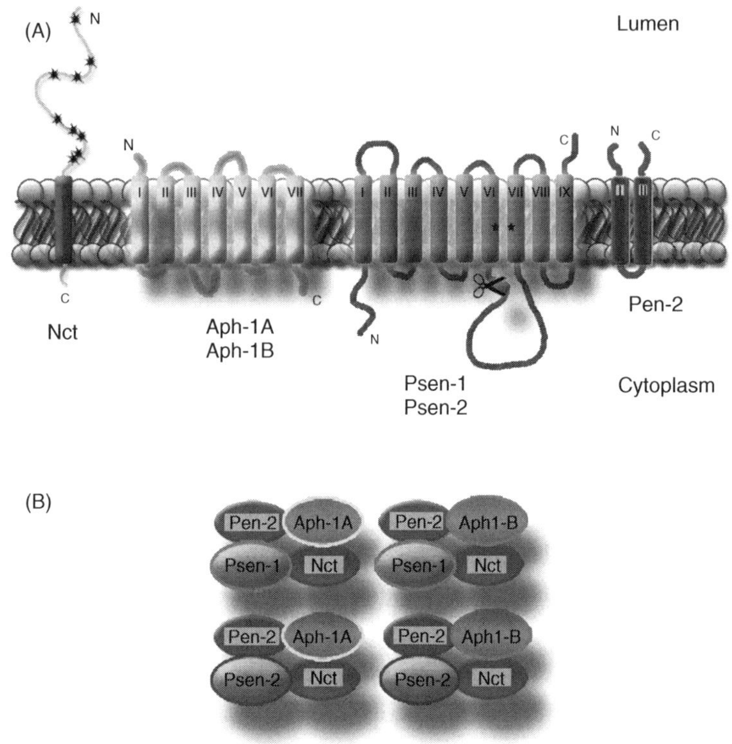

Figure 6.1 The architecture of the γ-secretase complex. (A) Subunit topology: γ-secretase consists of four proteins: Psen, Nct, Aph-1, and Pen-2. Psen forms the catalytic subunit (the catalytic aspartyl residues are indicated with stars). The endoproteolytic cleavage site in the cytoplasmic loop of Psen is indicated. Stars on the luminal domain of Nct indicate complex glycosylation of the mature form of Nct. (B) Schematic representation of the four different γ-secretase complexes. The heterogeneity is based on the existence of two different Psen genes, Psen-1 and Psen-2, and two different Aph-1 genes, Aph-1A and Aph-1B. (See Color Insert.)

complex (Anderton *et al.*, 2000; Nyabi *et al.*, 2003). Psen reveals weak amino acid sequence similarities to bacterial type-4 prepilin proteases, a class of membrane-embedded aspartyl proteases (Steiner *et al.*, 2000). Psen-1 KO mice revealed a Notch phenotype (De Strooper *et al.*, 1999; Shen *et al.*, 1997; Wong *et al.*, 1997) and resulted in a reduced γ-site cleavage of APP (De Strooper *et al.*, 1998). The remaining activity turned out to be contributed by Psen-2 (Herreman *et al.*, 2000; Zhang *et al.*, 2000). This provides evidence that Psen is the catalytic active core component of γ-secretase as was proposed by three groups at the same time, for various reasons (De Strooper *et al.*, 1999; Wolfe *et al.*, 1999; Struhl and Adachi, 1999). Final proof that Psen is the catalytic subunit of γ-secretase came from the observation that transition state protease inhibitors blocking γ-secretase bound to this protein (Esler *et al.*, 2000; Li *et al.*, 2000).

Psen is endoproteolytically cleaved into stable 30-kDa N- and 20-kDa C-terminal fragment (NTF & CTF, respectively) that associate to form a heterodimer (Thinakaran *et al.*, 1996). This cleavage occurs within the large cytoplasmic loop between TM6 and TM7 within a short hydrophobic domain (Edbauer *et al.*, 2003; Ratovitski *et al.*, 1997). The cleavage protease, although unknown, has been called presenilinase. The facts that presenilinase is inhibited by transition state compounds targeting aspartic proteases and that catalytically inactive Psen mutants do not undergo endoproteolytic cleavage, have led to the proposal that Psen itself is presenilinase (Xia, 2008). However, the potencies of several well-characterized γ-secretase inhibitors (GSIs) blocking Psen and Presenilinase do not correlate, suggesting that the two activities are pharmacologically distinct (Xia and Wolfe, 2003). Endoproteolysis is not an absolute requirement for γ-secretase activity (Steiner *et al.*, 1999b), in contrast to what was initially suggested (Wolfe *et al.*, 1999). The familial AD-linked Psen-1 delta exon9 deletion mutant (Steiner *et al.*, 1999a), as well as mutants of the endoproteolysis site (Jacobsen *et al.*, 1999), blocks Psen endoproteolysis, but allows γ-secretase activity. On the other hand the cleaved Psen-NTF and Psen-CTF fragments are the predominant forms of Psen in the cell, whereas the Psen holoprotein is rapidly degraded (Thinakaran *et al.*, 1996, 1997). Simultaneously co-expression of Psen-NTF and Psen-CTF in the absence of endogenous Psen was able to produce active γ-secretase (Laudon *et al.*, 2004; Levitan *et al.*, 2001). An interesting model supported by fluorescence lifetime-imaging microscopy experiments assumes two conformations for Psen: an inactive conformation characterized by less interference of fluorophores on the NTF and CTF domains of presenilin, and a second conformation after proteolysis, resulting in the mature γ-secretase conformation in which the N and C terminus of Psen come close together. FAD-linked Psen-1 mutations change the conformation to an even closer state of the NTF and CTF, suggesting that the molecular conformation of Psen-1 is linked to the precision of γ-cleavage of APP to yield Aβ species of different lengths (Berezovska *et al.*, 2005; Uemura *et al.*, 2009). Recent data show that this hydrophobic domain in the Psen loop can alternate between positions that are water accessible and not, supporting its conformational flexibility (Berezovska *et al.*, 2005; Bergman *et al.*, 2004; Sato *et al.*, 2008; Tolia *et al.*, 2008; Uemura *et al.*, 2009).

Among the more intriguing questions on the entire family of intramembrane-cleaving proteases is how they hydrolyze substrates within the hydrophobic environment of the lipid bilayer. Thus, the active site within Psen has to be part of a water-containing pore or channel, as was confirmed recently using cysteine-scanning mutagenesis (Sato *et al.*, 2006; Tolia *et al.*, 2006). However, the integral membrane substrates initially should interact on the surface of the protease before entering the internal active site.

Evidence for such a mechanism came from co-precipitation experiments of an endogenous γ-secretase substrate with the protease complex isolated with an immobilized transition-state analog, suggesting that the substrate can bind at some other site while the immobilized inhibitor occupies the active site (Annaert *et al.*, 2001; Esler *et al.*, 2002). Designed helical peptides based on the transmembrane domain of APP can potently inhibit γ-secretase, apparently by interacting with this docking site (Das *et al.*, 2003). Conversion of these helical peptide inhibitors into photo affinity probes localized the substrate-binding site on Psen near the active site (Kornilova *et al.*, 2005), which was also suggested by binding studies between APP CTF and fragments of Psen (Annaert *et al.*, 2001). Interestingly, extension of a 10-residue helical peptide inhibitor by just three additional residues resulted in a more potent inhibitor apparently capable of binding both the docking site and the active site (Bihel *et al.*, 2004; Kornilova *et al.*, 2005), suggesting that these two substrate binding sites are relatively close.

5.2. Nicastrin as gatekeeper?

When Psen was immunoprecipitated in non-denaturing conditions, Nicastrin (Nct) co-precipitated (Yu *et al.*, 2000). Nct was also identified as Aph-2 (anterior pharynx defective) in a screen for genetic modifiers of the GLP-1 pathway (Notch in *C. elegans*) and was found to be essential for γ-secretase processing of both APP and Notch (Goutte *et al.*, 2000; Yu *et al.*, 2000). Nct is a 130-kDa type-I transmembrane glycoprotein and has a large extracellular domain with a conserved (Fig. 6.1A), functionally important DYIGS motif and a relatively small ICD with little sequence similarity between species. Nct undergoes maturation (N-glycosylation) during trafficking through the Golgi/trans Golgi pathway, although glycosylation itself is apparently not required for γ-secretase assembly or activity (Edbauer *et al.*, 2002; Herreman *et al.*, 2003; Kimberly *et al.*, 2002; Tomita *et al.*, 2002; Yang *et al.*, 2002; Yu *et al.*, 2000).

The exact role of Nct in the γ-secretase complex remains an issue of debate. One group has suggested that Nct might play a role in substrate recognition. The Nct ectodomain can bind the new N-terminus that is generated upon shedding of the ectodomain of γ-secretase substrates (Shah *et al.*, 2005) Chemical- or antibody-mediated blocking of the free N-terminus, addition of purified Nct ectodomain, or mutations in the ectodomain of Nct markedly reduced the binding and cleavage of substrate by γ-secretase. One conserved glutamate residues (E333) within the highly conserved DYIGS motif was noted to be crucial for Nct–substrate interaction (Dries *et al.*, 2009; Shah *et al.*, 2005). These observations led to the interesting proposal that Nct is actually the gatekeeper for the γ-secretase complex, restricting access of substrates to the complex. However, recent work showed that the E333 residue is critical for the assembly and maturation of the γ-secretase complex rather than the recognition of the substrate. In fact it was shown that the little amount of γ-secretase complex that is still generated when the E333 mutant is expressed was as active as the wild-type counterpart on a per mole basis, challenging the proposed model that Nct is a receptor for γ-secretase substrates in mature, active enzyme (Chavez-Gutierrez *et al.*, 2008). A recent study confirmed and extended this in Nct-deficient mouse embryonic fibroblasts where the trimeric Psen-1/Pen-2/Aph-1a complex is an active but highly unstable enzyme. Thus, Nct acts to stabilize γ-secretase but is dispensable for γ-secretase activity (Zhao *et al* 2010).

5.3. Aph-1 and Pen-2

The genes encoding for Aph-1 and Pen-2 were identified via a genetic screen for Psen modifiers in *C. elegans* (Francis *et al.*, 2002; Goutte *et al.*, 2002). Aph-1 has a seven TM structure with the C-terminus located in the cytoplasm (Fig. 6.1A) (Fortna *et al.*, 2004). Two homologs of Aph-1, Aph-1a, and Aph-1b have been identified in humans, and they are in addition alternatively spliced. In rodents, gene duplication of Aph-1b has given rise to a third gene Aph-1c. Aph-1a has 55% sequence similarity with Aph-1b/Aph-1c, whereas Aph-1b and Aph-1c share 96% similarity at the nucleotide level. Aph-1a exists as a long (Aph-1aL) and a short (Aph-1aS) splice variant differing by the addition of 18 residues on the C- terminus of Aph1aL (Gu *et al.*, 2003; Lee *et al.*, 2002). Two highly conserved histidine residues in TM5 and TM6 contribute to Aph-1 function and can affect Psen catalytic activity: mutations in these residues affected Psen-1 cleavage and altered binding to other γ-secretase components, resulting in decreased Aβ-generating activity (Pardossi-Piquard *et al.*, 2009). The conserved GxxxG motif in TM4 of Aph-1a is necessary for intermolecular interactions with Psen and Pen-2, but not with Nct (Araki *et al.*, 2006; Lee *et al.*, 2004; Niimura *et al.*, 2005).

Pen-2 is a small hairpin protein with two transmembrane domains and both termini located at the luminal side (Fig. 6.1A) (Crystal *et al.*, 2003). The N-terminus of Pen-2 plays a role in the interaction with Psen (Crystal *et al.*, 2003), whereas the C-terminus and TM1 are necessary for the endoproteolysis and subsequent activation and stabilization of Psen in the complex, by a yet unknown mechanism (Hasegawa *et al.*, 2004; Kim and Sisodia, 2005).

6. γ-Secretases Are Tetrameric Complexes

Co-expression of all four components (Psen, Nct, Aph-1, and Pen-2) increased γ-secretase activity in both *Drosophila* and mammalian cells and reconstituted γ-secretase activity in Sf9 insect cells and in the budding yeast *Saccharomyces cerevisiae* (Edbauer *et al.*, 2003; Hayashi *et al.*, 2004; Kimberly *et al.*, 2003; Takasugi *et al.*, 2003; Zhang *et al.*, 2005). The latter experiment was particularly compelling since yeast does not possess such protease activity and contains no orthologs of these four metazoan proteins. That these four proteins are necessary and sufficient to reconstitute γ-secretase was subsequently confirmed through purification of the protease complex (Fraering *et al.*, 2004) and numerous other experiments. Co-immunoprecipitation of the individual γ-secretase components Aph-1, Nct, Psen-1, or Pen-2 pulled down each of the other three factors (Baulac *et al.*, 2003; Kimberly *et al.*, 2003; Steiner *et al.*, 2002; Takasugi *et al.*, 2003). The Nct KO and Aph-1 KO mice have phenotypes similar to Notch1 or double Psen-1&2 KO, demonstrating that both Nct and Aph-1 are essential parts of the γ-secretase complex (Li *et al.*, 2003a). Absence of functional Nct in cells abolished NICD and Aβ production, further implicating Nct in γ-secretase activity (Chen *et al.*, 2001; Chung and Struhl, 2001; Hu *et al.*, 2002; Li *et al.*, 2003b; Lopez-Schier and St Johnston, 2002). Additionally, RNA interference (RNAi)-mediated inactivation of Aph-1, Pen-2, or Nct in cultured *C. elegans, Drosophila*, and mammalian cells inhibited Psen endoproteolysis and deficient γ-secretase activity (Edbauer *et al.*, 2002; Francis *et al.*, 2002; Hu *et al.*, 2002; Lee *et al.*, 2002; Yu *et al.*, 2000). Thus, downregulation of any of the four proteins using either RNAi or classical genetic KO results in instability and lack of maturation of the other members of the complex and loss of γ-secretase function (reviewed in (De Strooper, 2003)).

When the four proteins are overexpressed together, the increase in total γ-secretase activity is not increasing linearly with protein expression, suggesting that additional factors are involved in γ-secretase regulation (see below).

Over the last few years it is realized that in mammalian cells at least four different γ-secretase complexes exist (Hebert *et al.*, 2004; Shirotani *et al.*, 2004) with likely different biological functions. This heterogeneity is based on the fact that two different Psen genes and two different Aph-1 genes exist and that the incorporation of the encoded proteins is mutually exclusive (Fig. 6.1B) (De Strooper, 2003). In overexpression paradigms all four different complexes are able to cleave similar substrates (our unpublished results), but one should not jump from this to conclusions that the four enzymes have identical activities. More careful kinetic studies are likely needed to analyze this question. Some initial data suggest indeed that complexes containing Psen-1 are more efficient in cleaving APP than those containing Psen-2 (Bentahir *et al.*, 2006; Mastrangelo *et al.*, 2005). The phenotype of the single Psen-1 KO mouse is also far more severe than that of the Psen-2 KO, suggesting that Psen-2 is dispensable for normal Notch signaling during embryogenesis (Donoviel *et al.*, 1999; Shen *et al.*, 1997; Wong *et al.*, 1997). A series of more recently published work indicates that also the different Aph-1 proteins contribute to tissue and substrate specificity of the γ-secretase complex. Aph-1a is essential for Notch mediated signaling during mammalian development while Aph-1b/c γ-secretase plays a role in normal brain function and also appears to have a brain-specific function in Neuregulin-dependent developmental pathways (Dejaegere *et al.*, 2008; Ma *et al.*, 2005; Serneels *et al.*, 2005). While Notch signaling appeared to be conserved in the Aph-1b/c KO mice, the total amount of Aβ peptides in Aph-1b KO brains was decreased (Serneels *et al.*, 2009), dissociating Aβ generation from Notch signaling in this model.

7. Structure and Assembly of the Complex

Cysteine-scanning mutagenesis of Psen-1 shows that TM6 and TM7 that harbor the two catalytic aspartates, delineate a water-containing cavity inside the membrane (Sato *et al.*, 2006; Tolia *et al.*, 2006). Recently, it was demonstrated that TM9 and the hydrophobic domain in the large cytoplasmic loop of Psen (between TM6 and TM7) are dynamic parts of the water-containing cavity; also the conserved PAL motif in TM9 contributes to the catalytic center because it can be cross-linked to the active aspartate located within TM6 (Sato *et al.*, 2008; Tolia *et al.*, 2008). However, the regulation of water entry and substrate translocation to the catalytic site are important issues that still need to be resolved. Because of the tremendous difficulty in obtaining high amounts of pure γ-secretase, high-resolution structural information is not yet available. However, the first structural studies of recombinant human γ-secretase complex expressed in CHO cells using electron microscopy (EM) showed a large spherical structure with a low-density interior, which was suggested to be a water-accessible proteolytic chamber (Lazarov *et al.*, 2006). A second EM structural study of an γ-secretase preparation from Sf9 cells reported images at lower resolution (Ogura *et al.*, 2006) and suggested a Y-shaped 3D structure with a large pore. These structures obtained with negative staining are limited in value as they only provide the contours of the complex. Recently a higher resolution structure of γ-secretase was determined using cryo-EM at 12 Å. In principle, such approach can provide real structural information if the resolution can be pushed below 10 Å or lower (Osenkowski *et al.*, 2009), but at this moment we have only a very general view of the complex. Instead of the single chamber, the new cryo-EM structure reveals three smaller low-density interior regions and provides a better definition of the transmembrane regions and extracellular density domains. Moreover, a vertically oriented groove on the surface of the transmembrane portion was speculated to provide potentially the initial substrate-docking sites (Kornilova *et al.*, 2005; Osenkowski *et al.*, 2009). The electrolucent cavities seemed open to the cytoplasm or extracellular space, but do not cross the membrane. Other crystals from intramembrane cleaving proteases have been published (Ha, 2007; Urban and Shi, 2008) suggesting that these proteins create cavities in the membrane which harbor the active sites of the proteases. However, γ-secretases consist of four proteins, while rhomboids and other intramembrane cleaving proteases usual consist of only one protein, suggesting that it will be very difficult to extrapolate what is known of the other proteases toward γ-secretase.

Ongoing work using cryo-EM and/or 2D crystallization might further improve the resolution and will continue to play an important role in our understanding of the molecular basis of γ-secretase and the development of mechanism-based GSIs. Three-dimensional crystallization of the entire complex, with its 19 transmembrane domains, is a particular challenge.

8. Consecutive Cleavage Model for γ-Secretase

Recent biochemical studies have indicated that γ-secretase cleaves its substrates at multiple positions in a stepwise manner within their membrane domain (Qi-Takahara *et al.*, 2005; Zhao *et al.*, 2005) (Fig. 6.2). APP has

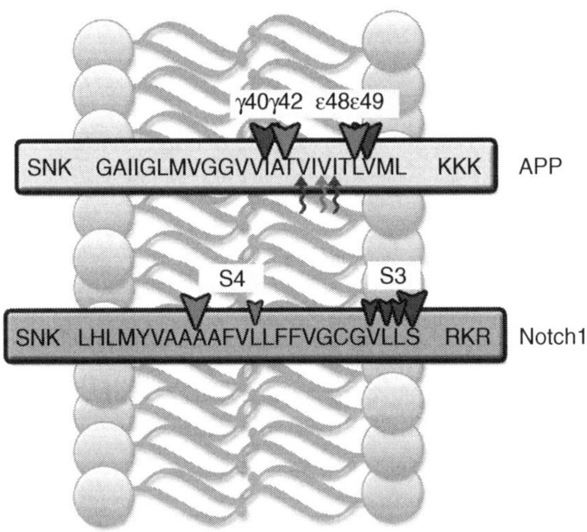

Figure 6.2 γ-Cleavage sites in APP and Notch. Schematic representation of the amino acid sequences of the TMs of APP and Notch; the cleavage sites are indicated with arrows. For APP the ε-cleavage of the βCTF generates $A\beta_{49}$ and $A\beta_{48}$. γ-Secretase cleaves then in the direction from the ε-cleavage to γ-cleavage sites by releasing tripeptides, finally producing $A\beta_{40}$ and $A\beta_{42}$, respectively. Notch NEXT fragment with arrowheads showing the S3 and S4 proteolytic sites.

been best investigated in that regard. The current model suggests that proteolysis of APP occurs first close to the cytoplasmic border of the membrane at the ε-site, releasing the AICD in the cytoplasm and leaving long Aβ (Aβ48,Aβ49) species in the membrane. Further cleavages then occur every third residue (every alpha-helical turn) until the γ-site is reached so that the remaining Aβ peptide becomes short enough to be released from the membrane. The cleavages at the ε-site can occur at two main sites which determine largely the consecutive cleavages. Therefore two different product lines are proposed: Aβ40 is generated through Aβ43/46 from Aβ49, and Aβ42 is generated through Aβ45 from Aβ48, which is generated by ε-cleavage (Kakuda *et al.*, 2006; Qi-Takahara *et al.*, 2005; Takami *et al.*, 2009; Wolfe, 2007; Zhao *et al.*, 2007). The model has gained considerable support from the recent finding that the tripeptides predicted by this consecutive cleavage model can actually be detected in a cell-free γ-secretase assay (Takami *et al.*, 2009). It remains a highly interesting question how such sequential processing occurs mechanistically and whether unwinding of the α-helix, or movement of the active site of the γ-secretase, is responsible for the progressive cutting of the transmembrane domain.

Sequential proteolysis of Notch-1 by γ-secretase has been proposed, as well (Fig. 6.2) (Chandu *et al.*, 2006). The S4-cleavage near the middle of the transmembrane domain, and generating Nβ, depends on prior cleavage at the S3 site, a position equivalent to the ε-site in APP and close to the cytoplasmic border of the membrane (Chandu *et al.*, 2006; Okochi *et al.*, 2002; Tagami *et al.*, 2008). Also for Notch the initial S3 or ε-cleavage is heterogenous. Interestingly the NICD fragments have therefore either a Val or a Ser/Leu at their aminoterminus. The latter two fragments have a shorter half-life than the NICD starting with Val (Tagami *et al.*, 2008), in accordance with the N-end rule for protein stability (Bachmair *et al.*, 1986; Gonda *et al.*, 1989). "Unstable" NICD is apparently mainly generated in the endosomes, while "stable" NICD is produced at the cell surface (Tagami *et al.*, 2008). Interestingly also the AICDs have either a Val or a Leu N-terminal amino acid residue, but it is not known whether the half-life of these two products is different. In any event, the observations with NICD suggests at least the interesting possibility that Notch signaling is partially regulated at this level, coupling the subcellular localization of the proteolytic process to the strength (duration) of the resulting Notch signal.

9. Regulation of γ-Secretase Activity

The regulation of γ-secretase activity is indeed potentially a very interesting field of research, both from a practical point of view as insight in the regulation of this protease might yield novel drug targets and from a fundamental scientific point of view as we might expect to identify unusual and unexpected novel regulatory mechanisms. Indeed, the above-discussed differential S3 cleavage of Notch supports the concept of γ-secretase being regulated to a certain extent by specific subcellular compartmentalization. Also the association of γ-secretase and substrates with specific detergent extraction-resistant microdomains, e.g., tetraspanin web domains or rafts (Cheng *et al.*, 2009; Vetrivel *et al.*, 2004, 2005; Wakabayashi *et al.*, 2009), suggests unexpected and novel ways of regulating a crucial protease activity like γ-secretase. A recent study using tandem purification of proteins associated with the active γ-secretase complex confirmed association of the protease with a whole series of proteins that likely transiently interact and are possibly involved in complex maturation, membrane trafficking (Rab11, annexin-2,...), and tetraspanin web (CD81, CD9, EWI-2). Modulation of the tetraspanin web using antibodies affected γ-secretase activity (Wakabayashi *et al.*, 2009). The tetraspanin web consists of tetraspanin proteins that associate with sphingolipids, cholesterol, and other proteins and form platforms for proteolytic activity and signaling (Boucheix and Rubinstein, 2001; Hemler, 2008).

Another level of regulation of γ-secretase by cell biological mechanisms has been demonstrated. The endoplasmic reticulum retrieval receptor Rer1p was found to bind to immature Nct and to compete with Aph-1. In this way, Rer1p trafficked back Nct to the ER until binding to Aph-1 removes Rer1p which was then the sign for the complex to leave the ER. Modulation of Rer1p or Aph-1 expression could regulate γ-secretase activity (Spasic *et al.*, 2007). Similarly, Rer1p also appears to interact with unassembled Pen-2 (Kaether *et al.*, 2007). These initial reports suggest clearly that the assembly of the complex is finely tuned, and further work into that direction could also greatly enhance our understanding of γ-secretase activity and potential specificity.

10. γ-Secretase as a Drug Target: AD and Cancer

The central role of γ-secretase in Aβ generation makes it a drug target for AD. The interference with Notch signaling is on the other hand an important concern with regard to potential side effects, including gastrointestinal bleeding (van Es *et al.*, 2005), skin cancer (Demehri *et al.*, 2009; Nicolas *et al.*, 2003), and (auto)-immune problems (Hadland *et al.*, 2001; Tournoy *et al.*, 2004). We will discuss below how the field has developed GSIs with Notch-sparing properties. However, Notch inhibition might be an advantage for the treatment of certain cancers. Excessive Notch signaling has been indeed implicated in several types of neoplastic diseases (Shih Ie and Wang, 2007). For instance about half of the human T-cell acute lymphoblastic leukemias (T-ALL) display activating mutations in the Notch gene (Grabher *et al.*, 2006; Weng *et al.*, 2004). Changes of Notch expression levels are observed in subsets of lung cancer (Dang *et al.*, 2000), ovarian cancer (Park *et al.*, 2006), and breast cancer (Gallahan and Callahan, 1997). Experimental treatments with Notch inhibitors in animal models were very promising. In leukemia mouse models, GSI increased the survival (Cullion *et al.*, 2009) and could reduce osteosarcomas in mice (Engin *et al.*, 2009). GSIs had a synergistic effect with other chemotherapy in human colon adenocarcinoma cell lines (Meng *et al.*, 2009). However, as these GSIs block also Notch processing in the gastrointestinal crypts (van Es *et al.*, 2005), they resulted in serious gastrointestinal side effects by the transformation of proliferative intestinal crypt cells into postmitotic goblet cells (DeAngelo *et al.*, 2006; Milano *et al.*, 2004; Searfoss *et al.*, 2003; van Es *et al.*, 2005). In mouse models they also affected maturation of B and T lymphocytes, causing immunosuppression (Wong *et al.*, 2004). The hope is that a therapeutic window exists or that GSIs could be given for shorter periods of time and in smaller doses, in combination with other antineoplastic agents so that the levels of γ-GSI needed for therapeutic benefit are small enough to avoid side effects. Inhibition of Notch1 signaling with GSIs in glucocorticoid-resistant T-ALL restored corticoid sensitivity and co-treatment with glucocorticoids inhibited GSI-induced gut toxicity *in vitro* and *in vivo* (Real *et al.*, 2009). The mechanism of this effect is unclear. Although this opens perspectives for the treatment of cancer, this insight is probably not helpful for the treatment of AD patients, as glucocorticoids have severe side effects if taken chronically.

For the treatment of AD the trick is to make inhibitors that block APP processing to decrease Aβ generation, while maintaining Notch signaling as much as possible. Classical transition state compounds against aspartic proteases, which target the catalytic site, inhibit similarly all cleavages of all substrates and are likely not useful for clinical development. However non-transition state inhibitors, such as peptide-based inhibitors (DAPT), sulfonamides and benzodiazepines, when used at low concentrations, do not affect cleavage at the ε-site of APP and Notch, but are potent inhibitors of abeta40/42 (γ-site) (Yagishita *et al.*, 2006). The first reported *in vivo* testing of a GSI involved the compound DAPT, which reduced Aβ peptide levels in brain (Dovey *et al.*, 2001). A sulfonamide inhibitor, inhibiting Aβ production in HEK293 cells stably overexpressing APP, is reported to be selective for inhibiting the processing of APP over Notch. Administration of this compound into APP transgenic mice lowered brain Aβ and plasma Aβ (Anderson *et al.*, 2005; Barten *et al.*, 2005). Benzodiazepine analog LY-411575 and benzolactam LY-450139 are highly potent GSI that have been tested extensively *in vivo* (Best *et al.*, 2005; Siemers *et al.*, 2005, 2006). LY-411575 lowered brain, CSF, and plasma Aβ in an APP transgenic mouse model, although reduction in brain Aβ levels lagged behind that of CSF and plasma. However, treatment of mice with GSI LY-411575 at high doses also caused severe gastrointestinal toxicity and interfered with the maturation of B and T lymphocytes, due to the inhibition of Notch processing (Searfoss *et al.*, 2003; Wong *et al.*, 2004). Benzolactam LY450139 is currently in a phase III clinical study (Wong *et al.*, 2004). A single dose of this compound has been shown by stable-isotope labeling to lower CSF Aβ production in healthy men, but a substantial decrease in total Aβ was evident only in the treatment group receiving the highest dose of the compound (Bateman *et al.*, 2009; Siemers *et al.*, 2006).

To overcome these toxicity issues, pharmaceutical companies have been trying to develop "Notch-sparing" GSIs. Such compounds have been identified from kinase inhibitor collections. The discovery of such agents emerged from the finding that ATP and other nucleotides can stimulate Aβ production (Netzer *et al.*, 2003), even in purified γ-secretase preparations (Fraering *et al.*, 2005), suggesting a nucleotide-binding site on the enzyme complex and/or the APP substrate that serves to allosterically regulate substrate selectivity. However the original compounds were not particularly potent.

Two new Notch-sparing inhibitors, with unknown working mechanism, have recently been described, i.e., begacestat (Mayer *et al.*, 2008), for which no human data have yet been disclosed, and BMS-708,163, which has been shown in phase I clinical trials to lower plasma and CFS Aβ levels (Fagan, 2008). Some authors have proposed that these GSIs work by binding to the substrate-docking sites on γ-secretase that are distinct between Notch and APP or that they target selectively different γ-secretase complexes (Fagan, 2008).

Hope remains that a GSI might lower Aβ production in the brain enough to prevent Aβ oligomerization and fibril formation while leaving enough Notch signaling intact to avoid toxic effects.

An unexpected finding some years ago led to the concept of γ-secretase modulators (GSM) to alter Aβ production with little or no effect on Notch signaling. A subset of nonsteroidal anti-inflammatory drugs, such as ibuprofen and sulindac sulfide, as well as enantiomers of flurbiprofen, were shown to change the profiles of the different Aβ species in cell cultures and mice *in vivo*, decreasing the relative amounts of aggregation-prone Aβ42, independently of their ability to inhibit cyclooxygenases (Lleo *et al.*, 2004; Weggen *et al.*, 2001). These compounds shift the cleavages of γ-secretase toward the production of shorter Aβ38 peptides, without affecting Aβ40, AICD, or NICD generation (Eriksen *et al.*, 2003; Weggen *et al.*, 2003). The mechanism of action is not understood, and effects on the conformation of γ-secretase (Beher *et al.*, 2004) or binding to APP have been proposed (Kukar *et al.*, 2008; Munter *et al.*, 2007), as biotinylated photoactivable derivatives of some of these compounds were shown to directly bind APP (Kukar *et al.*, 2008). This supported a hypothesis that these compounds affect the positioning of the substrate in γ-secretase, thus altering the initial cleavage site by the enzyme. It remains to be proven whether this is a general mechanism for GSMs and how it is connected to conformational changes of the enzyme itself.

Tarenflurbil (R-fluriprofen), the first GSM evaluated in humans, had an excellent safety profile, but failed recently in phase III clinical trials (Galasko *et al.*, 2007; Green *et al.*, 2009). The major problem with this trial was that the levels of R-fluriprofen in the brain were not assessed and the doses used were probably not high enough to reach a critical concentration in the brain. Indeed one of the problems with this type of drugs is the high concentrations needed to see the effects on APP processing.

More theoretical avenues of research are currently also considered. Specific inactivation of the Aph-1b γ-secretase complexes in a mouse AD model led to significant improvements of AD-relevant phenotypic features without any Notch-related side effects. The Aph-1b complex contributes to total γ-secretase activity in the human brain, and thus specific targeting of Aph-1b-containing γ-secretase complexes may help generate less-toxic therapies for AD (Serneels *et al.*, 2009). Recent reports have also shown that alteration of cell signaling, especially G protein-coupled receptor (GPCR) signaling, is related to abnormal Aβ production and AD pathogenesis (Cai *et al.*, 2006; Ni *et al.*, 2006; Nitsch *et al.*, 1992; Weggen *et al.*, 2001; Xu *et al.*, 1998). The activation of β2-adrenergic receptor or δ-opioid receptor (DOR) directly enhances γ-secretase activity and accelerates Aβ production (Ni *et al.*, 2006). These DOR receptors seem regulating amyloidogenic APP processing by GPCR-mediated endocytic sorting of BACE1 and γ-secretase in cells (Teng *et al.*, 2010). Recently, Thathiah *et al.* (2009) have reported that expression of GPR3 led to increased formation and cell-surface localization of the mature γ-secretase complex in the absence of an effect on Notch processing. These findings, together with other studies, suggest that cellular membrane receptors, especially GPCRs, might be potential targets for modulating APP processing. This work also opens interesting perspectives on the regulation of γ-secretase by changing its subcellular localization as discussed above.

11. Conclusion

It is clear that work on γ-secretase is a blooming area of research with both fundamental and clinical importance. Over the next years we will see an increasing understanding of the cell biology of this complex, trying to unravel the assembly and regulation of these fascinating enzymes, and also a progressive better understanding of the role of the different γ-secretases in different physiological functions. On the longer run, one hopes that the crystallization of the different complexes will lead to a real understanding of the working mechanisms of these proteases. This would help tremendously in rationale drug design.

ACKNOWLEDGMENTS

The research in the authors' laboratory is supported by the Fund for Scientific Research Flanders, KULeuven, VIB, the Federal Office for Scientific Affairs, Belgium (IUAP P6/43/), a Methusalem grant of the KULeuven and the Flemish Government and Memosad (FZ-2007-200611) of the European union. EJ was supported by IWT and a short-term fellowship from EMBO.

REFERENCES

Alzheimer's_Disease_Collaborative_Group (1995). The structure of the presenilin 1 (S182) gene and identification of six novel mutations in early onset AD families. *Nat. Genet.* **11**, 219–222.

Anderson, J. J., Holtz, G., Baskin, P. P., Turner, M., Rowe, B., Wang, B., Kounnas, M. Z., Lamb, B. T., Barten, D., Felsenstein, K., McDonald, I., Srinivasan, K., *et al.* (2005). Reductions in beta-amyloid concentrations in vivo by the gamma-secretase inhibitors BMS-289948 and BMS-299897. *Biochem. Pharmacol.* **69**, 689–698.

Anderton, B. H., Dayanandan, R., Killick, R., and Lovestone, S. (2000). Does dysregulation of the Notch and wingless/Wnt pathways underlie the pathogenesis of Alzheimer's disease? *Mol. Med. Today* **6**, 54–59.

Annaert, W., and De Strooper, B. (1999). Presenilins: molecular switches between proteolysis and signal transduction. *Trends. Neurosci.* **22**, 439–443.

Annaert, W. G., Esselens, C., Baert, V., Boeve, C., Snellings, G., Cupers, P., Craessaerts, K., and De Strooper, B. (2001). Interaction with telencephalin and the amyloid precursor protein predicts a ring structure for presenilins. *Neuron* **32**, 579–589.

Araki, W., Saito, S., Takahashi-Sasaki, N., Shiraishi, H., Komano, H., and Murayama, K. S. (2006). Characterization of APH-1 mutants with a disrupted transmembrane GxxxG motif. *J. Mol. Neurosci.* **29**, 35–43.

Bachmair, A., Finley, D., and Varshavsky, A. (1986). In vivo half-life of a protein is a function of its amino-terminal residue. *Science* **234**, 179–186.

Barten, D. M., Guss, V. L., Corsa, J. A., Loo, A., Hansel, S. B., Zheng, M., Munoz, B., Srinivasan, K., Wang, B., Robertson, B. J., Polson, C. T., Wang, J., *et al.* (2005). Dynamics of {beta}-amyloid reductions in brain, cerebrospinal fluid, and plasma of {beta}-amyloid precursor protein transgenic mice treated with a {gamma}-secretase inhibitor. *J. Pharmacol. Exp. Ther.* **312**, 635–643.

Bateman, R. J., Siemers, E. R., Mawuenyega, K. G., Wen, G., Browning, K. R., Sigurdson, W. C., Yarasheski, K. E., Friedrich, S. W., Demattos, R. B., May, P. C., Paul, S. M., and Holtzman, D. M. (2009). A gamma-secretase inhibitor decreases amyloid-beta production in the central nervous system. *Ann. Neurol.* **66**, 48–54.

Baulac, S., LaVoie, M. J., Kimberly, W. T., Strahle, J., Wolfe, M. S., Selkoe, D. J., and Xia, W. (2003). Functional gamma-secretase complex assembly in Golgi/trans-Golgi network: interactions among presenilin, nicastrin, Aph1, Pen-2, and gamma-secretase substrates. *Neurobiol. Dis.* **14**, 194–204.

Beher, D., Clarke, E. E., Wrigley, J. D., Martin, A. C., Nadin, A., Churcher, I., and Shearman, M. S. (2004). Selected non-steroidal anti-inflammatory drugs and their derivatives target gamma-secretase at a novel site. Evidence for an allosteric mechanism. *J. Biol. Chem.* **279**, 43419–43426.

Bentahir, M., Nyabi, O., Verhamme, J., Tolia, A., Horre, K., Wiltfang, J., Esselmann, H., and De Strooper, B. (2006). Presenilin clinical mutations can affect gamma-secretase activity by different mechanisms. *J. Neurochem.* **96**, 732–742.

Berechid, B. E., Thinakaran, G., Wong, P. C., Sisodia, S. S., and Nye, J. S. (1999). Lack of requirement for presenilin1 in Notch1 signaling. *Curr. Biol.* **9**, 1493–1496.

Berezovska, O., Jack, C., McLean, P., Aster, J. C., Hicks, C., Xia, W., Wolfe, M. S., Kimberly, W. T., Weinmaster, G., Selkoe, D. J., and Hyman, B. T. (2000). Aspartate mutations in presenilin and gamma-secretase inhibitors both impair notch1 proteolysis and nuclear translocation with relative preservation of notch1 signaling. *J. Neurochem.* **75**, 583–593.

Berezovska, O., Lleo, A., Herl, L. D., Frosch, M. P., Stern, E. A., Bacskai, B. J., and Hyman, B. T. (2005). Familial Alzheimer's disease presenilin 1 mutations cause alterations in the conformation of presenilin and interactions with amyloid precursor protein. *J. Neurosci.* **25**, 3009–3017.

Bergman, A., Laudon, H., Winblad, B., Lundkvist, J., and Naslund, J. (2004). The extreme C terminus of presenilin 1 is essential for gamma-secretase complex assembly and activity. *J. Biol. Chem.* **279**, 45564–45572.

Best, J. D., Jay, M. T., Otu, F., Ma, J., Nadin, A., Ellis, S., Lewis, H. D., Pattison, C., Reilly, M., Harrison, T., Shearman, M. S., Williamson, T. L., *et al.* (2005). Quantitative measurement of changes in amyloid-beta(40) in the rat brain and cerebrospinal fluid following treatment with the gamma-secretase inhibitor LY-411575 [N2-[(2S)-2-(3,5-difluorophenyl)-2-hydroxyethanoyl]-N1-[(7S)-5-methyl-6-ox o-6,7-dihydro-5H-dibenzo[b,d]azepin-7-yl]-L-alaninamide]. *J. Pharmacol. Exp. Ther.* **313**, 902–908.

Bihel, F., Das, C., Bowman, M. J., and Wolfe, M. S. (2004). Discovery of a Subnanomolar helical D-tridecapeptide inhibitor of gamma-secretase. *J. Med. Chem.* **47**, 3931–3933.

Blaumueller, C. M., Qi, H., Zagouras, P., and Artavanis-Tsakonas, S. (1997). Intracellular cleavage of Notch leads to a heterodimeric receptor on the plasma membrane. *Cell* **90**, 281–291.

Boucheix, C., and Rubinstein, E. (2001). Tetraspanins. *Cell Mol. Life Sci.* **58**, 1189–1205.

Bozkulak, E. C., Weinmaster, G. (2009). Selective use of ADAM10 and ADAM17 in activation of Notch1 signaling. *Mol Cell Biol.* **29**(21), 5679–5695.

Brou, C., Logeat, F., Gupta, N., Bessia, C., LeBail, O., Doedens, J. R., Cumano, A., Roux, P., Black, R. A., and Israel, A. (2000). A novel proteolytic cleavage involved in Notch signaling: the role of the disintegrin-metalloprotease TACE. *Mol. Cell* **5**, 207–216.

Brown, M. S., Ye, J., Rawson, R. B., and Goldstein, J. L. (2000). Regulated intramembrane proteolysis: a control mechanism conserved from bacteria to humans. *Cell* **100**, 391–398.

Cagavi Bozkulak, E., and Weinmaster, G. (2009). Selective use of Adam10 and Adam17 in activation of Notch1 signaling. *Mol. Cell. Biol.* **29**(21), 5679–5695.

Cai, D., Netzer, W. J., Zhong, M., Lin, Y., Du, G., Frohman, M., Foster, D. A., Sisodia, S. S., Xu, H., Gorelick, F. S., and Greengard, P. (2006). Presenilin-1 uses phospholipase D1 as a negative regulator of beta-amyloid formation. *Proc. Natl. Acad. Sci. U.S.A.* **103**, 1941–1946.

Chandu, D., Huppert, S. S., and Kopan, R. (2006). Analysis of transmembrane domain mutants is consistent with sequential cleavage of Notch by gamma-secretase. *J. Neurochem.* **96**, 228–235.

Chavez-Gutierrez, L., Tolia, A., Maes, E., Li, T., Wong, P. C., and de Strooper, B. (2008). Glu(332) in the Nicastrin ectodomain is essential for gamma-secretase complex maturation but not for its activity. *J. Biol. Chem.* **283**, 20096–20105.

Chen, F., Yu, G., Arawaka, S., Nishimura, M., Kawarai, T., Yu, H., Tandon, A., Supala, A., Song, Y. Q., Rogaeva, E., Milman, P., Sato, C., *et al.* (2001). Nicastrin binds to membrane-tethered Notch. *Nat. Cell Biol.* **3**, 751–754.

Cheng, H., Vetrivel, K. S., Drisdel, R. C., Meckler, X., Gong, P., Leem, J. Y., Li, T., Carter, M., Chen, Y., Nguyen, P., Iwatsubo, T., Tomita, T., *et al.* (2009). S-palmitoylation of gamma-secretase subunits nicastrin and APH-1. *J. Biol. Chem.* **284**, 1373–1384.

Chung, H. M., and Struhl, G. (2001). Nicastrin is required for Presenilin-mediated transmembrane cleavage in Drosophila. *Nat. Cell Biol.* **3**, 1129–1132.

Cole, S. L., and Vassar, R. (2008). BACE1 structure and function in health and Alzheimer's disease. *Curr. Alzheimer Res.* **5**, 100–120.

Crystal, A. S., Morais, V. A., Pierson, T. C., Pijak, D. S., Carlin, D., Lee, V. M., and Doms, R. W. (2003). Membrane topology of gamma-secretase component PEN-2. *J. Biol. Chem.* **278**, 20117–20123.

Cullion, K., Draheim, K. M., Hermance, N., Tammam, J., Sharma, V. M., Ware, C., Nikov, G., Krishnamoorthy, V., Majumder, P. K., and Kelliher, M. A. (2009). Targeting the Notch1 and mTOR pathways in a mouse T-ALL model. *Blood* **113**, 6172–6181.

Dang, T. P., Gazdar, A. F., Virmani, A. K., Sepetavec, T., Hande, K. R., Minna, J. D., Roberts, J. R., and Carbone, D. P. (2000). Chromosome 19 translocation, overexpression of Notch3, and human lung cancer. *J. Natl. Cancer Inst.* **92**, 1355–1357.

Das, C., Berezovska, O., Diehl, T. S., Genet, C., Buldyrev, I., Tsai, J. Y., Hyman, B. T., and Wolfe, M. S. (2003). Designed helical peptides inhibit an intramembrane protease. *J. Am. Chem. Soc.* **125**, 11794–5.

de la Pompa, J. L., Wakeham, A., Correia, K. M., Samper, E., Brown, S., Aguilera, R. J., Nakano, T., Honjo, T., Mak, T. W., Rossant, J., and Conlon, R. A. (1997). Conservation of the Notch signalling pathway in mammalian neurogenesis. *Development* **124**, 1139–1148.

De Strooper, B. (2003). Aph-1, Pen-2, and Nicastrin with Presenilin generate an active gamma-Secretase complex. *Neuron* **38**, 9–12.

De Strooper, B., and Annaert, W. (2001). Where Notch and Wnt signaling meet. The presenilin hub. *J. Cell Biol.* **152**, F17–F20.

De Strooper, B., Annaert, W., Cupers, P., Saftig, P., Craessaerts, K., Mumm, J. S., Schroeter, E. H., Schrijvers, V., Wolfe, M. S., Ray, W. J., Goate, A., and Kopan, R. (1999). A presenilin-1-dependent gamma-secretase-like protease mediates release of Notch intracellular domain. *Nature* **398**, 518–522.

De Strooper, B., Saftig, P., Craessaerts, K., Vanderstichele, H., Guhde, G., Annaert, W., Von Figura, K., and Van Leuven, F. (1998). Deficiency of presenilin-1 inhibits the normal cleavage of amyloid precursor protein. *Nature* **391**, 387–390.

DeAngelo, D. J., Stone, R. M., Heaney, M. L., Nimer, S. D., Paquette, R. L., Klisovic, R. B., Caligiuri, M. A., Cooper, M. R., Lecerf, J. M., Karol, M. D., Sheng, S., Holford, N., *et al.* (2006). Phase 1 clinical results with tandutinib (MLN518), a novel FLT3 antagonist, in patients with acute myelogenous leukemia or high-risk myelodysplastic syndrome: safety, pharmacokinetics, and pharmacodynamics. *Blood* **108**, 3674–3681.

Dejaegere, T., Serneels, L., Schafer, M. K., Van Biervliet, J., Horre, K., Depboylu, C., Alvarez-Fischer, D., Herreman, A., Willem, M., Haass, C., Hoglinger, G. U., D'Hooge, R., *et al.* (2008). Deficiency of Aph1B/C-gamma-secretase disturbs Nrg1 cleavage and sensorimotor gating that can be reversed with antipsychotic treatment. *Proc. Natl. Acad. Sci. U.S.A.* **105**, 9775–9780.

Demehri, S., Turkoz, A., and Kopan, R. (2009). Epidermal Notch1 loss promotes skin tumorigenesis by impacting the stromal microenvironment. *Cancer Cell.* **16**, 55–66.

Donoviel, D. B., Hadjantonakis, A. K., Ikeda, M., Zheng, H., Hyslop, P. S., and Bernstein, A. (1999). Mice lacking both presenilin genes exhibit early embryonic patterning defects. *Genes Dev.* **13**, 2801–2810.

Dovey, H. F., John, V., Anderson, J. P., Chen, L. Z., de Saint Andrieu, P., Fang, L. Y., Freedman, S. B., Folmer, B., Goldbach, E., Holsztynska, E. J., Hu, K. L., Johnson-Wood, K. L., *et al.* (2001). Functional gamma-secretase inhibitors reduce beta-amyloid peptide levels in brain. *J. Neurochem.* **76**, 173–181.

Dries, D. R., Shah, S., Han, Y. H., Yu, C., Yu, S., Shearman, M. S., and Yu, G. (2009). Glu-333 of nicastrin directly participates in gamma-secretase activity. *J. Biol. Chem.* **284**, 29714–29724.

D'Souza, B., Miyamoto, A., and Weinmaster, G. (2008). The many facets of Notch ligands. *Oncogene* **27**, 5148–5167.

Edbauer, D., Winkler, E., Haass, C., and Steiner, H. (2002). Presenilin and nicastrin regulate each other and determine amyloid beta-peptide production via complex formation. *Proc. Natl. Acad. Sci. U.S.A.* **99**, 8666–8671.

Edbauer, D., Winkler, E., Regula, J. T., Pesold, B., Steiner, H., and Haass, C. (2003). Reconstitution of gamma-secretase activity. *Nat. Cell Biol.* **5**, 486–488.

Engin, F., Bertin, T., Ma, O., Jiang, M. M., Wang, L., Sutton, R. E., Donehower, L. A., and Lee, B. (2009). Notch signaling contributes to the pathogenesis of human osteosarcomas. *Hum. Mol. Genet.* **18**, 1464–1470.

Eriksen, J. L., Sagi, S. A., Smith, T. E., Weggen, S., Das, P., McLendon, D. C., Ozols, V. V., Jessing, K. W., Zavitz, K. H., Koo, E. H., and Golde, T. E. (2003). NSAIDs and enantiomers of flurbiprofen target gamma-secretase and lower Abeta 42 in vivo. *J. Clin. Invest.* **112**, 440–449.

Esler, W. P., Kimberly, W. T., Ostaszewski, B. L., Diehl, T. S., Moore, C. L., Tsai, J. Y., Rahmati, T., Xia, W., Selkoe, D. J., and Wolfe, M. S. (2000). Transition-state analogue inhibitors of gamma-secretase bind directly to presenilin-1. *Nat. Cell Biol.* **2**, 428–434.

Esler, W. P., Kimberly, W. T., Ostaszewski, B. L., Ye, W., Diehl, T. S., Selkoe, D. J., and Wolfe, M. S. (2002). Activity-dependent isolation of the presenilin- gamma -secretase complex reveals nicastrin and a gamma substrate. *Proc. Natl. Acad. Sci. U.S.A.* **99**, 2720–2725.

Fagan, T. (2008) DC: New γ Secretase Inhibitors Hit APP, Spare Notch. *The AlzGene Database. Alzheimer Research Forum [online]*, http://www.alzforum.org/new/detail.asp?id=1992

Fortna, R. R., Crystal, A. S., Morais, V. A., Pijak, D. S., Lee, V. M., and Doms, R. W. (2004). Membrane topology and nicastrin-enhanced endoproteolysis of APH-1, a component of the gamma-secretase complex. *J. Biol. Chem.* **279**, 3685–3693.

Fraering, P. C., Ye, W., LaVoie, M. J., Ostaszewski, B. L., Selkoe, D. J., and Wolfe, M. S. (2005). gamma-Secretase substrate selectivity can be modulated directly via interaction with a nucleotide-binding site. *J. Biol. Chem.* **280**, 41987–41996.

Fraering, P. C., Ye, W., Strub, J. M., Dolios, G., LaVoie, M. J., Ostaszewski, B. L., van Dorsselaer, A., Wang, R., Selkoe, D. J., and Wolfe, M. S. (2004). Purification and characterization of the human gamma-secretase complex. *Biochemistry* **43**, 9774–9789.

Francis, R., McGrath, G., Zhang, J., Ruddy, D. A., Sym, M., Apfeld, J., Nicoll, M., Maxwell, M., Hai, B., Ellis, M. C., Parks, A. L., Xu, W., *et al.* (2002). aph-1 and pen-2 are required for Notch pathway signaling, gamma-secretase cleavage of betaAPP, and presenilin protein accumulation. *Dev. Cell* **3**, 85–97.

Freeman, M. (2008). Rhomboid proteases and their biological functions. *Annu. Rev. Genet.* **42**, 191–210.

Galasko, D. R., Graff-Radford, N., May, S., Hendrix, S., Cottrell, B. A., Sagi, S. A., Mather, G., Laughlin, M., Zavitz, K. H., Swabb, E., Golde, T. E., Murphy, M. P., *et al.* (2007). Safety, tolerability, pharmacokinetics, and Abeta levels after short-term administration of R-flurbiprofen in healthy elderly individuals. *Alzheimer Dis. Assoc. Disord.* **21**, 292–299.

Gallahan, D., and Callahan, R. (1997). The mouse mammary tumor associated gene INT3 is a unique member of the NOTCH gene family (NOTCH4). *Oncogene* **14**, 1883–1890.

Gonda, D. K., Bachmair, A., Wunning, I., Tobias, J. W., Lane, W. S., and Varshavsky, A. (1989). Universality and structure of the N-end rule. *J. Biol. Chem.* **264**, 16700–16712.

Gordon, W. R., Vardar-Ulu, D., Histen, G., Sanchez-Irizarry, C., Aster, J. C., and Blacklow, S. C. (2007). Structural basis for autoinhibition of Notch. *Nat. Struct. Mol. Biol.* **14**, 295–300.

Goutte, C., Hepler, W., Mickey, K. M., and Priess, J. R. (2000). aph-2 encodes a novel extracellular protein required for GLP-1-mediated signaling. *Development* **127**, 2481–2492.

Goutte, C., Tsunozaki, M., Hale, V. A., and Priess, J. R. (2002). APH-1 is a multipass membrane protein essential for the Notch signaling pathway in Caenorhabditis elegans embryos. *Proc. Natl. Acad. Sci. U.S.A.* **99**, 775–779.

Grabher, C., von Boehmer, H., and Look, A. T. (2006). Notch 1 activation in the molecular pathogenesis of T-cell acute lymphoblastic leukaemia. *Nat. Rev. Cancer.* **6**, 347–359.

Green, R. C., Schneider, L. S., Amato, D. A., Beelen, A. P., Wilcock, G., Swabb, E. A., and Zavitz, K. H. (2009). Effect of tarenflurbil on cognitive decline and activities of daily living in patients with mild Alzheimer disease: a randomized controlled trial. *J. Am. Med. Assoc.* **302**, 2557–2564.

Gu, Y., Chen, F., Sanjo, N., Kawarai, T., Hasegawa, H., Duthie, M., Li, W., Ruan, X., Luthra, A., Mount, H. T., Tandon, A., Fraser, P. E., *et al.* (2003). APH-1 interacts with mature and immature forms of presenilins and nicastrin and may play a role in maturation of presenilin.nicastrin complexes. *J. Biol. Chem.* **278**, 7374–7380.

Ha, Y. (2007). Structural principles of intramembrane proteases. *Curr. Opin. Struct. Biol.* **17**, 405–411.

Haass, C. (2004). Take five–BACE and the gamma-secretase quartet conduct Alzheimer's amyloid beta-peptide generation. *EMBO J.* **23**, 483–488.

Haass, C., and Selkoe, D. J. (1993). Cellular processing of beta-amyloid precursor protein and the genesis of amyloid beta-peptide. *Cell* **75**, 1039–1042.

Hadland, B. K., Manley, N. R., Su, D., Longmore, G. D., Moore, C. L., Wolfe, M. S., Schroeter, E. H., and Kopan, R. (2001). Gamma -secretase inhibitors repress thymocyte development. *Proc. Natl. Acad. Sci. U.S.A.* **98**, 7487–7491.

Hartmann, D., De Strooper, B., and Saftig, P. (1999). Presenilin-1 deficiency leads to loss of Cajal-Retzius neurons and cortical dysplasia similar to human type 2 lissencephaly. *Curr. Biol.* **9**, 719–727.

Hartmann, D., de Strooper, B., Serneels, L., Craessaerts, K., Herreman, A., Annaert, W., Umans, L., Lubke, T., Lena Illert, A., von Figura, K., and Saftig, P. (2002). The disintegrin/metalloprotease ADAM 10 is essential for Notch signalling but not for alpha-secretase activity in fibroblasts. *Hum. Mol. Genet.* **11**, 2615–2624.

Hasegawa, H., Sanjo, N., Chen, F., Gu, Y. J., Shier, C., Petit, A., Kawarai, T., Katayama, T., Schmidt, S. D., Mathews, P. M., Schmitt-Ulms, G., Fraser, P. E., *et al.* (2004). Both the sequence and length of the C terminus of PEN-2 are critical for intermolecular interactions and function of presenilin complexes. *J. Biol. Chem.* **279**, 46455–46463.

Hayashi, I., Urano, Y., Fukuda, R., Isoo, N., Kodama, T., Hamakubo, T., Tomita, T., and Iwatsubo, T. (2004). Selective reconstitution and recovery of functional gamma-secretase complex on budded baculovirus particles. *J. Biol. Chem.* **279**, 38040–38046.

Hebert, S. S., Serneels, L., Dejaegere, T., Horre, K., Dabrowski, M., Baert, V., Annaert, W., Hartmann, D., and De Strooper, B. (2004). Coordinated and widespread expression of gamma-secretase in vivo: evidence for size and molecular heterogeneity. *Neurobiol. Dis.* **17**, 260–272.

Hemler, M. E. (2008). Targeting of tetraspanin proteins–potential benefits and strategies. *Nat. Rev. Drug Discov.* **7**, 747–758.

Henricson, A., Kall, L., and Sonnhammer, E. L. (2005). A novel transmembrane topology of presenilin based on reconciling experimental and computational evidence. *FEBS J.* **272**, 2727–2733.

Herreman, A., Hartmann, D., Annaert, W., Saftig, P., Craessaerts, K., Serneels, L., Umans, L., Schrijvers, V., Checler, F., Vanderstichele, H., Baekelandt, V., Dressel, R., *et al.*

(1999). Presenilin 2 deficiency causes a mild pulmonary phenotype and no changes in amyloid precursor protein processing but enhances the embryonic lethal phenotype of presenilin 1 deficiency. *Proc. Natl. Acad. Sci. U.S.A.* **96**, 11872–11877.

Herreman, A., Serneels, L., Annaert, W., Collen, D., Schoonjans, L., and De Strooper, B. (2000). Total inactivation of gamma-secretase activity in presenilin-deficient embryonic stem cells. *Nat. Cell Biol.* **2**, 461–2.

Herreman, A., Van Gassen, G., Bentahir, M., Nyabi, O., Craessaerts, K., Mueller, U., Annaert, W., and De Strooper, B. (2003). gamma-Secretase activity requires the presenilin-dependent trafficking of nicastrin through the Golgi apparatus but not its complex glycosylation. *J. Cell. Sci.* **116**, 1127–1136.

Hu, Y., Ye, Y., and Fortini, M. E. (2002). Nicastrin is required for gamma-secretase cleavage of the Drosophila Notch receptor. *Dev. Cell* **2**, 69–78.

Ikeuchi, T., and Sisodia, S. S. (2003). The Notch ligands, Delta1 and Jagged2, are substrates for presenilin-dependent "gamma-secretase" cleavage. *J. Biol. Chem.* **278**, 7751–7754.

Jacobsen, H., Reinhardt, D., Brockhaus, M., Bur, D., Kocyba, C., Kurt, H., Grim, M. G., Baumeister, R., and Loetscher, H. (1999). The influence of endoproteolytic processing of familial Alzheimer's disease presenilin 2 on abeta42 amyloid peptide formation. *J. Biol. Chem.* **274**, 35233–35239.

Jorissen, E., Prox, J., Bernreuther, C., Weber, S., Schwanbeck, R., Serneels, L., Snellinx, A., Craessaerts, K., Thathiah, A., Tesseur, I., Bartsch, U., Weskamp, G., *et al.* (2010). *J. Neurosci.* **30**(14), 4833–4844.

Kaether, C., Scheuermann, J., Fassler, M., Zilow, S., Shirotani, K., Valkova, C., Novak, B., Kacmar, S., Steiner, H., and Haass, C. (2007). Endoplasmic reticulum retention of the gamma-secretase complex component Pen2 by Rer1. *EMBO Rep.* **8**, 743–748.

Kakuda, N., Funamoto, S., Yagishita, S., Takami, M., Osawa, S., Dohmae, N., and Ihara, Y. (2006). Equimolar production of amyloid beta-protein and amyloid precursor protein intracellular domain from beta-carboxyl-terminal fragment by gamma-secretase. *J. Biol. Chem.* **281**, 14776–14786.

Kim, S. H., and Sisodia, S. S. (2005). A sequence within the first transmembrane domain of PEN-2 is critical for PEN-2-mediated endoproteolysis of presenilin 1. *J. Biol. Chem.* **280**, 1992–2001.

Kimberly, W. T., LaVoie, M. J., Ostaszewski, B. L., Ye, W., Wolfe, M. S., and Selkoe, D. J. (2002). Complex N-linked glycosylated nicastrin associates with active gamma-secretase and undergoes tight cellular regulation. *J. Biol. Chem.* **277**, 35113–35117.

Kimberly, W. T., LaVoie, M. J., Ostaszewski, B. L., Ye, W., Wolfe, M. S., and Selkoe, D. J. (2003). Gamma-secretase is a membrane protein complex comprised of presenilin, nicastrin, Aph-1, and Pen-2. *Proc. Natl. Acad. Sci. U.S.A.* **100**, 6382–6387.

Kimberly, W. T., Xia, W., Rahmati, T., Wolfe, M. S., and Selkoe, D. J. (2000). The transmembrane aspartates in presenilin 1 and 2 are obligatory for gamma-secretase activity and amyloid beta-protein generation. *J. Biol. Chem.* **275**, 3173–3178.

Kopan, R., and Ilagan, M. X. (2004). Gamma-secretase: proteasome of the membrane?. *Nat. Rev. Mol. Cell Biol.* **5**, 499–504.

Kopan, R., and Ilagan, M. X. (2009). The canonical Notch signaling pathway: unfolding the activation mechanism. *Cell* **137**, 216–233.

Kornilova, A. Y., Bihel, F., Das, C., and Wolfe, M. S. (2005). The initial substrate-binding site of gamma-secretase is located on presenilin near the active site. *Proc. Natl. Acad. Sci. U.S.A.* **102**, 3230–3235.

Kukar, T. L., Ladd, T. B., Bann, M. A., Fraering, P. C., Narlawar, R., Maharvi, G. M., Healy, B., Chapman, R., Welzel, A. T., Price, R. W., Moore, B., Rangachari, V., *et al.* (2008). Substrate-targeting gamma-secretase modulators. *Nature* **453**, 925–929.

LaVoie, M. J., and Selkoe, D. J. (2003). The Notch ligands, Jagged and Delta, are sequentially processed by alpha-secretase and presenilin/gamma-secretase and release signaling fragments. *J. Biol. Chem.* **278**, 34427–34437.

Lammich, S., Okochi, M., Takeda, M., Kaether, C., Capell, A., Zimmer, A. K., Edbauer, D., Walter, J., Steiner, H., and Haass, C. (2002). Presenilin-dependent intramembrane proteolysis of CD44 leads to the liberation of its intracellular domain and the secretion of an Abeta-like peptide. *J. Biol. Chem.* **277**, 44754–44759.

Laudon, H., Hansson, E. M., Melen, K., Bergman, A., Farmery, M. R., Winblad, B., Lendahl, U., von Heijne, G., and Naslund, J. (2005). A nine-transmembrane domain topology for presenilin 1. *J. Biol. Chem.* **280**, 35352–35360.

Laudon, H., Mathews, P. M., Karlstrom, H., Bergman, A., Farmery, M. R., Nixon, R. A., Winblad, B., Gandy, S. E., Lendahl, U., Lundkvist, J., and Naslund, J. (2004). Co-expressed presenilin 1 NTF and CTF form functional gamma-secretase complexes in cells devoid of full-length protein. *J. Neurochem.* **89**, 44–53.

Lazarov, O., Peterson, L. D., Peterson, D. A., and Sisodia, S. S. (2006). Expression of a familial Alzheimer's disease-linked presenilin-1 variant enhances perforant pathway lesion-induced neuronal loss in the entorhinal cortex. *J. Neurosci.* **26**, 429–434.

Lee, S. F., Shah, S., Li, H., Yu, C., Han, W., and Yu, G. (2002). Mammalian APH-1 interacts with presenilin and nicastrin and is required for intramembrane proteolysis of amyloid-beta precursor protein and Notch. *J. Biol. Chem.* **277**, 45013–45019.

Lee, S. F., Shah, S., Yu, C., Wigley, W. C., Li, H., Lim, M., Pedersen, K., Han, W., Thomas, P., Lundkvist, J., Hao, Y. H., and Yu, G. (2004). A conserved GXXXG motif in APH-1 is critical for assembly and activity of the gamma-secretase complex. *J. Biol. Chem.* **279**, 4144–4152.

Levitan, D., and Greenwald, I. (1995). Facilitation of lin-12-mediated signalling by sel-12, a Caenorhabditis elegans S182 Alzheimer's disease gene. *Nature* **377**, 351–354.

Levitan, D., Lee, J., Song, L., Manning, R., Wong, G., Parker, E., and Zhang, L. (2001). PS1 N- and C-terminal fragments form a complex that functions in APP processing and Notch signaling. *Proc. Natl. Acad. Sci. U.S.A.* **98**, 12186–12190.

Levy-Lahad, E., Wasco, W., Poorkaj, P., Romano, D. M., Oshima, J., Pettingell, W. H., Yu, C. E., Jondro, P. D., Schmidt, S. D., Wang, K., *et al.* (1995). Candidate gene for the chromosome 1 familial Alzheimer's disease locus [see comments]. *Science* **269**, 973–977.

Li, J., Fici, G. J., Mao, C. A., Myers, R. L., Shuang, R., Donoho, G. P., Pauley, A. M., Himes, C. S., Qin, W., Kola, I., Merchant, K. M., and Nye, J. S. (2003a). Positive and negative regulation of the gamma-secretase activity by nicastrin in a murine model. *J. Biol. Chem.* **278**, 33445–33449.

Li, X., and Greenwald, I. (1997). HOP-1, a Caenorhabditis elegans presenilin, appears to be functionally redundant with SEL-12 presenilin and to facilitate LIN-12 and GLP-1 signaling. *Proc. Natl. Acad. Sci. U.S.A.* **94**, 12204–12209.

Li, T., Ma, G., Cai, H., Price, D. L., and Wong, P. C. (2003b). Nicastrin is required for assembly of presenilin/gamma-secretase complexes to mediate Notch signaling and for processing and trafficking of beta-amyloid precursor protein in mammals. *J. Neurosci.* **23**, 3272–3277.

Li, Y. M., Xu, M., Lai, M. T., Huang, Q., Castro, J. L., DiMuzio-Mower, J., Harrison, T., Lellis, C., Nadin, A., Neduvelil, J. G., Register, R. B., Sardana, M. K., *et al.* (2000). Photoactivated gamma-secretase inhibitors directed to the active site covalently label presenilin 1. *Nature* **405**, 689–694.

Lleo, A., Berezovska, O., Herl, L., Raju, S., Deng, A., Bacskai, B. J., Frosch, M. P., Irizarry, M., and Hyman, B. T. (2004). Nonsteroidal anti-inflammatory drugs lower Abeta42 and change presenilin 1 conformation. *Nat. Med.* **10**, 1065–1066.

Logeat, F., Bessia, C., Brou, C., LeBail, O., Jarriault, S., Seidah, N. G., and Israel, A. (1998). The Notch1 receptor is cleaved constitutively by a furin-like convertase. *Proc. Natl. Acad. Sci. U.S.A.* **95**, 8108–8112.

Lopez-Schier, H., and St Johnston, D. (2002). Drosophila nicastrin is essential for the intramembranous cleavage of notch. *Dev. Cell* **2**, 79–89.

Ma, G., Li, T., Price, D. L., and Wong, P. C. (2005). APH-1a is the principal mammalian APH-1 isoform present in gamma-secretase complexes during embryonic development. *J. Neurosci.* **25**, 192–198.

Marambaud, P., Shioi, J., Serban, G., Georgakopoulos, A., Sarner, S., Nagy, V., Baki, L., Wen, P., Efthimiopoulos, S., Shao, Z., Wisniewski, T., and Robakis, N. K. (2002). A presenilin-1/gamma-secretase cleavage releases the E-cadherin intracellular domain and regulates disassembly of adherens junctions. *EMBO J.* **21**, 1948–1956.

Mastrangelo, P., Mathews, P. M., Chishti, M. A., Schmidt, S. D., Gu, Y., Yang, J., Mazzella, M. J., Coomaraswamy, J., Horne, P., Strome, B., Pelly, H., Levesque, G., *et al.* (2005). Dissociated phenotypes in presenilin transgenic mice define functionally distinct gamma-secretases. *Proc. Natl. Acad. Sci. U.S.A.* **102**, 8972–8977.

Mayer, S. C., Kreft, A. F., Harrison, B., Abou-Gharbia, M., Antane, M., Aschmies, S., Atchison, K., Chlenov, M., Cole, D. C., Comery, T., Diamantidis, G., Ellingboe, J., *et al.* (2008). Discovery of begacestat, a Notch-1-sparing gamma-secretase inhibitor for the treatment of Alzheimer's disease. *J Med Chem.* **51**, 7348–7351.

McCarthy, J. V., Twomey, C., and Wujek, P. (2009). Presenilin-dependent regulated intramembrane proteolysis and gamma-secretase activity. *Cell Mol. Life Sci.* **66**, 1534–1555.

Meng, R. D., Shelton, C. C., Li, Y. M., Qin, L. X., Notterman, D., Paty, P. B., and Schwartz, G. K. (2009). gamma-Secretase inhibitors abrogate oxaliplatin-induced activation of the Notch-1 signaling pathway in colon cancer cells resulting in enhanced chemosensitivity. *Cancer Res.* **69**, 573–582.

Milano, J., McKay, J., Dagenais, C., Foster-Brown, L., Pognan, F., Gadient, R., Jacobs, R. T., Zacco, A., Greenberg, B., and Ciaccio, P. J. (2004). Modulation of notch processing by gamma-secretase inhibitors causes intestinal goblet cell metaplasia and induction of genes known to specify gut secretory lineage differentiation. *Toxicol. Sci.* **82**, 341–358.

Mumm, J. S., and Kopan, R. (2000). Notch signaling: from the outside in. *Dev. Biol.* **228**, 151–165.

Munter, L. M., Voigt, P., Harmeier, A., Kaden, D., Gottschalk, K. E., Weise, C., Pipkorn, R., Schaefer, M., Langosch, D., and Multhaup, G. (2007). GxxxG motifs within the amyloid precursor protein transmembrane sequence are critical for the etiology of Abeta42. *EMBO J.* **26**, 1702–1712.

Netzer, W. J., Dou, F., Cai, D., Veach, D., Jean, S., Li, Y., Bornmann, W. G., Clarkson, B., Xu, H., and Greengard, P. (2003). Gleevec inhibits beta-amyloid production but not Notch cleavage. *Proc. Natl. Acad. Sci. U.S.A.* **100**, 12444–12449.

Ni, Y., Zhao, X., Bao, G., Zou, L., Teng, L., Wang, Z., Song, M., Xiong, J., Bai, Y., and Pei, G. (2006). Activation of beta2-adrenergic receptor stimulates gamma-secretase activity and accelerates amyloid plaque formation. *Nat. Med.* **12**, 1390–1396.

Nicolas, M., Wolfer, A., Raj, K., Kummer, J. A., Mill, P., van Noort, M., Hui, C. C., Clevers, H., Dotto, G. P., and Radtke, F. (2003). Notch1 functions as a tumor suppressor in mouse skin. *Nat. Genet.* **33**, 416–421.

Niimura, M., Isoo, N., Takasugi, N., Tsuruoka, M., Ui-Tei, K., Saigo, K., Morohashi, Y., Tomita, T., and Iwatsubo, T. (2005). Aph-1 contributes to the stabilization and trafficking of the gamma-secretase complex through mechanisms involving intermolecular and intramolecular interactions. *J. Biol. Chem.* **280**, 12967–12975.

Nitsch, R. M., Blusztajn, J. K., Pittas, A. G., Slack, B. E., Growdon, J. H., and Wurtman, R. J. (1992). Evidence for a membrane defect in Alzheimer disease brain. *Proc. Natl. Acad. Sci. U.S.A.* **89**, 1671–1675.

Nyabi, O., Bentahir, M., Horre, K., Herreman, A., Gottardi-Littell, N., Van Broeckhoven, C., Merchiers, P., Spittaels, K., Annaert, W., and De Strooper, B. (2003). Presenilins mutated at Asp-257 or Asp-385 restore Pen-2 expression and Nicastrin glycosylation but remain catalytically inactive in the absence of wild type Presenilin. *J. Biol. Chem.* **278**, 43430–43436.

Ogura, T., Mio, K., Hayashi, I., Miyashita, H., Fukuda, R., Kopan, R., Kodama, T., Hamakubo, T., Iwatsubo, T., Tomita, T., and Sato, C. (2006). Three-dimensional structure of the gamma-secretase complex. *Biochem. Biophys. Res. Commun.* **343**, 525–534.

Oh, Y. S., and Turner, R. J. (2005). Topology of the C-terminal fragment of human presenilin 1. *Biochemistry* **44**, 11821–11828.

Oka, C., Nakano, T., Wakeham, A., de la Pompa, J. L., Mori, C., Sakai, T., Okazaki, S., Kawaichi, M., Shiota, K., Mak, T. W., and Honjo, T. (1995). Disruption of the mouse RBP-J kappa gene results in early embryonic death. *Development* **121**, 3291–3301.

Okochi, M., Steiner, H., Fukumori, A., Tanii, H., Tomita, T., Tanaka, T., Iwatsubo, T., Kudo, T., Takeda, M., and Haass, C. (2002). Presenilins mediate a dual intramembranous gamma-secretase cleavage of Notch-1. *EMBO J.* **21**, 5408–5416.

Osenkowski, P., Li, H., Ye, W., Li, D., Aeschbach, L., Fraering, P. C., Wolfe, M. S., and Selkoe, D. J. (2009). Cryoelectron microscopy structure of purified gamma-secretase at 12 A resolution. *J. Mol. Biol.* **385**, 642–652.

Pardossi-Piquard, R., Yang, S. P., Kanemoto, S., Gu, Y., Chen, F., Bohm, C., Sevalle, J., Li, T., Wong, P. C., Checler, F., Schmitt-Ulms, G., St George-Hyslop, P., *et al.* (2009). APH1 polar transmembrane residues regulate the assembly and activity of presenilin complexes. *J. Biol. Chem.* **284**, 16298–16307.

Park, J. T., Li, M., Nakayama, K., Mao, T. L., Davidson, B., Zhang, Z., Kurman, R. J., Eberhart, C. G., Shih Ie, M., and Wang, T. L. (2006). Notch3 gene amplification in ovarian cancer. *Cancer Res.* **66**, 6312–6318.

Qi-Takahara, Y., Morishima-Kawashima, M., Tanimura, Y., Dolios, G., Hirotani, N., Horikoshi, Y., Kametani, F., Maeda, M., Saido, T. C., Wang, R., and Ihara, Y. (2005). Longer forms of amyloid beta protein: implications for the mechanism of intramembrane cleavage by gamma-secretase. *J. Neurosci.* **25**, 436–445.

Rand, M. D., Grimm, L. M., Artavanis-Tsakonas, S., Patriub, V., Blacklow, S. C., Sklar, J., and Aster, J. C. (2000). Calcium depletion dissociates and activates heterodimeric notch receptors. *Mol. Cell. Biol.* **20**, 1825–1835.

Ratovitski, T., Slunt, H. H., Thinakaran, G., Price, D. L., Sisodia, S. S., and Borchelt, D. R. (1997). Endoproteolytic processing and stabilization of wild-type and mutant presenilin. *J. Biol. Chem.* **272**, 24536–24541.

Rawson, R. B. (2003). The SREBP pathway–insights from Insigs and insects. *Nat. Rev. Mol. Cell Biol.* **4**, 631–640.

Real, P. J., Tosello, V., Palomero, T., Castillo, M., Hernando, E., de Stanchina, E., Sulis, M. L., Barnes, K., Sawai, C., Homminga, I., Meijerink, J., Aifantis, I., *et al.* (2009). Gamma-secretase inhibitors reverse glucocorticoid resistance in T cell acute lymphoblastic leukemia. *Nat. Med.* **15**, 50–58.

Rogaev, E. I., Sherrington, R., Rogaeva, E. A., Levesque, G., Ikeda, M., Liang, Y., Chi, H., Lin, C., Holman, K., Tsuda, T., *et al.* (1995). Familial Alzheimer's disease in kindreds with missense mutations in a gene on chromosome 1 related to the Alzheimer's disease type 3 gene. *Nature* **376**, 775–778.

Sato, C., Morohashi, Y., Tomita, T., and Iwatsubo, T. (2006). Structure of the catalytic pore of gamma-secretase probed by the accessibility of substituted cysteines. *J. Neurosci.* **26**, 12081–12088.

Sato, C., Takagi, S., Tomita, T., and Iwatsubo, T. (2008). The C-terminal PAL motif and transmembrane domain 9 of presenilin 1 are involved in the formation of the catalytic pore of the gamma-secretase. *J. Neurosci.* **28**, 6264–6271.

Saxena, M. T., Schroeter, E. H., Mumm, J. S., and Kopan, R. (2001). Murine notch homologs (N1-4) undergo presenilin-dependent proteolysis. *J. Biol. Chem.* **276**, 40268–40273.

Schroeter, E. H., Kisslinger, J. A., and Kopan, R. (1998). Notch-1 signalling requires ligand-induced proteolytic release of intracellular domain. *Nature* **393**, 382–386.

Searfoss, G. H., Jordan, W. H., Calligaro, D. O., Galbreath, E. J., Schirtzinger, L. M., Berridge, B. R., Gao, H., Higgins, M. A., May, P. C., and Ryan, T. P. (2003). Adipsin, a biomarker of gastrointestinal toxicity mediated by a functional gamma-secretase inhibitor. *J. Biol. Chem.* **278**, 46107–46116.

Seiffert, D., Bradley, J. D., Rominger, C. M., Rominger, D. H., Yang, F., Meredith, J. E., Jr., Wang, Q., Roach, A. H., Thompson, L. A., Spitz, S. M., Higaki, J. N., Prakash, S. R., *et al.* (2000). Presenilin-1 and -2 are molecular targets for gamma-secretase inhibitors. *J. Biol. Chem.* **275**, 34086–34091.

Serneels, L., Dejaegere, T., Craessaerts, K., Horre, K., Jorissen, E., Tousseyn, T., Hebert, S., Coolen, M., Martens, G., Zwijsen, A., Annaert, W., Hartmann, D., *et al.* (2005). Differential contribution of the three Aph1 genes to gamma-secretase activity in vivo. *Proc. Natl. Acad. Sci. U.S.A.* **102**, 1719–1724.

Serneels, L., Van Biervliet, J., Craessaerts, K., Dejaegere, T., Horre, K., Van Houtvin, T., Esselmann, H., Paul, S., Schafer, M. K., Berezovska, O., Hyman, B. T., Sprangers, B., *et al.* (2009). gamma-Secretase heterogeneity in the Aph1 subunit: relevance for Alzheimer's disease. *Science* **324**, 639–642.

Shah, S., Lee, S. F., Tabuchi, K., Hao, Y. H., Yu, C., LaPlant, Q., Ball, H., Dann, C. E.3rd, Sudhof, T., and Yu, G. (2005). Nicastrin functions as a gamma-secretase-substrate receptor. *Cell* **122**, 435–447.

Shen, J., Bronson, R. T., Chen, D. F., Xia, W., Selkoe, D. J., and Tonegawa, S. (1997). Skeletal and CNS defects in Presenilin-1-deficient mice. *Cell* **89**, 629–639.

Sherrington, R., Rogaev, E. I., Liang, Y., Rogaeva, E. A., Levesque, G., Ikeda, M., Chi, H., Lin, C., Li, G., Holman, K., *et al.* (1995). Cloning of a gene bearing missense mutations in early-onset familial Alzheimer's disease [see comments]. *Nature* **375**, 754–760.

Shih Ie, M., and Wang, T. L. (2007). Notch signaling, gamma-secretase inhibitors, and cancer therapy. *Cancer Res.* **67**, 1879–1882.

Shirotani, K., Edbauer, D., Prokop, S., Haass, C., and Steiner, H. (2004). Identification of distinct gamma-secretase complexes with different APH-1 variants. *J. Biol. Chem.* **279**, 41340–41345.

Siemers, E. R., Quinn, J. F., Kaye, J., Farlow, M. R., Porsteinsson, A., Tariot, P., Zoulnouni, P., Galvin, J. E., Holtzman, D. M., Knopman, D. S., Satterwhite, J., Gonzales, C., *et al.* (2006). Effects of a gamma-secretase inhibitor in a randomized study of patients with Alzheimer disease. *Neurology* **66**, 602–604.

Siemers, E., Skinner, M., Dean, R. A., Gonzales, C., Satterwhite, J., Farlow, M., Ness, D., and May, P. C. (2005). Safety, tolerability, and changes in amyloid beta concentrations after administration of a gamma-secretase inhibitor in volunteers. *Clin. Neuropharmacol.* **28**, 126–132.

Song, W., Nadeau, P., Yuan, M., Yang, X., Shen, J., and Yankner, B. A. (1999). Proteolytic release and nuclear translocation of Notch-1 are induced by presenilin-1 and impaired by pathogenic presenilin-1 mutations. *Proc. Natl. Acad. Sci. U.S.A.* **96**, 6959–6963.

Spasic, D., Raemaekers, T., Dillen, K., Declerck, I., Baert, V., Serneels, L., Fullekrug, J., and Annaert, W. (2007). Rer1p competes with APH-1 for binding to nicastrin and regulates gamma-secretase complex assembly in the early secretory pathway. *J. Cell Biol.* **176**, 629–640.

Spasic, D., Tolia, A., Dillen, K., Baert, V., De Strooper, B., Vrijens, S., and Annaert, W. (2006). Presenilin-1 maintains a nine-transmembrane topology throughout the secretory pathway. *J. Biol. Chem.* **281**, 26569–26577.

Steiner, H., Kostka, M., Romig, H., Basset, G., Pesold, B., Hardy, J., Capell, A., Meyn, L., Grim, M. L., Baumeister, R., Fechteler, K., and Haass, C. (2000). Glycine 384 is required for presenilin-1 function and is conserved in bacterial polytopic aspartyl proteases. *Nat. Cell Biol.* **2**, 848–851.

Steiner, H., Romig, H., Grim, M. G., Philipp, U., Pesold, B., Citron, M., Baumeister, R., and Haass, C. (1999a). The biological and pathological function of the presenilin-1 Deltaexon 9 mutation is independent of its defect to undergo proteolytic processing. *J. Biol. Chem.* **274**, 7615–7618.

Steiner, H., Romig, H., Pesold, B., Philipp, U., Baader, M., Citron, M., Loetscher, H., Jacobsen, H., and Haass, C. (1999a). Amyloidogenic function of the Alzheimer's disease-associated presenilin 1 in the absence of endoproteolysis. *Biochemistry* **38**, 14600–14605.

Steiner, H., Winkler, E., Edbauer, D., Prokop, S., Basset, G., Yamasaki, A., Kostka, M., and Haass, C. (2002). PEN-2 is an integral component of the gamma-secretase complex required for coordinated expression of presenilin and nicastrin. *J. Biol. Chem.* **277**, 39062–39065.

Struhl, G., and Adachi, A. (2000). Requirements for presenilin-dependent cleavage of notch and other transmembrane proteins. *Mol Cell.* **6**, 625–636.

Struhl, G., and Greenwald, I. (1999). Presenilin is required for activity and nuclear access of Notch in Drosophila. *Nature* **398**, 522–525.

Tagami, S., Okochi, M., Yanagida, K., Ikuta, A., Fukumori, A., Matsumoto, N., Ishizuka-Katsura, Y., Nakayama, T., Itoh, N., Jiang, J., Nishitomi, K., Kamino, K., *et al.* (2008). Regulation of Notch signaling by dynamic changes in the precision of S3 cleavage of Notch-1. *Mol. Cell. Biol.* **28**, 165–176.

Takami, M., Nagashima, Y., Sano, Y., Ishihara, S., Morishima-Kawashima, M., Funamoto, S., and Ihara, Y. (2009). gamma-Secretase: successive tripeptide and tetrapeptide release from the transmembrane domain of beta-carboxyl terminal fragment. *J. Neurosci.* **29**, 13042–13052.

Takasugi, N., Tomita, T., Hayashi, I., Tsuruoka, M., Niimura, M., Takahashi, Y., Thinakaran, G., and Iwatsubo, T. (2003). The role of presenilin cofactors in the gamma-secretase complex. *Nature* **422**, 438–441.

Teng, L., Zhao, J., Wang, F., Ma, L., and Pei, G. (2010). A GPCR/secretase complex regulates beta- and gamma-secretase specificity for Abeta production and contributes to AD pathogenesis. *Cell Res.* **20**, 138–153.

Thathiah, A., Spittaels, K., Hoffmann, M., Staes, M., Cohen, A., Horre, K., Vanbrabant, M., Coun, F., Baekelandt, V., Delacourte, A., Fischer, D. F., Pollet, D., *et al.* (2009). The orphan G protein-coupled receptor 3 modulates amyloid-beta peptide generation in neurons. *Science* **323**, 946–951.

Thinakaran, G., Borchelt, D. R., Lee, M. K., Slunt, H. H., Spitzer, L., Kim, G., Ratovitsky, T., Davenport, F., Nordstedt, C., Seeger, M., Hardy, J., Levey, A. I., *et al.* (1996). Endoproteolysis of presenilin 1 and accumulation of processed derivatives in vivo. *Neuron* **17**, 181–190.

Thinakaran, G., Harris, C. L., Ratovitski, T., Davenport, F., Slunt, H. H., Price, D. L., Borchelt, D. R., and Sisodia, S. S. (1997). Evidence that levels of presenilins (PS1 and PS2) are coordinately regulated by competition for limiting cellular factors. *J. Biol. Chem.* **272**, 28415–28422.

Tolia, A., Chavez-Gutierrez, L., and De Strooper, B. (2006). Contribution of presenilin transmembrane domains 6 and 7 to a water-containing cavity in the gamma-secretase complex. *J. Biol. Chem.* **281**, 27633–27642.

Tolia, A., Horre, K., and De Strooper, B. (2008). Transmembrane domain 9 of presenilin determines the dynamic conformation of the catalytic site of gamma-secretase. *J. Biol. Chem.* **283**, 19793–19803.

Tomita, T., Katayama, R., Takikawa, R., and Iwatsubo, T. (2002). Complex N-glycosylated form of nicastrin is stabilized and selectively bound to presenilin fragments. *FEBS Lett.* **520**, 117–121.

Tournoy, J., Bossuyt, X., Snellinx, A., Regent, M., Garmyn, M., Serneels, L., Saftig, P., Craessaerts, K., De Strooper, B., and Hartmann, D. (2004). Partial loss of presenilins causes seborrheic keratosis and autoimmune disease in mice. *Hum. Mol. Genet.* **13**, 1321–1331.

Uemura, K., Lill, C. M., Li, X., Peters, J. A., Ivanov, A., Fan, Z., DeStrooper, B., Bacskai, B. J., Hyman, B. T., and Berezovska, O. (2009). Allosteric modulation of PS1/gamma-secretase conformation correlates with amyloid beta(42/40) ratio. *PLoS One* **4**, e7893.

Urban, S. (2009). Making the cut: central roles of intramembrane proteolysis in pathogenic microorganisms. *Nat. Rev. Microbiol.* **7**, 411–423.

Urban, S., and Shi, Y. (2008). Core principles of intramembrane proteolysis: comparison of rhomboid and site-2 family proteases. *Curr. Opin. Struct. Biol.* **18**, 432–441.

van Es, J. H., van Gijn, M. E., Riccio, O., van den Born, M., Vooijs, M., Begthel, H., Cozijnsen, M., Robine, S., Winton, D. J., Radtke, F., and Clevers, H. (2005). Notch/gamma-secretase inhibition turns proliferative cells in intestinal crypts and adenomas into goblet cells. *Nature* **435**, 959–963.

van Tetering, G., van Diest, P., Verlaan, I., van der Wall, E., Kopan, R., and Vooijs, M. (2009). The metalloprotease ADAM10 is required for notch1 S2 cleavage. *J. Biol. Chem.*

Vetrivel, K. S., Cheng, H., Kim, S. H., Chen, Y., Barnes, N. Y., Parent, A. T., Sisodia, S. S., and Thinakaran, G. (2005). Spatial segregation of gamma-secretase and substrates in distinct membrane domains. *J. Biol. Chem.* **280**, 25892–25900.

Vetrivel, K. S., Cheng, H., Lin, W., Sakurai, T., Li, T., Nukina, N., Wong, P. C., Xu, H., and Thinakaran, G. (2004). Association of gamma-secretase with lipid rafts in post-Golgi and endosome membranes. *J. Biol. Chem.* **279**, 44945–44954.

Wakabayashi, T., Craessaerts, K., Bammens, L., Bentahir, M., Borgions, F., Herdewijn, P., Staes, A., Timmerman, E., Vandekerckhove, J., Rubinstein, E., Boucheix, C., Gevaert, K., *et al.* (2009). Analysis of the gamma-secretase interactome and validation of its association with tetraspanin-enriched microdomains. *Nat. Cell Biol.* **11**, 1340–1346.

Wakabayashi, T., and De Strooper, B. (2008). Presenilins: members of the gamma-secretase quartets, but part-time soloists too. *Physiology (Bethesda)* **23**, 194–204.

Weggen, S., Eriksen, J. L., Das, P., Sagi, S. A., Wang, R., Pietrzik, C. U., Findlay, K. A., Smith, T. E., Murphy, M. P., Bulter, T., Kang, D. E., Marquez-Sterling, N., *et al.* (2001). A subset of NSAIDs lower amyloidogenic Abeta42 independently of cyclooxygenase activity. *Nature* **414**, 212–216.

Weggen, S., Eriksen, J. L., Sagi, S. A., Pietrzik, C. U., Ozols, V., Fauq, A., Golde, T. E., and Koo, E. H. (2003). Evidence that nonsteroidal anti-inflammatory drugs decrease amyloid beta 42 production by direct modulation of gamma-secretase activity. *J. Biol. Chem.* **278**, 31831–31837.

Weng, A. P., Ferrando, A. A., Lee, W., Morris, J.Pt., Silverman, L. B., Sanchez-Irizarry, C., Blacklow, S. C., Look, A. T., and Aster, J. C. (2004). Activating mutations of NOTCH1 in human T cell acute lymphoblastic leukemia. *Science* **306**, 269–271.

Westlund, B., Parry, D., Clover, R., Basson, M., and Johnson, C. D. (1999). Reverse genetic analysis of Caenorhabditis elegans presenilins reveals redundant but unequal roles for sel-12 and hop-1 in Notch-pathway signaling. *Proc. Natl. Acad. Sci. U.S.A.* **96**, 2497–2502.

Wolfe, M. S. (2007). gamma-Secretase modulators. *Cur.r Alzheimer Res.* **4**, 571–573.

Wolfe, M. S., and Kopan, R. (2004). Intramembrane proteolysis: theme and variations. *Science* **305**, 1119–1123.

Wolfe, M. S., Xia, W., Ostaszewski, B. L., Diehl, T. S., Kimberly, W. T., and Selkoe, D. J. (1999). Two transmembrane aspartates in presenilin-1 required for presenilin endoproteolysis and gamma-secretase activity. *Nature* **398**, 513–517.

Wong, G. T., Manfra, D., Poulet, F. M., Zhang, Q., Josien, H., Bara, T., Engstrom, L., Pinzon-Ortiz, M., Fine, J. S., Lee, H. J., Zhang, L., Higgins, G. A., *et al.* (2004). Chronic treatment with the gamma-secretase inhibitor LY-411,575 inhibits beta-amyloid peptide production and alters lymphopoiesis and intestinal cell differentiation. *J. Biol. Chem.* **279**, 12876–12882.

Wong, P. C., Zheng, H., Chen, H., Becher, M. W., Sirinathsinghji, D. J., Trumbauer, M. E., Chen, H. Y., Price, D. L., Van der Ploeg, L. H., and Sisodia, S. S. (1997). Presenilin 1 is required for Notch1 and Dll1 expression in the paraxial mesoderm. *Nature* **387**, 288–292.

Xia, W. (2008). From presenilinase to gamma-secretase, cleave to capacitate. *Curr. Alzheimer Res.* **5**, 172–178.

Xia, W., and Wolfe, M. S. (2003). Intramembrane proteolysis by presenilin and presenilin-like proteases. *J. Cell. Sci.* **116**, 2839–2844.

Xu, H., Lu, Y. F., Partilla, J. S., Pinto, J., Calderon, S. N., Matecka, D., Rice, K. C., Lai, J., Porreca, F., Ananthan, S., and Rothman, R. B. (1998). Opioid peptide receptor studies. 8. One of the mouse brain deltaNCX binding sites is similar to the cloned mouse opioid delta receptor: further evidence for heterogeneity of delta opioid receptors. *Peptides* **19**, 343–350.

Yagishita, S., Morishima-Kawashima, M., Tanimura, Y., Ishiura, S., and Ihara, Y. (2006). DAPT-induced intracellular accumulations of longer amyloid beta-proteins: further implications for the mechanism of intramembrane cleavage by gamma-secretase. *Biochemistry* **45**, 3952–3960.

Yanagida, K., Okochi, M., Tagami, S., Nakayama, T., Kodama, T. S., Nishitomi, K., Jiang, J., Mori, K., Tatsumi, S., Arai, T., Ikeuchi, T., Kasuga, K., *et al.* (2009). The 28-amino acid form of an APLP1-derived Abeta-like peptide is a surrogate marker for Abeta42 production in the central nervous system. *EMBO Mol. Med.* **1**, 223–235.

Yang, D. S., Tandon, A., Chen, F., Yu, G., Yu, H., Arawaka, S., Hasegawa, H., Duthie, M., Schmidt, S. D., Ramabhadran, T. V., Nixon, R. A., Mathews, P. M., *et al.* (2002). Mature glycosylation and trafficking of nicastrin modulate its binding to presenilins. *J. Biol. Chem.* **277**, 28135–28142.

Ye, Y., and Fortini, M. E. (1999). Apoptotic activities of wild-type and Alzheimer's disease-related mutant presenilins in Drosophila melanogaster. *J. Cell Biol.* **146**, 1351–1364.

Yu, G., Nishimura, M., Arawaka, S., Levitan, D., Zhang, L., Tandon, A., Song, Y. Q., Rogaeva, E., Chen, F., Kawarai, T., Supala, A., Levesque, L., *et al.* (2000). Nicastrin modulates presenilin-mediated notch/glp-1 signal transduction and betaAPP processing. *Nature* **407**, 48–54.

Zhang, L., Lee, J., Song, L., Sun, X., Shen, J., Terracina, G., and Parker, E. M. (2005). Characterization of the reconstituted gamma-secretase complex from Sf9 cells co-expressing presenilin 1, nicastrin [correction of nacastrin], aph-1a, and pen-2. *Biochemistry* **44**, 4450–4457.

Zhang, Z., Nadeau, P., Song, W., Donoviel, D., Yuan, M., Bernstein, A., and Yankner, B. A. (2000). Presenilins are required for gamma-secretase cleavage of beta-APP and transmembrane cleavage of Notch-1. *Nat. Cell Biol.* **2**, 463–465.

Zhao, G., Cui, M. Z., Mao, G., Dong, Y., Tan, J., Sun, L., and Xu, X. (2005). gamma-Cleavage is dependent on zeta-cleavage during the proteolytic processing of amyloid precursor protein within its transmembrane domain. *J. Biol. Chem.* **280**, 37689–37697.

Zhao, G., Liu, Z., Ilagan, M.X., and Kopan, R. (2010) {gamma}-Secretase Composed of PS1/Pen2/Aph1a Can Cleave Notch and Amyloid Precursor Protein in the Absence of Nicastrin. *J. Neurosci.* **30**, 1648–1656.

Zhao, G., Tan, J., Mao, G., Cui, M. Z., and Xu, X. (2007). The same gamma-secretase accounts for the multiple intramembrane cleavages of APP. *J. Neurochem.* **100**, 1234–1246.

CHAPTER SEVEN

Two Opposing Roles of RBP-J in Notch Signaling

Kenji Tanigaki* *and* Tasuku Honjo†

Contents

Abstract

RBP-J/Su(H)/Lag1, the main transcriptional mediator of Notch signaling, binds DNA with the consensus sequence YRTGDGAD. Notch target genes can be controlled by two opposing activities of RBP-J. The interaction of the Notch intracellular domain with RBP-J induces a weak transcriptional activation and requires an additional tissue-specific transcriptional activator such as bHLH proteins or GATA to mediate strong target gene expression. For example, during *Drosophila* sensory organ precursor (SOP) cell development, proneural bHLH interacts with Da, a *Drosophila* orthologue of *E2A*, to form a tissue-specific activator of Su(H), the *Drosophila* orthologue of *RBP-J*. This complex and Su(H) act synergistically to promote the epidermal cell fate. In contrast, a complex of Su(H) with Hairless, a *Drosophila* functional homologue of *MINT*, has transcriptional repression activity that promotes SOP differentiation to neurons. Recent conditional loss-of-function studies demonstrated that transcriptional networks involving RBP-J, MINT, and E2A are conserved in mammalian cell differentiation, including multiple steps of lymphocyte development, and probably also in neuronal maturation in adult neurogenesis. During neurogenesis, Notch–RBP-J signaling was thought historically to be

* Research Institute, Shiga Medical Center, Moriyama, Shiga, Japan
† Department of Immunology and Genomic Medicine, Kyoto University, Sakyo-ku, Kyoto, Japan

Current Topics in Developmental Biology, Volume 92
ISSN 0070-2153, DOI 10.1016/S0070-2153(10)92007-3

involved mainly in the maintenance of undifferentiated neural progenitors. However, the identification of a tissue-specific transcriptional activator of RBP-J-Notch has revealed new roles of RBP-J in the promotion of neuronal maturation. Finally, the Notch-independent function of RBP-J was recently discovered and will be reviewed here.

1. The Identification of RBP-J and its Connection to Notch Signaling

RBP-J, also known as *CSL*, reflecting other names of this gene in vertebrates (*CBF1*), fly *Su(H)*, and worm (*Lag1*) was originally purified from nuclear extracts of a mouse pre-B cell line as a protein that binds a DNA fragment carrying the *immunoglobulin Jk* recombination signal sequence (CACTGTG) (Hamaguchi *et al.*, 1989, 1992; Kawaichi *et al.*, 1992). Subsequent studies using recombinant RBP-J protein showed that the core recognition sequence of RBP-J is YRTGDGAD (Barolo *et al.*, 2000; Henkel *et al.*, 1994; Tun *et al.*, 1994). The biological function of RBP-J began to be elucidated by genetic mapping of the *Drosophila* orthologue of *RBP-J* to the *Suppressor of Hairless Su(H)* locus (Furukawa *et al.*, 1991, 1992). Schweisguth and Posakony independently isolated the *Su(H)* gene by chromosome walking and demonstrated its identity with *RBP-J* (Schweisguth and Posakony, 1992). Genetic analyses showed that Su(H) functions downstream of Notch signaling (Fortini and Artavanis, 1994; Furukawa *et al.*, 1994). The RBP-J/Su(H) binding sequence was identified in the promoter region of a target gene of Notch signaling, *enhancer of split (E(spl)) m8*; transcriptional activation of *m8* by Notch was shown to be directly regulated by Su(H) (Furukawa *et al.*, 1995; Lecourtois and Schweisguth, 1995).

Notch is a membrane-bound receptor and interacts with ligands such as Delta and Jagged, which are also expressed on the cytoplasmic membrane. Notch interacts directly with RBP-J, mainly through Notch's RAM (RBP-J-associated molecule) domain and weakly through its ANK (ankyrin) repeats (Kato *et al.*, 1997; Tamura *et al.*, 1995). In parallel, it was shown that the intracellular domain of Notch can transactivate the promoter of *HES1*, the mammalian homologue of *Hairy* and *E(spl)*, which also contains RBP-J-binding sites (Honjo, 1996; Jarriault *et al.*, 1995). Subsequent studies showed that ligand binding to the Notch receptor leads to proteolytic processing of Notch within its transmembrane domain, resulting in the release of Notch's intracellular domain (Mizutani *et al.*, 2001; Schroeter *et al.*, 1998; Struhl and Adachi, 1998). The released intracellular domain translocates to the nucleus to interact with RBP-J in the nucleus (Sakai *et al.*, 1995).

2. RBP-J as a Transcription Factor

RBP-J was also identified as a cellular factor binding to a viral promoter (Henkel *et al.*, 1994) and was initially considered to be a transcriptional repressor (Dou *et al.*, 1994; Hsieh and Hayward, 1995). RBP-J binding to the promoter of the adenovirus capsid protein polypeptideIX (pIX) perturbs the interaction between the transcription factors TFIIA and TFIID (Olave *et al.*, 1998). RBP-J was also shown to interact with corepressors, such as SMRT (silencing mediator for retinoid and thyroid receptor)/N-CoR (nuclear receptor corepressor) and MINT (Msx2-interacting nuclear target protein, also known as Sharp/SPEN), which, like *Drosophila* Hairless, interact with C-terminal binding protein (CtBP) or other global corepressor complexes, and recruit histone deacetylases to suppress the transcription of RBP-J's target genes (Fig. 7.1) (Heitzler and Simpson, 1991; Kao *et al.*, 1998; Kuroda *et al.*, 2003; Kurooka *et al.*, 1998; Morel *et al.*, 2001; Oswald *et al.*, 2005; Zhou and Hayward, 2001). An isoform of *FHL1C/KyoT* gene, KyoT2, is another RBP-J-associated molecule that has recently been reported to recruit RING1, a polycomb group (PcG) protein (Qin *et al.*, 2004; Tani *et al.*, 1999; Taniguchi *et al.*, 1998). This finding suggested that RBP-J might be involved in the regulation of chromosome accessibility through PcG-mediated transcriptional silencing.

Studies on Epstein–Barr virus (EBV) were also important in describing the role of RBP-J as a transcriptional activator. EBV infects B lymphocytes and induces their immortalization. EBV nuclear antigen 2 (EBNA2) is essential for this immortalization. EBNA2 has been shown to mask RBP-J's corepressor interaction domain to convert RBP-J from a transcriptional repressor to an activator (Hsieh and Hayward, 1995; Zimber-Strobl *et al.*, 1994). The intracellular domain of Notch has an analogous function (Hsieh *et al.*, 1996, 1997; Sakai *et al.*, 1998) because the RAM domain and ANK repeats of Notch displace corepressors from RBP-J (Kato *et al.*, 1997; Kurooka *et al.*, 1998). In addition, the Notch intracellular domain actively recruits global coactivators (CoAs), Mastermind, PCAF, and GCN5 (Kitagawa *et al.*, 2001; Kurooka and Honjo, 2000; Kurooka *et al.*, 1998; Wu *et al.*, 2000) to enhance the RBP-J-dependent transcription of target genes such as *Hes-1* and *Hes-5*.

Mutational and truncational analyses of RBP-J revealed that the central domain of the RBP-J protein is important for its interactions with DNA and the RAM domain of Notch and that its N- and C-terminal domains are important for its interaction with the ANK repeats of Notch. All these interactions are indispensable for RBP-J-mediated transactivation by the Notch intracellular domain (Chung *et al.*, 1994; Hsieh *et al.*, 1996; Kao *et al.*, 1998; Tani *et al.*, 2001). These findings were recently

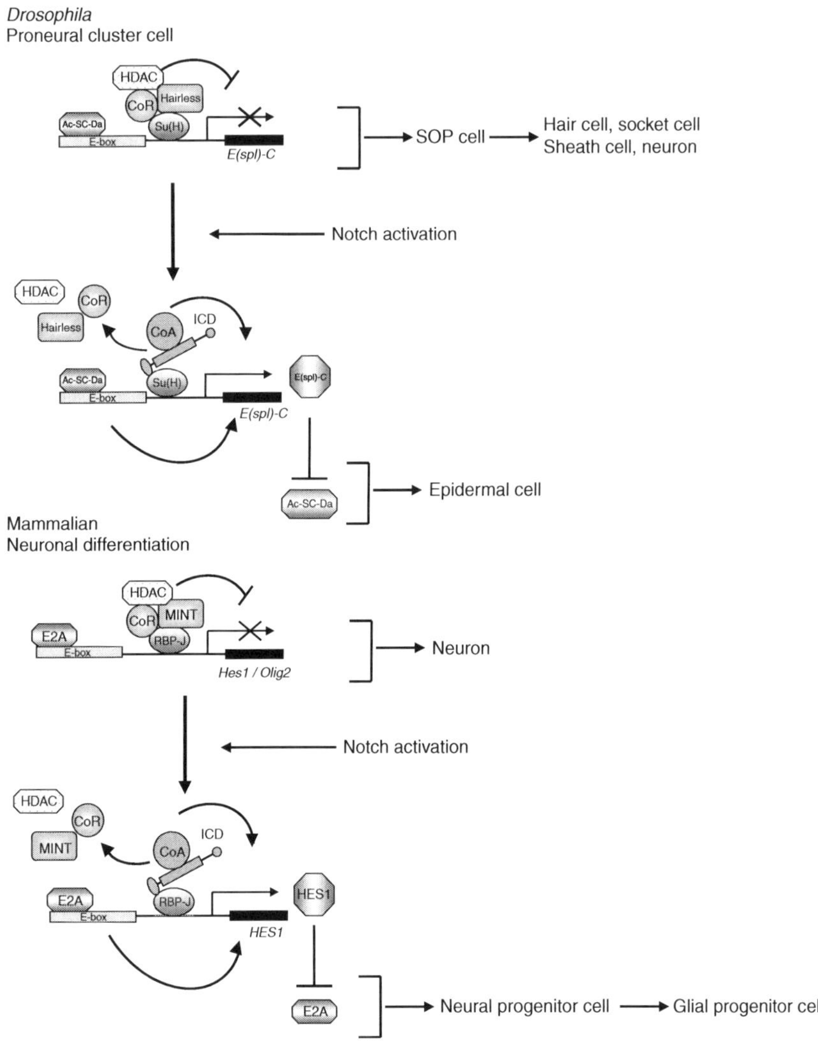

Figure 7.1 The dual roles of RBP-J in cell-lineage commitment. In the absence of Notch signaling, RBP-J/Su(H) recruits corepressors (CoR) through Hairless/MINT and represses target gene expression. After Notch activation, the intracellular domain of Notch (ICD) translocates to the nucleus, cooperates with the Da/E2A complex, and transactivates RBP-J-mediated transcription by displacing CoR and recruiting coactivators (CoAs). Next, target genes such as *E(spl)* and *HES1* inhibit the functions of the Da/E2A complex and terminate their synergistic transactivation. This negative feedback loop involving Notch, RBP-J, MINT, and E2A is conserved from *Drosophila* to mammals. The repressive activity of Su(H) and Hairless is required for the differentiation of neurons from SOP cells. A similar mechanism is observed for the regulation of *Olig2* in differentiating mammalian neurons.

confirmed by structural studies of the complex of the Notch intracellular domain and RBP-J.

The interaction of the central portion of RBP-J with the RAM domain of Notch induces a drastic conformational change in the N-terminal domain of RBP-J. This conformational change at the N-terminal domain increases the accessibility of the ANK repeats of Notch to the C-terminal domain of RBP-J. The subsequent binding of the Notch ANK repeats to the C-terminal domain of RBP-J forms a complete docking site for the coactivator Mastermind (Friedmann *et al.*, 2008; Kovall and Hendrickson, 2004; Wilson and Kovall, 2006).

3. Biological Functions of RBP-J in *Drosophila*

Notch signaling regulates various developmental processes in *Drosophila*, including lateral inhibition, binary cell-lineage commitment, and the formation of tissue boundaries (Artavanis *et al.*, 1995; Bray, 1998). The analysis of Notch signaling in the development of the sensory organ precursors (SOPs) of *Drosophila* has provided a useful model for understanding the molecular mechanisms of Notch signaling in cell-lineage commitment (Bang *et al.*, 1995). Each SOP divides, and its progeny differentiate into a hair cell, a socket cell, a sheath cell, and a neuron, to generate a mechanosensory bristle. Notch signaling selects out a single cell from a cluster of cells [a proneural cluster (PNC)], all of which have the same potential to become SOP cells. This selection process, called lateral inhibition, is explained as follows: one of the PNC cells expresses a higher level of Notch ligand than its neighbors, which activates the Notch receptors of the surrounding PNC cells, and induces them to decide the epidermal cell fate by inhibiting their differentiation into SOP cells. In the absence of Notch signaling, all PNC cells commit to the SOP fate.

Notch activation leads to the Su(H)-mediated transcription of *E(spl)* and suppresses SOP cell differentiation (Fig. 7.1). E-box-mediated transcription of *E(spl)* is promoted by Achaete/Scute (Ac/Sc) –Daughterless (Da), but E(spl) inhibits its own E-box-mediated transcription, and this feedback inhibition is required for SOP cell differentiation. In the absence of Notch activation, Su(H) recruits Hairless, CtBP, and Groucho to repress the transcription of E(spl) (Castro *et al.*, 2005; Kao *et al.*, 1998; Morel *et al.*, 2001). Whether a cell commits to the SOP or the epidermal fate is determined by the activity balance between the Ac/Sc–Da complex and Notch–Su(H) signaling (Fig. 7.2A) (Castro *et al.*, 2005). The repressive activity of the Su(H) –Hairless complex is required for the determination of the SOP cell fate. This complex completely suppresses the expression of

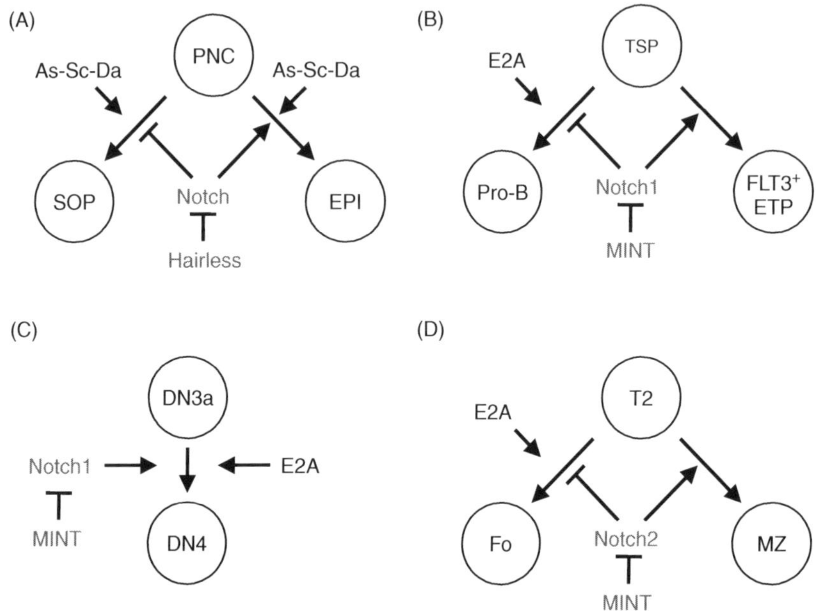

Figure 7.2 Notch signaling regulates binary cell-lineage commitments cooperatively with the Da–E2A complex. SOP, sensory organ precursor; PNC, proneural cluster; EPI, epidermal cell; TSP, thymus seeding precursor; Pro-B, pro-B cell; ETP, early T-cell precursor; DN, double-negative cell; T2, type2 transitional B cell; Fo, follicular B cell; MZ, marginal zone B cell.

E(spl) and maintains a high activity of Ac/Sc–Da in SOP cells. The loss of *Su(H)* causes the derepression of *E(spl)* and leads to defects in the neuronal maturation of SOP cell progeny (Koelzer and Klein, 2003). Thus, the transcriptional repression activity of RBP-J is indispensable for the induction of neuronal differentiation.

In contrast, the negative feedback regulation of the Ac/Sc–Da complex by E(spl) is indispensable for epidermal cell differentiation. In epidermal cells, the Ac/Sc–Da complex and Notch intracellular domain–Su(H) complex synergistically transactivate *E(spl)* expression, which in turn inhibits the Ac/Sc–Da-mediated transcription. Mutations in the E box of the *E(spl)* promoter, even when the Su(H)-binding sites are intact, eliminate transcription from the promoter (Castro *et al.*, 2005). Thus, Su(H)–Notch alone is a weak transcriptional activator, whereas the combination of the Su(H)-binding sites and the E-box *cis*-regulatory element provides the robust and specific upregulation of *E(spl)* gene expression. *Su(H)* promotes the epidermal cell fate as a transcriptional activator and the neuronal cell fate as a transcriptional repressor. Thus, the dual roles of Su(H) are pivotal in SOP development.

4. Regulation of Mammalian Neuronal Development by RBP-J

The biological functions of Notch signaling are conserved in mammalian development, in which they regulate various cell fate specifications such as myogenesis, neurogenesis, gliogenesis, skin/hair formation, and lymphoid development (Table 7.1) (Artavanis *et al.*, 1995; Blanpain *et al.*, 2006; Demehri and Kopan, 2009; de la Pompa *et al.*, 1997; Han *et al.*, 2002; Kuroda *et al.*, 1999; Tanigaki and Honjo, 2007; Tanigaki *et al.*, 2001, 2003, 2004; Yamamoto *et al.*, 2003). In mammalian systems, the *cis*-regulatory elements of the RBP-J-binding sites and E box are also conserved in the upstream region of *HES1*, a homologue of *E(spl)*, which is regulated by E2A, a homologue of *Drosophila Da*, and by Notch–RBP-J (Ikawa *et al.*, 2006). HES1 plays a critical role in the negative feedback regulation of HES1 by inhibiting the E-box-mediated transcription and N-box-mediated repression (Kuroda *et al.*, 1999; Sasai *et al.*, 1992) (Fig. 7.1). MINT, a functional homologue of *Drosophila Hairless*, binds to RBP-J in the absence of Notch activation and negatively regulates *HES1* expression in a manner similar to Hairless in *Drosophila* (Kuroda *et al.*, 2003). The collaboration of Notch, RBP-J, MINT, and E2A during binary cell fate decision operates in many different cell-lineage commitment systems, such as neuronal development and lymphocyte differentiation, including B-cell versus T-cell-lineage commitment, intrathymic T-cell differentiation, and B-cell differentiation.

Initially, Notch–RBP-J signaling was thought to suppress neuronal differentiation and maintain undifferentiated neural progenitors through the regulation of its targets, *HES1* and *HES5* (de la Pompa *et al.*, 1997; Ohtsuka *et al.*, 1999), which enabled the neural progenitors to differentiate into glial progenitors in later developmental stages. Notch activation induces gliogenesis from both neural multipotent progenitors and neural crest stem cells and loss of RBP-J delayed astrocyte differentiation *in vitro* (Ge *et al.*, 2002; Morrison *et al.*, 2000; Tanigaki *et al.*, 2001). However, it was difficult to distinguish whether Notch signaling directly induces gliogenesis or only inhibits premature depletion of undifferentiated multipotent neural progenitors before gliogenesis.

It has been recently demonstrated that conditional knockout of *RBP-J* at a late developmental stage severely impaired gliogenesis which was independent of its effect on the maintenance of undifferentiated neural progenitors, because clonal analysis clearly shows RBP-J-deficient late neural progenitors still maintain multipotency (Taylor *et al.*, 2007). In addition, loss-of-function analysis of RBP-J also demonstrated RBP-J-mediated Notch signaling induces *nuclear factor I (NFI) and Sox9*, which are indispensable for glial development (Namihira *et al.*, 2009; Taylor *et al.*, 2007). These findings

Table 7.1 Functions of Notch/RBP-J/MINT confirmed by loss-of-function analysis of RBP-J or MINT

Skin/hair	Epidermis/hair fate determination of hair follicular stem cells (Demehri and Kopan, 2009; Yamamoto *et al.*, 2003), regulation of spinous cell differentiation (Blanpain *et al.*, 2006; Demehri and Kopan, 2009), regulation of melanoblast differentiation and maintenance (Aubin-Houzelstein *et al.*, 2008; Moriyama *et al.*, 2006)
Central nervous system	Maintenance of neuronal progenitors, promotion of neuronal differentiation (de la Pompa *et al.*, 1997; Fujimoto *et al.*, 2009; Komine *et al.*, 2007; Yabe *et al.*, 2007; Zhu *et al.*, 2006), promotion of glial specification (Komine *et al.*, 2007; Namihira *et al.*, 2009; Taylor *et al.*, 2007), memory formation (Costa *et al.*, 2003)
Inner ear	Maintenance of supporting cells (Yamamoto *et al.*, 2006)
Retina	Maintenance of retinal progenitor cells and suppression of retinal ganglion and cone cell fate (Riesenberg *et al.*, 2009; Zheng *et al.*, 2009)
Mammary gland	Regulation of the composition of luminal and basal cells during pregnancy (Buono *et al.*, 2006)
Lens	Maintenance of epithelial cells (Jia *et al.*, 2007), regulation of fiber cell growth and differentiation (Rowan *et al.*, 2008)
Somite	Regulation of somitogenesis (Oka *et al.*, 1995)
Bone	Regulation of proliferation and differentiation of chondrocytes (Mead and Yutzey, 2009)
Muscle	Maintenance of muscle progenitor cells (Vasyutina *et al.*, 2007)
Heart	Regulation of trabeculae formation and cardiomyocyte proliferation and differentiation (Grego-Bessa *et al.*, 2007; Schroeder *et al.*, 2003)
Hematopoietic cells	Definitive hematopoietic progenitor cell generation (Robert-Moreno *et al.*, 2005), T versus B cell commitment (Han *et al.*, 2002), T-cell development (Amsen *et al.*, 2004; Ong *et al.*, 2008; Tanaka *et al.*, 2006; Tanigaki *et al.*, 2004; Tsuji *et al.*, 2007), regulatory T-cell generation (Ou-Yang *et al.*, 2009), MZ/FO B cell-lineage commitment regulation (Kuroda *et al.*, 2003; Tanigaki *et al.*, 2002; Yabe *et al.*, 2007), splenic dendritic cell generation (Caton *et al.*, 2007)

Vascular structure	Regulation of endothelial cell differentiation and function (Krebs *et al.*, 2004; Wang *et al.*, 2009b) and regulation of angiogenesis (Dou *et al.*, 2008)
Lung	Regulation of differentiation and mobilization of myofibroblasts (Xu *et al.*, 2010). Regulation of ciliated cell and Clara cell differentiation (Morimoto *et al.*, 2010; Tsao *et al.*, 2009)
Kidney	Regulation of proximal nephron differentiation (Cheng *et al.*, 2007; Surendran *et al.*, 2009)
Liver	Regulation of sinusoid endothelial cell proliferation (Wang *et al.*, 2009a)
Intestine	Goblet cell generation(van Es *et al.*, 2005)

suggest that Notch–RBP-J signaling not only maintains undifferentiated progenitors at early developmental stages but also might promote gliogenesis. In contrast with the functions of Notch–RBP-J signaling in undifferentiated neural progenitors, its functions after neuronal lineage commitment have not been well demonstrated.

It has recently been reported that the repressor activity of RBP-J was shown to play pivotal roles in neuronal maturation in mammalian adult neurogenesis, as was shown for *Su(H)* in *Drosophila* (Fujimoto *et al.*, 2009). Olfactory bulb interneurons are generated continuously in adulthood. In this system, astroglial-like stem cells divide slowly and give rise to rapidly dividing, transiently amplifying multipotent precursors (TA precursors). The TA precursors in turn differentiate into mature interneurons in the olfactory bulb. The loss of *Notch1* or *Jagged1* disrupts the self-replication of neural progenitors, but no abnormality in the neuronal cell-lineage commitment was observed (Nyfeler *et al.*, 2005). However, in the absence of RBP-J, neuronal maturation is affected, and *Olig2* is ectopically expressed in differentiating neurons. The *Olig2* promoter contains both the E box and the RBP-J-binding site, which are conserved from mice to humans (Kenji Tanigaki unpublished data) (Fujimoto *et al.*, 2009). *Olig2* is known to be indispensable for oligodendroglial differentiation (Takebayashi *et al.*, 2002; Zhou and Anderson, 2002) and to inhibit neuronal differentiation (Buffo *et al.*, 2005; Hack *et al.*, 2005; Marshall *et al.*, 2005). Reporter analyses showed that RBP-J can repress the expression of *Olig2*, suggesting that the derepression of *Olig2* might be one of the causes of the maturation defects of RBP-J-deficient neurons (Fujimoto *et al.*, 2009) (Figs. 7.1 and 7.3). Notch signaling promotes the commitment of oligodendroglial progenitors in the embryonic spinal cord (Park and Appel, 2003). In the spinal cord of mice with a Notch1 deficiency specifically in neural progenitors, neuronal differentiation is enhanced, and the number of $Olig2^{+}$

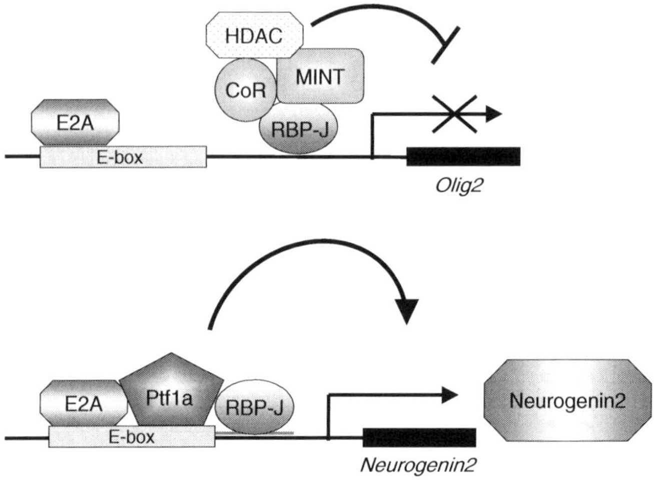

Figure 7.3 Notch-independent functions of RBP-J in differentiating neurons. The repressive activity of RBP-J is essential for neuronal maturation. Loss of RBP-J causes a derepression of *Olig2*, which leads to defects in neuronal differentiation. In the mammalian spinal cord, Ptf1a, a tissue-specific activator, competes with Notch ICD for RBP-J and promotes GABAergic interneuron differentiation by inducing neuronal differentiation-promoting genes such as *Neurogenin2*.

oligodendroglial progenitors is decreased at E11.5 (Yang *et al.*, 2006), whereas a neural progenitor-specific RBP-J deficiency has no effect on the number of Olig2^{+} cells in the E11.5 spinal cord, but enhances the generation of Olig2^{+} cells in later developmental stages (Taylor *et al.*, 2007). These differences in the effects on *Olig2* expression caused by Notch1 deficiency versus RBP-J deficiency can be reconciled by the dual functions of RBP-J. The coordination of transcriptional regulation by Notch, RBP-J, MINT, and E2A might be also important for the oligodendroglial lineage commitment and neuronal maturation in adult neurogenesis, at least in part through the regulation of *Olig2*. The collaborative regulation of *Olig2* by Notch, RBP-J, and E2A was also reported in hematopoietic cells (Ikawa *et al.*, 2006).

5. Regulation of Mammalian Hematopoietic Cell Development by RBP-J

The lymphoid developmental process also shares many similarities with the SOP development in *Drosophila*. As in neuronal development, a transcriptional network consisting of Notch, RBP-J, MINT, and E2A plays pivotal roles in several binary cell-lineage commitments in lymphoid cell development. T-cell differentiation occurs in the thymus, which is seeded by precursors released from the bone marrow to the bloodstream. The most immature T-cell progenitors in the thymus, the Flt3$^+$ early T-cell progenitors (ETPs), retain a weak, delayed B-lineage potential *in vivo*, although their B-lineage potential cannot be detected in OP9 stromal cell cultures (Allman *et al.*, 2003; Sambandam *et al.*, 2005). Flt3$^+$ ETPs differentiate to CD4$^-$ CD8$^-$ double-negative (DN) two cells via a Flt3$^-$ ETP stage. DN2 cells differentiate to αβ T cells via DN3, four stages.

The B-cell versus T-cell lineage commitment occurs before the ETP developmental stages. The loss of Notch1 or RBP-J in adult bone marrow cells leads to impairments in T-cell development and ectopic B-cell differentiation in the thymus (Han *et al.*, 2002; Radtke *et al.*, 1999; Wilson *et al.*, 2001). MINT deficiency results in an increase in ETPs, which is consistent with the requirement for high Notch signaling to generate ETPs. E2A is indispensable for B-cell differentiation (Ordentlich *et al.*, 1998). The overexpression of HES1 in hematopoietic progenitor cells partially perturbs B-cell development (Hoebeke *et al.*, 2006). Notch–RBP-J-MINT signaling provides an inductive signal for the T-cell fate determination from T/B precursors and inhibits B-lineage commitment, probably through the inhibition of E2A function (Fig. 7.2B).

The establishment of a Delta-like1(DLL1)-expressing OP9 stromal cell culture system has made it possible to perform detailed analyses of T-cell differentiation (Schmitt and Zuniga-Pflucker, 2002). Continuous Notch activation is essential for DN T-cell differentiation. Notch signaling is required for the DN1-to-DN2 transition, the maintenance of CD25 expression on the DN2 and DN3 cells (Sambandam *et al.*, 2005), and the survival of DN2, DN3, and DN4 cells (Ciofani and Zuniga-Pflucker, 2005; Taghon *et al.*, 2005). E2A is critical for the DN1-to-DN2 transition (Bain *et al.*, 1997), and MINT deficiency also leads to an impaired DN1-to-DN2 transition (Tsuji *et al.*, 2007). This again illustrates collaborative functions of Notch, RBP-J, MINT, and E2A in the mammalian T-cell development system.

A T-cell-specific conditional knockout of Notch1 or RBP-J results in the developmental arrest of αβ T cells at the DN3 stage (Tanigaki *et al.*, 2004; Wolfer *et al.*, 2002), and the DN3-to-DN4 transition is enhanced by MINT deficiency (Tsuji *et al.*, 2007). Notch signaling is essential for the β selection of pre-T cells in various ways. First, Notch1 is indispensable for the upregulation of the Vβ germ line transcript and T cell receptor (TCR) β-gene rearrangement (Hoflinger *et al.*, 2004; Wolfer *et al.*, 2002). Second, Notch signaling leads to the activation of phosphatidylinositol-3-OH kinase and Akt signaling and supports the survival of pre-T cells at the β-selection checkpoint (Ciofani and Zuniga-Pflucker, 2005). Third, E2A and Notch signaling cooperatively induces the expression of *pTα* (Ikawa *et al.*, 2006). Then, Notch signaling collaborates with pTα to suppress the activity of E2A, which induces the proliferation of β-selected pre-T cells (Engel and Murre, 2002; Nie *et al.*, 2003; Talora *et al.*, 2003). This negative feedback of the transcriptional network is in perfect agreement with that of SOP development in *Drosophila* (Fig. 7.2C).

B lymphocytes differentiate from common lymphoid progenitors in the bone marrow through pro-B and pre-B stages. Immature B cells migrate to the secondary lymphoid organs. Two types of transitional precursors for mature B cells exist in the spleen (Loder *et al.*, 1999). Immature B cells that have just arrived at the spleen from the bone marrow are type 1 transitional (T1) B cells. T1 B cells differentiate into type 2 transitional (T2) B cells in the spleen and then further develop into two types of mature B cells, follicular (Fo) and marginal zone (MZ) B cells, in the spleen. An RBP-J deficiency in B cells results in the complete loss of MZ B cells, with a concomitant increase in Fo B cells, suggesting that Notch–RBP-J signaling regulates the MZ/Fo B-cell fate determination from a common T2 precursor (Kuroda *et al.*, 2003; Tanigaki *et al.*, 2002). The inactivation of Notch2 in hematopoietic cells also causes a loss of MZ B cells (Saito *et al.*, 2003). The loss of Delta1 in hematopoietic cells results in the disappearance of MZ cells, whereas B-cell-specific Dll1-null mice have normal MZ B cells (Hozumi *et al.*, 2004).

Dll1 is expressed on dendritic cells and not on lymphocytes (Kuroda *et al.*, 2003), indicating that the Notch2 on B cells may interact with the Delta1 on dendritic cells to induce MZ B cells. MINT, a specific negative regulator of Notch–RBP-J signaling, and E2A are expressed in Fo and transitional B cells, but less so in MZ B cells. Mice with a MINT deficiency or E2A heterozygous loss show an increase in MZ B cells and decrease in Fo B cells (Kuroda *et al.*, 2003; Quong *et al.*, 2004). These findings indicate that the strict regulation of Notch and E2A activation levels is indispensable for MZ/Fo B cell fate determination (Fig. 7.2D).

6. Notch-Independent RBP-J Functions

Apparent discrepancies between the phenotypes of Notch deficiency and RBP-J deficiency led to the speculation that RBP-J might have Notch-independent functions. In many cases, such discrepancies can be explained by the transcriptional repressive activities of RBP-J (Barolo *et al.*, 2000; Fujimoto *et al.*, 2009; Koelzer and Klein, 2003). In mammals, most of the phenotypes observed in Notch conditional knockout mice, i.e., the abnormalities in the T/B cell lineages, marginal zone B cells, and αβ T-cell development, are also observed in RBP-J conditional knockout mice (Han *et al.*, 2002; Radtke *et al.*, 1999; Saito *et al.*, 2003; Tanigaki *et al.*, 2002; Witt *et al.*, 2003; Wolfer *et al.*, 2002). However, it was recently reported that the phenotype of pancreas-specific RBP-J-deficient mice is drastically different from that of pancreas-specific Notch1/2-deficient mice (Nakhai *et al.*, 2008). The loss of RBP-J in pancreatic development leads to endocrine and exocrine pancreatic hypoplasia (Fujikura *et al.*, 2006, 2007). In contrast to this severe phenotype, the loss of Notch1 and Notch2 causes a moderate reduction in the proliferation of pancreatic epithelial cells during an early embryonic stage (Nakhai *et al.*, 2008).

A breakthrough in the elucidation of RBP-J's Notch-independent functions came from the two-hybrid screening for novel RBP-J-binding molecules, which showed that RBP-J interacts with Ptf1a/p48, a transcription factor indispensable for pancreas development (Obata *et al.*, 2001). Ptf1a requires interactions with both RBP-J and an E-protein such as E2A to transactivate its target genes. The RBP-J–Ptf1a complex recognizes a juxtaposed E box and RBP-J-binding site (Beres *et al.*, 2006) (Fig. 7.3). The domain of RBP-J that interacts with Ptf1a is consistent with the domain that interacts with Notch RAM, and the Notch intracellular domain competes with Ptf1a for RBP-J (Beres *et al.*, 2006). The binding of Ptf1a and the Notch intracellular domain to RBP-J is mutually exclusive.

Ptf1a-mediated transcription is also enhanced by RBP-J-like (RBP-L), which is a paralogue of RBP-J that is specifically expressed in the lung and pancreas (Minoguchi *et al.*, 1999). RBP-L binds to a DNA sequence identical to the RBP-J-binding site, but it does not interact with the Notch intracellular domain (Minoguchi *et al.*, 1997). The severe pancreatic defect of RBP-J-deficient mice mirrors that of Ptf1a-null mice (Masui *et al.*, 2007). Ptf1a, RBP-J, and RBP-L are essential for pancreas growth and pancreas-specific gene transcription (Beres *et al.*, 2006; Masui *et al.*, 2007, 2008). RBP-L might have evolved to specialize in Notch-independent functions that promote pancreatic exocrine cell differentiation and insulate Ptf1a from further interference by Notch signaling competing for RBP-J. Indeed, recent findings demonstrate that Ngn3-positive progenitors depend on Notch2 to titrate RBP-J away from Ptf1a and protect their endocrine choice; reduction in Notch2 dose permits Ptf1a to recruit RBP-J and this complex outcompetes Ngn3 for E2A, diverting the Ngn3 linage to the acinar fate (Cras-Meneur *et al.*, 2009).

The interaction of Ptf1a and RBP-J was also shown to promote GABAergic inhibitory neurons in the spinal cord in a manner similar to the pancreas (Hori *et al.*, 2008). RBP-J and Ptf1a suppress the glutamatergic neuronal fate and induce GABAergic neurons. Neurogenin2, a neuronal differentiation transcriptional factor, was shown to be a direct target of the RBP-J–Ptf1a transcriptional complex (Fig. 7.3) (Henke *et al.*, 2009). These findings demonstrated new roles for RBP-J in neuronal maturation. Thus, RBP-J can promote neuronal differentiation in at least two ways. First, RBP-J/Su(H) actively represses the expression of genes that are inhibitory for neuronal differentiation, such as *E(spl)* and *Olig2*. Second, RBP-J recruits a tissue-specific activator such as Ptf1a to promote a GABAergic neuronal fate. This mechanism may be conserved in *Drosophila* because the *Drosophila* homologue of PTf1a, Fer1, can interact with Su(H) and is extensively expressed in the embryonic central nervous system (Beres *et al.*, 2006; Masui *et al.*, 2007).

The RBP-J/Su(H)/Notch intracellular domain complex is a weak transcriptional activator and needs to synergize with other tissue-specific transcriptional activators to have its greatest effects (Cooper *et al.*, 2000; Furriols and Bray, 2001; Neves *et al.*, 2007). However, few transcriptional activators of RBP-J-Su(H) have been identified except for proneural bHLH proteins (Ac/Sc) and Ptf1a. Genome-wide studies are likely to elucidate direct target genes of RBP-J/Su(H) and lead to the identification of other tissue-specific transcriptional activators of RBP-J (Krejci *et al.*, 2009).

ACKNOWLEDGMENTS

This research was supported by a Center for Excellence Grant and Grant-in-Aid for Specially Promoted Research 17002015, Young Scientists (A) 17689014 and (B) 20790244 of the Ministry of Education, Culture, Sports, Science and Technology (MEXT), Japan.

REFERENCES

Allman, D., Sambandam, A., Kim, S., Miller, J. P., Pagan, A., Well, D., Meraz, A., and Bhandoola, A. (2003). Thymopoiesis independent of common lymphoid progenitors. *Nat. Immunol.* **4**, 168–174.

Amsen, D., Blander, J. M., Lee, G. R., Tanigaki, K., Honjo, T., and Flavell, R. A. (2004). Instruction of distinct CD4 T helper cell fates by different notch ligands on antigen-presenting cells. *Cell* **117**, 515–526.

Artavanis, T. S., Matsuno, K., and Fortini, M. E. (1995). Notch signaling. *Science* **268**, 225–232.

Aubin-Houzelstein, G., Djian-Zaouche, J., Bernex, F., Gadin, S., Delmas, V., Larue, L., and Panthier, J. J. (2008). Melanoblasts' proper location and timed differentiation depend on Notch/RBP-J signaling in postnatal hair follicles. *J. Invest. Dermatol.* **128**, 2686–2695.

Bain, G., Engel, I., Robanus Maandag, E. C., te Riele, H. P., Voland, J. R., Sharp, L. L., Chun, J., Huey, B., Pinkel, D., and Murre, C. (1997). E2A deficiency leads to abnormalities in alphabeta T-cell development and to rapid development of T-cell lymphomas. *Mol. Cell. Biol.* **17**, 4782–4791.

Bang, A. G., Bailey, A. M., and Posakony, J. W. (1995). Hairless promotes stable commitment to the sensory organ precursor cell fate by negatively regulating the activity of the Notch signaling pathway. *Dev. Biol.* **172**, 479–494.

Barolo, S., Walker, R. G., Polyanovsky, A. D., Freschi, G., Keil, T., and Posakony, J. W. (2000). A notch-independent activity of suppressor of hairless is required for normal mechanoreceptor physiology. *Cell* **103**, 957–969.

Beres, T. M., Masui, T., Swift, G. H., Shi, L., Henke, R. M., and MacDonald, R. J. (2006). PTF1 is an organ-specific and Notch-independent basic helix-loop-helix complex containing the mammalian Suppressor of Hairless (RBP-J) or its paralogue, RBP-L. *Mol. Cell. Biol.* **26**, 117–130.

Blanpain, C., Lowry, W. E., Pasolli, H. A., and Fuchs, E. (2006). Canonical notch signaling functions as a commitment switch in the epidermal lineage. *Genes Dev.* **20**, 3022–3035.

Bray, S. (1998). Notch signalling in *Drosophila*: three ways to use a pathway. *Semin. Cell Dev. Biol.* **9**, 591–597.

Buffo, A., Vosko, M. R., Erturk, D., Hamann, G. F., Jucker, M., Rowitch, D., and Gotz, M. (2005). Expression pattern of the transcription factor Olig2 in response to brain injuries: implications for neuronal repair. *Proc. Natl. Acad. Sci. U.S.A.* **102**, 18183–18188.

Buono, K. D., Robinson, G. W., Martin, C., Shi, S., Stanley, P., Tanigaki, K., Honjo, T., and Hennighausen, L. (2006). The canonical Notch/RBP-J signaling pathway controls the balance of cell lineages in mammary epithelium during pregnancy. *Dev. Biol.* **293**, 565–580.

Castro, B., Barolo, S., Bailey, A. M., and Posakony, J. W. (2005). Lateral inhibition in proneural clusters: cis-regulatory logic and default repression by Suppressor of Hairless. *Development* **132**, 3333–3344.

Caton, M. L., Smith-Raska, M. R., and Reizis, B. (2007). Notch-RBP-J signaling controls the homeostasis of CD8- dendritic cells in the spleen. *J. Exp. Med.* **204**, 1653–1664.

Cheng, H. T., Kim, M., Valerius, M. T., Surendran, K., Schuster-Gossler, K., Gossler, A., McMahon, A. P., and Kopan, R. (2007). Notch2, but not Notch1, is required for proximal fate acquisition in the mammalian nephron. *Development* **134**, 801–811.

Chung, C. N., Hamaguchi, Y., Honjo, T., and Kawaichi, M. (1994). Site-directed mutagenesis study on DNA binding regions of the mouse homologue of Suppressor of Hairless, RBP-J kappa. *Nucleic Acids Res.* **22**, 2938–2944.

Ciofani, M., and Zuniga-Pflucker, J. C. (2005). Notch promotes survival of pre-T cells at the beta-selection checkpoint by regulating cellular metabolism. *Nat. Immunol.* **6**, 881–888.

Cooper, M. T., Tyler, D. M., Furriols, M., Chalkiadaki, A., Delidakis, C., and Bray, S. (2000). Spatially restricted factors cooperate with notch in the regulation of Enhancer of split genes. *Dev. Biol.* **221**, 390–403.

Costa, R. M., Honjo, T., and Silva, A. J. (2003). Learning and memory deficits in Notch mutant mice. *Curr. Biol.* **13**, 1348–1354.

Cras-Meneur, C., Li, L., Kopan, R., and Permutt, M. A. (2009). Presenilins, Notch dose control the fate of pancreatic endocrine progenitors during a narrow developmental window. *Genes Dev.* **23**, 2088–2101.

de la Pompa, J. L., Wakeham, A., Correia, K. M., Samper, E., Brown, S., Aguilera, R. J., Nakano, T., Honjo, T., Mak, T. W., Rossant, J., and Conlon, R. A. (1997). Conservation of the Notch signalling pathway in mammalian neurogenesis. *Development* **124**, 1139–1148.

Demehri, S., and Kopan, R. (2009). Notch signaling in bulge stem cells is not required for selection of hair follicle fate. *Development* **136**, 891–896.

Dou, G. R., Wang, Y. C., Hu, X. B., Hou, L. H., Wang, C. M., Xu, J. F., Wang, Y. S., Liang, Y. M., Yao, L. B., Yang, A. G., and Han, H. (2008). RBP-J, the transcription factor downstream of Notch receptors, is essential for the maintenance of vascular homeostasis in adult mice. *FASEB J.* **22**, 1606–1617.

Dou, S., Zeng, X., Cortes, P., Erdjument, B. H., Tempst, P., Honjo, T., and Vales, L. D. (1994). The recombination signal sequence-binding protein RBP-2N functions as a transcriptional repressor. *Mol. Cell. Biol.* **14**, 3310–3319.

Engel, I., and Murre, C. (2002). Disruption of pre-TCR expression accelerates lymphomagenesis in E2A-deficient mice. *Proc. Natl. Acad. Sci. U.S.A.* **99**, 11322–11327.

Fortini, M. E., and Artavanis, T. S. (1994). The suppressor of hairless protein participates in notch receptor signaling. *Cell* **79**, 273–282.

Friedmann, D. R., Wilson, J. J., and Kovall, R. A. (2008). RAM-induced allostery facilitates assembly of a notch pathway active transcription complex. *J. Biol. Chem.* **283**, 14781–14791.

Fujikura, J., Hosoda, K., Iwakura, H., Tomita, T., Noguchi, M., Masuzaki, H., Tanigaki, K., Yabe, D., Honjo, T., and Nakao, K. (2006). Notch/Rbp-j signaling prevents premature endocrine and ductal cell differentiation in the pancreas. *Cell Metab.* **3**, 59–65.

Fujikura, J., Hosoda, K., Kawaguchi, Y., Noguchi, M., Iwakura, H., Odori, S., Mori, E., Tomita, T., Hirata, M., Ebihara, K., Masuzaki, H., Fukuda, A., *et al.* (2007). Rbp-j regulates expansion of pancreatic epithelial cells and their differentiation into exocrine cells during mouse development. *Dev. Dyn.* **236**, 2779–2791.

Fujimoto, M., Takagi, Y., Muraki, K., Nozaki, K., Yamamoto, N., Tsuji, M., Hashimoto, N., Honjo, T., and Tanigaki, K. (2009). RBP-J promotes neuronal differentiation and inhibits oligodendroglial development in adult neurogenesis. *Dev. Biol.* **332**, 339–350.

Furriols, M., and Bray, S. (2001). A model Notch response element detects Suppressor of Hairless-dependent molecular switch. *Curr. Biol.* **11**, 60–64.

Furukawa, T., Kawaichi, M., Matsunami, N., Ryo, H., Nishida, Y., and Honjo, T. (1991). The *Drosophila* RBP-J kappa gene encodes the binding protein for the immunoglobulin J kappa recombination signal sequence. *J. Biol. Chem.* **266**, 23334–23340.

Furukawa, T., Kimura, K., Kobayakawa, Y., Tamura, K., Kawaichi, M., Tanimura, T., and Honjo, T. (1994). Genetic characterization of *Drosophila* RBP-J kappa (suppressor of hairless) as a neurogenic gene in adult PNS development. *Jpn. J. Genet.* **69**, 701–711.

Furukawa, T., Kobayakawa, Y., Tamura, K., Kimura, K., Kawaichi, M., Tanimura, T., and Honjo, T. (1995). Suppressor of hairless, the *Drosophila* homologue of RBP-J kappa, transactivates the neurogenic gene E(spl)m8. *Jpn. J. Genet.* **70**, 505–524.

Furukawa, T., Maruyama, S., Kawaichi, M., and Honjo, T. (1992). The *Drosophila* homolog of the immunoglobulin recombination signal-binding protein regulates peripheral nervous system development. *Cell* **69**, 1191–1197.

Ge, W., Martinowich, K., Wu, X., He, F., Miyamoto, A., Fan, G., Weinmaster, G., and Sun, Y. E. (2002). Notch signaling promotes astrogliogenesis via direct CSL-mediated glial gene activation. *J. Neurosci. Res.* **69**, 848–860.

Grego-Bessa, J., Luna-Zurita, L., del Monte, G., Bolos, V., Melgar, P., Arandilla, A., Garratt, A. N., Zang, H., Mukouyama, Y. S., Chen, H., Shou, W., Ballestar, E., *et al.* (2007). Notch signaling is essential for ventricular chamber development. *Dev. Cell* **12**, 415–429.

Hack, M. A., Saghatelyan, A., de Chevigny, A., Pfeifer, A., Ashery-Padan, R., Lledo, P. M., and Gotz, M. (2005). Neuronal fate determinants of adult olfactory bulb neurogenesis. *Nat. Neurosci.* **8**, 865–872.

Hamaguchi, Y., Matsunami, N., Yamamoto, Y., and Honjo, T. (1989). Purification and characterization of a protein that binds to the recombination signal sequence of the immunoglobulin J kappa segment. *Nucleic Acids Res.* **17**, 9015–9026.

Hamaguchi, Y., Yamamoto, Y., Iwanari, H., Maruyama, S., Furukawa, T., Matsunami, N., and Honjo, T. (1992). Biochemical and immunological characterization of the DNA binding protein (RBP-J kappa) to mouse J kappa recombination signal sequence. *J. Biochem. (Tokyo)* **112**, 314–320.

Han, H., Tanigaki, K., Yamamoto, N., Kuroda, K., Yoshimoto, M., Nakahata, T., Ikuta, K., and Honjo, T. (2002). Inducible gene knockout of transcription factor recombination signal binding protein-J reveals its essential role in T versus B lineage decision. *Int. Immunol.* **14**, 637–645.

Heitzler, P., and Simpson, P. (1991). The choice of cell fate in the epidermis of *Drosophila*. *Cell* **64**, 1083–1092.

Henke, R. M., Savage, T. K., Meredith, D. M., Glasgow, S. M., Hori, K., Dumas, J., MacDonald, R. J., and Johnson, J. E. (2009). Neurog2 is a direct downstream target of the Ptf1a-Rbpj transcription complex in dorsal spinal cord. *Development* **136**, 2945–2954.

Henkel, T., Ling, P. D., Hayward, S. D., and Peterson, M. G. (1994). Mediation of Epstein-Barr virus EBNA2 transactivation by recombination signal-binding protein J kappa. *Science* **265**, 92–95.

Hoebeke, I., De Smedt, M., Van de Walle, I., Reynvoet, K., De Smet, G., Plum, J., and Leclercq, G. (2006). Overexpression of HES-1 is not sufficient to impose T-cell differentiation on human hematopoietic stem cells. *Blood* **107**, 2879–2881.

Hoflinger, S., Kesavan, K., Fuxa, M., Hutter, C., Heavey, B., Radtke, F., and Busslinger, M. (2004). Analysis of Notch1 function by in vitro T cell differentiation of Pax5 mutant lymphoid progenitors. *J. Immunol.* **173**, 3935–3944.

Honjo, T. (1996). The shortest path from the surface to the nucleus: RBP-J kappa/Su(H) transcription factor. *Genes Cells* **1**, 1–9.

Hori, K., Cholewa-Waclaw, J., Nakada, Y., Glasgow, S. M., Masui, T., Henke, R. M., Wildner, H., Martarelli, B., Beres, T. M., Epstein, J. A., Magnuson, M. A., Macdonald, R. J., *et al.* (2008). A nonclassical bHLH Rbpj transcription factor complex is required for specification of GABAergic neurons independent of Notch signaling. *Genes Dev.* **22**, 166–178.

Hozumi, K., Negishi, N., Suzuki, D., Abe, N., Sotomaru, Y., Tamaoki, N., Mailhos, C., Ish-Horowicz, D., Habu, S., and Owen, M. J. (2004). Delta-like 1 is necessary for the generation of marginal zone B cells but not T cells in vivo. *Nat. Immunol.* **5**, 638–644.

Hsieh, J. J., and Hayward, S. D. (1995). Masking of the CBF1/RBPJ kappa transcriptional repression domain by Epstein-Barr virus EBNA2. *Science* **268**, 560–563.

Hsieh, J. J., Henkel, T., Salmon, P., Robey, E., Peterson, M. G., and Hayward, S. D. (1996). Truncated mammalian Notch1 activates CBF1/RBPJk-repressed genes by a mechanism resembling that of Epstein-Barr virus EBNA2. *Mol. Cell. Biol.* **16**, 952–959.

Hsieh, J. J., Nofziger, D. E., Weinmaster, G., and Hayward, S. D. (1997). Epstein-Barr virus immortalization: Notch2 interacts with CBF1 and blocks differentiation. *J. Virol.* **71**, 1938–1945.

Ikawa, T., Kawamoto, H., Goldrath, A. W., and Murre, C. (2006). E proteins and Notch signaling cooperate to promote T cell lineage specification and commitment. *J. Exp. Med.* **203**, 1329–1342.

Jarriault, S., Brou, C., Logeat, F., Schroeter, E. H., Kopan, R., and Israel, A. (1995). Signalling downstream of activated mammalian Notch. *Nature* **377**, 355–358.

Jia, J., Lin, M., Zhang, L., York, J. P., and Zhang, P. (2007). The Notch signaling pathway controls the size of the ocular lens by directly suppressing p57Kip2 expression. *Mol. Cell. Biol.* **27**, 7236–7247.

Kao, H. Y., Ordentlich, P., Koyano-Nakagawa, N., Tang, Z., Downes, M., Kintner, C. R., Evans, R. M., and Kadesch, T. (1998). A histone deacetylase corepressor complex regulates the Notch signal transduction pathway. *Genes Dev.* **12**, 2269–2277.

Kato, H., Taniguchi, Y., Kurooka, H., Minoguchi, S., Sakai, T., Nomura-Okazaki, S., Tamura, K., and Honjo, T. (1997). Involvement of RBP-J in biological functions of mouse Notch1 and its derivatives. *Development* **124**, 4133–4141.

Kawaichi, M., Oka, C., Shibayama, S., Koromilas, A. E., Matsunami, N., Hamaguchi, Y., and Honjo, T. (1992). Genomic organization of mouse J kappa recombination signal binding protein (RBP-J kappa) gene. *J. Biol. Chem.* **267**, 4016–4022.

Kitagawa, M., Oyama, T., Kawashima, T., Yedvobnick, B., Kumar, A., Matsuno, K., and Harigaya, K. (2001). A human protein with sequence similarity to *Drosophila* mastermind coordinates the nuclear form of notch and a CSL protein to build a transcriptional activator complex on target promoters. *Mol. Cell. Biol.* **21**, 4337–4346.

Koelzer, S., and Klein, T. (2003). A Notch-independent function of Suppressor of Hairless during the development of the bristle sensory organ precursor cell of *Drosophila. Development* **130**, 1973–1988.

Komine, O., Nagaoka, M., Watase, K., Gutmann, D. H., Tanigaki, K., Honjo, T., Radtke, F., Saito, T., Chiba, S., and Tanaka, K. (2007). The monolayer formation of Bergmann glial cells is regulated by Notch/RBP-J signaling. *Dev. Biol.* **311**, 238–250.

Kovall, R. A., and Hendrickson, W. A. (2004). Crystal structure of the nuclear effector of Notch signaling, CSL, bound to DNA. *EMBO J.* **23**, 3441–3451.

Krebs, L. T., Shutter, J. R., Tanigaki, K., Honjo, T., Stark, K. L., and Gridley, T. (2004). Haploinsufficient lethality and formation of arteriovenous malformations in Notch pathway mutants. *Genes Dev.* **18**, 2469–2473.

Krejci, A., Bernard, F., Housden, B. E., Collins, S., and Bray, S. J. (2009). Direct response to Notch activation: signaling crosstalk and incoherent logic. *Sci. Signal.* **2**, ra1.

Kuroda, K., Han, H., Tani, S., Tanigaki, K., Tun, T., Furukawa, T., Taniguchi, Y., Kurooka, H., Hamada, Y., Toyokuni, S., and Honjo, T. (2003). Regulation of marginal zone B cell development by MINT, a suppressor of Notch/RBP-J signaling pathway. *Immunity* **18**, 301–312.

Kuroda, K., Tani, S., Tamura, K., Minoguchi, S., Kurooka, H., and Honjo, T. (1999). Delta-induced Notch signaling mediated by RBP-J inhibits MyoD expression and myogenesis. *J. Biol. Chem.* **274**, 7238–7244.

Kurooka, H., and Honjo, T. (2000). Functional interaction between the mouse notch1 intracellular region and histone acetyltransferases PCAF and GCN5. *J. Biol. Chem.* **275**, 17211–17220.

Kurooka, H., Kuroda, K., and Honjo, T. (1998). Roles of the ankyrin repeats and C-terminal region of the mouse notch1 intracellular region. *Nucleic Acids Res.* **26**, 5448–5455.

Lecourtois, M., and Schweisguth, F. (1995). The neurogenic suppressor of hairless DNA-binding protein mediates the transcriptional activation of the enhancer of split complex genes triggered by Notch signaling. *Genes Dev.* **9**, 2598–2608.

Loder, F., Mutschler, B., Ray, R. J., Paige, C. J., Sideras, P., Torres, R., Lamers, M. C., and Carsetti, R. (1999). B cell development in the spleen takes place in discrete steps and is determined by the quality of B cell receptor-derived signals. *J. Exp. Med.* **190**, 75–89.

Marshall, C. A., Novitch, B. G., and Goldman, J. E. (2005). Olig2 directs astrocyte and oligodendrocyte formation in postnatal subventricular zone cells. *J. Neurosci.* **25**, 7289–7298.

Masui, T., Long, Q., Beres, T. M., Magnuson, M. A., and MacDonald, R. J. (2007). Early pancreatic development requires the vertebrate Suppressor of Hairless (RBPJ) in the PTF1 bHLH complex. *Genes Dev.* **21**, 2629–2643.

Masui, T., Swift, G. H., Hale, M. A., Meredith, D. M., Johnson, J. E., and Macdonald, R. J. (2008). Transcriptional autoregulation controls pancreatic Ptf1a expression during development and adulthood. *Mol. Cell. Biol.* **28**, 5458–5468.

Mead, T. J., and Yutzey, K. E. (2009). Notch pathway regulation of chondrocyte differentiation and proliferation during appendicular and axial skeleton development. *Proc. Natl. Acad. Sci. U.S.A.* **106**, 14420–14425.

Minoguchi, S., Ikeda, T., Itohara, S., Kaneko, T., Okaichi, H., and Honjo, T. (1999). Studies on the cell-type specific expression of RBP-L, a RBP-J family member, by replacement insertion of beta-galactosidase. *J. Biochem. (Tokyo)* **126**, 738–747.

Minoguchi, S., Taniguchi, Y., Kato, H., Okazaki, T., Strobl, L. J., Zimber-Strobl, U., Bornkamm, G. W., and Honjo, T. (1997). RBP-L, a transcription factor related to RBP-Jkappa. *Mol. Cell. Biol.* **17**, 2679–2687.

Mizutani, T., Taniguchi, Y., Aoki, T., Hashimoto, N., and Honjo, T. (2001). Conservation of the biochemical mechanisms of signal transduction among mammalian Notch family members. *Proc. Natl. Acad. Sci. U.S.A.* **98**, 9026–9031.

Morel, V., Lecourtois, M., Massiani, O., Maier, D., Preiss, A., and Schweisguth, F. (2001). Transcriptional repression by suppressor of hairless involves the binding of a hairless-dCtBP complex in *Drosophila*. *Curr. Biol.* **11**, 789–792.

Morimoto, M., Liu, Z., Cheng, H. T., Winters, N., Bader, D., and Kopan, R. (2010). Canonical Notch signaling in the developing lung is required for determination of arterial smooth muscle cells and selection of Clara versus ciliated cell fate. *J. Cell. Sci.* **123**, 213–224.

Moriyama, M., Osawa, M., Mak, S. S., Ohtsuka, T., Yamamoto, N., Han, H., Delmas, V., Kageyama, R., Beermann, F., Larue, L., and Nishikawa, S. (2006). Notch signaling via Hes1 transcription factor maintains survival of melanoblasts and melanocyte stem cells. *J. Cell Biol.* **173**, 333–339.

Morrison, S. J., Perez, S. E., Qiao, Z., Verdi, J. M., Hicks, C., Weinmaster, G., and Anderson, D. J. (2000). Transient Notch activation initiates an irreversible switch from neurogenesis to gliogenesis by neural crest stem cells. *Cell* **101**, 499–510.

Nakhai, H., Siveke, J. T., Klein, B., Mendoza-Torres, L., Mazur, P. K., Algul, H., Radtke, F., Strobl, L., Zimber-Strobl, U., and Schmid, R. M. (2008). Conditional ablation of Notch signaling in pancreatic development. *Development* **135**, 2757–2765.

Namihira, M., Kohyama, J., Semi, K., Sanosaka, T., Deneen, B., Taga, T., and Nakashima, K. (2009). Committed neuronal precursors confer astrocytic potential on residual neural precursor cells. *Dev. Cell* **16**, 245–255.

Neves, A., English, K., and Priess, J. R. (2007). Notch-GATA synergy promotes endoderm-specific expression of ref-1 in *C. elegans*. *Development* **134**, 4459–4468.

Nie, L., Xu, M., Vladimirova, A., and Sun, X. H. (2003). Notch-induced E2A ubiquitination and degradation are controlled by MAP kinase activities. *EMBO J.* **22**, 5780–5792.

Nyfeler, Y., Kirch, R. D., Mantei, N., Leone, D. P., Radtke, F., Suter, U., and Taylor, V. (2005). Jagged1 signals in the postnatal subventricular zone are required for neural stem cell self-renewal. *EMBO J.* **24**, 3504–3515.

Obata, J., Yano, M., Mimura, H., Goto, T., Nakayama, R., Mibu, Y., Oka, C., and Kawaichi, M. (2001). p48 subunit of mouse PTF1 binds to RBP-Jkappa/CBF-1, the intracellular mediator of Notch signalling, and is expressed in the neural tube of early stage embryos. *Genes Cells* **6**, 345–360.

Ohtsuka, T., Ishibashi, M., Gradwohl, G., Nakanishi, S., Guillemot, F., and Kageyama, R. (1999). Hes1 and Hes5 as notch effectors in mammalian neuronal differentiation. *EMBO J.* **18**, 2196–2207.

Oka, C., Nakano, T., Wakeham, A., de la Pompa, J. L., Mori, C., Sakai, T., Okazaki, S., Kawaichi, M., Shiota, K., Mak, T. W., and Honjo, T. (1995). Disruption of the mouse RBP-J kappa gene results in early embryonic death. *Development* **121**, 3291–3301.

Olave, I., Reinberg, D., and Vales, L. D. (1998). The mammalian transcriptional repressor RBP (CBF1) targets TFIID and TFIIA to prevent activated transcription. *Genes Dev.* **12**, 1621–1637.

Ong, C. T., Sedy, J. R., Murphy, K. M., and Kopan, R. (2008). Notch and presenilin regulate cellular expansion and cytokine secretion but cannot instruct Th1/Th2 fate acquisition. *PLoS One* **3**, e2823.

Ordentlich, P., Lin, A., Shen, C. P., Blaumueller, C., Matsuno, K., Artavanis-Tsakonas, S., and Kadesch, T. (1998). Notch inhibition of E47 supports the existence of a novel signaling pathway. *Mol. Cell. Biol.* **18**, 2230–2239.

Oswald, F., Winkler, M., Cao, Y., Astrahantseff, K., Bourteele, S., Knochel, W., and Borggrefe, T. (2005). RBP-Jkappa/SHARP recruits CtIP/CtBP corepressors to silence Notch target genes. *Mol. Cell. Biol.* **25**, 10379–10390.

Ou-Yang, H. F., Zhang, H. W., Wu, C. G., Zhang, P., Zhang, J., Li, J. C., Hou, L. H., He, F., Ti, X. Y., Song, L. Q., Zhang, S. Z., Feng, L., *et al.* (2009). Notch signaling regulates the FOXP3 promoter through RBP-J- and Hes1-dependent mechanisms. *Mol. Cell. Biochem.* **320**, 109–114.

Park, H. C., and Appel, B. (2003). Delta-Notch signaling regulates oligodendrocyte specification. *Development* **130**, 3747–3755.

Qin, H., Wang, J., Liang, Y., Taniguchi, Y., Tanigaki, K., and Han, H. (2004). RING1 inhibits transactivation of RBP-J by Notch through interaction with LIM protein KyoT2. *Nucleic Acids Res.* **32**, 1492–1501.

Quong, M. W., Martensson, A., Langerak, A. W., Rivera, R. R., Nemazee, D., and Murre, C. (2004). Receptor editing and marginal zone B cell development are regulated by the helix-loop-helix protein, E2A. *J. Exp. Med.* **199**, 1101–1112.

Radtke, F., Wilson, A., Stark, G., Bauer, M., van Meerwijk, J., MacDonald, H. R., and Aguet, M. (1999). Deficient T cell fate specification in mice with an induced inactivation of Notch1. *Immunity* **10**, 547–558.

Riesenberg, A. N., Liu, Z., Kopan, R., and Brown, N. L. (2009). Rbpj cell autonomous regulation of retinal ganglion cell and cone photoreceptor fates in the mouse retina. *J. Neurosci.* **29**, 12865–12877.

Robert-Moreno, A., Espinosa, L., de la Pompa, J. L., and Bigas, A. (2005). RBPjkappa-dependent Notch function regulates Gata2 and is essential for the formation of intra-embryonic hematopoietic cells. *Development* **132**, 1117–1126.

Rowan, S., Conley, K. W., Le, T. T., Donner, A. L., Maas, R. L., and Brown, N. L. (2008). Notch signaling regulates growth and differentiation in the mammalian lens. *Dev. Biol.* **321**, 111–122.

Saito, T., Chiba, S., Ichikawa, M., Kunisato, A., Asai, T., Shimizu, K., Yamaguchi, T., Yamamoto, G., Seo, S., Kumano, K., Nakagami-Yamaguchi, E., Hamada, Y., *et al.* (2003). Notch2 is preferentially expressed in mature B cells and indispensable for marginal zone B lineage development. *Immunity* **18**, 675–685.

Sakai, T., Furukawa, T., Iwanari, H., Oka, C., Nakano, T., Kawaichi, M., and Honjo, T. (1995). Loss of immunostaining of the RBP-J kappa transcription factor upon F9 cell differentiation induced by retinoic acid. *J. Biochem. (Tokyo)* **118**, 621–628.

Sakai, T., Taniguchi, Y., Tamura, K., Minoguchi, S., Fukuhara, T., Strobl, L. J., Zimber-Strobl, U., Bornkamm, G. W., and Honjo, T. (1998). Functional replacement of the intracellular region of the Notch1 receptor by Epstein-Barr virus nuclear antigen 2. *J. Virol.* **72**, 6034–6039.

Sambandam, A., Maillard, I., Zediak, V. P., Xu, L., Gerstein, R. M., Aster, J. C., Pear, W. S., and Bhandoola, A. (2005). Notch signaling controls the generation and differentiation of early T lineage progenitors. *Nat. Immunol.* **6**, 663–670.

Sasai, Y., Kageyama, R., Tagawa, Y., Shigemoto, R., and Nakanishi, S. (1992). Two mammalian helix-loop-helix factors structurally related to *Drosophila* hairy and Enhancer of split. *Genes Dev.* **6**, 2620–2634.

Schmitt, T. M., and Zuniga-Pflucker, J. C. (2002). Induction of T cell development from hematopoietic progenitor cells by delta-like-1 in vitro. *Immunity* **17**, 749–756.

Schroeder, T., Fraser, S. T., Ogawa, M., Nishikawa, S., Oka, C., Bornkamm, G. W., Nishikawa, S., Honjo, T., and Just, U. (2003). Recombination signal sequence-binding protein J kappa alters mesodermal cell fate decisions by suppressing cardiomyogenesis. *Proc. Natl. Acad. Sci. U.S.A.* **100**, 4018–4023.

Schroeter, E. H., Kisslinger, J. A., and Kopan, R. (1998). Notch-1 signalling requires ligand-induced proteolytic release of intracellular domain [see comments]. *Nature* **393**, 382–386.

Schweisguth, F., and Posakony, J. W. (1992). Suppressor of Hairless, the *Drosophila* homolog of the mouse recombination signal-binding protein gene, controls sensory organ cell fates. *Cell* **69**, 1199–1212.

Struhl, G., and Adachi, A. (1998). Nuclear access and action of notch in vivo. *Cell* **93**, 649–660.

Surendran, K., Boyle, S., Barak, H., Kim, M., Stromberski, C., McCright, B., and Kopan, R. (2010). The contribution of Notch1 to nephron segmentation in the developing kidney is revealed in a sensitized Notch2 background and can be augmented by reducing Mint dosage. *Dev. Biol.* **337**, 386–395.

Taghon, T. N., David, E. S., Zuniga-Pflucker, J. C., and Rothenberg, E. V. (2005). Delayed, asynchronous, and reversible T-lineage specification induced by Notch/Delta signaling. *Genes Dev.* **19**, 965–978.

Takebayashi, H., Nabeshima, Y., Yoshida, S., Chisaka, O., Ikenaka, K., and Nabeshima, Y. (2002). The basic helix-loop-helix factor olig2 is essential for the development of motoneuron and oligodendrocyte lineages. *Curr. Biol.* **12**, 1157–1163.

Talora, C., Campese, A. F., Bellavia, D., Pascucci, M., Checquolo, S., Groppioni, M., Frati, L., von Boehmer, H., Gulino, A., and Screpanti, I. (2003). Pre-TCR-triggered ERK signalling-dependent downregulation of E2A activity in Notch3-induced T-cell lymphoma. *EMBO Rep.* **4**, 1067–1072.

Tamura, K., Taniguchi, Y., Minoguchi, S., Sakai, T., Tun, T., Furukawa, T., and Honjo, T. (1995). Physical interaction between a novel domain of the receptor Notch and the transcription factor RBP-J kappa/Su(H). *Curr. Biol.* **5**, 1416–1423.

Tanaka, S., Tsukada, J., Suzuki, W., Hayashi, K., Tanigaki, K., Tsuji, M., Inoue, H., Honjo, T., and Kubo, M. (2006). The interleukin-4 enhancer CNS-2 is regulated by Notch signals and controls initial expression in NKT cells and memory-type CD4 T cells. *Immunity* **24**, 689–701.

Tani, S., Kurooka, H., Aoki, T., Hashimoto, N., and Honjo, T. (2001). The N- and C-terminal regions of RBP-J interact with the ankyrin repeats of Notch1 RAMIC to activate transcription. *Nucleic Acids Res.* **29**, 1373–1380.

Tani, S., Taniwaki, M., Taniguchi, Y., Minoguchi, S., Kuroda, K., Han, H., Aoki, T., Miyatake, S., Hashimoto, N., and Honjo, T. (1999). Chromosomal mapping of two RBP-J-related genes: Kyo-T and RBP-L. *J. Hum. Genet.* **44**, 73–75.

Tanigaki, K., Han, H., Yamamoto, N., Tashiro, K., Ikegawa, M., Kuroda, K., Suzuki, A., Nakano, T., and Honjo, T. (2002). Notch-RBP-J signaling is involved in cell fate determination of marginal zone B cells. *Nat. Immunol.* **3**, 443–450.

Tanigaki, K., and Honjo, T. (2007). Regulation of lymphocyte development by Notch signaling. *Nat. Immunol.* **8**, 451–456.

Tanigaki, K., Kuroda, K., Han, H., and Honjo, T. (2003). Regulation of B cell development by Notch/RBP-J signaling. *Semin. Immunol.* **15**, 113–119.

Tanigaki, K., Nogaki, F., Takahashi, J., Tashiro, K., Kurooka, H., and Honjo, T. (2001). Notch1 and Notch3 instructively restrict bFGF-responsive multipotent neural progenitor cells to an astroglial fate. *Neuron* **29**, 45–55.

Tanigaki, K., Tsuji, M., Yamamoto, N., Han, H., Tsukada, J., Inoue, H., Kubo, M., and Honjo, T. (2004). Regulation of alphabeta/gammadelta T cell lineage commitment and peripheral T cell responses by Notch/RBP-J signaling. *Immunity* **20**, 611–622.

Taniguchi, Y., Furukawa, T., Tun, T., Han, H., and Honjo, T. (1998). LIM protein KyoT2 negatively regulates transcription by association with the RBP-J DNA-binding protein. *Mol. Cell. Biol.* **18**, 644–654.

Taylor, M. K., Yeager, K., and Morrison, S. J. (2007). Physiological Notch signaling promotes gliogenesis in the developing peripheral and central nervous systems. *Development* **134**, 2435–2447.

Tsao, P. N., Vasconcelos, M., Izvolsky, K. I., Qian, J., Lu, J., and Cardoso, W. V. (2009). Notch signaling controls the balance of ciliated and secretory cell fates in developing airways. *Development* **136**, 2297–2307.

Tsuji, M., Shinkura, R., Kuroda, K., Yabe, D., and Honjo, T. (2007). Msx2-interacting nuclear target protein (Mint) deficiency reveals negative regulation of early thymocyte differentiation by Notch/RBP-J signaling. *Proc. Natl. Acad. Sci. U.S.A.* **104**, 1610–1615.

Tun, T., Hamaguchi, Y., Matsunami, N., Furukawa, T., Honjo, T., and Kawaichi, M. (1994). Recognition sequence of a highly conserved DNA binding protein RBP-J kappa. *Nucleic Acids Res.* **22**, 965–971.

van Es, J. H., van Gijn, M. E., Riccio, O., van den Born, M., Vooijs, M., Begthel, H., Cozijnsen, M., Robine, S., Winton, D. J., Radtke, F., and Clevers, H. (2005). Notch/gamma-secretase inhibition turns proliferative cells in intestinal crypts and adenomas into goblet cells. *Nature* **435**, 959–963.

Vasyutina, E., Lenhard, D. C., Wende, H., Erdmann, B., Epstein, J. A., and Birchmeier, C. (2007). RBP-J (Rbpsuh) is essential to maintain muscle progenitor cells and to generate satellite cells. *Proc. Natl. Acad. Sci. U.S.A.* **104**, 4443–4448.

Wang, L., Wang, C. M., Hou, L. H., Dou, G. R., Wang, Y. C., Hu, X. B., He, F., Feng, F., Zhang, H. W., Liang, Y. M., Dou, K. F., and Han, H. (2009a). Disruption of the transcription factor recombination signal-binding protein-Jkappa (RBP-J) leads to veno-occlusive disease and interfered liver regeneration in mice. *Hepatology* **49**, 268–277.

Wang, L., Wang, Y. C., Hu, X. B., Zhang, B. F., Dou, G. R., He, F., Gao, F., Feng, F., Liang, Y. M., Dou, K. F., and Han, H. (2009b). Notch-RBP-J signaling regulates the mobilization and function of endothelial progenitor cells by dynamic modulation of CXCR4 expression in mice. *PLoS One* **4**, e7572.

Wilson, J. J., and Kovall, R. A. (2006). Crystal structure of the CSL-Notch-Mastermind ternary complex bound to DNA. *Cell* **124**, 985–996.

Wilson, A., MacDonald, H. R., and Radtke, F. (2001). Notch 1-deficient common lymphoid precursors adopt a B cell fate in the thymus. *J. Exp. Med.* **194**, 1003–1012.

Witt, C. M., Won, W. J., Hurez, V., and Klug, C. A. (2003). Notch2 haploinsufficiency results in diminished B1 B cells and a severe reduction in marginal zone B cells. *J. Immunol.* **171**, 2783–2788.

Wolfer, A., Wilson, A., Nemir, M., MacDonald, H. R., and Radtke, F. (2002). Inactivation of Notch1 impairs VDJbeta rearrangement and allows pre-TCR-independent survival of early alpha beta Lineage Thymocytes. *Immunity* **16**, 869–879.

Wu, L., Aster, J. C., Blacklow, S. C., Lake, R., Artavanis-Tsakonas, S., and Griffin, J. D. (2000). MAML1, a human homologue of *Drosophila* mastermind, is a transcriptional co-activator for NOTCH receptors. *Nat. Genet.* **26**, 484–489.

Xu, K., Nieuwenhuis, E., Cohen, B. L., Wang, W., Canty, A. J., Danska, J. S., Coultas, L., Rossant, J., Wu, M. Y., Piscione, T. D., Nagy, A., Gossler, A., *et al.* (2010). Lunatic Fringe-mediated Notch signaling is required for lung alveogenesis. *Am. J. Physiol. Lung Cell Mol. Physiol.* **298**, L45–L56.

Yabe, D., Fukuda, H., Aoki, M., Yamada, S., Takebayashi, S., Shinkura, R., Yamamoto, N., and Honjo, T. (2007). Generation of a conditional knockout allele for mammalian Spen protein Mint/SHARP. *Genesis* **45**, 300–306.

Yamamoto, N., Tanigaki, K., Han, H., Hiai, H., and Honjo, T. (2003). Notch/RBP-J signaling regulates epidermis/hair fate determination of hair follicular stem cells. *Curr. Biol.* **13**, 333–338.

Yamamoto, N., Tanigaki, K., Tsuji, M., Yabe, D., Ito, J., and Honjo, T. (2006). Inhibition of Notch/RBP-J signaling induces hair cell formation in neonate mouse cochleas. *J. Mol. Med.* **84**, 37–45.

Yang, X., Tomita, T., Wines-Samuelson, M., Beglopoulos, V., Tansey, M. G., Kopan, R., and Shen, J. (2006). Notch1 signaling influences v2 interneuron and motor neuron development in the spinal cord. *Dev. Neurosci.* **28**, 102–117.

Zheng, M. H., Shi, M., Pei, Z., Gao, F., Han, H., and Ding, Y. Q. (2009). The transcription factor RBP-J is essential for retinal cell differentiation and lamination. *Mol. Brain* **2**, 38.

Zhou, Q., and Anderson, D. J. (2002). The bHLH transcription factors OLIG2 and OLIG1 couple neuronal and glial subtype specification. *Cell* **109**, 61–73.

Zhou, S., and Hayward, S. D. (2001). Nuclear localization of CBF1 is regulated by interactions with the SMRT corepressor complex. *Mol. Cell. Biol.* **21**, 6222–6232.

Zhu, X., Zhang, J., Tollkuhn, J., Ohsawa, R., Bresnick, E. H., Guillemot, F., Kageyama, R., and Rosenfeld, M. G. (2006). Sustained Notch signaling in progenitors is required for sequential emergence of distinct cell lineages during organogenesis. *Genes Dev.* **20**, 2739–2753.

Zimber-Strobl, U., Strobl, L. J., Meitinger, C., Hinrichs, R., Sakai, T., Furukawa, T., Honjo, T., and Bornkamm, G. W. (1994). Epstein-Barr virus nuclear antigen 2 exerts its transactivating function through interaction with recombination signal binding protein RBP-J kappa, the homologue of *Drosophila* Suppressor of Hairless. *EMBO J.* **13**, 4973–4982.

CHAPTER EIGHT

Notch Targets and Their Regulation

Sarah Bray *and* Fred Bernard

Contents

Abstract

The proteolytic cleavages elicited by activation of the Notch receptor release an intracellular fragment, Notch intracellular domain, which enters the nucleus to activate the transcription of targets. Changes in transcription are therefore a major output of this pathway. However, the Notch outputs clearly differ from cell type to cell type. In this review we discuss current understanding of Notch targets, the mechanisms involved in their transcriptional regulation, and what might underlie the activation of different sets of targets in different cell types.

1. Introduction

Notch signaling has widespread roles in development and adult homeostasis, as well as a pathogenic role, when misregulated in human disease. The transcription factor CSL (CBF1-Suppressor of Hairless) plays a central role in transducing Notch signals into transcriptional outputs. Following activation, the formation of a ternary complex containing CSL, the Notch intracellular domain (NICD) and Mastermind (Mam), is essential for upregulating transcription from Notch target genes (Bray, 2006; Kopan and Ilagan, 2009). This underscores the importance of transcriptional regulation

Department of Physiology Development and Neuroscience, University of Cambridge, Cambridge, UK

Current Topics in Developmental Biology, Volume 92
ISSN 0070-2153, DOI 10.1016/S0070-2153(10)92008-5

in the Notch pathway. Here we consider our current understanding about the transcriptional response to Notch, both the types of genes that are regulated and the mechanisms underlying this regulation. The focus is on the direct targets of NICD/CSL, using the criterion that they contain validated CSL binding sites. Although our examples draw heavily from studies of Notch function in *Drosophila,* because of familiarity, our aim is to illustrate mechanisms that are generally relevant to Notch signaling in all species. However, for simplicity we refer to all Notch receptors as "Notch" and we do not discuss the added implications of the different Notch paralogues that are present in many species including humans (Kopan and llagan, 2009). We have also not discussed in detail the partners that interact with CSL, which have been well summarized in a recent review (Borggrefe and Oswald, 2009).

2. Number and Diversity of Notch Targets

The best-characterized Notch targets are the bHLH genes of the HES/HEY families, exemplified by the *E(spl)* genes in *Drosophila* and *HES1* in mouse. These were the first genes whose transcription was shown to change following Notch activation and provided a key paradigm for unraveling Notch pathway activity (Fischer and Gessler, 2007). Induction of *E(spl)* genes can be detected within 20-30 min of Notch activation (Krejci and Bray, 2007). Their expression is usually transient and reflects the dynamic nature of Notch signaling. In addition there is evidence for autoregulation such that oscillations of HES expression have been observed and are thought to contribute to clocks that regulate somitogenesis, limb segmentation, and neural progenitor maintenance (Brend and Holley, 2009; Kageyama *et al.*, 2007; Lewis *et al.*, 2009; Pascoal *et al.*, 2007; Shimojo *et al.*, 2008). Altogether HES/HEY have now been shown to function downstream of Notch in many critical processes and to contribute to oncogenesis. For example, in tumor cells *HES1* may participate in the regulatory circuitry sustaining cell growth by repressing expression of PTEN (Palomero *et al.*, 2008).

All HES/HEY proteins appear to function as transcriptional repressors. For example, they share a C-terminal tetrapeptide motif WRPW/Y, which is sufficient to recruit transcriptional corepressors of the Groucho family (Fisher *et al.*, 1996; Paroush *et al.*, 1994), but note interacts less well with Groucho and may recruit alternative factors (Fischer *et al.*, 2002). Interactions with Sir2 class of proteins (Rosenberg and Parkhurst, 2002; Takata and Ishikawa, 2003) and with CtBP have also been demonstrated, the latter requiring a PLSLV/PVNLA motif (Poortinga *et al.*, 1998; Zhang and Levine, 1999). Indeed, on a genome-wide scale it appears that that binding of the HES protein Hairy overlaps to a larger extent with CtBP and Sir2 than with Groucho (Bianchi-Frias *et al.*, 2004). Similarly, the closest related Notch targets in nematodes, the Ref-1 family, also appear to function as repressors by recruiting CtBP (Neves and Priess, 2005). Although highly diverged from the HES family, Ref-1 and relatives contain two bHLH domains, which have moderate similarity to the basic regions of HES proteins, and a terminal FRPWE motif shown to be a weak CtBP binding domain (Alper and Kenyon, 2001). Thus, the regulation by Notch of E(spl)/HES/HEY/Ref1 bHLH repressors (which we will refer to collectively as "HESR") appears to be an ancient phenomenon and these proteins are essential in many Notch-dependent processes where they repress key cell fate determinants and cell cycle regulators (Fischer and Gessler, 2007).

Although HESR genes fulfill multiple pivotal roles in Notch-dependent processes, it is evident that they are not sufficient to explain all Notch functions. For example, elimination of *E(spl)* genes in the *Drosophila* wing fails to mimic the classic wing "notching" caused by reductions in Notch function. Here and elsewhere other targets are essential. Initially a relatively small number of other direct targets were identified. These included *vestigial* (Kim *et al.*, 1996), required for wing development in *Drosophila, single-minded* (Morel and Schweisguth, 2000), a midline determinant in *Drosophila, GATA* 3, required for physiological Th2 responses to parasite in mammals (Amsen *et al.*, 2007; Fang *et al.*, 2007), and *egl-43*, an EVI1 homologue with crucial roles in the *Caenorhabditis.elegans* reproductive system (Hwang *et al.*, 2007). More recently genome-wide studies in human T-ALL cells and in *Drosophila* myogenic precursor-related cells have revealed that, even within these specific cell types, Notch regulates a diverse array of direct targets (Krejci *et al.*, 2009; Palomero *et al.*, 2006).

Apart from the HESR genes, so far there are relatively few genes that have been found to be Notch regulated in both vertebrates and invertebrates. This may be because studies have not focused on the same processes but it may also reflect species divergence in the outputs. Nevertheless, several consistent messages have emerged (Fig. 8.1). First, Notch has been found to directly regulate genes involved in proliferation and apoptosis. For example, the *myc* gene is a direct target of Notch in several types of cancer cells and in *Drosophila* cells (Klinakis *et al.*, 2006; Krejci *et al.*, 2009; Palomero *et al.*, 2006; Weng *et al.*, 2006). Knock down of *myc* in these contexts compromised the extent of proliferation, arguing that *myc* is an important intermediate in the proliferative response to Notch activation. Other direct targets involved in promoting proliferation include CyclinD (Jeffries *et al.*, 2002; Joshi *et al.*, 2009; Ronchini and Capobianco, 2001), *string/CDC25* (Krejci *et al.*, 2009; Palomero *et al.*, 2006), and *CDK5* (Palomero *et al.*, 2006). Although Notch activates these proproliferative genes in several contexts, in others it activates cell cycle inhibitors like *p21* (Rangarajan *et al.*, 2001) reflecting the differing consequences on

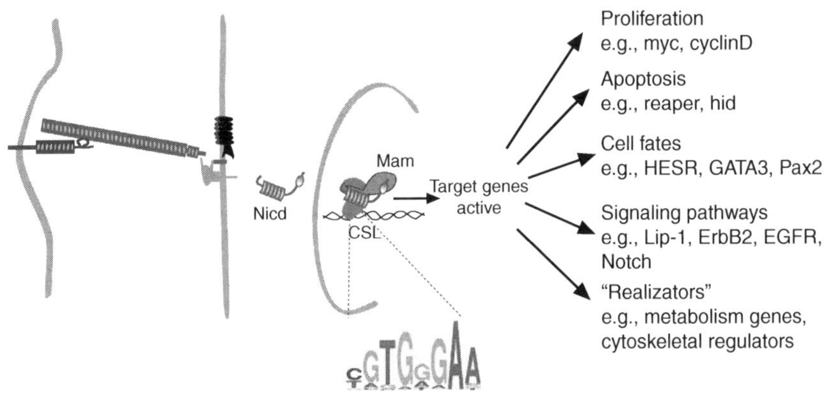

Figure 8.1 Diversity in Notch outputs. Simplified diagram of the Notch pathway. Interaction between the ligand (green) and the Notch (purple) leads to cleavage by ADAM metaiioproteases (yellow) and gamma secretase complex (brown) to release the NICD. In the nucleus, NICD binds to CSL (orange) and recruits Mam (green) to activate target genes. Arrows indicate different types of output that have been observed, with examples of some of the direct targets identified. CSL consensus binding site is depicted below, relative sizes are indicative of frequency for a given base occupying that position in the 56 validated Su(H) sites used to compile the logo. (See Color Insert.)

proliferation (Koch and Radtke, 2007). Notch has also been shown to directly control apoptosis effector genes. Hence *reaper* and *Wrinkled/hid* in *Drosophila* have been found as direct targets (Krejci *et al.*, 2009). Similarly *bcl-2* in mammals has been reported to respond rapidly to Notch activation consistent with being a direct target (Deftos *et al.*, 1998), but direct CSL binding to its promoter remains to be proved. Finding out what underlies the selection of apoptotic and proliferative targets is of major importance for understanding the diverse roles of Notch in development and cancers.

Second, many components of the Notch pathway are themselves direct targets. *DELTEX1,* encoding a ubiquitin ligase that regulates Notch trafficking, was first shown to be positively regulated by Notch in C2C12 cells (Kishi *et al.*, 2001) and has subsequently emerged as a target in multiple vertebrate tissues but not yet in invertebrates [e.g., Campese *et al.* (2006), Chang *et al.* (2000), Deftos *et al.* (1998), and Deftos *et al.* (2000)]. *NRARP,* a Notch inhibitor, appears to be a target in a range of vertebrate cell types [e.g., Krebs *et al.* (2001), Lamar *et al.* (2001), Phng *et al.* (2009), Pirot *et al.* (2004) and Weerkamp *et al.* (2006)]. Other pathway members have so far only emerged as direct targets in invertebrates [e.g., *Serrate,* (Martinez *et al.*, 2009; Yan *et al.*, 2004); *Su(H),* (Barolo *et al.*, 2000; Christensen *et al.*, 1996); *neuralized, numb, Kuzbanian/Adam10,* (Krejci *et al.*, 2009), although indirect evidence suggest that some are also targets in mammalian processes [e.g., Cheng *et al.* (2003, 2007)]. In addition Notch autoregulates its own expression in some mammalian (Weng *et al.*, 2006; Yashiro-Ohtani *et al.*, 2009) and *Drosophila* cells (Krejci *et al.*, 2009) as well as in *C. elegans* (Christensen *et al.*, 1996), providing a feedback mechanism that reinforces signaling (Christensen *et al.*, 1996).

Third, common targets include components of other signaling pathways. Multiple Ras pathway regulators were identified through bioinformatics and genetic screens in *C. elegans,* where the MAP kinase phosphatase (MKP) *lip-1* is a direct target along with five other negative regulators of the RAS-MAPK pathway (Berset *et al.*, 2001; Yoo *et al.*, 2004). A similar elaborate cross talk with EGF receptor signaling network and with other signaling pathways is evident in *Drosophila* and direct Notch targets include positive as well as negative regulators (Hurlbut *et al.*, 2009; Krejci *et al.*, 2009). Hints at similar cross talk in mammalian cells are seen with the identification of *ErbB-2* as a direct target (Chen *et al.*, 1997), with upregulation of MAPK regulators in hematopoietic progenitors (Weerkamp *et al.*, 2006) and with the oscillatory network related to Notch signaling in somitogenesis (although in this case there is as yet no proof that the cross talk involves direct regulation). The precise nature of the Notch targets and the consequences for the cross-regulation of signaling pathways are likely to differ depending on the context of the cell. In the *C. elegans* vulva and *Drosophila* wing veins the consequences on Ras signaling are inhibitory (Berset *et al.*, 2001; Yoo *et al.*, 2004), but elsewhere Notch can cooperate with Ras (e.g., R7 development in *Drosophila* eye) suggesting a requirement for different cohorts of targets (Hurlbut *et al.*, 2009; Mittal *et al.*, 2009; Sundaram, 2005). As more studies of direct targets are carried out, it may prove possible to extract underlying rules.

Fourth, it is evident that Notch also directly regulates expression of genes encoding proteins that actually implement cell functions ("realizator" genes). For example, in T-ALL cells many of the direct targets are involved in metabolism (Margolin *et al.*, 2009; Palomero *et al.*, 2006). And in several developmental contexts direct targets include cytoskeletal regulators such as cytoskeletal crosslinkers *Short stop* and *Gas2* and the genes encoding Ig cell adhesion receptors Roughest and Hibris (Apitz *et al.*, 2005; Artero *et al.*, 2003; Fuss *et al.*, 2004; Krejci *et al.*, 2009; Pines *et al.*, 2010). Likewise, *Tenascin-C* is a target of Notch2 in glioblastoma cells, where it may contribute to invasiveness of the tumor cells (Sivasankaran *et al.*, 2009).

Finally, several regulatory motifs are beginning to emerge from systematic studies of Notch targets. These include positive feed-forward loops, exemplified by the role of Myc in T-ALL cells (Palomero *et al.*, 2006), and incoherent (IFL), characteristic of the response in *Drosophila* myogenic precursors (Krejci *et al.*, 2009). In this type of IFL, the stimulus (Notch) regulates both a gene and a repressor of the gene. Classic examples involve members of the HESR family. For example, PTEN, *atonal* and *twist* are all directly responsive to CSL/Notch, and in each case these genes can also be repressed by HESR proteins (Ligoxygakis *et al.*, 1998; Palomero *et al.*, 2008; Tapanes-Castillo and Baylies, 2004; Whelan *et al.*, 2007). Genome-wide studies revealed further targets that form IFL independent of HESR members including *String/CDC25-hindsight* and *myc-brat* (Krejci et al., 2009). The overall output of IFL is difficult to predict since it is dependent on several criteria such as the rate of synthesis and the thresholds required for activation and repression, but in some conditions it has been shown to create pulse of target activities (Alon, 2007) and it is proposed to render the response proportional to the fold change in the input signal (Goentoro *et al.*, 2009)

3. How Does the Notch Switch Work?

Binding of NICD to the DNA-binding CSL mediates the "transcriptional switch" to activate gene expression from the target promoters. CSL binds to DNA as a monomer and initial studies identified high-affinity binding sites for both *Drosophila* and mammalian CSL proteins with the core consensus YGTGRGAA (Bailey and Posakony, 1995; Lecourtois and Schweisguth, 1995; Tun *et al.*, 1994). The verification of more target binding sites implied a less stringent consensus [e.g., Nellesen *et al.* (1999)] as illustrated by the logo in Fig 8.1. Matches to the CSL consensus are detected throughout the genome: one estimate places a high affinity site in the vicinity of ~40% of *Drosophila* genes (Rebeiz *et al.*, 2002). It is unclear how many such sites are functional and what determines functionality. Certainly in one cell type only ~260 genes (<2%) were directly responsive to Notch demonstrating that at any one time only a subset of binding sites are utilized (Krejci *et al.*, 2009).

One factor contributing to target selection could be the arrangement of sites. In the best-characterized targets, *E(spl)* genes and *Hes* 1, there is a specific site architecture comprising two CSL binding sites arranged in a head-to-head manner with an approximately 16 base pair A/T-rich spacer sequence (SPS motif). It is proposed that SPS could confer the ability to respond at lower levels of NICD, explaining their presence in the strongly responding *HES1* and *E(spl)* genes. Cooperative interactions between NICD-containing complexes have been detected when SPS have the appropriate spacing (15–22 nucleotides), suggesting a mechanism that would ensure a sensitive and tight response at promoters at such targets (Nam *et al.*, 2006; 2007). However, relatively few targets contain SPS motifs implying that additional mechanisms contribute to binding site selection and activity, as discussed further.

The interaction of NICD with CSL creates an interface that is recognized by Mam, a critical adaptor in the activation of targets. Recent structural analysis revealed that there is a stepwise assembly of the complex, with binding by the N-terminal part of NICD (RAM domain) providing a tether for the interaction and causing a conformational change that also favors Mam recruitment [reviewed in Gordon *et al.* (2008)]. Mam in turn can recruit histone acetyltransferase (HAT) complexes such as p300-PCAF and GCN5 (Kurooka and Honjo, 2000; Oswald *et al.*, 2001). Mam is required for p300-dependent acetylation of nucleosomes at a minimal Notch enhancer in vitro (Fryer *et al.*, 2002) and enhances p300 acetylation (Hansson *et al.*, 2009). Additional HAT complexes, such as the Tip60 complex containing TRRAP/Nipped-A, may also facilitate target gene transcription in some contexts (Gause *et al.*, 2006), although Tip60 has also been reported to suppress Notch activity (Kim *et al.*, 2007). The recruitment of HAT complexes explains the increased H4 acetylation seen at actively transcribed Notch targets in *Drosophila* cells (Krejci and Bray, 2007).

Other histone modifications, such as ubiquitination of H2B and associated methylation of H3K4, are important for expression of Notch targets in *Drosophila* (Bray *et al.*, 2005; Buszczak *et al.*, 2009; Moshkin *et al.*, 2009; Tenney *et al.*, 2006). Mutations in the H2B ubiquitinating enzyme Bre1 result in loss of target gene expression (Bray *et al.*, 2005) while mutations in *scrawny,* encoding an ubiquitin-specific protease that deubiquitinates H2B, lead to premature expression of key differentiation genes, including Notch targets, in stem cells (Buszczak *et al.*, 2009). While these histone modifications are likely important generally for transcription, Notch-regulated genes appear particularly susceptible. This may be because the modifications at the CSL binding site are critical for activity (Liefke *et al.*, 2010) and/or because NICD, and hence the activation complex, is present only transiently. In transfected HeLa cells Mam was found to promote phosphorylation of NICD by CDK8, rendering it a substrate for proteasomal degradation by E3 ubiquitin ligases that include Sel10/Fbw7 (Fryer *et al.*, 2004; Gupta-Rossi *et al.*, 2001; Tsunematsu *et al.*, 2004). This suggests that target gene activation is coupled to a mechanism that down regulates the signal (Fryer *et al.*, 2004). Given the dynamic requirements for Notch signaling during development, and the oncogenic effects of mutations that interfere with NICD turnover (O'Neil and Look, 2007; Welcker and Clurman, 2008), this aspect of Notch regulation is of major importance.

The role of CSL in mediating transcriptional activation of Notch targets appeared initially at odds with its preceding characterization as a repressor in mammals (Dou *et al.*, 1994). Several different corepressors were identified in mammalian cells, including CIR, SMRT, and SHARP (Hsieh *et al.*, 1999; Oswald *et al.*, 2005; Zhou and Hayward, 2001). These interacted directly with CSL and, when added in increasing amounts in cell transfection assays, antagonized the activation by NICD. From these data, a model emerged where NICD displaced corepressors to convert DNA-bound CSL to an activator (Borggrefe and Oswald, 2009; Bray, 2006; Kopan and llagan, 2009). This elegant "switch" model helped to explain many complex observations, such as the fact that some target genes are still expressed in

Su(H) mutants, albeit at lower levels and with a broader domain (Koelzer and Klein, 2003; Li and Baker, 2001; Morel and Schweisguth, 2000). There is now increasing *in vivo* evidence that supports the repressive role of CSL at Notch targets in flies (Bardin *et al.*, 2010; Castro *et al.*, 2005; Furriols and Bray, 2001; Koelzer and Klein, 2006; Nagel *et al.*, 2005), where it is sometimes referred to as "default repression" (Barolo *et al.*, 2002), but there has been less definitive *in vivo* evidence from mammalian studies so far (Kopan and llagan, 2009).

A key component of the repressor complex in *Drosophila* is the protein Hairless, which serves as a platform or adaptor to recruit the corepressors Groucho and CtBP (Barolo *et al.*, 2002; Morel *et al.*, 2001; Nagel *et al.*, 2005). These proteins contribute to Hairless-mediated repression to differing extents, and it remains to be determined whether both act in combination or whether they are part of discrete complexes. Similarly, SHARP/MINT may function in an analogous manner in mammals (Kuroda *et al.*, 2003; Oswald *et al.*, 2002; Yabe *et al.*, 2007) and has been shown to bind several different corepressors including CtBP and SMRT/NCoR (Oswald *et al.*, 2005; Borggrefe and Oswald 2009). However, the diversity of CSL corepressors identified in mammals (Borggrefe and Oswald, 2009) and emerging examples also in *Drosophila* (Tsuda *et al.*, 2006) raise questions whether CSL is associated with distinct classes of repressor complex and how this would impact on its relationship with Notch. It is also unclear whether CSL has functions in gene repression where it is insensitive to Notch. Such a possibility has emerged from studies of *hlh-6* gene in *C. elegans* (Ghai and Gaudet, 2008). Detailed analysis of its regulation demonstrated a requirement for CSL (Lag-1) but its expression was unaffected by changes in *C. elegans* Notch gene function *(lin-12* or *glp-1)*. And in mammalian cells, CSL (RBPjk) is a potent repressor of the HIV-LTR promoter (Tyagi and Karn, 2007). It remains to be determined whether Notch-insensitive CSL repression is more widespread and, if so, what renders the targets Notch insensitive.

In both flies and mammals one function of the CSL corepressor complex is thought to be the recruitment of histone deacetylases (HDACs). SMRT, CtBP, and Gro have all been shown to interact directly with class 1 HDACs (Chen *et al.*, 1999; Nagy *et al.*, 1998; Subramanian and Chinnadurai, 2003). Furthermore, elevated levels of HESR (*ESR-1*) gene expression were detected following treatment of *Xenopus* caps with the HDAC inhibitor (Kao *et al.*, 1998). Likewise, the Notch-responsive HESR genes *her6* and *her4* were ectopically expressed at distinct sites within the developing nervous system in zebrafish *hdacl* mutant embryos (Cunliffe, 2004; Yamaguchi *et al.*, 2005). However, this does not appear to be the whole story as other chromatin modifications appear to be important (Borggrefe and Oswald, 2009). For example, the Hairless/CSL repressor complex was found to associate with large protein complexes, containing histone chaperones and the histone H3K4 demethylase Lid/KDM5A, that contribute to target gene repression (Goodfellow *et al.*, 2007; Moshkin *et al.*, 2009). Similarly specific interactions were also detected between CSL (RBPjK) and KDM5A in mammalian cells, where methylation of histone H3K4 was erased at CSL sites upon Notch inhibition (Liefke *et al.*, 2010). Thus a combination of histone-modifying and remodeling activities are likely to contribute to the silencing of targets in the absence of Notch activation. Transcription elongation may also be regulated; some Notch pathway genes have paused polymerase at their promoters in *Drosophila* embryos and may be affected by mutations in elongation factors (Chopra *et al.*, 2009; Zeitlinger *et al.*, 2007).

The most commonly depicted model of the switch suggests that CSL is statically bound to DNA while regulating transcription from Notch target genes (Fig. 8.2). More recent studies, including work from our lab, suggested that CSL binding to DNA is dynamic rather than static and it is

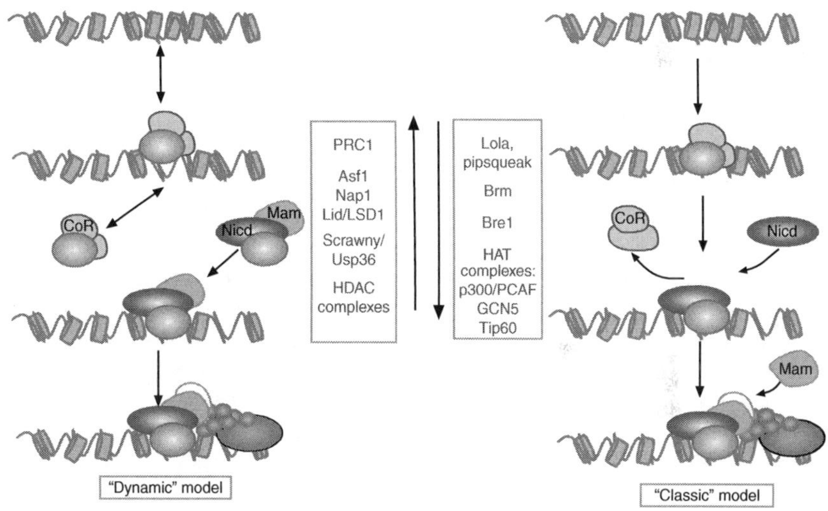

Figure 8.2 Alternative models for the transcriptional switch and factors that enhance or suppress Notch-mediated activation. Both models are based on the fundamental principal that (1) CSL (orange) bound to corepressors (CoR, grey) contribute to target gene repression (2) NICD (purple) interacts with CSL (orange) and recruits Mam (green) and coactivators to activate transcription (red arrow). The boxes list other factors known to suppress (left) or promote (right) activation of targets. The two models differ in how stably the CSL is bound to the DNA. In the dynamic model we propose that there is equilibrium between bound and unbound CSL repression complexes, that NICD containing complexes can form off the DNA, and that the exchange is between different CSL-containing complexes (repression and activation). The activation complex becomes stabilized by interactions with the basal transcription machinery. In the classic model, CSL remains bound to the DNA and exchange occurs between NICD and co-repressors on a DNA-bound CSL. This implies that CSL has high affinity for DNA and no other interactions are needed to stabilize this interaction. (See Color Insert.)

notable that, in both *Drosophila* and human cells, CSL occupancy on promoters was enhanced when NICD was present (Fryer *et al.*, 2004; Joshi *et al.*, 2009; Krejci and Bray, 2007; Zhou and Hayward, 2001). This raises the possibility that complexes may be forming/exchanging in the nucleoplasm and indeed makes it possible that there are distinct pools of CSL complexes (Fig. 8.2). Recent measurements of the affinity of CSL for DNA support this more dynamic model (Friedmann and Kovall, 2010). If correct, this model also suggests that cooperative mechanisms will be required to recruit and/or stabilize CSL to sites on the DNA. One mechanism is likely to involve interactions between NICD-containing complexes at appropriately spaced CSL sites (e.g., SPS motif), which would increase the stability of the activation complex (Cave *et al.*, 2005; Gordon *et al.*, 2008; Ong *et al.*, 2006). However, few enhancers contain the optimal pairing of sites, suggesting that other mechanisms are also likely to be important.

4. Different Enhancer Logics

One prediction of the switch model is that target genes will be de-repressed in the absence of CSL, as seen for a number of targets in *Drosophila* which are ectopically expressed in *Su(H)* mutants [e.g., Bardin *et al.* (2010), Koelzer and Klein(2006), and Morel and Schweisguth(2000)]. This ectopic expression is limited and often weaker than normal but contributes to phenotypic differences in Su(H) and Notch signaling mutants. For this reason, defects in Notch signaling may in some cases be alleviated by mutations in the co-repressors. For example, conditional inactivation of the corepressor SHARP/MINT in developing nephrons resulted in a moderate rescue of the Notch2 mutant phenotype (Surendran *et al.*, 2010). CSL-mediated repression is clearly essential in some processes (Bardin *et al.*, 2010; Koelzer and Klein, 2006), but in others there has been no clear evidence for target gene de-repression contributing to the phenotypes of CSL-knockouts (Han *et al.*, 2002; Oka *et al.*, 1995; Shen *et al.*, 1997). It thus remains to be resolved to what extent CSL-mediated repression is important at all target enhancers.

A second prediction is that targets will have reduced expression in the absence of CSL in the places where they are normally responsive. This would seem to be a fundamental expectation for Notch-regulated targets and is certainly the case for many of the best characterized: in most cases HESR gene expression is compromised by the loss of CSL binding [e.g., Bailey and Posakony (1995), Lamar and Kintner (2005), Lecourtois and Schweisguth (1995), and Nellesen *et al.* (1999)]. However, there are examples of direct targets in *Drosophila* that show no loss of activation in *Su(H)* mutants, including *atonal* in the eye imaginal disc and *sox15* in the sensory organ lineage (Li and Baker, 2001; Miller *et al.*, 2009). Despite their continued expression in Su(H) mutants, *atonal* and *Sox15* are both Notch regulated and require CSL for this regulation. Such observations have led to the proposal that the function of NICD at some targets, referred to as Notch-permissive targets, is primarily to alleviate the repression function of CSL complexes (Bray and Furriols, 2001). Permissive targets would not require the activation function of NICD and would achieve high levels of expression through other factors bound to the enhancer. In contrast, so-called Notch-inductive targets, such as HESR genes, require the activation function of NICD for high levels of expression, and their expression is compromised by mutations in the CSL-binding sites (Bailey and Posakony, 1995; Flores *et al.*, 2000; Lecourtois and Schweisguth, 1995; Nellesen *et al.*, 1999; Neves and Priess, 2005). It is unclear what mechanisms might underlie these differences in the target responses; it may depend on whether the promoters are in a poised conformation, what histone modifications are present and what type or amounts of cooperating factors are already bound.

5. Context Dependence of Notch Responses

The context corresponds to the mechanisms that make a gene responsive when the Notch pathway is activated. Thus, while most Notch-dependent processes are associated with expression of HESR genes, the specific HESR gene(s) activated varies according to the context, illustrating that even these common targets acquire additional specificity-conferring inputs. For example, in Drosophila there are seven closely related *E(spl) bHLH* genes that have arisen through recent gene duplications and are clustered on the chromosome (Schlatter and Maier, 2005). Despite their relatively recent origins and close proximity, the different genes have distinct patterns of expression (especially during post embryonic stages) and can only be activated by Notch in limited territories [e.g., Cooper *et al.* (2000), de Celis *et al.* (1996),and Nellesen *et al.*, (1999)]. Similar spatial restricted patterns occur for the *Hes* and *Hey* genes in mouse, the *Her* genes in zebrafish (Kageyama *et al.*, 2007) and *ESR* genes in *Xenopus* (Lamar and Kintner, 2005). Thus despite the fact that many target genes contain multiple CSL binding sites, they are only able to respond in a subset of the places where Notch is activated. Clearly there is other information that restricts where the specific gene targets are responsive. Further complexity arises with targets that show Notch-dependent and Notch-independent expression [e.g., Yeo *et al.* (2007)] and with targets that cross-regulate each other (Fior and Henrique, 2005; Hatakeyama *et al.*, 2004).

Combinatorial regulation with patterning transcription factors is one way that genes acquire specificity in their response to Notch. An obligate integration at enhancers of signaling inputs, such as Notch, with patterning protein inputs was proposed from studies in the *Drosophila* wing and eye (Flores *et al.*, 2000; Guss *et al.*, 2001). Synergy between Notch and the proneural bHLH proteins emerged as critical in the regulation of HESR genes during neurogenesis (Bailey and Posakony, 1995; Castro *et al.*, 2005; Cooper *et al.*, 2000; Kramatschek and Campos-Ortega, 1994; Lamar and Kintner, 2005; Singson *et al.*, 1994). Other examples of combinatorial factors include GATA factors, which synergize in regulating *ref-1* in *C. elegans* endoderm (Neves and Priess, 2005), NFκB family members, which co-regulate *HES, Deltex*- 1, and *cyclinD3* (Joshi *et al.*, 2009; Moran *et al.*, 2007), the AML1 homologue Lozenge, which combines with Notch on the *Pax2* enhancer (Flores *et al.*, 2000) and Twist, which coregulates many targets in muscle progenitors (Bernard et al., 2010).

At present the mechanisms underlying the combinatorial and synergistic interactions between Notch and other factors are not fully understood. One possibility is that direct interactions between synergizing factors and CSL complexes could lead to their mutual stabilization on the target enhancer, so favoring activation (Fig. 8.3A). Both ELT-2/GATA and Daughterless/E2A have been shown to interact directly with CSL in support of this model (Cave *et al.*, 2005; Neves and Priess, 2005). NFκB has also been found to augment binding of NICD to *cyclinD3* (Joshi *et al.*, 2009). Such interactions may be facilitated by specific arrangements of binding sites. For example, Daughterless/E2A combines with proneural proteins to bind a target site that is found associated with paired CSL sites (SPS) in *E(spl)* gene enhancers (Cave *et al.*, 2005). Synergy is no longer observed if the geometry of these sites is altered, arguing that their configuration is important in enabling interactions. Although, many enhancers do not appear to contain a stereotypic site architecture [e.g., Guss *et al.*, (2001), and Lamar and Kintner (2005)] a recent analysis of the *Drosophila Pax2 "sparkling"* enhancer reveals that the organization of CSL and other regulatory sites is important to determine the correct expression pattern (Swanson *et al.*, 2010). The constraints on spacing and organization suggest that expression is dependent on short-range regulatory interactions that could be compatible with direct protein–protein interactions.

A second possibility to explain synergy between Notch and cooperating factors is that they make independent interactions with targets within the transcriptional machinery such that the readout is an integration of their inputs (Fig. 8.3B). Despite the importance of site organization in the *sparkling* enhancer, the composition and distribution of sites are not maintained in a functionally conserved enhancer from distantly related species (Swanson *et al.*, 2010). This implies greater flexibility than is likely to be feasible for direct interactions between specific transcription factors. Such binding site flexibility could result from NICD/CSL and coregulating transcription factors being able to contact the basal machinery or chromatin-modifying cofactors, from many different configurations (Arnosti and Kulkarni, 2005).

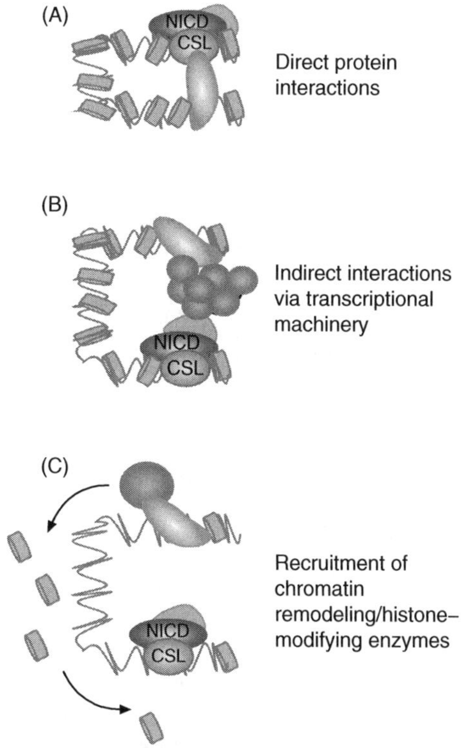

Figure 8.3 Possible mechanisms underlying combinatorial regulation at Notch targets. Context-conferring factors (blue) could act by (A) directly contacting CSL (or NICD) to stabilize interactions at the enhancer; (B) contacting intermediate targets, such as the basal machinery (grey), allowing greater diversity in the configurations of sites; (C) recruiting chromatin-modifying cofactors (grey) that allow accessibility to NICD/CSL by altering the chromatin conformation (CSL, orange; NICD, purple; Mam, green). (See Color Insert.)

A third possibility is that context-conferring factors alter the chromatin at Notch enhancers, making it accessible to CSL/NICD, a mechanism that would not necessitate simultaneous binding. Instead these factors could result in altered nucleosome placement or histone modifications (Fig. 8.3C). There is mounting evidence that chromatin modifications are important for Notch outputs. For example, two BTB/POZ proteins, Lola and Pipsqueak, synergize with Notch activity in *Drosophila* through a mechanism that appears to involve changes in histone and DNA methylation at critical targets (Ferres-Marco *et al.*, 2006). Similarly, the Brm/Brahma ATPase component of the SWI/SNF chromatin-remodeling complex interacts with the NICD in C2C12 cells and shows functional interactions with the Notch pathway in *Drosophila* (Armstrong *et al.*, 2005; Kadam and Emerson, 2003). Conversely, recent studies suggest that Notch pathway is repressed by PRC1, one of the Polycomb chromatin-silencing complexes (or at least by proteins that are part of PRC1) (Martinez *et al.*, 2009; Tolhuis *et al.*, 2006). It has been suggested that repression by the PcG could raise the threshold Notch has to overcome to activate certain genes (Merdes *et al.*, 2004). However, it remains possible that PcG may regulate transcription of the receptor and ligands, rather than altering the accessibility of targets, especially since other studies reported different consequences on target gene activation in the PRC1 mutant cells (Classen *et al.*, 2009). Further studies will be needed to unravel the contribution of these and other epigenetic regulators at Notch targets.

6. Concluding Comments

Starting with an initial trickle and increasing to the current deluge from genome-wide studies, the number of direct Notch-regulated targets has risen exponentially since the initial discovery that NICD is a transcriptional activator. Here we have summarized some of the general conclusions that have emerged from studies of Notch targets and their regulation so far. One of the challenges in future will be to extract fundamental messages from the repertoire of targets identified in different tissues and diseases, in order to determine whether there are specific signatures and conserved patterns in the responses. A second challenge will be to unravel the mechanisms that confer different Notch responses: the cell context is fundamental to target gene activation and the identification of factors conferring specificity remains of primary importance. Possible mechanisms contributing to this specificity include chromatin accessibility and, although it is evident that epigenetic factors contribute to target gene activation, the critical changes that make enhancers accessible to CSL/Notch complexes remain to be established. Also unclear is how different the mechanisms of regulation are at individual targets, for example, whether there are different modes of CSL repression complexes. Finally, another factor that has largely been overlooked is whether the level or duration of the signal could impact on the sets of genes activated. In *Drosophila,* use of a thermosensitive Notch allele suggests that some genes are more sensitive to a slight decrease of signal than others (Becam and Milan, 2008) and studies with hematopoietic progenitors indicate that quantitative aspects of Notch signaling are relevant for cell fate outcomes (Delaney *et al.*, 2005). Clearly this will be another important question for the future.

References

Alon, U. (2007). Network motifs: theory and experimental approaches. *Nat. Rev. Genet.* **8**, 450–461.

Alper, S., and Kenyon, C. (2001). REF-1, a protein with two bHLH domains, alters the pattern of cell fusion in *C. elegans* by regulating Hox protein activity. *Development* **128**, 1793–1804.

Amsen, D., Antov, A., Jankovic, D., Sher, A., Radtke, F., Souabni, A., Busslinger, M., McCright, B., Gridley, T., and Flavell, R.A. (2007). Direct regulation of Gata3 expression determines the T helper differentiation potential of Notch. *Immunity* **27**, 89–99.

Apitz, H., Strunkelnberg, M., de Couet, H. G., and Fischbach, K.F. (2005). Single-minded, Dmef2, Pointed, and Su(H) act on identified regulatory sequences of the roughest gene in *Drosophila* melanogaster. *Dev. Genes Evol.* **215**, 460–469.

Armstrong, J. A., Sperling, A. S., Deuring, R., Manning, L., Moseley, S. L., Papoulas, O., Piatek, C. I., Doe, C. Q., and Tamkun, J. W. (2005). Genetic screens for enhancers of brahma reveal functional interactions between the BRM chromatin-remodeling complex and the delta-notch signal transduction pathway in *Drosophila. Genetics* **170**, 1761–1774.

Arnosti, D. N., and Kulkarni, M. M. (2005). Transcriptional enhancers: intelligent enhanceosomes or flexible billboards? *J. Cell. Biochem.* **94**, 890–898.

Artero, R., Furlong, E. E., Beckett, K., Scott, M. P., and Baylies, M. (2003). Notch and Ras signaling pathway effector genes expressed in fusion competent and founder cells during *Drosophila* myogenesis. *Development* **130**, 6257–6272.

Bailey, A. M., and Posakony, J. W. (1995). Suppressor of hairless directly activates transcription of enhancer of split complex genes in response to Notch receptor activity. *Genes Dev.* **9**, 2609–2622.

Bardin, A. J., Perdigoto, C. N., Southall, T. D., Brand, A. H., and Schweisguth, F. (2010). Transcriptional control of stem cell maintenance in the *Drosophila* intestine. *Development* **137**, 705–714.

Barolo, S., Stone, T., Bang, A. G., and Posakony, J. W. (2002). Default repression and Notch signaling: hairless acts as an adaptor to recruit the corepressors Groucho and dCtBP to suppressor of hairless. *Genes Dev.* **16**, 1964–1976.

Barolo, S., Walker, R. G., Polyanovsky, A. D., Freschi, G., Keil, T., and Posakony, J. W. (2000). A notch-independent activity of suppressor of hairless is required for normal mechanoreceptor physiology. *Cell* **103**, 957–969.

Becam, I., and Milan, M. (2008). A permissive role of Notch in maintaining the DV affinity boundary of the *Drosophila* wing. *Dev. Biol.* **322**, 190–198.

Bernard, F., Krejci, A., Housden, B., Adryan, B., and Bray, S. (2010).Specificity of Notch pathway activation: twist controls the transcriptional output in adult muscle progenitors. *Development.* In Press.

Berset, T., Hoier, E. F., Battu, G., Canevascini, S., and Hajnal, A. (2001). Notch inhibition of RAS signaling through MAP kinase phosphatase LIP-1 during *C. elegans* vulval development *Science* **291**, 1055–1058.

Bianchi-Frias, D., Orian, A., Delrow, J. J., Vazquez, J., Rosales-Nieves, A. E., and Parkhurst, S. M. (2004). Hairy transcriptional repression targets and cofactor recruitment in *Drosophila. PLoS Biol.* **2**, E178.

Borggrefe, T., and Oswald, F. (2009). The Notch signaling pathway: transcriptional regulation at Notch target genes. *Cell Mol. Life Sci.* **66**, 1631–1646.

Bray, S., and Furriols, M. (2001). Notch pathway: making sense of suppressor of hairless. *Curr. Biol.* **11**, R217–R221.

Bray, S., Musisi, H., and Bienz, M. (2005). Bre1 is required for Notch signaling and histone modification. *Dev. Cell.* **8**, 279–286.

Bray, S. J. (2006). Notch signalling: a simple pathway becomes complex. *Nat. Rev. Mol. Cell. Biol.* **7**, 678–689.

Brend, T., and Holley, S. A. (2009). Expression of the oscillating gene her1 is directly regulated by Hairy/Enhancer of Split, T-box, and suppressor of hairless proteins in the zebrafish segmentation clock. *Dev. Dyn.* **238**, 2745–2759.

Buszczak, M., Paterno, S., and Spradling, A. C. (2009). *Drosophila* stem cells share a common requirement for the histone H2B ubiquitin protease scrawny. *Science* **323**, 248–251.

Campese, A. F., Garbe, A. I., Zhang, F., Grassi, F., Screpanti, I., and von Boehmer, H. (2006). Notch1-dependent lymphomagenesis is assisted by but does not essentially require pre-TCR signaling. *Blood* **108**, 305–310.

Castro, B., Barolo, S., Bailey, A. M., and Posakony, J. W. (2005). Lateral inhibition in proneural clusters: cis-regulatory logic and default repression by suppressor of hairless. *Development* **132**, 3333–3344.

Cave, J. W., Loh, F., Surpris, J. W., Xia, L., and Caudy, M. A. (2005). A DNA transcription code for cell-specific gene activation by notch signaling. *Curr. Biol.* **15**, 94–104.

Chang, D., Valdez, P., Ho, T., and Robey, E. (2000). MHC recognition in thymic development: distinct, parallel pathways for survival and lineage commitment. *J. Immunol.* **165**, 6710–6715.

Chen, G., Fernandez, J., Mische, S., and Courey, A. J. (1999). A functional interaction between the histone deacetylase Rpd3 and the corepressor groucho in *Drosophila* development *Genes Dev.* **13**, 2218–2230.

Chen, Y., Fischer, W. H., and Gill, G. N. (1997). Regulation of the ERBB-2 promoter by RBPJkappa and NOTCH. *J. Biol. Chem.* **272**, 14110–14114.

Cheng, H. T., Kim, M., Valerius, M. T., Surendran, K., Schuster-Gossler, K., Gossler, A., McMahon, A. P., and Kopan, R. (2007). Notch2, but not Notch1, is required for proximal fate acquisition in the mammalian nephron. *Development* **134**, 801–811.

Cheng, H. T., Miner, J. H., Lin, M., Tansey, M. G., Roth, K., and Kopan, R. (2003). Gamma-secretase activity is dispensable for mesenchyme-to-epithelium transition but required for podocyte and proximal tubule formation in developing mouse kidney. *Development* **130**, 5031–5042.

Chopra, V. S., Hong, J. W., and Levine, M. (2009). Regulation of Hox gene activity by transcriptional elongation in *Drosophila. Curr. Biol.* **19**, 688–693.

Christensen, S., Kodoyianni, V., Bosenberg, M., Friedman, L., and Kimble, J. (1996). lag-1, a gene required for lin-12 and glp-1 signaling in *Caenorhabditis elegans*, is homologous to human CBF1 and *Drosophila* Su(H). *Development* **122**, 1373–1383.

Classen, A. K., Bunker, B. D., Harvey, K. F., Vaccari, T., and Bilder, D. (2009). A tumor suppressor activity of *Drosophila* Polycomb genes mediated by JAK-STAT signaling. *Nat. Genet.* **41**, 1150–1155.

Cooper, M. T., Tyler, D. M., Furriols, M., Chalkiadaki, A., Delidakis, C., and Bray, S. (2000). Spatially restricted factors cooperate with notch in the regulation of Enhancer of split genes. *Dev. Biol.* **221**, 390–403.

Cunliffe, V. T. (2004). Histone deacetylase 1 is required to repress Notch target gene expression during zebrafish neurogenesis and to maintain the production of motoneurones in response to hedgehog signalling. *Development* **131**, 2983–2995.

de Celis, J. F., de Celis, J., Ligoxygakis, P., Preiss, A., Delidakis, C., and Bray, S. (1996). Functional relationships between Notch, Su(H) and the bHLH genes of the E(spl) complex: the E(spl) genes mediate only a subset of Notch activities during imaginal development. *Development* **122**, 2719–2728.

Deftos, M. L., He, Y. W., Ojala, E. W., and Bevan, M. J. (1998). Correlating notch signaling with thymocyte maturation. *Immunity* **9**, 777–786.

Deftos, M. L., Huang, E., Ojala, E. W., Forbush, K. A., and Bevan, M. J. (2000). Notch1 signaling promotes the maturation of CD4 and CD8 SP thymocytes. *Immunity* **13**, 73–84.

Delaney, C., Varnum-Finney, B., Aoyama, K., Brashem-Stein, C., and Bernstein, I. D. (2005). Dose-dependent effects of the Notch ligand Delta1 on ex vivo differentiation and in vivo marrow repopulating ability of cord blood cells. *Blood* **106**, 2693–2699.

Dou, S., Zeng, X., Cortes, P., Erdjument-Bromage, H., Tempst, P., Honjo, T., and Vales, L. D. (1994). The recombination signal sequence-binding protein RBP-2N functions as a transcriptional repressor. *Mol. Cell. Biol.* **14**, 3310–3319.

Fang, T. C., Yashiro-Ohtani, Y., Del Bianco, C., Knoblock, D. M., Blacklow, S. C., and Pear, W. S. (2007). Notch directly regulates Gata3 expression during T helper 2 cell differentiation. *Immunity* **27**, 100–110.

Ferres-Marco, D., Gutierrez-Garcia, I., Vallejo, D. M., Bolivar, J., Gutierrez-Avino, F. J., and Dominguez, M. (2006). Epigenetic silencers and Notch collaborate to promote malignant tumours by Rb silencing. *Nature* **439**, 430–436.

Fior, R., and Henrique, D. (2005). A novel hes5/hes6 circuitry of negative regulation controls Notch activity during neurogenesis. *Dev. Biol.* **281**, 318–333.

Fischer, A., and Gessler, M. (2007). Delta-Notch–and then? Protein interactions and proposed modes of repression by Hes and Hey bHLH factors. *Nucleic Acids Res.* **35**, 4583–4596.

Fischer, A., Leimeister, C., Winkler, C., Schumacher, N., Klamt, B., Elmasri, H., Steidl, C., Maier, M., Knobeloch, K. P., Amann, K., Helisch, A., Sendtner, M., and Gessler, M. (2002). Hey bHLH factors in cardiovascular development. *Cold Spring Harb. Symp. Quant. Biol.* **67**, 63–70.

Fisher, A. L., Ohsako, S., and Caudy, M. (1996). The WRPW motif of the hairy-related basic helix-loop-helix repressor proteins acts as a 4-amino-acid transcription repression and protein-protein interaction domain. *Mol. Cell. Biol.* **16**, 2670–2677.

Flores, G. V., Duan, H., Yan, H., Nagaraj, R., Fu, W., Zou, Y., Noll, M., and Banerjee, U. (2000). Combinatorial signaling in the specification of unique cell fates. *Cell* **103**, 75–85.

Friedmann, D. R., and Kovall, R. A. (2010). Thermodynamic and structural insights into CSL-DNA complexes. *Protein Sci.* **19**, 34–46.

Fryer, C. J., Lamar, E., Turbachova, I., Kintner, C., and Jones, K. A. (2002). Mastermind mediates chromatin-specific transcription and turnover of the Notch enhancer complex. *Genes Dev.* **16**, 1397–1411.

Fryer, C. J., White, J. B., and Jones, K. A. (2004). Mastermind recruits CycC:CDK8 to phosphorylate the Notch ICD and coordinate activation with turnover. *Mol. Cell.* **16**, 509–520.

Furriols, M., and Bray, S. (2001). A model Notch response element detects suppressor of hairless-dependent molecular switch. *Curr. Biol.* **11**, 60–64.

Fuss, B., Josten, F., Feix, M., and Hoch, M. (2004). Cell movements controlled by the Notch signalling cascade during foregut development in *Drosophila*. *Development* **131**, 1587–1595.

Gause, M., Eissenberg, J. C., Macrae, A. F., Dorsett, M., Misulovin, Z., and Dorsett, D. (2006). Nipped-A, the Tra1/TRRAP subunit of the *Drosophila* SAGA and Tip60 complexes, has multiple roles in Notch signaling during wing development. *Mol. Cell. Biol.* **26**, 2347–2359.

Ghai, V., and Gaudet, J. (2008). The CSL transcription factor LAG-1 directly represses hlh-6 expression in *C. elegans*. *Dev. Biol.* **322**, 334–344.

Goentoro, L., Shoval, O., Kirschner, M. W., and Alon, U. (2009). The incoherent feedforward loop can provide fold-change detection in gene regulation. *Mol. Cell.* **36**, 894–899.

Goodfellow, H., Krejci, A., Moshkin, Y., Verrijzer, C. P., Karch, F., and Bray, S. J. (2007). Gene-specific targeting of the histone chaperone asf1 to mediate silencing. *Dev. Cell.* **13**, 593–600.

Gordon, W. R., Arnett, K. L., and Blacklow, S. C. (2008). The molecular logic of Notch signaling—a structural and biochemical perspective. *J. Cell. Sci.* **121**, 3109–3119.

Gupta-Rossi, N., Le Bail, O., Gonen, H., Brou, C., Logeat, F., Six, E., Ciechanover, A., and Israel, A. (2001). Functional interaction between SEL-10, an F-box protein, and the nuclear form of activated Notch1 receptor. *J. Biol. Chem.* **276**, 34371–34378.

Guss, K. A., Nelson, C. E., Hudson, A., Kraus, M. E., and Carroll, S. B. (2001). Control of a genetic regulatory network by a selector gene. *Science* **292**, 1164–1167.

Han, H., Tanigaki, K., Yamamoto, N., Kuroda, K., Yoshimoto, M., Nakahata, T., Ikuta, K., and Honjo, T. (2002). Inducible gene knockout of transcription factor recombination signal binding protein-J reveals its essential role in T versus B lineage decision. *Int. Immunol.* **14**, 637–645.

Hansson, M. L., Popko-Scibor, A. E., Saint Just Ribeiro, M., Dancy, B. M., Lindberg, M. J., Cole, P. A., and Wallberg, A. E. (2009). The transcriptional coactivator MAML1 regulates p300 autoacetylation and HAT activity. *Nucleic Acids Res.* **37**, 2996–3006.

Hatakeyama, J., Bessho, Y., Katoh, K., Ookawara, S., Fujioka, M., Guillemot, F., and Kageyama, R. (2004). Hes genes regulate size, shape and histogenesis of the nervous system by control of the timing of neural stem cell differentiation. *Development* **131**, 5539–5550.

Hsieh, J. J., Zhou, S., Chen, L., Young, D. B., and Hayward, S. D. (1999). CIR, a corepressor linking the DNA binding factor CBF1 to the histone deacetylase complex. *Proc. Natl. Acad. Sci. U.S.A.* **96**, 23–28.

Hurlbut, G. D., Kankel, M. W., and Artavanis-Tsakonas, S. (2009). Nodal points and complexity of Notch-Ras signal integration. *Proc. Natl. Acad. Sci. U.S.A.* **106**, 2218–2223.

Hwang, B. J., Meruelo, A. D., and Sternberg, P. W. (2007). *C. elegans* EVI1 proto-oncogene, EGL-43, is necessary for Notch-mediated cell fate specification and regulates cell invasion. *Development* **134**, 669–679.

Jeffries, S., Robbins, D. J., and Capobianco, A. J. (2002). Characterization of a high-molecular-weight Notch complex in the nucleus of Notch(ic)-transformed RKE cells and in a human T-cell leukemia cell line. *Mol. Cell. Biol.* **22**, 3927–3941.

Joshi, I., Minter, L. M., Telfer, J., Demarest, R. M., Capobianco, A. J., Aster, J. C., Sicinski, P., Fauq, A., Golde, T. E., and Osborne, B. A. (2009). Notch signaling mediates G1/S cell-cycle progression in T cells via cyclin D3 and its dependent kinases. *Blood* **113**, 1689–1698.

Kadam, S., and Emerson, B. M. (2003). Transcriptional specificity of human SWI/SNF BRG1 and BRM chromatin remodeling complexes. *Mol. Cell.* **11**, 377–389.

Kageyama, R., Ohtsuka, T., and Kobayashi, T. (2007). The Hes gene family: repressors and oscillators that orchestrate embryogenesis. *Development* **134**, 1243–1251.

Kao, H. Y., Ordentlich, P., Koyano-Nakagawa, N., Tang, Z., Downes, M., Kintner, C. R., Evans, R. M., and Kadesch, T. (1998). A histone deacetylase corepressor complex regulates the Notch signal transduction pathway. *Genes Dev.* **12**, 2269–2277.

Kim, J., Sebring, A., Esch, J. J., Kraus, M. E., Vorwerk, K., Magee, J., and Carroll, S. B. (1996). Integration of positional signals and regulation of wing formation and identity by *Drosophila* vestigial gene. *Nature* **382**, 133–138.

Kim, M. Y., Ann, E. J., Kim, J. Y., Mo, J. S., Park, J. H., Kim, S. Y., Seo, M. S., and Park, H. S. (2007). Tip60 histone acetyltransferase acts as a negative regulator of Notch1 signaling by means of acetylation. *Mol. Cell. Biol.* **27**, 6506–6519.

Kishi, N., Tang, Z., Maeda, Y., Hirai, A., Mo, R., Ito, M., Suzuki, S., Nakao, K., Kinoshita, T., Kadesch, T., Hui, C., Artavanis-Tsakonas, S., Okano, H., and Matsuno, K. (2001). Murine homologs of deltex define a novel gene family involved in vertebrate Notch signaling and neurogenesis. *Int. J. Dev. Neurosci.* **19**, 21–35.

Klinakis, A., Szabolcs, M., Politi, K., Kiaris, H., Artavanis-Tsakonas, S., and Efstratiadis, A. (2006). Myc is a Notch1 transcriptional target and a requisite for Notch1-induced mammary tumorigenesis in mice. *Proc. Natl. Acad. Sci. U.S.A.* **103**, 9262–9267.

Koch, U., and Radtke, F. (2007). Notch and cancer: a double-edged sword. *Cell Mol. Life Sci.* **64**, 2746–2762.

Koelzer, S., and Klein, T. (2003). A Notch-independent function of suppressor of hairless during the development of the bristle sensory organ precursor cell of *Drosophila*. *Development* **130**, 1973–1988.

Koelzer, S., and Klein, T. (2006). Regulation of expression of Vg and establishment of the dorsoventral compartment boundary in the wing imaginal disc by suppressor of hairless. *Dev. Biol.* **289**, 77–90.

Kopan, R., and Ilagan, M. X. (2009). The canonical Notch signaling pathway: unfolding the activation mechanism. *Cell* **137**, 216–233.

Kramatschek, B., and Campos-Ortega, J. A. (1994). Neuroectodermal transcription of the *Drosophila* neurogenic genes E(spl) and HLH-m5 is regulated by proneural genes. *Development* **120**, 815–826.

Krebs, L. T., Deftos, M. L., Bevan, M. J., and Gridley, T. (2001). The Nrarp gene encodes an ankyrin-repeat protein that is transcriptionally regulated by the notch signaling pathway. *Dev. Biol.* **238**, 110–119.

Krejci, A., Bernard, F., Housden, B. E., Collins, S., and Bray, S. J. (2009). Direct response to Notch activation: signaling crosstalk and incoherent logic. *Sci. Signal* **2**, ra1.

Krejci, A., and Bray, S. (2007). Notch activation stimulates transient and selective binding of Su(H)/CSL to target enhancers. *Genes Dev.* **21**, 1322–1327.

Kuroda, K., Han, H., Tani, S., Tanigaki, K., Tun, T., Furukawa, T., Taniguchi, Y., Kurooka, H., Hamada, Y., Toyokuni, S., and Honjo, T. (2003). Regulation of marginal zone B cell development by MINT, a suppressor of Notch/RBP-J signaling pathway. *Immunity* **18**, 301–312.

Kurooka, H., and Honjo, T. (2000). Functional interaction between the mouse notch1 intracellular region and histone acetyltransferases PCAF and GCN5. *J. Biol. Chem.* **275**, 17211–17220.

Lamar, E., Deblandre, G., Wettstein, D., Gawantka, V., Pollet, N., Niehrs, C., and Kintner, C. (2001). Nrarp is a novel intracellular component of the Notch signaling pathway. *Genes Dev.* **15**, 1885–1899.

Lamar, E., and Kintner, C. (2005). The Notch targets Esr1 and Esr10 are differentially regulated in Xenopus neural precursors. *Development* **132**, 3619–3630.

Lecourtois, M., and Schweisguth, F. (1995). The neurogenic suppressor of hairless DNA-binding protein mediates the transcriptional activation of the enhancer of split complex genes triggered by Notch signaling. *Genes Dev.* **9**, 2598–2608.

Lewis, J., Hanisch, A., and Holder, M. (2009). Notch signaling, the segmentation clock, and the patterning of vertebrate somites. *J. Biol.* **8**, 44.

Li, Y., and Baker, N. E. (2001). Proneural enhancement by Notch overcomes Suppressor-of-Hairless repressor function in the developing *Drosophila* eye. *Curr. Biol.* **11**, 330–338.

Liefke, R., Oswald, F., Alvarado, C., Ferres-Marco, D., Mittler, G., Rodriguez, P., Dominguez, M., and Borggrefe, T. (2010). Histone demethylase KDM5A is an integral part of the core Notch-RBP-J repressor complex. *Genes Dev.* **24**, 590–601.

Ligoxygakis, P., Yu, S. Y., Delidakis, C., and Baker, N. E. (1998). A subset of notch functions during *Drosophila* eye development require Su(H) and the E(spl) gene complex. *Development* **125**, 2893–2900.

Margolin, A. A., Palomero, T., Sumazin, P., Califano, A., Ferrando, A. A., and Stolovitzky, G. (2009). ChIP-on-chip significance analysis reveals large-scale binding and regulation by human transcription factor oncogenes. *Proc. Natl. Acad. Sci. U.S.A.* **106**, 244–249.

Martinez, A. M., Schuettengruber, B., Sakr, S., Janic, A., Gonzalez, C., and Cavalli, G. (2009). Polyhomeotic has a tumor suppressor activity mediated by repression of Notch signaling. *Nat. Genet.* **41**, 1076–1082.

Merdes, G., Soba, P., Loewer, A., Bilic, M. V., Beyreuther, K., and Paro, R. (2004). Interference of human and *Drosophila* APP and APP-like proteins with PNS development in *Drosophila*. *EMBO J.* **23**, 4082–4095.

Miller, S. W., Avidor-Reiss, T., Polyanovsky, A., and Posakony, J. W. (2009). Complex interplay of three transcription factors in controlling the tormogen differentiation program of *Drosophila* mechanoreceptors. *Dev. Biol.* **329**, 386–399.

Mittal, S., Subramanyam, D., Dey, D., Kumar, R. V., and Rangarajan, A. (2009). Cooperation of Notch and Ras/MAPK signaling pathways in human breast carcinogenesis. *Mol. Cancer.* **8**, 128.

Moran, S. T., Cariappa, A., Liu, H., Muir, B., Sgroi, D., Boboila, C., and Pillai, S. (2007). Synergism between NF-kappa B1/p50 and Notch2 during the development of marginal zone B lymphocytes. *J. Immunol.* **179**, 195–200.

Morel, V., Lecourtois, M., Massiani, O., Maier, D., Preiss, A., and Schweisguth, F. (2001). Transcriptional repression by suppressor of hairless involves the binding of a hairless-dCtBP complex in *Drosophila. Curr. Biol.* **11**, 789–792.

Morel, V., and Schweisguth, F. (2000). Repression by suppressor of hairless and activation by Notch are required to define a single row of single-minded expressing cells in the *Drosophila* embryo. *Genes Dev.* **14**, 377–388.

Moshkin, Y. M., Kan, T. W., Goodfellow, H., Bezstarosti, K., Maeda, R. K., Pilyugin, M., Karch, F., Bray, S. J., Demmers, J. A., and Verrijzer, C. P. (2009). Histone chaperones ASF1 and NAP1 differentially modulate removal of active histone marks by LID-RPD3 complexes during NOTCH silencing. *Mol. Cell.* **35**, 782–793.

Nagel, A. C., Krejci, A., Tenin, G., Bravo-Patino, A., Bray, S., Maier, D., and Preiss, A. (2005). Hairless-mediated repression of notch target genes requires the combined activity of Groucho and CtBP corepressors. *Mol. Cell. Biol.* **25**, 10433–10441.

Nagy, L., Tontonoz, P., Alvarez, J. G., Chen, H., and Evans, R. M. (1998). Oxidized LDL regulates macrophage gene expression through ligand activation of PPARgamma. *Cell* **93**, 229–240.

Nam, Y., Sliz, P., Pear, W. S., Aster, J. C., and Blacklow, S. C. (2007). Cooperative assembly of higher-order Notch complexes functions as a switch to induce transcription. *Proc. Natl. Acad. Sci. U.S.A.* **104**, 2103–2108.

Nam, Y., Sliz, P., Song, L., Aster, J. C., and Blacklow, S. C. (2006). Structural basis for cooperativity in recruitment of MAML coactivators to Notch transcription complexes. *Cell* **124**, 973–983.

Nellesen, D. T., Lai, E. C., and Posakony, J. W. (1999). Discrete enhancer elements mediate selective responsiveness of enhancer of split complex genes to common transcriptional activators. *Dev. Biol.* **213**, 33–53.

Neves, A., and Priess, J. R. (2005). The REF-1 family of bHLH transcription factors pattern *C. elegans* embryos through Notch-dependent and Notch-independent pathways. *Dev. Cell.* **8**, 867–879.

O'Neil, J., and Look, A. T. (2007). Mechanisms of transcription factor deregulation in lymphoid cell transformation. *Oncogene* **26**, 6838–6849.

Oka, C., Nakano, T., Wakeham, A., de la Pompa, J. L., Mori, C., Sakai, T., Okazaki, S., Kawaichi, M., Shiota, K., Mak, T. W., and Honjo, T. (1995). Disruption of the mouse RBP-J kappa gene results in early embryonic death. *Development* **121**, 3291–3301.

Ong, C. T., Cheng, H. T., Chang, L. W., Ohtsuka, T., Kageyama, R., Stormo, G. D., and Kopan, R. (2006). Target selectivity of vertebrate notch proteins. Collaboration between discrete domains and CSL-binding site architecture determines activation probability. *J. Biol. Chem.* **281**, 5106–5119.

Oswald, F., Kostezka, U., Astrahantseff, K., Bourteele, S., Dillinger, K., Zechner, U., Ludwig, L., Wilda, M., Hameister, H., Knochel, W., Liptay, S., and Schmid, R. M. (2002). SHARP is a novel component of the Notch/RBP-Jkappa signalling pathway. *EMBO J.* **21**, 5417–5426.

Oswald, F., Tauber, B., Dobner, T., Bourteele, S., Kostezka, U., Adler, G., Liptay, S., and Schmid, R. M. (2001). p300 acts as a transcriptional coactivator for mammalian Notch-1. *Mol. Cell. Biol.* **21**, 7761–7774.

Oswald, F., Winkler, M., Cao, Y., Astrahantseff, K., Bourteele, S., Knochel, W., and Borggrefe, T. (2005). RBP-Jkappa/SHARP recruits CtIP/CtBP corepressors to silence Notch target genes. *Mol. Cell. Biol.* **25**, 10379–10390.

Palomero, T., Dominguez, M., and Ferrando, A. A. (2008). The role of the PTEN/AKT Pathway in NOTCH1-induced leukemia. *Cell Cycle* **7**, 965–970.

Palomero, T., Lim, W. K., Odom, D. T., Sulis, M. L., Real, P. J., Margolin, A., Barnes, K. C., O'Neil, J., Neuberg, D., Weng, A. P., Aster, J. C., Sigaux, F., Soulier, J., Look, A. T., Young, R. A., Califano, A., and Ferrando, A. A. (2006). NOTCH1 directly regulates c-MYC and activates a feed-forward-loop transcriptional network promoting leukemic cell growth. *Proc. Natl. Acad. Sci. U.S.A.* **103**, 18261–18266.

Paroush, Z., Finley, R. L., Jr., Kidd, T., Wainwright, S. M., Ingham, P. W., Brent, R., and Ish-Horowicz, D. (1994). Groucho is required for *Drosophila* neurogenesis, segmentation, and sex determination and interacts directly with hairy-related bHLH proteins. *Cell* **79**, 805–815.

Pascoal, S., Carvalho, C. R., Rodriguez-Leon, J., Delfini, M. C., Duprez, D., Thorsteinsdottir, S., and Palmeirim, I. (2007). A molecular clock operates during chick autopod proximal-distal outgrowth. *J. Mol. Biol.* **368**, 303–309.

Phng, L. K., Potente, M., Leslie, J. D., Babbage, J., Nyqvist, D., Lobov, I., Ondr, J. K., Rao, S., Lang, R. A., Thurston, G., and Gerhardt, H. (2009). Nrarp coordinates endothelial Notch and Wnt signaling to control vessel density in angiogenesis. *Dev. Cell.* **16**, 70–82.

Pines, M. K., Housden, B. E., Bernard, F., Bray, S. J., and Roper, K. (2010). The cytolinker Pigs is a direct target and a negative regulator of Notch signalling. *Development* **137**, 913–922.

Pirot, P., van Grunsven, L. A., Marine, J. C., Huylebroeck, D., and Bellefroid, E. J. (2004). Direct regulation of the Nrarp gene promoter by the Notch signaling pathway. *Biochem. Biophys. Res. Commun.* **322**, 526–534.

Poortinga, G., Watanabe, M., and Parkhurst, S. M. (1998). *Drosophila* CtBP: a hairy-interacting protein required for embryonic segmentation and hairy-mediated transcriptional repression. *EMBO J.* **17**, 2067–2078.

Rangarajan, A., Talora, C., Okuyama, R., Nicolas, M., Mammucari, C., Oh, H., Aster, J. C., Krishna, S., Metzger, D., Chambon, P., Miele, L., Aguet, M., Radtke, F., and Dotto, G. P. (2001). Notch signaling is a direct determinant of keratinocyte growth arrest and entry into differentiation. *EMBO J.* **20**, 3427–3436.

Rebeiz, M., Reeves, N. L., and Posakony, J. W. (2002). SCORE: a computational approach to the identification of cis-regulatory modules and target genes in whole-genome sequence data. Site clustering over random expectation. *Proc. Natl. Acad. Sci. U.S.A.* **99**, 9888–9893.

Ronchini, C., and Capobianco, A. J. (2001). Induction of cyclin D1 transcription and CDK2 activity by Notch(ic): implication for cell cycle disruption in transformation by Notch(ic). *Mol. Cell. Biol.* **21**, 5925–5934.

Rosenberg, M. I., and Parkhurst, S. M. (2002). *Drosophila* Sir2 is required for heterochromatic silencing and by euchromatic Hairy/E(Spl) bHLH repressors in segmentation and sex determination. *Cell* **109**, 447–458.

Schlatter, R., and Maier, D. (2005). The Enhancer of split and Achaete-Scute complexes of Drosophilids derived from simple ur-complexes preserved in mosquito and honeybee. *BMC Evol. Biol.* **5**, 67.

Shen, J., Bronson, R. T., Chen, D. F., Xia, W., Selkoe, D. J., and Tonegawa, S. (1997). Skeletal and CNS defects in Presenilin-1-deficient mice. *Cell* **89**, 629–639.

Shimojo, H., Ohtsuka, T., and Kageyama, R. (2008). Oscillations in notch signaling regulate maintenance of neural progenitors. *Neuron* **58**, 52–64.

Singson, A., Leviten, M. W., Bang, A. G., Hua, X. H., and Posakony, J. W. (1994). Direct downstream targets of proneural activators in the imaginal disc include genes involved in lateral inhibitory signaling. *Genes Dev.* **8**, 2058–2071.

Sivasankaran, B., Degen, M., Ghaffari, A., Hegi, M. E., Hamou, M. F., Ionescu, M. C., Zweifel, C., Tolnay, M., Wasner, M., Mergenthaler, S., Miserez, A. R., Kiss, R., Lino, M. M., Merlo, A., Chiquet-Ehrismann, R., and Boulay, J. L. (2009). Tenascin-C is a novel RBPJkappa-induced target gene for Notch signaling in gliomas. *Cancer Res.* **69**, 458–465.

Subramanian, T., and Chinnadurai, G. (2003). Association of class I histone deacetylases with transcriptional corepressor CtBP. *FEBS Lett.* **540**, 255–258.

Sundaram, M. V. (2005). The love-hate relationship between Ras and Notch. *Genes Dev.* **19**, 1825–1839.

Surendran, K., Boyle, S., Barak, H., Kim, M., Stomberski, C., McCright, B., and Kopan, R. (2010). The contribution of Notch1 to nephron segmentation in the developing kidney is revealed in a sensitized Notch2 background and can be augmented by reducing Mint dosage. *Dev. Biol.* **337**, 386–395.

Swanson, C. I., Evans, N. C., and Barolo, S. (2010). Structural rules and complex regulatory circuitry constrain expression of a Notch- and EGFR-regulated eye enhancer. *Dev. Cell.* **18**, 359–370.

Takata, T., and Ishikawa, F. (2003). Human Sir2-related protein SIRT1 associates with the bHLH repressors HES1 and HEY2 and is involved in HES1- and HEY2-mediated transcriptional repression. *Biochem. Biophys. Res. Commun.* **301**, 250–257.

Tapanes-Castillo, A., and Baylies, M. K. (2004). Notch signaling patterns *Drosophila* mesodermal segments by regulating the bHLH transcription factor twist. *Development* **131**, 2359–2372.

Tenney, K., Gerber, M., Ilvarsonn, A., Schneider, J., Gause, M., Dorsett, D., Eissenberg, J. C., and Shilatifard, A. (2006). *Drosophila* Rtf1 functions in histone methylation, gene expression, and Notch signaling. *Proc. Natl. Acad. Sci. U.S.A.* **103**, 11970–11974.

Tolhuis, B., de Wit, E., Muijrers, I., Teunissen, H., Talhout, W., van Steensel, B., and van Lohuizen, M. (2006). Genome-wide profiling of PRC1 and PRC2 Polycomb chromatin binding in *Drosophila melanogaster*. *Nat. Genet.* **38**, 694–699.

Tsuda, L., Kaido, M., Lim, Y. M., Kato, K., Aigaki, T., and Hayashi, S. (2006). An NRSF/REST-like repressor downstream of Ebi/SMRTER/Su(H) regulates eye development in *Drosophila*. *EMBO J.* **25**, 3191–3202.

Tsunematsu, R., Nakayama, K., Oike, Y., Nishiyama, M., Ishida, N., Hatakeyama, S., Bessho, Y., Kageyama, R., Suda, T., and Nakayama, K. I. (2004). Mouse Fbw7/Sel-10/Cdc4 is required for notch degradation during vascular development. *J. Biol. Chem.* **279**, 9417–9423.

Tun, T., Hamaguchi, Y., Matsunami, N., Furukawa, T., Honjo, T., and Kawaichi, M. (1994). Recognition sequence of a highly conserved DNA binding protein RBP-J kappa. *Nucleic Acids Res.* **22**, 965–971.

Tyagi, M., and Karn, J. (2007). CBF-1 promotes transcriptional silencing during the establishment of HIV-1 latency. *EMBO J.* **26**, 4985–4995.

Weerkamp, F., Luis, T. C., Naber, B. A., Koster, E. E., Jeannotte, L., van Dongen, J. J., and Staal, F. J. (2006). Identification of Notch target genes in uncommitted T-cell progenitors: no direct induction of a T-cell specific gene program. *Leukemia* **20**, 1967–1977.

Welcker, M., and Clurman, B. E. (2008). FBW7 ubiquitin ligase: a tumour suppressor at the crossroads of cell division, growth and differentiation. *Nat. Rev. Cancer* **8**, 83–93.

Weng, A. P., Millholland, J. M., Yashiro-Ohtani, Y., Arcangeli, M. L., Lau, A., Wai, C., Del Bianco, C., Rodriguez, C. G., Sai, H., Tobias, J., Li, Y., Wolfe, M. S., Shachaf, C., Felsher, D., Blacklow, S. C., Pear, W. S., and Aster, J. C. (2006). c-Myc is an important direct target of Notch1 in T-cell acute lymphoblastic leukemia/lymphoma. *Genes Dev.* **20**, 2096–2109.

Whelan, J. T., Forbes, S. L., and Bertrand, F. E. (2007). CBF-1 (RBP-J kappa) binds to the PTEN promoter and regulates PTEN gene expression. *Cell Cycle* **6**, 80–84.

Yabe, D., Fukuda, H., Aoki, M., Yamada, S., Takebayashi, S., Shinkura, R., Yamamoto, N., and Honjo, T. (2007). Generation of a conditional knockout allele for mammalian Spen protein Mint/SHARP. *Genesis* **45**, 300–306.

Yamaguchi, M., Tonou-Fujimori, N., Komori, A., Maeda, R., Nojima, Y., Li, H., Okamoto, H., and Masai, I. (2005). Histone deacetylase 1 regulates retinal neurogenesis in zebrafish by suppressing Wnt and Notch signaling pathways. *Development* **132**, 3027–3043.

Yan, S. J., Gu, Y., Li, W. X., and Fleming, R. J. (2004). Multiple signaling pathways and a selector protein sequentially regulate *Drosophila* wing development. *Development* **131**, 285–298.

Yashiro-Ohtani, Y., He, Y., Ohtani, T., Jones, M. E., Shestova, O., Xu, L., Fang, T. C., Chiang, M. Y., Intlekofer, A. M., Blacklow, S. C., Zhuang, Y., and Pear, W. S. (2009). Pre-TCR signaling inactivates Notch1 transcription by antagonizing E2A. *Genes Dev.* **23**, 1665–1676.

Yeo, S. Y., Kim, M., Kim, H. S., Huh, T. L., and Chitnis, A. B. (2007). Fluorescent protein expression driven by her4 regulatory elements reveals the spatiotemporal pattern of Notch signaling in the nervous system of zebrafish embryos. *Dev. Biol.* **301**, 555–567.

Yoo, A. S., Bais, C., and Greenwald, I. (2004). Crosstalk between the EGFR and LIN-12/Notch pathways in *C. elegans* vulval development. *Science* **303**, 663–666.

Zeitlinger, J., Stark, A., Kellis, M., Hong, J. W., Nechaev, S., Adelman, K., Levine, M., and Young, R. A. (2007). RNA polymerase stalling at developmental control genes in the *Drosophila melanogaster* embryo. *Nat. Genet.* **39**, 1512–1516.

Zhang, H., and Levine, M. (1999). Groucho and dCtBP mediate separate pathways of transcriptional repression in the *Drosophila* embryo. *Proc. Natl. Acad. Sci. U.S.A.* **96**, 535–540.

Zhou, S., and Hayward, S. D. (2001). Nuclear localization of CBF1 is regulated by interactions with the SMRT corepressor complex. *Mol. Cell. Biol.* **21**, 6222–6232.

CHAPTER NINE

Notch Signaling in the Vasculature

Thomas Gridley

Contents

The Jackson Laboratory, Bar Harbor, ME, USA

Current Topics in Developmental Biology, Volume 92
ISSN 0070-2153, DOI 10.1016/S0070-2153(10)92009-7

Abstract

Notch signaling is an evolutionarily conserved, intercellular signaling mechanism that plays myriad roles during vascular development and physiology in vertebrates. These roles include the regulation of arteriovenous specification and differentiation in both endothelial cells and vascular smooth muscle cells, regulation of blood vessel sprouting and branching during normal and pathological angiogenesis, and the physiological responses of vascular smooth muscle cells. Defects in Notch signaling also cause inherited vascular diseases, such as the degenerative vascular disorder cerebral autosomal dominant arteriopathy with subcortical infarcts and leukoencephalopathy. This review summarizes recent studies that highlight the multiple roles the Notch signaling pathway plays during vascular development and physiology.

1. Introduction

The Notch signaling pathway plays myriad roles during vascular development and physiology in vertebrates. This review will summarize recent work highlighting the multiple roles the Notch pathway plays during vascular development, physiology, and disease, emphasizing its role in mammals. Additional views on the role of Notch signaling during these processes can be found in a number of previously published reviews (Gridley, 2007; Hofmann and Iruela-Arispe, 2007; Kume, 2009; Phng and Gerhardt, 2009; Roca and Adams, 2007). Notch signaling also plays an important role in the development of the vertebrate heart, which is not addressed here. Details on the roles of Notch signaling in heart development can be found in several recent reviews (High and Epstein, 2008; Nemir and Pedrazzini, 2008; Niessen and Karsan, 2008).

Notch family receptors are large single-pass type I transmembrane proteins (Fig. 9.1). In mammals, four Notch family receptors have been described: NOTCH1 through NOTCH4. The extracellular domain of Notch family proteins contains up to 36 tandemly repeated copies of an epidermal growth factor (EGF)-like motif. A Notch family receptor exists at the cell surface as a proteolytically cleaved heterodimer consisting of a large ectodomain and a membrane-tethered intracellular domain. Notch receptors interact with ligands that are also single-pass type I transmembrane proteins. This restricts the Notch pathway to regulating signaling interactions between physically adjacent cells (which has been termed juxtacrine signaling). In mammals, the Notch ligands are encoded by the Jagged (*JAG1* and *JAG2*) and Delta-like (*DLL1, DLL3*, and *DLL4*) gene families.

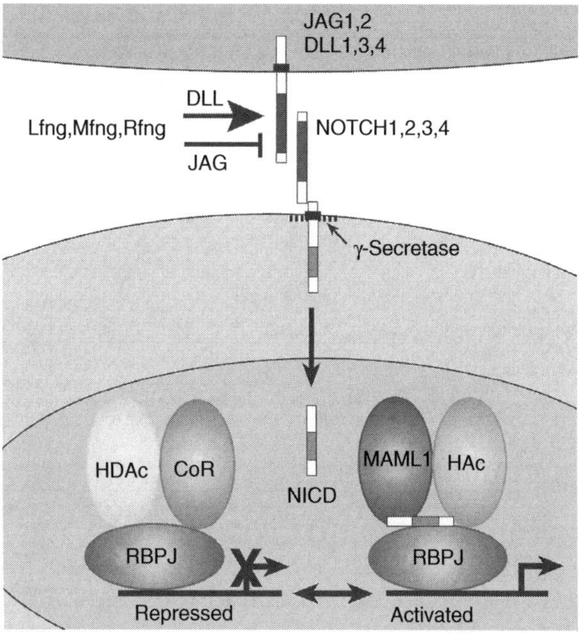

Figure 9.1 Core components of the canonical Notch signaling pathway. Ligands of the Jagged (JAG1, JAG2) and Delta-like (DLL1, DLL3, DLL4) families (upper cell) interact with Notch family receptors (NOTCH1 through NOTCH4) on an adjacent cell. The Notch receptor exists at the cell surface as a proteolytically cleaved heterodimer consisting of a large ectodomain and a membrane-tethered intracellular domain. The Fringe proteins (Lfng, Mfng, and Rfng) glycosylate Notch family receptors, potentiating signaling from DLL family ligands and suppressing signaling from JAG family ligands. The receptor/ligand interaction induces additional proteolytic cleavages in the membrane-tethered intracellular domain. The final cleavage, catalyzed by the γ-secretase complex, frees the NICD from the cell membrane. NICD translocates to the nucleus, where it forms a complex with the RBPJ protein, displacing a histone deacetylase (HDAc)/corepressor (CoR) complex from the RBPJ protein. Components of an activation complex, such as MAML1 and histone acetyltransferases (HAc), are recruited to the NICD/RBPJ complex, leading to the transcriptional activation of Notch target genes.

The signal induced by ligand binding is transmitted intracellularly by a process involving proteolytic cleavage of the receptor and nuclear translocation of the intracellular domain of the Notch family protein. The receptor/ligand interaction induces two additional proteolytic cleavages in the membrane-tethered fragment of the Notch heterodimer. The final cleavage, catalyzed by the γ-secretase complex, frees the intracellular domain of the Notch receptor from the cell membrane. The cleaved fragment translocates to the nucleus due to the presence of nuclear localization signals located in the Notch intracellular domain (NICD). Once in the nucleus, NICD forms a complex with the RBPJ protein, a sequence-specific DNA-binding protein (also known in mammals as CSL and CBF1). In the absence of NICD, the RBPJ protein binds to specific DNA sequences in the regulatory elements of various target genes and represses transcription by recruiting histone deacetylases and other components of a corepressor complex. Nuclear translocation of the NICD displaces the histone deacetylase/corepressor complex from the RBPJ protein. The NICD/RBPJ complex recruits other proteins, such as MAML1 and histone acetyltransferases, leading to the transcriptional activation of Notch target genes. Among the most commonly induced Notch target genes are basic helix-loop-helix (bHLH) transcriptional repressors of the Hes/Hey family (Borggrefe and Oswald, 2009). For additional details on the biochemistry of the Notch signaling pathway and references to the primary literature, please refer to a number of excellent review articles (Bray, 2006; Ehebauer *et al.*, 2006a, b; Fortini, 2009; Ilagan and Kopan, 2007; Kageyama *et al.*, 2007; Kopan and Ilagan, 2009; Le Borgne, 2006).

2. Arteriovenous Differentiation

2.1. Notch signaling is downstream of the Vegf pathway during vascular development

A role for the Notch pathway in regulating vascular development was demonstrated initially from analysis of targeted mouse mutants in Notch pathway components. Mouse mutants for which targeted mutagenesis and transgenic studies have demonstrated a role in embryonic vascular development include the receptors Notch1 (Huppert *et al.*, 2000; Krebs *et al.*, 2000; Limbourg *et al.*, 2005) and Notch4 (Carlson *et al.*, 2005; Krebs *et al.*, 2000; Uyttendaele *et al.*, 2001), the ligands Jag1 (Xue *et al.*, 1999) and Dll4 (Duarte *et al.*, 2004; Gale *et al.*, 2004; Krebs *et al.*, 2004), the Notch transcriptional regulator Rpbj (Krebs *et al.*, 2004), the E3 ubiquitin ligase Mib1 (Barsi *et al.*, 2005; Koo *et al.*, 2005), components of the γ-secretase complex such as nicastrin (Li *et al.*, 2003) and presenilin 1 and 2 (Herreman *et al.*, 1999), and the Notch pathway downstream effector bHLH proteins Hey1 and Hey2 (Fischer *et al.*, 2004; Kokubo *et al.*, 2005). Most of these mutants exhibit a similar phenotype characterized by the absence of angiogenic vascular remodeling in the extraembryonic yolk sac, placenta, and embryo proper.

However, analysis of zebra fish embryos with reduced Notch signaling gave the first clues that a primary function of the Notch pathway during vascular development was to regulate the specification of arterial fate in endothelial cells. It had long been believed that the primary factor regulating differentiation of arteries and veins was blood flow. Endothelial cells lining arteries experience higher blood pressures, higher rates of hemodynamic flow, and higher oxygen tensions than endothelial cells lining veins. However, it has recently become clear that genetic prepatterning, mediated

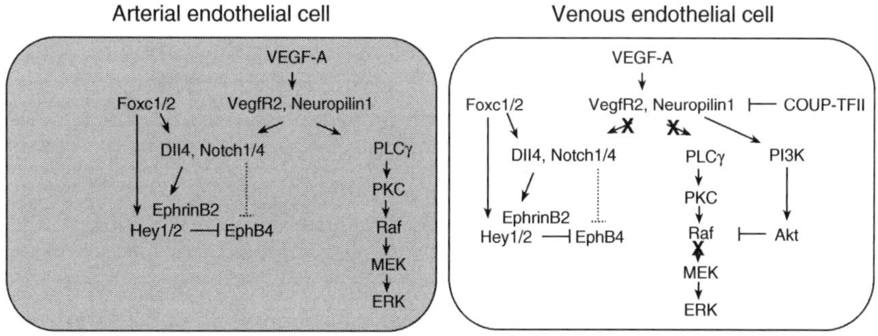

Figure 9.2 Model for genetic regulation of artery–vein differentiation by the Notch and PLCγ/MAPK pathways. An arterial endothelial cell is shown on the left, while a venous endothelial cell is depicted on the right. Two main signaling pathways operate downstream of VEGF-A to induce arterial differentiation: the Notch pathway and the phospholipase Cγ (PLCγ)/mitogen-activated protein kinase (MAPK) pathway. The transcription factors Foxc1 and Foxc2 directly induce *Dll4* and *Hey2* transcription. VEGF-A signaling, by an unknown mechanism, augments Foxc1/Foxc2 induction of *Dll4* and *Hey2* gene expression. During venous differentiation, two different mechanisms inhibit artery differentiation. The orphan nuclear receptor COUP-TFII suppresses neuropilin1 expression, thereby suppressing reception of the VEGF-A signal and activation of Notch signaling. In addition, activation of PI3K/Akt signaling antagonizes promotion of arterial cell differentiation by blocking ERK activation.

in large part by the Notch pathway, plays a primary role in regulating arteriovenous differentiation. The role of the Notch pathway in regulating early embryonic vascular development is intertwined with that of another major regulator of vascular development and physiology, the vascular endothelial growth factor-A (VEGF-A) pathway (Fig. 9.2). VEGF-A is a secreted glycoprotein that is a potent inducer of angiogenesis that also regulates multiple other aspects of blood vessel homeostasis (Byrne *et al.*, 2005; Coultas *et al.*, 2005; Shibuya and Claesson-Welsh, 2006). The roles and interdependence of the Notch and VEGF-A pathways in regulating formation of the large axial blood vessels of the trunk, the dorsal aorta, and the posterior cardinal vein was studied first in zebra fish (Lawson *et al.*, 2001, 2002). Notch signaling-deficient embryos exhibited a poorly formed dorsal aorta and posterior cardinal vein with accompanying arteriovenous malformations (the fusion of arteries and veins without an intervening capillary bed). These embryos exhibited loss of expression of arterial markers such as ephrinB2 from arterial vessels with an accompanying expansion of venous markers into normally arterial domains. Embryos in which Notch signaling had been ectopically activated exhibited the reverse phenotype: suppression of vein-specific markers with ectopic expression of arterial markers in venous vessels (Lawson *et al.*, 2001). A similar phenotype was observed in embryos mutant for some Notch target genes, such as the bHLH transcriptional repressor Hey2 (referred to in zebra fish as the gridlock gene) (Zhong *et al.*, 2000, 2001).

This analysis of formation of the major trunk vessels of the zebra fish embryo revealed a signaling cascade responsible for determining arterial and venous cell fates in these vessels (Lawson *et al.*, 2002). Reduction of *Vegf* activity resulted in loss of expression of arterial markers from the dorsal aorta and ectopic arterial expression of vein markers. Injection of *Vegf* mRNA induced ectopic expression of the arterial marker *ephrinB2* in the posterior cardinal vein. *Vegf* expression was regulated by expression of the secreted morphogen Sonic hedgehog (*Shh*) along the axial midline. Similar to what was observed in *Vegf*-deficient embryos, *Shh* mutant zebra fish embryos also exhibited loss of arterial differentiation, while injection of *Shh* mRNA caused ectopic expression of arterial markers. *Shh* acted upstream of *Vegf*, since injection of *Vegf* mRNA into the *Shh* mutant embryos rescued arterial differentiation. This work also demonstrated that the Notch pathway acted downstream of the Vegf pathway. While injection of *Vegf* mRNA into Notch signaling-deficient zebra fish embryos could not rescue arterial marker gene expression, expression of an activated *Notch1* transgene in *Vegf*-deficient embryos could rescue expression of arterial markers (Lawson *et al.*, 2002). A new study has demonstrated by high-resolution *in vivo* imaging analysis that formation of the major trunk vessels of the zebra fish embryo takes place by a novel mechanism (Herbert *et al.*, 2009). Angioblasts coalesced along the embryonic midline to form a single vascular cord in the position of the future dorsal aorta, which then lumenized. A subset of angioblasts then sprouted ventrally from this progenitor vessel to form the cardinal vein. Sprouting behaviors were regulated in the embryos by coordinated Vegf, Notch, and ephrinB2-EphB4 signaling.

Studies in mammalian cells in culture have also placed the Notch pathway downstream of the Vegf pathway. VEGF-A administration induced expression of mRNA for the Notch1 receptor and the Dll4 ligand in human arterial endothelial cells, but not in venous endothelial cells (Liu *et al.*, 2003). Targeted mutagenesis studies have demonstrated that VEGF-A is essential for vascular development in mice. Embryos heterozygous for a *Vegfa*-targeted mutation exhibited lethal haploinsufficiency (Carmeliet *et al.*, 1996; Ferrara *et al.*, 1996). Blood vessels formed in these embryos, but were severely constricted or atretic. It is not known whether artery-vein differentiation is affected in *Vegfa+/−* embryos. However, gain of function transgenic experiments have demonstrated a role for *Vegfa* in regulating arterial endothelial cell differentiation in mice. Alternative splicing of the *Vegfa* gene results in production of several different protein isoforms (VEGF-A 120, VEGF-A 164, and VEGF-A 188). Genetically engineered mice expressing only the VEGF-A 164 isoform exhibited normal retinal vascular development. However, mice expressing only VEGF-A 120 exhibited severe defects in vascular outgrowth, while mice expressing only VEGF-A 188 exhibited impaired retinal arterial development, but normal venous development (Stalmans *et al.*, 2002).

Overexpression of the VEGF-A 164 isoform in cardiac muscle increased the number of ephrinB2-positive capillaries in the heart while reducing the number of EphB4-positive venules (Visconti *et al.*, 2002). VEGF-A could induce *ephrinB2* expression in mouse primary embryonic endothelial cells, and VEGF-A derived from sensory neurons, motor neurons, and Schwann cells was required for arterial differentiation of small diameter nerve-associated vessels in mice (Mukouyama *et al.*, 2002, 2005).

2.2. DLL4 is a key regulator of early embryonic vascular development

In mice, DLL4 is the key Notch ligand required for vascular development. Similarly to *Vegfa*+/− heterozygous embryos, *Dll4*+/− heterozygous embryos exhibited embryonic lethal haploinsufficiency due to vascular defects on inbred genetic backgrounds (Duarte *et al.*, 2004; Gale *et al.*, 2004; Krebs *et al.*, 2004). However, some *Dll4*+/− mice were viable on an outbred background, permitting the examination of *Dll4*−/− embryos. The phenotype of the *Dll4*−/− homozygotes was similar, although more severe, than that of the *Dll4*+/− heterozygous embryos (Duarte *et al.*, 2004; Gale *et al.*, 2004). Similar to what was observed in Notch signaling-deficient zebra fish embryos, both *Dll4*-deficient embryos and other types of Notch signaling-deficient mouse embryos such as *Rbpj* mutant and *Hey1*/*Hey2* double mutant embryos did not express arterial markers (Duarte *et al.*, 2004; Fischer *et al.*, 2004; Gale *et al.*, 2004; Kokubo *et al.*, 2005; Krebs *et al.*, 2004). Supporting a direct role for Notch signaling in regulating expression of important arterially expressed genes, Dll4-mediated Notch signaling induced *ephrinB2* expression in cultured endothelial cells (Iso *et al.*, 2006), and the *ephrinB2* gene was demonstrated to be a direct Notch target (Grego-Bessa *et al.*, 2007).

Little is known of the transcriptional regulation of genes that exhibit arterially restricted expression in early embryos. The winged helix/forkhead (Fox) proteins are a large family of evolutionarily conserved transcription factors (Kaestner *et al.*, 2000). Mouse embryos with compound mutations of the *Foxc1* and *Foxc2* genes, two related Fox family transcription factors, exhibited defects in vascular remodeling in the yolk sac and embryo (Kume *et al.*, 2001), accompanied by reduced or absent expression of arterial markers and arteriovenous malformations (Hayashi and Kume, 2008; Seo *et al.*, 2006). The mechanism for this failure of arterial specification was likely through disrupted regulation of *Dll4* transcription (Fig. 9.2). The Foxc1 and Foxc2 proteins directly activate *Dll4* transcription through a Foxc-binding element in the upstream region of the *Dll4* gene (Hayashi and Kume, 2008; Seo *et al.*, 2006). The Foxc1 and Foxc2 proteins also bind directly to two Foxc-binding elements to activate transcription of the Notch target gene *Hey2*, and the Foxc2 protein can form a complex with the RBPJ and NICD proteins (Hayashi and Kume, 2008). In bovine aortic endothelial cells, VEGF-A treatment augmented Foxc-induced activity *Dll4* and *Hey2* luciferase reporter constructs. These results demonstrate that the Foxc proteins are key transcriptional regulators that act upstream of the Notch pathway during arteriovenous differentiation (Hayashi and Kume, 2008; Seo *et al.*, 2006).

2.3. DLL1 is required for both fetal and postnatal arterial development

During embryogenesis in mice, *Dll1* expression in the vasculature is detected at approximately 13 days of gestation in arterial, but not venous, endothelial cells (Sorensen *et al.*, 2009). In *Dll1* loss of function mutant embryos (*Dll1* hypomorphs, or embryos with endothelial cell-specific *Dll1* gene deletion), generation of the Notch1 intracellular domain and expression of arterial markers such as neuropilin1, VEGF receptor 2, and ephrinB2 were lost, despite the fact that DLL4 and JAG1 continued to be expressed in the arterial endothelium. These results established DLL1 as a critical Notch ligand required for maintaining arterial identity of endothelial cells during mouse fetal development and suggested context-dependent cross-regulation of the VEGF-A and Notch signaling pathways (Sorensen *et al.*, 2009).

Dll1 function is also required for arteriogenesis postnatally. Arterial growth is required for restoration of blood flow following ischemia. During neovascularization induced by a mouse hindlimb ischemia model, DLL1 expression was strongly induced in arterial endothelial cells, and neovascularization was impaired in *Dll1+/−* heterozygous mutant mice (Limbourg *et al.*, 2007). Blood flow recovery and postnatal neovascularization in response to hindlimb ischemia were also impaired in *Notch1+/−* heterozygous mice and in mice heterozygous for an endothelial cell-specific deletion of the *Notch1* gene (Takeshita *et al.*, 2007). However, postnatal arteriogenesis and recovery from hindlimb ischemia were normal in *Notch4−/−* mice, indicating that expression of the NOTCH1 protein in endothelial cells was critical for these processes (Takeshita *et al.*, 2007).

2.4. Formation of arteriovenous malformations in Notch pathway mutants

Arteries normally connect to veins only through an intervening capillary bed. An aberrant direct communication between an artery and vein is termed an arteriovenous malformation. One model for the formation of arteriovenous malformations is an inability to establish or maintain distinct arterial and venous vascular beds. Both zebra fish (Lawson *et al.*, 2001) and mouse (Duarte *et al.*, 2004; Gale *et al.*, 2004; Krebs *et al.*, 2004) embryos deficient in Notch signaling formed arteriovenous malformations. Notch pathway gain of function mutations can also cause formation of arteriovenous malformations. Ectopic *Notch4* (Carlson *et al.*, 2005; Kim *et al.*, 2008; Uyttendaele *et al.*, 2001) or *Notch1* (Krebs *et al.*, 2010) activation in endothelial cells as well as conditional *Dll4* overexpression (Trindade *et al.*, 2008) all resulted in formation of arteriovenous malformations and embryonic vascular remodeling defects. Arteriovenous malformations that formed in Notch pathway gain of function mutants were distinct from those exhibited by Notch pathway loss of function mutants (Krebs *et al.*, 2010). Formation of arteriovenous malformations in both Notch pathway gain and loss of function mutants is likely due to an inability of the vascular beds to maintain distinct arterial and venous identities in these mutants.

EphrinB2 and *EphB4* loss of function mutants also formed arteriovenous malformations (Kim *et al.*, 2008; Krebs *et al.*, 2010). *EphrinB2* is a direct Notch target gene whose expression is induced by Notch signal reception (Grego-Bessa *et al.*, 2007). Surprisingly, arteriovenous malformations in the *ephrinB2* and *EphB4* loss of function mutants phenocopied the arteriovenous malformations present in embryos with conditional *Notch1* activation in endothelial cells, rather than the arteriovenous malformations exhibited by Notch pathway loss of function mutants. This result is contrary to the phenotype expected if the ephrinB2/EphB4 pathway was simply acting downstream of the Notch pathway and suggests independent mechanisms for formation of arteriovenous malformations in Notch pathway gain of function mutant embryos and in *ephrinB2* and *EphB4* loss of function mutant embryos. This idea is supported by the finding that *ephrinB2* and *EphB4* loss of function mutant embryos contain venous endothelial cells mislocalized to the aorta, whereas the aortas of embryos with *Notch4* gain of function mutations do not contain these venous endothelial cells (Kim *et al.*, 2008). These data suggest that ephrinB2/EphB4 signaling functions distinctly from Notch signaling, by sorting arterial and venous endothelial cells into their respective vessels.

Inducible expression of an activated *Notch4* transgene in adult mice resulted in vessel arterialization, such as induction of venous expression of ephrinB2 and caused arteriovenous malformations in several organs, including liver, uterus, and skin (Carlson *et al.*, 2005). Surprisingly, these malformations were reversible if activated *Notch4* transgene expression was repressed. These studies demonstrate that the ability of Notch signaling to arterialize blood vessels is not confined to the embryonic period. This inducible *Notch4* transgenic line has been utilized to model arteriovenous shunts and malformations in the lung (Miniati *et al.*, 2010) and brain (Murphy *et al.*, 2008). These mouse models have direct clinical relevance, as recent studies have established that *NOTCH1* signaling is activated in brain arteriovenous malformations in humans (Murphy *et al.*, 2009; ZhuGe *et al.*, 2009).

2.5. Arterial specification of vascular smooth muscle cells

In addition to regulating arterial specification of endothelial cells, Notch signaling also regulates arterial specification of vascular smooth muscle cells. The *Notch3* gene is expressed in vascular smooth muscle cells of arteries, but not veins. *Notch3-/-* mice exhibited marked arterial defects, including enlarged arteries with a thinner vascular smooth muscle cell coat than wild-type arteries (Domenga *et al.*, 2004). These defects arose postnatally as a consequence of defects in arterial vessel maturation. Morphologically, vascular smooth muscle cells in arteries of *Notch3-/-* mice resembled the vascular smooth muscle cells surrounding veins in wild-type mice. Only a few markers are expressed predominantly in arterial vascular smooth muscle cells, rather than venous cells. These include smoothelin (van der Loop *et al.*, 1997) and a transgenic line expressing the β-galactosidase protein from arterial-specific regulatory elements of the SM22α promoter (Moessler *et al.*, 1996). Expression of both these markers was markedly downregulated in arteries of *Notch3-/-* mice. Combined with the morphological data, this indicates that vascular smooth muscle cells surrounding arteries in *Notch3-/-* mice have acquired a venous fate. Notably, arteries in *Notch3-/-* mice, which did not express arterial markers for vascular smooth muscle cells, exhibited normal expression of several endothelial cell arterial markers (Domenga *et al.*, 2004). These results demonstrate that arterial identity of endothelial cells and the vascular smooth muscle cells surrounding them is specified independently.

3. Endothelial Tip Cell Differentiation

3.1. Notch signaling is a key regulator of endothelial tip cell formation and function

During angiogenesis, new capillaries sprout from previously existing blood vessels. Tip cells are specialized endothelial cells situated at the tips of vascular sprouts (Fig. 9.3). These cells extend filopodia that sense the local extracellular environment and guide growth of these sprouts along VEGF-A gradients (Gerhardt *et al.*, 2003, 2004). The Notch pathway has a primary role in regulating formation and function of endothelial tip cells. Such a role was initially described by Hughes and colleagues (Sainson *et al.*, 2005). In an *in vitro* angiogenesis culture system utilizing human umbilical vein endothelial cells (HUVECs), Notch signaling suppressed branching at the tip of developing angiogenic sprouts. Suppression of Notch signaling led to tip cell division, with both daughter cells being specified as tip cells. This led to increased branching through vessel bifurcation.

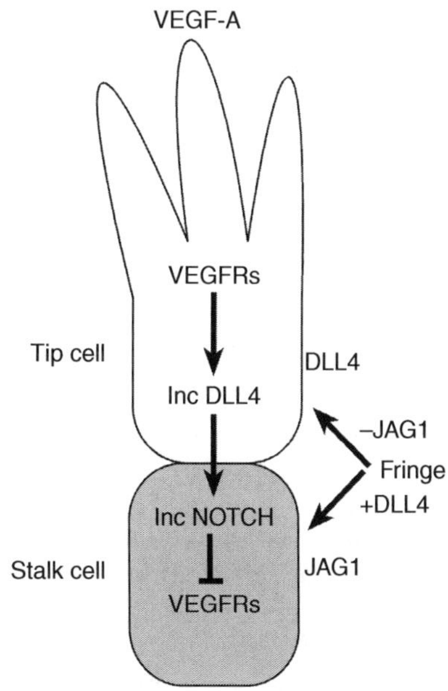

Figure 9.3 Antagonistic roles of the DLL4 and JAG1 proteins in tip cell selection during sprouting angiogenesis. An endothelial tip cell (white) at the leading edge of a vascular sprout extends filopodia and migrates toward a VEGF-A gradient. Signal reception by VEGF receptors (VEGFRs) in the tip cell leads to upregulation of DLL4 expression (Inc DLL4). The tip cell signals via DLL4 to the adjacent Notch receptor-expressing cell, downregulating VEGF receptor expression. This suppresses the tip cell fate and promotes differentiation of the adjacent endothelial cell as a stalk cell (gray). Glycosylation of Notch receptors (likely NOTCH1) by Fringe proteins enhances DLL4-Notch signaling and suppresses JAG1-Notch signaling. The DLL4 and JAG1 ligands act in an antagonistic fashion during endothelial tip cell selection and angiogenic sprouting. DLL4, expressed more highly on tip cells, suppresses adoption of the tip cell fate and angiogenic sprouting. JAG1, which is expressed more highly on stalk cells, promotes tip cell differentiation and sprouting behavior by antagonizing DLL4-Notch signaling between stalk cells.

More recent work has confirmed and extended our understanding of the role that Notch signaling plays in tip cell formation. Dll4/Notch signaling regulates tip cell numbers, filopodia extension, and branching of angiogenic sprouts in several model systems in addition to HUVECs: the mouse retina and hindbrain (Hellstrom *et al.*, 2007; Lobov *et al.*, 2007; Ridgway *et al.*, 2006; Suchting *et al.*, 2007), the zebra fish embryo (Leslie *et al.*, 2007; Siekmann and Lawson, 2007), and xenograft tumor models (Noguera-Troise *et al.*, 2006; Ridgway *et al.*, 2006; Scehnet *et al.*, 2007). In all of these studies, Notch signal inhibition led to increased sprouting and branching of blood vessels. The Notch pathway regulates sprouting and branching behaviors by influencing the differentiation, migration, and proliferation of vascular tip cells. The Notch ligand DLL4 is most highly expressed in endothelial tip cells, where it signals to adjacent Notch receptor-expressing endothelial cells, causing them to adopt the stalk cell fate. Reduced Notch signaling leads to increases in tip cell numbers, filopodia extension, and vessel branching. Suppression of tip cell formation and angiogenic sprouting by Notch signaling was downstream of the VEGF-A signal, since pharmacological or genetic manipulations that blocked VEGF-A function reduced both Dll4 expression and blood vessel sprouting.

Several studies have assessed the effects of modulating Notch signaling on differentiation of the postnatal retinal vasculature in mice (Hellstrom *et al.*, 2007; Lobov *et al.*, 2007; Ridgway *et al.*, 2006; Suchting *et al.*, 2007). The mouse retina possesses distinct advantages for analysis of developmental angiogenesis (Dorrell and Friedlander, 2006; Gariano and Gardner, 2005; Uemura *et al.*, 2006). Development of the vascular system of the mouse retina occurs postnatally in a highly reproducible spatial and temporal pattern. The retinal vascular system emerges first in the region of the optic nerve head and then grows radially toward the periphery. The primitive vascular plexus that forms initially is remodeled into large and small arterial and venous vessels. During these stages, the retinal vasculature is accessible both for observation and for experimental administration of exogenous agents.

The *Dll4* gene is highly expressed in the developing retinal vasculature. Reduced DLL4/Notch signaling leads to striking defects in the early postnatal retinal vasculature. The observed defects are concordant whether DLL4/Notch signaling is reduced genetically, by assessing *Dll4+/−* heterozygous mice (Hellstrom *et al.*, 2007; Lobov *et al.*, 2007; Suchting *et al.*, 2007) or mice with temporally regulated *Notch1* deletion in the retinal vasculature (Hellstrom *et al.*, 2007), or by administering anti-DLL4 blocking reagents (Lobov *et al.*, 2007; Ridgway *et al.*, 2006) or γ-secretase inhibitors (Hellstrom *et al.*, 2007; Suchting *et al.*, 2007). The retinal vasculature in these mice exhibited severe patterning defects. The vascular plexus had increased capillary density and diameter, with increased filopodial extensions both at the growing vascular front and in the interior of the plexus. Portions of the vascular plexus fused to form syncytial sinuses. Markers specific for tip cells, such as *pdgfb* and *unc5b*, were also upregulated in mice with reduced DLL4/Notch signaling. These data indicate that DLL4/Notch signaling restricts acquisition of endothelial tip cell fate in angiogenic sprouts, causing adjacent Notch receptor-expressing endothelial cells to acquire the stalk cell fate.

Mathematical and computational modeling of angiogenesis has a long history, and many types of computational approaches have been utilized to model different aspects of angiogenesis (Peirce, 2008). Recently, agent-based modeling of angiogenesis has been utilized to more accurately model endothelial tip cell selection and capillary sprouting and the roles of VEGF-A and DLL4/Notch signaling in these processes (Bentley *et al.*,

2008, 2009; Qutub and Popel, 2009). Agent-based modeling, utilized initially in fields such as ecology and the social sciences, has been employed to study a variety of multicellular morphogenic processes occurring during embryonic development (Thorne *et al.*, 2007). The new models (Bentley *et al.*, 2008, 2009; Qutub and Popel, 2009) specifically incorporate features such as VEGF-A concentration gradients and DLL4/Notch signaling during endothelial tip cell selection and are leading to novel insights amenable to experimental verification.

3.2. Antagonistic roles of the Notch ligands DLL4 and JAG1 during tip cell formation and function

Recent studies of the postnatal retinal vasculature have demonstrated that the Notch ligands DLL4 and JAG1 play antagonistic roles during tip cell selection and sprouting angiogenesis (Benedito *et al.*, 2009). Contrary to the phenotypes observed by modulation of DLL4-mediated Notch signaling, JAG1 loss of function mutants in the retina reduced sprouting angiogenesis, while JAG1 overexpression enhanced angiogenesis and resulted in increased number of tip cells. JAG1 expression therefore acted as a proangiogenic signal during postnatal retinal angiogenesis. JAG1 expression antagonized DLL4-mediated Notch signaling and required expression of Fringe family glycosyltransferases to do so. Glycosylation of Notch family receptors by the three mammalian Fringe proteins (Lfng, Lunatic fringe; Mfng, Manic fringe; Rfng, Radical fringe) potentiates Notch signal transmission by DLL family ligands but suppresses signaling by JAG family ligands (Fig. 9.3). During retinal angiogenesis, Fringe-mediated glycosylation of Notch receptors enhanced DLL4-Notch signaling, while simultaneously diminishing JAG1 signaling ability (Benedito *et al.*, 2009). This work demonstrated that the equilibrium between two Notch ligands with distinct spatial expression patterns and opposing functional roles regulates postnatal angiogenesis in the retina.

3.3. Tip cell function in zebra fish

Additional insights into endothelial tip cell formation, migration, and behavior have been obtained from analysis of zebra fish embryos (Leslie *et al.*, 2007; Siekmann and Lawson, 2007). The optical clarity of these embryos and the availability of transgenic lines expressing fluorescent proteins in endothelial cells makes this model system ideal for high-resolution fluorescent microscopy, including time-lapse confocal microscopy on living embryos. Mosaic analysis revealed that transplanted cells lacking Rbpj did not contribute to the dorsal aorta but were preferentially located in the posterior cardinal vein or the most dorsal position of the segmental arteries.

Conversely, transplanted cells expressing activated Notch1 preferentially located in the dorsal aorta or the base of developing sprouts (Siekmann and Lawson, 2007). These results indicate that Notch signaling is required cell-autonomously for determination of endothelial cell fate in segmental artery sprouts. Utilization of time-lapse confocal microscopy on living embryos revealed that in both *Dll4* morphant and *Rbpj* morphant embryos, segmental artery sprouts contained more cells than controls. These additional cells were incorporated via both increased migration of endothelial cells into the initial sprout and proliferation of normally quiescent stalk cells. Interestingly, vascular defects in *Rbpj* morphant embryos were more severe than those in *Dll4* morphant embryos, suggesting that additional Notch ligands play important roles during early vascular development (Leslie *et al.*, 2007; Siekmann and Lawson, 2007). Blocking Vegf signaling with a small molecule inhibitor blocked both normal endothelial sprouting, as well as the ectopic sprouting observed in *Dll4* morphant embryos (Leslie *et al.*, 2007). In addition, reducing levels of Vegf receptor 3 in *Rbpj* morphant embryos partially rescued the *Rbpj* knockdown phenotype, suggesting that Notch activation might normally repress Vegf receptor 3 to limit angiogenic cell behavior in developing segmental artery sprouts (Siekmann and Lawson, 2007). Taken together, the studies in both the mouse retina and the zebra fish embryo indicate that Notch signaling acts as a negative regulator of VEGF-A-induced angiogenesis and is essential for proper vascular morphogenesis.

4. Tumor Angiogenesis

4.1. Use of DLL4 blocking reagents to disrupt tumor angiogenesis

The maintenance, growth, and metastasis of solid tumors require the recruitment of host blood vessels into the tumor. Many solid tumors express VEGF-A, and therapies utilizing anti-VEGF-A antibodies or other blocking reagents are effective in inhibiting solid tumor growth in preclinical rodent models (Ferrara and Kerbel, 2005; Jain *et al.*, 2006). Given the prominent role of the Notch pathway in regulating vascular development, components of the Notch pathway may provide novel drug targets during tumor angiogenesis (Dufraine *et al.*, 2008; Li and Harris, 2009). The Notch ligand DLL4 is expressed at high levels in tumor vasculature (Gale *et al.*, 2004; Hainaud *et al.*, 2006; Mailhos *et al.*, 2001; Patel *et al.*, 2005), and recent studies have identified the DLL4 protein as a potential drug target (Noguera-Troise *et al.*, 2006; Ridgway *et al.*, 2006; Scehnet *et al.*, 2007). Systemic administration of neutralizing anti-DLL4 antibodies (Noguera-Troise *et al.*, 2006; Ridgway *et al.*, 2006) and systemic

(Noguera-Troise *et al.*, 2006) or localized (Scehnet *et al.*, 2007) administration of recombinant forms of the DLL4 protein that had been modified to block DLL4/Notch signaling inhibited growth of several different solid tumors in mice. Similar to the findings in zebra fish embryos and mouse retinas, anti-DLL4 treatment (also termed DLL4 blockade) increased blood vessel sprouting and branching and led to a marked increase in tumor blood vessel density in the treated tumors. Paradoxically, tumor growth was inhibited despite the increased blood vessel density. Testing of the vascular network in the anti-DLL4-treated tumors by perfusion assays with fluorescent lectins or assessment of hypoxic regions in the tumors revealed that the newly induced vessels functioned inefficiently. Many of these vessels were not connected to the vascular network in the tumors, leading to poor perfusion, increased hypoxia, and an overall inhibition of tumor growth. This mechanism of inhibition of tumor growth has been termed nonproductive angiogenesis (Dufraine *et al.*, 2008; Sainson and Harris, 2007; Thurston *et al.*, 2007; Yan and Plowman, 2007). One complication of anti-VEGF-A therapies is acquired resistance of tumors that initially respond to the anti-VEGF-A treatments (Azam *et al.*, 2010). Importantly, anti-DLL4 therapies were effective against tumors that were resistant to anti-VEGF-A treatments and could provide synergistic effects against certain tumors when combined with anti-VEGF-A therapies (Noguera-Troise *et al.*, 2006; Ridgway *et al.*, 2006).

An important new study has dissected mechanisms involved in inhibition of tumor growth by DLL4 blockade in xenograft mouse models (Hoey *et al.*, 2009). Specific and selective anti-human DLL4 and anti-mouse DLL4 antibodies were generated, enabling the selective targeting of DLL4 expression in the tumor, in the host vasculature and stromal cells, or both. These studies demonstrated that blocking DLL4 signaling inhibited tumor growth through multiple mechanisms. Administration of anti-mouse DLL4 antibodies inhibited tumor growth in a similar fashion to the studies described above (through nonproductive angiogenesis resulting in increased density of poorly perfusing vessels). However, administration of anti-human DLL4 antibodies also inhibited tumor growth, but by different mechanisms. Anti-human DLL4 administration inhibited tumor growth and the expression of Notch pathway target genes and reduced proliferation of tumor cells. Significantly, anti-human DLL4 administration reduced the frequency of tumor-initiating cancer stem cells. Combined anti-human and anti-mouse DLL4 antibody treatment had an additive effect on inhibiting tumor growth, as did combined treatment with the chemotherapeutic agent irinotecan (Hoey *et al.*, 2009).

4.2. Safety concerns regarding anti-DLL4 therapies

While the studies described above are quite promising, a number of issues remain before anti-DLL4 therapies reach the clinic. For example, despite the efficacy of anti-VEGF-A therapies in treatment of xenograft tumor models in rodents, in clinical trials anti-VEGF-A antibody treatment of several types of cancer only provided an overall survival benefit for patients when it was combined with conventional chemotherapy treatment (Ferrara and Kerbel, 2005; Jain *et al.*, 2006). Similar issues may arise as anti-DLL4 treatments progress from preclinical models into clinical trials. Phase 1 clinical trials of the use of two different anti-DLL4 human monoclonal antibodies in treatment of solid tumors are currently in progress (A multiple-ascending-dose study of the safety and tolerability of REGN421 in patients with advanced solid malignancies, http://clinicaltrials.gov/ct2/show/NCT00871559; A Phase 1 dose escalation study of OMP-21M18 in subjects with solid tumors, http://clinicaltrials.gov/ct2/show/NCT00744562).

A potentially much more serious problem was revealed in a recent study that demonstrated chronic DLL4 blockade caused pathological activation of endothelial cells, resulted in histopathological changes in several organs, and induced the formation of vascular neoplasms (Yan *et al.*, 2010). Affected organs included liver and thymus, and deleterious effects of chronic DLL4 blockade were observed in mouse, rat, and cynomolgus monkey models. Chronic DLL4 blockade in male rats resulted in a dose-dependent increase in ulcerating subcutaneous tumors that exhibited histopathological features characteristic of vascular neoplasms, although the tumors did not appear to be malignant. These studies raise concerns about the safety of chronic DLL4 blockade and suggest that refined strategies and treatment regimens may be required to safely utilize anti-DLL4 reagents. Despite these issues, anti-DLL4 treatments remain a novel therapeutic approach for cancer treatment and retain much promise, particularly for treatment of tumors unresponsive to anti-VEGF-A therapies.

5. Notch Signaling and Vascular Smooth Muscle Cells

5.1. Notch signaling is a key regulator of vascular smooth muscle cell differentiation and response to vascular injury

Notch signaling plays an important role in the differentiation, physiology, and function of vascular smooth muscle cells (Morrow *et al.*, 2008). However, contradictory results suggest that its role may be context, temporally, or cell line dependent. Several groups have published studies indicating that Notch signaling represses smooth muscle cell differentiation during *in vitro* culture (Doi *et al.*, 2005; Morrow *et al.*, 2005a;

Proweller *et al.*, 2005) and that this repressive effect is likely mediated through induction of the Hey2 protein (a demonstrated Notch target gene). More recent studies, however, have indicated that Notch signaling induces smooth muscle cell differentiation (Doi *et al.*, 2006; High *et al.*, 2007). Jag1-mediated Notch signaling promoted smooth muscle cell differentiation in both human aortic smooth muscle cells and a murine embryonic fibroblast cell line (Doi *et al.*, 2006). Both smooth muscle myosin heavy chain (Doi *et al.*, 2006) and smooth muscle α-actin (Noseda *et al.*, 2006; Tang *et al.*, 2008) have been demonstrated to be direct Notch target genes. *In vivo* studies in which Notch signaling was inactivated specifically in mouse neural crest cells demonstrated that Notch signaling plays an essential role in differentiation of cardiac neural crest cells into smooth muscle cells (High *et al.*, 2007).

Analysis of an *in vitro* angiogenesis model involving coculture of human vascular endothelial cells and mural cells (progenitors for vascular smooth muscle cells) revealed that expression of the *NOTCH3* gene was strongly induced in mural cells by coculture (Lilly and Kennard, 2009; Liu *et al.*, 2009). Knockdown by small interfering RNA revealed that *NOTCH3* expression was required for endothelial-dependent mural cell differentiation, whereas *NOTCH3* overexpression promoted smooth muscle gene expression. NOTCH3 promoted its own expression, as well as that of the ligand JAG1 in mural cells. These findings suggested that NOTCH3 has the capacity to establish and maintain a differentiated phenotype in vascular mural cells through a positive-feedback loop that includes both NOTCH3 autoregulation and induction of *JAG1* expression (Liu *et al.*, 2009). Such a model is consistent with the finding that endothelial cell-specific *Jag1* gene deletion in mice leads to a deficiency of vascular smooth muscle cell recruitment and differentiation, causing hemorrhages and early embryonic lethality (High *et al.*, 2008).

Several studies have characterized expression of Notch pathway genes during the response to vascular injury (Campos *et al.*, 2002; Doi *et al.*, 2005; Lindner *et al.*, 2001; Wang *et al.*, 2002). After vascular injury such as carotid artery ligation, smooth muscle cells in the vascular wall of the injured area proliferate and form a thickened layer of smooth muscle cells termed the neointima. Expression of several Notch pathway components, including Notch1, Notch3, Jag1, Jag2, Hey1, and Hey2, is modulated after experimentally induced vascular injury. Expression of these genes is downregulated within the first 2 days following vascular injury but is upregulated compared to uninjured contralateral control vessels by 7–14 days after injury. Supporting a functional role for the modulation of Notch pathway components during the response to vascular injury, neointima formation after vascular injury was significantly decreased in *Hey2−/−* mice (Sakata *et al.*, 2004). Culture of primary aortic vascular smooth muscle cells from *Hey2−/−* mice revealed that the mutant cells exhibited a reduced proliferation rate compared to wild-type cells. Overexpression of Hey1 (Wang *et al.*, 2003) or Hey2 (Havrda *et al.*, 2006) in vascular smooth muscle cells led to increased vascular smooth muscle cell proliferation associated with reduced levels of the cyclin-dependent kinase inhibitors $p21^{waf1/cip1}$ (Wang *et al.*, 2003) or $p27^{kip1}$ (Havrda *et al.*, 2006). The Hey2 protein interacted directly with the $p27^{kip1}$ promoter to repress transcription (Havrda *et al.*, 2006). The Hey1 and Hey2 proteins suppress NICD/RBPJ binding to the smooth muscle α-actin promoter, providing further evidence that temporally regulated induction of Hey1/Hey2 expression by Notch signaling may constitute a negative feedback mechanism involved in the regulation of vascular smooth muscle cell differentiation (Tang *et al.*, 2008).

Neointima formation after carotid artery ligation also was decreased in heterozygous *Notch1+/−* mice, as well as in mice heterozygous for a conditional deletion of the *Notch1* gene in smooth muscle cells (Li *et al.*, 2009). Smooth muscle cells explanted from the aortas of these mice exhibited decreased migration and proliferation and increased apoptosis compared to control littermate mice. Surprisingly, neointima formation after carotid artery ligation in *Notch3−/−* mice was unaffected. These data indicate that *Notch1* function in smooth muscle cells of the vascular wall, rather than *Notch3* function, mediates smooth muscle cell proliferation and neointima formation subsequent to vascular injury.

5.2. Notch signaling and mechanical stress

Mechanical forces are one of a number of factors implicated in regulating vascular smooth muscle cell differentiation and physiology. Adult vascular smooth muscle cells are not terminally differentiated and can exhibit substantial plasticity in their phenotype in response to changes in the local environment (Owens *et al.*, 2004). Exposure of primary human or rat vascular smooth muscle cells to cyclic strain during *in vitro* culture led to significant reductions in Notch1 and Notch3 receptor expression, concomitant with an increase in expression of vascular smooth muscle cell differentiation markers (Morrow *et al.*, 2005a, b). Strain-exposed vascular smooth muscle cells also exhibited reduced proliferation and increased apoptosis. These changes could be reversed by overexpression of the Notch1 or Notch3 intracellular domains. These results indicate that cyclic strain inhibits vascular smooth muscle cell growth while increasing apoptosis and that these effects are mediated at least in part by modulation of Notch signaling. Interestingly, cyclic strain led to an upregulation of Notch receptor expression in endothelial cells (Morrow *et al.*, 2007). Cyclic strain also caused increased endothelial cell vascular network formation in Matrigel, indicating that effects of mechanical forces on Notch signaling in the vasculature are not restricted to vascular smooth muscle cells.

6. Medical Consequences of Aberrant NOCH3 Signaling in Vascular Smooth Muscle Cells: CADASIL

6.1. Cerebral Autosomal Dominant Arteriopathy with Subcortical Infarcts and Leukoencephalopathy

Mutations in the *NOTCH3* gene cause an inherited degenerative vascular disease that affects vascular smooth muscle cells. Cerebral autosomal dominant arteriopathy with subcortical infarcts and leukoencephalopathy (CADASIL) is the most common genetic form of stroke and vascular dementia (Chabriat *et al.*, 2009; Kalaria *et al.*, 2004). Affected individuals exhibit a variety of symptoms, including migraines, mood disorders, recurrent subcortical ischemic strokes, progressive cognitive decline, dementia, and premature death. CADASIL is characterized by progressive degeneration of vascular smooth muscle cells with accumulation of granular osmiophilic material (GOM) within the smooth muscle cell basement membrane (Chabriat *et al.*, 2009; Kalaria *et al.*, 2004). GOM accumulation in vascular smooth muscle cells is one of the hallmark features of this disease.

CADASIL is caused by mutations in the *NOTCH3* gene (Joutel *et al.*, 1996). All mutations associated with CADASIL result in a gain or loss of a cysteine residue in one of the 34 EGF-like repeats in the extracellular domain of the NOTCH3 protein. The characteristic nature of these mutations, in addition to the absence of any examples of obviously inactivating mutations or deletions of the *NOTCH3* gene of CADASIL patients, strongly suggests that mutations causing CADASIL do not create *NOTCH3*-null alleles. In CADASIL patients, the ectodomain of the NOTCH3 protein accumulates in the cerebral microvasculature (Joutel *et al.*, 2000). The NOTCH3 ectodomain accumulates at the cytoplasmic membrane of vascular smooth muscle cells, although it is controversial whether the NOTCH3 ectodomain is a constituent of the GOM deposits (Ishiko *et al.*, 2006; Joutel *et al.*, 2000).

6.2. Mouse models of CADASIL

Mice homozygous for a targeted null mutation of the *Notch3* gene are viable and fertile (Krebs *et al.*, 2003). *Notch3−/−* mice exhibited marked arterial defects, including enlarged arteries with a thinner vascular smooth muscle cell coat than wild-type arteries (Domenga *et al.*, 2004) (Table 9.1). These defects arose postnatally as a consequence of defects in arterial vessel maturation. However, *Notch3−/−* mice did not exhibit age-dependent

Table 9.1 Phenotypes of *Notch3* mutant and CADASIL mouse models

***Notch3* mutation**	**Allele type**	**Species of *Notch3* coding sequences**	**Phenotypes**	**References**
Null	Knockout	NA	Vascular smooth muscle cells surrounding arteries acquire venous fate; decreased myogenic tone in tail, and cerebral resistance arteries; no CADASIL-related brain pathology	Domenga *et al.* (2004) and Belin de Chantemele *et al.* (2008)
Null	Gene trap	NA	Enhanced susceptibility in middle cerebral artery ischemia model	Arboleda-Velasquez *et al.* (2008)
Arg142Cys	Knockin	Mouse	No phenotype or CADASIL-related brain pathology in mice heterozygous or homozygous for the knockin allele	Lundkvist *et al.* (2005)
Arg90Cys	Transgenic	Human	Age-dependent accumulation of NOTCH3 ectodomain and of GOM deposits in vascular smooth muscle cells; impaired cerebral vasoreactivity; impaired vascular mechanotransduction	Ruchoux *et al.* (2003), Lacombe *et al.* (2005), and Dubroca *et al.* (2005)

Cys428Ser	Transgenic	Human	Age-dependent accumulation of NOTCH3 ectodomain and of GOM deposits in vascular smooth muscle cells; mild dominant-negative activity	Monet-Lepretre *et al.* (2009)
Arg169Cys	PAC transgenic	Rat	NOTCH3 extracellular domain aggregates and GOM deposits in brain vessels; progressive degeneration of the white matter; reduced cerebral blood flow; most accurate CADASIL model reported to date	Joutel *et al.* (2010)

NA, not applicable; PAC, P1-derived artificial chromosome; CADASIL, cerebral autosomal dominant arteriopathy with subcortical infarcts and leukoencephalopathy; GOM, granular osmiophilic material.

accumulation of the NOTCH3 ectodomain and GOM deposits in vascular smooth muscle cells, or any CADASIL-related brain pathology. The *Notch3−/−* mice did, however, exhibit a reduction in pressure-induced myogenic tone that was associated with a higher flow-mediated dilation of tail and cerebral resistance arteries (Belin de Chantemele *et al.*, 2008). Another group analyzed an independently generated *Notch3* gene trap-null allele (Mitchell *et al.*, 2001). Mice homozygous for this *Notch3*-null allele exhibited enhanced susceptibility in a proximal middle cerebral artery ischemia model (Arboleda-Velasquez *et al.*, 2008). *Notch3−/−* mice developed ischemic lesions approximately twice as large as those observed in heterozygous or wild-type littermates and developed a 60% larger area of severe cerebral blood flow deficit than wild-type mice. The results of these studies indicate that while *Notch3*-null mice exhibit altered vascular maturation and function in at least some arteries, they do not develop CADASIL-related pathologies.

Several mouse models expressing NOTCH3 proteins with CADASIL mutations have been developed (Table 9.1). An Arg142Cys knock-in mutation was introduced into the endogenous mouse *Notch3* gene (Lundkvist *et al.*, 2005). These mice did not exhibit any CADASIL-like morphological or behavioral phenotypes, even when homozygous for the *Notch3* Arg142Cys mutant allele. Another model more successfully recapitulated the early, preclinical phase of CADASIL. In this model, transgenic mice were generated that expressed the human NOTCH3 cDNA containing a different CADASIL mutation, the Arg90Cys mutation, in vascular smooth muscle cells (Ruchoux *et al.*, 2003). These mice demonstrated age-dependent accumulation of the NOTCH3 ectodomain and of GOM deposits in vascular smooth muscle cells of both cerebral and peripheral arterioles. However, despite GOM accumulation, these mice did not exhibit any evidence of damage to the brain parenchyma. Physiological studies of these NOTCH3 Arg90Cys transgenic mice revealed impaired cerebral vasoreactivity that suggested either decreased relaxation or increased resistance of cerebral blood vessels (Lacombe *et al.*, 2005) and increased pressure-induced contraction and decreased flow-induced dilation of tail arteries (Dubroca *et al.*, 2005). Transgenic mice expressing either the wild-type NOTCH3 protein or the NOTCH3 Arg90Cys mutation were equally effective in rescuing arterial defects of *Notch3−/−* mice, and the mutant NOTCH3 Arg90Cys protein exhibited normal activity in regulating *in vivo* expression of a Notch signaling reporter (Monet *et al.*, 2007). These data suggested a novel pathogenic role for the NOTCH3 Arg90Cys protein, rather than compromised NOTCH3 signaling activity, as the primary defect leading to the CADASIL phenotype.

However, it was not clear whether all CADASIL mutations retained normal NOTCH3 signaling activity, or whether genotype–phenotype correlations were observable in CADASIL patients with different NOTCH3 mutations. In particular, CADASIL mutations in the ligand-binding domain, located in EGF-like repeats 10 and 11 of the NOTCH3 protein, abrogated Notch signal transduction in cell-based assays (Joutel *et al.*, 2004; Peters *et al.*, 2004). To assess the characteristics of NOTCH3 mutations in the ligand-binding domain *in vivo*, transgenic mice expressing a human NOTCH3 protein with the Cys428Ser CADASIL mutation were constructed and analyzed (Monet-Lepretre *et al.*, 2009). The NOTCH3 Cys428Ser mice, like those with the more common NOTCH3 Arg90Cys mutation, developed characteristic arterial accumulation of the NOTCH3 extracellular domain and GOM deposits upon aging. However, introducing the mutant Cys428Ser NOTCH3 transgene into a *Notch3−/−*-null background revealed that, in contrast to the Arg90Cys mutant protein, the Cys428Ser mutant protein had lost wild-type NOTCH3 activity and exhibited mild dominant-negative activity. This study also revealed genotype–phenotype correlations in patients with EGF repeat 10–11 mutations. From a prospectively recruited cohort of 176 CADASIL patients, 10 patients from 5 distinct pedigrees were identified that carried a NOTCH3 mutation in EGF repeat 10 or 11 (including 6 patients with the Cys428Ser mutation). These patients had higher cognitive function than patients with mutations in EGF repeats 2–5. The patients with EGF repeat 10–11 mutations also had a distinctive presentation on magnetic resonance imaging analyses. These results revealed distinctive functional and phenotypic features of EGF repeat 10–11 mutations, relative to the common CADASIL mutations (Monet-Lepretre *et al.*, 2009).

In the most recently described mouse CADASIL model (Joutel *et al.*, 2010), bacterial recombineering was used to introduce the Arg169Cys CADASIL mutation into *Notch3* coding sequences of a rat P1-derived artificial chromosome (PAC) clone. Transgenic mice containing the wild-type rat *Notch3* PAC construct and two independent lines containing the Arg169Cys CADASIL mutation were analyzed. Expression of high levels of the mutant Arg169Cys NOTCH3 transgene reproduced the endogenous NOTCH3 expression pattern and the main pathological features of CADASIL, including NOTCH3 extracellular domain aggregates and GOM deposits in brain vessels, progressive degeneration of the white matter, and reduced cerebral blood flow. The mutant mice exhibited a number of neuropathological changes that occurred in the absence of either histologically detectable alterations in cerebral artery structure or blood–brain barrier breakdown. These findings provide evidence for cerebrovascular dysfunction and microcirculatory failure as the earliest detectable consequences of the expression of the Arg169Cys NOTCH3 mutant protein. The continued development and improvement of these mouse CADASIL models are leading to valuable new insights into the onset and progression of CADASIL.

6.3. Mechanistic insights into the pathogenesis of CADASIL mutations

Recent work has yielded insights into the cellular and molecular mechanisms involved in processing and clearance of the wild-type NOTCH3 protein and NOTCH3 proteins containing CADASIL mutations. Human embryonic kidney (HEK) 293 cell lines with inducible expression of either the wild-type NOTCH3 protein or the NOTCH3 proteins containing Arg133Cys and Cys185Arg mutations were analyzed (Takahashi *et al.*, 2010). Both NOTCH3 mutant proteins were prone to aggregation and were retained within the endoplasmic reticulum. The turnover rates of the NOTCH3 proteins containing CADASIL mutations were strikingly slow, with half lives greater than 6 days, whereas the wild-type NOTCH3 protein was rapidly degraded, with a half-life of 0.7 days. Expression of the mutant NOTCH3 proteins also impaired cell proliferation compared with expression of wild-type NOTCH3. Cell lines expressing mutant NOTCH3 proteins also were more sensitive than cell lines expressing wild-type NOTCH3 to proteasome inhibition, resulting in cell death. These findings suggest that prolonged retention of mutant NOTCH3 aggregates in the endoplasmic reticulum can decrease cell growth and increase sensitivity to other cellular stresses. The finding that cell lines expressing mutant NOTCH3 proteins exhibited increased sensitivity to proteasome inhibitors (Takahashi *et al.*, 2010) stands in contrast to another recent study, which concluded that the NOTCH3 protein in transfected HEK293 cells was degraded primarily in lysosomes (Jia *et al.*, 2009).

Another study demonstrated that the extracellular domains of both wild-type NOTCH3 protein and NOTCH3 proteins containing CADASIL mutations spontaneously formed oligomers and higher order multimers *in vitro* and that multimerization was mediated by disulfide bonds (Opherk *et al.*, 2009). Three CADASIL mutant proteins (Arg133Cys, Cys183Arg, and Cys455Arg) were tested in these studies, with concordant results. Using single-molecule analysis techniques, it was shown that CADASIL-associated mutations significantly enhanced multimerization compared to the wild-type NOTCH3 extracellular domain. These results provide experimental evidence for spontaneous NOTCH3 self-association and are consistent with a neomorphic effect of CADASIL mutations in disease pathogenesis.

One caveat to the studies described in this section is that they all utilized HEK293 cells, rather than vascular smooth muscle cells. One of the studies reported that expression of either wild-type or mutant NOTCH3 proteins in cultured human aortic smooth muscle cells resulted in cell death as early as 2 days after transfection (Takahashi *et al.*, 2010). Further studies will be required to assess whether these mechanisms also operate in more physiologically relevant situations, such as the *in vivo* mouse models.

7. Perspectives and Conclusions

I last reviewed the field of Notch signaling in vascular biology 3 years ago (Gridley, 2007), and in the intervening period great advances have been made. We know substantially more about the roles and requirements for Notch ligands during both embryonic and postnatal arterial development and during endothelial tip cell selection. Great progress has been made in developing mouse models that have important clinical implications (e.g., models for lung and brain arteriovenous malformations, tumor angiogenesis, and CADASIL models that more accurately reproduce the pathology of the human disease) and in developing computational models that accurately predict novel cellular behaviors that can then be tested experimentally.

However, much remains to be learned. We need to learn much more of the role of Notch signaling during vascular physiology in adults. Areas in which there will continue to be advances in upcoming years include the development and evaluation of anti-DLL4 therapies (as well as other types of "anti-Notch" therapies) for tumor angiogenesis and other vascular diseases, such as macular degeneration. We will also see the continued development and characterization of improved animal models for CADASIL and the utilization of these models to gain insights into the poorly understood mechanisms underlying the pathogenesis of these NOTCH3 mutations. Another area that will see significant advances will be determining the mechanisms for cross talk between the Notch pathway and other signaling pathways such as the TGFβ, Wnt, ephrin/Eph receptor and PI3K/Akt pathways, and the developmental and physiological decisions in which such cross talk is operative.

Acknowledgments

The work in my laboratory on these topics has been supported by the National Institutes of Health.

References

Arboleda-Velasquez, J. F., Zhou, Z., Shin, H. K., Louvi, A., Kim, H. H., Savitz, S. I., Liao, J. K., Salomone, S., Ayata, C., Moskowitz, M. A., et al. (2008). Linking Notch signaling to ischemic stroke. *Proc. Natl. Acad. Sci. U.S.A.* **105**, 4856–4861.

Azam, F., Mehta, S., and Harris, A. L. (2010). Mechanisms of resistance to antiangiogenesis therapy. *Eur. J. Cancer.* **46**, 1323–1332.

Barsi, J. C., Rajendra, R., Wu, J. I., and Artzt, K. (2005). Mind bomb1 is a ubiquitin ligase essential for mouse embryonic development and Notch signaling. *Mech. Dev.* **122**, 1106–1117.

Belin de Chantemele, E. J., Retailleau, K., Pinaud, F., Vessieres, E., Bocquet, A., Guihot, A. L., Lemaire, B., Domenga, V., Baufreton, C., Loufrani, L., et al. (2008). Notch3 is a major regulator of vascular tone in cerebral and tail resistance arteries. *Arterioscler. Thromb. Vasc. Biol.* **28**, 2216–2224.

Benedito, R., Roca, C., Sorensen, I., Adams, S., Gossler, A., Fruttiger, M., and Adams, R. H. (2009). The notch ligands Dll4 and Jagged1 have opposing effects on angiogenesis. *Cell* **137**, 1124–1135.

Bentley, K., Gerhardt, H., and Bates, P. A. (2008). Agent-based simulation of notch-mediated tip cell selection in angiogenic sprout initialisation. *J. Theor. Biol.* **250**, 25–36.

Bentley, K., Mariggi, G., Gerhardt, H., and Bates, P. A. (2009). Tipping the balance: robustness of tip cell selection, migration and fusion in angiogenesis. *PLoS Comput. Biol.* **5**, e1000549.

Borggrefe, T., and Oswald, F. (2009). The Notch signaling pathway: transcriptional regulation at Notch target genes. *Cell Mol. Life Sci.* **66**, 1631–1646.

Bray, S. J. (2006). Notch signalling: a simple pathway becomes complex. *Nat. Rev. Mol. Cell Biol.* **7**, 678–689.

Byrne, A. M., Bouchier-Hayes, D. J., and Harmey, J. H. (2005). Angiogenic and cell survival functions of vascular endothelial growth factor (VEGF). *J. Cell Mol. Med.* **9**, 777–794.

Campos, A. H., Wang, W., Pollman, M. J., and Gibbons, G. H. (2002). Determinants of Notch-3 receptor expression and signaling in vascular smooth muscle cells: implications in cell-cycle regulation. *Circ. Res.* **91**, 999–1006.

Carlson, T. R., Yan, Y., Wu, X., Lam, M. T., Tang, G. L., Beverly, L. J., Messina, L. M., Capobianco, A. J., Werb, Z., and Wang, R. (2005). Endothelial expression of constitutively active Notch4 elicits reversible arteriovenous malformations in adult mice. *Proc. Natl. Acad. Sci. U.S.A.* **102**, 9884–9889.

Carmeliet, P., Ferreira, V., Breier, G., Pollefeyt, S., Kieckens, L., Gertsenstein, M., Fahrig, M., Vandenhoeck, A., Harpal, K., Eberhardt, C., et al. (1996). Abnormal blood vessel development and lethality in embryos lacking a single VEGF allele. *Nature* **380**, 435–439.

Chabriat, H., Joutel, A., Dichgans, M., Tournier-Lasserve, E., and Bousser, M. G. (2009). Cadasil. *Lancet Neurol.* **8**, 643–653.

Coultas, L., Chawengsaksophak, K., and Rossant, J. (2005). Endothelial cells and VEGF in vascular development. *Nature* **438**, 937–945.

Doi, H., Iso, T., Sato, H., Yamazaki, M., Matsui, H., Tanaka, T., Manabe, I., Arai, M., Nagai, R., and Kurabayashi, M. (2006). Jagged1-selective notch signaling induces smooth muscle differentiation via a RBP-Jkappa-dependent pathway. *J. Biol. Chem.* **281**, 28555–28564.

Doi, H., Iso, T., Yamazaki, M., Akiyama, H., Kanai, H., Sato, H., Kawai-Kowase, K., Tanaka, T., Maeno, T., Okamoto, E., et al. (2005). HERP1 inhibits myocardin-induced vascular smooth muscle cell differentiation by interfering with SRF binding to CArG box. *Arterioscler. Thromb. Vasc. Biol.* **25**, 2328–2334.

Domenga, V., Fardoux, P., Lacombe, P., Monet, M., Maciazek, J., Krebs, L. T., Klonjkowski, B., Berrou, E., Mericskay, M., Li, Z., et al. (2004). Notch3 is required for arterial identity and maturation of vascular smooth muscle cells. *Genes Dev.* **18**, 2730–2735.

Dorrell, M. I., and Friedlander, M. (2006). Mechanisms of endothelial cell guidance and vascular patterning in the developing mouse retina. *Prog. Retin. Eye Res.* **25**, 277–295.

Duarte, A., Hirashima, M., Benedito, R., Trindade, A., Diniz, P., Bekman, E., Costa, L., Henrique, D., and Rossant, J. (2004). Dosage-sensitive requirement for mouse Dll4 in artery development. *Genes Dev.* **18**, 2474–2478.

Dubroca, C., Lacombe, P., Domenga, V., Maciazek, J., Levy, B., Tournier-Lasserve, E., Joutel, A., and Henrion, D. (2005). Impaired vascular mechanotransduction in a transgenic mouse model of CADASIL arteriopathy. *Stroke* **36**, 113–117.

Dufraine, J., Funahashi, Y., and Kitajewski, J. (2008). Notch signaling regulates tumor angiogenesis by diverse mechanisms. *Oncogene* **27**, 5132–5137.

Ehebauer, M., Hayward, P., and Martinez-Arias, A. (2006). Notch signaling pathway. *Sci. STKE* **2006**, cm7.

Ehebauer, M., Hayward, P., and Martinez-Arias, A. (2006). Notch, a universal arbiter of cell fate decisions. *Science* **314**, 1414–5.

Ferrara, N., Carver-Moore, K., Chen, H., Dowd, M., Lu, L., O'Shea, K. S., Powell-Braxton, L., Hillan, K. J., and Moore, M. W. (1996). Heterozygous embryonic lethality induced by targeted inactivation of the VEGF gene. *Nature* **380**, 439–442.

Ferrara, N., and Kerbel, R. S. (2005). Angiogenesis as a therapeutic target. *Nature* **438**, 967–974.

Fischer, A., Schumacher, N., Maier, M., Sendtner, M., and Gessler, M. (2004). The Notch target genes Hey1 and Hey2 are required for embryonic vascular development. *Genes Dev.* **18**, 901–911.

Fortini, M. E. (2009). Notch signaling: the core pathway and its posttranslational regulation. *Dev. Cell* **16**, 633–647.

Gale, N. W., Dominguez, M. G., Noguera, I., Pan, L., Hughes, V., Valenzuela, D. M., Murphy, A. J., Adams, N. C., Lin, H. C., Holash, J., et al. (2004). Haploinsufficiency of delta-like 4 ligand results in embryonic lethality due to major defects in arterial and vascular development. *Proc. Natl. Acad. Sci. U.S.A.* **101**, 15949–15954.

Gariano, R. F., and Gardner, T. W. (2005). Retinal angiogenesis in development and disease. *Nature* **438**, 960–966.

Gerhardt, H., Golding, M., Fruttiger, M., Ruhrberg, C., Lundkvist, A., Abramsson, A., Jeltsch, M., Mitchell, C., Alitalo, K., Shima, D., et al. (2003). VEGF guides angiogenic sprouting utilizing endothelial tip cell filopodia. *J. Cell Biol.* **161**, 1163–1177.

Gerhardt, H., Ruhrberg, C., Abramsson, A., Fujisawa, H., Shima, D., and Betsholtz, C. (2004). Neuropilin-1 is required for endothelial tip cell guidance in the developing central nervous system. *Dev. Dyn.* **231**, 503–509.

Grego-Bessa, J., Luna-Zurita, L., Del Monte, G., Bolos, V., Melgar, P., Arandilla, A., Garratt, A. N., Zang, H., Mukouyama, Y. S., Chen, H., et al. (2007). Notch signaling is essential for ventricular chamber development. *Dev. Cell* **12**, 415–429.

Gridley, T. (2007). Notch signaling in vascular development and physiology. *Development* **134**, 2709–2718.

Hainaud, P., Contreres, J. O., Villemain, A., Liu, L. X., Plouet, J., Tobelem, G., and Dupuy, E. (2006). The role of the vascular endothelial growth factor-Delta-like 4 ligand/Notch4-ephrin B2 cascade in tumor vessel remodeling and endothelial cell functions. *Cancer Res.* **66**, 8501–8510.

Havrda, M. C., Johnson, M. J., O'Neill, C. F., and Liaw, L. (2006). A novel mechanism of transcriptional repression of p27kip1 through Notch/HRT2 signaling in vascular smooth muscle cells. *Thromb. Haemost.* **96**, 361–370.

Hayashi, H., and Kume, T. (2008). Foxc transcription factors directly regulate Dll4 and Hey2 expression by interacting with the VEGF-Notch signaling pathways in endothelial cells. *PLoS ONE* **3**, e2401.

Hellstrom, M., Phng, L. K., Hofmann, J. J., Wallgard, E., Coultas, L., Lindblom, P., Alva, J., Nilsson, A. K., Karlsson, L., Gaiano, N., et al. (2007). Dll4 signalling through Notch1 regulates formation of tip cells during angiogenesis. *Nature* **445**, 776–780.

Herbert, S. P., Huisken, J., Kim, T. N., Feldman, M. E., Houseman, B. T., Wang, R. A., Shokat, K. M., and Stainier, D. Y. (2009). Arterial-venous segregation by selective cell sprouting: an alternative mode of blood vessel formation. *Science* **326**, 294–298.

Herreman, A., Hartmann, D., Annaert, W., Saftig, P., Craessaerts, K., Serneels, L., Umans, L., Schrijvers, V., Checler, F., Vanderstichele, H., et al. (1999). Presenilin 2 deficiency causes a mild pulmonary phenotype and no changes in amyloid precursor protein processing but enhances the embryonic lethal phenotype of presenilin 1 deficiency. *Proc. Natl. Acad. Sci. U.S.A.* **96**, 11872–11877.

High, F. A., and Epstein, J. A. (2008). The multifaceted role of Notch in cardiac development and disease. *Nat. Rev. Genet.* **9**, 49–61.

High, F. A., Lu, M. M., Pear, W. S., Loomes, K. M., Kaestner, K. H., and Epstein, J. A. (2008). Endothelial expression of the Notch ligand Jagged1 is required for vascular smooth muscle development. *Proc. Natl. Acad. Sci. U.S.A.* **105**, 1955–1959.

High, F. A., Zhang, M., Proweller, A., Tu, L., Parmacek, M. S., Pear, W. S., and Epstein, J. A. (2007). An essential role for Notch in neural crest during cardiovascular development and smooth muscle differentiation. *J. Clin. Invest.* **117**, 353–363.

Hoey, T., Yen, W. C., Axelrod, F., Basi, J., Donigian, L., Dylla, S., Fitch-Bruhns, M., Lazetic, S., Park, I. K., Sato, A., et al. (2009). DLL4 blockade inhibits tumor growth and reduces tumor-initiating cell frequency. *Cell Stem Cell* **5**, 168–177.

Hofmann, J. J., and Iruela-Arispe, M. L. (2007). Notch signaling in blood vessels: who is talking to whom about what? *Circ. Res.* **100**, 1556–1568.

Huppert, S. S., Le, A., Schroeter, E. H., Mumm, J. S., Saxena, M. T., Milner, L. A., and Kopan, R. (2000). Embryonic lethality in mice homozygous for a processing-deficient allele of Notch1. *Nature* **405**, 966–970.

Ilagan, M. X., and Kopan, R. (2007). SnapShot: notch signaling pathway. *Cell* **128**, 1246.

Ishiko, A., Shimizu, A., Nagata, E., Takahashi, K., Tabira, T., and Suzuki, N. (2006). Notch3 ectodomain is a major component of granular osmiophilic material (GOM) in CADASIL. *Acta Neuropathol. (Berl)* **112**, 333–339.

Iso, T., Maeno, T., Oike, Y., Yamazaki, M., Doi, H., Arai, M., and Kurabayashi, M. (2006). Dll4-selective Notch signaling induces ephrinB2 gene expression in endothelial cells. *Biochem. Biophys. Res. Commun.* **341**, 708–714.

Jain, R. K., Duda, D. G., Clark, J. W., and Loeffler, J. S. (2006). Lessons from phase III clinical trials on anti-VEGF therapy for cancer. *Nat. Clin. Pract. Oncol.* **3**, 24–40.

Jia, L., Yu, G., Zhang, Y., and Wang, M. M. (2009). Lysosome-dependent degradation of Notch3. *Int. J. Biochem. Cell Biol.* **41**, 2594–2598.

Joutel, A., Andreux, F., Gaulis, S., Domenga, V., Cecillon, M., Battail, N., Piga, N., Chapon, F., Godfrain, C., and Tournier-Lasserve, E. (2000). The ectodomain of the Notch3 receptor accumulates within the cerebrovasculature of CADASIL patients. *J. Clin. Invest.* **105**, 597–605.

Joutel, A., Corpechot, C., Ducros, A., Vahedi, K., Chabriat, H., Mouton, P., Alamowitch, S., Domenga, V., Cecillion, M., Marechal, E., et al. (1996). Notch3 mutations in CADASIL, a hereditary adult-onset condition causing stroke and dementia. *Nature* **383**, 707–710.

Joutel, A., Monet, M., Domenga, V., Riant, F., and Tournier-Lasserve, E. (2004). Pathogenic mutations associated with cerebral autosomal dominant arteriopathy with subcortical infarcts and leukoencephalopathy differently affect Jagged1 binding and Notch3 activity via the RBP/JK signaling Pathway. *Am. J. Hum. Genet.* **74**, 338–347.

Joutel, A., Monet-Lepretre, M., Gosele, C., Baron-Menguy, C., Hammes, A., Schmidt, S., Lemaire-Carrette, B., Domenga, V., Schedl, A., Lacombe, P., et al. (2010). Cerebrovascular dysfunction and microcirculation rarefaction precede white matter lesions in a mouse genetic model of cerebral ischemic small vessel disease. *J. Clin. Invest.* **120**, 433–445.

Kaestner, K. H., Knochel, W., and Martinez, D. E. (2000). Unified nomenclature for the winged helix/forkhead transcription factors. *Genes Dev.* **14**, 142–146.

Kageyama, R., Ohtsuka, T., and Kobayashi, T. (2007). The Hes gene family: repressors and oscillators that orchestrate embryogenesis. *Development* **134**, 1243–1251.

Kalaria, R. N., Viitanen, M., Kalimo, H., Dichgans, M., and Tabira, T. (2004). The pathogenesis of CADASIL: an update. *J. Neuropsychiatr.* **226**, 35–39.

Kim, Y. H., Hu, H., Guevara-Gallardo, S., Lam, M. T., Fong, S. Y., and Wang, R. A. (2008). Artery and vein size is balanced by Notch and ephrin B2/EphB4 during angiogenesis. *Development* **135**, 3755–3764.

Kokubo, H., Miyagawa-Tomita, S., Nakazawa, M., Saga, Y., and Johnson, R. L. (2005). Mouse hesr1 and hesr2 genes are redundantly required to mediate Notch signaling in the developing cardiovascular system. *Dev. Biol.* **278**, 301–309.

Koo, B. K., Lim, H. S., Song, R., Yoon, M. J., Yoon, K. J., Moon, J. S., Kim, Y. W., Kwon, M. C., Yoo, K. W., Kong, M. P., et al. (2005). Mind bomb 1 is essential for generating functional Notch ligands to activate Notch. *Development* **132**, 3459–3470.

Kopan, R., and Ilagan, M. X. (2009). The canonical Notch signaling pathway: unfolding the activation mechanism. *Cell* **137**, 216–233.

Krebs, L. T., Shutter, J. R., Tanigaki, K., Honjo, T., Stark, K. L., and Gridley, T. (2004). Haploinsufficient lethality and formation of arteriovenous malformations in Notch pathway mutants. *Genes Dev.* **18**, 2469–2473.

Krebs, L. T., Starling, C., Chervonsky, A. V., and Gridley, T. (2010). Notch1 activation in mice causes arteriovenous malformations phenocopied by ephrinB2 and EphB4 mutants. *Genesis* **48**, 146–150.

Krebs, L. T., Xue, Y., Norton, C. R., Shutter, J. R., Maguire, M., Sundberg, J. P., Gallahan, D., Closson, V., Kitajewski, J., Callahan, R., et al. (2000). Notch signaling is essential for vascular morphogenesis in mice. *Genes Dev.* **14**, 1343–1352.

Krebs, L. T., Xue, Y., Norton, C. R., Sundberg, J. P., Beatus, P., Lendahl, U., Joutel, A., and Gridley, T. (2003). Characterization of Notch3-deficient mice: normal embryonic development and absence of genetic interactions with a Notch1 mutation. *Genesis* **37**, 139–143.

Kume, T. (2009). Novel insights into the differential functions of Notch ligands in vascular formation. *J. Angiogenes. Res.* **1**, 8.

Kume, T., Jiang, H., Topczewska, J. M., and Hogan, B. L. (2001). The murine winged helix transcription factors, Foxc1 and Foxc2, are both required for cardiovascular development and somitogenesis. *Genes Dev.* **15**, 2470–2482.

Lacombe, P., Oligo, C., Domenga, V., Tournier-Lasserve, E., and Joutel, A. (2005). Impaired cerebral vasoreactivity in a transgenic mouse model of cerebral autosomal dominant arteriopathy with subcortical infarcts and leukoencephalopathy arteriopathy. *Stroke* **36**, 1053–1058.

Lawson, N. D., Scheer, N., Pham, V. N., Kim, C. H., Chitnis, A. B., Campos-Ortega, J. A., and Weinstein, B. M. (2001). Notch signaling is required for arterial-venous differentiation during embryonic vascular development. *Development* **128**, 3675–3683.

Lawson, N. D., Vogel, A. M., and Weinstein, B. M. (2002). Sonic hedgehog and vascular endothelial growth factor act upstream of the Notch pathway during arterial endothelial differentiation. *Dev. Cell* **3**, 127–136.

Le Borgne, R. (2006). Regulation of Notch signalling by endocytosis and endosomal sorting. *Curr. Opin. Cell Biol.* **18**, 213–222.

Leslie, J. D., Ariza-McNaughton, L., Bermange, A. L., McAdow, R., Johnson, S. L., and Lewis, J. (2007). Endothelial signalling by the Notch ligand Delta-like 4 restricts angiogenesis. *Development* **134**, 839–844.

Li, J. L., and Harris, A. L. (2009). Crosstalk of VEGF and Notch pathways in tumour angiogenesis: therapeutic implications. *Front. Biosci.* **14**, 3094–3110.

Li, T., Ma, G., Cai, H., Price, D. L., and Wong, P. C. (2003). Nicastrin is required for assembly of presenilin/gamma-secretase complexes to mediate Notch signaling and for processing and trafficking of beta-amyloid precursor protein in mammals. *J. Neurosci.* **23**, 3272–3277.

Li, Y., Takeshita, K., Liu, P. Y., Satoh, M., Oyama, N., Mukai, Y., Chin, M. T., Krebs, L., Kotlikoff, M. I., Radtke, F., et al. (2009). Smooth muscle Notch1 mediates neointimal formation after vascular injury. *Circulation* **119**, 2686–2692.

Lilly, B., and Kennard, S. (2009). Differential gene expression in a coculture model of angiogenesis reveals modulation of select pathways and a role for Notch signaling. *Physiol. Genomics* **36**, 69–78.

Limbourg, A., Ploom, M., Elligsen, D., Sorensen, I., Ziegelhoeffer, T., Gossler, A., Drexler, H., and Limbourg, F. P. (2007). Notch ligand Delta-like 1 is essential for postnatal arteriogenesis. *Circ. Res.* **100**, 363–371.

Limbourg, F. P., Takeshita, K., Radtke, F., Bronson, R. T., Chin, M. T., and Liao, J. K. (2005). Essential role of endothelial Notch1 in angiogenesis. *Circulation* **111**, 1826–1832.

Lindner, V., Booth, C., Prudovsky, I., Small, D., Maciag, T., and Liaw, L. (2001). Members of the Jagged/Notch gene families are expressed in injured arteries and regulate cell phenotype via alterations in cell matrix and cell-cell interaction. *Am. J. Pathol.* **159**, 875–883.

Liu, H., Kennard, S., and Lilly, B. (2009). NOTCH3 expression is induced in mural cells through an autoregulatory loop that requires endothelial-expressed JAGGED1. *Circ. Res.* **104**, 466–475.

Liu, Z. J., Shirakawa, T., Li, Y., Soma, A., Oka, M., Dotto, G. P., Fairman, R. M., Velazquez, O. C., and Herlyn, M. (2003). Regulation of Notch1 and Dll4 by vascular endothelial growth factor in arterial endothelial cells: implications for modulating arteriogenesis and angiogenesis. *Mol. Cell. Biol.* **23**, 14–25.

Lobov, I. B., Renard, R. A., Papadopoulos, N., Gale, N. W., Thurston, G., Yancopoulos, G. D., and Wiegand, S. J. (2007). Delta-like ligand 4 (Dll4) is induced by VEGF as a negative regulator of angiogenic sprouting. *Proc. Natl. Acad. Sci. U.S.A.* **104**, 3219–3224.

Lundkvist, J., Zhu, S., Hansson, E. M., Schweinhardt, P., Miao, Q., Beatus, P., Dannaeus, K., Karlstrom, H., Johansson, C. B., Viitanen, M., et al. (2005). Mice carrying a R142C Notch 3 knock-in mutation do not develop a CADASIL-like phenotype. *Genesis* **41**, 13–22.

Mailhos, C., Modlich, U., Lewis, J., Harris, A., Bicknell, R., and Ish-Horowicz, D. (2001). Delta4, an endothelial specific notch ligand expressed at sites of physiological and tumor angiogenesis. *Differentiation* **69**, 135–144.

Miniati, D., Jelin, E. B., Ng, J., Wu, J., Carlson, T. R., Wu, X., Looney, M. R., and Wang, R. A. (2010). Constitutively active endothelial Notch4 causes lung arteriovenous shunts in mice. *Am. J. Physiol. Lung Cell Mol. Physiol.* **298**, L169–L177.

Mitchell, K. J., Pinson, K. I., Kelly, O. G., Brennan, J., Zupicich, J., Scherz, P., Leighton, P. A., Goodrich, L. V., Lu, X., Avery, B. J., et al. (2001). Functional analysis of secreted and transmembrane proteins critical to mouse development. *Nat. Genet.* **28**, 241–249.

Moessler, H., Mericskay, M., Li, Z., Nagl, S., Paulin, D., and Small, J. V. (1996). The SM 22 promoter directs tissue-specific expression in arterial but not in venous or visceral smooth muscle cells in transgenic mice. *Development* **122**, 2415–2425.

Monet, M., Domenga, V., Lemaire, B., Souilhol, C., Langa, F., Babinet, C., Gridley, T., Tournier-Lasserve, E., Cohen-Tannoudji, M., and Joutel, A. (2007). The archetypal R90C CADASIL-NOTCH3 mutation retains NOTCH3 function in vivo. *Hum. Mol. Genet.* **16**, 982–992.

Monet-Lepretre, M., Bardot, B., Lemaire, B., Domenga, V., Godin, O., Dichgans, M., Tournier-Lasserve, E., Cohen-Tannoudji, M., Chabriat, H., and Joutel, A. (2009). Distinct phenotypic and functional features of CADASIL mutations in the Notch3 ligand binding domain. *Brain* **132**, 1601–1612.

Morrow, D., Cullen, J. P., Cahill, P. A., and Redmond, E. M. (2007). Cyclic strain regulates the Notch/CBF-1 signaling pathway in endothelial cells: role in angiogenic activity. *Arterioscler. Thromb. Vasc. Biol.* **27**, 1289–1296.

Morrow, D., Guha, S., Sweeney, C., Birney, Y., Walshe, T., O'Brien, C., Walls, D., Redmond, E. M., and Cahill, P. A. (2008). Notch and vascular smooth muscle cell phenotype. *Circ. Res.* **103**, 1370–1382.

Morrow, D., Scheller, A., Birney, Y. A., Sweeney, C., Guha, S., Cummins, P. M., Murphy, R., Walls, D., Redmond, E. M., and Cahill, P. A. (2005a). Notch-mediated CBF-1/

RBP-J{kappa}-dependent regulation of human vascular smooth muscle cell phenotype in vitro. *Am. J. Physiol. Cell Physiol.* **289**, C1188–C1196.

Morrow, D., Sweeney, C., Birney, Y. A., Cummins, P. M., Walls, D., Redmond, E. M., and Cahill, P. A. (2005b). Cyclic strain inhibits Notch receptor signaling in vascular smooth muscle cells in vitro. *Circ. Res.* **96**, 567–575.

Mukouyama, Y. S., Gerber, H. P., Ferrara, N., Gu, C., and Anderson, D. J. (2005). Peripheral nerve-derived VEGF promotes arterial differentiation via neuropilin 1-mediated positive feedback. *Development* **132**, 941–952.

Mukouyama, Y. S., Shin, D., Britsch, S., Taniguchi, M., and Anderson, D. J. (2002). Sensory nerves determine the pattern of arterial differentiation and blood vessel branching in the skin. *Cell* **109**, 693–705.

Murphy, P. A., Lam, M. T., Wu, X., Kim, T. N., Vartanian, S. M., Bollen, A. W., Carlson, T. R., and Wang, R. A. (2008). Endothelial Notch4 signaling induces hallmarks of brain arteriovenous malformations in mice. *Proc. Natl. Acad. Sci. U.S. A.* **105**, 10901–10906.

Murphy, P. A., Lu, G., Shiah, S., Bollen, A. W., and Wang, R. A. (2009). Endothelial Notch signaling is upregulated in human brain arteriovenous malformations and a mouse model of the disease. *Lab. Invest.* **89**, 971–982.

Nemir, M., and Pedrazzini, T. (2008). Functional role of Notch signaling in the developing and postnatal heart. *J. Mol. Cell Cardiol.* **45**, 495–504.

Niessen, K., and Karsan, A. (2008). Notch signaling in cardiac development. *Circ. Res.* **102**, 1169–1181.

Noguera-Troise, I., Daly, C., Papadopoulos, N. J., Coetzee, S., Boland, P., Gale, N. W., Lin, H. C., Yancopoulos, G. D., and Thurston, G. (2006). Blockade of Dll4 inhibits tumour growth by promoting non-productive angiogenesis. *Nature* **444**, 1032–1037.

Noseda, M., Fu, Y., Niessen, K., Wong, F., Chang, L., McLean, G., and Karsan, A. (2006). Smooth Muscle alpha-actin is a direct target of Notch/CSL. *Circ. Res.* **98**, 1468–1470.

Opherk, C., Duering, M., Peters, N., Karpinska, A., Rosner, S., Schneider, E., Bader, B., Giese, A., and Dichgans, M. (2009). CADASIL mutations enhance spontaneous multimerization of NOTCH3. *Hum. Mol. Genet.* **18**, 2761–2767.

Owens, G. K., Kumar, M. S., and Wamhoff, B. R. (2004). Molecular regulation of vascular smooth muscle cell differentiation in development and disease. *Physiol. Rev.* **84**, 767–801.

Patel, N. S., Li, J. L., Generali, D., Poulsom, R., Cranston, D. W., and Harris, A. L. (2005). Up-regulation of delta-like 4 ligand in human tumor vasculature and the role of basal expression in endothelial cell function. *Cancer Res.* **65**, 8690–8697.

Peirce, S. M. (2008). Computational and mathematical modeling of angiogenesis. *Microcirculation* **15**, 739–751.

Peters, N., Opherk, C., Zacherle, S., Capell, A., Gempel, P., and Dichgans, M. (2004). CADASIL-associated Notch3 mutations have differential effects both on ligand binding and ligand-induced Notch3 receptor signaling through RBP-Jk. *Exp. Cell Res.* **299**, 454–464.

Phng, L. K., and Gerhardt, H. (2009). Angiogenesis: a team effort coordinated by notch. *Dev. Cell* **16**, 196–208.

Proweller, A., Pear, W. S., and Parmacek, M. S. (2005). Notch signaling represses myocardin-induced smooth muscle cell differentiation. *J. Biol. Chem.* **280**, 8994–9004.

Qutub, A. A., and Popel, A. S. (2009). Elongation, proliferation & migration differentiate endothelial cell phenotypes and determine capillary sprouting. *BMC Syst. Biol.* **3**, 13.

Ridgway, J., Zhang, G., Wu, Y., Stawicki, S., Liang, W. C., Chanthery, Y., Kowalski, J., Watts, R. J., Callahan, C., Kasman, I., et al. (2006). Inhibition of Dll4 signalling inhibits tumour growth by deregulating angiogenesis. *Nature* **444**, 1083–1087.

Roca, C., and Adams, R. H. (2007). Regulation of vascular morphogenesis by Notch signaling. *Genes Dev.* **21**, 2511–2524.

Ruchoux, M. M., Domenga, V., Brulin, P., Maciazek, J., Limol, S., Tournier-Lasserve, E., and Joutel, A. (2003). Transgenic mice expressing mutant Notch3 develop vascular alterations characteristic of cerebral autosomal dominant arteriopathy with subcortical infarcts and leukoencephalopathy. *Am. J. Pathol.* **162**, 329–342.

Sainson, R. C., Aoto, J., Nakatsu, M. N., Holderfield, M., Conn, E., Koller, E., and Hughes, C. C. (2005). Cell-autonomous notch signaling regulates endothelial cell branching and proliferation during vascular tubulogenesis. *FASEB J.* **19**, 1027–1029.

Sainson, R. C., and Harris, A. L. (2007). Anti-Dll4 therapy: can we block tumour growth by increasing angiogenesis? *Trends Mol. Med.* **13**, 389–395.

Sakata, Y., Xiang, F., Chen, Z., Kiriyama, Y., Kamei, C. N., Simon, D. I., and Chin, M. T. (2004). Transcription factor CHF1/Hey2 regulates neointimal formation in vivo and vascular smooth muscle proliferation and migration in vitro. *Arterioscler. Thromb. Vasc. Biol.* **24**, 2069–2074.

Scehnet, J. S., Jiang, W., Kumar, S. R., Krasnoperov, V., Trindade, A., Benedito, R., Djokovic, D., Borges, C., Ley, E. J., Duarte, A., et al. (2007). Inhibition of Dll4 mediated signaling induces proliferation of immature vessels and results in poor tissue perfusion. *Blood.* **109**, 4753–4760.

Seo, S., Fujita, H., Nakano, A., Kang, M., Duarte, A., and Kume, T. (2006). The forkhead transcription factors, Foxc1 and Foxc2, are required for arterial specification and lymphatic sprouting during vascular development. *Dev. Biol.* **294**, 458–470.

Shibuya, M., and Claesson-Welsh, L. (2006). Signal transduction by VEGF receptors in regulation of angiogenesis and lymphangiogenesis. *Exp. Cell Res.* **312**, 549–560.

Siekmann, A. F., and Lawson, N. D. (2007). Notch signalling limits angiogenic cell behaviour in developing zebrafish arteries. *Nature* **445**, 781–784.

Sorensen, I., Adams, R. H., and Gossler, A. (2009). DLL1-mediated Notch activation regulates endothelial identity in mouse fetal arteries. *Blood* **113**, 5680–5688.

Stalmans, I., Ng, Y. S., Rohan, R., Fruttiger, M., Bouche, A., Yuce, A., Fujisawa, H., Hermans, B., Shani, M., Jansen, S., et al. (2002). Arteriolar and venular patterning in retinas of mice selectively expressing VEGF isoforms. *J. Clin. Invest.* **109**, 327–336.

Suchting, S., Freitas, C., le Noble, F., Benedito, R., Breant, C., Duarte, A., and Eichmann, A. (2007). The Notch ligand Delta-like 4 negatively regulates endothelial tip cell formation and vessel branching. *Proc. Natl. Acad. Sci. U.S.A.* **104**, 3225–3230.

Takahashi, K., Adachi, K., Yoshizaki, K., Kunimoto, S., Kalaria, R. N., and Watanabe, A. (2010). Mutations in NOTCH3 cause the formation and retention of aggregates in the endoplasmic reticulum, leading to impaired cell proliferation. *Hum. Mol. Genet.* **19**, 79–89.

Takeshita, K., Satoh, M., Ii, M., Silver, M., Limbourg, F. P., Mukai, Y., Rikitake, Y., Radtke, F., Gridley, T., Losordo, D. W., et al. (2007). Critical role of endothelial Notch1 signaling in postnatal angiogenesis. *Circ. Res.* **100**, 70–78.

Tang, Y., Urs, S., and Liaw, L. (2008). Hairy-related transcription factors inhibit Notch-induced smooth muscle alpha-actin expression by interfering with Notch intracellular domain/CBF-1 complex interaction with the CBF-1-binding site. *Circ. Res.* **102**, 661–668.

Thorne, B. C., Bailey, A. M., DeSimone, D. W., and Peirce, S. M. (2007). Agent-based modeling of multicell morphogenic processes during development. *Birth Defects Res. C Embryo Today* **81**, 344–353.

Thurston, G., Noguera-Troise, I., and Yancopoulos, G. D. (2007). The Delta paradox: DLL4 blockade leads to more tumour vessels but less tumour growth. *Nat. Rev. Cancer* **7**, 327–331.

Trindade, A., Kumar, S. R., Scehnet, J. S., Lopes-da-Costa, L., Becker, J., Jiang, W., Liu, R., Gill, P. S., and Duarte, A. (2008). Overexpression of delta-like 4 induces arterialization and attenuates vessel formation in developing mouse embryos. *Blood* **112**, 1720–1729.

Uemura, A., Kusuhara, S., Katsuta, H., and Nishikawa, S. (2006). Angiogenesis in the mouse retina: a model system for experimental manipulation. *Exp. Cell Res.* **312**, 676–683.

Uyttendaele, H., Ho, J., Rossant, J., and Kitajewski, J. (2001). Vascular patterning defects associated with expression of activated Notch4 in embryonic endothelium. *Proc. Natl. Acad. Sci. U.S.A.* **98**, 5643–5648.

van der Loop, F. T., Gabbiani, G., Kohnen, G., Ramaekers, F. C., and van Eys, G. J. (1997). Differentiation of smooth muscle cells in human blood vessels as defined by smoothelin, a novel marker for the contractile phenotype. *Arterioscler. Thromb. Vasc. Biol.* **17**, 665–671.

Visconti, R. P., Richardson, C. D., and Sato, T. N. (2002). Orchestration of angiogenesis and arteriovenous contribution by angiopoietins and vascular endothelial growth factor (VEGF). *Proc. Natl. Acad. Sci. U.S.A.* **99**, 8219–8224.

Wang, W., Campos, A. H., Prince, C. Z., Mou, Y., and Pollman, M. J. (2002). Coordinate Notch3-hairy-related transcription factor pathway regulation in response to arterial injury. Mediator role of platelet-derived growth factor and ERK. *J. Biol. Chem.* **277**, 23165–23171.

Wang, W., Prince, C. Z., Hu, X., and Pollman, M. J. (2003). HRT1 modulates vascular smooth muscle cell proliferation and apoptosis. *Biochem. Biophys. Res. Commun.* **308**, 596–601.

Xue, Y., Gao, X., Lindsell, C. E., Norton, C. R., Chang, B., Hicks, C., Gendron-Maguire, M., Rand, E. B., Weinmaster, G., and Gridley, T. (1999). Embryonic lethality and vascular defects in mice lacking the Notch ligand Jagged1. *Hum. Mol. Genet.* **8**, 723–730.

Yan, M., Callahan, C. A., Beyer, J. C., Allamneni, K. P., Zhang, G., Ridgway, J. B., Niessen, K., and Plowman, G. D. (2010). Chronic DLL4 blockade induces vascular neoplasms. *Nature* **463**, E6–E7.

Yan, M., and Plowman, G. D. (2007). Delta-like 4/Notch signaling and its therapeutic implications. *Clin. Cancer Res.* **13**, 7243–7246.

Zhong, T. P., Childs, S., Leu, J. P., and Fishman, M. C. (2001). Gridlock signalling pathway fashions the first embryonic artery. *Nature* **414**, 216–220.

Zhong, T. P., Rosenberg, M., Mohideen, M. A., Weinstein, B., and Fishman, M. C. (2000). Gridlock, an HLH gene required for assembly of the aorta in zebrafish. *Science* **287**, 1820–1824.

ZhuGe, Q., Zhong, M., Zheng, W., Yang, G. Y., Mao, X., Xie, L., Chen, G., Chen, Y., Lawton, M. T., Young, W. L., et al. (2009). Notch-1 signalling is activated in brain arteriovenous malformations in humans. *Brain* **132**, 3231–3241.

CHAPTER TEN

Ultradian Oscillations in Notch Signaling Regulate Dynamic Biological Events

Ryoichiro Kageyama, Yasutaka Niwa, Hiromi Shimojo, Taeko Kobayashi, *and* Toshiyuki Ohtsuka

Contents

Abstract

Notch signaling regulates many dynamic processes; accordingly, expression of genes in this pathway is also dynamic. In mouse embryos, one dynamic process regulated by Notch is somite segmentation, which occurs with a 2-h periodicity. This periodic event is regulated by a biological clock called the segmentation clock, which involves cyclic expression of the Notch effector gene *Hes7*. Loss of Hes7 expression and sustained expression of Hes7 result in identical and severe somite defects, suggesting that Hes7 oscillation is required for proper somite segmentation. Mathematical models of this oscillator have been used to generate and test hypothesis, helping to uncover the role of negative feedback in regulating the oscillator. Oscillations of another Notch effector gene, *Hes1*, plays an important role in maintenance of neural stem cells. Hes1 expression

Institute for Virus Research, Kyoto University; and Japan Science and Technology Agency, CREST, Kyoto, Japan

Current Topics in Developmental Biology, Volume 92
ISSN 0070-2153, DOI 10.1016/S0070-2153(10)92010-3

oscillates with a period of about 2–3 h in neural stem cells, whereas sustained Hes1 expression inhibits proliferation and differentiation of these cells, suggesting that Hes1 oscillations are important for their proper activities. Hes1 inhibits its own expression as well as the expression of the proneural gene *Neurogenin2* and the Notch ligand *Delta1*, driving oscillations of these two genes. *Delta1* oscillations in turn maintain neural stem cells by mutual activation of Notch signaling, which re-activates Hes1 to close the cycle. Hes1 expression also oscillates in embryonic stem (ES) cells. Cells expressing low and high levels of Hes1 tend to differentiate into neural and mesodermal cells, respectively. Furthermore, *Hes1*-null ES cells display early and uniform neural differentiation, indicating that *Hes1* oscillations act to promote multipotency by generating heterogeneity in both the differentiation timing and the fate choice. Taken together, these results suggest that Notch signaling can drive short-period oscillatory expression of *Hes7* and *Hes1* (ultradian oscillation) and that ultradian oscillations are important for many biological events.

1. Introduction

Upon activation of Notch signaling, the Notch intracellular domain (NICD) is released from the membrane region and is transferred to the nucleus, where NICD forms a complex with the DNA-binding protein RBPj (Fig. 10.1A) (Honjo, 1996; Kopan and Ilagan, 2009). The NICD–RBPj complex recruits additional transcriptional activators and induces downstream genes such as *Hes* and *Hey* genes, forming the Notch–RBPj–Hes axis, often called the canonical pathway (Fig. 10.1A). *Hes* genes are mammalian homologues of *Drosophila hairy* and *Enhancer of split* and encode basic helix-loop-helix (bHLH)-type transcriptional repressors (Fig. 10.1B) (Sasai *et al.*, 1992; Kageyama *et al.*, 2007). There are seven members in the *Hes* family (*Hes1* to *Hes7*), although *Hes4* is not present in the mouse genome. Through their bHLH domain, Hes factors form homodimers or heterodimers with Hes-related bHLH factors such as Hey1 and bind to the DNA sequences called the N box (CACNAG) or the class C site (CACG(C/A)G) (Iso *et al.*, 2003; Kageyama *et al.*, 2007). In addition, Hes factors contain an orange domain located just C-terminal to the bHLH domain followed by the WRPW sequence at the carboxyl terminus, and both the orange domain and the WRPW sequence recruit several co-repressors. Thus, Hes proteins act as transcriptional repressors by binding to the target sequences. The orange domain also regulates the selection of bHLH heterodimer partners (Dawson *et al.*, 1995; Taelman *et al.*, 2004), while the WRPW sequence acts as a polyubiquitination signal, controlling the half-life of Hes protein by promotion of proteasome-mediated degradation (Kang *et al.*, 2005).

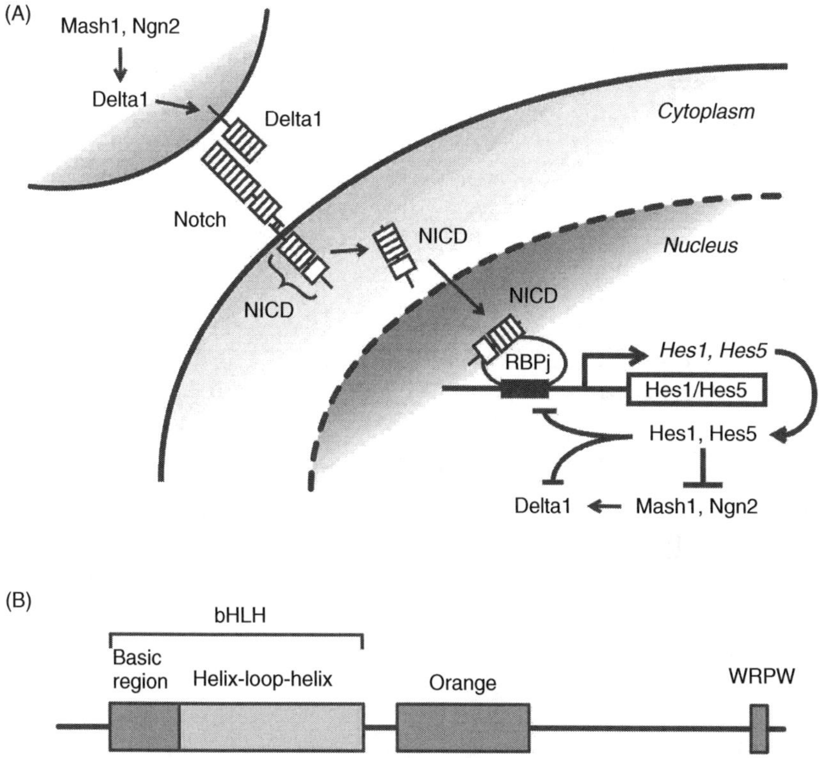

Figure 10.1 The Notch–RBPj–Hes pathway and lateral inhibition. (A) Proneural genes such as *Mash1* and *Ngn2* induce neuronal differentiation and *Delta1* expression. Delta1 activates Notch signaling of neighboring cells. Upon activation of Notch, NICD is released and transferred to the nucleus, where it forms a complex with RBPj. The NICD–RBPj complex induces expression of Hes1 and Hes5, which repress proneural gene and *Delta1* expression. Thus, differentiating neurons inhibit their neighboring cells from neuronal differentiation (lateral inhibition). (B) The conserved domains of Hes factors. Hes factors form dimers through the bHLH domain, bind to the DNA through the basic region, and recruit co-repressors through the orange domain and the WRPW sequence. Thus, Hes factors function as transcriptional repressors.

It is well established that the Notch–RBPj–Hes pathway regulates many biological events by repressing target gene expression (Honjo, 1996; Kageyama *et al.*, 2007). Major Hes target genes include proneural genes such as *Mash1* and *Neurogenin2* (*Ngn2*), themselves bHLH proteins acting as transcriptional activators (Fig. 10.1A) (Bertrand *et al.*, 2002; Ross *et al.*, 2003). Proneural genes promote differentiation of neural stem cells into neurons, while *Hes* genes repress proneural gene expression and maintain neural stem cells (Kageyama *et al.*, 2007). Other major Hes target genes are *Hes* genes themselves (Fig. 10.1A) (Takebayashi *et al.*, 1994; Bessho *et al.*, 2003). For example, Hes1 and Hes7 can repress their own expression by directly binding to their own promoters, thereby forming negative feedback loops. This negative feedback is sufficient to produce oscillating gene expression (Hirata *et al.*, 2002; Bessho *et al.*, 2003). In this chapter, we describe the mechanism and significance of oscillating Hes expression in somite formation, neural development, and embryonic stem (ES) cell differentiation and discuss that not only the expression level but also the expression mode (oscillating versus sustained) is very important for many biological events.

2. Hes7 Oscillations Regulate Somite Segmentation

2.1. Hes7 oscillations by negative feedback

Somites are segmental axial structures of vertebrate embryos that give rise to vertebral column, ribs, skeletal muscles, and subcutaneous tissues. A bilateral pair of somites forms periodically at the anterior ends of the presomitic mesoderm (PSM), located at the caudal part of embryos (Fig. 10.2A). During this process, mesenchymal cells of the PSM are transformed into

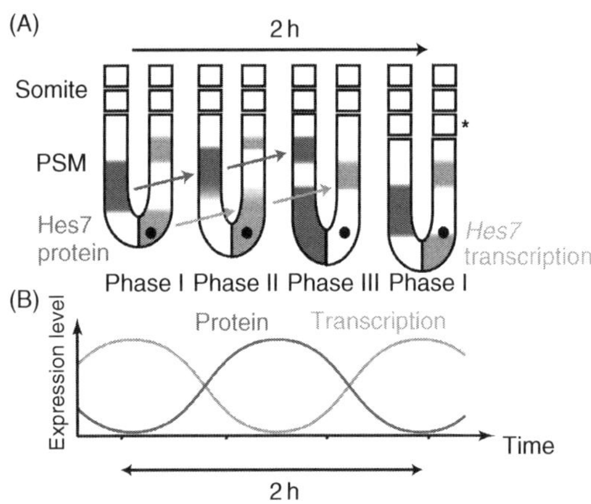

Figure 10.2 Hes7 oscillations in the PSM during somite segmentation. (A) The anterior ends of the PSM are segmented every 2 h in mouse embryos, forming a bilateral pair of somites (asterisk). *Hes7* gene transcription is initiated in the posterior end of the PSM (phase I) and is propagated into the anterior region (phase II), stopping near the anterior end (phase III). *Hes7* gene transcription and Hes7 protein expression occur in a mutually exclusive manner in all three phases, indicating that *Hes7* gene transcription is repressed by Hes7 protein. (B) Dynamic Hes7 expression is caused by oscillation in individual cells [indicated by a dot in (A)]. *Hes7* gene transcription and Hes7 protein expression oscillate in an antiphase manner. (See Color Insert.)

the epithelial sheet (mesenchymal–epithelial transition) at each somite border, which segments a block of somitic cells from the PSM (segmentation). A bilateral pair of somites is formed once every 2 h in mice, every 90 min in chick, and every 30 min in zebrafish. Thus, the segmentation period differs from species to species. In addition, the period becomes longer at lower temperatures (Jiang *et al.*, 2000), indicating that it is also temperature dependent. However, within each species, the period length remains precise during development, so that the number of somites is used to identify the embryonic stages. The biological clock that regulates this periodic segmentation is called the segmentation clock (Pourquié, 2003); its molecular mechanisms have been analyzed intensively in zebrafish, chick, and mice (Dequéant and Pourquié, 2008; Mara and Holley, 2007).

In mouse embryos, both *Hes1* and *Hes7* are expressed dynamically in the PSM (Bessho *et al.*, 2001a; Jouve *et al.*, 2000). The expression cycle is initiated in the posterior tip of the PSM and is propagated toward the anterior, ending near the anterior boundary of the PSM, after which segmentation of one bilateral pair of somites occurs (Fig. 10.2A). This dynamic expression of Hes1 and Hes7 is caused by synchronized oscillation in PSM cells (Fig. 10.2B). Of these two genes, *Hes7* is functionally more important than *Hes1* for somite segmentation: somites fail to segment and thus are severely fused in the absence of *Hes7* but not in the absence of *Hes1* (Bessho *et al.*, 2001b). Interestingly, sustained expression of Hes7 also leads to severe somite fusion, suggesting that the oscillating expression of Hes7 is the key for maintaining periodic somite segmentation (Niwa *et al.*, 2007).

What is the mechanism producing *Hes7* oscillations? In the PSM, *Hes7* gene transcription and translation are mutually exclusive (Fig. 10.2A,B), and in the absence of a functional Hes7 protein, the *Hes7* gene is continuously transcribed in the PSM (Bessho *et al.*, 2003). Activation of *Hes7* promoter induces synthesis of *Hes7* mRNA which is then translated to generate Hes7 protein; Hes7 protein levels reach a peak within about 1 h. Hes7 protein binds to the multiple N box sequences located in the *Hes7* promoter, repressing its own expression (Fig. 10.3). This repression leads to disappearance of *Hes7* mRNA, and then the Hes7 proteins disappear within an hour by the ubiquitin–proteasome-mediated degradation. The disappearance of Hes7 protein relieves autorepression, allowing *Hes7* transcription to restart. As a result, Hes7 expression oscillates with a period of about 2 h. Thus, the negative regulation of Hes7 expression by Hes7 protein forms a negative feedback loop that is critical for maintaining oscillations.

2.2. Mathematical simulation of Hes7 oscillations

Based on the negative feedback mechanism, Hes7 oscillations as well as other oscillatory expression patters have been mathematically simulated

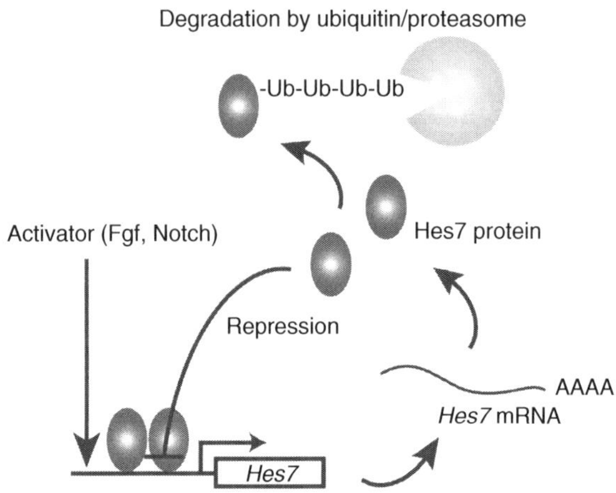

Figure 10.3 Hes7 oscillations are regulated by negative feedback and rapid degradation of gene products. Activation of *Hes7* promoter induces synthesis of both *Hes7* mRNA and Hes7 protein. Hes7 protein then binds to multiple N box sequences of *Hes7* promoter and represses its own expression. This repression leads to disappearance of *Hes7* mRNA and Hes7 protein because they are extremely unstable. Hes7 protein is polyubiquitinated and degraded by proteasome. Disappearance of Hes7 protein relieves negative autoregulation, allowing the next round of expression. As a result, Hes7 expression oscillates in the PSM.

with the following differential equations (Hirata *et al.*, 2004; Jensen *et al.*, 2003; Lewis, 2003; Monk, 2003):

$$\mathrm{d}p(t)/\mathrm{d}t = am(t-Tp) - bp(t)$$

$$\mathrm{d}m(t)/\mathrm{d}t = k/[1 + \{p(t-Tm)\}^2/p_0^2] - cm(t)$$

where $p(t)$ and $m(t)$ are the quantities of functional Hes7 protein and *Hes7* mRNA per cell at time t, respectively, and p_0 is the amount of protein that shows half-maximal inhibition. a is the rate constant for translation, while b and c are the degradation rate constants for Hes7 protein and *Hes7* mRNA, respectively. Confirming that all the important variables were captured by the model, these equations successfully simulated 2-h cycle oscillation of Hes7 in the PSM (Fig. 10.4A). According to this mathematical model, the half-lives of gene products must be sufficiently short (or the degradation rate constants must be sufficiently large) to sustain oscillatory expression, predicting that even a slight increase in the half-life of the gene products should dampen the oscillations (Fig. 10.4A) (Hirata *et al.*, 2004). To evaluate this prediction, mice expressing a stabilized Hes7 protein (with a slightly reduced degradation rate) were produced by introducing a point mutation into the Hes7 locus (changing the 14th amino acid residue from lysine to

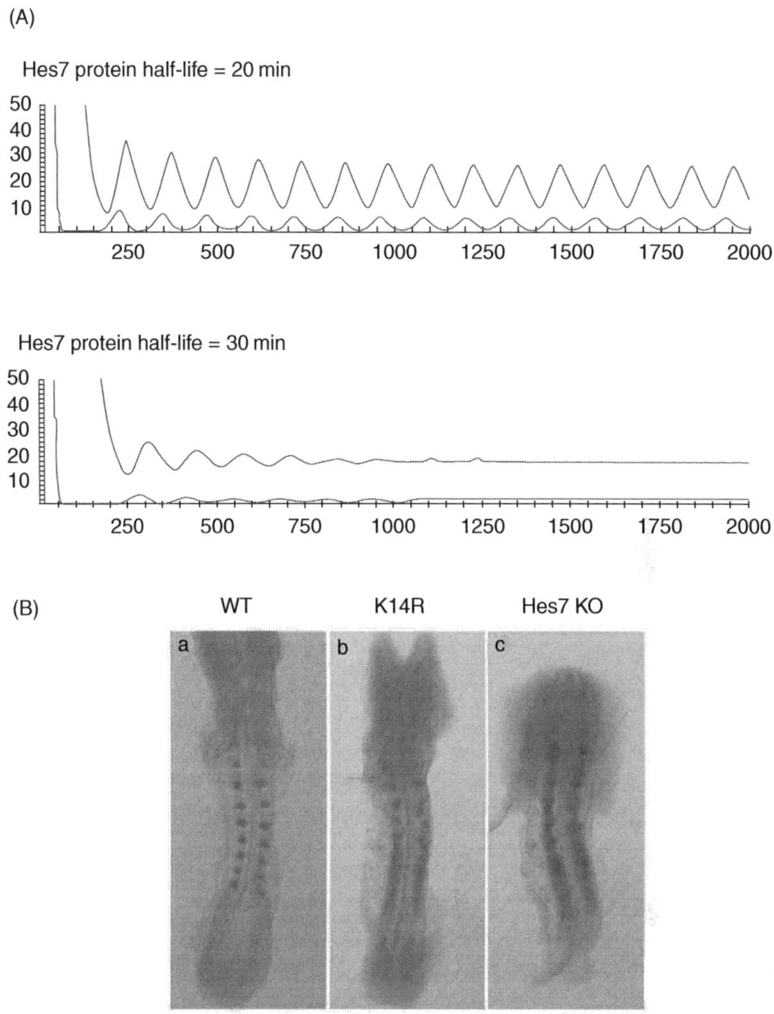

Figure 10.4 Hes7 oscillations depend on the instability of Hes7 protein. (A) Mathematical simulation of Hes7 oscillations. According to the mathematical simulations (Hirata *et al.*, 2004), Hes7 oscillations continue when the half-life of Hes7 protein is 20 min, but not when the half-life of Hes7 protein is 30 min. In the latter case, Hes7 oscillations are damped after three or four cycles. (B) *Uncx4.1* expression was analyzed by *in situ* hybridization with wild-type (a), K14R mutant (b), and *Hes7* knock-out mice (c) at a eight-somite stage. The half-life of wild-type Hes7 protein is about 22 min, while that of Hes7 protein with a K14R mutation (the 14th amino acid residue lysine was changed to arginine) is about 30 min. In mice expressing Hes7 with a K14R mutation (b) instead of wild-type Hes7 protein (a), Hes7 oscillations were damped after three or four cycles. As a result, although the initial three or four pairs of somites were segmented, the following somites were severely fused. These results suggest that sustained Hes7 oscillations depend on the instability of gene products. All somites were segmented in wild-type mice (a) but were severely fused in *Hes7* KO mice (c). This figure is adapted from Hirata *et al.*, 2004. (See Color Insert.)

arginine, K14R). In these mice, Hes7 oscillations stopped after three or four cycles. As a result, although the initial three or four pairs of somites were segmented properly, all following somites were fused (Fig. 10.4B) (Hirata *et al.*, 2004). These results confirmed the prediction from a mathematical model that sustained Hes7 oscillations depend on the instability of gene products.

2.3. Synchronized oscillations underlie the segmentation clock

Expression of Lunatic fringe (Lfng), β1,3-*N*-acetylglucosaminyl-transferase of Notch, also oscillates in the mouse PSM. As seen with Hes7, both loss of expression and sustained expression of Lfng result in severe somite fusion, suggesting that oscillating expression of Lfng is also required for periodic somite segmentation (Evrard *et al.*, 1998; Zhang and Gridley, 1998; Serth *et al.*, 2003). Lfng is constitutively expressed throughout the PSM in the absence of Hes7 (Bessho *et al.*, 2001b), indicating that Lfng oscillations are regulated by Hes7 oscillations (Fig. 10.5A). Lfng acts as an inhibitor of Notch signaling in the PSM, and therefore Lfng oscillations result in oscillating accumulation of NICD (Fig. 10.5A) (Dale *et al.*, 2003; Huppert *et al.*, 2005; Morimoto *et al.*, 2005). NICD levels peak in the anterior where it periodically induces expression of Mesp2, an essential regulator of segment border formation (Morimoto *et al.*, 2005).

Fgf signaling activates extracellular signal-regulated kinase (ERK) by phosphorylation, and then phosphorylated ERK (pERK) induces Dusp4 expression. Dusp4, a phosphatase, can inactivate pERK by dephosphorylation. As a result, the amounts of pERK and Dusp4 oscillate in antiphase (Fig. 10.5A). In *Hes7*-null mice, Dusp4 expression does not oscillate, indicating that Dusp4 oscillations depend on Hes7 oscillations, as Lfng and NICD oscillations do (Niwa *et al.*, 2007). Thus, Hes7 oscillations regulate coupled oscillations of pERK–Dusp4 and Lfng–NICD (Fig. 10.5A) (Niwa *et al.*, 2007).

As discussed above, Hes7 expression is controlled by Notch signaling. However, although Hes7 expression is severely down-regulated in the absence of Notch signaling, oscillating expression still occurs at a low level in the posterior region of the PSM (Niwa *et al.*, 2007). It is likely that Notch signaling regulates the amplification of Hes7 expression and the anterior propagation of Hes7 oscillations but is not involved in initiating Hes7 expression in the posterior region (Fig. 10.5B). Fgf8 is highly expressed in the posterior region of the PSM and forms the posterior-to-anterior gradient. In the absence of Fgf signaling, Hes7 expression disappears completely, suggesting that Fgf signaling initiates Hes7 oscillation, while Notch signaling maintains them (Fig. 10.5B) (Niwa *et al.*, 2007). These results indicate that Fgf and Notch signaling cooperatively regulate Hes7 oscillations, and conversely, Hes7 oscillations couple and synchronize the oscillations of Fgf and Notch signaling (Fig. 10.5A). It is likely that these

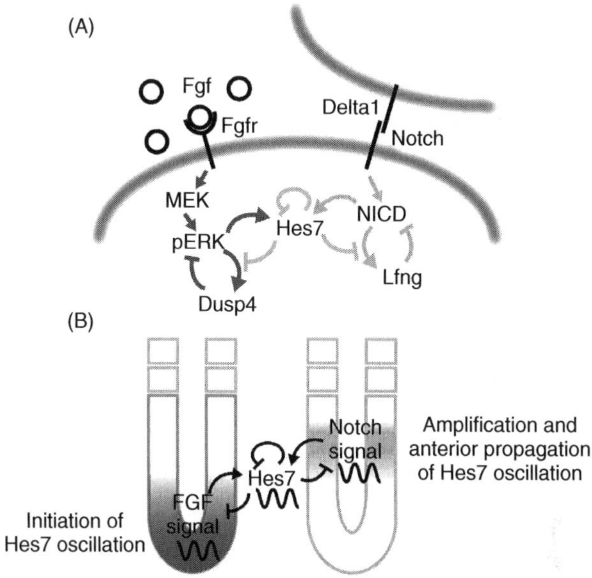

Figure 10.5 Hes7-mediated coupled oscillations in Fgf and Notch signaling underlie the segmentation clock. (A) Hes7 oscillations periodically repress Dusp4, a phosphatase induced by Fgf signaling, and induce Dusp4 oscillation. Fgf signaling activates ERK by phosphorylation, and pERK induces Dusp4 expression, whereas pERK is periodically inactivated by Dusp4 oscillation. As a result, the amounts of pERK and Dusp4 oscillate in antiphase. Hes7 oscillations also periodically repress Lfng. Lfng acts as an inhibitor of Notch signaling in the PSM, and therefore Lfng oscillations lead to oscillating formation of NICD. Thus, pERK–Dusp4 and Lfng–NICD oscillations are coupled by Hes7 oscillations. (B) Hes7 oscillations are initiated by Fgf signaling in the posterior PSM and then amplified and propagated into the anterior PSM by Notch signaling. The coupled oscillator networks comprising Hes7, pERK–Dusp4 and Lfng–NICD underlie the segmentation clock.

interdependent yet coupled oscillations may be sufficient to produce the segmentation clock (Fig. 10.5A, B); however, many other molecules oscillate in the PSM (Dequéant *et al.*, 2006), suggesting that other oscillators, some of which are independent of Hes7, could also regulate the gene networks of the segmentation clock (Aulehla *et al.*, 2008; Ferjentsik *et al.*, 2009).

Oscillations in the PSM are stable in their amplitude and periodicity, but become unstable when PSM cells are dissociated. This result indicates that individual PSM cells cannot maintain a stable oscillator without cell–cell communication (Maroto *et al.*, 2005; Masamizu *et al.*, 2006). It has been suggested that Notch signaling is required to synchronize oscillations between PSM cells because oscillations become desynchronized in the absence of Notch signaling (Horikawa *et al.*, 2006; Jiang *et al.*, 2000; Riedel-Kruse *et al.*, 2007). However, whether Notch is assisted by additional molecular mechanisms involving cell–cell communication remains to be determined. Reconstitution of the synchronized oscillators from nonsynchronized (non-PSM) cells by introducing a minimal set of genes will be required to understand this process.

3. Hes1 Oscillations Regulate Neural Stem Cells

3.1. The role of bHLH genes in neural development

In the developing nervous system, neuroepithelial cells proliferate extensively by repeated replicative, or symmetric, cell division. In such division, each dividing neuroepithelial cell will produce two neuroepithelial cells (Fig. 10.6). As neural development proceeds, neuroepithelial cells become gradually elongated and adopt the fate of radial glial cells. Each radial glial cell has its cell body in an inner layer lining the ventricle, thus called the ventricular zone; each cell extends a long radial fiber that reaches the pial surface (Fig. 10.6). Radial glial cells sequentially give rise to different types of neurons by a different mode of replication called asymmetric cell division, where each radial glial cell divides into one daughter that remains radial glia and another that differentiates into a neuron or a basal progenitor (Fig. 10.6). Basal progenitors further divide to produce more neurons. After neurogenesis is completed, radial glial cells produce other differentiated cell types such as astrocytes (Fig. 10.6). Both neuroepithelial cells and radial glial cells are considered embryonic neural stem cells (Alvarez-Buylla *et al.*, 2001; Fishell and Kriegstein, 2003; Fujita, 2003).

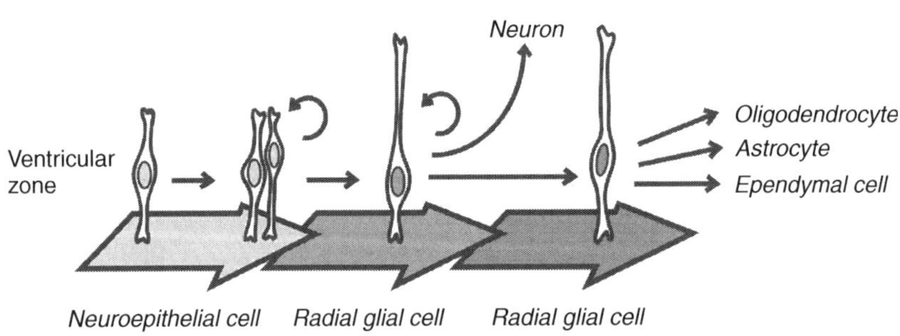

Figure 10.6 Neural stem cells and their proliferation and differentiation. Neuroepithelial cells proliferate extensively by repeating symmetric cell division. As neural development proceeds, neuroepithelial cells are gradually elongated and become radial glial cells. Radial glial cells sequentially give rise to different types of neurons by repeating asymmetric cell division. After neurogenesis, radial glial cells differentiate into oligodendrocytes, ependymal cells, and astrocytes. Both neuroepithelial cells and radial glial cells are considered embryonic neural stem cells.

It has been shown that proneural genes such as *Mash1* and *Neurogenin2* (*Ngn2*), bHLH transcriptional activator genes, are required for and can induce neuronal differentiation (Bertrand *et al.*, 2002; Ross *et al.*, 2003). They also induce the expression of Notch ligands such as Delta1, which activate Notch signaling in the neighboring cells (Fig. 10.1A). Activation of Notch signaling leads to upregulation of *Hes1* and *Hes5*, which repress proneural gene expression, thereby inhibiting neuronal differentiation (Fig. 10.1A) (Ohtsuka *et al.*, 1999, 2001). Thus, differentiating neurons inhibit their neighbors from engaging in neuronal differentiation through Notch signaling (a process known as lateral inhibition, Fig. 10.1A). In the absence of *Hes* genes, proneural genes cannot be repressed after Notch activation, and neural stem cells differentiate prematurely into neurons at the expense of later-born cell types (Hatakeyama *et al.*, 2004; Ishibashi *et al.*, 1995; Tomita *et al.*, 1996). Thus, *Hes* genes are essential for the maintenance of neural stem cells, and thanks to Notch signaling-dependent lateral inhibition, maintenance of neural stem cells and formation of neurons are well balanced in the developing nervous system.

Hes genes are required for the maintenance of neural stem cells, but their expression depends on Delta-expressing, differentiating neurons capable of activating Notch on stem cells. This mechanism raises one important question: during the early stages of neural development, before any mature neurons are formed, how was Notch activated, and how were neural stem cells maintained? The answer emerged from developmental studies in the mouse brain. In the dorsal telencephalon of mouse embryos, formation of neurons occurs from around embryonic day 11 (E11) onward, and expression analysis indicates that proneural genes, the Notch ligand *Delta1* and the Notch target *Hes1* are all expressed at variable levels in a salt-and-pepper pattern by dividing neuroepithelial cells beginning as early as E8.5. When Hes1 expression is high, expression of Delta1 and proneural genes is low, and when Hes1 expression is low, proneural genes and Notch ligands are highly expressed. This indicates that an inverse correlation exists between Hes1 and proneural/Delta1 expression levels and suggests that Notch signaling is active without any mature neurons. Why proneural gene expression is insufficient to induce neuronal formation before E11 is unknown.

3.2. Notch signaling oscillations in embryonic neural stem cells

Hes1 expression oscillates with a period of about 2 h in many cell types, and this oscillation is regulated by negative feedback, like Hes7 oscillations (Hirata *et al.*, 2002). Real-time imaging analysis demonstrated that Hes1 expression oscillates with a period of about 2–3 h in embryonic neural stem cells, although the period and amplitude are variable from cell to cell and from cycle to cycle (Shimojo *et al.*, 2008), suggesting that Hes1 oscillations are not synchronized and consistent with the hypothesis that Hes1 is important to generate heterogeneity among cells rather than to function as a neuronal biological clock. Real-time imaging analysis demonstrated that *Ngn2* and *Delta1* expression also oscillates in neural stem cells, but not in differentiating neurons that lost Hes1 expression where expression levels are sustained (Shimojo *et al.*, 2008). These results suggest that Hes1 oscillation drives *Ngn2* and *Delta1* oscillations in neural stem cells (Fig. 10.7). Sustained expression of Ngn2 seems to be required for neuronal differentiation, because many downstream genes respond rather slowly to Ngn2 expression (Heng *et al.*, 2008). It is likely that when *Ngn2* expression is oscillating, only early-response genes such as *Delta1* are induced, and that *Delta1* oscillations activate Notch signaling between neural stem cells (Fig. 10.7A), activate Hes1, and prevent late targets of Ngn2 from being induced. Thus, *Ngn2* oscillation contributes to the maintenance of neural stem cells without induction of neuronal formation. These results suggest that depending on *Hes1, Ngn2* can lead to two opposite outcomes: when

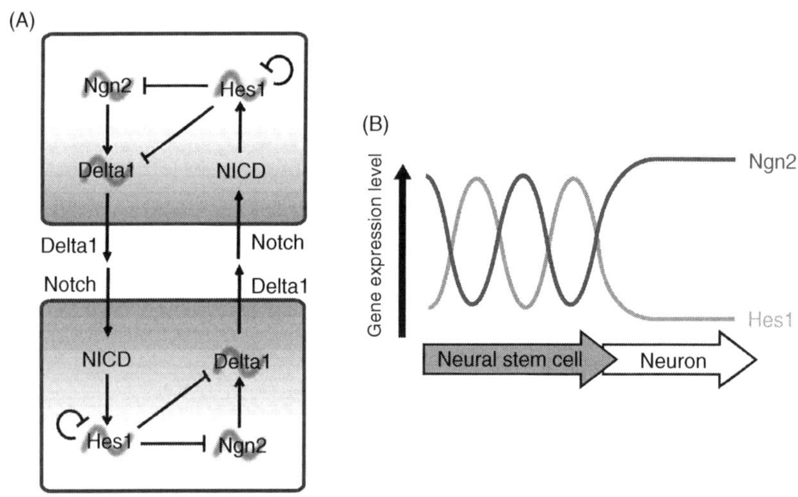

Figure 10.7 Oscillations in Notch signaling regulate maintenance of neural stem cells. (A) Hes1 expression oscillates in neural stem cells, and Hes1 oscillations drive *Ngn2* and *Delta1* oscillations. Delta1 oscillations lead to mutual activation of Notch signaling in neural stem cells. (B) Hes1 expression oscillates in neural stem cells. When Hes1 is repressed, Ngn2 expression becomes sustained. Sustained expression of Ngn2 seems to be required for neuronal differentiation, because many downstream genes respond rather slowly to Ngn2 expression. When Ngn2 expression is oscillating, only early-response genes such as *Delta1* are selectively induced, which is good for maintenance of neural stem cells. Thus, depending on the expression mode, *Ngn2* can lead to two opposite outcomes: when the expression is oscillating, *Ngn2* can maintain neural stem cells, whereas when the expression is sustained, *Ngn2* can induce neuronal differentiation.

the expression is oscillating, *Ngn2* can maintain neural stem cells by activating Delta1 and indirectly Hes1; when *Ngn2* expression is sustained, Ngn2 protein will induce neuronal differentiation (Fig. 10.7B).

The above findings raise a question about the interpretation of salt-and-pepper patterns of proneural genes and *Delta1* expression in the developing nervous system. According to the classical view, certain cells express a higher level of Delta1 due to stochastic variations, activating Notch signaling more in their neighboring cells. As a result, the latter cells express less Delta1, while the former cells receive less Notch activation and express more Delta1, amplifying a stochastic difference in the level of Notch signaling. As Delta1-high cells express even more Delta1, their neighbors express less Delta1 and accumulate more NICD and Hes proteins. As a result, Delta1-high cells have become the first neurons, while Delta1-low cells will remain neural stem cells. This view suggests that the salt-and-pepper pattern reflects the neuronal selection process (Artavanis-Tsakonas *et al.*, 1999). However, the results of real-time imaging revealed that this salt-and-pepper pattern reflects oscillating gene expression, with positive cells becoming negative and negative cells becoming positive at the next moment. Thus, positive and negative cells seem to be equivalent to each other (Kageyama *et al.*, 2008), and none appears to be selected for a particular fate at this stage.

3.3. Sustained Hes1 expression in boundary regions

The developing nervous system is partitioned into many compartments by boundaries such as the isthmus and zona limitans intrathalamica (Zli, Fig. 10.8A). The nervous system is also divided into the right and left halves by the roof plate and the floor plate (Fig. 10.8A). Very few cells can cross these boundary regions into different compartments, and therefore each compartment develops as a unit. Boundary cells express morphogens such as Fgf8 and regulate regional specification of adjacent compartments (Kiecker and Lumsden, 2005). Cells in compartments proliferate intensively and divide asymmetrically, maintaining a progenitor population and differentiating into neurons, whereas cells in boundary regions proliferate poorly and do not give rise to neurons. In boundary regions, Hes1 is highly and stably expressed, and this expression does not oscillate (Fig. 10.8B). Forced, sustained expression of Hes1 in compartmental neural stem cells inhibits both proliferation and neuronal differentiation by repressing the expression of cell cycle regulator genes such as cyclins and proneural genes such as *Mash1* and *Ngn2* (Baek *et al.*, 2006; Shimojo *et al.*, 2008). Conversely, in mice lacking *Hes* genes, cells in boundary regions gain expression of proneural genes and differentiate into neurons, suggesting that persistent Hes1 expression is important for maintenance of boundary cells (Baek *et al.*, 2006). Thus, it is likely that the expression mode (oscillating versus

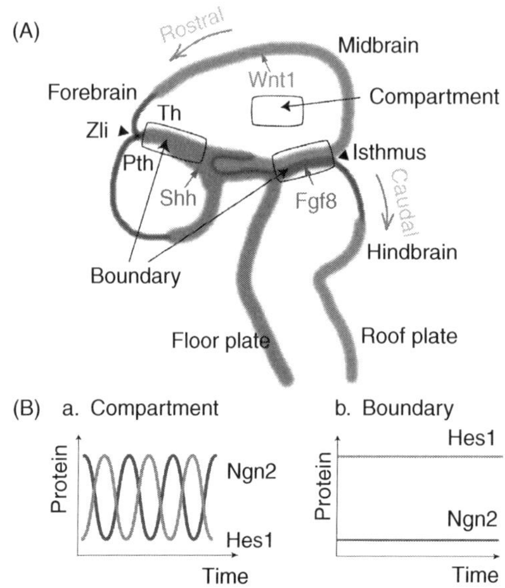

Figure 10.8 Different mode of Hes1 expression in the developing nervous system. (A) The developing nervous system is partitioned into many compartments by boundaries such as the isthmus and zona limitans intrathalamica (Zli). The nervous system is also divided into the right and left halves by the roof plate and the floor plate. These boundary cells express morphogens such as Fgf8 and regulate regional specification of adjacent compartments. Cells in boundary regions proliferate poorly and do not give rise to neurons, whereas cells in compartments proliferate intensively and differentiate into neurons. Pth, prethalamus; Th, thalamus. (B) Hes1 expression oscillates in compartment cells but is sustained in boundary cells. As a result, expression of proneural genes such as *Ngn2* oscillates in compartment cells but is repressed in boundary cells. The expression mode (oscillating versus sustained) of Hes1 regulates different characteristics of cell proliferation and differentiation between compartment and boundary cells. (See Color Insert.)

sustained) of Hes1 is involved in defining the characteristics of cell proliferation and differentiation in compartment and boundary cells: oscillating Hes1 expression is required in compartment cells, while sustained Hes1 expression is required in boundary cells (Fig. 10.8B).

The precise mechanism of how oscillating versus sustained Hes1 expression is regulated remains to be determined, but it seems to involve yet another feedback loop. In fibroblasts, Jak-Stat signaling is involved in Hes1 oscillation. Jak activates Stat3 by phosphorylation, and phosphorylated Stat3 (pStat3) induces expression of downstream genes such as Socs3. Socs3 dephosphorylates pStat3, thereby inactivating Stat3. Due to this negative feedback, the amounts of pStat3 and Socs3 oscillate in antiphase (Yoshiura *et al.*, 2007). It seems that pStat3 oscillations periodically destabilize Hes1 protein, contributing to Hes1 oscillation. Inhibition of Jak activity dampens Hes1 oscillation not only in fibroblasts but also in neural stem cells

(Yoshiura *et al.*, 2007; Shimojo *et al.*, 2008), suggesting that Jak-Stat signaling regulates Hes1 oscillations in these cells. In neural stem cells, Hes1 protein associates with both Jak and Stat and promotes phosphorylation of Stat by Jak (Kamakura *et al.*, 2004). Thus, Hes1 oscillation and Jak-Stat signaling seem to depend on each other in neural stem cells.

The above results indicate that oscillating gene expression is important for proliferation and differentiation of neural stem cells and that a gene expression mode is involved in specification of compartment and boundary cells. Whether these oscillators regulate the timing of differentiation in the nervous system as a clock remains to be determined.

4. Hes1 Oscillations Regulate ES Cell Differentiation

ES cells have the ability to differentiate into cell types of all three germ layers (pluripotency), and application of these cells to regenerative medicine is greatly anticipated. However, ES cells are heterogeneous in response to differentiation cues, and it is very difficult to direct their differentiation into a pure population of desired cell types. The molecular mechanism of such heterogeneity of ES cells in differentiation responses is not well understood. Recent studies have shown that expression of the homeodomain factor Nanog changes over several days in ES cells and that ES cells expressing different levels of Nanog display different differentiation competency (Chambers *et al.*, 2007). However, even Nanog-positive ES cells are not homogeneous, suggesting that expression of other genes also changes dynamically.

Hes1 is highly expressed by ES cells under the control of BMP and LIF (Fig. 10.9), and real-time imaging revealed that Hes1 expression oscillates in ES cells with a period of about 3–5 h, which is much faster than Nanog fluctuation (Kobayashi *et al.*, 2009). Furthermore, Hes1 oscillations lead to dynamic changes in expression of downstream genes such as *Delta1* and the cell cycle inhibitor *Gadd45g* (Fig. 10.9) (Kobayashi *et al.*, 2009). To examine whether Hes1 oscillations impact differentiation competency, ES cells were generated in which the Venus fluorescence gene was knocked in frame to the first Hes1 exon. These ES cells express Venus–Hes1 fusion protein, allowing Hes1-high and Hes1-low ES cells to be separated according to the fluorescence intensity. Following sorting, Hes1-high and Hes1-low ES cells were subjected to differentiation conditions, and it was found that whereas Hes1-high ES cells tended to differentiate into mesodermal cells, Hes1-low ES cells tended to differentiate into neural cells (Fig. 10.10). This suggested that Hes1 oscillations contribute to the heterogeneity of ES cell differentiation (Kobayashi *et al.*, 2009). In agreement with this idea, *Hes1*-null ES cells

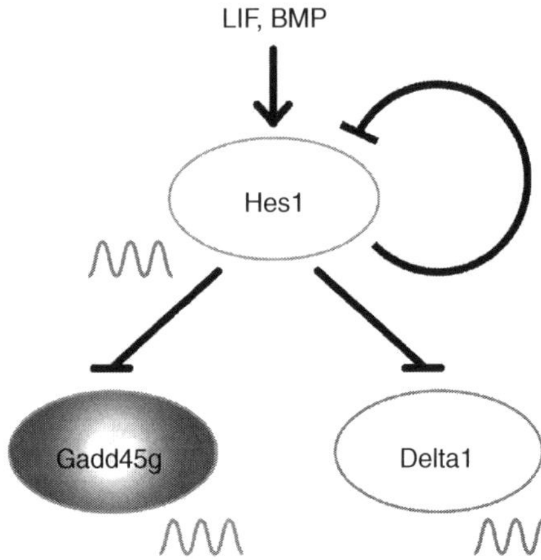

Figure 10.9 Hes1 oscillations regulate the differentiation and proliferation of ES cells. Hes1 expression oscillates in ES cells with a period of about 3–5 h under the control of BMP and LIF. Hes1 oscillations lead to dynamic changes in expression of downstream genes such as the Notch ligand *Delta1* and the cell cycle inhibitor *Gadd45g*. Thus, Hes1 oscillations affect the differentiation and proliferation of ES cells by regulating the expression of downstream genes.

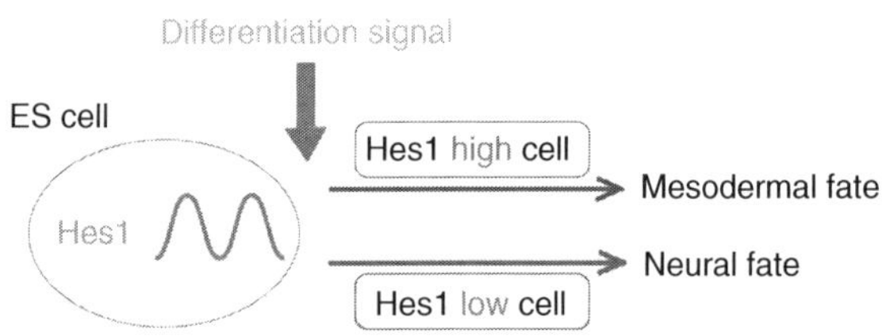

Figure 10.10 Hes1 oscillations contribute to the heterogeneity of ES cell differentiation. Hes1 expression oscillates in ES cells. When differentiation cues come, Hes1-high ES cells tend to differentiate into mesodermal cells, while Hes1-low ES cells tend to differentiate into neural cells. Thus, Hes1 oscillations contribute to the heterogeneity of ES cell differentiation.

differentiated into neural cells earlier and more uniformly than wild-type ES cells (Kobayashi *et al.*, 2009).

It was previously reported that inactivation of Notch signaling in ES cells favors cardiac mesoderm differentiation (Jang *et al.*, 2008; Nemir *et al.*, 2006; Schroeder *et al.*, 2003), whereas forced activation of Notch signaling advances neural differentiation (Lowell *et al.*, 2006). Interestingly, upregulation of Hes1 inactivates Notch signaling by repressing Delta1 expression, while downregulation of Hes1 activates Notch signaling by inducing Delta1 expression in ES cells (Kobayashi and Kageyama, 2010). Thus, it is likely that cyclic expression of Hes1 dynamically changes the activity of Notch signaling and thereby regulates the fate preference of ES cells.

5. Conclusions

The above results suggest that oscillatory expression with short periods (ultradian oscillation) is important for many biological events. During somite segmentation, oscillations occur in a synchronized manner between neighboring cells in the PSM, and each cycle leads to segmentation of a bilateral pair of somites. In this case, oscillators function as a biological clock. However, unlike the circadian clock, the oscillation period of the segmentation clock is species specific and temperature dependent: the period is about 2 h in mice at 37°C but different at different temperatures and in different species. In contrast, the circadian rhythm is always about 24 h irrespective of temperature, cell types, and species. A precise mechanism of how the period of the segmentation clock is determined in a species-specific manner remains to be analyzed. Another feature is that except for the PSM, Hes1 oscillations are not synchronized between neighboring cells and that the period is variable from cycle to cycle and from cell to cell (Masamizu *et al.*, 2006). Why only PSM cells can synchronize with each other remains to be determined. Nonsynchronized oscillations in non-PSM cells make heterogeneous Hes1 expression levels even in clonal cell populations and is therefore advantageous to make different responses to environmental conditions. Similarly, Hes1 expression oscillates in ES cells, and variable Hes1 levels contribute to multipotency of ES cells. Importantly, when Hes1 expression becomes steady instead of oscillatory, neural stem cells become dormant, and ES cells display less heterogeneity in the differentiation timing and the fate choice. Thus, not only the expression but also the expression mode (oscillatory versus sustained) is very important for many biological events.

Oscillating gene expression is not unique to Notch signaling. For example, Jak-Stat signaling and Fgf signaling exhibit oscillating responses (Yoshiura *et al.*, 2007; Nakayama *et al.*, 2008). In addition, p53 induces Mdm2, an E3 ubiquitin ligase, which leads to degradation of p53, and this negative feedback makes p53 oscillations (Lahav *et al.*, 2004). It has been suggested that the number of p53 pulses is important for the choice of subsequent events, DNA repair or apoptosis. Another example is NF-κB. This factor induces IκB, which traps NF-κB in the cytoplasm, and this negative feedback makes oscillatory transition of NF-κB between the nucleus and the cytoplasm (Hoffmann *et al.*, 2002). Thus, it is likely that oscillating gene expression occurs more widely than previously thought. Because the expression mode and the number of pulses are important for the choice of subsequent outcomes, further analysis on the dynamics of gene expression will be required to understand the complex regulatory mechanism of many biological events.

REFERENCES

Alvarez-Buylla, A., Garcia-Verdugo, J. M., and Tramontin, A. D. (2001). A unified hypothesis on the lineage of neural stem cells. *Nat. Rev. Neurosci.* **2**, 287–293.

Artavanis-Tsakonas, S., Rand, M. D., and Lake, R. J. (1999). Notch signaling: cell fate control and signal integration in development. *Science* **284**, 770–776.

Aulehla, A., Wiegraebe, W., Baubet, V., Wahl, M. B., Deng, C., Taketo, M., Lewandoski, M., and Pourquié, O. (2008). A beta-catenin gradient links the clock and wavefront systems in mouse embryo segmentation. *Nat. Cell Biol.* **10**, 186–193.

Baek, J. H., Hatakeyama, J., Sakamoto, S., Ohtsuka, T., and Kageyama, R. (2006). Persistent and high levels of Hes1 expression regulate boundary formation in the developing central nervous system. *Development* **133**, 2467–2476.

Bertrand, N., Castro, D. S., and Guillemot, F. (2002). Proneural genes and the specification of neural cell types. *Nat. Rev. Neurosci.* **3**, 517–530.

Bessho, Y., Hirata, H., Masamizu, Y., and Kageyama, R. (2003). Periodic repression by the bHLH factor Hes7 is an essential mechanism for the somite segmentation clock. *Genes Dev.* **17**, 1451–1456.

Bessho, Y., Miyoshi, G., Sakata, R., and Kageyama, R. (2001a). *Hes7*: a bHLH-type repressor gene regulated by Notch and expressed in the presomitic mesoderm. *Genes Cells* **6**, 175–185.

Bessho, Y., Sakata, R., Komatsu, S., Shiota, K., Yamada, S., and Kageyama, R. (2001b). Dynamic expression and essential functions of *Hes7* in somite segmentation. *Genes Dev.* **15**, 2642–2647.

Chambers, I., Silva, J., Colby, D., Nichols, J., Nijmeijer, B., Robertson, M., Vrana, J., Jones, K., Grotewold, L., and Smith, A. (2007). Nanog safeguards pluripotency and mediates germline development. *Nature* **450**, 1230–1234.

Dale, J. K., Maroto, M., Dequeant, M. L., Malapert, P., McGrew, M., and Pourquié, O. (2003). Periodic Notch inhibition by Lunatic Fringe underlies the chick segmentation clock. *Nature* **421**, 275–278.

Dawson, S. R., Turner, D. L., Weintraub, H., and Parkhurst, S. M. (1995). Specificity for the hairy/enhancer of split basic helix-loop-helix (bHLH) proteins maps outside the bHLH domain and suggests two separable modes of transcriptional repression. *Mol. Cell. Biol.* **15**, 6923–6931.

Dequéant, M.-L., Glynn, E., Gaudenz, K., Wahl, M., Chen, J., Mushegian, A., and Pourquié, O. (2006). A complex oscillating network of signaling genes underlies the mouse segmentation clock. *Science* **314**, 1595–1598.

Dequéant, M. L., and Pourquié, O. (2008). Segmental patterning of the vertebrate embryonic axis. *Nat. Rev. Genet.* **9**, 370–382.

Evrard, Y. A., Lun, Y., Aulehla, A., Gan, L., and Johnson, R. L. (1998). *Lunatic fringe* is an essential mediator of somite segmentation and patterning. *Nature* **394**, 377–381.

Ferjentsik, Z., Hayashi, S., Dale, J. K., Bessho, Y., Herreman, A., De Strooper, B., del Monte, G., de la Pompa, J. L., and Maroto, M. (2009). Notch is a critical component of the mouse somitogenesis oscillator and is essential for the formation of the somites. *PLoS Genet.* **5**, e1000662.

Fishell, G., and Kriegstein, A. R. (2003). Neurons from radial glia: the consequences of asymmetric inheritance. *Curr. Opin. Neurobiol.* **13**, 34–41.

Fujita, S. (2003). The discovery of the matrix cell, the identification of the multipotent neural stem cell and the development of the central nervous system. *Cell Struct. Funct.* **28**, 205–228.

Hatakeyama, J., Bessho, Y., Katoh, K., Ookawara, S., Fujioka, M., Guillemot, F., and Kageyama, R. (2004). Hes genes regulate size, shape and histogenesis of the nervous system by control of the timing of neural stem cell differentiation. *Development* **131**, 5539–5550.

Heng, J. I., Nguyen, L., Castro, D. S., Zimmer, C., Wildner, H., Armant, O., Skowronska-Krawczyk, D., Bedogni, F., Matter, J. M., Hevner, R., and Guillemot, F. (2008). Neurogenin 2 controls cortical neuron migration through regulation of *Rnd2*. *Nature* **455**, 114–118.

Hirata, H., Bessho, Y., Kokubu, H., Masamizu, Y., Yamada, S., Lewis, J., and Kageyama, R. (2004). Instability of Hes7 protein is crucial for the somite segmentation clock. *Nat. Genet.* **36**, 750–754.

Hirata, H., Yoshiura, S., Ohtsuka, T., Bessho, Y., Harada, T., Yoshikawa, K., and Kageyama, R. (2002). Oscillatory expression of the bHLH factor Hes1 regulated by a negative feedback loop. *Science* **298**, 840–843.

Hoffmann, A., Levchenko, A., Scott, M. L., and Baltimore, D. (2002). The IκB—NF-κB signaling module: temporal control and selective gene activation. *Science* **298**, 1241–1245.

Honjo, T. (1996). The shortest path from the surface to the nucleus: RBP-J κ/Su(H) transcription factor. *Genes Cells* **1**, 1–9.

Horikawa, K., Ishimatsu, K., Yoshimoto, E., Kondo, S., and Takeda, H. (2006). Noise-resistant and synchronized oscillation of the segmentation clock. *Nature* **441**, 719–723.

Huppert, S. S., Ilagan, M.X.G., De Strooper, B., and Kopan, R. (2005). Analysis of Notch function in presomitic mesoderm suggests a γ-secretase-independent role for presenilins in somite differentiation. *Dev. Cell* **8**, 677–688.

Ishibashi, M., Ang, S. L., Shiota, K., Nakanishi, S., Kageyama, R., and Guillemot, F. (1995). Targeted disruption of mammalian hairy and Enhancer of split homolog-1 (HES-1) leads to up-regulation of neural helix-loop-helix factors, premature neurogenesis, and severe neural tube defects. *Genes Dev.* **9**, 3136–3148.

Iso, T., Kedes, L., and Hamamori, Y. (2003). HES and HERP families: multiple effectors of the Notch signaling pathway. *J. Cell. Physiol.* **194**, 237–255.

Jang, J., Ku, S. Y., Kim, J. E., *et al.* (2008). Notch inhibition promotes human embryonic stem cell-derived cardiac mesoderm differentiation. *Stem Cells* **26**, 2782–2790.

Jensen, M. H., Sneppen, K., and Tiana, G. (2003). Sustained oscillations and time delays in gene expression of protein Hes1. *FEBS Lett.* **541**, 176–177.

Jiang, Y. -J., Aerne, B. L., Smithers, L., Haddon, C., Ish-Horowicz, D., and Lewis, J. (2000). Notch signaling and the synchronization of the somite segmentation clock. *Nature* **408**, 475–479.

Jouve, C., Palmeirim, I., Henrique, D., Beckers, J., Gossler, A., Ish-Horowicz, D., and Pourquié, O. (2000). Notch signalling is required for cyclic expression of the hairy-like gene *HES1* in the presomitic mesoderm. *Development* **127**, 1421–1429.

Kageyama, R., Ohtsuka, T., and Kobayashi, T. (2007). The Hes gene family: repressors and oscillators that orchestrate embryogenesis. *Development* **134**, 1243–1251.

Kageyama, R., Ohtsuka, T., Shimojo, H., and Imayoshi, I. (2008). Dynamic Notch signaling in neural progenitor cells and a revised view of lateral inhibition. *Nat. Neurosci.* **11**, 1247–1251.

Kamakura, S., Oishi, K., Yoshimatsu, T., Nakafuku, M., Masuyama, N., and Gotoh, Y. (2004). Hes binding to STAT3 mediates crosstalk between Notch and JAK-STAT signaling. *Nat. Cell Biol.* **6**, 547–554.

Kang, S. A., Seol, J. H., and Kim, J. (2005). The conserved WRPW motif of Hes6 mediates proteasomal degradation. *Biochem. Biophys. Res. Commun.* **332**, 33–36.

Kiecker, C., and Lumsden, A. (2005). Compartments and their boundaries in vertebrate brain development. *Nat. Rev. Neurosci.* **6**, 553–564.

Kobayashi, T., Mizuno, H., Imayoshi, I., Furusawa, C., Shirahige, K., and Kageyama, R. (2009). The cyclic gene Hes1 contributes to diverse differentiation responses of embryonic stem cells. *Genes Dev.* **23**, 1870–1875.

Kobayashi, T., and Kageyama, R. (2010). Hes1 regulates embryonic stem cell differentiation by suppressing Notch signaling. *Genes Cells* in press.

Kopan, R., and Ilagan, M.X.G. (2009). The canonical Notch signaling pathway: unfolding the activation mechanism. *Cell* **137**, 216–233.

Lahav, G., Rosenfeld, N., Sigal, A., Geva-Zatorsky, N., Levine, A. J., Elowitz, M. B., and Alon, U. (2004). Dynamics of the p53-Mdm2 feedback loop in individual cells. *Nat. Genet.* **36**, 147–150.

Lewis, J. (2003). Autoinhibition with transcriptional delay: a simple mechanism for the zebrafish somitogenesis oscillator. *Curr. Biol.* **13**, 1398–1408.

Lowell, S., Benchoua, A., Heavey, B., and Smith, A. G. (2006). Notch promotes neural lineage entry by pluripotent embryonic stem cells. *PLoS Biol.* **4**, 805–818.

Mara, A., and Holley, S. A. (2007). Oscillators and the emergence of tissue organization during zebrafish somitogenesis. *Trends Cell Biol.* **17**, 593–599.

Maroto, M., Dale, J. K., Dequéant, M.-L., Petit, A.-C., and Pourquié, O. (2005). Synchronised cycling gene oscillations in presomitic mesoderm cells require cell-cell contact. *Int. J. Dev. Biol.* **49**, 309–315.

Masamizu, Y., Ohtsuka, T., Takashima, Y., Nagahara, H., Takenaka, Y., Yoshikawa, K., Okamura, H., and Kageyama, R. (2006). Real-time imaging of the somite segmentation clock: revelation of unstable oscillators in the individual presomitic mesoderm cells. *Proc. Natl. Acad. Sci. U.S.A.* **103**, 1313–1318.

Monk, N.A.M. (2003). Oscillatory expression of Hes1, p53, and NF-κB driven by transcriptional time delays. *Curr. Biol.* **13**, **1409–1413**.

Morimoto, M., Takahashi, Y., Endo, M., and Saga, Y. (2005). The Mesp2 transcription factor establishes segmental borders by suppressing Notch activity. *Nature* **435**, 354–359.

Nakayama, K., Satoh, T., Igari, A., Kageyama, R., and Nishida, E. (2008). FGF induces oscillations of Hes1 expression and Ras/ERK activation. *Curr. Biol.* **18**, R332–R334.

Nemir, M., Croquelois, A., Pedrazzini, T., and Radtke, F. (2006). Induction of cardiogenesis in embryonic stem cells via downregulation of Notch1 signaling. *Circ. Res.* **98**, 1471–1478.

Niwa, Y., Masamizu, Y., Liu, T., Nakayama, R., Deng, C.-X., and Kageyama, R. (2007). The initiation and propagation of Hes7 oscillation are cooperatively regulated by Fgf and Notch signaling in the somite segmentation clock. *Dev. Cell* **13**, 298–304.

Ohtsuka, T., Ishibashi, M., Gradwohl, G., Nakanishi, S., Guillemot, F., and Kageyama, R. (1999). *Hes1* and *Hes5* as Notch effectors in mammalian neuronal differentiation. *EMBO J.* **18**, 2196–2207.

Ohtsuka, T., Sakamoto, M., Guillemot, F., and Kageyama, R. (2001). Roles of the basic helix-loop-helix genes *Hes1* and *Hes5* in expansion of neural stem cells of the developing brain. *J. Biol. Chem.* **276**, 30467–30474.

Pourquié, O. (2003). The ,segmentation clock: converting embryonic time into spatial pattern. *Science* **301**, 328–330.

Riedel-Kruse, I. H., Müller, C., and Oates, A. C. (2007). Synchrony dynamics during initiation, failure, and rescue of the segmentation clock. *Science* **317**, 1911–1915.

Ross, S. E., Greenberg, M. E., and Stiles, C. D. (2003). Basic helix-loop-helix factors in cortical development. *Neuron* **39**, 13–25.

Sasai, Y., Kageyama, R., Tagawa, Y., Shigemoto, R., and Nakanishi, S. (1992). Two mammalian helix-loop-helix factors structurally related to *Drosophila hairy* and *Enhancer of split*. *Genes Dev.* **6**, 2620–2634.

Schroeder, T., Fraser, S. T., Ogawa, M., Nishikawa, S., Oka, C., Bornkamm, G. W., Nishikawa, S.-I., Honjo, T., and Just, U. (2003). Recombination signal sequence-binding protein Jk alters mesodermal cell fate decisions by suppressing cardiomyogenesis. *Proc. Natl. Acad. Sci. U.S.A.* **100**, 4018–4023.

Serth, K., Schuster-Gossler, K., Cordes, R., and Gossler, A. (2003). Transcriptional oscillation of Lunatic fringe is essential for somitogenesis. *Genes Dev.* **17**, 912–925.

Shimojo, H., Ohtsuka, T., and Kageyama, R. (2008). Oscillations in notch signaling regulate maintenance of neural progenitors. *Neuron* **58**, 52–64.

Taelman, V., Van Wayenbergh, R., Solter, M., Pichon, B., Pieler, T., Christophe, D., and Bellefroid, E. J. (2004). Sequences downstream of the bHLH domain of the *Xenopus* hairy-related transcription factor-1 act as an extended dimerization domain that contributes to the selection of the partners. *Dev. Biol.* **276**, 47–63.

Takebayashi, K., Sasai, Y., Sakai, Y., Watanabe, T., Nakanishi, S., and Kageyama, R. (1994). Structure, chromosomal locus, and promoter analysis of the gene encoding the mouse helix-loop-helix factor HES-1: negative autoregulation through the multiple N box elements. *J. Biol. Chem.* **269**, 5150–5156.

Tomita, K., Ishibashi, M., Nakahara, K., Ang, S.-L., Nakanishi, S., Guillemot, F., and Kageyama, R. (1996). Mammalian *hairy* and *Enhancer of split* homolog 1 regulates differentiation of retinal neurons and is essential for eye morphogenesis. *Neuron* **16**, 723–734.

Yoshiura, S., Ohtsuka, T., Takenaka, Y., Nagahara, H., Yoshikawa, K., and Kageyama, R. (2007). Ultradian oscillations of Stat, Smad, and Hes1 expression in response to serum. *Proc. Natl. Acad. Sci. U.S.A.* **104**, 11292–11297.

Zhang, N., and Gridley, T. (1998). Defects in somite formation in lunatic fringe-deficient mice. *Nature* **394**, 374–377.

CHAPTER ELEVEN

Notch Signaling in Cardiac Development and Disease

Donal MacGrogan, Meritxell Nus, *and* José Luis de la Pompa

Contents

Abstract

The Notch-signaling pathway is involved in multiple processes during vertebrate cardiac development. Cardiomyocyte differentiation, patterning of the different cardiac regions, valve development, ventricular trabeculation, and outflow tract development have all been shown to depend on the activity of specific Notch-signaling elements. From these studies, it becomes obvious that Notch regulates in a cell autonomous or non-cell autonomous manner different signaling pathways, pointing to a role for Notch as a signal coordinator during cardiogenesis. While most of the research has concentrated on Notch signaling in the myocardium, the importance of Notch activity in the cardiac endothelium (endocardium) must not be overlooked. Endocardial Notch activity is crucial for valve and ventricular trabeculae development, two processes that illustrate the role of Notch as a signal coordinator. The importance of Notch signaling in human disease is evident from the discovery that many mutations in components of this pathway segregate in several inherited and acquired disorders.

Laboratorio de Biología Celular y del Desarrollo, **Dpto. de Biología del Desarrollo Cardiovascular, Centro Nacional de Investigaciones Cardiovasculares (CNIC), Madrid, Spain**

Current Topics in Developmental Biology, Volume 92
ISSN 0070-2153, DOI 10.1016/S0070-2153(10)92011-5

This reflects the fundamental roles that Notch performs during cardiac ontogeny. This review examines the experimental evidence supporting a role for Notch in cardiac development and adult heart homeostasis, and how dysregulated Notch signaling may lead to cardiac disease in the newborn and in the adult.

1. Outline of Cardiac Development

The heart is the first organ to develop and function during embryogenesis, arising shortly after gastrulation from the splanchnic mesoderm at the cranial end of the embryo. The cranial mesoderm distributes and forms a crescent that comprises at least two distinct populations of precardiac cells, the so-called first and second heart fields (Fig. 11.1A). Both heart fields contain the progenitors of the earliest cardiac tissues, the myocardium, and the endocardium. As development proceeds, the bilateral cardiac primordia coalesce and fuse at the embryonic midline to form a single, straight heart tube (Fig. 11.1B). This primary outline of the cardiac organ constitutes cells derived from the first heart field. The heart tube contains two tissues, the inner endocardium, and the outer myocardium which are separated by a thick extracellular matrix termed cardiac jelly. Shortly after the straight heart tube is formed it initiates a characteristic R-looping (Fig. 11.1C; right heart tube bending) that will realign cardiac domains, allowing for the initiation of the complex processes leading to cardiac septation and valve formation. This process takes place while secondary heart field (SHF) cells incorporate into both the anterior (arterial) and posterior (venous) poles of the heart tube. The recruitment of cells to the myocardium and endocardium substantially increases the length of the primary heart tube during cardiac looping. Once a third layer (the outermost cardiac tissue layer or epicardium) forms over the myocardium, the morphological differentiation of the different cardiac regions becomes apparent (Fig. 11.1D). Notch makes crucial contributions to multiple processes during cardiac development both in endocardium and in myocardium. These include chamber cardiomyocyte differentiation, patterning of the different cardiac regions, heart valve development, ventricular trabeculation, and outflow tract (OFT) development. The expression pattern of Notch-signaling elements in the developing heart is very complex and in the adult heart is still under study. Figure 11.1E shows the expression of Notch ligands and receptors in the E10.5 heart. In the following sections we will discuss the experimental data supporting a role for Notch in cardiac development and disease.

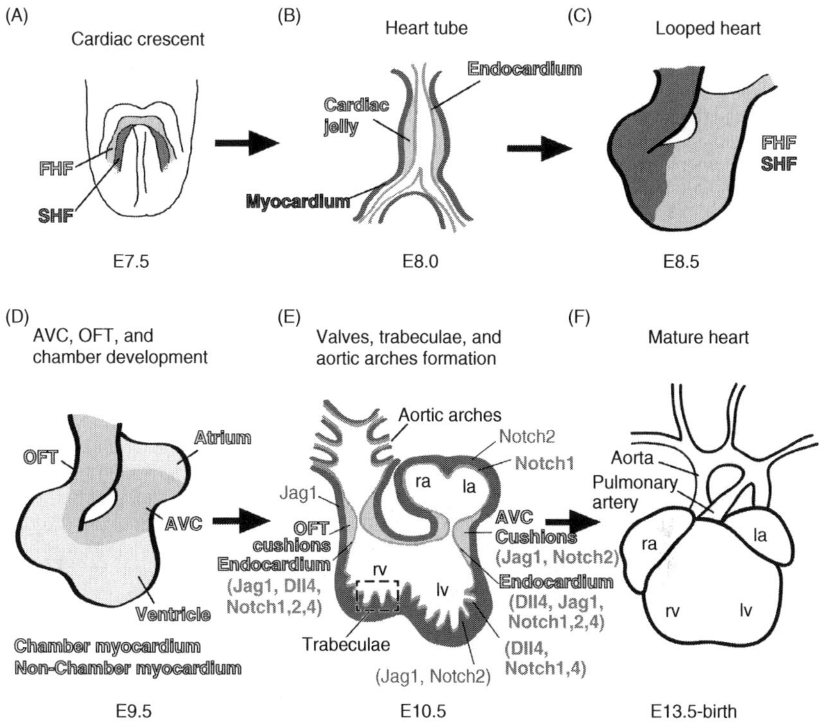

Figure 11.1 Stages of cardiac development and Notch ligands/receptors expression at midgestation. (A) Cardiac crescent. The E7.5 mouse heart is formed by two populations: the first- and second-heart fields (FHF and SHF) that have fused in the midline of the embryo and form the so-called cardiac crescent. (B) At E8.0 the heart looks like an endothelial tube (the endocardium, blue), surrounded by a layer of myocardial cells (red). Between endocardium and myocardium there is an extracellular matrix, the cardiac jelly (grey). (C) At E8.5, the heart is elongated and begins to loop rightwards, giving rise to a complex structure with four chambers. (D) E9.5, formation of the atrio-ventricular canal (AVC) is crucial to separate the two chambers' territories, the developing atria and ventricles. (E) Valve primordia formation occurs between E9.5 and E10.5, together with ventricular chamber development that begins with trabeculae formation in the developing left and right ventricles (LV and RV). Tissue expression of Notch ligands and receptors is color-coded: Endocardium, blue; myocardium, red and endocardial cushions, grey. (F) From E10.5 to E13.5, the OFT remodels to form the aorta and pulmonary arteries. Developmental stages are indicated in embryonic days. In all panels the heart is viewed from its ventral side and anterior is to the top. (See Color Insert.)

2. Cardiomyocyte Specification and Differentiation

Studies in different model systems show that during early cardiogenesis Notch inhibits cardiomyocyte cell fate specification (Nemir *et al.*, 2006; Rones *et al.*, 2000; Schroeder *et al.*, 2003). Activation or inhibition of Notch signaling in *Xenopus* provided evidence for a role of Notch in suppressing cardiomyogenesis within the heart field (Rones *et al.*, 2000). Embryonic stem (ES) cells allowed to differentiate and undergo embryoid body (EB) formation differentiate into precursors of the three germ layers, and then give rise to areas with beating cardiomyocytes that resemble those of the developing heart (Kehat *et al.*, 2001). Consistent with the data from *Xenopus*, ES cells deficient for the Notch effector RBPJK/CBF1 show increased cardiomyogenic differentiation in EB cultures (Schroeder *et al.*, 2003). Elegant gain- and loss-of-function (GOF and LOF) experiments in ES cells have refined this notion and supported an inhibitory role for Notch during mesodermal and cardiac differentiation (Nemir *et al.*, 2006). However, it is not clear how to extrapolate from the ES data how Notch might regulate the process *in vivo* (Hadland *et al.*, 2004). In addition, the molecular mechanism by which Notch inhibits cardiomyogenesis is not well understood. An early study showed that NICD inhibits skeletal muscle myogenesis by blocking the DNA-binding ability of the myocyte enhancer factor 2 (MEF2c) transcription factor (Wilson-Rawls *et al.*, 1999) or by sequestering mastermind-like 1 (MAML1), a transcriptional coactivator for both Notch and MEF2c *in vitro* (Shen *et al.*, 2006). Moreover, a recent study indicates that the Notch target Hey1/HRT1/Hesr1 represses myogenesis by binding to the *MEF2c* promoter and inhibiting its activity (Buas *et al.*, 2010). Interestingly, in contrast to the inhibitory cardiogenic function of Notch in early development, Notch is subsequently required for cardiomyocyte differentiation during ventricular chamber development, which will be discussed below.

3. Atrioventricular Canal Development

During cardiac development, soon after the linear heart tube loops, different cardiac domains can be identified as regions separated by external myocardial groves. Some of these will later contribute to the formation of the cardiac chambers (atria and ventricles, see Fig. 11.1E). The region between the atria and the ventricles is called the atrioventricular canal (AVC), and its development depends on the activity of the T-box transcription factor Tbx2. Biochemical and functional studies show that Tbx2 inhibits chamber-specific gene expression within the AVC myocardium (Habets *et al.*, 2002; Harrelson *et al.*, 2004). Manipulation of Bmp2 signaling in chicken and mice indicates that it regulates *Tbx2* expression in the AVC (Yamada *et al.*, 2000). Studies in chicken (Rutenberg *et al.*, 2006) and mouse (Kokubo *et al.*, 2007) embryos have implicated Notch signaling restricting *Bmp2* and *Tbx2* to the AVC region (Fig. 11.2A). In zebrafish, chicken, and mouse embryos, myocardial expression of the Notch target genes *Hey1* and *Hey2/HRT2/Hesr2* is absent from the AVC but present in the flanking ventricular and atrial chamber tissue. GOF of *Notch2, Hey1,* or *Hey2* in the chicken heart results in down-regulation of *Bmp2* in the AVC (Rutenberg *et al.*, 2006), suggesting that Notch activity may suppress *Bmp2* expression in chamber tissue thus limiting its expression to AVC and OFT regions. On the other hand, *Tbx2* misexpression inhibits expression of both *Hey1* and *Hey2*, suggesting the existence of a negative feedback loop between Notch and Bmp2 acting to limit *Hey* expression to prospective chamber territories (Rutenberg *et al.*, 2006) and *Tbx2* to the AVC (Fig. 11.2A). Consistent with this hypothesis, ectopic expression of *Hey1* or *Hey2* in the mouse cardiac lineage using the *Mesp1-Cre* driver results in *Bmp2* and *Tbx2* down-regulation and severe reduction in the AVC territory (Kokubo *et al.*, 2007). Also consistent is the finding that ectopic expression of *N1ICD* in the myocardium leads to *Bmp2* and *Tbx2* down-regulation and expansion of chamber territory at the expense of the AVC (Watanabe *et al.*, 2006). The combined deletion of *Hey1* and *Hey2* leads to an expansion of AVC territory and *Bmp2* expression, which does not invade the atria or ventricles (Watanabe *et al.*, 2006), indicating that other factors must restrict *Bmp2* expression to AVC tissue. It is unlikely that *Hey1* and *Hey2* are the only Notch targets involved in chamber and AVC patterning, and *Hey* genes are likely to have additional activators besides Notch. For example, *Hey2* expression in chamber myocardium does not depend on Notch activity (Timmerman *et al.*, 2004). Nonetheless, *Hey1-3* are expressed in AVC endocardium, where Notch1 activity is detected (Del Monte *et al.*, 2007), suggesting they participate in the EMT leading to valve primordia formation (see below). Figure 11.2B shows *Hey* genes expression pattern in the developing heart.

An unanswered question is how the AVC expression of *Bmp2* is regulated temporally to allow valve formation. The careful dissection of Notch

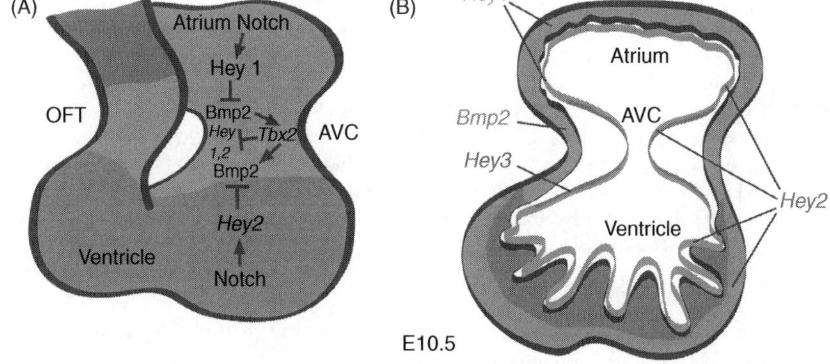

Figure 11.2 *(Continued)*

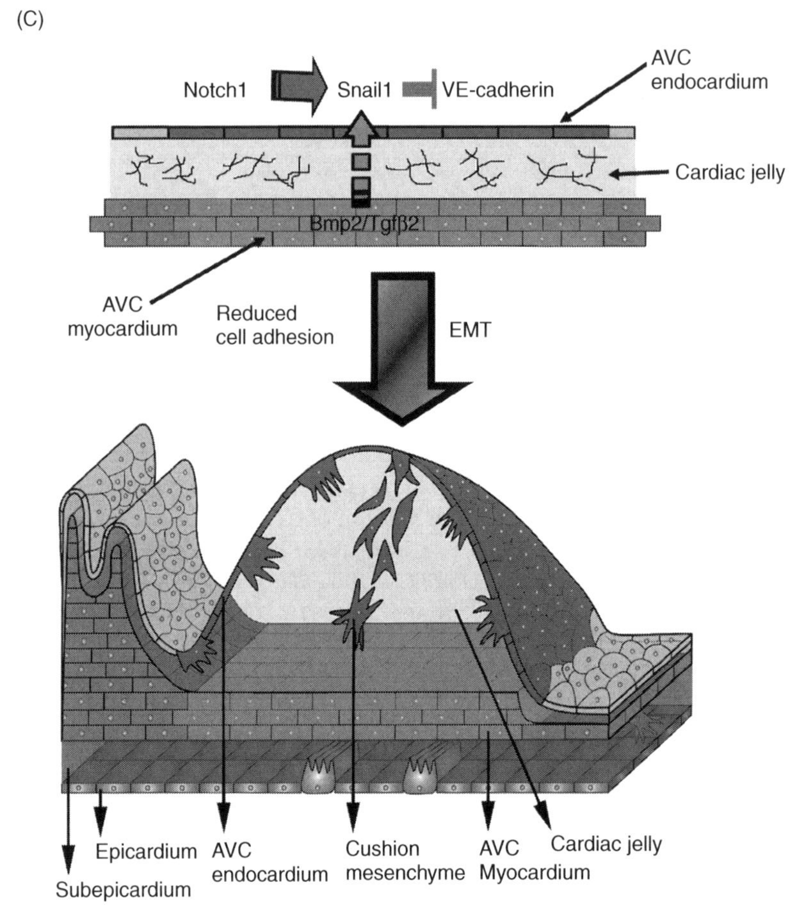

Figure 11.2 Notch signaling in atrioventricular canal (AVC) development and cardiac cushion formation. (A) View of the mouse heart at the time of AVC development. The dorsal side of the embryo is at the top. Hey1 is expressed in developing atrium and Hey2 in the forming ventricles. These transcription factors repress *Bmp2* expression in chamber tissue. In the presumptive AVC Bmp2 represses via Tbx2 the expression of *Hey* and chamber-specific genes, so that chamber (brown and blue) and non-chamber AVC and OFT tissues (green) are generated. (B) *Hey* and *Bmp2* expression at E10.5. Their tissue distribution is color-coded: *Hey1*, blue; *Hey2*, orange; *Hey3*, pink; *Bmp2*, green. For details see text. (C) Detail of the AVC region at E9.5. Endocardial Notch1/RBPJK activates *Snail1* expression in AVC endocardium that in turn represses *VE-cadherin* allowing EMT initiation and valve primordia formation (bottom). Myocardial signals (Bmp2, Tgfβ2) also converge on *Snail1* expression to elicit EMT. (See Color Insert.)

function in AVC development using tissue-specific (myocardium vs. endocardium) drivers and biochemistry will help us to fully understand the underlying molecular mechanisms and their functional integration.

4. Cardiac Valve Development

Cardiac valve development begins in mice at E9.5, when signals from the AVC and OFT myocardium instruct adjacent endocardial cells to undergo an EMT (Person *et al.*, 2005). These cells detach from each other, acquire a mesenchymal phenotype, and finally invade the underlying extracellular matrix to generate the heart valve primordia (endocardial cushions) (Markwald *et al.*, 1977). Through poorly understood processes of cell proliferation, programmed cell death and tissue morphogenesis, the valve primordia will give rise to the atrioventricular (AV) valves (tricuspid and mitral) and the OFT valves (pulmonic and aortic).

The nature of the EMT-inducing signal that triggers valve formation has been studied extensively, initially in chicken and later in mice. Earlier studies with chicken AVC tissue using collagen gel explant assays suggested that Tgfβ2 or Tgfβ3 were the main myocardial EMT inducers (Boyer *et al.*, 1999). Later studies indicated that *Tgfβ2*- or *Tgfβ3*-deficient mice did not display any endocardial EMT defect, suggesting that the induction of EMT in mammals was triggered by other signals (Camenisch *et al.*, 2002). More recent studies with targeted mutagenesis in mice showed that Bmp2 is a critical signal in EMT induction and cushion formation in the AVC, acting upstream of Tgfß2 (Ma *et al.*, 2005) whereas Bmp2 alone can specify a field of cardiac progenitors as a heart valve-inducing region (Rivera-Feliciano and Tabin, 2006; Fig. 11.2C)

Within the responsive endocardial tissue, EMT is primarily regulated by Notch. The Delta4 Dll4 ligand is expressed in murine AVC and OFT endocardium prior to valve formation (Timmerman *et al.*, 2004), anti-N1ICD antibody staining delineates the pre-valve endocardium of the E9.0 heart (Del Monte *et al.*, 2007), and lineage tracing of cells deploying Notch1 activity labels the valve from E10.5 onwards (Vooijs *et al.*, 2007). Targeted deletion of *RBPJk* and *Notch1* causes severe defects in valve formation because the AVC of such mutants lacks mesenchymal cushion cells (Timmerman *et al.*, 2004).

Although endocardial cells in E9.5 *RBPJk*-deficient AVC show features of activated premigratory endocardium, they remain in close association, maintain their adherens junctions, and do not invade the cardiac jelly. This behavior correlates with reduced transcription of *Snail1,* which is a Notch target gene expressed in the AVC and OFT endocardium in wild-type (WT) embryos, to repress *Vascular Endothelial Cadherin* expression (Timmerman *et al.*, 2004; Fig. 11.2C). Loss of *Snail1* expression thus appears to block EMT by preventing down-regulation of endocardial cell adhesion. *RBPjk* or *Notch1* mutants also show reduced *Tgfβ2* AVC and OFT expression (Timmerman *et al.*, 2004), suggesting that Notch exerts a non-cell autonomous effect in myocardial Tgfβ2 signaling. AVC explant assays with *Notch1* and *RBPJk* mutants show that Notch is required for endocardial EMT, a finding confirmed by Notch inhibition in WT mouse AVC explants and zebrafish embryos (Timmerman *et al.*, 2004), and by the ventricular septal defects seen in mice lacking *RBPJk* in the *Twist 2* expression domain (Morimoto *et al.*, 2010).

The cardiac expression of Notch target genes is highly dynamic (see *Hey* genes expression pattern in Fig. 11.2B). *Hey1* is transcribed throughout the endocardium until E9.0 (Timmerman *et al.*, 2004) but becomes restricted to the atrial myocardium and endocardium at E9.5 (Kokubo *et al.*, 2005a; Nakagawa *et al.*, 1999). It appears again in the AVC cushion tissue at E11.5 (Fischer *et al.*, 2007). At E9.5, *Hey2* is expressed in the ventricles, in the myocardium of the compact zone, in the trabeculae at lower levels (Nakagawa *et al.*, 1999), throughout the endocardium (our unpublished data), and is also found in the cushion tissue at E11.5 (Fischer *et al.*, 2007). As noted previously, *Hey1* and *Hey2* expression in the myocardium of atria/ventricles, respectively, confines *Bmp2* expression to the AVC myocardium. *Hey3* is expressed in AVC endocardium between E9.5 (unpublished data) and E11.5 (Fischer *et al.*, 2007). Functional studies have shown that *Hey2* deletion leads to AVC defects that cause valve dysfunction, ventricular septal defect, and tetralogy of Fallot leading to early death of mutant mice (Donovan *et al.*, 2002; Gessler *et al.*, 2002; Kokubo *et al.*, 2004). Double *Hey1;Hey2* mutants show among other alterations a severe defect in EMT that causes death at E11.5 (Kokubo *et al.*, 2005b). Lastly, *Hey1; Hey3* mutants show a dysplastic AVC and pulmonary valves and a ventricular septal defect at E15.5 that leads to congestive heart failure and lethality at birth (Fischer *et al.*, 2007). The data from mutants in Notch-signaling pathway, the AVC endocardium-restricted expression of active Notch1 (Del Monte *et al.*, 2007) and *Notch2* (Fischer *et al.*, 2007), and the Notch1 lineage tracing (Vooijs *et al.*, 2007), all indicate that Notch signaling in the AVC endocardium is critical for valve formation and morphogenesis. The nature of the interaction between AVC endocardium and myocardium remains unclear. The AVC endocardium may be a passive tissue that responds to inductive myocardial signals (Bmp2, etc.) or it does play an active role in EMT, as the initial work suggested (Timmerman *et al.*, 2004). With regard to these questions, recent data appear contradictory. Elegant work has shown that *N1ICD* overexpression in zebrafish AVC endocardium inhibits EMT (Beis *et al.*, 2005), but *Mesp1-Cre*-driven *N1ICD* activation in the mouse cardiac lineage impairs AVC myocardial differentiation without affecting EMT (Watanabe *et al.*, 2006). Thus the precise role of Notch in the regulation of endocardial EMT and its downstream molecular mechanism and relationship with the other activators of this process (e.g. Tgfβs and Bmps) remains to be defined. Another issue is the role of Notch signaling in the final stages of valve development. N1ICD expression is observed in valve endocardium until the latest stages of embryogenesis (Del Monte *et al.*, 2007), suggesting that Notch may participate in valve morphogenesis.

5. Ventricular Chamber Development: Trabeculation

Ventricular trabeculae are highly organized sheets of cardiomyocytes forming muscular ridges lined by endocardial cells. Trabecular development relies on the interaction between endocardial and myocardial cells and they are the first morphological sign of ventricular chamber development (Ben-Shachar *et al.*, 1985). Trabeculae form a sponge-like structure protruding toward the ventricular lumen that substantially increases the endocardial surface area (Vuillemin and Pexieder, 1989). Trabecular myocardium soon stops proliferating and differentiates into the fast-conducting peripheral ventricular conduction system. In contrast, the outer myocardial layer becomes highly proliferative and forms the compact layer of the ventricular wall.

Notch-signaling activity is detected in trabeculae since early developmental stages (Del Monte *et al.*, 2007). At E9.5, expression of *Delta4* and N1ICD activity is predominant in the endocardium at the base of developing trabeculae, suggesting that endocardial Delta-Notch expression is activated by a signal from the underlying myocardium (Del Monte *et al.*, 2007; Grego-Bessa *et al.*, 2007; Fig. 11.3A). The trabeculation-defective phenotype of *Notch1* and *RBPJk* mutants supports a role for Notch in the development of ventricular myocardium, and molecular analysis indicates that Notch1-dependent signaling is required for ventricular myocardial differentiation. Since ES cells lacking Notch1 contribute exclusively to cardiomyocytes (Vooijs *et al.*, 2007), this requirement appears to be non-cell autonomous. Notch mutants show defective expression of the three signaling pathways that are essential for trabeculation: Bmp10, ephrinB2, and Nrg1 (Fig. 11.3B).

Bmp10 is a signaling molecule expressed in trabecular myocardium and is required for cardiomyocyte proliferation (Chen *et al.*, 2004). *Bmp10* expression and signaling is reduced in *RBPJk* mutants suggesting that Bmp10 and Notch interact during chamber development (Grego-Bessa *et al.*, 2007). Loss of Bmp10 signaling strongly correlates with a reduction in cardiomyocyte proliferation in *RBPJk* mutants, a defect which is rescued by culturing *RBPJk* mutants in Bmp10-conditioned media, suggesting that in trabecular myocardium Notch modulates proliferation via Bmp10.

The Nrg1/ErbB-signaling pathway is essential for trabeculation (Gassmann *et al.*, 1995; Lee *et al.*, 1995; Meyer and Birchmeier, 1995). Nrg1 is expressed in ventricular endocardium and activates dimerization and tyrosine kinase-signaling activity of ErbB2–4 receptors in the myocardium (Falls, 2003). Nrg1 expression and activity is reduced in *Notch1* and *RBPJk* embryos, indicating that Notch and Nrg1/ErbB signaling interact during trabecular development. Nrg1 promotes the *in vitro* proliferation of neonatal

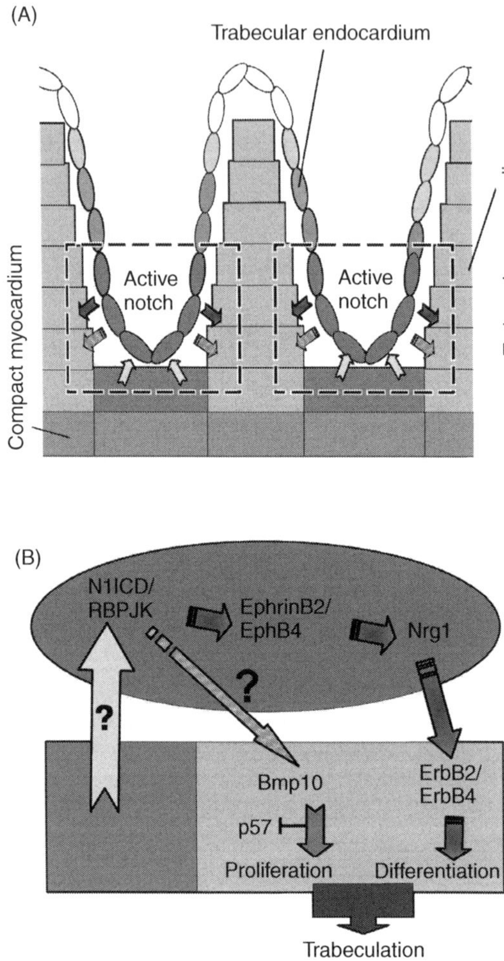

Figure 11.3 Notch and ventricular trabeculation. (A) Detail of developing trabeculae. Active Notch1 signaling (red) is predominant in endocardial cells at the base of the forming trabeculae, suggesting that a myocardial signal (yellow arrow) may activate Notch1 in this region. Notch activity in turn signals to the trabecular myocardium (blue and red arrows) to promote cardiomyocyte proliferation and differentiation. (B) Detail of the endocardium-myocardium molecular interplay at the base of trabeculae. Notch1 activity induced from the myocardium, activates ephrinB2/EphB4 signaling in endocardium which in turn is required for *Nrg1* expression in this tissue. Nrg1 activates ErbB2/ErbB4 signaling in myocardium to promote cardiomyocyte differentiation. In addition, Notch1 signaling in endocardium is required for *Bmp10* expression and therefore proliferation of trabecular cardiomyocytes. Modified from (Grego-Bessa *et al.*, 2007). (See Color Insert.)

cardiomyocytes (Zhao *et al.*, 1998), but in cultured embryos Nrg1 induces trabeculation (Hertig *et al.*, 1999), or cardiac conduction system development (Rentschler *et al.*, 2002), without affecting proliferation. Consistent with these data, the myocardial differentiation defect of *RBPJk* mutants is rescued by culturing them in Nrg1-containing media. Together, these data indicate that in Notch mutants, cardiomyocytes stop dividing prematurely and do not progress in their differentiation toward trabecular muscle because they are deprived of trophic factors.

The ephrinB2/EphB4-signaling system is required for vascular and trabecular development (Wang *et al.*, 1998). Cardiac *ephrinB2 ligand* and *EphB4* receptor expression and activity are impaired in Notch mutants and ChIP and luciferase reporter assays demonstrated that *EphrinB2* is a direct transcriptional target of N1ICD/RBPJK (Grego-Bessa *et al.*, 2007). Expression analysis in *ephrinB2, Nrg1*, and *Bmp10* mutants indicates that cardiac *Nrg1* transcription is reduced in *ephrinB2* mutants while *Bmp10* mutants show normal *ephrinB2* and *Nrg1* expression. These results indicate that during trabeculation endocardial *ephrinB2* acts upstream of Nrg1, and that both molecules act independently of Bmp10.

We have proposed that during trabeculation Notch links endocardial and myocardial signaling via a direct effect on the ephrinB2/EphB4 pathway, which is required for endocardial Nrg1 production and subsequent ErbB2/B4 receptors activation in myocardial cells. Notch1 is active in endocardium and *Bmp10* is expressed in myocardium, suggesting that Notch1 is required for production of a soluble endocardial signal, which induces myocardial *Bmp10* expression in a non-cell autonomous manner. Our hypothesis is that Notch mediates an endocardium–myocardium interaction essential for trabeculation and ventricular chamber development via two Notch-dependent processes, (i) the transition of primitive myocardial epithelium to trabecular and compact myocardium (ephrinB2- and Nrg1-dependent) and (ii) the maintenance of a proliferating trabecular cardiomyocyte population during this transition (Bmp10-dependent; (Grego-Bessa *et al.*, 2007); see Fig. 11.3B). An unsolved question is what is the mechanism that restricts Notch-signaling activity to the base of the developing trabeculae? A recent report shows that ectopic and supra-physiological Notch1 activation using the *Mesp1-Cre* driver results in ectopic trabeculation in the AVC (Watanabe *et al.*, 2006), suggesting that the heart of *Mesp1-Cre;N1ICD* mice might have lost AVC features at the expense of ventricular ones. This is somehow paradoxical taking into account the different N1ICD expression patterns in the AVC and the ventricles: N1ICD is quite uniformly expressed in AVC endocardium but is mostly restricted to the base of trabecular endocardium in the ventricles (Del Monte *et al.*, 2007 and unpublished data). One would expect to generate a homogenous N1ICD expression throughout the heart upon *Mesp1-Cre* driver activation, resembling that of the *Avc* and not of ventricles.

The involvement of other Notch receptors in chamber development is suggested by the defective trabecular phenotype and myocardial hypoplasia of hypomorphic *Notch2* mutants (McCright *et al.*, 2001) and the defective trabecular myocytes of *Hey1;Hey2* mice (Kokubo *et al.*, 2005a). Although the ventricular chamber containing both the compact and the trabecular zones forms initially, subsequent apoptosis of trabecular zone myocytes leads to poor trabecular formation. Clearly Hey2 is required for cardiomyocyte development and the interaction between *Hey1*-expressing endocardial and epicardial cells, and *Hey2*-expressing cells in the compact layer of the ventricle is also required to produce and/or maintain trabecular myocytes (Kokubo *et al.*, 2005b).

6. Outflow Tract Development

Development of the OFT is a complex process, involving cardiac progenitor cells from the second heart field, cardiac neural crest, and endothelial cells. The OFT includes the arterial pole of the heart, where the right and left ventricles empty into the pulmonary artery and aorta. The relationship between Notch signaling and OFT development was established originally from studies of Alagille syndrome patients (see below). A number of Notch ligands and receptors are expressed at E12.5 in remodeling OFT and aortic arch arteries (High *et al.*, 2007). The Jagged1 (Jag1) and Jagged2 (Jag2) ligands are expressed in neural crest-derived tissues including vascular smooth muscle cells, whereas Delta 1 (Dll1) and Delta 4 (Dll4) are expressed in endocardium. The receptors Notch1 and Notch4 are expressed in the endothelium (High *et al.*, 2007) and, in the case of Notch1, also in the OFT endocardium (Del Monte *et al.*, 2007). Notch2 is expressed in pharyngeal mesenchyme and together with Notch3, in neural crest-derived cells surrounding the aortic arch arteries (High *et al.*, 2007). Notch-signaling inhibition in cardiac neural crest derivatives was achieved using a silent (stop cassette containing) dominant negative version of MAML (*DNMAML*), activated using two different neural crest-specific Cre drivers (*Pax3* and *Wnt1*). DNMAML expression resulted in aortic arch-patterning defects, pulmonary artery stenosis, and ventricular septal defects. Lineage analysis of DNMAML-GFP expressing neural crest cells indicated that DNMAML expression does not affect neural crest proliferation, migration into the OFT, or their contribution to the aortic arch. However, Notch signals are essential for the differentiation of cardiac neural crest precursors into smooth muscle cells (High *et al.*, 2007).

As stated above, the Jag1 ligand is expressed in OFT myocardium (High *et al.*, 2007) Inhibition of Notch signaling in the SHF was achieved by activating *DNMAML* with two different Cre drivers (*Islet1* and *Mef2c*). The same Cre lines were also used to inactivate a conditional *Jag1* mutant. In all cases, loss of Notch signaling resulted in OFT and aortic artery defects resembling those seen in patients with congenital heart disease (CHD), indicating that Jag1 is an essential Notch ligand in the SHF (High *et al.*, 2009).

To identify downstream effectors of Notch signaling in the SHF, E9.5 WT explants composed of pharynx and distal OFT were treated with the γ-secretase inhibitor DAPT to inhibit Notch signaling. Quantitative reverse transcriptase polymerase chain reaction (RT-PCR) analysis of a set of genes involved in OFT development revealed that Fgf8 expression was strongly down-regulated in the explants. This observation was validated by quantitative RT-PCR analysis of mRNA obtained from pharyngeal and the OFT region of *Islet1*$^{Cre/+}$*;DNMAML* mutant embryos at E10.5 and by *in situ* hybridization. Further analysis revealed that expression and signaling activity of Bmp4, a downstream effector of Fgf8 in the OFT (Park *et al.*, 2008), was down-regulated in OFT cushions and myocardium of *Islet1*$^{Cre/+}$*;DNMAML* mutant embryos. Similar results were obtained with *Islet1*$^{Cre/+}$*;Jag1*$^{flox/flox}$ mice. OFT explant assays showed that EMT was drastically impaired in *Islet1*$^{Cre/+}$*;DNMAML* mutants, as very few mutant endocardial-derived cells migrated away from the myocardium and those that migrated were unable to invade the gel matrix. This phenotype could be rescued by adding Fgf8 in the explant media. These results indicate that Notch signaling regulates *Fgf8* expression in the SHF myocardium and thus modulates EMT in adjacent endothelial cells via activation of the secreted factor Fgf8 and its downstream mediator Bmp4 (High *et al.*, 2009). In the context of CHD these data link SHF, neural crest, and endocardial cushion development and suggest that abnormal Jag1-Notch signaling within the SHF could account for some of CHD involving these tissues.

The relevance of Notch signaling in OFT development is also indicated by the analysis of *presenilin 1* (*Psen1*) mutant mice, which show ventricular septal defect, double outlet right ventricle, and pulmonary artery stenosis (Nakajima *et al.*, 2004). Recent work shows that the expression of *Hes1* in the SHF is required for proper OFT extension, neural crest proliferation, and OFT alignment. Transgenic reporter insertion upstream of the *Hes1* gene reveals expression in pharyngeal endoderm and mesoderm including the SHF. Analysis of targeted *Hes1* mutant mice indicates that at embryonic day 15.5 the OFT shows alignment defects, including ventricular septal defects and overriding aorta. At earlier developmental stages, *Hes1* mutant embryos display defects in SHF-cell proliferation, reduction in cardiac neural crest cell migration, and failure to completely extend the OFT. Thus, *Hes1* is required for development of the arterial pole of the heart (Rochais *et al.*, 2009).

Table 11.1 summarizes the cardiac phenotypes of Notch LOF and GOF mutations described in the previous sections.

CHD are the most common birth defects worldwide with a prevalence of 5–10 per 1000 live births (Christianson and Modell, 2004), and anomalies of the heart valves account for over 30% of all CHD (Loffredo, 2000). In light of the essential function of Notch in OFT development, it seems logical to examine the impact of defective Notch signaling in disorders of the left and right ventricular OFTs (LVOT and RVOT, respectively),

Table 11.1 LOF and GOF mutations in the murine Notch pathway causing a cardiac phenotype

Mutations	Lethality	Cardiac phenotype	References
Loss of function			
$Dll4^{-/-}$	E9.5	Reduced atrial and ventricular chambers, defective ventricular trabeculation.	(Duarte *et al.*, 2004)
$Jag1^{+/-};Notch2^{+/del1}$	Postnatal	Right ventricular hypoplasia, atrial and ventricular septal defects, pulmonary artery stenosis	(McCright *et al.*, 2002)
$Notch1^{-/-}$	E10.5	Hypoplastic cardiac cushions and impaired trabeculation	(Grego-Bessa *et al.*, 2007; Timmerman *et al.*, 2004)
$Notch2^{del1/del1}$	E11.5 to birth	Myocardial hypoplasia, and reduced trabeculation	(McCright *et al.*, 2001)
Tie2-Cre; $Notch1^{flox/flox}$	E10.5	Impaired trabeculation	(Grego-Bessa *et al.*, 2007)
$RBPJk^{-/-}$	E10.5	Defective cardiac looping, hypoplastic cardiac cushions and impaired trabeculation	(Grego-Bessa *et al.*, 2007; Oka *et al.*, 1995; Timmerman *et al.*, 2004)
Tie2-Cre; $RBPJk^{flox/flox}$	E10.5	Impaired trabeculation	(Grego-Bessa *et al.*, 2007)
Twist2-Cre; $RPJk^{flox/flox}$	Perinatal	Ventricular septal defect	(Morimoto *et al.*, 2010)
$Hey2^{-/-}$	Postnatal	Cardiomyopathy, atrial and ventricular septal defects, AV valve disease, tricuspid valve atresia	(Donovan *et al.*, 2002; Gessler *et al.*, 2002; Kokubo *et al.*, 2004; Sakata *et al.*, 2002; 2006)

Hey1$^{-/-}$;*Hey2*$^{-/-}$	E11.5	Looping defective, impaired trabeculation, hypoplastic cardiac cushions, AVC expansion	(Fischer *et al.*, 2004; Kokubo *et al.*, 2005b; 2007)
Hey1$^{-/-}$;*Hey3*$^{-/-}$	Postnatal	Ventricular septal defect, aortic valve defects	(Fischer *et al.*, 2007)
Hes1$^{-/-}$	Perinatal	Ventricular septal defect and overriding aorta	(Rochais *et al.*, 2009)
Psen1$^{-/-}$	Perinatal	Ventricular septal defect, double outlet right ventricle, pulmonary artery stenosis	(Nakajima *et al.*, 2004)
Pax3-Cre; DNMAML	Perinatal	OFT and aortic arch abnormalities,ventricular septal defect	(High *et al.*, 2007)
Wnt1-Cre; DNMAML	Postnatal	OFT and aortic arch abnormalities	(High *et al.*, 2007)
Islet1-Cre; DNMAML	Perinatal	OFT and aortic arch abnormalities, ventricular septal defect, tricuspid valve atresia	(High *et al.*, 2009)
Mef2c-AHF-Cre; DNMAML	Postnatal	OFT and aortic arch abnormalities, ventricular septal defect	(High *et al.*, 2009)
Islet1-Cre; Jag1$^{flox/flox}$	Perinatal	Aortic arch abnormalities, ventricular and atrial septal defect	(High *et al.*, 2009)
Mef2c-AHF-Cre; Jag1$^{flox/flox}$	Postnatal	Aortic arch abnormalities, ventricular and atrial septal defect	(High *et al.*, 2009)
Gain of function			
Mesp1-Cre; N1ICD ME	E10.5–11.5	Impaired ventricular myocardial differentiation and maturation, ectopic trabeculation in AVC	(Watanabe *et al.*, 2006)
Mesp1-Cre; Hey1 ME	E11.5	Reduced AVC extension, cardiac cushions are formed	(Kokubo *et al.*, 2007)
Mesp1-Cre; Hey2 ME	E10.5	Absent AVC, no cardiac cushions,defective trabeculation	(Kokubo *et al.*, 2007)

which represent specialized realms of Notch action in the developing heart (Towbin and Belmont, 2000).

7. Left Ventricular Outflow Tract Obstruction (LVOTO)

Bicuspid aortic valve (BAV) (OMIM database #108730), coarctation of the aorta (COA OMIM #120000), and hypoplastic left heart syndrome (HLHS #241550) form a spectrum of cardiovascular malformations causing obstruction of the left ventricular OFT (LVOTO). The structures affected in LVOTO, including the aortic valve, aorta, and/or left ventricle, are abnormally narrowed or underdeveloped, creating an impediment to systemic blood flow. If LVOTO were a "continuum" of disease at birth then the order from mild to severe would be BAV > COA > HLHS consistent with a dysmorphogenic "hierarchy" established earlier in the embryonic life. The BAV is the mildest and commonest CHD form found in 1–2% of the population, and paradoxically associated with the highest morbidity and mortality because complications, including aortic valve stenosis (AS), regurgitation, dilation and dissection arise later in life (Braverman *et al.*, 2005). A bicuspid valve therefore constitutes a predisposing LVOTO lesion, and additional environmental or lifestyle factors are important for the development of disease.

Familial clustering and inheritance analyses have supported a significant inherited component in LVOTO (Cripe *et al.*, 2004; Hinton *et al.*, 2007; McBride *et al.*, 2005a; Probst *et al.*, 2006; Wessels *et al.*, 2005). A high correlation was found for the presence of a CHD among families with LVOTO compared with the general population, with particularly high concordance for BAV in the relatives of severely affected children (Huntington *et al.*, 1997; Lewin *et al.*, 2004; Loffredo *et al.*, 2004). Locus heterogeneity for familial BAV has been established in several studies and genome-wide association studies in families with BAV and/or LVOTO demonstrate linkage to multiple chromosomal loci (Hinton *et al.*, 2007; Martin *et al.*, 2007; McBride *et al.*, 2009). Based on these findings, a complex multigenic etiology of BAV and other left-sided cardiac malformations has been proposed in conjunction with environmental or stochastic modifying influences, but the contributing genes remain largely unknown.

A significant finding by Garg *et al.* was that of mutated *NOTCH1* alleles in families with AS in the context of BAV and isolated left- or right-side heart malformations (Garg *et al.*, 2005). In one family a nonsense heterozygous mutation was found within EGF-like repeat (EGFR) encoding sequences in the extracellular domain of NOTCH1, predicting a premature stop codon. In a second unrelated family, a frameshift-causing deletion resulted in a premature stop codon within the NOTCH1 ectodomain.

The three affected BAV family members all inherited this mutation. Both mutations were predicted to generate an unstable ribonucleotide message or the expression of a truncated polypeptide lacking the intracellular domain. These observations, implicating mutated *NOTCH1* alleles in familial BAV, lead to the suggestion that non-familial LVOTO lesions may be explained by other loss-of-function Notch1 mutations (Fig. 11.4). Mohamed *et al.*

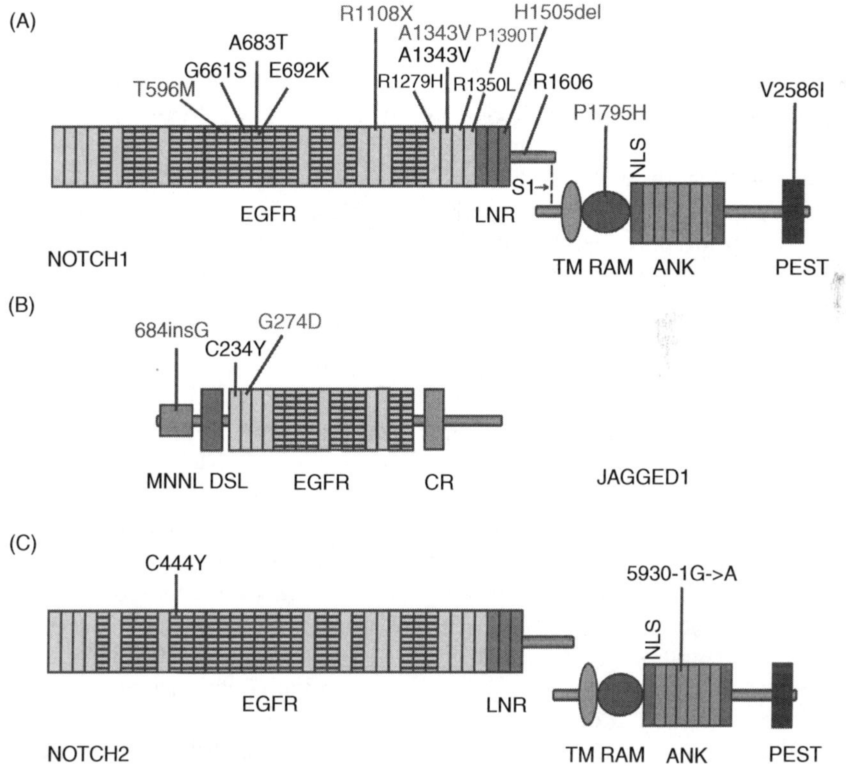

Figure 11.4 Topography of *NOTCH* mutations in left and right ventricular outflow tract malformations. (A) Schematic representation of NOTCH1 protein indicates the location of missense, nonsense and frameshift mutations found in LVOTO malformations. Color code for identified mutations: Garg *et al.* (blue), Mohamed *et al.* (green), McKellar *et al.* (red), McBride *et al.* (black). (B) Schematic representation of JAG1 protein shows the location of missense and frameshift mutations found in TOF patients with subclinical ALGS or isolated disease. Color code: Krantz *et al.* (green), LeCaignec *et al.* (black), Eldadah *et al.* (red). (C) Schematic representation of NOTCH2 protein indicating the location of the missense and splicing mutations in mutation-negative JAG1 ALGS families identified by McDaniell *et al.* Abbreviations for protein domains: EGFR: EGF-like repeat (shaded repeats correspond to calcium-binding consensus sequence); LNR: Lin/NOTCH repeat; TM: transmembrane domain; RAM: juxtramembrane RAM domain; ANK: ankyrin repeat S1: S1 cleavage site; MNNL: N-terminus of Notch ligand; DSL: Delta, Serrate, Lag2; CR: cysteine-rich. (See Color Insert.)

evaluated 48 sporadic BAV cases and found two polymorphic alleles with amino acid substitutions at evolutionary conserved positions that were not present in control population. One variant (T596M) is located in EGFR 15 near a potential GluNac site (Matsuura *et al.*, 2008); the other (P1795H), was located near the RAM domain (Mohamed *et al.*, 2006) but outside the region involved in RBPJK binding (Johnson *et al.*, 2010). McKellar *et al.* analyzed by PCR, DHPLC profiling and direct DNA sequencing four *NOTCH1* exons (11, 20, 25 and 29) that have been previously implicated in BAV (Garg *et al.*, 2005; Mohamed *et al.*, 2006). 48 patients with sporadic BAV and thoracic aortic aneurysms were analyzed (McKellar *et al.*, 2007). Two novel polymorphisms were found, one at A1343V and another at P1390T, in EGFR 34 and 36, respectively (Fig. 11.4A). Both showed association with BAVs and tricuspid aortic aneurysms and were absent from the control population (McKellar *et al.*, 2007). In the most comprehensive study published to date, McBride *et al.* examined 91 patients with LVOTO including BAV, COA and HLHS (McBride *et al.*, 2008). They identified eight rare polymorphisms of which six were novel and five were completely absent or under-represented in their patient control group. Six highly non-synonymous mutations including three amino acids completely conserved across nine species (G661S, R1279H, and A1343V) mapped to the EGF repeats, whereas one was found in the extreme terminal C-terminal region of the ectodomain, and another in the intracellular PEST motif (Fig. 11.4A). The latter substituted an isoleucine for a poorly conserved valine. Whereas some may reduce function, others may be functionally "silent" since all these substitutions are tolerated in other Notch1 EGF repeats.

Examining the putative contribution of *NOTCH1* mutations to LVOTO in all these studies suggests that 4.2–6.6% of patients with a BAV or other LVOTO lesion have a rare *NOTCH1* allele (McBride *et al.*, 2008). This low frequency suggests either that *NOTCH1* mutation contributes only a small fraction of LVOTO or that these particular mutations are linked to another disease causing mutations nearby. Linkage analysis of familial LVOTO identified up to six genes that may be involved (McBride *et al.*, 2005b). However, there was no clear association between the genotype and phenotype; for example, the same polymorphism (G661S; Fig. 11.4A) was found in the full LVOTO spectrum including AS, BAV, COA and HLHS, but the new Ser residue does not form a known glycosylation prone sequence (see chapter 4) and is often found in other EGF repeats in a similar position. It is hard to imagine that this amino acid substation can be responsible for all these diverse phenotypic outcomes. Together, the findings of infrequent *NOTCH1* mutations and variable phenotypical expressivity are suggestive. Since the mechanism(s) by which *NOTCH1* mutations promote LVOTO remain largely unknown, the connection remains inconclusive.

The six missense mutations in the extracellular domain of NOTCH1 appear to cluster in two regions corresponding to EGFR 15–18 and EGFR

33–36 (Fig. 4A, see chapter 2). Functional analyses in *Drosophila* have shown that EGFR 11–12 is necessary and sufficient to mediate binding of the DSL domain of Delta and Serrate (Rebay *et al.*, 1991). EGFR 15–18 are predicted to bind calcium (Gordon *et al.*, 2008; Xu *et al.*, 2005), but these polymorphic alleles are not predicted to change calcium binding.

The functional implications of *NOTCH1* missense mutations in LVOTO patients were examined by McBride *et al.* They suggested that some *NOTCH1* mutations associated with cardiac disease resulted in mis-processing in the Golgi (McBride *et al.*, 2008). They used cell-based and biochemical assays to examine the ability of soluble JAG1 ligand to activate signaling through the WT receptor and several over expressed mutants including G661S, A683T, A1343V, and R1279H. NOTCH1 signaling was marginally reduced in the G661S and A683T mutants compared with the WT, consistent with the notion that these polypeptides may be inadequate substrates for processing and maturation in the Golgi. However, all these mutant proteins produce a signal, and in heterozygote individuals will thus produce a stronger Notch signal than the haploinsufficient alleles described by Garg *et al.* Importantly, G661S and A683T do not disrupt any important features in Notch EGF repeats, i.e., ASP hydroxylation, Ca^{++} binding, or glycosylation sites. No known new sites were added, either, and these amino acid substitutions are well tolerated in other EGF repeats. For example, the sequence $GxxC_4xC_5$ appears in 10 EGF repeats and $SxxC_4xC_5$ appears in 9 repeats. Since Ser is tolerated at equivalent positions, it is very unlikely that G661S represents a misfolding or loss-of-function mutation. However, as detailed in chapter 4, much remains to be learned about the possible modifications at Ser and Thr residues, and many LVOTO mutations either add such residues or alter their immediate environment.

8. Right Ventricular Outflow Tract Obstruction

Branch pulmonic stenosis, peripheral pulmonary hyperplasia, pulmonic valve stenosis, and tetralogy of Fallot (TOF) (OMIM #187500) form a spectrum of cardiac diseases with right ventricular OFT obstruction (RVOTO). The RVOTO malformations have been studied in the context of Allagile Syndrome (ALGS, OMIM #118450) a dominant multisystem disorder defined primarily by liver disease in combination with cardiac, skeletal, ocular, and facial manifestations. Deletions or mutations of *JAG1* have been found in over 94% of ALGS patients with about 60–70% shown to arise *de novo* and the remaining inherited (Li *et al.*, 1997; Oda *et al.*, 1997; Spinner *et al.*, 2001; Warthen *et al.*, 2006). The majority of over 400 *JAG1* mutations identified to date have been found to result in protein truncation with about

10% being missense mutations located mainly in the N-terminal portion of the JAG1 extracellular domain. In the absence of obvious phenotypic differences between complete gene deletion and mutation, haploinsufficiency of *JAG1* was postulated as the pathogenetic mechanism causing ALGS (Crosnier *et al.*, 1999; Krantz *et al.*, 1998).

The cardiac disease ALGS phenotype is highly penetrant occurring in 94% of ALGS individuals and presents variably from mild pulmonary stenosis to severe forms of TOF and infrequent left-side anomalies (McElhinney *et al.*, 2002). The severe end of this spectrum is dominated by TOF, a complex cardiac phenotype that in addition to the obstruction of right ventricular outflow is characterized by a ventricular septal defect, aortic dextroposition, and right ventricular hypertrophy. The TOF accounts for up to one-tenth of congenital cardiac lesions and occurs in one of every 3000 live births (Bailliard and Anderson, 2009). The classic form of TOF has been observed as a component of other heritable syndromes including the branchial arch disorders but less is known about the etiology of sporadic and isolated TOF.

JAG1 mutations have been found in a subset of patients with TOF and pulmonary stenosis that did not meet the clinical criteria for ALGS (Fig. 11.4B). Krantz *et al.* identified two *JAG1* mutations in subdiagnostic or "microform" ALGS patients (Krantz *et al.*, 1999). One was an insertion 684insG predicted to cause a frame-shift and result in a truncated JAG1 protein and the other was a hemizygous deletion of *JAG1*. LeCaignec *et al.* described a family segregating a unique phenotype with highly penetrant deafness, ocular defects, and TOF and identified a missense mutation (C234Y) in the first of the sixteen EGFR present in the extracellular portion of JAG1 (Le Caignec *et al.*, 2002). Perhaps more relevant, Eldadah *et al.* (2001) identified a heterozygous missense mutation (G274D) in the second EGFR in JAG1 in a large family segregating TOF as an autosomal dominant trait with reduced penetrance and absence of any other ALGS phenotype.

The C234Y and G274D mutations were found in the first and second EGFR, respectively, and with the N-terminal MNNL and DSL domains (Fig. 11.4B) have been shown to be important for establishing a high-affinity complex with Notch (Parks *et al.*, 2006; Shimizu *et al.*, 1999). X-ray crystallography analysis of the DSL domain and EGFR 1–3 suggests it adopts an extended structure reminiscent of the elongated conformation seen in EGFR 11–13 of NOTCH1 (Cordle *et al.*, 2008). The receptor-binding portion of the ligand, encoded by the first six exons of the *JAG1* gene, harbors the majority of ALGS missense mutations known to date highlighting its functional significance (Spinner *et al.*, 2001; Warthen *et al.*, 2006).

The substitution of tyrosine for cysteine in C234Y predicting the removal of one of the three intramolecular disulfide bonds responsible for EGFR domain stability is considered to be highly structurally disruptive, and it impacts the receptor-binding surface of JAG1 (see chapter 2 and Kopan and Ilagan, 2009). The functional implications of the G274D substitution are less clear. Detailed *in vitro* oxidative folding of synthetic domains encompassing the second EGFR suggested that substituting the glycine caused focal abnormalities of polypeptide folding that could not be rescued by compensatory mutations (Guarnaccia *et al.*, 2009). The refolding studies suggested that even if the misfolded region in JAG1 were minimal as would be expected in a non-cysteine substitution, it might still be disruptive for DSL–ligand–Notch interactions (see Chapter 2 and Kopan and Ilagan, 2009). Indeed most if not all substitutions within the Notch-interacting domain of JAG1 directed ligand misfolding, delayed trafficking, and even when present at the cell surface resulted in the loss of Notch-signaling activity relative to WT JAG1 (Lu *et al.*, 2003; Morrissette *et al.*, 2001).

The paucity of findings of *JAG1* mutations in isolated TOF suggests a mainly syndromic disease involvement but it is probable that a systematic targeted search would uncover new *JAG1* mutations. Indeed recent studies performed a genome-wide survey of 114 subjects with sporadic TOF and their unaffected parents and identified eleven *de novo* copy number variants that were absent or less than 0.1% in more than 2000 controls (Greenway *et al.*, 2009). They found loss of chromosomal arms 9q34 and 20p12, which contain *NOTCH1* and *JAG1*, respectively, identifying *NOTCH1* and as a novel candidate TOF-associated gene and confirming *JAG1* mutations in sporadic TOF.

Several mouse models of inactivation of the Hairy/Enhancer-of-split gene homologs *HEY* developed RVOTO defects. A mutation that alters the carboxy terminus of gridlock, the zebrafish homolog of *Hey2*, was found to be associated with a localized defect in the aorta resembling human COA (Zhong *et al.*, 2000). *Hey2* inactivation in mouse resulted in a spectrum of cardiac malformations including TOF, ventricular septal defects, and tricuspid atresia, that resembled those associated with mutations of human *JAG1* (Donovan *et al.*, 2002), whereas LOF of both *Hey1* and *Hey3* caused AV dysplasia and membranous ventricular septal defects during cardiogenesis (Fischer *et al.*, 2007). These animal models suggested an important role for *hey* genes in cardiac morphogenesis and it was hypothesized on that basis that *HEY2* could be a candidate gene for ALGS and sporadic TOF. Sequencing of *HEY2* in 61 patients with congenital heart defects or ALGS and with no *JAG1* mutations did not reveal any disease-related polymorphism (Fischer *et al.*, 2004), although *HEY2* mutations were found in severely malformed hearts in conjunction with mutated *TBX5* and *GATA4* genes (Reamon-Buettner and Borlak, 2006). ALGS is characterized by highly variable expressivity and the absence of a clear relationship between genotype and phenotype, indicating the existence of modifying loci (Slavotinek and Biesecker, 2003). Accordingly, the type or location of

JAG1 mutation does not seem to be predictive of the severity of cardiac manifestations, which can be associated with anomalies in other organ systems (McElhinney *et al.*, 2002).

Studies in mice uncovered *Notch2* as genetic ALGS modifier (McCright *et al.*, 2002) and they are predicted to alter the clinical manifestations of ALGS1 in patients. Mice heterozygous for a *Jag1* null mutation exhibited anterior chamber eye defects but did not mimic the other phenotypes typical of ALGS in humans (Xue *et al.*, 1999). However compound heterozygotes for the *Jag1* null mutation and a *Notch2* hypomorphic allele reproduced most of the human ALGS-associated phenotypes, including typical heart and liver anomalies (McCright *et al.*, 2002). As anticipated from the mice studies, screening of eleven *JAG1* mutation-negative ALGS families found *NOTCH2* mutations segregating in two families and identified five affected individuals (Fig. 11.4C). Renal defects seen in the mutant mice but are a minor feature of ALGS were also present in the affected individuals (McDaniell *et al.*, 2006). Mutations of *NOTCH2* identify a clinical entity distinct from the common *JAG1*-linked ALGS1, characterized by kidney manifestations seen also in mice with loss of *Notch2* (Cheng *et al.*, 2007; Kopan *et al.*, 2007; Surendran *et al.*, 2010) and called ALGS2 (OMIM # 610205).

The *FRINGE* genes were identified as potential genetic modifiers of the hepatic phenotype associated with *JAG1* mutation in ALGS. They encode glycosylation enzymes responsible for mediating sugar elongation of O-fucose residues in the EGFRs located in the extracellular domain of the Notch receptors (Stanley, 2007; Tien *et al.*, 2009). This process was shown to affect the binding affinity of Notch receptor to DSL ligands to reduce or enhance S2 cleavage of various Notch paralogs thereby modulating the strength of the signal output (see Chapter 4). Mice heterozygous for mutations in *Jag1* and *Fringe* genes displayed bile duct proliferation and remodeling, phenotypes not normally associated in either single heterozygous animals or in *Fringe* homozygous mice (Ryan *et al.*, 2008). Given their ability to modulate Notch signaling, the *FRINGE* genes might be reasonable candidates for screening as modifiers of cardiac phenotypes in ALGS and isolated cardiac disease.

A converging theme in cardiac malformations is that mutations in the Notch pathway that result in reduced Notch signaling underlie a pathology involving heart, liver, and kidney. Heterozygous insertion/deletion of genomic sequences or nonsense mutations would be expected to lead to a 50% reduction in protein and may cause haploinsufficiency. Missense point mutations may affect protein folding, delay processing, or target misfolded proteins for degradation. Moreover LOF might be achieved in some mutations by modifying the strength or affinity of Notch and DSL–ligand interactions or due to a "dominant-negative" effect, but this remains to be demonstrated for polymorphisms associated with LVOTO. These mechanisms of disease are consistent with the notion that Notch-signaling levels must be tightly regulated during cardiogenesis and that even subtle alterations of dosage or function may promote LVOTO and RVOTO malformations.

9. Acquired Disease

Acquired forms of heart disease develop in adult life because of chronic hypertension, myocardial ischemia, AS, and unrecognized congenital defects. With the exception of AS there is little evidence implicating predisposing *NOTCH* mutations or polymorphisms in acquired heart disease in adulthood. Murine models of genetic GOF or LOF have implicated the activation of Notch signaling as an important regulator of adaptive repair mechanisms in the damaged myocardium.

Calcific aortic valve disease (CAVD) is the third leading cause of adult heart disease and the most common form of acquired valve disease in high-income communities (Leggett and Otto, 1996; Ramaraj and Sorrell, 2008). A BAV and aging are the risk factors most closely associated with CAVD. Incomplete leaflet separation between two of the three valve leaflets in BAV causes a hemodynamic disturbance predisposing to precocious valvular calcification, while age-related wear-and-tear may lead to long-term calcification in tricuspid valves (Butcher *et al.*, 2008). Obstruction of blood flow and regurgitation caused by an improper valve function results in increased workload on the left ventricle and ultimately heart failure. Pathological examination of the leaflets of calcified aortic valves provides evidence for a systemic disease process involving lipid deposition, inflammation, endothelial cell activation, bone formation, and calcium deposition akin to atherosclerosis (Otto, 2009; Rajamannan *et al.*, 2007; Xu *et al.*, 2009).

Valve calcification is thought to proceed via the trans-differentiation of valve myofibroblasts into osteoblastic cells, which can spontaneously form calcific nodules in culture. NOTCH1 signaling appears to play a key role in preventing osteoblastogenic transformation and aortic valve calcification by repressing pro-osteogenic signaling pathways (Yang *et al.*, 2009), and these pathways are activated in the valves of mutated *NOTCH1* CAVD patients. NOTCH1 was shown to block activation of the master osteogenic regulator RUNX2 via physical interaction of HEY2 with RUNX2, suggesting that derepression of *RUNX2* follows *NOTCH1* inactivation in AS patients (Garg *et al.*, 2005). Consistent with an anti-osteogenic function in the aortic valves, Notch1 was shown to repress valvular Bmp2 expression in aortic valve leaflets and derepression of Bmp2 following *Notch1* inhibition caused by the formation of calcific nodules in valve interstitial cell cultures (Nigam and Srivastava, 2009). In accordance with these *in vitro* findings, heterozygous null *Notch1* mutant mice were predisposed to aortic valve

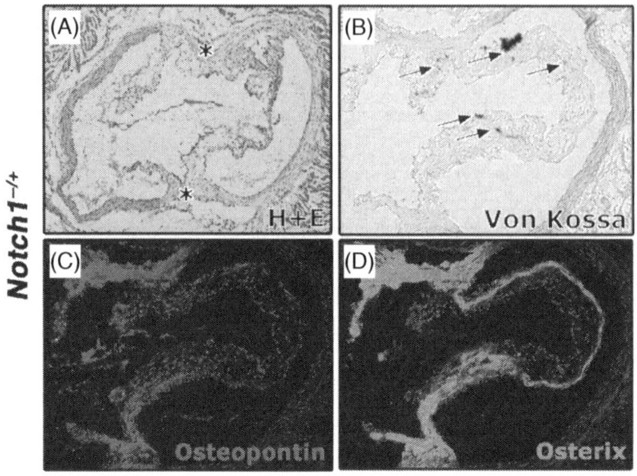

Figure 11.5 A bicuspid aortic valve predisposes to calcification. (A) Eosin-hematoxylin staining of a BAV from a Notch1 heterozygous null mutant mouse suggests the presence of two instead of the normal three leaflets (*). (B) Von Kossa staining indicates the presence of calcific nodules in valve leaflets, at the base or attachments of the leaflets and also in the leaflet proper (arrows). (C, D) Double indirect immunofluorescence staining of osteopontin and osterix indicating that an active process of osteoblastic differentiation is taking place in the valve leaflets. (See Color Insert.)

calcification when fed a western diet for over 10 months (Nigam and Srivastava, 2009). We have recently established a diet-induced model of aortic valve calcification with stenosis in heterozygous *RBPJk*-targeted mice (M. Nus *et al.* unpublished data). When *RBPJk*$^{+/-}$ mice are fed with a hypercholesterolemic diet they develop valvular dysfunction after 4 months, associated with leaflet thickening, osteogenic protein expression, and calcification. Parallel examination of heterozygous mutant *Notch1* mice indicated that they developed calcification but not valve dysfunction (Fig. 11.5), suggesting that other Notch receptors may contribute to maintaining valve homeostasis.

Cardiac failure occurs when the pumping activity of the heart fails to meet the metabolic demands of the body (Jessup and Brozena, 2003; McMurray and Pfeffer, 2005). This is one of the most common medical conditions affecting between 1 and 2 % of the adult population in high-income communities. The principal risk factors are advanced age (6–10 % of people over the age of 65 years are affected), hypertension, myocardial infarction, and valve disease. After suffering myocardial infarction, the heart undergoes a series of morphological changes affecting the left ventricular chamber including increased sphericity, dilatation, and wall thinning (Lorell and Carabello, 2000; McMullen and Jennings, 2007). This pathological remodeling is associated with cellular and molecular changes that include myocyte hypertrophy, apoptosis and fibrosis, and the activation of a fetal genetic program (Chien *et al.*, 1991; Dorn *et al.*, 2003; Hein *et al.*, 2003).

The loss of net muscle mass and thinning of the left ventricular myocardium results in functional decompensation and systolic or diastolic failure (Dorn, 2007).

Cardiac myocytes divide actively during embryogenesis but shortly after birth they exit the cell cycle, stop proliferating, and begin to terminally differentiate by forming binucleated syncytium and sarcomerization. The incapacity of adult cardiomyocytes to divide is related to the complex process of myofibrillar disassembly and reassembly (Ahuja *et al.*, 2004) and greatly impairs the ability of cardiac muscle to regenerate significantly after injury (see chapter 12). Postnatally, endogenous Notch signaling is down-regulated as cardiomyocyte differentiate and exit the cell cycle (Nemir *et al.*, 2006; Rones *et al.*, 2000). Sustained Notch activation *in vitro* can induce cell cycle reentry in quiescent immature cardiac myocytes by activation of the cell cycle regulator cyclin D1 (Campa *et al.*, 2008; Collesi *et al.*, 2008); but cell-cycle withdrawal is irreversible in terminally differentiated cardio myocytes.

Several studies have shown that under conditions of chronic cardiac stress induced by myocardial infarction or decompensated hypertrophy, Notch signaling may be reactivated to control or limit the extent of pathological cardiac remodeling. Crosstalk between Notch and hepatocyte growth factor and its cognate receptor c-Met was shown to regulate survival mechanisms in cardiomyocytes via activation of the pro-survival Akt kinase pathway (Gude *et al.*, 2008). Genetically engineered *Notch1* mutant mice specifically inactivated in the myocardium demonstrated increased apoptotic cell death and worsened function compared with WT mice during hemodynamic overload consistent with a protective anti-apoptotic role (Croquelois *et al.*, 2008). In a transient undifferentiated cell compartment, reactivated Notch signaling prevented cardiogenic differentiation and facilitated the expansion of a myogenic precusor pool. Injecting a viral vector expressing the N1ICD into infarcted mouse hearts (Gude *et al.*, 2008) or over-expressing the N1ICD specifically in the myocardium (Kratsios *et al.*, 2009) improved hemodynamic performance via mechanisms promoting cell survival and angiogenesis. In summary, the Notch pathway is not normally active in the adult myocardium but under conditions of chronic cellular stress reactivation of Notch signaling contributes to cardiac survival and it has been speculated that it may control the maturation of newly formed cardiac myocytes.

10. Conclusions

This review has covered the data implicating the Notch pathway in multiple aspects of cardiac development and demonstrating that mutations in the NOTCH-signaling pathway can cause human heart disease. Open questions remain regarding the mechanism of Notch activation during patterning, morphogenesis, and repair of different cardiac structures in development and disease and how downstream and upstream signals interact with Notch during cardiac development. Future studies with carefully engineered mouse models to understand the underlying molecular mechanism of disease, combined with cell culture and genetic and epidemiologic studies in patients, will give us a more clear understanding of the contribution of altered NOTCH signaling to neonatal and adult cardiac disease.

ACKNOWLEDGMENTS

We wish to thank to J. M. Pérez-Pomares for helpful suggestions and critical reading of this MS.

Donal MacGrogan is supported by a postdoctoral contract associated to grant RD06/0014/0038 (Spanish Ministry of Health, Carlos III Health Institute) and Meritxell Nus by a Sara Borrell postdoctoral contract (Carlos III Health Institute). The work described here was funded by grants SAF2007-62445 (Spanish Ministry of Science and Innovation), P-2006/BIO-194 (Regional Government of Madrid), RD06/0014/0038 and RD06/0010/1013 (Spanish Ministry of Health, Carlos III Health Institute), and LSHM-CT-2005-018630 (Heart Repair, EU FP6) to J. L.d.l.P. The CNIC is supported by the Spanish Ministry of Science and Innovation and the Pro-CNIC Foundation.

REFERENCES

Ahuja, P., Perriard, E., Perriard, J. C., and Ehler, E. (2004). Sequential myofibrillar breakdown accompanies mitotic division of mammalian cardiomyocytes. *J. Cell. Sci.* **117**, 3295–3306.

Bailliard, F., and Anderson, R. H. (2009). Tetralogy of Fallot. *Orphanet J. Rare Dis.* **4**, 2.

Beis, D., Bartman, T., Jin, S. W., Scott, I. C., D'Amico, L. A., Ober, E. A., Verkade, H., Frantsve, J., Field, H. A., Wehman, A., Baier, H., Tallafuss, A., *et al.* (2005). Genetic and cellular analyses of zebrafish atrioventricular cushion and valve development. *Development* **132**, 4193–4204.

Ben-Shachar, G., Arcilla, R. A., Lucas, R. V., and Manasek, F. J. (1985). Ventricular trabeculations in the chick embryo heart and their contribution to ventricular and muscular septal development. *Circ. Res.* **57**, 759–766.

Boyer, A. S., Ayerinskas, I. I., Vincent, E. B., McKinney, L. A., Weeks, D. L., and Runyan, R. B. (1999). TGFbeta2 and TGFbeta3 have separate and sequential activities during epithelial-mesenchymal cell transformation in the embryonic heart. *Dev. Biol.* **208**, 530–545.

Braverman, A. C., Guven, H., Beardslee, M. A., Makan, M., Kates, A. M., and Moon, M. R. (2005). The bicuspid aortic valve. *Curr. Probl. Cardiol.* **30**, 470–522.

Butcher, J. T., Simmons, C. A., and Warnock, J. N. (2008). Mechanobiology of the aortic heart valve. *J. Heart Valve Dis.* **17**, 62–73.

Camenisch, T. D., Molin, D. G., Person, A., Runyan, R. B., Gittenberger-de Groot, A. C., McDonald, J. A., and Klewer, S. E. (2002). Temporal and distinct TGFbeta ligand requirements during mouse and avian endocardial cushion morphogenesis. *Dev. Biol.* **248**, 170–181.

Campa, V. M., Gutierrez-Lanza, R., Cerignoli, F., Diaz-Trelles, R., Nelson, B., Tsuji, T., Barcova, M., Jiang, W., and Mercola, M. (2008). Notch activates cell cycle reentry and progression in quiescent cardiomyocytes. *J. Cell Biol.* **183**, 129–141.

Chen, H., Shi, S., Acosta, L., Li, W., Lu, J., Bao, S., Chen, Z., Yang, Z., Schneider, M. D., Chien, K. R., Conway, S. J., Yoder, M. C., *et al.* (2004). BMP10 is essential for maintaining cardiac growth during murine cardiogenesis. *Development* **131**, 2219–2231.

Cheng, H. T., Kim, M., Valerius, M. T., Surendran, K., Schuster-Gossler, K., Gossler, A., McMahon, A. P., and Kopan, R. (2007). Notch2, but not Notch1, is required for proximal fate acquisition in the mammalian nephron. *Development* **134**, 801–811.

Chien, K. R., Knowlton, K. U., Zhu, H., and Chien, S. (1991). Regulation of cardiac gene expression during myocardial growth and hypertrophy: molecular studies of an adaptive physiologic response. *FASEB J.* **5**, 3037–3046.

Christianson, A., and Modell, B. (2004). Medical genetics in developing countries. *Annu. Rev. Genomics Hum. Genet.* **5**, 219–265.

Collesi, C., Zentilin, L., Sinagra, G., and Giacca, M. (2008). Notch1 signaling stimulates proliferation of immature cardiomyocytes. *J. Cell Biol.* **183**, 117–128.

Cordle, J., Johnson, S., Tay, J. Z., Roversi, P., Wilkin, M. B., de Madrid, B. H., Shimizu, H., Jensen, S., Whiteman, P., Jin, B., Redfield, C., Baron, M., *et al.* (2008). A conserved face of the Jagged/Serrate DSL domain is involved in Notch trans-activation and cis-inhibition. *Nat. Struct. Mol. Biol.* **15**, 849–857.

Cripe, L., Andelfinger, G., Martin, L. J., Shooner, K., and Benson, D. W. (2004). Bicuspid aortic valve is heritable. *J. Am. Coll. Cardiol.* **44**, 138–143.

Croquelois, A., Domenighetti, A. A., Nemir, M., Lepore, M., Rosenblatt-Velin, N., Radtke, F., and Pedrazzini, T. (2008). Control of the adaptive response of the heart to stress via the Notch1 receptor pathway. *J. Exp. Med.* **205**, 3173–3185.

Crosnier, C., Driancourt, C., Raynaud, N., Dhorne-Pollet, S., Pollet, N., Bernard, O., Hadchouel, M., and Meunier-Rotival, M. (1999). Mutations in JAGGED1 gene are predominantly sporadic in Alagille syndrome. *Gastroenterology* **116**, 1141–1148.

Del Monte, G., Grego-Bessa, J., Gonzalez-Rajal, A., Bolos, V., and De La Pompa, J. L. (2007). Monitoring Notch1 activity in development: evidence for a feedback regulatory loop. *Dev. Dyn.* **236**, 2594–2614.

Donovan, J., Kordylewska, A., Jan, Y. N., and Utset, M. F. (2002). Tetralogy of fallot and other congenital heart defects in Hey2 mutant mice. *Curr. Biol.* **12**, 1605–1610.

Dorn, G. W., 2nd (2007). The fuzzy logic of physiological cardiac hypertrophy. *Hypertension* **49**, 962–970.

Dorn, G. W., 2nd, Robbins, J., and Sugden, P. H. (2003). Phenotyping hypertrophy: eschew obfuscation. *Circ. Res.* **92**, 1171–1175.

Duarte, A., Hirashima, M., Benedito, R., Trindade, A., Diniz, P., Bekman, E., Costa, L., Henrique, D., and Rossant, J. (2004). Dosage-sensitive requirement for mouse Dll4 in artery development. *Genes Dev.* **18**, 2474–2478.

Eldadah, Z. A., Hamosh, A., Biery, N. J., Montgomery, R. A., Duke, M., Elkins, R., and Dietz, H. C. (2001). Familial Tetralogy of Fallot caused by mutation in the jagged1 gene. *Hum. Mol. Genet.* **10**, 163–169.

Falls, D. L. (2003). Neuregulins: functions, forms, and signaling strategies. *Exp. Cell Res.* **284**, 14–30.

Fischer, A., Schumacher, N., Maier, M., Sendtner, M., and Gessler, M. (2004). The Notch target genes Hey1 and Hey2 are required for embryonic vascular development. *Genes Dev.* **18**, 901–911.

Fischer, A., Steidl, C., Wagner, T. U., Lang, E., Jakob, P. M., Friedl, P., Knobeloch, K. P., and Gessler, M. (2007). Combined loss of Hey1 and HeyL causes congenital heart defects because of impaired epithelial to mesenchymal transition. *Circ. Res.* **100**, 856–863.

Garg, V., Muth, A. N., Ransom, J. F., Schluterman, M. K., Barnes, R., King, I. N., Grossfeld, P. D., and Srivastava, D. (2005). Mutations in NOTCH1 cause aortic valve disease. *Nature* **437**, 270–274.

Gassmann, M., Casagranda, F., Orioli, D., Simon, H., Lai, C., Klein, R., and Lemke, G. (1995). Aberrant neural and cardiac development in mice lacking the ErbB4 neuregulin receptor. *Nature* **378**, 390–394.

Gessler, M., Knobeloch, K. P., Helisch, A., Amann, K., Schumacher, N., Rohde, E., Fischer, A., and Leimeister, C. (2002). Mouse gridlock: no aortic coarctation or deficiency, but fatal cardiac defects in Hey2 -/- mice. *Curr. Biol.* **12**, 1601–1604.

Gordon, W. R., Arnett, K. L., and Blacklow, S. C. (2008). The molecular logic of Notch signaling – a structural and biochemical perspective. *J. Cell Sci.* **121**, 3109–3119.

Greenway, S. C., Pereira, A. C., Lin, J. C., DePalma, S. R., Israel, S. J., Mesquita, S. M., Ergul, E., Conta, J. H., Korn, J. M., McCarroll, S. A., Gorham, J. M., Gabriel, S., *et al.* (2009). De novo copy number variants identify new genes and loci in isolated sporadic tetralogy of Fallot. *Nat. Genet.* **41**, 931–935.

Grego-Bessa, J., Luna-Zurita, L., del Monte, G., Bolos, V., Melgar, P., Arandilla, A., Garratt, A. N., Zang, H., Mukouyama, Y. S., Chen, H., Shou, W., Ballestar, E., *et al.* (2007). Notch signaling is essential for ventricular chamber development. *Dev. Cell* **12**, 415–429.

Guarnaccia, C., Dhir, S., Pintar, A., and Pongor, S. (2009). The tetralogy of Fallot-associated G274D mutation impairs folding of the second epidermal growth factor repeat in Jagged-1. *FEBS J.* **276**, 6247–6257.

Gude, N. A., Emmanuel, G., Wu, W., Cottage, C. T., Fischer, K., Quijada, P., Muraski, J. A., Alvarez, R., Rubio, M., Schaefer, E., and Sussman, M. A. (2008). Activation of Notch-mediated protective signaling in the myocardium. *Circ. Res.* **102**, 1025–1035.

Habets, P. E., Moorman, A. F., Clout, D. E., van Roon, M. A., Lingbeek, M., van Lohuizen, M., Campione, M., and Christoffels, V. M. (2002). Cooperative action of Tbx2 and Nkx2.5 inhibits ANF expression in the atrioventricular canal: implications for cardiac chamber formation. *Genes Dev.* **16**, 1234–1246.

Hadland, B. K., Huppert, S. S., Kanungo, J., Xue, Y., Jiang, R., Gridley, T., Conlon, R. A., Cheng, A. M., Kopan, R., and Longmore, G. D. (2004). A requirement for Notch1 distinguishes 2 phases of definitive hematopoiesis during development. *Blood* **104**, 3097–3105.

Harrelson, Z., Kelly, R. G., Goldin, S. N., Gibson-Brown, J. J., Bollag, R. J., Silver, L. M., and Papaioannou, V. E. (2004). Tbx2 is essential for patterning the atrioventricular canal and for morphogenesis of the outflow tract during heart development. *Development* **131**, 5041–5052.

Hein, S., Arnon, E., Kostin, S., Schonburg, M., Elsasser, A., Polyakova, V., Bauer, E. P., Klovekorn, W. P., and Schaper, J. (2003). Progression from compensated hypertrophy to failure in the pressure-overloaded human heart: structural deterioration and compensatory mechanisms. *Circulation* **107**, 984–991.

Hertig, C. M., Kubalak, S. W., Wang, Y., and Chien, K. R. (1999). Synergistic roles of neuregulin-1 and insulin-like growth factor-I in activation of the phosphatidylinositol 3-kinase pathway and cardiac chamber morphogenesis. *J. Biol. Chem.* **274**, 37362–37369.

High, F. A., Jain, R., Stoller, J. Z., Antonucci, N. B., Lu, M. M., Loomes, K. M., Kaestner, K. H., Pear, W. S., and Epstein, J. A. (2009). Murine Jagged1/Notch signaling in the second heart field orchestrates Fgf8 expression and tissue-tissue interactions during outflow tract development. *J. Clin. Invest.* **119**, 1986–1996.

High, F. A., Zhang, M., Proweller, A., Tu, L., Parmacek, M. S., Pear, W. S., and Epstein, J. A. (2007). An essential role for Notch in neural crest during cardiovascular development and smooth muscle differentiation. *J. Clin. Invest.* **117**, 353–363.

Hinton, R. B., Jr., Martin, L. J., Tabangin, M. E., Mazwi, M. L., Cripe, L. H., and Benson, D. W. (2007). Hypoplastic left heart syndrome is heritable. *J. Am. Coll. Cardiol.* **50**, 1590–1595.

Huntington, K., Hunter, A. G., and Chan, K. L. (1997). A prospective study to assess the frequency of familial clustering of congenital bicuspid aortic valve. *J. Am. Coll. Cardiol.* **30**, 1809–1812.

Jessup, M., and Brozena, S. (2003). Heart failure. *N. Engl. J. Med.* **348**, 2007–2018.

Johnson, S.E., Ilagan, M.X., Kopan, R., and Barrick, D.Thermodynamic analysis of the CSL x Notch interaction: distribution of binding energy of the Notch RAM region to the CSL beta-trefoil domain and the mode of competition with the viral transactivator EBNA2. *J. Biol. Chem.* **285**, 6681–6692.

Kehat, I., Kenyagin-Karsenti, D., Snir, M., Segev, H., Amit, M., Gepstein, A., Livne, E., Binah, O., Itskovitz-Eldor, J., and Gepstein, L. (2001). Human embryonic stem cells can differentiate into myocytes with structural and functional properties of cardiomyocytes. *J. Clin. Invest.* **108**, 407–414.

Kokubo, H., Miyagawa-Tomita, S., and Johnson, R. L. (2005a). Hesr, a mediator of the Notch signaling, functions in heart and vessel development. *Trends Cardiovasc. Med.* **15**, 190–194.

Kokubo, H., Miyagawa-Tomita, S., Nakazawa, M., Saga, Y., and Johnson, R. L. (2005b). Mouse hesr1 and hesr2 genes are redundantly required to mediate Notch signaling in the developing cardiovascular system. *Dev. Biol.* **278**, 301–309.

Kokubo, H., Miyagawa-Tomita, S., Tomimatsu, H., Nakashima, Y., Nakazawa, M., Saga, Y., and Johnson, R. L. (2004). Targeted disruption of hesr2 results in atrioventricular valve anomalies that lead to heart dysfunction. *Circ. Res.* **95**, 540–547.

Kokubo, H., Tomita-Miyagawa, S., Hamada, Y., and Saga, Y. (2007). Hesr1 and Hesr2 regulate atrioventricular boundary formation in the developing heart through the repression of Tbx2. *Development* **134**, 747–755.

Kopan, R., Cheng, H. T., and Surendran, K. (2007). Molecular insights into segmentation along the proximal-distal axis of the nephron. *J. Am. Soc. Nephrol.* **18**, 2014–2020.

Kopan, R., and Ilagan, M. X. (2009). The canonical Notch signaling pathway: unfolding the activation mechanism. *Cell* **137**, 216–233.

Krantz, I. D., Colliton, R. P., Genin, A., Rand, E. B., Li, L., Piccoli, D. A., and Spinner, N. B. (1998). Spectrum and frequency of jagged1 (JAG1) mutations in Alagille syndrome patients and their families. *Am. J. Hum. Genet.* **62**, 1361–1369.

Krantz, I. D., Smith, R., Colliton, R. P., Tinkel, H., Zackai, E. H., Piccoli, D. A., Goldmuntz, E., and Spinner, N. B. (1999). Jagged1 mutations in patients ascertained with isolated congenital heart defects. *Am. J. Med. Genet.* **84**, 56–60.

Kratsios, P., Catela, C., Salimova, E., Huth, M., Berno, V., Rosenthal, N., and Mourkioti, F. (2010). Distinct roles for cell-autonomous Notch signaling in cardiomyocytes of the embryonic and adult heart. *Circ. Res.* **106**, 559–572.

Le Caignec, C., Lefevre, M., Schott, J. J., Chaventre, A., Gayet, M., Calais, C., and Moisan, J. P. (2002). Familial deafness, congenital heart defects, and posterior embryotoxon caused by cysteine substitution in the first epidermal-growth-factor-like domain of jagged 1. *Am. J. Hum. Genet.* **71**, 180–186.

Lee, K. F., Simon, H., Chen, H., Bates, B., Hung, M. C., and Hauser, C. (1995). Requirement for neuregulin receptor erbB2 in neural and cardiac development. *Nature* **378**, 394–398.

Leggett, M., and Otto, C. M. (1996). Aortic valve disease. *Curr. Opin. Cardiol.* **11**, 120–125.

Lewin, M. B., McBride, K. L., Pignatelli, R., Fernbach, S., Combes, A., Menesses, A., Lam, W., Bezold, L. I., Kaplan, N., Towbin, J. A., and Belmont, J. W. (2004). Echocardiographic evaluation of asymptomatic parental and sibling cardiovascular anomalies associated with congenital left ventricular outflow tract lesions. *Pediatrics* **114**, 691–696.

Li, L., Krantz, I. D., Deng, Y., Genin, A., Banta, A. B., Collins, C. C., Qi, M., Trask, B. J., Kuo, W. L., Cochran, J., Costa, T., Pierpont, M. E., *et al.* (1997). Alagille syndrome is caused by mutations in human Jagged1, which encodes a ligand for Notch1. *Nat. Genet.* **16**, 243–251.

Loffredo, C. A. (2000). Epidemiology of cardiovascular malformations: prevalence and risk factors. *Am. J. Med. Genet.* **97**, 319–325.

Loffredo, C. A., Chokkalingam, A., Sill, A. M., Boughman, J. A., Clark, E. B., Scheel, J., and Brenner, J. I. (2004). Prevalence of congenital cardiovascular malformations among relatives of infants with hypoplastic left heart, coarctation of the aorta, and d-transposition of the great arteries. *Am. J. Med. Genet. A* **124**, 225–230.

Lorell, B. H., and Carabello, B. A. (2000). Left ventricular hypertrophy: pathogenesis, detection, and prognosis. *Circulation* **102**, 470–479.

Lu, F., Morrissette, J. J., and Spinner, N. B. (2003). Conditional JAG1 mutation shows the developing heart is more sensitive than developing liver to JAG1 dosage. *Am. J. Hum. Genet.* **72**, 1065–1070.

Ma, L., Lu, M. F., Schwartz, R. J., and Martin, J. F. (2005). Bmp2 is essential for cardiac cushion epithelial-mesenchymal transition and myocardial patterning. *Development* **132**, 5601–5611.

Markwald, R. R., Fitzharris, T. P., and Manasek, F. J. (1977). Structural development of endocardial cushions. *Am. J. Anat.* **148**, 85–119.

Martin, L. J., Ramachandran, V., Cripe, L. H., Hinton, R. B., Andelfinger, G., Tabangin, M., Shooner, K., Keddache, M., and Benson, D. W. (2007). Evidence in favor of linkage to human chromosomal regions 18q, 5q and 13q for bicuspid aortic valve and associated cardiovascular malformations. *Hum. Genet.* **121**, 275–284.

Matsuura, A., Ito, M., Sakaidani, Y., Kondo, T., Murakami, K., Furukawa, K., Nadano, D., Matsuda, T., and Okajima, T. (2008). O-linked N-acetylglucosamine is present on the extracellular domain of notch receptors. *J. Biol. Chem.* **283**, 35486–35495.

McBride, K. L., Pignatelli, R., Lewin, M., Ho, T., Fernbach, S., Menesses, A., Lam, W., Leal, S. M., Kaplan, N., Schliekelman, P., Towbin, J. A., and Belmont, J. W. (2005). Inheritance analysis of congenital left ventricular outflow tract obstruction malformations: Segregation, multiplex relative risk, and heritability. *Am. J. Med. Genet. A* **134A**, 180–186.

McBride, K. L., Riley, M. F., Zender, G. A., Fitzgerald-Butt, S. M., Towbin, J. A., Belmont, J. W., and Cole, S. E. (2008). NOTCH1 mutations in individuals with left ventricular outflow tract malformations reduce ligand-induced signaling. *Hum. Mol. Genet.* **17**, 2886–2893.

McBride, K. L., Zender, G. A., Fitzgerald-Butt, S. M., Koehler, D., Menesses-Diaz, A., Fernbach, S., Lee, K., Towbin, J. A., Leal, S., and Belmont, J. W. (2009). Linkage analysis of left ventricular outflow tract malformations (aortic valve stenosis, coarctation of the aorta, and hypoplastic left heart syndrome. *Eur. J. Hum. Genet.* **17**, 811–819.

McCright, B., Gao, X., Shen, L., Lozier, J., Lan, Y., Maguire, M., Herzlinger, D., Weinmaster, G., Jiang, R., and Gridley, T. (2001). Defects in development of the kidney, heart and eye vasculature in mice homozygous for a hypomorphic Notch2 mutation. *Development* **128**, 491–502.

McCright, B., Lozier, J., and Gridley, T. (2002). A mouse model of Alagille syndrome: Notch2 as a genetic modifier of Jag1 haploinsufficiency. *Development* **129**, 1075–1082.

McDaniell, R., Warthen, D. M., Sanchez-Lara, P. A., Pai, A., Krantz, I. D., Piccoli, D. A., and Spinner, N. B. (2006). NOTCH2 mutations cause Alagille syndrome, a heterogeneous disorder of the notch signaling pathway. *Am. J. Hum. Genet.* **79**, 169–173.

McElhinney, D. B., Krantz, I. D., Bason, L., Piccoli, D. A., Emerick, K. M., Spinner, N. B., and Goldmuntz, E. (2002). Analysis of cardiovascular phenotype and genotype-phenotype correlation in individuals with a JAG1 mutation and/or Alagille syndrome. *Circulation* **106**, 2567–2574.

McKellar, S. H., Tester, D. J., Yagubyan, M., Majumdar, R., Ackerman, M. J., and Sundt, T. M., 3rd. (2007). Novel NOTCH1 mutations in patients with bicuspid aortic valve disease and thoracic aortic aneurysms. *J. Thorac. Cardiovasc. Surg.* **134**, 290–296.

McMullen, J. R., and Jennings, G. L. (2007). Differences between pathological and physiological cardiac hypertrophy: novel therapeutic strategies to treat heart failure. *Clin. Exp. Pharmacol. Physiol.* **34**, 255–262.

McMurray, J. J., and Pfeffer, M. A. (2005). Heart failure. *Lancet* **365**, 1877–1889.

Meyer, D., and Birchmeier, C. (1995). Multiple essential functions of neuregulin in development. *Nature* **378**, 386–390.

Mohamed, S. A., Aherrahrou, Z., Liptau, H., Erasmi, A. W., Hagemann, C., Wrobel, S., Borzym, K., Schunkert, H., Sievers, H. H., and Erdmann, J. (2006). Novel missense mutations (p.T596M and p.P1797H) in NOTCH1 in patients with bicuspid aortic valve. *Biochem. Biophys. Res. Commun.* **345**, 1460–1465.

Morimoto, M., Liu, Z., Cheng, H. T., Winters, N., Bader, D., and Kopan, R. (2010). Canonical Notch signaling in the developing lung is required for determination of arterial smooth muscle cells and selection of Clara versus ciliated cell fate. *J. Cell. Sci.* **123**, 213–224.

Morrissette, J. D., Colliton, R. P., and Spinner, N. B. (2001). Defective intracellular transport and processing of JAG1 missense mutations in Alagille syndrome. *Hum. Mol. Genet.* **10**, 405–413.

Nakagawa, O., Nakagawa, M., Richardson, J. A., Olson, E. N., and Srivastava, D. (1999). HRT1, HRT2, and HRT3: a new subclass of bHLH transcription factors marking specific cardiac, somitic, and pharyngeal arch segments. *Dev. Biol.* **216**, 72–84.

Nakajima, M., Moriizumi, E., Koseki, H., and Shirasawa, T. (2004). Presenilin 1 is essential for cardiac morphogenesis. *Dev. Dyn.* **230**, 795–799.

Nemir, M., Croquelois, A., Pedrazzini, T., and Radtke, F. (2006). Induction of cardiogenesis in embryonic stem cells via downregulation of Notch1 signaling. *Circ. Res.* **98**, 1471–1478.

Nigam, V., and Srivastava, D. (2009). Notch1 represses osteogenic pathways in aortic valve cells. *J. Mol. Cell. Cardiol.* **47**, 828–834.

Oda, T., Elkahloun, A. G., Pike, B. L., Okajima, K., Krantz, I. D., Genin, A., Piccoli, D. A., Meltzer, P. S., Spinner, N. B., Collins, F. S., Chandrasekharappa, S. C., Washburn, T., *et al.* (1997). Mutations in the human Jagged1 gene are responsible for Alagille syndrome. *Nat. Genet.* **16**, 235–242.

Oka, C., Nakano, T., Wakeham, A., de la Pompa, J. L., Mori, C., Sakai, T., Okazaki, S., Kawaichi, M., Shiota, K., Mak, T. W., and Honjo, T. (1995). Disruption of the mouse RBP-J kappa gene results in early embryonic death. *Development* **121**, 3291–3301.

Otto, C. M. (2009). Calcific aortic valve disease: outflow obstruction is the end stage of a systemic disease process. *Eur. Heart J.* **30**, 1940–1942.

Park, E. J., Watanabe, Y., Smyth, G., Miyagawa-Tomita, S., Meyers, E., Klingensmith, J., Camenisch, T., Buckingham, M., and Moon, A. M. (2008). An FGF autocrine loop initiated in second heart field mesoderm regulates morphogenesis at the arterial pole of the heart. *Development* **135**, 3599–3610.

Parks, A. L., Stout, J. R., Shepard, S. B., Klueg, K. M., Dos Santos, A. A., Parody, T. R., Vaskova, M., and Muskavitch, M. A. (2006). Structure-function analysis of delta trafficking, receptor binding and signaling in Drosophila. *Genetics* **174**, 1947–1961.

Person, A. D., Klewer, S. E., and Runyan, R. B. (2005). Cell biology of cardiac cushion development. *Int. Rev. Cytol.* **243**, 287–335.

Probst, V., Allouis, M., Sacher, F., Pattier, S., Babuty, D., Mabo, P., Mansourati, J., Victor, J., Nguyen, J. M., Schott, J. J., Boisseau, P., Escande, D., *et al.* (2006). Progressive cardiac conduction defect is the prevailing phenotype in carriers of a Brugada syndrome SCN5A mutation. *J. Cardiovasc. Electrophysiol.* **17**, 270–275.

Rajamannan, N. M., Bonow, R. O., and Rahimtoola, S. H. (2007). Calcific aortic stenosis: an update. *Nat. Clin. Pract. Cardiovasc. Med.* **4**, 254–262.

Ramaraj, R., and Sorrell, V. L. (2008). Degenerative aortic stenosis. *BMJ* **336**, 550–555.

Reamon-Buettner, S. M., and Borlak, J. (2006). Somatic mutations in cardiac malformations. *J. Med. Genet.* **43**, e45.

Rebay, I., Fleming, R. J., Fehon, R. G., Cherbas, L., Cherbas, P., and Artavanis-Tsakonas, S. (1991). Specific EGF repeats of Notch mediate interactions with Delta and Serrate: implications for Notch as a multifunctional receptor. *Cell* **67**, 687–699.

Rentschler, S., Zander, J., Meyers, K., France, D., Levine, R., Porter, G., Rivkees, S. A., Morley, G. E., and Fishman, G. I. (2002). Neuregulin-1 promotes formation of the murine cardiac conduction system. *Proc. Natl. Acad. Sci. U.S.A.* **99**, 10464–10469.

Rivera-Feliciano, J., and Tabin, C. J. (2006). Bmp2 instructs cardiac progenitors to form the heart-valve-inducing field. *Dev. Biol.* **295**, 580–588.

Rochais, F., Dandonneau, M., Mesbah, K., Jarry, T., Mattei, M. G., and Kelly, R. G. (2009). Hes1 is expressed in the second heart field and is required for outflow tract development. *PLoS One* **4**, e6267.

Rones, M. S., McLaughlin, K. A., Raffin, M., and Mercola, M. (2000). Serrate and Notch specify cell fates in the heart field by suppressing cardiomyogenesis. *Development* **127**, 3865–3876.

Rutenberg, J. B., Fischer, A., Jia, H., Gessler, M., Zhong, T. P., and Mercola, M. (2006). Developmental patterning of the cardiac atrioventricular canal by Notch and Hairy-related transcription factors. *Development* **133**, 4381–4390.

Ryan, M. J., Bales, C., Nelson, A., Gonzalez, D. M., Underkoffler, L., Segalov, M., Wilson-Rawls, J., Cole, S. E., Moran, J. L., Russo, P., Spinner, N. B., Kusumi, K., *et al.* (2008). Bile duct proliferation in Jag1/fringe heterozygous mice identifies candidate modifiers of the Alagille syndrome hepatic phenotype. *Hepatology* **48**, 1989–1997.

Sakata, Y., Kamei, C. N., Nakagami, H., Bronson, R., Liao, J. K., and Chin, M. T. (2002). Ventricular septal defect and cardiomyopathy in mice lacking the transcription factor CHF1/Hey2. *Proc. Natl. Acad. Sci. U.S.A.* **99**, 16197–16202.

Sakata, Y., Koibuchi, N., Xiang, F., Youngblood, J. M., Kamei, C. N., and Chin, M. T. (2006). The spectrum of cardiovascular anomalies in CHF1/Hey2 deficient mice reveals roles in endocardial cushion, myocardial and vascular maturation. *J. Mol. Cell. Cardiol.* **40**, 267–273.

Schroeder, T., Fraser, S. T., Ogawa, M., Nishikawa, S., Oka, C., Bornkamm, G. W., Nishikawa, S., Honjo, T., and Just, U. (2003). Recombination signal sequence-binding protein Jkappa alters mesodermal cell fate decisions by suppressing cardiomyogenesis. *Proc. Natl. Acad. Sci. U.S.A.* **100**, 4018–4023.

Shen, H., McElhinny, A. S., Cao, Y., Gao, P., Liu, J., Bronson, R., Griffin, J. D., and Wu, L. (2006). The Notch coactivator, MAML1, functions as a novel coactivator for MEF2C-mediated transcription and is required for normal myogenesis. *Genes Dev.* **20**, 675–688.

Shimizu, K., Chiba, S., Kumano, K., Hosoya, N., Takahashi, T., Kanda, Y., Hamada, Y., Yazaki, Y., and Hirai, H. (1999). Mouse jagged1 physically interacts with notch2 and other notch receptors. Assessment by quantitative methods. *J. Biol. Chem.* **274**, 32961–32969.

Slavotinek, A., and Biesecker, L. G. (2003). Genetic modifiers in human development and malformation syndromes, including chaperone proteins. *Hum. Mol. Genet.* **12(Spec No 1)**, R45–R50.

Spinner, N. B., Colliton, R. P., Crosnier, C., Krantz, I. D., Hadchouel, M., and Meunier-Rotival, M. (2001). Jagged1 mutations in alagille syndrome. *Hum. Mutat.* **17**, 18–33.

Stanley, P. (2007). Regulation of Notch signaling by glycosylation. *Curr. Opin. Struct. Biol.* **17**, 530–535.

Surendran, K., Boyle, S., Barak, H., Kim, M., Stomberski, C., McCright, B., and Kopan, R. The contribution of Notch1 to nephron segmentation in the developing kidney is revealed in a sensitized Notch2 background and can be augmented by reducing Mint dosage. *Dev. Biol.* **337**, 386–395.

Tien, A. C., Rajan, A., and Bellen, H. J. (2009). A Notch updated. *J. Cell Biol.* **184**, 621–629.

Timmerman, L. A., Grego-Bessa, J., Raya, A., Bertran, E., Perez-Pomares, J. M., Diez, J., Aranda, S., Palomo, S., McCormick, F., Izpisua-Belmonte, J. C., and de la Pompa, J. L. (2004). Notch promotes epithelial-mesenchymal transition during cardiac development and oncogenic transformation. *Genes Dev.* **18**, 99–115.

Towbin, J. A., and Belmont, J. (2000). Molecular determinants of left and right outflow tract obstruction. *Am. J. Med. Genet.* **97**, 297–303.

Vooijs, M., Ong, C. T., Hadland, B., Huppert, S., Liu, Z., Korving, J., van den Born, M., Stappenbeck, T., Wu, Y., Clevers, H., and Kopan, R. (2007). Mapping the consequence of Notch1 proteolysis in vivo with NIP-CRE. *Development* **134**, 535–544.

Vuillemin, M., and Pexieder, T. (1989). Normal stages of cardiac organogenesis in the mouse: II. Development of the internal relief of the heart. *Am. J. Anat.* **184**, 114–128.

Wang, H. U., Chen, Z. F., and Anderson, D. J. (1998). Molecular distinction and angiogenic interaction between embryonic arteries and veins revealed by ephrin-B2 and its receptor Eph-B4 [see comments]. *Cell* **93**, 741–753.

Warthen, D. M., Moore, E. C., Kamath, B. M., Morrissette, J. J., Sanchez, P., Piccoli, D. A., Krantz, I. D., and Spinner, N. B. (2006). Jagged1 (JAG1) mutations in Alagille syndrome: increasing the mutation detection rate. *Hum. Mutat.* **27**, 436–443.

Watanabe, Y., Kokubo, H., Miyagawa-Tomita, S., Endo, M., Igarashi, K., Aisaki, K. I., Kanno, J., and Saga, Y. (2006). Activation of Notch1 signaling in cardiogenic mesoderm induces abnormal heart morphogenesis in mouse. *Development* **133**, 1625–1634.

Wessels, M. W., Berger, R. M., Frohn-Mulder, I. M., Roos-Hesselink, J. W., Hoogeboom, J. J., Mancini, G. S., Bartelings, M. M., Krijger, R., Wladimiroff, J. W., Niermeijer, M. F., Grossfeld, P., and Willems, P. J. (2005). Autosomal dominant inheritance of left ventricular outflow tract obstruction. *Am. J. Med. Genet. A* **134A**, 171–179.

Wilson-Rawls, J., Molkentin, J. D., Black, B. L., and Olson, E. N. (1999). Activated notch inhibits myogenic activity of the MADS-Box transcription factor myocyte enhancer factor 2C. *Mol. Cell. Biol.* **19**, 2853–2862.

Xu, A., Lei, L., and Irvine, K. D. (2005). Regions of *Drosophila* Notch that contribute to ligand binding and the modulatory influence of Fringe. *J. Biol. Chem.* **280**, 30158–30165.

Xu, S., Liu, A. C., and Gotlieb, A. I. (2010). Common pathogenic features of atherosclerosis and calcific aortic stenosis: role of transforming growth factor-beta. *Cardiovasc. Pathol.* **19**, 236–247.

Xue, Y., Gao, X., Lindsell, C. E., Norton, C. R., Chang, B., Hicks, C., Gendron-Maguire, M., Rand, E. B., Weinmaster, G., and Gridley, T. (1999). Embryonic lethality and vascular defects in mice lacking the Notch ligand Jagged1. *Hum. Mol. Genet.* **8**, 723–730.

Yamada, M., Revelli, J. P., Eichele, G., Barron, M., and Schwartz, R. J. (2000). Expression of chick Tbx-2, Tbx-3, and Tbx-5 genes during early heart development: evidence for BMP2 induction of Tbx2. *Dev. Biol.* **228**, 95–105.

Yang, X., Meng, X., Su, X., Mauchley, D. C., Ao, L., Cleveland, J. C., Jr., and Fullerton, D. A. (2009). Bone morphogenic protein 2 induces Runx2 and osteopontin expression in human aortic valve interstitial cells: role of Smad1 and extracellular signal-regulated kinase 1/2. *J. Thorac. Cardiovasc. Surg.* **138**, 1008–1015.

Zhao, Y. Y., Sawyer, D. R., Baliga, R. R., Opel, D. J., Han, X., Marchionni, M. A., and Kelly, R. A. (1998). Neuregulins promote survival and growth of cardiac myocytes. Persistence of ErbB2 and ErbB4 expression in neonatal and adult ventricular myocytes. *J. Biol. Chem.* **273**, 10261–10269.

Zhong, T. P., Rosenberg, M., Mohideen, M. A., Weinstein, B., and Fishman, M. C. (2000). gridlock, an HLH gene required for assembly of the aorta in zebrafish. *Science* **287**, 1820–1824.

CHAPTER TWELVE

Notch Signaling in the Regulation of Stem Cell Self-Renewal and Differentiation

Jianing Liu,* Chihiro Sato,† Massimiliano Cerletti,* *and* Amy Wagers*

Contents

Abstract

Stem cells are rare and unique precursor cells that participate in the building and rebuilding of tissues and organs during embryogenesis, postnatal growth, and injury repair. Stem cells are distinctively endowed with the ability to both self-renew and differentiate, such that they can replenish the stem cell pool while continuing to produce the differentiated daughter cells that are essential for tissue function. Stem cell self-renewal/differentiation decisions must

* Joslin Diabetes Center, One Joslin Place, Boston, MA, USA
† Washington University, St. Louis, MO, USA

Current Topics in Developmental Biology, Volume 92
ISSN 0070-2153, DOI 10.1016/S0070-2153(10)91012-7

be carefully controlled during organogenesis, tissue homeostasis, and regeneration, as failure in stem cell maintenance or activation can lead to progressive tissue wasting, while unchecked self-renewal is a hallmark of many cancers. Here, we review evidence implicating the Notch signaling pathway, an evolutionarily conserved cell fate determinant with widespread roles in a variety of tissues and organisms, as a crucial regulator of stem cell behavior. As discussed below, this pathway plays varied and critical roles at multiple stages of organismal development, in lineage-specific differentiation of pluripotent embryonic stem cells, and in controlling stem cell numbers and activity in the context of age-related tissue degeneration, injury-induced tissue repair, and malignancy.

1. Introduction to Stem Cells and Stem Cell Biology

Stem cells are a rare population of cells that possess the ability to self-renew to preserve the stem cell pool and to differentiate to produce progeny cells needed for the physiological functions of tissues and organs. Stem cells exist in many different organisms and have been identified in both the plant and the animal kingdoms. Embryonic stem (ES) cells (isolated from early-stage embryos) are able to differentiate into all the cell types required to form an entire organism and thus may be regarded as a fundamental building block of life. Similarly, in multicellular organisms, the establishment, maintenance, and repair of highly specialized tissues often depend upon rare tissue-resident stem cells (also called adult stem cells).

Stem cells from various developmental stages and organs share many common features, and all possess the ability to self-renew and differentiate, but these cells differ to some degree with regard to their developmental potency. *Totipotent* mammalian stem cells are found only in early embryos and are able to form complete organisms. *Pluripotent* stem cells can be found in, and cultured from, the inner cell mass (ICM) of the blastocyst and can form any cell type found in the adult body. As discussed further in later sections of this chapter, recent exciting advances in reprogramming technology have made possible also the derivation of pluripotent stem cells from adult somatic cells, through induced expression of transcriptional "reprogramming" factors (Park *et al.*, 2008b; Takahashi and Yamanaka, 2006; Takahashi *et al.*, 2007).

Tissue-specific adult stem cells, and stem cells derived from fetal tissues and cord blood, may be *multipotent, oligopotent*, or even *unipotent*. These cells are responsible for organogenesis, tissue maturation, repair, and rejuvenation. It is still controversial whether *every* mammalian tissue and organ possesses an adult stem cell, but many tissue-specific adult stem cells have been successfully identified and isolated, e.g., hematopoietic stem cells (HSCs), germ line stem cell (GSC), epithelial stem cells, muscle stem cells (satellite cells), intestinal stem cells (ISCs), and mesenchymal stem cells (MSCs) (see Table 12.1 and Fig. 12.1). Notch signaling plays varied and important roles in regulating many types of stem cells, as well as the signals they receive from their microenvironment, or "niche." These roles will be reviewed here, with particular emphasis on how this conserved signaling pathway has been broadly adapted to play differing yet important roles in the functions of pluripotent and tissue-specific stem cells in developing and adult animals.

1.1. Cellular properties of stem cells: cell division, cell cycle, and life span

Stem cells may undergo both symmetric and asymmetric divisions, instructed by diverse molecular, cellular, and environmental cues at discrete developmental stages. Both symmetric and asymmetric cell divisions can support stem cell self-renewal. Symmetric cell division enables stem cells to generate two daughter cells, each with properties that are indistinguishable from the mother cell. This mode of division is critical for expanding stem cell reservoirs, and also can produce two differentiated daughter cells having less potency than the parental stem cell, leading to rapid production of tissue "effector" cells but potential depletion of the stem cell pool.

Asymmetric cell division gives rise to two daughter cells with distinct cell fates: one daughter maintains stem cell properties and functions, while the other daughter loses these characteristics. In all likelihood, both symmetric and asymmetric cell divisions are employed *in vivo* to maintain the fine balance between self-renewal and differentiation of stem cells (Molofsky *et al.*, 2004). The particular roles of Notch in asymmetric cell divisions within the fly intestine and the developing nervous system are described later in this chapter.

Different types of stem cells have different life spans, linked in part to chromosome stability (or instability), oxidative stress, telomere length, and DNA damage repair activity (Chen *et al.*, 2006; Nijnik *et al.*, 2007; Nussenzweig *et al.*, 1996; Rai *et al.*, 2009; Rossi *et al.*, 2007). ES cells, and perhaps cancer stem cells (Reya *et al.*, 2001; Rossi *et al.*, 2008), may be relatively more resistant to DNA damage and maintain a stable telomere length; these cells thus appear to be able to undergo an infinite number of cell divisions, demonstrated by their extensive passage in culture. In contrast, most adult stem cells possess a more limited self-renewal and proliferative potential, which declines with advancing age of the organism (Rossi *et al.*, 2008). This limited longevity is exemplified experimentally by the functional exhaustion of HSCs that typically occurs after four to five rounds

Table 12.1 Summary of Notch signaling in stem cells within various tissues of the fly, worm, fish, chicken, and mouse. For each organism and tissue, the table summarizes stem cell populations present, the known Notch receptors and ligands that are expressed, and the major functions of Notch signaling. Data from a large number of publications are summarized; see text for relevant references

Species	Tissue compartments	Stem and progenitor cells	Notch receptor known to be expressed	Notch ligand known to be expressed	Major functions	Reference
Fly	Midgut	ISC	Notch	Dl	Inhibit ISC proliferation and enteroblast (EB) to secretory ee differentiation	Mathur, *et al.*, Science, 2010; reviewed by Wang *et al.*, 2009, JCP
	Ovary	GSC	Notch (in cap cells)	Dl	Induce formation and maintenance of GSC niche cells to support GSCs	reviewed by Xie *et al.*, 2008, CSHSQB
	Gonad	GSC	Notch	Dl	Induce formation and maintenance of GSC niche cells to support GSCs	reviewed by Xie *et al.*, 2008, CSHSQB
	External sensory organ	SOP	Notch	Dl	Inhibit SOP cell-specification and control daughter cell fate	D.F. Lyman, *et al.*, Genetics, 1995;
Worm	Gonad	GSC	Notch1	GLP-1 in distal tip cell	Maintain stem cell, promote mitotic division	Byrd DT, *et al.*, Semin. Cell. Dev. Biol., 2009
Zebrafish	Blood	HSC	Notch	?	Expand HSC in AGM region; specify self-renewing HSCs	Burns CE, *et al.*, Blood, 2009; Burns CE, Genes Dev., 2005.
	Gut	Gut stem cell	Notch	DeltaD	Binary cell fate decision toward absorptive versus secretory cells	Crosnier C, *et al.*, Development, 2005

(Continued)

Chicken	Embryo	SMC and BC/EC	Notch	?	Promote SMC progenitor formation and mediates separation of SMC and BC/EC common progenitors	Shin M, *et al.*, Development, 2009
	Pancreas	Pancreatic progenitor cells	Notch	Delta	Inhibit endocrine development, stimulate progenitor cell proliferation	Ahnfelt-Ronne, *et al.*, BMC Dev. Biol., 2007
	Retina	RPC	Notch	Dll1, Dll4	Coordinate retinogenesis	Nelson BR, *et al.*, Dev. Dyn, 2008.
Mouse		ES cells	Notch1	?	Induce mesoderm differentiation, cardiomyogenesis, maintain the balance between endothelial cell versus vascular smooth cells of blood vessels	Schroeder, *et al.*, 2006; Schroeder, *et al.*, 2003a.
	Small intestinal epithelium	+4 cell, CBC	Notch1, Notch2	?	Involved in daughter cell fate decision: skewing toward absorptive enterocyte cell than secretory cells	See text
	Neuron system	NPC, NSC	Notch1	?	Maintain neural progenitor cells quiescence, inhibit neuronal differentiation	See text

(Continued)

Table 12.1 (*Continued*)

Species	Tissue compartments	Stem and progenitor cells	Notch receptor known to be expressed	Notch ligand known to be expressed	Major functions	Reference
	Skin	Epidermal stem cell	Notch1	Jagged1	Inhibit apoptosis of melanocyte precursor cell, Mb, to maintain the population, promote spinous cell differentiation, and exit from niche. Potential tumor suppressor	See text
	Hair follicle	Melanocyte stem cells		Jagged1	Support survival of immature Mbs	See text
	Hematopoietic system	HSC	Notch1, etc.	Jagged1, Jagged2, Delta1, and Delta4 in niche cells	Maintain HSC and progenitor cell function, and interaction with osteoblast; inhibit myeloid differentiation, enhance T-cell lineage differentiation over B lineage. Potential T-ALL oncogene	See text
	Bone	MSC	Notch1, etc.	?	Maintain mesenchymal progenitor cells, osteogenesis, and bone formation	See text

(*Continued*)

Skeletal muscle	Satellite cells, or SMP cell	Notch1, Notch2, Notch3	Dll	Maintain SMP pool and progenitor properties and regeneration ability, enhance proliferation of SMPs, and inhibit differentiation to myoblasts	reviewed by Farniw G., *et al.*, Stem Cell Rev., 2007
Mammary gland	MaSCs	Notch1, Notch4		Support proliferation and differentiation of MSCs, potential breast cancer oncogene	See text

ISC, adult intestinal stem cells; Dl, Delta; EB, enteroblast; GSC, germ line stem cells; SOP, sensory organ precursors; HSC, hematopoietic stem cells; SMC, smooth muscle progenitors; BC, blood common progenitors; EC, endothelial common progenitors; RPC, retinal progenitor cells; ES cells, embryonic stem cells; CBC, crypt base columnar cells; NPC, neural progenitor cells; NSC, neural stem cells; MSC, mesenchymal stem cells; SMP, skeletal muscle progenitors; AGM, aorta-gonad-mesonephros; Mb, melanoblasts; ee cells, enteroendocrine cells

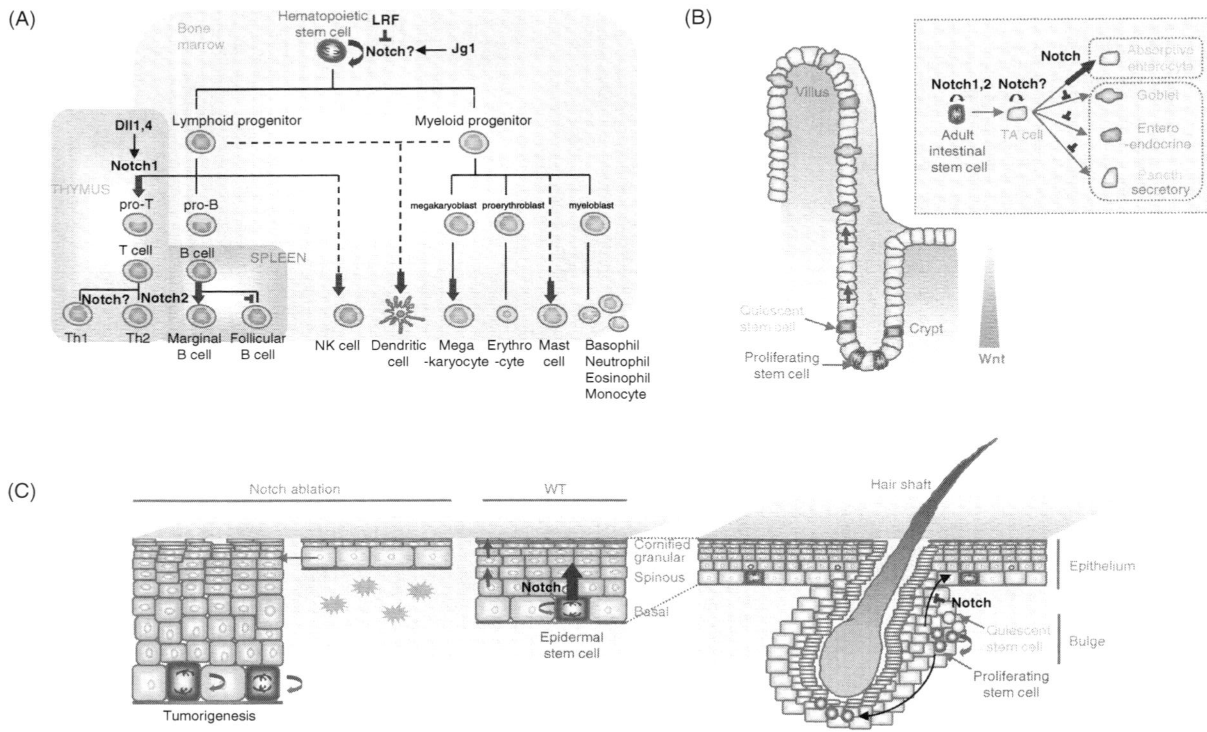

Figure 12.1 *(Continued)*

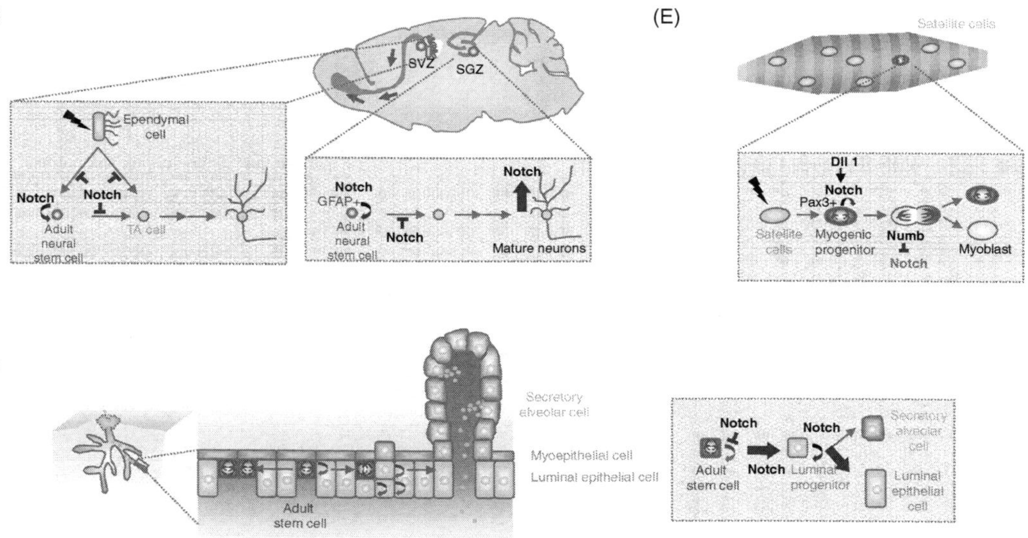

Figure 12.1 Notch signaling in stem cell niches. Notch signaling is utilized in multiple organs for cell renewal and tissue maintenance in the adult. (A) hematopoiesis, (B) intestine, (C) skin and hair follicle, (D) nervous system, (E) muscle, and (F) mammary gland. See text for details. (See Color Insert.)

of serial transplantation, despite the fact that the HSC telomere length exceeds that of their progenitor cell progeny and differentiated cells (Allsopp *et al.*, 2003a; Lansdorp, 2008).

The longer telomere length of HSCs may reflect maintenance of telomerase expression and activity specifically in this cell population, as telomerase activity is barely detectable in most mature human hematopoietic cells (Wang *et al.*, 2005). It is generally believed that when cells fail to repair accumulated DNA damage or to restore telomere length as they replicate, they eventually will enter into either senescence or apoptosis. Supporting this notion, mutations in hTERC, which impair the enzymatic function of the telomerase complex, were found in primary samples of aplastic anemia and bone marrow failure syndromes and correlate well with shortened telomere length in patient blood cells (Ly *et al.*, 2005). However, definitive conclusions are still elusive, since contradictory data derived from studies of transgenic TERT-overexpressing mice suggest that telomerase-independent factors also restrain the self-renewal ability of adult stem cells; HSCs from such mutant mice show no detectable improvement of transplantation ability *in vivo*, despite maintenance of telomere length (Allsopp *et al.*, 2003b). A combination of other stress factors may also need to be taken into consideration in future experiments.

To prevent or delay rapid exhaustion of the stem cell pool, adult stem cells may be maintained in quiescence under optimal conditions. For most of these cells, quiescence is a reversible cellular state closely linked to self-renewal capacity (Foudi *et al.*, 2009; Wilson *et al.*, 2008, 2009) and helps to ensure maintenance of stem cell reserves throughout life. Indeed, levels of the proliferation indicator Ki67 and of telomerase transcriptase could barely be detected in the resting mammary gland, where mammary stem cells are localized (Kolquist *et al.*, 1998). Interestingly, regulation of the proliferation and differentiation of mammary stem cells, which harbor the potential to generate both luminal and myoepithelial lineages of the mammary gland (Shackleton *et al.*, 2006; Stingl *et al.*, 2006), appears to be controlled in large part through the activity of the Notch pathway. Inhibition of Notch signaling enhances mammary stem cell self-renewal, whereas ectopic activation of Notch signaling drives commitment of these cells to the luminal lineage and further enhances proliferation of luminal progenitor cells, leading ultimately to their transformation (Bouras *et al.*, 2008) (Fig. 12.1F). These data suggest that controlled signaling through the Notch pathway is critical in mammary stem cells, not only for appropriately balancing the production of their differentiated daughters but also for regulating cell cycle progression in mammary lineage cells to protect from tumorigenesis (see accompanying Chapter 13).

The ability of stem cells to reenter the cell cycle from dormancy is critical to their ability to execute physiological functions by producing terminally differentiated, functional effector cells. This activity is essential to their normal regenerative role in response to tissue injury and is also a routine function for stem cell populations that seed rapidly recycling tissues (e.g., epithelial stem cells within the gut and skin, see below). Major cell cycle regulators, such as p21, p27, p53, Ras, Cyclins, and cyclin-dependent kinases (CDKs), participate in control of stem cell proliferation. In addition, as discussed in more detail below, it is increasingly clear that, as in the mammary gland (Bouras *et al.*, 2008), Notch plays a central role in these processes in a number of different types of stem cells (Table 12.1).

1.2. The stem cell microenvironment, or "niche," regulates stem cell function

The fate of stem cells is regulated concomitantly by cell-intrinsic and cell-extrinsic mechanisms. While the identification of core transcriptional networks and signaling pathways within various stem cell populations has provided significant insights into how cell identities are maintained, increasing attention has been paid to the microenvironment surrounding stem cells, which provides diverse external cues to instruct stem cell activities. The concept of the "niche," an optimal physiological location for stem cells, was proposed first by Schofield (1978) as a means of understanding cell non-autonomous regulation of hematopoietic precursor cells (Schofield, 1978). However, this concept subsequently has proven relevant to many different stem cell systems, and the definition of the niche has been expanded further to include functional regulation of stem cells by both cellular and acellular (extracellular matrix, ECM) components of the niche (Jones and Wagers, 2008).

"Niche cells" are specialized cells in the microenvironment that provide both physical signals to specify the correct location of stem cells and molecular signals to maintain their stem cell-specific activities while preventing both rapid depletion and aberrant tumorigenic expansion. In the *Drosophila* testis, for example, GSCs are localized adjacent to a cluster of postmitotic somatic hub cells, which activate the JAK-STAT and BMP signaling pathways (Kawase *et al.*, 2004; Song *et al.*, 2004; Tran *et al.*, 2000). Likewise, in the *Drosophila* ovary, GSCs are maintained next to the inner germarial sheath cells and cap cells through E-cadherin-mediated cell adhesion (Song *et al.*, 2002; Xie and Spradling, 1998). Studies of the *Drosophila* ovary indicate that activated Notch signaling is critical in specifying the number of GSC-supportive cap cells and thereby the size of the GSC niche. Indeed, enhanced Notch signaling results in a larger niche and a concomitant increase in the number of GSCs, while impaired Notch signaling either during development or in adulthood reduces niche size and causes significant reductions in GSC number (Song *et al.*, 2007).

Stem cell niches have also been identified for many other stem cell populations, including hair follicle stem cells (the dermal papilla in the bulge region), intestinal crypt stem cells (mesenchymal pericryptal fibroblasts), and HSCs (osteoblasts, endothelial cells, and stromal reticular cells), and as in the *Drosophila* gonad, Notch signaling is implicated in the interactions of stem cells with their niches in each of these tissues as well (Jones and Wagers, 2008). In the hair follicle, Notch regulates, in an injury-dependent fashion, the availability to differentiating stem cells of particular cell fates (Demehri and Kopan, 2009), and in the mammalian small intestine, Notch promotes stem cell proliferation and regulates alternative cell fate decisions between absorptive and secretory cells (Fre *et al.*, 2005). In the fly intestine, Notch signaling mediates asymmetric cell division of ISCs, which normally generate both enterocytes and enteroendocrine cells. Unique expression of the Notch ligand Delta by ISCs that remain in contact with the basement membrane (which forms the ISC niche) allows these cells to activate Notch targets in their daughters, which are displaced away from the niche and differentiate to form enteroblasts (Ohlstein and Spradling, 2007). Finally, as discussed in further detail below, Notch signaling may be involved in the maintenance of marrow-resident HSCs through interactions with bone-lining osteoblasts (Calvi *et al.*, 2003), although the precise requirement for Notch in regulating HSC function remains somewhat controversial (see below). Thus, Notch signaling plays varied yet crucial roles in the responses of a number of tissue stem cells to extrinsic cues provided by their specialized niches and therefore represents an attractive target for directly manipulating stem cell activity in both physiological and pathological conditions.

2. The Notch Pathway in Stem Cell Regulation and Function

The role of Notch signaling in both embryonic development and adult life has been the primary focus of many laboratories. Like many other signaling pathways, e.g., Sonic Hedgehog and Wnt/β-catenin, Notch signaling is evolutionarily conserved from invertebrates to vertebrates. Recent advances in inducible Cre-loxP targeting technology have greatly facilitated the *in vivo* dissection of the role of Notch in adult mammalian tissues. As alluded to above, it has been demonstrated that Notch signaling is critical in tissue renewal and maintenance in many organs, including, but not limited to, the skin, blood, intestine, liver, kidney, central nervous system, bone, and muscle. In stem cell biology, Notch signaling is highly context dependent, and the biological consequences of pathway activation can vary from stem cell maintenance or expansion to promotion of stem cell differentiation (Table 12.1). Below, we highlight some of the key roles of Notch signaling in rapidly renewing tissues, such as the hematopoietic system, intestine, and skin, and in highly proliferative ES cells, which can be propagated indefinitely *in vitro*, as well as in the renewal and regeneration of tissue systems with slower turnover, such as the brain and skeletal muscle (Fig. 12.1).

2.1. Notch signaling in pluripotent stem cells

ES cells are pluripotent cells derived from an early embryonic stage (Thomson *et al.*, 1998). They express Notch1, even though cells of the ICM, from which ES cells are derived, do not (Hadland *et al.*, 2004). Transient activation of Notch signaling during discrete stages of ES cell differentiation has been proposed to enhance and/or direct the generation of particular, therapeutically relevant tissue precursor cells (Chen *et al.*, 2008; Kobayashi *et al.*, 2009). Indeed, timed activation of Notch/RBP-J signaling at 1, 2, or 3 days after induction of ES cell differentiation into mesodermal cell lineages showed that production of Flk1^{+} mesodermal cells was reduced by activated Notch, suggesting that Notch/RBP-J signaling may block the generation of Flk1^{+} cells at several stages of mesoderm induction (Schroeder *et al.*, 2006). Transduced signals from the Flk1 receptor are critical for induction of primary and ES-derived mesodermal cells and for proper generation of their progeny, including definitive hematopoietic cells (Hidaka *et al.*, 1999). Activated Notch signaling in mesodermal cells blocks the generation of cardiac muscle, endothelial, and hematopoietic cells at the expense of vascular smooth muscle cells and pericytes, and inhibition of Notch signaling in ES cells, by deletion of the Notch downstream transducer RBP-J, directs differentiation along the cardiomyocyte lineage (Schroeder *et al.*, 2003a). On the other hand, activated Notch appears to promote neural commitment of ES cells when cultured in the absence of self-renewal factors (Lowell *et al.*, 2006). Together, these findings might suggest that Notch signaling plays a role in mesodermal development, in cardiomyogenesis, and in balancing the generation of endothelial cells versus vascular smooth muscle cells of blood vessels (Schroeder *et al.*, 2006). However, *in vivo*, mouse embryos deficient in RBPjk (Oka *et al.*, 1995), Notch1 and 2 (Huppert *et al.*, 2005), presenilin 1 and 2 (Donoviel *et al.*, 1999; Herreman *et al.*, 1999), Nicastrin (Li *et al.*, 2003), and Aph proteins (Serneels *et al.*, 2005) all complete gastrulation and progress to day 9 postcoitum (see accompanying Chapter 9). Therefore, although Notch signaling can modulate the outcome of ES cell differentiation, there appears to be no *in vivo* requirement for this pathway until after all three germ layers have formed.

2.2. Notch signaling in hematopoietic development

Blood is a critical component in the bodies of vertebrates. Blood cells circulate throughout all organs, transporting oxygen and nutrients, supplying immune cells to guard against infection and promote tissue repair, and carrying away waste and metabolites. In adult mammals, rare HSCs reside predominantly in the bone marrow and constantly give rise to lineage-specific progenitor cells and effector blood cells that perform the physiological functions of the hematopoietic system.

Blood formation begins early during embryogenesis and persists into adult life. Based on the location and timing of blood cell formation, mammalian hematopoietic development is believed to occur in two waves. The first wave of so-called primitive hematopoiesis is initiated in the extraembryonic yolk sac between embryonic days 7 and 11 (E 7–11) in mice. The second, "definitive" wave of hematopoiesis initiates in the aorta-gonad-mesonephros (AGM) region, which contains hemogenic endothelium that "buds" newborn hematopoietic cells into the aortic lumen. Recent studies have found that HSCs reside also in the placenta during the same period of time as AGM hematopoiesis initiates and are present there until day 13 (Gekas *et al.*, 2005; Ottersbach and Dzierzak, 2005). These data suggest that there are multiple sites of hematopoietic origin during development.

Beginning from days 10 to 11 in mice, hematopoiesis migrates to the fetal liver and eventually to all adult hematopoietic compartments, including the spleen, thymus, and bone marrow, which continues to support definitive blood cell production after birth. Some evidence from amphibians, birds, and even mice supports the notion that the primitive hematopoietic sites in the yolk sac may also support or seed definitive hematopoiesis (Samokhvalov *et al.*, 2007; Turpen *et al.*, 1997), although no universally accepted conclusions have been reached on this controversial issue (Cumano *et al.*, 1996; Medvinsky and Dzierzak, 1996). The major function of early, primitive hematopoiesis is thought to be a "rapid production" phase of hematopoiesis, which provides red blood cells needed to oxygenate rapidly growing embryonic tissues. Definitive hematopoiesis, in contrast, produces the HSC pool, which will generate the full spectrum of functional blood and immune cells (Orkin and Zon, 2008). Interestingly, studies of zebrafish *mindbomb* mutants, which lack functional Notch ligands, indicate that Notch signaling is essential for specification of self-renewing HSCs, although it appears to be dispensable for the formation and function of nonself-renewing erythromyeloid progenitor cells, which are formed prior to multipotential HSCs and give rise only transiently during embryogenesis to a more limited subset of mature blood cell lineages (Bertrand *et al.*, 2010a, b).

Throughout development, HSCs are regulated by complicated intrinsic and extrinsic signals. The size of the HSC pool in the adult bone marrow is determined by a delicate balance of self-renewal and differentiation, although at any given time a majority of HSCs appear to exist in a deeply quiescent state. Indeed, it has been estimated that the most primitive HSCs enter cell cycle only five times in a mouse's entire lifetime (Wilson *et al.*,

2008; 2009), a number significantly lower than the cycling rate of MPP cells (Christensen and Weissman, 2001; Morrison and Weissman, 1994) and oligopotent progenitor cells (Akashi *et al.*, 2000; Kondo *et al.*, 1997; Morrison and Weissman, 1994; Ogawa *et al.*, 1993).

The inherent differences between long-term HSCs, MPPs, and oligopotent progenitor cells may be attributed in part to an uneven distribution of cell fate determinants, perhaps established during asymmetric cell division, such that each progenitor possesses a distinct transcriptional program. In addition, microenvironmental input from bone marrow stroma and from other hematopoietic sites appears also to be essential for maintaining cell identity and cell fate. During development, movement of HSCs and progenitor cells from one anatomical site to another is tightly regulated. However, even after HSCs arrive at the bone marrow, some stem cells will reenter circulation and migrate to other organs and tissues, such as the spleen (Massberg *et al.*, 2007; Wright *et al.*, 2001). It has been estimated that approximately 100–400 stem cells are circulating in the peripheral blood at any given time (Wright *et al.*, 2001), and only a fraction of these cells will relocalize to hematopoietic sites, highlighting the dynamic state of the hematopoietic system.

Understanding the mechanisms of retention, mobilization, and migration of HSCs/progenitor cells has been instrumental to improving the therapeutic application of these cells in bone marrow (BM) transplantation. Yet, relatively little information is currently available describing the molecular and cellular interplay between HSC/progenitor cells and various niche cells. Intriguingly, some published studies indicate that Notch is particularly involved in maintaining the HSC pool and its capacity for self-renewal and differentiation. Notch receptors are widely expressed in human and mouse hematopoietic cells, including stem cells, progenitor cells, and mature cells (Duncan *et al.*, 2005; Jonsson *et al.*, 2001; Milner *et al.*, 1994). This broad expression pattern suggests widespread participation of the Notch pathway in blood cell development and function. Importantly, Notch receptors, ligands, and signaling components exhibit differential expression patterns within the hematopoietic hierarchy (Jonsson *et al.*, 2001) and in different hematopoietic compartments (Han *et al.*, 2000; Jones *et al.*, 1998), reflecting the complex temporal and spatial regulation of the blood system by the Notch signaling network.

The process of hematopoiesis is closely related to angiogenesis in the embryo, which suggests the existence of a common ancestor for HSCs and endothelial cells. This shared ancestry has been corroborated by the identification of potential hemangioblast and hemogenic endothelial cells (Bertrand *et al.*, 2010a; Choi *et al.*, 1998; Dzierzak and Speck, 2008; Eilken *et al.*, 2009; Lancrin *et al.*, 2009; Li *et al.*, 2006). Significant hematopoietic and angiogenic defects, as well as decreased HSC activity in the AGM, are observed in Notch1-deficient mice, but not in Notch2-deficient mice, implying that Notch1 signaling is particularly crucial for the generation of definitive HSCs, although it appears dispensable for primitive hematopoiesis (Kumano *et al.*, 2003). A late developmental requirement of Notch1 is supported also by studies of chimeric mice derived from Notch1$^{-/-}$ ES cells, which show proper colony-forming ability—the ability of committed hematopoietic progenitor cells to differentiate into blood cells *in vitro*—among yolk sac derived and fetal liver derived hematopoietic cells, but loss of long-term hematopoietic reconstitution ability *in vivo* (Hadland *et al.*, 2004). It has been suggested that the supporting role of Notch1 in blood development is mediated through binding with the transcriptional cofactor RBPjκ to activate expression of the key hematopoietic transcription factor GATA2 (Robert-Moreno *et al.*, 2005). The Notch ligand Jagged1 also is required for initiating the definitive hematopoietic program in the embryo, and coculturing with Jagged1-expressing stromal cells or overexpression of GATA2 can rescue the blood formation defect observed in Jagged1-null AGM cells (Robert-Moreno *et al.*, 2008).

Notch1 signaling pathways also have been suggested to affect the self-renewal, proliferation, and differentiation of adult HSCs *in vitro* and *in vivo*, but existing data are somewhat contradictory (Fig. 12.2A). Some studies report that activated Notch signaling enhances proliferation and numerically expands hematopoietic progenitor cell lines and mouse hematopoietic stem/progenitor cells, while inhibiting differentiation in response to various cytokines, mostly under myeloid promoting conditions (Carlesso *et al.*, 1999; Kumano *et al.*, 2001; Milner *et al.*, 1996; Varnum-Finney *et al.*, 1998). Recent data using Notch reporter strains further suggest a predominant Notch signal in hematopoietic progenitor cells and indicates that Notch expression may correlate with the capacity for both symmetric and asymmetric cell division, the balance of which could be altered by oncogene overexpression (Wu *et al.*, 2007). In line with these data, human primitive blood cells expressing ectopic Notch1 receptor (Vercauteren and Sutherland, 2004) or treated with soluble or immobilized extracellular Notch ligands (Jagged1, Delta1 and Delta4) also exhibit increased HSC expansion and reconstitution ability *in vitro* and *in vivo* (Karanu *et al.*, 2000, 2001; Ohishi *et al.*, 2002). Furthermore, mouse hematopoietic progenitor cells (the Lin$^-$Sca1$^+$cKit$^+$ compartment) could be immortalized by ectopic expression of activated Notch1 while still maintaining their capacity for multilineage hematopoietic reconstitution upon *in vivo* transplantation (Varnum-Finney *et al.*, 2000). Although the molecular mechanisms underlying Notch1 function have yet to be fully understood, Notch-mediated inhibition of cytokine-induced differentiation has been suggested to occur via decreased expression of GATA2 and increased expression of HES1 (Kumano *et al.*, 2001), which may facilitate self-renewal of hematopoietic stem/progenitor cells both *in vitro* and *in vivo* (Kunisato *et al.*, 2003).

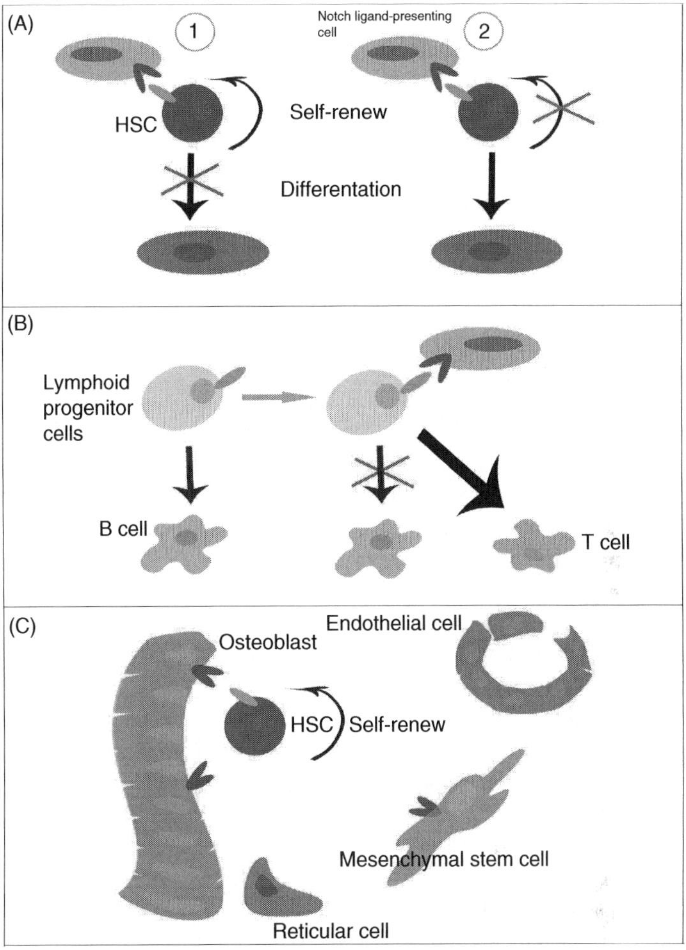

Figure 12.2 Models of Notch regulating hematopoietic stem/progenitor cells. (A) Two models depicting role of Notch signaling regulating the proliferation, self-renewal, and differentiation of HSCs; (left) model 1. Activation of Notch pathway enhances HSC proliferation and self-renewal, while inhibiting differentiation, especially that of myeloid lineage; (right) model 2. Notch signaling inhibits HSC self-renewal but promotes differentiation. (B) Activation of the Notch pathway directs lymphoid progenitor cells to differentiate toward the T-cell lineage at the expense of B-cell lineage differentiation, the default route in the absence of Notch signals. (C) Binding between Notch receptor on HSC and Notch ligand, e.g., Jagged1, from niche cells, e.g., osteoblasts, contributes to preservation of HSC pool, migration, and retention in the bone marrow compartment.

The studies discussed above clearly support the notion that Notch1 positively regulates HSC/progenitor cell maintenance and proliferation; however, in dramatic contrast, additional reports provide equally compelling data that argue *against* an important role for Notch in maintaining HSCs.

For example, inducible Notch1 expression in the 32D cell line actually *decreased* proliferation and *accelerated* myeloid differentiation (Schroeder and Just, 2000a). Similar results were obtained in 32D cells ectopically expressing the active form of RBP-j or exposed extracellularly to the Notch ligand Jagged1 (Schroeder and Just, 2000b; Vas *et al.*, 2004). These effects were confirmed using additional hematopoietic progenitor cell lines, in which Notch activation drove expression of PU.1, a transcription factor well known to direct myeloid differentiation (Schroeder *et al.*, 2003b). Furthermore, cell cycle progression in Notch1-expressing human cord blood progenitor cells was greatly inhibited, due to Notch-induced transcriptional elevation of the cell cycle inhibitor P21 (Chadwick *et al.*, 2007).

It is possible that the discrepancies among these conclusions about the impact of Notch signaling on hematopoietic differentiation, which are based largely on overexpression and ligand exposure studies, arise from the systematic differences in experimental design that render it difficult to accurately control the quantitative expression levels of Notch pathway components. To alleviate this concern, several groups have employed gene-targeted "knockout" mouse models to investigate the role of Notch signaling components in hematopoietic development and HSC function. Surprisingly, conditional inactivation of RBP-j (a DNA-binding protein that is commonly utilized for signal transduction by all four Notch receptors) (Han *et al.*, 2002), activation of a dominant-negative Mastermind-like1 (MAML1, a potent Notch inhibitor) (Maillard *et al.*, 2008), or conditional deletion in the hematopoietic compartment of Notch1 and Notch2 receptors (singly or in combination) did not result in any adult HSC phenotype or myeloid differentiation alterations. Thus, although it remains possible that the lack of HSC phenotype in these animals could be explained by redundant functions of other Notch-related receptors, or by the existence of ligand isoforms that trigger RBP-j-independent transcriptional networks, these genetic data argue against a critical requirement for Notch signaling in embryonic or adult HSCs. Perhaps the most parsimonious interpretation of currently existing data is that Notch signaling is dispensable for adult mammalian HSC functions but may support or assist in processes that maintain the HSC pool and promote HSC activity. In this regard, Notch's role in hematopoiesis appears quite complicated and highly dependent on signal strength and cellular context. Further clarification of the molecular mechanisms underlying the transduction of Notch signals in various cell types will certainly help to illuminate the particular role of this pathway in the normal physiology of HSCs.

2.3. Role of Notch in hematopoietic progenitor cells

Downstream of HSCs, Notch signaling clearly plays an important role in cell fate decisions of a variety of oligopotent progenitor cells in the hematopoietic system. Recent data indicate that Delta1-mediated Notch1 activation stimulates megakaryocytic development from HSCs *in vivo* (Mercher *et al.*, 2008), and among lymphoid cells, an extensive collection of data has confirmed the importance of Notch in binary cell fate determination of lymphoid lineages (Fig. 12.2B). Notch gain-of-function, induced by either ectopic expression of Notch1 (Hozumi *et al.*, 2003; Jaleco *et al.*, 2001; Pui *et al.*, 1999) or stimulation with the extracellular ligands Delta1 and Delta4 (de La Coste *et al.*, 2005; Jaleco *et al.*, 2001; La Motte-Mohs *et al.*, 2005), leads to aberrant proliferation and differentiation of T-cell progenitors at the expense of B-cell populations. Conversely, loss-of-function studies employing expression of inactivated Notch1 (Wilson *et al.*, 2001), a DNA-binding dead mutant of RBP-j (Han *et al.*, 2002), or addition of γ-secretase inhibitor (GSI) (Hadland *et al.*, 2001), produced analogous phenotypes. These data suggest that the physiological function of Notch during lymphopoiesis is to instruct differentiation along the T-cell, rather than B-cell, lineage. The choice to favor T-cell over B-cell differentiation executed by Notch signaling likely occurs at the early lymphoid progenitor stage (Han *et al.*, 2002), and similar effects persist to the later maturation step of functional double-positive T cells (Hadland *et al.*, 2001), independent of the microenvironment (Allman *et al.*, 2001; Hozumi *et al.*, 2003; Wilson *et al.*, 2001). Importantly, in the absence of Notch1 signaling, lymphoid progenitors adopt a default B-lymphopoiesis program, as shown by multiple transgenic mice with ectopically inhibited Notch pathways, including those expressing the dominant-negative form of MAML1 (Maillard *et al.*, 2008) (Fig. 12.2B). Thus, the Notch pathway functions as an inhibitory signal for maturation of B-cell lineages but acts as a positive stimulating factor for T-cell development, possibly to ensure a balanced production of immune cells required to properly support the body's immune defenses.

2.4. Notch signaling in the hematopoietic microenvironment

The central paradigm of interaction between signal-sending cells and signal-receiving cells suggests that Notch signaling may be crucial for interactions of developing blood cells with their surrounding microenvironment (Fig. 12.2C). The interaction between HSCs/progenitor cells and various nonhematopoietic cell populations within hematopoietic sites has long been recognized to influence the dynamics of blood production. Within the bone marrow, particular nonhematopoietic cells have been proposed to form specialized HSC "niches," providing physiologically appropriate environmental cues to HSCs and progenitors in response to stress or injury in hematopoietic compartments. Potential HSC "niche cells" currently include bone-forming osteoblasts (Calvi *et al.*, 2003; Mayack and Wagers, 2008; Visnjic *et al.*, 2004; Xie *et al.*, 2009; Zhang *et al.*, 2003), sinusoidal endothelial cells (Kiel *et al.*, 2005), SDF-1-producing reticular cells (Sugiyama *et al.*, 2006), mural cells (Frenette *et al.*, 1998), and multilineage

MSCs that can differentiate into adipocytes, chondrocytes, and osteoblasts (Garrett and Emerson, 2009). Notch ligands are widely expressed in many of these hematopoietic niche cells. For example, Jagged1 is expressed in the 3T3 cell line, murine hematopoietic stromal cell lines, murine fetal liver stroma, and cultured murine BM stroma cells (Varnum-Finney *et al.*, 1998). Jagged2 is also expressed by human BM endothelial cell lines and primary endothelial cells (Fernandez *et al.*, 2008), while Delta1 and Delta4 have been shown to be expressed in both human BM stromal cells and endothelial cells (Karanu *et al.*, 2001), although their physiological functions have not been fully investigated.

Notch signaling appears to be required for proper long bone formation, a process crucial for supporting adult hematopoietic development. Skeletogenic mesenchyme-specific deletion of the Notch pathway components presenilin1 and presenilin2, the two catalytic components of the γ-secretase complex that mediates the cleavage and activation of Notch receptor, resulted in increased bone mass with a dramatic loss of mesenchymal progenitors, and osteopenia in mice *in vivo*, suggesting that Notch signaling positively regulates *in vivo* bone formation (Hilton *et al.*, 2008). Coincidently, osteoblast-specific activation of the parathyroid hormone receptor (under the regulation of Colα1 promoter) leads to elevated numbers of osteoblast cells *in vivo*, which support an expanded HSC pool and enhance transplantation efficiency (Adams *et al.*, 2006; Calvi *et al.*, 2003). This effect appears to be mediated by increased secretion of Jagged1 from osteoblastic cells (Calvi *et al.*, 2003; Weber *et al.*, 2006; Whitfield, 2005), which in turn activates Notch receptors on stem/progenitor cells through adenylate cyclase/protein kinase A (Weber *et al.*, 2006). Moreover, it has been suggested that the supportive roles of MSCs for HSC/progenitor cells and their modulating effect on lymphoid differentiation also depend on the Jagged1–Notch–Hes1 axis (Fujita *et al.*, 2008; Li *et al.*, 2008b). These data suggest a role for Notch in HSC/progenitor cell retention in the marrow environment, and possibly also in the maintenance of stem/progenitor cell identity; however, current information about the role of this pathway in regulating HSC mobilization and migration is still quite limited. Some recent *in vitro* data suggest that the Notch ligand, Delta1, can interact with Dlg1 (a human homolog of *Drosophila* discs large tumor suppressor) on the cell surface of the 3T3 cells and thus reduce the mobility of these cells (Six *et al.*, 2004). Whether this effect holds true *in vivo* for stem and progenitor cells, and which "niche" cells are involved, remains unknown.

To date, the most striking evidence attesting to the extensive role of Notch signaling in the bone marrow microenvironment comes from the *de novo* development of myeloid proliferative disease (MPD) in transgenic Mind bomb-1 (Mib1)-null mice. These animals delete Mib1 under the regulation of either the MMTV1 promoter, which deletes the gene in both BM and stromal cells, or the Mx1 promoter, which deletes the gene predominantly (though not exclusively) in hematopoietic compartments. Loss of Mib1 perturbs Notch ligand endocytosis processes in these cells. The MPD disease induced in Mib1-null mice is reproduced when wild-type bone marrow cells are transplanted into lethally irradiated mutant-recipient mice, but not when mutant bone marrow cells are transplanted into wild-type recipients (Kim *et al.*, 2008), suggesting a microenvironmental rather than hematopoietic origin for the disease. This microenvironmental effect most likely reflects presence of the cytokine thymic stromal lymphopoietin (TSLP), expressed by epithelial cells deficient in Notch signaling (Demehri *et al.*, 2008; Dumortier *et al.*, 2010). Still, more specific genetic and biochemical studies are needed to fully understand the specific ligand and receptor combination(s) that are recruited by each niche cell population to regulate hematopoiesis. The role of Notch signaling in MPD, as well as several other hematopoietic malignancies, is further discussed in the following section. In-depth understanding of the complex roles that this pathway plays in the development of hematopoietic hierarchy will clearly be enlightening both for better understanding physiological cell fate specification in hematopoietic tissues and for facilitating therapeutic targeting under specific leukemic contexts.

2.5. Notch in hematologic malignancy and leukemia stem cells

Given the crucial involvement of the Notch pathway in hematopoiesis, it is perhaps not surprising that alterations in Notch expression are found in diverse arrays of leukemias. The tumorigenic potential of Notch mutation was first observed in human T-cell acute lymphoblastic leukemia (T-ALL), where a rare t(7;9) translocation generates a constitutively active, truncated form of the Notch1 receptor *in vivo*. This translocation is associated with less than 1% of all T-ALL patient cases. Interestingly, ectopic expression of activated Notch1 in mouse bone marrow cells phenocopies these human T-cell malignancies *in vivo* (Kawamata *et al.*, 2002), which may be caused by aberrant proliferation and differentiation of T-lymphoid progenitors before they complete T-cell receptor (TCR)-α rearrangement (Li *et al.*, 2008a). Studies using T-ALL cell lines also indicate that tumorgenesis in some of these models may be a consequence of an activated PI3K/AKT pathway in the absence of PTEN (Calzavara *et al.*, 2008). Additionally, mutations of the Notch1 receptor may function as a "second hit" in leukemogenesis that facilitates leukemic initiation and progression. For example, Notch1-activating mutations have been found in SAP (signal-induced proliferation associated gene-1)-null hematopoietic progenitor cell-derived T-ALL cell lines (Wang *et al.*, 2008) and some rare C/EBPα-related acute myelogenous leukemia (AML) cell lines (Wouters *et al.*, 2007). Aberrantly high expression of Notch receptor, stimulated by its ligand Jagged1, has been detected in B- and T-cell-derived tumor cells of Hodgkin's and anaplastic large-cell lymphoma (Jundt *et al.*, 2002), while deregulated Notch expression has been detected in a number of AML cell lines and patient samples, typically in conjunction with decreased PU.1 expression, which contributes to myeloid leukemogenesis (Chen *et al.*, 2008). Thus, dysregulated Notch signaling may function as a diagnostic marker and potential drug target for multiple types of leukemias. Excitingly, γ-secretase inhibitor (GSI) has been employed in various studies, either alone or in combination with other chemotherapy agents, to treat T-ALL cell lines or mouse models, and has shown a significant anti-proliferation effect that is possibly mediated through both cell cycle arrest and induction of apoptosis (Cullion *et al.*, 2009; De Keersmaecker *et al.*, 2008; Kindler *et al.*, 2008; Lewis *et al.*, 2007; Masuda *et al.*, 2009; Rao *et al.*, 2009; Tammam *et al.*, 2009). These data support the possible therapeutic promise of targeting the Notch pathway in T-ALL.

Indirect data also suggest that the dysregulation of Notch signaling is critical for leukemogenesis. Cocultures of bone marrow-nucleated cells (BMNCs) from Myelodysplastic syndrome (MDS) patient samples with normal marrow stromal cells (MSCs), and of normal BMNCs with MDS patient MSCs, showed a reduced frequency of both early and late cobblestone area-forming cells (CAFC, an empirical assay that measures the ability of HSC/progenitor cells to form cobblestone-like cell clusters when cultured atop supporting stromal cells and correlates with *in vivo* multi-lineage repopulating capability). Incubation with soluble Notch ligand, Jagged1, could inhibit the late CAFC of normal BMNCs on both MDS and normal MSCs, but not that of MDS BMNCs, indicating perturbed Notch signaling in the BMNCs from MDS patients (Varga *et al.*, 2007). Similarly, the Notch ligand Delta-like1 (Dlk1) selectively exhibits high-level expression in hematopoietic stem/progenitor cells from MDS patient samples compared with AML samples (Miyazato *et al.*, 2001; Qi *et al.*, 2008), suggesting a potential role of Dlk1 as a diagnostic marker for MDS. Evidence from studies using myeloid leukemia cell lines indicates that ectopic expression of Dlk1 upregulates HES1 expression (Qi *et al.*, 2008) and inhibits myeloid differentiation and proliferation, which may provide a plausible explanation for its association with MDS (Li *et al.*, 2005). Jagged2, another Notch ligand, is also significantly expressed in $CD34^{+}CD38^{-}$ populations of AML cells, enriched for leukemic stem cells (LSCs, an often rare subset of tumor cells that can self-renew, highly proliferate, and generate all the heterogeneous cell populations in the tumor). *In vitro* growth of LSCs can be selectively impaired by treating these cells with GSI (Gal *et al.*, 2006). Finally, consistent with the aforementioned requirement for Notch1 in megakaryocyte development (Mercher *et al.*, 2008), in a mouse model of acute megakaryoblastic leukemia (generated by combinatorial transformation using OTT-MAL (one-twenty-two-megakaryocytic acute leukemia) and an activating mutant of thrombopoietin receptor MPL (myeloproliferative leukemia virus oncogene), transcriptional activation of RBP-j appears to be an indispensible concomitant step that transforms hematopoietic cells for further leukemic events (Mercher *et al.*, 2009).

In conclusion, Notch pathways are highly involved in both normal HSC/progenitor cell functions and in leukemic transformation. Further clarification of the complicated roles that this pathway plays in blood cell development and transformation will certainly advance its possible targeting under particular leukemic contexts.

2.6. Notch signaling in the intestine

The epithelium of the intestine renews rapidly—every 4–5 days—and failure to maintain this fast paced homeostasis can result in premature death. Mild disruptions in the rate of cell replacement in the intestine can cause malnutrition, infection, or cancer (see accompanying Chapter 13). Intestinal cell replacement relies heavily on a highly proliferative, LGR5-positive stem cell population which resides at the bottom of the crypt, and a relatively quiescent, Bmi-1-positive stem cell population found at the +4 position (Li and Clevers, 2010). $LGR5^{+}$ cells generate transit-amplifying (TA) cells, which upon leaving the TA compartment at the crypt–villus junction, subsequently give rise to all four terminally differentiated cell types in the gut: absorptive enterocyte, secretory goblet cells, enteroendocrine, and Paneth cells (Fig. 12.1B) (Barker *et al.*, 2008; Casali and Batlle, 2009; Crosnier *et al.*, 2006; Wang and Hou, 2010).

Notch signaling and Wnt signaling cooperate to regulate cell renewal and binary fate decisions in the adult intestine. Wnt functions as the master switch promoting cell proliferation and suppressing differentiation (Casali and Batlle, 2009; Chiba, 2006; Crosnier *et al.*, 2006; Scoville *et al.*, 2008; Van der Flier *et al.*, 2007; Wang and Hou, 2010). Ablation of Notch signaling in the intestine, using RBPjk conditional knockout mice, GSI, dibenzazepine, or double knockout of Notch 1/Notch2 (Riccio *et al.*, 2008) specifically increases the number of secretory goblet cells at the expense of ISCs and absorptive enterocytes. Notch1-null ISCs can self-renew, but they produce an excess of goblet cells (Vooijs *et al.*, 2007). These data suggest that Notch receptors are redundant in the intestinal niche and that Notch signaling regulates cell fate decisions controlling the relative production of secretory versus enterocyte cell lineages (van der Flier and Clevers, 2009; van Es and Clevers, 2005; van Es *et al.*, 2005; Yang *et al.*, 2001). Consistent with this notion, forced expression of Notch intracellular domain (NICD) in the intestine leads to reduction in secretory cell numbers and increased cell proliferation (Fre *et al.*, 2005). In zebrafish, lateral inhibition through Delta-mediated Notch signaling has been implicated in binary cell fate decisions (Crosnier *et al.*, 2005), but in mammals this type of regulation remains unproven. In summary, it appears that Notch signaling functions in two ways in the adult intestine: first, Notch promotes proliferation in the stem cell and/or TA cell compartment and second it regulates binary fate decisions between absorptive and secretory cells. While this first role of Notch signaling was shown to be Wnt dependent, the second is independent of Wnt signaling (Fre *et al.*, 2009).

In addition to its role in normal cell turnover, Notch activation in the intestine also can promote tumorigenesis in sensitized $Apc^{min/+}$ mice, which have a propensity for developing intestinal adenomas (Fre *et al.*, 2009; Vooijs *et al.*, 2007). Treatment with GSIs promoted goblet cell differentiation and reduced proliferation in $Apc^{min/+}$-associated adenomas (van Es *et al.*, 2005), suggesting that GSI may be useful in the treatment of neoplastic diseases in the gut. However, severe gut toxicity following GSI treatment has been observed in multiple independent studies (Milano *et al.*, 2004; Searfoss *et al.*, 2003; van Es *et al.*, 2005; Wong *et al.*, 2004), reinforcing the need to carefully control dosage or tissue-specific access of γ-secretase inhibition for *in vivo* therapeutic approaches. Encouragingly, however, a recent report showed that combination therapy employing GSI and glucocorticoids reduced GSI-associated intestinal toxicity and improved its antileukemic effects (Real *et al.*, 2009).

2.7. Notch signaling in skin stem cells

The epidermis consists of four layers: basal (innermost), spinous, granular, and cornified layers, each expressing different molecular markers (Fig. 12.2C). Like the gut, the skin maintains robust cell replacement throughout life, and dysfunction in skin cell differentiation can result in dehydration, infection, atopic disease, or cancer (Demehri *et al.*, 2009a, b; Zhang *et al.*, 2009). Adult stem cells appear to reside in two places in the skin. A proliferative unipotential stem cell population is found in the basal layer of the epidermis, and as these cells commit to terminal differentiation, they detach and migrate outward (Blanpain and Fuchs, 2009; Fuchs and Raghavan, 2002). A second, distinct population of multipotential stem cells resides in the bulge region of the hair follicle (Blanpain and Fuchs, 2006; Blanpain *et al.*, 2007; Fuchs *et al.*, 2004; Li and Clevers, 2010); these cells are typically quiescent but can enter a proliferative state in response to traumatic injury or during the normal course of the hair cycle (Blanpain and Fuchs, 2009).

Notch signaling regulates the differentiation and proliferation of adult epidermal stem cells (Ambler and Maatta, 2009; Okuyama *et al.*, 2008; Watt *et al.*, 2008) (see also accompanying Chapter 13) (Fig. 12.2C). Conditional gain or loss of Notch function (via NICD overexpression, or $Notch1^{-/-}$ or $RBPj^{-/-}$, respectively) in the epidermis established that canonical Notch signaling promotes spinous cell differentiation and exit from the niche *in vivo* (Blanpain *et al.*, 2006; Rangarajan *et al.*, 2001). However, while embryonic ablation of RBPjk in the epidermis causes epidermal hypoplasia (Blanpain *et al.*, 2006), after birth the skin compensates for loss of Notch signaling by hyperplasia, creating a tumor-promoting environment (Demehri *et al.*, 2008, b; Dotto, 2008; Nicolas *et al.*, 2003). This Notch "tumor suppressor" phenotype invokes both cell autonomous (Dotto, 2008; Nicolas *et al.*, 2003) and noncell autonomous (Demehri *et al.*, 2009b) mechanisms and highlights the complex role played by Notch signaling in regulating exit from the proliferative stem cell niche and controlling epidermal differentiation.

Even moderate reduction of Notch signaling in the skin can increase susceptibility to tumorigenesis in a dose-dependent manner. While conditional knockouts of either Notch 2 or 3 alone in the postnatal epidermis exhibit no phenotype, stepwise deletion of Notch paralogs accelerate skin carcinogenesis, suggesting that Notch 1, 2, and 3 are not redundant but instead exhibit additive functions (Demehri *et al.*, 2009b). Similarly, moderate reduction of γ-secretase activity (either $PS1^{+/-}$, $PS1^{+/-}$;$PS2^{-/-}$, $Nicastrin^{+/-}$, or $Aph1a^{+/-}$) increased the risk of squamous cell carcinoma (Li *et al.*, 2007; Tournoy *et al.*, 2004). Finally, skin barrier defects can induce systemic B-lymphoproliferative disorders in newborn mice by dose-dependent induction of TSLP secretion, which eventually can lead to atopic dermatitis and asthma in adult animals (Demehri *et al.*, 2009a; Dumortier *et al.*, 2010). Importantly, while TSLP is useful as a biomarker for skin differentiation defects, these results raise concern that even moderate reduction of Notch signaling can increase cancer susceptibility and simultaneously add risks of atopic dermatitis, asthma, and MPD. Indeed, $PS1^{+/-}PS2^{-/-}$ mice with reduced dosage of γ-secretase *in vivo* develop autoimmunity (Qyang *et al.*, 2004; Tournoy *et al.*, 2004), most likely in synergy with skin barrier defects.

In the quiescent bulge stem cell population, no evidence has yet emerged to support a specific role for Notch signaling in stem cell maintenance or instruction of lineage choice during differentiation. Instead, Notch signaling appears to play a "gate-keeper" role. In the absence of Notch, stem cells or their immediate descendents can select either epidermal or follicular differentiation (Demehri and Kopan, 2009). In its presence, only the follicular fate is selected unless an injury has occurred. The differences observed in the specific activities of Notch signaling in discrete epidermal stem cell populations with overlapping developmental potentials highlight a key point of this review—that Notch signaling plays complex, context-dependent, and cell-type specific roles in stem cell biology and tissue regeneration and that the functions of this pathway should not be overgeneralized across tissue systems, or even within a single tissue.

2.8. Notch signaling in adult neurogenesis and synaptic plasticity

Unlike the intestine, the hematopoietic system, or the epidermis, where cell replacement occurs continuously throughout life, neurogenesis takes place predominantly during embryonic development, with only a few specialized regions of the brain [notably, the olfactory bulb, the subgranular zone (SGZ) in the hippocampus, and the subventicular zone (SVZ) in the lateral ventricle] maintaining stem cell activity in adult life (Fig. 12.2D). Adult neurogenesis can be enhanced upon injury, exercise, or in an enriched environment. Although the functional significance of adult neurogenesis is not fully understood, it has been implicated in olfaction, in some types of learning and memory, and in neurological disorders such as epilepsy and Alzheimer's disease (Lledo *et al.*, 2006; Suh *et al.*, 2009).

During development, Notch signaling has been shown to maintain neural progenitors and inhibit neuronal differentiation (Louvi and Artavanis-Tsakonas, 2006) (also see accompanying Chapter 10). Interestingly, Notch1 protein is asymmetrically inherited during division of mammalian cortical neuronal progenitors (Chenn and McConnell, 1995), and ectopic induction of Notch targets inhibits neuronal development (Ishibashi *et al.*, 1994; Sakamoto *et al.*, 2003). Current views hold that while Notch signaling may inhibit neural fates in early progenitors, it may function later as an instructive or permissive signal, regulating fate choices between different neural cell subtypes (Louvi and Artavanis-Tsakonas, 2006). For example, Notch activation appears to promote neuronal differentiation over gliogenesis (Grandbarbe *et al.*, 2003) and to favor the differentiation of radial glia, Muller cells, and astrocytes, at the expense of oligodendrocyte formation (Grandbarbe *et al.*, 2003; Wang *et al.*, 1998). Thus, as in the gut and skin, Notch activity in the developing nervous system may serve as a binary switch of cell fate determination (Cau and Blader, 2009).

Emerging evidence also suggests multiple roles for Notch signaling in adult neurogenesis (Johnson *et al.*, 2009) (Fig. 12.2D). In the SGZ, conditional deletion of *Notch1* in adult, GFAP (glial fibrillary acidic protein)-positive neural stem cells, as well as reciprocal experiments activating NICD1 overexpression in these same cells, demonstrate that Notch1 signaling inhibits exit from the cell cycle and promotes proliferation of adult neural stem cells. Notch signaling also enhances maturation and survival of the newborn neurons (Breunig *et al.*, 2007). In the SVZ, a study using Nestin-CreERT-induced deletion of a conditional RBPjk allele suggested that Notch signaling maintains a quiescent neural stem cell pool, preventing differentiation into TA descendents. Notch is therefore critical for long-term maintenance of adult neural stem cells (Imayoshi *et al.*, 2010). Furthermore, although not formally considered stem cells, forebrain ependymal cells have been shown to give rise to neuroblasts and astrocytes after stroke, and Notch signaling plays an active role in maintaining their quiescence (Carlen *et al.*, 2009). Thus, in the nervous system, Notch signaling appears to promote stem cell maintenance and may influence cell fate decisions of differentiating neural precursor cells.

2.9. Notch signaling in skeletal muscle and muscle satellite cells

Skeletal muscle is composed of multinucleated fibers bundled together by tendons that anchor the muscle to the skeleton. Striated muscles contract to generate force and movement and play a major role in regulating metabolism. Muscle growth and repair depends on a specialized subset of muscle fiber-associated mononuclear cells called "satellite cells" (Mauro, 1961), named for their very close association with mature muscle fibers (Fig. 12.2E).

Satellite cells are small (~8 μm) mononuclear precursor cells with scant cytoplasm and are located between the basement membrane and the sarcolemma (cell membrane surrounding the muscle cytoplasm, or sarcoplasm) of individual muscle fibers. Satellite cells express a number of distinct genetic markers, such as Pax7 and Pax3 (Relaix *et al.*, 2006), which distinguish them from nonmyogenic cells that may also reside in the muscle. Upon muscle damage, satellite cells proliferate and differentiate into fusion-competent myoblasts to regenerate the muscle (Hawke and Garry, 2001; Wagers and Conboy, 2005). The dynamics of satellite cell activation and quiescence, and induction of the myogenic program, invokes a cascade of myogenic regulatory factors, including myf5, myoD, myogenin, and MRF4, although a complete description of this process has yet to be established. Once activated, skeletal muscle satellite cells must decide their fate, undergoing either myogenic differentiation or self-renewal (hallmark properties of tissue stem cells). Because the outcome of this decision determines the efficiency of muscle repair, understanding the molecular inputs upon which this choice is made is critical to enhancing muscle repair activity and maintaining adequate muscle function throughout life. In addition, such knowledge has important implications for the understanding and treatment of congenital muscle disease, as the inability to maintain a sufficient reservoir of satellite cells during postnatal development, as occurs in Pax7 mutant mice, leads to impaired muscle regeneration (Oustanina *et al.*, 2004). Likewise, in human congenital myotonic dystrophy, satellite cells exhibit reduced proliferation and impaired myogenic differentiation, leading to reduced muscle repair capacity (Furling *et al.*, 2001).

As discussed above, Notch signaling is required in several developmental processes, and myogenesis is also crucially regulated by Notch signaling, both during vertebrate somitogenesis and during postnatal muscle repair and regeneration (Buas *et al.* 2010; Conboy and Rando, 2002; Huppert *et al.*, 2005; Kopan *et al.*, 1994; Rida *et al.*, 2004; Vasyutina *et al.*, 2007) (Fig. 12.2E). Overexpression of the Notch ligand delta-like 1 (Dll1) in signal-sending cells, or constitutive expression of Notch1 in satellite cells (Conboy and Rando, 2002; Conboy *et al.*, 2003; Kopan *et al.*, 1994; Sun *et al.*, 2008), inhibits myogenic differentiation. In contrast, overexpression of Numb, a negative regulator of Notch (Conboy and Rando, 2002), or inhibition of γ-secretase activity (Kitzmann *et al.*, 2006; Sun *et al.*, 2008) promotes myoblast differentiation and stimulates formation of larger myotubes (Fig. 12.2E). These data indicate that Notch acts in proliferating myogenic precursor cells as a negative regulator of terminal differentiation.

Interestingly, complementing its role as an inhibitor of differentiation, Notch also acts as a critical positive regulator of muscle precursor cells (Fig. 12.2E). For example, Dll1 modulates both differentiation and maintenance of myogenic precursor cells during development of the mouse embryo, and Dll1 hypomorphic mutant fetuses exhibit accelerated myogenic differentiation associated with an excess of myotomal muscle fibers and a loss of progenitor cells, leading to reduced muscle growth and severe muscle hypotrophy (Schuster-Gossler *et al.*, 2007). Likewise, conditional mutagenesis of RBP-J results in uncontrolled myogenic differentiation, which is associated with depletion of the myogenic precursor pool and causes severe muscle hypotrophy during fetal development (Tanigaki *et al.*, 2002; Vasyutina *et al.*, 2007). Thus, Notch signaling initiated by Dll1 ligand and transduced by RBP-J is essential for establishing and maintaining the satellite cell compartment during development and acts at least in part by blocking satellite cell differentiation to mature myoblasts.

In addition to its key importance in developing skeletal muscle, Notch signaling plays a continuing and essential role in satellite cell proliferation during muscle regeneration. Moreover, age-related impairment in the upregulation of Dll1 contributes significantly to the loss of muscle regeneration in older animals (Carlson and Conboy, 2007; Conboy *et al.*, 2005). To analyze the impact of circulating factors on aged satellite cells, Conboy *et al.* utilized a heterochronic parabiosis system, joining aged mice to young partners. Interestingly, satellite cells isolated from aged heterochronic pairs exhibited restored upregulation of Dll1 and enhanced cell activation and proliferation. The levels of Dll1 upregulation were comparable to levels found in their young partners and in young isochronic pairs, whereas Dll1 induction was lacking in the aged, isochronic parabionts—typical of the response of aged muscle. Furthermore, exposure of satellite cells isolated from aged mice to young serum promoted expression of the Notch ligand Dll1, increased Notch activation, and enhanced satellite cell proliferation *in vitro* (Conboy *et al.*, 2005). These findings suggest that systemic factors may modulate Notch activation locally within the muscle in an age-dependent manner. Moreover, they suggest that Notch signaling is critical to maintaining appropriate activity of muscle stem cells throughout life.

In order to control muscle stem and progenitor cell activity, Notch signals must be integrated with a host of other intrinsic and extrinsic inputs, which ultimately determine cell fate. Indeed, genetic and pharmacological "epistasis" analyses indicate significant cross talk between this pathway and several other key regulators of muscle development and regeneration. Interestingly, Notch signals can either reinforce or counteract these additional tissue regulators in a developmental and tissue-dependent manner.

Such complex, combinatorial control may have considerable ramifications for cell fate determination and regenerative medicine approaches in the skeletal muscle.

Similar to Notch, induction of BMP signaling appears to block differentiation of myogenic cells (Kopan *et al.*, 1994; Kuroda *et al.*, 1999). Addition of BMP4 during induction of myogenic differentiation dramatically reduces the number of differentiated myoblasts formed from satellite cells *in vitro* and simultaneously induces Notch responsive genes (Hey1 and Hes1), suggesting that BMP4 may inhibit myogenic differentiation through upregulation of Notch signaling (Dahlqvist *et al.*, 2003). Consistent with this notion, concomitant blockade of Notch signaling in BMP4-treated cell cultures, either by addition of GSI or by introduction of a dominant-negative version of CSL, can restore myogenic differentiation. Thus, functional Notch signaling appears to act in concert with BMP4 to restrict myogenic differentiation and promote a more primitive, stem cell fate among muscle satellite cells.

Like BMP4, transforming growth factor-beta (TGF-β), another member of the TGF/BMP superfamily, also initiates a signaling cascade that ultimately intersects with Notch pathway mediators. However, in contrast to BMP4, TGF-β appears to negatively regulate myogenic differentiation. For example, in addition to loss of Notch activation, aged muscle also produces excessive TGF-β, which induces unusually high levels of phosphorylated Smad3 (the active signal transducer of TGF-β signals) in muscle satellite cells. It appears that these high levels of Smad3 activity impair muscle regenerative capacity through direct antagonism of endogenous Notch signals. Whereas in young satellite cells, activated Notch inhibits expression of the CDK inhibitors p15, p16, p21, and p27, which can restrict satellite cell proliferation, in aged cells, reduced Notch signaling permits TGF-β-dependent upregulation of these CDKs. Thus, inhibition of TGF-β/Smad3 or, conversely, activation of Notch signaling in the injured muscle of aged mice can restore muscle regenerative potential (Carlson *et al.*, 2008; Derynck and Zhang, 2003; Massague and Wotton, 2000).

3. Conclusions and Perspective

Regardless of their origins, both embryonic and adult stem cells have shown immense promise for the treatment of human disease, and recent developments in the stem cell field have helped to rapidly push forward the potential applications of these cells in regenerative medicine. The ultimate goal of regenerative medicine is to utilize inherent biological mechanisms to either stimulate tissue regeneration inside the human body (a power comparable to that of mythical Greek titan Prometheus), or, if internal healing fails, to grow a healthy organ *ex vivo* and then safely transplant it into the body. If successful, it is envisaged that such strategies could one day diminish or even eliminate injury-induced and aging-related adult stem cell exhaustion and thereby transform our ability to treat the many tissue degenerative diseases that currently afflict the human population.

Although clearly a long way in the future, many studies already have demonstrated that tissue repair and regeneration can be achieved using adult stem cells, both *in vitro* and *in vivo*. Such approaches are epitomized by the extensive application of skin grafts containing epidermal stem cells to the treatment of burn patients (Brouard and Barrandon, 2003; Gambardella and Barrandon, 2003) and of bone marrow and mobilized peripheral blood cell grafts, containing HSCs, to the treatment of patients with a variety of hematopoietic malignancies and insufficiencies (Weissman and Shizuru, 2008). Although clinical approaches employing pluripotent stem cell derivatives have lagged behind in comparison to these particular adult stem cell populations (in part due to ethical and technical considerations), recent progress on the generation of both murine- and human-induced pluripotent stem cells (iPS) from adult cells, have helped to overcome previous obstacles and dramatically accelerated the application of human pluripotent stem cells toward therapeutic purposes (Takahashi and Yamanaka, 2006; Yu *et al.*, 2007). Indeed, because iPS cells can be generated from almost any somatic cell by introduction of a small number of pluripotency-associated transcription factors (including *Oct4, Sox2*, c-myc, and *Klf4*) (Hockemeyer *et al.*, 2008; Ichida *et al.*, 2009; Kim *et al.*, 2009; Takahashi *et al.*, 2007; Takahashi and Yamanaka, 2006; Xu *et al.*, 2004; Yu *et al.*, 2007), iPS technology allows for the direct generation of pluripotent stem cells without the use of embryos or embryonic tissues. Already, several patient-specific pluripotent cell lines have been generated using such approaches (Dimos *et al.*, 2008; Ebert *et al.*, 2009; Lengerke and Daley, 2009; Maehr *et al.*, 2009; Park *et al.*, 2008a; Park and Daley, 2009), and these cells serve both as novel *in vitro* models of human genetic disorders and as a potential source of replacement cells for future transplantation strategies. Yet, for either of these strategies to succeed, it will be essential to dissect the key regulatory pathways that specify stem cell self-renewal and differentiation and limit tumor-forming potential. In this regard, insights from studies of Notch signaling in both embryonic and adult tissues have helped to reveal critical insights into the cell fate decisions that impact the establishment, growth, and regenerative potential of many of the body's tissues. Both in stem cells themselves, as well as in their niches, Notch signaling is repeatedly called upon, in a cell-type and context-dependent manner, to promote or inhibit self-renewal, to enhance or restrict proliferation, and to influence lineage decisions in a wide variety of tissues, organs, and malignancies. That the Notch signaling pathway cannot be categorized into a single cellular activity, e.g., stem cell maintenance, highlights its remarkable versatility and correlates well with its broad conservation throughout evolution. Ongoing efforts to define the full network of Notch regulators and effectors thus will have substantial implications for the treatment of a number of human diseases, and more specifically refined methods of targeting Notch in discrete tissues and cells will be essential to realizing its full therapeutic potential. Through such studies, this unique signaling pathway undoubtedly will continue to provide us with novel insights into the mechanisms regulating stem cell self-renewal and differentiation and the application of these mechanisms to regenerative medicine.

REFERENCES

Adams, G. B., Chabner, K. T., Alley, I. R., Olson, D. P., Szczepiorkowski, Z. M., Poznansky, M. C., Kos, C. H., Pollak, M. R., Brown, E. M., and Scadden, D. T. (2006). Stem cell engraftment at the endosteal niche is specified by the calcium-sensing receptor. *Nature* **439**, 599–603.

Akashi, K., Traver, D., Miyamoto, T., and Weissman, I. L. (2000). A clonogenic common myeloid progenitor that gives rise to all myeloid lineages. *Nature* **404**, 193–197.

Allman, D., Karnell, F. G., Punt, J. A., Bakkour, S., Xu, L., Myung, P., Koretzky, G. A., Pui, J. C., Aster, J. C., and Pear, W. S. (2001). Separation of Notch1 promoted lineage commitment and expansion/transformation in developing T cells. *J. Exp. Med.* **194**, 99–106.

Allsopp, R. C., Morin, G. B., DePinho, R., Harley, C. B., and Weissman, I. L. (2003a). Telomerase is required to slow telomere shortening and extend replicative lifespan of HSCs during serial transplantation. *Blood* **102**, 517–520.

Allsopp, R. C., Morin, G. B., Horner, J. W., DePinho, R., Harley, C. B., and Weissman, I. L. (2003a). Effect of TERT over-expression on the long-term transplantation capacity of hematopoietic stem cells. *Nat. Med.* **9**, 369–371.

Ambler, C. A., and Maatta, A. (2009). Epidermal stem cells: location, potential and contribution to cancer. *J. pathol.* **217**, 206–216.

Barker, N., van de Wetering, M., and Clevers, H. (2008). The intestinal stem cell. *Genes Dev.* **22**, 1856–1864.

Bertrand, J. Y., Chi, N. C., Santoso, B., Teng, S., Stainier, D. Y., and Traver, D. (2010a). Haematopoietic stem cells derive directly from aortic endothelium during development. *Nature* **464**, 108–111.

Bertrand, J. Y., Cisson, J. L., Stachura, D. L., and Traver, D. (2010b). Notch signaling distinguishes two waves of definitive hematopoiesis in the zebrafish embryo. *Blood.* **115**(14), 2777–2783.

Blanpain, C., and Fuchs, E. (2006). Epidermal Stem Cells of the Skin. *Annu. Rev. Cell Dev. Biol.* **22**, 339–373.

Blanpain, C., and Fuchs, E. (2009). Epidermal homeostasis: a balancing act of stem cells in the skin. *Nat. Rev. Mol. Cell Biol.* **10**, 207–217.

Blanpain, C., Horsley, V., and Fuchs, E. (2007). Epithelial stem cells: turning over new leaves. *Cell* **128**, 445–458.

Blanpain, C., Lowry, W. E., Pasolli, H. A., and Fuchs, E. (2006). Canonical notch signaling functions as a commitment switch in the epidermal lineage. *Genes Dev.* **20**, 3022–3035.

Bouras, T., Pal, B., Vaillant, F., Harburg, G., Asselin-Labat, M. L., Oakes, S. R., Lindeman, G. J., and Visvader, J. E. (2008). Notch signaling regulates mammary stem cell function and luminal cell-fate commitment. *Cell stem cell* **3**, 429–441.

Breunig, J. J., Silbereis, J., Vaccarino, F. M., Sestan, N., and Rakic, P. (2007). Notch regulates cell fate and dendrite morphology of newborn neurons in the postnatal dentate gyrus. *Proc. Natl. Acad. Sci. U.S.A* **104**, 20558–20563.

Brouard, M., and Barrandon, Y. (2003). Controlling skin morphogenesis: hope and despair. *Curr. Opin. Biotechnol.* **14**, 520–525.

Buas, M.F., Kabak, S., and Kadesch, T. The Notch effector Hey1 associates with myogenic target genes to repress myogenesis. *J. Biol. Chem.* **285**, 1249-1258.

Calvi, L. M., Adams, G. B., Weibrecht, K. W., Weber, J. M., Olson, D. P., Knight, M. C., Martin, R. P., Schipani, E., Divieti, P., Bringhurst, F. R., *et al.* (2003). Osteoblastic cells regulate the haematopoietic stem cell niche. *Nature* **425**, 841–846.

Calzavara, E., Chiaramonte, R., Cesana, D., Basile, A., Sherbet, G. V., and Comi, P. (2008). Reciprocal regulation of Notch and PI3K/Akt signalling in T-ALL cells in vitro. *J. Cell. Biochem.* **103**, 1405–1412.

Carlen, M., Meletis, K., Goritz, C., Darsalia, V., Evergren, E., Tanigaki, K., Amendola, M., Barnabe-Heider, F., Yeung, M. S., Naldini, L., *et al.* (2009). Forebrain ependymal cells are Notch-dependent and generate neuroblasts and astrocytes after stroke. *Nat. Neurosci.* **12**, 259–267.

Carlesso, N., Aster, J. C., Sklar, J., and Scadden, D. T. (1999). Notch1-induced delay of human hematopoietic progenitor cell differentiation is associated with altered cell cycle kinetics. *Blood* **93**, 838–848.

Carlson, M. E., and Conboy, I. M. (2007). Loss of stem cell regenerative capacity within aged niches. *Aging Cell* **6**, 371–382.

Carlson, M. E., Hsu, M., and Conboy, I. M. (2008). Imbalance between pSmad3 and Notch induces CDK inhibitors in old muscle stem cells. *Nature* **454**, 528–532.

Casali, A., and Batlle, E. (2009). Intestinal stem cells in mammals and Drosophila. *Cell stem cell* **4**, 124–127.

Cau, E., and Blader, P. (2009). Notch activity in the nervous system: to switch or not switch?. *Neural dev.* **4**, 36.

Chadwick, N., Nostro, M. C., Baron, M., Mottram, R., Brady, G., and Buckle, A. M. (2007). Notch signaling induces apoptosis in primary human CD34+ hematopoietic progenitor cells. *Stem Cells* **25**, 203–210.

Chen, M. F., Lin, C. T., Chen, W. C., Yang, C. T., Chen, C. C., Liao, S. K., Liu, J. M., Lu, C. H., and Lee, K. D. (2006). The sensitivity of human mesenchymal stem cells to ionizing radiation. *Int. J. Radiat. Oncol. Biol. Phys.* **66**, 244–253.

Chen, P. M., Yen, C. C., Wang, W. S., Lin, Y. J., Chu, C. J., Chiou, T. J., Liu, J. H., and Yang, M. H. (2008). Down-regulation of Notch-1 expression decreases PU.1-mediated myeloid differentiation signaling in acute myeloid leukemia. *Int. J. Oncol.* **32**, 1335–1341.

Chenn, A., and McConnell, S. K. (1995). Cleavage orientation and the asymmetric inheritance of Notch1 immunoreactivity in mammalian neurogenesis. *Cell* **82**, 631–641.

Chiba, S. (2006). Notch signaling in stem cell systems. *Stem cells (Dayton, Ohio)* **24**, 2437–2447.

Choi, K., Kennedy, M., Kazarov, A., Papadimitriou, J. C., and Keller, G. (1998). A common precursor for hematopoietic and endothelial cells. *Development* **125**, 725–732.

Christensen, J. L., and Weissman, I. L. (2001). Flk-2 is a marker in hematopoietic stem cell differentiation: a simple method to isolate long-term stem cells. *Proc. Natl. Acad. Sci. U.S.A* **98**, 14541–14546.

Conboy, I. M., Conboy, M. J., Smythe, G. M., and Rando, T. A. (2003). Notch-mediated restoration of regenerative potential to aged muscle. *Science* **302**, 1575–1577.

Conboy, I. M., Conboy, M. J., Wagers, A. J., Girma, E. R., Weissman, I. L., and Rando, T. A. (2005). Rejuvenation of aged progenitor cells by exposure to a young systemic environment. *Nature* **433**, 760–764.

Conboy, I. M., and Rando, T. A. (2002). The regulation of Notch signaling controls satellite cell activation and cell fate determination in postnatal myogenesis. *Dev. cell* **3**, 397–409.

Crosnier, C., Stamataki, D., and Lewis, J. (2006). Organizing cell renewal in the intestine: stem cells, signals and combinatorial control. *Nat. rev.* **7**, 349–359.

Crosnier, C., Vargesson, N., Gschmeissner, S., Ariza-McNaughton, L., Morrison, A., and Lewis, J. (2005). Delta-Notch signalling controls commitment to a secretory fate in the zebrafish intestine. *Development* **132**, 1093–1104.

Cullion, K., Draheim, K. M., Hermance, N., Tammam, J., Sharma, V. M., Ware, C., Nikov, G., Krishnamoorthy, V., Majumder, P. K., and Kelliher, M. A. (2009). Targeting the Notch1 and mTOR pathways in a mouse T-ALL model. *Blood* **113**, 6172–6181.

Cumano, A., Dieterlen-Lievre, F., and Godin, I. (1996). Lymphoid potential, probed before circulation in mouse, is restricted to caudal intraembryonic splanchnopleura. *Cell* **86**, 907–916.

Dahlqvist, C., Blokzijl, A., Chapman, G., Falk, A., Dannaeus, K., Ibanez, C. F., and Lendahl, U. (2003). Functional Notch signaling is required for BMP4-induced inhibition of myogenic differentiation. *Development* **130**, 6089–6099.

De Keersmaecker, K., Lahortiga, I., Mentens, N., Folens, C., Van Neste, L., Bekaert, S., Vandenberghe, P., Odero, M. D., Marynen, P., and Cools, J. (2008). In vitro validation of gamma-secretase inhibitors alone or in combination with other anti-cancer drugs for the treatment of T-cell acute lymphoblastic leukemia. *Haematologica* **93**, 533–542.

de La Coste, A., Six, E., Fazilleau, N., Mascarell, L., Legrand, N., Mailhe, M. P., Cumano, A., Laabi, Y., and Freitas, A. A. (2005). In vivo and in absence of a thymus, the enforced expression of the Notch ligands delta-1 or delta-4 promotes T cell development with specific unique effects. *J. Immunol.* **174**, 2730–2737.

Demehri, S., and Kopan, R. (2009). Notch signaling in bulge stem cells is not required for selection of hair follicle fate. *Development* **136**, 891–896.

Demehri, S., Liu, Z., Lee, J., Lin, M. H., Crosby, S. D., Roberts, C. J., Grigsby, P. W., Miner, J. H., Farr, A. G., and Kopan, R. (2008). Notch-deficient skin induces a lethal systemic B-lymphoproliferative disorder by secreting TSLP, a sentinel for epidermal integrity. *PLoS Biol.* **6**, e123.

Demehri, S., Morimoto, M., Holtzman, M. J., and Kopan, R. (2009a). Skin-derived TSLP triggers progression from epidermal-barrier defects to asthma. *PLoS Biol.* **7**, e1000067. doi:1000010.1001371/journal.pbio.1000067.

Demehri, S., Turkoz, A., and Kopan, R. (2009b). Epidermal Notch1 Loss Promotes Skin Tumorigenesis by Impacting the Stromal Microenvironment. *Cancer Cell* **16**, 55–66.

Derynck, R., and Zhang, Y. E. (2003). Smad-dependent and Smad-independent pathways in TGF-beta family signalling. *Nature* **425**, 577–584.

Dimos, J. T., Rodolfa, K. T., Niakan, K. K., Weisenthal, L. M., Mitsumoto, H., Chung, W., Croft, G. F., Saphier, G., Leibel, R., Goland, R., *et al.* (2008). Induced pluripotent stem cells generated from patients with ALS can be differentiated into motor neurons. *Science* **321**, 1218–1221.

Donoviel, D. B., Hadjantonakis, A. K., Ikeda, M., Zheng, H., Hyslop, P. S., and Bernstein, A. (1999). Mice lacking both presenilin genes exhibit early embryonic patterning defects. *Genes Dev.* **13**, 2801–2810.

Dotto, G. P. (2008). Notch tumor suppressor function. *Oncogene* **27**, 5115–5123.

Dumortier, A., Durham, A.-D., Di Piazza, M., Vauclair, S., Koch, U., Ferrand, G., Ferrero, I., Demehri, S., Song, L. L., Farr, A. G., *et al.* (2010). Atopic Dermatitis-Like Disease and Associated Lethal Myeloproliferative Disorder Arise from Loss of Notch Signaling in the Murine Skin. *PLoS ONE* **5**, e9258.

Duncan, A. W., Rattis, F. M., DiMascio, L. N., Congdon, K. L., Pazianos, G., Zhao, C., Yoon, K., Cook, J. M., Willert, K., Gaiano, N., *et al.* (2005). Integration of Notch and Wnt signaling in hematopoietic stem cell maintenance. *Nat. Immunol.* **6**, 314–322.

Dzierzak, E., and Speck, N. A. (2008). Of lineage and legacy: the development of mammalian hematopoietic stem cells. *Nat. Immunol.* **9**, 129–136.

Ebert, A. D., Yu, J., Rose, F. F., Jr., Mattis, V. B., Lorson, C. L., Thomson, J. A., and Svendsen, C. N. (2009). Induced pluripotent stem cells from a spinal muscular atrophy patient. *Nature* **457**, 277–280.

Eilken, H. M., Nishikawa, S., and Schroeder, T. (2009). Continuous single-cell imaging of blood generation from haemogenic endothelium. *Nature* **457**, 896–900.

Fernandez, L., Rodriguez, S., Huang, H., Chora, A., Fernandes, J., Mumaw, C., Cruz, E., Pollok, K., Cristina, F., Price, J. E., *et al.* (2008). Tumor necrosis factor-alpha and endothelial cells modulate Notch signaling in the bone marrow microenvironment during inflammation. *Exp. Hematol.* **36**, 545–558.

Foudi, A., Hochedlinger, K., Van Buren, D., Schindler, J. W., Jaenisch, R., Carey, V., and Hock, H. (2009). Analysis of histone 2B-GFP retention reveals slowly cycling hematopoietic stem cells. *Nat. Biotechnol.* **27**, 84–90.

Fre, S., Huyghe, M., Mourikis, P., Robine, S., Louvard, D., and Artavanis-Tsakonas, S. (2005). Notch signals control the fate of immature progenitor cells in the intestine. *Nature* **435**, 964–968.

Fre, S., Pallavi, S. K., Huyghe, M., Lae, M., Janssen, K. P., Robine, S., Artavanis-Tsakonas, S., and Louvard, D. (2009). Notch and Wnt signals cooperatively control cell proliferation and tumorigenesis in the intestine. *Proc. Natl. Acad. Sci. U.S.A* **106**, 6309–6314.

Frenette, P. S., Subbarao, S., Mazo, I. B., von Andrian, U. H., and Wagner, D. D. (1998). Endothelial selectins and vascular cell adhesion molecule-1 promote hematopoietic progenitor homing to bone marrow. *Proc. Natl. Acad. Sci. U.S.A* **95**, 14423–14428.

Fuchs, E., and Raghavan, S. (2002). Getting under the skin of epidermal morphogenesis. *Nat. rev.* **3**, 199–209.

Fuchs, E., Tumbar, T., and Guasch, G. (2004). Socializing with the neighbors: stem cells and their niche. *Cell* **116**, 769–778.

Fujita, S., Toguchida, J., Morita, Y., and Iwata, H. (2008). Clonal analysis of hematopoiesis-supporting activity of human mesenchymal stem cells in association with Jagged1 expression and osteogenic potential. *Cell Transplant.* **17**, 1169–1179.

Furling, D., Coiffier, L., Mouly, V., Barbet, J. P., St Guily, J. L., Taneja, K., Gourdon, G., Junien, C., and Butler-Browne, G. S. (2001). Defective satellite cells in congenital myotonic dystrophy. *Hum. mol. genet.* **10**, 2079–2087.

Gal, H., Amariglio, N., Trakhtenbrot, L., Jacob-Hirsh, J., Margalit, O., Avigdor, A., Nagler, A., Tavor, S., Ein-Dor, L., Lapidot, T., *et al.* (2006). Gene expression profiles of AML derived stem cells; similarity to hematopoietic stem cells. *Leukemia* **20**, 2147–2154.

Gambardella, L., and Barrandon, Y. (2003). The multifaceted adult epidermal stem cell. *Curr. Opin. Cell Biol.* **15**, 771–777.

Garrett, R. W., and Emerson, S. G. (2009). Bone and blood vessels: the hard and the soft of hematopoietic stem cell niches. *Cell Stem Cell* **4**, 503–506.

Gekas, C., Dieterlen-Lievre, F., Orkin, S. H., and Mikkola, H. K. (2005). The placenta is a niche for hematopoietic stem cells. *Dev. Cell* **8**, 365–375.

Grandbarbe, L., Bouissac, J., Rand, M., Hrabe de Angelis, M., Artavanis-Tsakonas, S., and Mohier, E. (2003). Delta-Notch signaling controls the generation of neurons/glia from neural stem cells in a stepwise process. *Development* **130**, 1391–1402.

Hadland, B. K., Huppert, S. S., Kanungo, J., Xue, Y., Jiang, R., Gridley, T., Conlon, R. A., Cheng, A. M., Kopan, R., and Longmore, G. D. (2004). A requirement for Notch1 distinguishes 2 phases of definitive hematopoiesis during development. *Blood* **104**, 3097–3105.

Hadland, B. K., Manley, N. R., Su, D., Longmore, G. D., Moore, C. L., Wolfe, M. S., Schroeter, E. H., and Kopan, R. (2001). Gamma -secretase inhibitors repress thymocyte development. *Proc. Natl. Acad. Sci. U.S.A* **98**, 7487–7491.

Han, H., Tanigaki, K., Yamamoto, N., Kuroda, K., Yoshimoto, M., Nakahata, T., Ikuta, K., and Honjo, T. (2002). Inducible gene knockout of transcription factor recombination signal binding protein-J reveals its essential role in T versus B lineage decision. *Int. Immunol.* **14**, 637–645.

Han, W., Ye, Q., and Moore, M. A. (2000). A soluble form of human Delta-like-1 inhibits differentiation of hematopoietic progenitor cells. *Blood* **95**, 1616–1625.

Hawke, T. J., and Garry, D. J. (2001). Myogenic satellite cells: physiology to molecular biology. *J. Appl. Physiol.* **91**, 534–551.

Herreman, A., Hartmann, D., Annaert, W., Saftig, P., Craessaerts, K., Serneels, L., Umans, L., Schrijvers, V., Checler, F., Vanderstichele, H., *et al.* (1999). Presenilin 2 deficiency causes a mild pulmonary phenotype and no changes in amyloid precursor protein processing but enhances the embryonic lethal phenotype of presenilin 1 deficiency. *Proc. Natl. Acad. Sci. U.S.A* **96**, 11872–11877.

Hidaka, M., Stanford, W. L., and Bernstein, A. (1999). Conditional requirement for the Flk-1 receptor in the in vitro generation of early hematopoietic cells. *Proc. Natl. Acad. Sci. U.S.A* **96**, 7370–7375.

Hilton, M. J., Tu, X., Wu, X., Bai, S., Zhao, H., Kobayashi, T., Kronenberg, H. M., Teitelbaum, S. L., Ross, F. P., Kopan, R., *et al.* (2008). Notch signaling maintains bone marrow mesenchymal progenitors by suppressing osteoblast differentiation. *Nat. Med.* **14**, 306–314.

Hockemeyer, D., Soldner, F., Cook, E. G., Gao, Q., Mitalipova, M., and Jaenisch, R. (2008). A drug-inducible system for direct reprogramming of human somatic cells to pluripotency. *Cell stem cell* **3**, 346–353.

Hozumi, K., Abe, N., Chiba, S., Hirai, H., and Habu, S. (2003). Active form of Notch members can enforce T lymphopoiesis on lymphoid progenitors in the monolayer culture specific for B cell development. *J. Immunol.* **170**, 4973–4979.

Huppert, S. S., Ilagan, M. X., De Strooper, B., and Kopan, R. (2005). Analysis of Notch function in presomitic mesoderm suggests a gamma-secretase-independent role for presenilins in somite differentiation. *Dev. cell* **8**, 677–688.

Ichida, J. K., Blanchard, J., Lam, K., Son, E. Y., Chung, J. E., Egli, D., Loh, K. M., Carter, A. C., Di Giorgio, F. P., Koszka, K., *et al.* (2009). A small-molecule inhibitor of tgf-Beta signaling replaces sox2 in reprogramming by inducing nanog. *Cell stem cell* **5**, 491–503.

Imayoshi, I., Sakamoto, M., Yamaguchi, M., Mori, K., and Kageyama, R. (2010). Essential roles of Notch signaling in maintenance of neural stem cells in developing and adult brains. *J. Neurosci.* **30**, 3489–3498.

Ishibashi, M., Moriyoshi, K., Sasai, Y., Shiota, K., Nakanishi, S., and Kageyama, R. (1994). Persistent expression of helix-loop-helix factor HES-1 prevents mammalian neural differentiation in the central nervous system. *EMBO J.* **13**, 1799–1805.

Jaleco, A. C., Neves, H., Hooijberg, E., Gameiro, P., Clode, N., Haury, M., Henrique, D., and Parreira, L. (2001). Differential effects of Notch ligands Delta-1 and Jagged-1 in human lymphoid differentiation. *J. Exp. Med.* **194**, 991–1002.

Johnson, M. A., Ables, J. L., and Eisch, A. J. (2009). Cell-intrinsic signals that regulate adult neurogenesis in vivo: insights from inducible approaches. *BMB rep.* **42**, 245–259.

Jones, D. L., and Wagers, A. J. (2008). No place like home: anatomy and function of the stem cell niche. *Nat. Rev. Mol. Cell Biol.* **9**, 11–21.

Jones, P., May, G., Healy, L., Brown, J., Hoyne, G., Delassus, S., and Enver, T. (1998). Stromal expression of Jagged 1 promotes colony formation by fetal hematopoietic progenitor cells. *Blood* **92**, 1505–1511.

Jonsson, J. I., Xiang, Z., Pettersson, M., Lardelli, M., and Nilsson, G. (2001). Distinct and regulated expression of Notch receptors in hematopoietic lineages and during myeloid differentiation. *Eur. J. Immunol.* **31**, 3240–3247.

Jundt, F., Anagnostopoulos, I., Forster, R., Mathas, S., Stein, H., and Dorken, B. (2002). Activated Notch1 signaling promotes tumor cell proliferation and survival in Hodgkin and anaplastic large cell lymphoma. *Blood* **99**, 3398–3403.

Karanu, F. N., Murdoch, B., Gallacher, L., Wu, D. M., Koremoto, M., Sakano, S., and Bhatia, M. (2000). The notch ligand jagged-1 represents a novel growth factor of human hematopoietic stem cells. *J. Exp. Med.* **192**, 1365–1372.

Karanu, F. N., Murdoch, B., Miyabayashi, T., Ohno, M., Koremoto, M., Gallacher, L., Wu, D., Itoh, A., Sakano, S., and Bhatia, M. (2001). Human homologues of Delta-1 and Delta-4 function as mitogenic regulators of primitive human hematopoietic cells. *Blood* **97**, 1960–1967.

Kawamata, S., Du, C., Li, K., and Lavau, C. (2002). Overexpression of the Notch target genes Hes in vivo induces lymphoid and myeloid alterations. *Oncogene* **21**, 3855–3863.

Kawase, E., Wong, M. D., Ding, B. C., and Xie, T. (2004). Gbb/Bmp signaling is essential for maintaining germline stem cells and for repressing bam transcription in the Drosophila testis. *Development* **131**, 1365–1375.

Kiel, M. J., Yilmaz, O. H., Iwashita, T., Terhorst, C., and Morrison, S. J. (2005). SLAM family receptors distinguish hematopoietic stem and progenitor cells and reveal endothelial niches for stem cells. *Cell* **121**, 1109–1121.

Kim, J. B., Sebastiano, V., Wu, G., Arauzo-Bravo, M. J., Sasse, P., Gentile, L., Ko, K., Ruau, D., Ehrich, M., van den Boom, D., *et al.* (2009). Oct4-induced pluripotency in adult neural stem cells. *Cell* **136**, 411–419.

Kim, Y. W., Koo, B. K., Jeong, H. W., Yoon, M. J., Song, R., Shin, J., Jeong, D. C., Kim, S. H., and Kong, Y. Y. (2008). Defective Notch activation in microenvironment leads to myeloproliferative disease. *Blood* **112**, 4628–4638.

Kindler, T., Cornejo, M. G., Scholl, C., Liu, J., Leeman, D. S., Haydu, J. E., Frohling, S., Lee, B. H., and Gilliland, D. G. (2008). K-RasG12D-induced T-cell lymphoblastic lymphoma/leukemias harbor Notch1 mutations and are sensitive to gamma-secretase inhibitors. *Blood* **112**, 3373–3382.

Kitzmann, M., Bonnieu, A., Duret, C., Vernus, B., Barro, M., Laoudj-Chenivesse, D., Verdi, J. M., and Carnac, G. (2006). Inhibition of Notch signaling induces myotube hypertrophy by recruiting a subpopulation of reserve cells. *J. Cell. Physiol.* **208**, 538–548.

Kobayashi, T., Mizuno, H., Imayoshi, I., Furusawa, C., Shirahige, K., and Kageyama, R. (2009). The cyclic gene Hes1 contributes to diverse differentiation responses of embryonic stem cells. *Genes Dev.* **23**, 1870–1875.

Kolquist, K. A., Ellisen, L. W., Counter, C. M., Meyerson, M., Tan, L. K., Weinberg, R. A., Haber, D. A., and Gerald, W. L. (1998). Expression of TERT in early premalignant lesions and a subset of cells in normal tissues. *Nat. Genet.* **19**, 182–186.

Kondo, M., Weissman, I. L., and Akashi, K. (1997). Identification of clonogenic common lymphoid progenitors in mouse bone marrow. *Cell* **91**, 661–672.

Kopan, R., Nye, J. S., and Weintraub, H. (1994). The intracellular domain of mouse Notch: a constitutively activated repressor of myogenesis directed at the basic helix-loop-helix region of MyoD. *Development* **120**, 2385–2396.

Kumano, K., Chiba, S., Kunisato, A., Sata, M., Saito, T., Nakagami-Yamaguchi, E., Yamaguchi, T., Masuda, S., Shimizu, K., Takahashi, T., *et al.* (2003). Notch1 but not Notch2 is essential for generating hematopoietic stem cells from endothelial cells. *Immunity* **18**, 699–711.

Kumano, K., Chiba, S., Shimizu, K., Yamagata, T., Hosoya, N., Saito, T., Takahashi, T., Hamada, Y., and Hirai, H. (2001). Notch1 inhibits differentiation of hematopoietic cells by sustaining GATA-2 expression. *Blood* **98**, 3283–3289.

Kunisato, A., Chiba, S., Nakagami-Yamaguchi, E., Kumano, K., Saito, T., Masuda, S., Yamaguchi, T., Osawa, M., Kageyama, R., Nakauchi, H., *et al.* (2003). HES-1 preserves purified hematopoietic stem cells ex vivo and accumulates side population cells in vivo. *Blood* **101**, 1777–1783.

Kuroda, K., Tani, S., Tamura, K., Minoguchi, S., Kurooka, H., and Honjo, T. (1999). Delta-induced Notch signaling mediated by RBP-J inhibits MyoD expression and myogenesis. *J. Biol. Chem.* **274**, 7238–7244.

La Motte-Mohs, R. N., Herer, E., and Zuniga-Pflucker, J. C. (2005). Induction of T-cell development from human cord blood hematopoietic stem cells by Delta-like 1 in vitro. *Blood* **105**, 1431–1439.

Lancrin, C., Sroczynska, P., Stephenson, C., Allen, T., Kouskoff, V., and Lacaud, G. (2009). The haemangioblast generates haematopoietic cells through a haemogenic endothelium stage. *Nature* **457**, 892–895.

Lansdorp, P. M. (2008). Telomeres, stem cells, and hematology. *Blood* **111**, 1759–1766.

Lengerke, C., and Daley, G. Q. (2009). Disease models from pluripotent stem cells. *Ann. N. Y. Acad. Sci.* **1176**, 191–196.

Lewis, H. D., Leveridge, M., Strack, P. R., Haldon, C. D., O'Neil, J., Kim, H., Madin, A., Hannam, J. C., Look, A. T., Kohl, N., *et al.* (2007). Apoptosis in T cell acute lymphoblastic leukemia cells after cell cycle arrest induced by pharmacological inhibition of notch signaling. *Chem. Biol.* **14**, 209–219.

Li, L., and Clevers, H. (2010). Coexistence of quiescent and active adult stem cells in mammals. *Science* **327**, 542–545.

Li, L., Forman, S. J., and Bhatia, R. (2005). Expression of DLK1 in hematopoietic cells results in inhibition of differentiation and proliferation. *Oncogene* **24**, 4472–4476.

Li, T., Ma, G., Cai, H., Price, D. L., and Wong, P. C. (2003). Nicastrin Is Required for Assembly of Presenilin/gamma -Secretase Complexes to Mediate Notch Signaling and for Processing and Trafficking of beta -Amyloid Precursor Protein in Mammals. *J. Neurosci.* **23**, 3272–3277.

Li, T., Wen, H., Brayton, C., Laird, F. M., Ma, G., Peng, S., Placanica, L., Wu, T. C., Crain, B. J., Price, D. L., *et al.* (2007). Moderate reduction of gamma-secretase attenuates amyloid burden and limits mechanism-based liabilities. *J. Neurosci.* **27**, 10849–10859.

Li, X., Gounari, F., Protopopov, A., Khazaie, K., and von Boehmer, H. (2008a). Oncogenesis of T-ALL and nonmalignant consequences of overexpressing intracellular NOTCH1. *J. Exp. Med.* **205**, 2851–2861.

Li, Y. P., Paczesny, S., Lauret, E., Poirault, S., Bordigoni, P., Mekhloufi, F., Hequet, O., Bertrand, Y., Ou-Yang, J. P., Stoltz, J. F., *et al.* (2008b). Human mesenchymal stem cells license adult CD34+ hemopoietic progenitor cells to differentiate into regulatory dendritic cells through activation of the Notch pathway. *J. Immunol.* **180**, 1598–1608.

Li, Z., Chen, M. J., Stacy, T., and Speck, N. A. (2006). Runx1 function in hematopoiesis is required in cells that express Tek. *Blood* **107**, 106–110.

Lledo, P. M., Alonso, M., and Grubb, M. S. (2006). Adult neurogenesis and functional plasticity in neuronal circuits. *Nat. Rev. Neurosci.* **7**, 179–193.

Louvi, A., and Artavanis-Tsakonas, S. (2006). Notch signalling in vertebrate neural development. *Nat. Rev. Neurosci.* **7**, 93–102.

Lowell, S., Benchoua, A., Heavey, B., and Smith, A. G. (2006). Notch promotes neural lineage entry by pluripotent embryonic stem cells. *PLoS biol.* **4**, e121.

Ly, H., Calado, R. T., Allard, P., Baerlocher, G. M., Lansdorp, P. M., Young, N. S., and Parslow, T. G. (2005). Functional characterization of telomerase RNA variants found in patients with hematologic disorders. *Blood* **105**, 2332–2339.

Maehr, R., Chen, S., Snitow, M., Ludwig, T., Yagasaki, L., Goland, R., Leibel, R. L., and Melton, D. A. (2009). Generation of pluripotent stem cells from patients with type 1 diabetes. *Proc. Natl. Acad. Sci. U.S.A* **106**, 15768–15773.

Maillard, I., Koch, U., Dumortier, A., Shestova, O., Xu, L., Sai, H., Pross, S. E., Aster, J. C., Bhandoola, A., Radtke, F., *et al.* (2008). Canonical notch signaling is dispensable for the maintenance of adult hematopoietic stem cells. *Cell Stem Cell* **2**, 356–366.

Massague, J., and Wotton, D. (2000). Transcriptional control by the TGF-beta/Smad signaling system. *EMBO J.* **19**, 1745–1754.

Massberg, S., Schaerli, P., Knezevic-Maramica, I., Kollnberger, M., Tubo, N., Moseman, E. A., Huff, I. V., Junt, T., Wagers, A. J., Mazo, I. B., *et al.* (2007). Immunosurveillance by hematopoietic progenitor cells trafficking through blood, lymph, and peripheral tissues. *Cell* **131**, 994–1008.

Masuda, S., Kumano, K., Suzuki, T., Tomita, T., Iwatsubo, T., Natsugari, H., Tojo, A., Shibutani, M., Mitsumori, K., Hanazono, Y., *et al.* (2009). Dual antitumor mechanisms of Notch signaling inhibitor in a T-cell acute lymphoblastic leukemia xenograft model. *Cancer Sci.* **100**(12), 2444–2450.

Mauro, A. (1961). Satellite cell of skeletal muscle fibers. *J. Biophys. Biochem. Cytol.* **9**, 493–495.

Mayack, S. R., and Wagers, A. J. (2008). Osteolineage niche cells initiate hematopoietic stem cell mobilization. *Blood* **112**, 519–531.

Medvinsky, A., and Dzierzak, E. (1996). Definitive hematopoiesis is autonomously initiated by the AGM region. *Cell* **86**, 897–906.

Mercher, T., Cornejo, M. G., Sears, C., Kindler, T., Moore, S. A., Maillard, I., Pear, W. S., Aster, J. C., and Gilliland, D. G. (2008). Notch signaling specifies megakaryocyte development from hematopoietic stem cells. *Cell Stem Cell* **3**, 314–326.

Mercher, T., Raffel, G. D., Moore, S. A., Cornejo, M. G., Baudry-Bluteau, D., Cagnard, N., Jesneck, J. L., Pikman, Y., Cullen, D., Williams, I. R., *et al.* (2009). The OTT-MAL fusion oncogene activates RBPJ-mediated transcription and induces acute megakaryoblastic leukemia in a knockin mouse model. *J. Clin. Invest.* **119**, 852–864.

Milano, J., McKay, J., Dagenais, C., Foster-Brown, L., Pognan, F., Gadient, R., Jacobs, R. T., Zacco, A., Greenberg, B., and Ciaccio, P. J. (2004). Modulation of notch processing by gamma-secretase inhibitors causes intestinal goblet cell metaplasia and induction of genes known to specify gut secretory lineage differentiation. *Toxicol. Sci.* **82**, 341–358.

Milner, L. A., Bigas, A., Kopan, R., Brashem-Stein, C., Bernstein, I. D., and Martin, D. I. (1996). Inhibition of granulocytic differentiation by mNotch1. *Proc. Natl. Acad. Sci. U.S.A* **93**, 13014–13019.

Milner, L. A., Kopan, R., Martin, D. I., and Bernstein, I. D. (1994). A human homologue of the Drosophila developmental gene, Notch, is expressed in CD34+ hematopoietic precursors. *Blood* **83**, 2057–2062.

Miyazato, A., Ueno, S., Ohmine, K., Ueda, M., Yoshida, K., Yamashita, Y., Kaneko, T., Mori, M., Kirito, K., Toshima, M., *et al.* (2001). Identification of myelodysplastic syndrome-specific genes by DNA microarray analysis with purified hematopoietic stem cell fraction. *Blood* **98**, 422–427.

Molofsky, A. V., Pardal, R., and Morrison, S. J. (2004). Diverse mechanisms regulate stem cell self-renewal. *Curr. Opin. Cell Biol.* **16**, 700–707.

Morrison, S. J., and Weissman, I. L. (1994). The long-term repopulating subset of hematopoietic stem cells is deterministic and isolatable by phenotype. *Immunity* **1**, 661–673.

Nicolas, M., Wolfer, A., Raj, K., Kummer, J. A., Mill, P., van Noort, M., Hui, C. C., Clevers, H., Dotto, G. P., and Radtke, F. (2003). Notch1 functions as a tumor suppressor in mouse skin. *Nat. genet.* **33**, 416–421.

Nijnik, A., Woodbine, L., Marchetti, C., Dawson, S., Lambe, T., Liu, C., Rodrigues, N. P., Crockford, T. L., Cabuy, E., Vindigni, A., *et al.* (2007). DNA repair is limiting for haematopoietic stem cells during ageing. *Nature* **447**, 686–690.

Nussenzweig, A., Chen, C., da Costa Soares, V., Sanchez, M., Sokol, K., Nussenzweig, M. C., and Li, G. C. (1996). Requirement for Ku80 in growth and immunoglobulin V(D)J recombination. *Nature* **382**, 551–555.

Ogawa, M., Nishikawa, S., Yoshinaga, K., Hayashi, S., Kunisada, T., Nakao, J., Kina, T., Sudo, T., and Kodama, H. (1993). Expression and function of c-Kit in fetal hemopoietic progenitor cells: transition from the early c-Kit-independent to the late c-Kit-dependent wave of hemopoiesis in the murine embryo. *Development* **117**, 1089–1098.

Ohishi, K., Varnum-Finney, B., and Bernstein, I. D. (2002). Delta-1 enhances marrow and thymus repopulating ability of human CD34(+)CD38(-) cord blood cells. *J. Clin. Invest.* **110**, 1165–1174.

Ohlstein, B., and Spradling, A. (2007). Multipotent Drosophila intestinal stem cells specify daughter cell fates by differential notch signaling. *Science (New York, NY)* **315**, 988–992.

Oka, C., Nakano, T., Wakeham, A., de la Pompa, J. A., Mori, C., Sakai, T., Okazaki, S., Kawaichi, M., Shiota, K., Mak, T. W., *et al.* (1995). Disruption of the mouse RBP-Jkappa gene results in early embryonic death. *Development* **121**, 3291–3301.

Okuyama, R., Tagami, H., and Aiba, S. (2008). Notch signaling: its role in epidermal homeostasis and in the pathogenesis of skin diseases. *J. Dermatol. Sci.* **49**, 187–194.

Orkin, S. H., and Zon, L. I. (2008). Hematopoiesis: an evolving paradigm for stem cell biology. *Cell* **132**, 631–644.

Ottersbach, K., and Dzierzak, E. (2005). The murine placenta contains hematopoietic stem cells within the vascular labyrinth region. *Dev. Cell* **8**, 377–387.

Oustanina, S., Hause, G., and Braun, T. (2004). Pax7 directs postnatal renewal and propagation of myogenic satellite cells but not their specification. *EMBO J.* **23**, 3430–3439.

Park, I. H., Arora, N., Huo, H., Maherali, N., Ahfeldt, T., Shimamura, A., Lensch, M. W., Cowan, C., Hochedlinger, K., and Daley, G. Q. (2008a). Disease-specific induced pluripotent stem cells. *Cell* **134**, 877–886.

Park, I. H., and Daley, G. Q. (2009). Human iPS cell derivation/reprogramming. *Curr. Protoc. Stem Cell Biol. Chapter 4*, Unit 4A 1.

Park, I. H., Zhao, R., West, J. A., Yabuuchi, A., Huo, H., Ince, T. A., Lerou, P. H., Lensch, M. W., and Daley, G. Q. (2008b). Reprogramming of human somatic cells to pluripotency with defined factors. *Nature* **451**, 141–146.

Pui, J. C., Allman, D., Xu, L., DeRocco, S., Karnell, F. G., Bakkour, S., Lee, J. Y., Kadesch, T., Hardy, R. R., Aster, J. C., *et al.* (1999). Notch1 expression in early lymphopoiesis influences B versus T lineage determination. *Immunity* **11**, 299–308.

Qi, X., Chen, Z., Liu, D., Cen, J., and Gu, M. (2008). Expression of Dlk1 gene in myelodysplastic syndrome determined by microarray, and its effects on leukemia cells. *Int. J. Mol. Med.* **22**, 61–68.

Qyang, Y., Chambers, S. M., Wang, P., Xia, X., Chen, X., Goodell, M. A., and Zheng, H. (2004). Myeloproliferative disease in mice with reduced presenilin gene dosage: effect of gamma-secretase blockage. *Biochemistry* **43**, 5352–5359.

Rai, P., Onder, T. T., Young, J. J., McFaline, J. L., Pang, B., Dedon, P. C., and Weinberg, R. A. (2009). Continuous elimination of oxidized nucleotides is necessary to prevent rapid onset of cellular senescence. *Proc. Natl. Acad. Sci. U.S.A* **106**, 169–174.

Rangarajan, A., Talora, C., Okuyama, R., Nicolas, M., Mammucari, C., Oh, H., Aster, J. C., Krishna, S., Metzger, D., Chambon, P., *et al.* (2001). Notch signaling is a direct determinant of keratinocyte growth arrest and entry into differentiation. *EMBO J.* **20**, 3427–3436.

Rao, S. S., O'Neil, J., Liberator, C. D., Hardwick, J. S., Dai, X., Zhang, T., Tyminski, E., Yuan, J., Kohl, N. E., Richon, V. M., *et al.* (2009). Inhibition of NOTCH signaling by gamma secretase inhibitor engages the RB pathway and elicits cell cycle exit in T-cell acute lymphoblastic leukemia cells. *Cancer Res.* **69**, 3060–3068.

Real, P. J., Tosello, V., Palomero, T., Castillo, M., Hernando, E., de Stanchina, E., Sulis, M. L., Barnes, K., Sawai, C., Homminga, I., *et al.* (2009). Gamma-secretase inhibitors reverse glucocorticoid resistance in T cell acute lymphoblastic leukemia. *Nat. Med.* **15**, 50–58.

Relaix, F., Montarras, D., Zaffran, S., Gayraud-Morel, B., Rocancourt, D., Tajbakhsh, S., Mansouri, A., Cumano, A., and Buckingham, M. (2006). Pax3 and Pax7 have distinct and overlapping functions in adult muscle progenitor cells. *J. Cell Biol.* **172**, 91–102.

Reya, T., Morrison, S. J., Clarke, M. F., and Weissman, I. L. (2001). Stem cells, cancer, and cancer stem cells. *Nature* **414**, 105–111.

Riccio, O., van Gijn, M. E., Bezdek, A. C., Pellegrinet, L., van Es, J. H., Zimber-Strobl, U., Strobl, L. J., Honjo, T., Clevers, H., and Radtke, F. (2008). Loss of intestinal crypt progenitor cells owing to inactivation of both Notch1 and Notch2 is accompanied by derepression of CDK inhibitors p27Kip1 and p57Kip2. *EMBO rep.* **9**, 377–383.

Rida, P. C., Le Minh, N., and Jiang, Y. J. (2004). A Notch feeling of somite segmentation and beyond. *Dev. Biol.* **265**, 2–22.

Robert-Moreno, A., Espinosa, L., de la Pompa, J. L., and Bigas, A. (2005). RBPjkappa-dependent Notch function regulates Gata2 and is essential for the formation of intra-embryonic hematopoietic cells. *Development* **132**, 1117–1126.

Robert-Moreno, A., Guiu, J., Ruiz-Herguido, C., Lopez, M. E., Ingles-Esteve, J., Riera, L., Tipping, A., Enver, T., Dzierzak, E., Gridley, T., *et al.* (2008). Impaired embryonic haematopoiesis yet normal arterial development in the absence of the Notch ligand Jagged1. *EMBO J.* **27**, 1886–1895.

Rossi, D. J., Bryder, D., Seita, J., Nussenzweig, A., Hoeijmakers, J., and Weissman, I. L. (2007). Deficiencies in DNA damage repair limit the function of haematopoietic stem cells with age. *Nature* **447**, 725–729.

Rossi, D. J., Jamieson, C. H., and Weissman, I. L. (2008). Stems cells and the pathways to aging and cancer. *Cell* **132**, 681–696.

Sakamoto, M., Hirata, H., Ohtsuka, T., Bessho, Y., and Kageyama, R. (2003). The basic helix-loop-helix genes Hesr1/Hey1 and Hesr2/Hey2 regulate maintenance of neural precursor cells in the brain. *J. Biol. Chem.* **278**, 44808–44815.

Samokhvalov, I. M., Samokhvalova, N. I., and Nishikawa, S. (2007). Cell tracing shows the contribution of the yolk sac to adult haematopoiesis. *Nature* **446**, 1056–1061.

Schofield, R. (1978). The relationship between the spleen colony-forming cell and the haemopoietic stem cell. *Blood Cells* **4**, 7–25.

Schroeder, T., Fraser, S. T., Ogawa, M., Nishikawa, S., Oka, C., Bornkamm, G. W., Nishikawa, S., Honjo, T., and Just, U. (2003a). Recombination signal sequence-binding protein Jkappa alters mesodermal cell fate decisions by suppressing cardiomyogenesis. *Proc. Natl. Acad. Sci. U.S.A* **100**, 4018–4023.

Schroeder, T., and Just, U. (2000a). mNotch1 signaling reduces proliferation of myeloid progenitor cells by altering cell-cycle kinetics. *Exp. Hematol.* **28**, 1206–1213.

Schroeder, T., and Just, U. (2000b). Notch signalling via RBP-J promotes myeloid differentiation. *EMBO J.* **19**, 2558–2568.

Schroeder, T., Kohlhof, H., Rieber, N., and Just, U. (2003b). Notch signaling induces multilineage myeloid differentiation and up-regulates PU.1 expression. *J. Immunol.* **170**, 5538–5548.

Schroeder, T., Meier-Stiegen, F., Schwanbeck, R., Eilken, H., Nishikawa, S., Hasler, R., Schreiber, S., Bornkamm, G. W., Nishikawa, S., and Just, U. (2006). Activated Notch1 alters differentiation of embryonic stem cells into mesodermal cell lineages at multiple stages of development. *Mech. dev.* **123**, 570–579.

Schuster-Gossler, K., Cordes, R., and Gossler, A. (2007). Premature myogenic differentiation and depletion of progenitor cells cause severe muscle hypotrophy in Delta1 mutants. *Proc. Natl. Acad. Sci. U.S.A* **104**, 537–542.

Scoville, D. H., Sato, T., He, X. C., and Li, L. (2008). Current view: intestinal stem cells and signaling. *Gastroenterology* **134**, 849–864.

Searfoss, G. H., Jordan, W. H., Calligaro, D. O., Galbreath, E. J., Schirtzinger, L. M., Berridge, B. R., Gao, H., Higgins, M. A., May, P. C., and Ryan, T. P. (2003). Adipsin, a biomarker of gastrointestinal toxicity mediated by a functional gamma-secretase inhibitor. *J. Biol. Chem.* **278**, 46107–46116.

Serneels, L., Dejaegere, T., Craessaerts, K., Horre, K., Jorissen, E., Tousseyn, T., Hebert, S., Coolen, M., Martens, G., Zwijsen, A., *et al.* (2005). Differential contribution of the three Aph1 genes to gamma-secretase activity in vivo. *Proc. Natl. Acad. Sci. U.S.A* **102**, 1719–1724.

Shackleton, M., Vaillant, F., Simpson, K. J., Stingl, J., Smyth, G. K., Asselin-Labat, M. L., Wu, L., Lindeman, G. J., and Visvader, J. E. (2006). Generation of a functional mammary gland from a single stem cell. *Nature* **439**, 84–88.

Six, E. M., Ndiaye, D., Sauer, G., Laabi, Y., Athman, R., Cumano, A., Brou, C., Israel, A., and Logeat, F. (2004). The notch ligand Delta1 recruits Dlg1 at cell-cell contacts and regulates cell migration. *J. Biol. Chem.* **279**, 55818–55826.

Song, X., Call, G. B., Kirilly, D., and Xie, T. (2007). Notch signaling controls germline stem cell niche formation in the Drosophila ovary. *Development* **134**, 1071–1080.

Song, X., Wong, M. D., Kawase, E., Xi, R., Ding, B. C., McCarthy, J. J., and Xie, T. (2004). Bmp signals from niche cells directly repress transcription of a differentiation-promoting gene, bag of marbles, in germline stem cells in the Drosophila ovary. *Development* **131**, 1353–1364.

Song, X., Zhu, C. H., Doan, C., and Xie, T. (2002). Germline stem cells anchored by adherens junctions in the Drosophila ovary niches. *Science* **296**, 1855–1857.

Stingl, J., Eirew, P., Ricketson, I., Shackleton, M., Vaillant, F., Choi, D., Li, H. I., and Eaves, C. J. (2006). Purification and unique properties of mammary epithelial stem cells. *Nature* **439**, 993–997.

Sugiyama, T., Kohara, H., Noda, M., and Nagasawa, T. (2006). Maintenance of the hematopoietic stem cell pool by CXCL12-CXCR4 chemokine signaling in bone marrow stromal cell niches. *Immunity* **25**, 977–988.

Suh, H., Deng, W., and Gage, F. H. (2009). Signaling in adult neurogenesis. *Annu. rev. cell dev. biol.* **25**, 253–275.

Sun, D., Li, H., and Zolkiewska, A. (2008). The role of Delta-like 1 shedding in muscle cell self-renewal and differentiation. *J. Cell. Sci.* **121**, 3815–3823.

Takahashi, K., Tanabe, K., Ohnuki, M., Narita, M., Ichisaka, T., Tomoda, K., and Yamanaka, S. (2007). Induction of pluripotent stem cells from adult human fibroblasts by defined factors. *Cell* **131**, 861–872.

Takahashi, K., and Yamanaka, S. (2006). Induction of pluripotent stem cells from mouse embryonic and adult fibroblast cultures by defined factors. *Cell* **126**, 663–676.

Tammam, J., Ware, C., Efferson, C., O'Neil, J., Rao, S., Qu, X., Gorenstein, J., Angagaw, M., Kim, H., Kenific, C., *et al.* (2009). Down-regulation of the Notch pathway mediated by a gamma-secretase inhibitor induces anti-tumour effects in mouse models of T-cell leukaemia. *Br. J. Pharmacol.* **158**, 1183–1195.

Tanigaki, K., Han, H., Yamamoto, N., Tashiro, K., Ikegawa, M., Kuroda, K., Suzuki, A., Nakano, T., and Honjo, T. (2002). Notch-RBP-J signaling is involved in cell fate determination of marginal zone B cells. *Nat. Immunol.* **3**, 443–450.

Thomson, J. A., Itskovitz-Eldor, J., Shapiro, S. S., Waknitz, M. A., Swiergiel, J. J., Marshall, V. S., and Jones, J. M. (1998). Embryonic stem cell lines derived from human blastocysts. *Science* **282**, 1145–1147.

Tournoy, J., Bossuyt, X., Snellinx, A., Regent, M., Garmyn, M., Serneels, L., Saftig, P., Craessaerts, K., De Strooper, B., and Hartmann, D. (2004). Partial loss of presenilins causes seborrheic keratosis and autoimmune disease in mice. *Hum. mol. genet.* **13**, 1321–1331.

Tran, J., Brenner, T. J., and DiNardo, S. (2000). Somatic control over the germline stem cell lineage during Drosophila spermatogenesis. *Nature* **407**, 754–757.

Turpen, J. B., Kelley, C. M., Mead, P. E., and Zon, L. I. (1997). Bipotential primitive-definitive hematopoietic progenitors in the vertebrate embryo. *Immunity* **7**, 325–334.

van der Flier, L. G., and Clevers, H. (2009). Stem cells, self-renewal, and differentiation in the intestinal epithelium. *Annu. rev. physiol.* **71**, 241–260.

Van der Flier, L. G., Sabates-Bellver, J., Oving, I., Haegebarth, A., De Palo, M., Anti, M., Van Gijn, M. E., Suijkerbuijk, S., Van de Wetering, M., Marra, G., *et al.* (2007). The Intestinal Wnt/TCF Signature. *Gastroenterology* **132**, 628–632.

van Es, J. H., and Clevers, H. (2005). Notch and Wnt inhibitors as potential new drugs for intestinal neoplastic disease. *Trends mol. med.* **11**, 496–502.

van Es, J. H., van Gijn, M. E., Riccio, O., van den Born, M., Vooijs, M., Begthel, H., Cozijnsen, M., Robine, S., Winton, D. J., Radtke, F., *et al.* (2005). Notch/gamma-secretase inhibition turns proliferative cells in intestinal crypts and adenomas into goblet cells. *Nature* **435**, 959–963.

Varga, G., Kiss, J., Varkonyi, J., Vas, V., Farkas, P., Paloczi, K., and Uher, F. (2007). Inappropriate Notch activity and limited mesenchymal stem cell plasticity in the bone marrow of patients with myelodysplastic syndromes. *Pathol. Oncol. Res.* **13**, 311–319.

Varnum-Finney, B., Purton, L. E., Yu, M., Brashem-Stein, C., Flowers, D., Staats, S., Moore, K. A., Le Roux, I., Mann, R., Gray, G., *et al.* (1998). The Notch ligand, Jagged-1, influences the development of primitive hematopoietic precursor cells. *Blood* **91**, 4084–4091.

Varnum-Finney, B., Xu, L., Brashem-Stein, C., Nourigat, C., Flowers, D., Bakkour, S., Pear, W. S., and Bernstein, I. D. (2000). Pluripotent, cytokine-dependent, hematopoietic stem cells are immortalized by constitutive Notch1 signaling. *Nat. Med.* **6**, 1278–1281.

Vas, V., Szilagyi, L., Paloczi, K., and Uher, F. (2004). Soluble Jagged-1 is able to inhibit the function of its multivalent form to induce hematopoietic stem cell self-renewal in a surrogate in vitro assay. *J. Leukoc. Biol.* **75**, 714–720.

Vasyutina, E., Lenhard, D. C., Wende, H., Erdmann, B., Epstein, J. A., and Birchmeier, C. (2007). RBP-J (Rbpsuh) is essential to maintain muscle progenitor cells and to generate satellite cells. *Proc. Natl. Acad. Sci. U.S.A* **104**, 4443–4448.

Vercauteren, S. M., and Sutherland, H. J. (2004). Constitutively active Notch4 promotes early human hematopoietic progenitor cell maintenance while inhibiting differentiation and causes lymphoid abnormalities in vivo. *Blood* **104**, 2315–2322.

Visnjic, D., Kalajzic, Z., Rowe, D. W., Katavic, V., Lorenzo, J., and Aguila, H. L. (2004). Hematopoiesis is severely altered in mice with an induced osteoblast deficiency. *Blood* **103**, 3258–3264.

Vooijs, M., Ong, C. T., Hadland, B., Huppert, S., Liu, Z., Korving, J., van den Born, M., Stappenbeck, T., Wu, Y., Clevers, H., *et al.* (2007). Mapping the consequence of Notch1 proteolysis in vivo with NIP-CRE. *Development* **134**, 535–544.

Wagers, A. J., and Conboy, I. M. (2005). Cellular and molecular signatures of muscle regeneration: current concepts and controversies in adult myogenesis. *Cell* **122**, 659–667.

Wang, J. C., Warner, J. K., Erdmann, N., Lansdorp, P. M., Harrington, L., and Dick, J. E. (2005). Dissociation of telomerase activity and telomere length maintenance in primitive human hematopoietic cells. *Proc. Natl. Acad. Sci. U.S.A* **102**, 14398–14403.

Wang, P., and Hou, S. X. (2010). Regulation of intestinal stem cells in mammals and Drosophila. *J. Cell. Physiol.* **222**, 33–37.

Wang, S., Sdrulla, A. D., DiSibio, G., Bush, G., Nofziger, D., Hicks, C., Weinmaster, G., and Barres, B. A. (1998). Notch receptor activation inhibits oligodendrocyte differentiation. *Neuron* **21**, 63–75.

Wang, S. F., Aoki, M., Nakashima, Y., Shinozuka, Y., Tanaka, H., Taniwaki, M., Hattori, M., and Minato, N. (2008). Development of Notch-dependent T-cell leukemia by deregulated Rap1 signaling. *Blood* **111**, 2878–2886.

Watt, F. M., Estrach, S., and Ambler, C. A. (2008). Epidermal Notch signalling: differentiation, cancer and adhesion. *Curr. opin. cell biol.* **20**, 171–179.

Weber, J. M., Forsythe, S. R., Christianson, C. A., Frisch, B. J., Gigliotti, B. J., Jordan, C. T., Milner, L. A., Guzman, M. L., and Calvi, L. M. (2006). Parathyroid hormone stimulates expression of the Notch ligand Jagged1 in osteoblastic cells. *Bone* **39**, 485–493.

Weissman, I. L., and Shizuru, J. A. (2008). The origins of the identification and isolation of hematopoietic stem cells, and their capability to induce donor-specific transplantation tolerance and treat autoimmune diseases. *Blood* **112**, 3543–3553.

Whitfield, J. F. (2005). Parathyroid hormone (PTH) and hematopoiesis: new support for some old observations. *J. Cell. Biochem.* **96**, 278–284.

Wilson, A., Laurenti, E., Oser, G., van der Wath, R. C., Blanco-Bose, W., Jaworski, M., Offner, S., Dunant, C. F., Eshkind, L., Bockamp, E., *et al.* (2008). Hematopoietic stem cells reversibly switch from dormancy to self-renewal during homeostasis and repair. *Cell* **135**, 1118–1129.

Wilson, A., Laurenti, E., and Trumpp, A. (2009). Balancing dormant and self-renewing hematopoietic stem cells. *Curr. opin. genet. dev.* **19**, 461–468.

Wilson, A., MacDonald, H. R., and Radtke, F. (2001). Notch 1-deficient common lymphoid precursors adopt a B cell fate in the thymus. *J. Exp. Med.* **194**, 1003–1012.

Wong, G. T., Manfra, D., Poulet, F. M., Zhang, Q., Josien, H., Bara, T., Engstrom, L., Pinzon-Ortiz, M., Fine, J. S., Lee, H. J., *et al.* (2004). Chronic treatment with the gamma-secretase inhibitor LY-411,575 inhibits beta-amyloid peptide production and alters lymphopoiesis and intestinal cell differentiation. *J. Biol. Chem.* **279**, 12876–12882.

Wouters, B. J., Jorda, M. A., Keeshan, K., Louwers, I., Erpelinck-Verschueren, C. A., Tielemans, D., Langerak, A. W., He, Y., Yashiro-Ohtani, Y., Zhang, P., *et al.* (2007). Distinct gene expression profiles of acute myeloid/T-lymphoid leukemia with silenced CEBPA and mutations in NOTCH1. *Blood* **110**, 3706–3714.

Wright, D. E., Wagers, A. J., Gulati, A. P., Johnson, F. L., and Weissman, I. L. (2001). Physiological migration of hematopoietic stem and progenitor cells. *Science* **294**, 1933–1936.

Wu, M., Kwon, H. Y., Rattis, F., Blum, J., Zhao, C., Ashkenazi, R., Jackson, T. L., Gaiano, N., Oliver, T., and Reya, T. (2007). Imaging hematopoietic precursor division in real time. *Cell Stem Cell* **1**, 541–554.

Xie, T., and Spradling, A. C. (1998). decapentaplegic is essential for the maintenance and division of germline stem cells in the Drosophila ovary. *Cell* **94**, 251–260.

Xie, Y., Yin, T., Wiegraebe, W., He, X. C., Miller, D., Stark, D., Perko, K., Alexander, R., Schwartz, J., Grindley, J. C., *et al.* (2009). Detection of functional haematopoietic stem cell niche using real-time imaging. *Nature* **457**, 97–101.

Xu, C., Jiang, J., Sottile, V., McWhir, J., Lebkowski, J., and Carpenter, M. K. (2004). Immortalized fibroblast-like cells derived from human embryonic stem cells support undifferentiated cell growth. *Stem cells (Dayton, Ohio)* **22**, 972–980.

Yang, Q., Bermingham, N. A., Finegold, M. J., and Zoghbi, H. Y. (2001). Requirement of Math1 for secretory cell lineage commitment in the mouse intestine. *Science* **294**, 2155–2158.

Yu, J., Vodyanik, M. A., Smuga-Otto, K., Antosiewicz-Bourget, J., Frane, J. L., Tian, S., Nie, J., Jonsdottir, G. A., Ruotti, V., Stewart, R., *et al.* (2007). Induced pluripotent stem cell lines derived from human somatic cells. *Science* **318**, 1917–1920.

Zhang, J., Niu, C., Ye, L., Huang, H., He, X., Tong, W. G., Ross, J., Haug, J., Johnson, T., Feng, J. Q., *et al.* (2003). Identification of the haematopoietic stem cell niche and control of the niche size. *Nature* **425**, 836–841.

Zhang, Z., Hener, P., Frossard, N., Kato, S., Metzger, D., Li, M., and Chambon, P. (2009). Thymic stromal lymphopoietin overproduced by keratinocytes in mouse skin aggravates experimental asthma. *Proc. Natl. Acad. Sci. U.S.A.* **106**, 1536–1541.

CHAPTER THIRTEEN

Notch Signaling in Solid Tumors

Ute Koch *and* Freddy Radtke

Contents

Abstract

In recent years a substantial body of evidence derived from not only preclinical but also clinical studies has accumulated in support of Notch signaling playing important oncogenic roles in several types of cancer. The finding that activating Notch mutations are frequently found in patients suffering from acute lymphoblastic leukemia is one of the best examples for a critical role of Notch signaling in cancer, a fact that motivated many researchers and clinicians to study the role of Notch also in solid tumors. Hence Notch signaling has gained increasing attention as a potential therapeutic target. In this book chapter we would like to discuss our current knowledge of Notch signaling within different types of solid cancers as well as advantages and disadvantages of potential new therapies that try to target the oncogenic properties of Notch signaling.

Ecole Polytechnique Fédérale de Lausanne (EPFL), Swiss Institute for Experimental Cancer Research (ISREC), Lausanne, Switzerland

Current Topics in Developmental Biology, Volume 92
ISSN 0070-2153, DOI 10.1016/S0070-2153(10)92013-9

1. Introduction

Thomas Hunt Morgan (1866–1945), an American geneticist and embryologist, was awarded the Nobel Prize in Physiology and Medicine in 1933 for his work on genetics using *Drosophila melanogaster* as a model system. In 1917, he and his colleagues described a strain of fruit flies with notches at the margin of their wing blades (Morgan, 1917). This notched wing phenotype is the result of a partial loss of function of the *Drosophila Notch* gene, which was cloned in the mid-1980s (Kidd *et al.*, 1986; Wharton *et al.*, 1985). It turned out that the *Notch* gene encodes for a transmembrane bound receptor that is part of a highly conserved signaling cascade that regulates cell-to cell-communication and is involved in important functions in higher organisms, both during development and tissue homeostasis. The first human homolog of the *Drososphila Notch* gene was identified by Jeff Sklar and colleagues in the late 1980s (Reynolds *et al.*, 1987) and beginning of the 1990s (Ellisen *et al.*, 1991) by cloning and sequencing of a chromosomal translocation within a small number of T cell lymphoblastic leukemia patient samples. In this original study, 4 out of 40 screened T cell leukemia patient samples carried a t(7;9)(q34;q34.3) chromosomal translocation, which results in the expression of an N-terminal truncated, dominant active, and ligand-independent human NOTCH1 receptor, which was named TAN1 for translocation-associated Notch homolog. This discovery represents the first link of the Notch cascade to human cancer. Nevertheless, it took additional 5 years to proof that TAN1 is indeed causative for disease development, which was shown by murine bone marrow (BM) reconstitution experiments. Mice transplanted with BM progenitors expressing TAN1 developed T cell neoplasms as early as 2 weeks after BM transplantation (Pear *et al.*, 1996). Although the association of Notch and T cell leukemia has thereafter been widely demonstrated in mouse models, the rare frequency (<1%) of the t(7;9) translocation in human T cell acute lymphoblstic leukemia (T-ALL) patients questioned the clinical importance of these findings. This changed dramatically in 2004, when Aster and colleagues were able to show that approximately 50% of all T-ALL patients had activating mutations within the human *NOTCH1* gene (Weng *et al.*, 2004). These milestones of Notch signaling in T cell leukemia paved the way for many researches to investigate a similar role of Notch in solid tumors. In the meantime developmental biologists have identified four mammalian Notch family members (Notch1–4), which are implicated in a broad range of functions during embryonic and post-natal development and different forms of cancer. In self-renewing tissues of vertebrates and during tumorigenesis, inhibition of differentiation, promotion of proliferation, lineage specification at developmental branch points and induction of differentiation are relevant functions of Notch signaling. As a result, there is growing interest to therapeutically block or activate Notch signaling in different contexts.

Before discussing examples of Notch signaling within solid tumors we briefly summarize the Notch cascade, to allow a better understanding of the different therapeutic strategies that are currently been exploited. The molecular and biochemical details of Notch signaling have recently been covered by excellent reviews (Gordon *et al.*, 2008; Kopan and Ilagan, 2009). In brief, mammals posses four Notch receptors (Notch1–4) that are bound by five ligands (Delta-like1, 3, and 4 and Jagged 1 and 2) (Bray, 2006). Newly synthesized Notch receptors are proteolytically cleaved in the Golgi (at site S1) during their transport to the cell surface by a furin-like protease. This cleavage generates a heterodimeric receptor consisting of an extracellular subunit (N^{EC}) that is non-covalently linked to a second subunit containing the extracellular heterodimerization domain and the transmembrane domain followed by the cytoplasmic region of the Notch receptor (N^{TM}). The extracellular part of the receptors contains between 29 and 36 epidermal growth factor-like repeats involved in ligand binding, followed by three cysteine-rich LIN12 repeats that prevent ligand-independent activation and a hydrophobic stretch of amino acids mediating heterodimerization between N^{EC} and N^{TM}. The cytoplasmic tail of the receptor harbors multiple conserved elements including nuclear localization signals, as well as protein–protein interaction and transactivation domains (Fig. 13.1A). Notch signaling is initiated by ligand-receptor interaction between neighboring cells, leading to two successive proteolytic cleavages of the receptor. The first is mediated by metalloproteases of the ADAM family, which cleave the receptors 12-13 amino acids external to the transmembrane domain (at site S2). The shedded extracellular domain is endocytosed by the ligand-expressing cell, a process that is dependent on monoubiquitinylation of the cytoplasmic tail of the ligands by E3-ubiquitin ligases of the mind bomb and neuralized family. Ligand binding to N^{EC} presumably induces a conformational change within the Notch receptors to expose the S2 cleavage site for proteolysis. After shedding of the extracellular domain, a second cleavage within the transmembrane domain (at site S3) is mediated by the γ-secretase activity of a multi-protein complex. This liberates the intracellular domain of Notch receptors (NICD), which subsequently traffics to the nucleus and heterodimerizes with the DNA binding transcription factor CSL in order to form a short-lived nuclear transcription complex. The transcription factor CSL is also known as CBF-1 in humans, Suppressor of hairless in *Drosophila*, Lag in Caenorhabditis elegans, and RBP-J in the mouse. Once bound to CSL, NICD recruits other coactivators including mastermind proteins (MAML1-3), which in turn recruit the MED8 mediator transcription activation complex in order to induce transcriptional expression of downstream target genes (Fig. 13.1B). Certain Notch target genes, including members of the *enhancer-of-split (Hes)* and *Hairy-related (Hrt* or *Hey)* families, are recurrently found in many tissues, while others seem to be tissue-specific target genes. In fact, recent gene expression studies combined with chromatin immunoprecipitation (ChIP) arrays revealed the existence of a large number of genes that can directly be regulated by Notch (Krejci *et al.*, 2009; Palomero *et al.*, 2006; Weng *et al.*, 2006). In the context of cancer the challenge will be to identify and distinguish driver target genes from the passengers. Moreover, Notch signaling can be regulated at multiple levels, including cell type specific and spatial expression of ligands and Notch receptors, or through glycosylation of the Notch extracellular domain by Fringe glycosyltransferase which influence the affinity of Notch receptors to certain ligands (Haines and Irvine, 2003; Haltiwanger and Stanley, 2002). An additional level of regulation is to ensure that a Notch signal is short lived, which is achieved by the PEST domain at the very C-terminus of the Notch receptors being responsible for rapid ubiquitin-mediated degradation (O'Neil *et al.*, 2007; Thompson *et al.*, 2007).

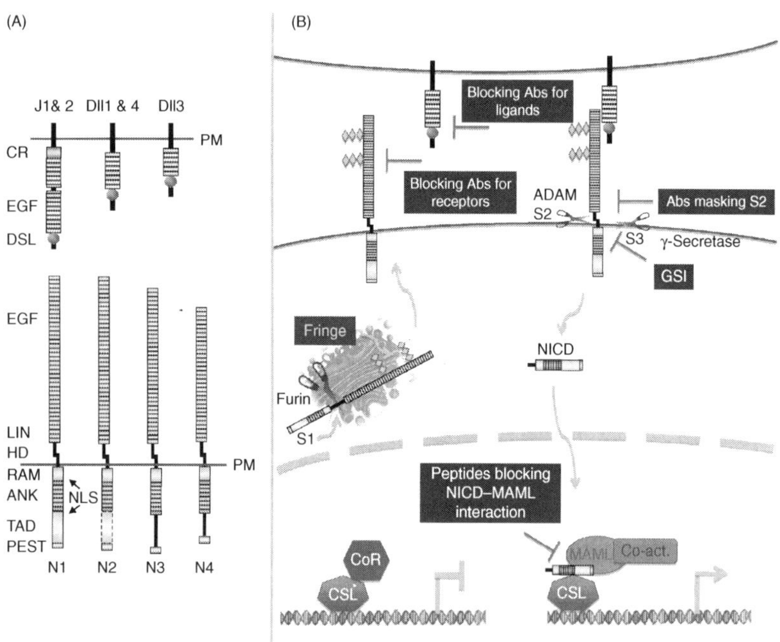

Figure 13.1 (*Continued*)

2. Notch and Breast Cancer

Breast cancer is one of the most common malignancies in women with a cumulative lifetime risk of developing the disease affecting one in 8 to one in 12 women over their lifetime. The present 5-year survival rate for women with breast cancer is 86%. The present 10-year survival rate is 76%. These rates consist of women at all stages or levels of harshness of breast cancer. The main risk factors are female sex, increasing age, and a genetic predisposition. Breast cancer most commonly arises in the ductal epithelium. DCIS or ductal carcinoma in-situ, and infiltrating ductal carcinoma accounts for up to 80% of all breast cancers. A major focus of breast cancer research has been to understand the genetic basis for the malignant transformation of breast epithelium.

Fig. 13.1 Therapeutic possibilities to interfere with Notch signaling. (A) *Notch receptor and ligands.* To date five conventional Notch ligands are known: Jagged1 (J1), Jaggged2 (J2), Delta-like1 (Dll1), Delta-like3 (Dll3), and Delta-like4 (Dll4). A common structural feature of all ligands is an aminoterminal domain called DSL (Delta, Serrate, and Lag-2) involved in receptor binding followed by EGF-like repeats. A cysteine rich domain (CR) is located downstream of the EGF-like repeats of J1 and J2 close to the plasma membrane (PM). Vertebrates have four Notch receptors (Notch1–Notch4; N1–N4). The extracellular domain of the receptors contains EGF-like repeats (36 in N1 and N2, 34 in N3, and 29 in N4) followed by three cysteine-rich LIN domains which prevent ligand independent activation and the heterodimerization domain (HD). The cytoplasmic domain contains a RAM domain followed by six ankyrin repeats (ANK) which bind to the CSL transcription factor, two nuclear localization signals (NLS), a transactivation domain (present in N1 and N2), and a PEST sequence involved in regulating protein stability. (B) *Notch signaling and therapeutic inhibitory strategies.* Notch proteins are synthesized as single precursor proteins, which are cleaved in the Golgi by a Furin-like convertase at site S1. Cleavage at S1 generates two subunits held together non-covalently by the N- and C-terminal subunits of the heterodimerization domains (HD). EGF-like repeats are glycosylated by Fringe proteins in the Golgi before receptors are transported to the cell surface. Notch signaling is initiated by ligand receptor interaction, which induces a second cleavage at site S2 (close to the transmembrane domain) mediated by ADAM-type metalloproteases followed by a third cleavage at S3 within the transmembrane domain mediated by the γ-secretase activity of a multi-protein complex containing presenilins. This last proteolytic cleavage liberates the cytoplasmic domain of Notch receptors (NICD), which translocates to the nucleus and binds to the transcription factor CSL (CBF1, Suppressor of hairless and Lag-1), converting it from a transcriptional repressor into a transcriptional activator by recruiting coactivators including mastermind like proteins (MAML). Multiple strategies to therapeutically interfere with Notch signaling are currently developed and tested in preclinical assays and clinical trials. Inhibitory strategies are highlighted in orange and include inhibitory antibodies against individual Notch receptor and ligands with the aim to block specific receptor–ligand interactions; receptor specific inhibitory antibodies masking the S2 cleavage site thereby blocking ADAM-protease-mediated cleavage of the receptors; divers forms of γ-secretase inhibitors (GSI) inhibiting the S3 cleavage and thereby activation of all Notch receptors and inhibitory peptides blocking the formation of a functional NICD–MAML transcription complex.

Historically, the first evidence describing a link between aberrant Notch signaling in solid tumors came from the observation in animal studies that integration of the mouse mammary tumor virus (MMTV) into the *Notch4* gene leads to the formation of mammary tumors (Gallahan *et al.*, 1987). MMTV integration into the *Notch4* locus results in the LTR-driven expression of transcripts encoding a truncated Notch4 mRNA species, named int3 (Uyttendaele *et al.*, 1996) that represents a gain-of-function mutation (Gallahan and Callahan, 1997; Kordon *et al.*, 1995; Raafat *et al.*, 2004; Robbins *et al.*, 1992). The expression of the truncated dominant active form of the *Notch4/Int3* gene (N4ICD) under control of mammary-specific regulatory elements, either the MMTV long terminal repeat or the whey acidic protein (WAP), in transgenic mice confirmed that activation of Notch signaling leads to the establishment of mammary tumors in 100% of female mice (Gallahan *et al.*, 1996; Jhappan *et al.*, 1992) (Fig. 13.2A). Microarray studies were performed on the Notch4/int3 induced mammary adenocarcinomas and revealed high levels of c-kit expression. Treatment of the tumor-bearing mice with the tyrosine kinase inhibitor Gleevec, targeting c-kit, PDGFRs, and c-Abl resulted in decreased proliferation and angiogenesis and the induction of apoptosis. Small interfering RNA (siRNA) to knock down c-kit, PDGFRs, and c-Abl revealed an oncogenic role for c-kit and PDGFR tyrosine kinases in the context of Int3 signaling (Raafat *et al.*, 2007). In addition, the expression of the N4ICD under the control of the *WAP* gene promoter resulted in an initial block of epithelial cell proliferation and differentiation during pregnancy (Gallahan *et al.*, 1996). This observation suggested a genuine role for Notch4 during normal breast development. However mice, which carry a gene-targeted homozygous disruption of the *Notch4* gene, are viable and fertile. They also revealed no defects upon histological and morphological analyses of mammary glands isolated from virgin, pregnant, and lactating females (Krebs *et al.*, 2000), suggesting that either Notch4 is dispensable for normal mammary development or that there is redundancy between the other Notch family members filling in the void in its absence. The role of Notch signaling during normal mammary gland development was assessed using Cre-Lox-mediated recombination to inactivate genes of the Notch signaling cascade using two independent approaches that address distinct aspects of the Notch signaling pathway (Fig. 13.2B). The conditional inactivation of *RBP-J*, the common downstream partner of all four Notch receptors, or *Pofut1*, a fucosyltransferase essential for efficient Notch receptor-ligand binding, revealed only an overt phenotype during pregnancy using the MMTV-Cre deleter line. The study by Buono *et al.* (2006) demonstrated that the canonical Notch signaling pathway through RBP-J is required for the expansion and maintenance of alveolar luminal cells during pregnancy and the suppression of basal cell proliferation. In the absence of RBP-J, luminal cells acquire basal characteristics and proliferate extensively, which

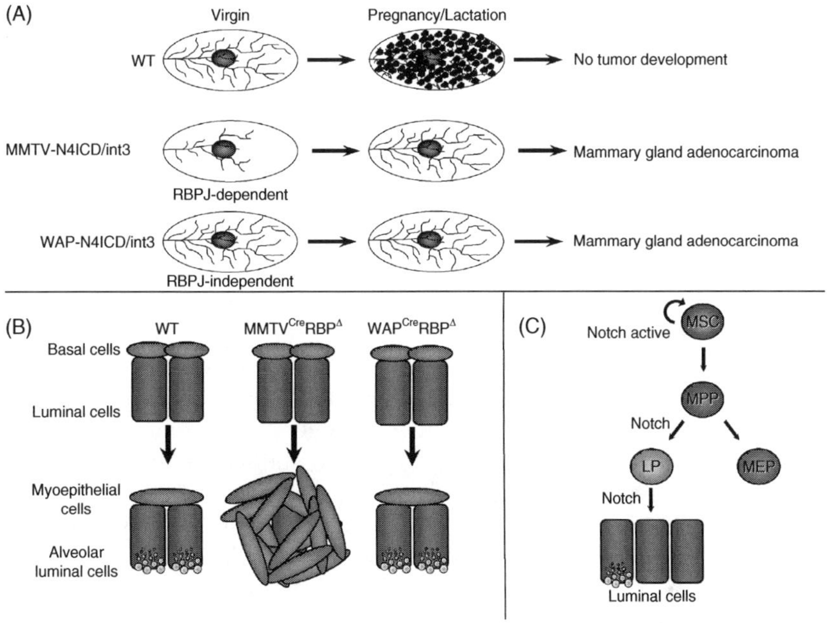

Figure 13.2 (A) *Constitutive Notch activation leads to mammary gland tumors.* The expression of the truncated dominant active form of the Notch4/int gene under the control of either the MMTV long terminal repeat or the whey acidic protein (WAP) in transgenic mice will lead to the establishment of mammary tumors with 100% penetrance. The ductal tree formation in MMTV-N4ICD/int3 mice is perturbed whereas the ductal tree in WAP-N4ICD/int3 mice develops normally. The deletion of RBP-J in these mice overexpression N4ICD showed that the development of the ductal tree structure in MMTV-Cre mice is RBP-J dependent, whereas it is independent in WAP-Cre mice. (B) *Role of RBP-J in mammary development during pregnancy.* The use of two different Cre deleter lines indicates different roles for RBP-J during pregnancy. Buono *et al.* (2006) used the MMTV-Cre deleter line and demonstrated that the canonical Notch signaling pathway through RBP-J is essential for the expansion and maintenance of alveolar luminal cells and the suppression of basal cell proliferation. Loss of RBP-J results in a fate switch toward the basal cell lineage and excessive proliferation of these cells. The study performed by Rafaat *et al.* (2009) employed the WAP-Cre deleter strain, which deletes around day 15 of pregnancy, whereas the MMTV-Cre line deletes much earlier during mammary gland development. In these mice mammary gland development occurs normally, which would support the notion that endogenous Notch/RBP-J signaling is required at an earlier time point to properly direct the development of alveolar/lobular structures. The $WAP^{Cre}RBP^{\Delta}$ mice do not develop any obvious phenotype. (C) *Notch regulates cell fate decisions in the mammary epithelium.* Mammary stem cells (MSC) give rise to MSC and multi-potent progenitors (MPP), which then develop along either the luminal (LP—luminal progenitor) or the myoepithelial (MEP—myoepithelial progenitor) lineage. A reduction in Notch signaling in the MSC will promote enhanced generation of MSC and MEP.

results in a disorganized compilation of myoepithelial-like cells. In addition, transplantation of mammary tissue from *RBP-J* deficient mice revealed that the development of the ductal tree was normal, suggesting that RBP-J is dispensable for the establishment of this structure (Fig. 13.2A). Whereas Buono *et al.* used the MMTV LTR-Cre deleter line to conditionally ablate *RBP-J* in the above-mentioned study, a more recent publication employed the WAP-Cre line to delete *RBP-J* in the mammary gland in *WAP-N4ICD/Int3* mice (Raafat *et al.*, 2009). While expression from the MMTV LTR begins early during mammary gland development, expression from the *WAP* promoter peaks around day 15 of pregnancy. This could explain the findings of Raafat *et al.* that mammary gland development occurs normally in *WAP-Cre RBP-J*$^{lox/lox}$ females arguing that endogenous Notch/RBP-J signaling is required at an earlier time point to promote the development of alveolar/lobular structures. Suggesting that by the time the WAP promoter is active at day 15 of pregnancy, endogenous Notch/RBP-J signaling was no longer required for mammary gland development. Although the deletion of *RBP-J* in the mammary gland suppresses the negative effect of N4ICD/Int3 signaling on alveolar development, there is little effect on mammary tumorigenesis. All *WAP-N4ICD/Int3 RBP-J* knockout female breeders develop mammary tumors that could be transplanted in nude mice, although with a longer latency than control animals. Taken together these findings imply that constitutive N4ICD/Int3 signaling blocks mammary alveolar development through an RBP-J dependent pathway, and that N4ICD/Int3-induced mammary tumorigenesis/growth occurs as a consequence of RBP-J independent N4ICD/Int3 signaling.

Besides MMTV-induced *Notch4/Int3* truncation found in mammary tumors, similar gain-of-function mutations truncating Notch1 were found in a study inducing insertional mutagenesis by MMTV integration. These accelerated the development of mammary tumors induced by the *MMTV-ErbB2* transgene (Dievart *et al.*, 1999) albeit at lower frequencies than the Notch4/Int3 induced tumors. In transgenic mice carrying an *MMTV-N1ICD* construct using human *NOTCH1* cDNA, all female mice developed by their third pregnancy lactation-dependent papillary tumors that were non-invasive and regressed upon gland involution (Klinakis *et al.*, 2006). However, after multiple pregnancies, non-regressing mammary carcinomas developed from remnants of N1ICD induced neoplasms, most likely through the occurrence of the accumulation of secondary tumorigenic events. These tumors revealed the same papillary histological architecture as their regressing counterparts, but were invasive (Klinakis *et al.*, 2006). In addition, using comparative microarray analysis N1ICD and c-myc induced tumors revealed a high degree of profile similarity and more conclusively it was shown that *c-myc* is a direct transcriptional target of aberrant Notch1 signaling in mammary tumors. Unexpectedly, when transgenic mice expressing a truncated mouse *Notch1* cDNA were generated (Hu *et al.*, 2006), ductal and lobuloalveolar mammary gland development was impaired in females of one line, resulting in a non-lactational phenotype. However, both virgin and parous animals eventually developed papillary carcinomas with long latency. Although the difference between MMTV-driven human and mouse *N1ICD* cDNAs remains unexplained, the tumorigenic process itself in the mammary glands seems identical and, in both cases, the invasive cancers are morphologically indistinguishable.

The significant insights gained from the mouse studies described above lead to the question whether aberrant Notch signaling and mammary tumorigenesis show any relevant parallel to human breast cancer development. Over the past years correlative evidence has accumulated implicating Notch signaling in human breast cancer. Activated forms of NOTCH1 and NOTCH4 have been identified in several human breast cancer cell lines (Imatani and Callahan, 2000; Stylianou *et al.*, 2006) and NOTCH3 has been shown to play an important role in the proliferation of ErbB2-negative breast tumor cell lines (Yamaguchi *et al.*, 2008). The first clue that Notch might be aberrantly expressed in primary human breast cancer came from a study demonstrating increased expression of Notch1 in four breast cancer tumors that overexpressed H-Ras identifying Notch1 as a downstream target of oncogenic H-Ras (Weijzen *et al.*, 2002). Although the Notch proteolysis inhibitor GSI1 used in this study was not selective for Notch (Han *et al.*, 2009; Monticone *et al.*, 2009), this *in vitro* analysis suggested that H-Ras activity increases Notch1 signaling activity, which implies that deregulated Notch1 activation is necessary to maintain the neoplastic phenotype in Ras-transformed cells. This is further supported by genetic evidence using double transgenic mice expressing human *DELTEX1*, which can negatively regulate Notch signaling, and *v-Ha-ras* under control of the MMTV promoter. The expression of the Notch antagonist Deltex inhibited strongly the oncogenic effects of Ha-ras expression in the mammary glands. In the majority of the double transgenic animals, palpable mammary tumors were not detected, even after allowing for a relatively long period of latency (Kiaris *et al.*, 2004). In a study of cultured primary breast cancer cells, the expression of Numb, a negative regulator of the Notch signaling pathway triggering endocytosis and degradation of Notch receptors was found to be inversely related to Notch expression and the growth inhibitory effects of Notch antagonists (Pece *et al.*, 2004). Pece *et al.* could show that NUMB-mediated control on Notch signaling was lost in approximately 50% of human mammary carcinomas. Mechanistically, Numb was suggested to operate as a tumor suppressor, as its ectopic expression in NUMB-negative, but not in Numb-positive, tumor cells inhibits proliferation. Interestingly, high levels of JAGGED1 and/or NOTCH1 expression correlate with poor survival and these molecules are independent prognostic indicators in human breast cancer (Reedijk *et al.*, 2005, 2008b). Increased accumulation of N1ICD and HES1 expression in DCIS compared with normal breast tissue predicts a reduced time to recurrence 5 years after surgery (Farnie *et al.*, 2007). This finding confirms that both the accumulation of NICD as a useful prognostic marker for the recurrence as well as changes in the Notch signaling pathway may be associated with the progression from DCIS to invasive disease. Consistent with these observations, activated Notch signaling (Pece *et al.*, 2004; Rizzo *et al.*, 2008; Stylianou *et al.*, 2006) and consequent upregulation of genes that promote tumor growth (Lee *et al.*, 2008; Leong *et al.*, 2007; Rustighi *et al.*, 2009) have been observed in breast cancer cell lines and primary breast cancers. In particular, NOTCH4 expression, as detected by immunohistochemistry, correlated with Ki67, a well-known proliferation marker in infiltrating breast carcinomas of ductal or lobular histologies (Rizzo *et al.*, 2008). Activation of Notch signaling in estrogen receptor (ER)-negative breast cancer cells results in direct transcriptional up-regulation of the apoptosis inhibitor, and cell-cycle regulator survivin (Lee *et al.*, 2008) and levels of Slug, a transcriptional repressor and Notch target, are elevated and correlate with increased expression of JAGGED1 in human breast cancers (Leong *et al.*, 2007). It is worth noting that the study by Lee *et al.* used the peptidyl GSI inhibitor Z-LLNle-CHO (GSI1), which suppressed survivin levels, induced apoptosis, and inhibited localized and metastatic tumor growth in mice. Rizzo and colleagues using the same GSI1 inhibitor showed also antineoplastic effects of the compound in breast cancer cells. Both groups argued that the reversion of the phenotype by N1ICD transfection indicated that GSI1 induced cell death through inhibition of the Notch signaling pathway. Recently, an alternative interpretation of this data was presented that N1ICD overexpression provided pro-survival signals that antagonize the pro-apoptotic effects of the GSI1 compound. It was shown by Han *et al.* (2009) that the induction of cell death in breast cancer cells by GSI1 is mediated by proteasome inhibition and not by γ-secretase inhibition. In human breast cancers elevated levels of Pin1 correlate with deregulated expression of activated NOTCH1 and HES1 (Rustighi *et al.*, 2009). The prolyl-isomerase Pin1 is a direct Notch target that amplified Notch signaling in a feed-forward loop. Rustighi *et al.* presented data that inhibition of Pin1 function together with GSI administration affects the growth of breast cancer cells with activated NOTCH1 signaling. Interestingly, NOTCH2 appears to antagonize signals by the other three Notch homologs in breast cancer cells (O'Neill *et al.*, 2007) and consistent with *in vitro* data, expression of Notch2 appears to perform a tumor-suppressive role in breast cancer and thus might have a positive prognostic significance (Parr *et al.*, 2004).

Adult stem cells (SC) are defined by two distinctive properties: self-renewal and the capacity to differentiate into multiple lineages. Normal SC in the adult organism are responsible for tissue renewal and repair of aged or damaged tissue. It is generally assumed that SC divide asymmetrically producing two daughter cells—one is a new SC and the second is progenitor cell, which has the ability for differentiation and proliferation, but not the capability for self-renewal (reviewed by Zou, 2007). Recently, there is increasing evidence supporting that certain types of cancer follow a hierarchical organization by harboring cells with tumor initiating activity, which are responsible for driving tumor growth. This concept is generally known as cancer stem cell hypothesis. Cancer stem cells (CSC) are in many aspects similar to normal SC in such a way that it has been shown that tumor cells are heterogeneous comprising rare tumor initiating cells and abundant non-tumor initiating cells. Just like the normal SC, CSC have the ability of self-renewal and proliferation and a large panel of studies shows that leukemia originates from a CSC (reviewed by Reya *et al.*, 2001). Over the past years the CSC hypothesis was extrapolated from the hematopoietic system to solid tumors. Cells with SC characteristics from human glioblastomas were first isolated using clonogenic neurosphere cultures (Ignatova *et al.*, 2002) and elegantly confirmed in a xenograft transplantation assay by the group of Peter Dirks identifying human brain tumor initiating cells *in vivo* (Singh *et al.*, 2003). Subsequently, CSC populations have also been found in prostate (Collins *et al.*, 2005; Lawson *et al.*, 2007), pancreas (Li *et al.*, 2007), colon (O'Brien *et al.*, 2007; Ricci-Vitiani *et al.*, 2007), and in a landmark study by Al-Hajj *et al.* (2003) in breast cancers. Whether Notch signaling plays a role in breast CSC and how one can therapeutically interfere with the Notch signaling cascade will be discussed below and in Section 8.

Two recent studies have demonstrated that, in both man and mice, the Notch pathway is important for promoting the commitment of mammary SC to the luminal lineage at the expense of the myoepithelial lineage (Bouras *et al.*, 2008; Raouf *et al.*, 2008) (Fig. 13.2C). The animal study by Bouras *et al.* assessed the role of Notch signaling in mammary stem and progenitor cells by either knocking down RBP-J using retrovirally expressed RBP-J shRNA constructs or constitutive activation of the pathway. The downregulation of Notch signaling in mammary SC led to a slightly increased repopulating activity *in vivo* and aberrant ductal morphogenesis. This finding provides some clues that Notch signaling plays a role in restricted expansion of the mammary SC pool. On the other hand the Notch pathway *in vivo* appears to be preferentially active in mammary luminal cells, with prominent expression of the active form of Notch1 and its target genes (*Hey1 and Hey2*) in luminal progenitor cells. Expression of N1ICD in mammary SC promoted luminal cell-fate specification at the expense of the myoepithelial lineage. The constitutive overexpression of N1ICD led to specific expansion of luminal progenitors and their self-renewal, finally leading to hyperplasia and tumorigenesis. In contrast to the mouse, where Notch1 is the key determinant of luminal fate selection, Raouf *et al.* performed transcriptome and functional analyses of human breast tissue and identified NOTCH3 as an important regulator of luminal cell-fate specification. Using human mammary progenitor cells and a shRNA knockdown approach they conclusively showed that NOTCH3 is critical for the restriction of bipotent progenitor cells to the luminal pathway and that the other Notch receptors could not substitute for this activity. However, they did not ask whether constitutive overexpression of any Notch in human mammary progenitor cells would lead to tumor development. Nevertheless, they showed that *NOTCH4* gene expression is highest in undifferentiated human clonogenic mammary progenitor cells, becoming markedly downregulated when these cells committed to the luminal lineage. This finding is interesting in light of a study by Harrison *et al.* (2010) assessing Notch signaling activity in human breast cancer cell lines and primary human tumor samples. They identified the cleaved/activated form of the NOTCH4 receptor in basal $CD44^+$ breast CSC cell lines and in primary samples, whereas N1ICD was observed at higher levels in the luminal cells of normal breast epithelium. The differential distribution of NOTCH1 and NOTCH4 in basal CSC and more differentiated cells would suggest different roles for each receptor. Therefore, targeting the NOTCH4 receptor specifically might be a feasible therapeutic approach. Taken together, the luminal progenitor cell can be implicated as a potential cell of origin for tumors in which the Notch pathway has been activated inappropriately, leading to hyperplasia and eventually tumorigenesis.

3. Notch Signaling in Cutaneous Melanoma

Melanoma is a highly aggressive neoplasma resistant to most conventional therapies. Melanomas arise from the transformation of melanocytes, which reside in the basal layer of the epidermis. Melanoma is one of the less common types of skin cancer, but causes the majority of skin cancer related deaths (75%). Long-term clinical and histopathologic observation of cutaneous melanoma has led to the definition of five major steps of tumor progression (Clark, 1991). (1) The common acquired nevus represents the earliest hyperplastic melanocytic lesion. (2) The dysplastic nevus with an increased level of architectural atypia is the candidate precursor for cutaneous melanoma. (3) The radial growth phase primary melanoma is the first recognizable malignant stage but does not show the capacity for rapid growth or metastasis. (4) In vertical growth phase primary melanoma lesions, melanoma cells infiltrate as an expanding mass into the dermis and the subcutaneous tissue with the associated risk of systemic dissemination. (5) Finally, metastatic melanoma represents the most advanced step of tumor progression.

Alterations of several signaling pathways, such as NRas (Demunter *et al.*, 2001), BRaf (Davies *et al.*, 2002), PTEN/PI3K/Akt (Stahl *et al.*, 2004), constitutive FGF-receptor signaling (Becker *et al.*, 1992), disregulated Wnt signaling (Bittner *et al.*, 2000; Weeraratna *et al.*, 2002), and p16/ARF (Cannon-Albright *et al.*, 1996) are found in melanomas and lead to acquisition of growth advantages, resistance to apoptosis, and invasion/metastatic behavior. A considerable effort has been put into the development of therapies aimed at inhibiting these protumorigenic functions as well as at identifying novel critical signaling pathways involved in melanoma.

Recent studies suggest that activation of the Notch signaling pathway is important to preserve the melanocyte SC (MSC) (see below) and may also play a role in melanoma progression extending the list of possible pathways involved in melanoma development. Microarray profiling comparing the gene expression pattern of normal melanocytes to human melanomas revealed up-regulation of Notch receptors, ligands, and downstream target genes. NOTCH2 and HEY1 mRNA were overexpressed in melanoma cells compared to nevi and normal melanocytes (Hoek *et al.*, 2004). Furthermore, JAGGED2 mRNA is upregulated in highly invasive melanoma cell lines (Gutgemann *et al.*, 2001). Massi *et al.* (2006) have performed an analysis of tissue and cellular distribution of Notch receptors and their ligands in a series of benign and malignant human melanocytic lesions. They found that the expression of NOTCH1 and NOTCH2, as well as Notch ligands, was upregulated in "dysplastic nevi" and melanomas as compared with common melanocytic nevi. These results suggest that the activation of Notch may represent an early event in melanocytic tumor growth leading to the hypothesis that upregulation of Notch signaling may sustain tumor progression. Inhibition of Notch signaling with a non-selective γ-secretase tripeptide inhibitor derived from MG132 (GSI1) induces apoptosis in malignant melanoma cells but only a G2-M arrest in melanocytes suggests that Notch signaling or the proteasome (Han *et al.*, 2009; Monticone *et al.*, 2009) might be required for malignant melanoma cell survival (Qin *et al.*, 2004). Conversely, constitutive activation of Notch signaling promotes the primary vertical growth phase of melanoma cells *in vitro* and *in vivo* and advances lung metastasis in adult mice under specific experimental conditions (Balint *et al.*, 2005; Liu *et al.*, 2006). However, although forced Notch signaling advances primary melanoma growth, it has little effect on metastatic melanoma cells and appears insufficient to transform melanocytes on its own. Furthermore, the oncogenic effect of NOTCH1 on primary melanoma cells was mediated by β-catenin, which was upregulated following Notch1 activation. Inhibiting β-catenin expression reversed NOTCH1-enhanced tumor growth and metastasis (Balint *et al.*, 2005). The oncogenic effect of activated Notch1 is at least partially mediated through regulation of the MAPK and PI3K-Akt pathways (Bedogni *et al.*, 2008; Liu *et al.*, 2006). Hyperactivated PI3K-Akt signaling results in the upregulation of Notch1 through NFκB activity, while constitutive low oxygen skin content increases Notch1 mRNA and protein via the stabilization of HIF-1α (Bedogni *et al.*, 2008), implicating that Notch1 contributes somehow to Akt-dependent melanocyte transformation in hypoxia and facilitates melanoma growth. In addition, Notch1 signaling enhances tumor cell adhesion and increases N-cadherin expression, a cell adhesion protein whose expression is highly correlated with melanoma progression and metastasis (Liu *et al.*, 2006). How Notch signaling interacts with these pathways in melanoma cells is currently unclear.

Given the role for Nodal in melanoma and the observation that perturbation of Notch signaling in the skin can result in malignancies including melanoma (Pinnix and Herlyn, 2007), it is possible that molecular cross-talk exists between Nodal and Notch signaling in human melanoma. Nodal, a member of the TGFβ family of proteins, is involved in SC maintenance and differentiation, and shown to be associated with cancer progression (Schier, 2003; Topczewska *et al.*, 2006). Evidence from developmental studies in mice and zebrafish describe a role for Notch in regulating expression of Nodal via two CSL binding sites in the Node-specific enhancer (Borggrefe and Oswald, 2009; Krebs *et al.*, 2003; Raya *et al.*, 2003). Nodal may be upregulated by Notch signaling in metastatic melanoma cells, and pharmacological inhibition of Notch signaling could result in decreased Nodal expression (Postovit *et al.*, 2007).

A recent study by Pinnix *et al.* (2009) provide good evidence that deregulation of Notch signaling activity plays a specific role in promoting a transformed phenotype in human melanocytes and has defined the importance of Notch signaling in human melanoma. Through analysis of a large panel of cell lines and patient lesions they could show that NOTCH receptors 1, 2, and 4 are overexpressed particularly when compared against primary melanocytes or normal human skin. Ectopic N1ICD expression resulted in the loss of E-cadherin expression and upregulation of MCAM, two well-characterized events in melanoma development. In addition, chromatin IP (ChIP) identified MCAM as a direct Notch target. The N1ICD oncoprotein conferred anchorage-independent growth, increased survival, and promoted loss of contact inhibition. To suppress Notch signaling, a dominant-negative (DN) *MAML* lentiviral vector was used. DNMAML inhibited growth of melanoma cell lines, whereas primary melanocytes were unaffected (Pinnix *et al.*, 2009). These findings suggest that Notch signaling plays a specific role in promoting the transformed phenotype in human melanocytes and acts as a driving force in melanocyte transformation. These studies provide convincing evidence that aberrant Notch signaling influences melanocyte transformation and stages of melanoma progression. Nonetheless, genetic loss-of-function analyses in established melanoma models need to be performed in order to convincingly demonstrate that Notch activation is an obligate event necessary for melanoma development and/or tumor progression.

A physiological role for Notch signaling in the melanocyte lineage has been demonstrated. Notch acts through Hes1 (hairy/enhancer of split 1) and plays an indispensable role in the maintenance of MSC and melanoblasts (MB) in the epidermis (Kumano *et al.*, 2008; Moriyama *et al.*, 2006; Pinnix and Herlyn, 2007; Schouwey *et al.*, 2007). Conditional deletion of *RBP-J* in the melanocyte lineage produces a diluted coat color at birth and accelerates hair graying after the first hair moult (Moriyama *et al.*, 2006). A similar phenotype is observed when Notch1 and/or Notch2 are ablated in the melanocyte lineage (Schouwey *et al.*, 2007). Overall the genetic ablation of Notch signaling in the mouse results in a dramatic reduction of embryonic MB (via apoptosis), which manifests itself as dilution of initial hair pigmentation (Moriyama *et al.*, 2006; Schouwey *et al.*, 2007). These animals also exhibit premature hair graying in subsequent hair cycles, suggesting an important role for Notch signaling in the maintenance of MSC. This function of Notch signaling has also been confirmed in adult mice using selective γ-secretase inhibitors (Kumano *et al.*, 2008). In addition, Kumano *et al.* (Kumano *et al.*, 2008) reported premature hair graying in $Notch1^{+/-}Notch2^{+/-}W^{+/v}$ mice, indicating a possibility of cross-talk between Notch and c-Kit signaling in regulation of the melanocyte lineage. This latter possibility was also noted in Notch1 deleted hair follicles (Lee *et al.*, 2007a). Employing an *in vivo* lineage tracing technique Aubin-Houzelstein *et al.* (2008) were able to show that in addition to the maintenance of MSC and MB, Notch signaling plays additional roles in maintaining the immature status of MB, promoting proper localization of MB and inducing proper differentiation in the hair matrix. As outlined above, melanomas emerge primarily within epidermal MB or melanocytes. Thus, it is reasonable to speculate that both normal and malignant MB may share key pathways that regulate biological homeostasis and maintenance. These similarities include the epithelial–mesenchymal transition and an invasive/migratory capacity that is central to both normal and malignant cells of the melanocyte lineage. Interestingly, as in normal MB development, Notch signaling participates in melanoma progression (see above—Balint *et al.*, 2005; Liu *et al.*, 2006; Pinnix *et al.*, 2009). Therefore, the elucidation of molecular mechanisms downstream of Notch is important not only for MB development but also for understanding the molecular mechanisms of melanomagenesis.

4. Notch and Colon Cancer

Colorectal cancer (CRC) is a common malignancy and with 655,000 deaths worldwide per year the third leading cause of cancer-related death in the Western world. CRCs arise most commonly from adenomatous polyps in the colon. These mushroom-shaped growths are usually benign, but some develop into cancer over time. Invasive cancers that are confined within the wall of the colon are curable with surgery. If untreated, they spread to regional lymph nodes, where up to 73% are curable by surgery and chemotherapy. Colon cancer that metastasizes to distant sites is not curable, although chemotherapy can extend survival and the 5-year survival rate is at most 8%. In comparison, women whose breast cancer has metastasized to other organs of the body have a 5-year survival rate of 21%.

The gut is a highly self-renewing tissue regenerating itself every 5 days. This turnover depends on the presence of SC and progenitor cells found in the crypts of the colon or between intestinal villi of the small intestine. The SC give rise to progenitor cells, which can in turn differentiate over successive divisions, to generate the various different cell types that populate and form the gut. The small intestinal epithelium comprises differentiated cells of four principal lineages—absorptive enterocytes and three secretory lineages consisting of goblet, enteroendocrine, and Paneth cells. The key factor for the continuous renewal of the intestinal epithelium is to maintain a balance between differentiation and proliferation of the epithelial SC and immature progenitors (reviewed by Radtke and Clevers, 2005; Radtke *et al.*, 2006) (Fig. 13.3A). Canonical Wnt signaling has long been regarded as the signaling pathway playing a central role in this epithelial cell fate determination (reviewed by Sancho *et al.*, 2004) and several findings support this notion. The proliferative cells at the bottom of the crypts of the small intestine and the colon accumulate nuclear β-catenin. Activating mutations in the WNT pathway are frequently associated with CRC in humans, as well as with adenomatous polyp formation in the murine intestine. Mutation of the intestinal-specific TCF/LEF family member TCF4 (TCF7L2) induces loss of the proliferative compartments in the small intestine (Korinek *et al.*, 1998). The inhibition of β-catenin/TCF4 activity in CRC cells induces these cells to switch from a crypt-like phenotype to a differentiated villus epithelial phenotype (Batlle *et al.*, 2002; van de Wetering *et al.*, 2002). Finally, the targeted expression of the soluble Wnt inhibitor DKK1 results in greatly reduced epithelial proliferation, coinciding with the loss of crypts (Pinto *et al.*, 2003).

However, more recent studies have shown that Notch signaling is also indispensable to maintain a balance between differentiation and proliferation of the epithelial SC and immature progenitors (Fre *et al.*, 2005; Riccio *et al.*, 2008; van Es *et al.*, 2005). Notch signaling pathway components are mainly expressed in the crypt region where SC and transient amplifying progenitors reside. In embryos and neonates increasing Notch signaling through the expression of N1ICD results in increased cell proliferation in the SC compartment and also shows a severe reduction of all secretory cells-goblet, enteroendocrine, and Paneth cells (Fre *et al.*, 2005). Conversely, inhibition of Notch signaling in the intestinal epithelium, either by deletion of *Hes1* (Jensen *et al.*, 2000), *RBP-J* (van Es *et al.*, 2005), or *Notch1*

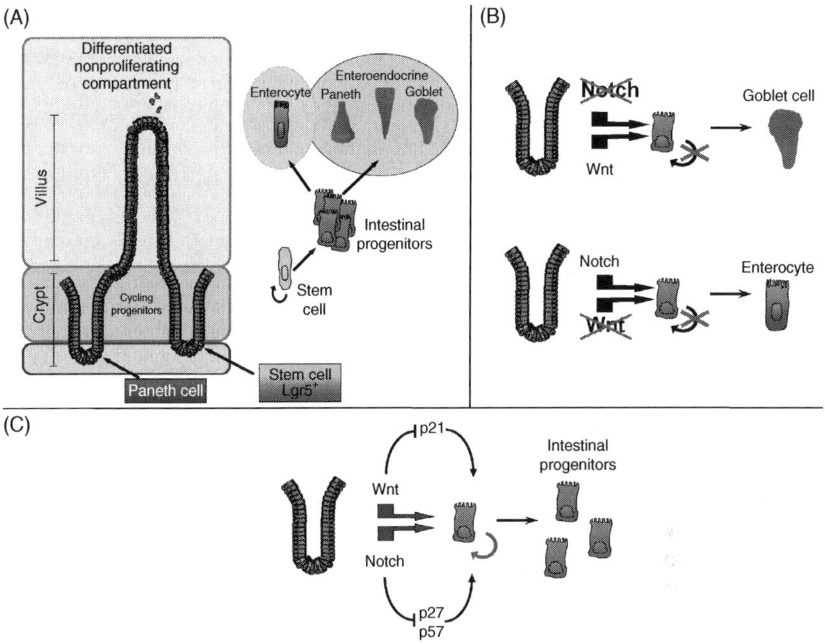

Figure 13.3 Notch signaling in the small intestine. (A) *Schematic representation of the crypt/villus of the small intestine.* Stem cells (yellow; Lgr5$^+$) and cycling progenitors (green) localize to the crypt compartment, which is maintained by Notch and Wnt signaling. All cycling progenitor cells with the exception of Paneth cells (purple), which localize to the bottom of the crypt, migrate upward and stop cycling at the crypt/villus boundary. The migration of non-proliferating differentiated cells continues toward the tip of the villus where they are shed into the lumen of the intestine. The villus is composed mostly of enterocytes (blue; absorptive lineage) with intermingled enteroendocrine and goblet cells (pink; secretory lineage). (B) *Notch and Wnt signaling act in concert to maintain stem/progenitor cell populations in the intestine and control binary cell fate decisions.* In the mammalian intestine, progenitor cells become committed toward one of the differentiated cell types as they migrate along the crypt-villus axis. This process implies at least one binary decision, i.e., to become an absorptive cell (enterocyte) or to differentiate toward a secretory phenotype (goblet, enteroendocrine, or Paneth cells). Loss of Notch signaling induces a massive conversion of intestinal progenitor cells types to the secretory lineages (mainly to goblet cells). Loss of Wnt signaling results in loss of progenitor cells, halts differentiation toward the goblet and enteroendocrine lineages, and drives intestinal progenitor cells into the absorptive lineage generating an abundance of enterocytes. Therefore, a combination of Notch and Wnt signaling specifies different cell types in the intestinal epithelium of mice: Wnt+ Notch− crypt cells are forced toward goblet or enteroendocrine differentiation, Wnt− Notch+ cells are converted to enterocytes, whereas Wnt+ Notch+ cells maintain an undifferentiated phenotype (see panel C). (C) Notch and Wnt signaling cascades work together to preserve the proliferative crypt compartment of the small intestine. Wnt signaling results in repression of the cyclin-dependent kinase inhibitor *p21*$^{CIP1/WAF1}$, whereas Notch signaling represses the cell-cycle regulators *p27*^{KIP1} and *p57*^{KIP2}. Although the Wnt and the Notch pathways do not show any apparent cross-talk in crypt progenitor cells, both show similar functions by inhibiting cyclin-dependent kinases. (See Color Insert.)

and *Notch2* concomitantly (Riccio *et al.*, 2008) or through pharmacological inhibition using γ-secretase inhibition (van Es *et al.*, 2005), results in loss of transient amplifying (TA) cells accompanied by goblet cell metaplasia. Math1 is a bHLH transcriptional activator required for differentiation of goblet, enteroendocrine, and paneth cells (Yang *et al.*, 2001) and is repressed by Hes1. The gene-targeted deletion of *Math1*, results in a largely reciprocal phenotype characterized by a fate switch toward the enterocytic lineage. Taken together, these complementary loss-of-function and gain-of-function studies specify two main physiological roles of Notch signaling in the intestine. One is to maintain the proliferating undifferentiated SC/progenitors acting as a gatekeeper of crypt cells and the other is to monitor binary cell fate decisions of the TA compartment controlling absorptive rather than secretory (i.e., goblet and Paneth) cell fate decisions in the intestinal epithelium (Fig. 13.3B and C). An additional function of Notch signaling has also been shown to provide an inductive role in driving the differentiation of post-mitotic cells into mature goblet cells using lineage-tracing approaches (Vooijs *et al.*, 2007; Zecchini *et al.*, 2005). Seminal papers by Micchelli (Micchelli and Perrimon, 2006) and Ohlstein (Ohlstein and Spradling, 2006) identified multi-potent SC in the adult *Drosophila* midgut epithelium and confirmed that the two types of differentiated intestinal cell lineages (enterocytes and enteroendocrine cells) arise from a single intestinal SC. The intestinal SC, however, could not be associated with a well-defined anatomical niche. In the fly the Notch signaling pathway is active in the adult intestinal epithelium and is required for normal intestinal SC function (Micchelli and Perrimon, 2006; Ohlstein and Spradling, 2006) in a similar way as in the mammalian system. Midgut intestinal SC express not only Notch but also high levels of Delta, whereas the differentiated progeny express only Notch. Thus, Delta expression serves as a specific SC marker in the *Drosophila* midgut. Similarly, Barker *et al.* (2007) have identified the Wnt target gene *Lgr5* as an adult SC marker in the murine intestine. However, it is important to note that Notch signaling in *Drosophila* seems to have different effects regarding self-renewal when compared to the mammalian intestinal epithelium. Whereas the inhibition of Notch signaling in mice will lead to the depletion of the progenitor compartment by promoting differentiation (van Es *et al.*, 2005), in the fruit fly the reduction of Notch signaling will lead to induction of overgrowth of intestinal SC and impaired differentiation (Micchelli and Perrimon, 2006; Ohlstein and Spradling, 2006). Conclusively, constitutive Notch expression in the mammalian crypts amplifies the progenitor cells (Fre *et al.*, 2005), albeit in *Drosophila* it induces differentiation of intestinal SC (Micchelli and Perrimon, 2006; Ohlstein and Spradling, 2006). Although there are distinct differences between fly and mouse intestinal SC, the *Drosophila* system might prove useful to understand the hierarchy between Notch and Wnt signaling pathways.

It has become evident that the accurate coordination of both the Notch and the Wnt signals controls intestinal epithelial cell fate decisions particularly through crosstalk at the tissue and cellular level (for a comprehensive review see Nakamura *et al.*, 2007). The coordination of these two signaling pathways as outlined above is essential in normal development and consequently it may play an important role in intestinal tumorigenesis. Indeed, tracing cells in which Notch1 was activated (Vooijs *et al.*, 2007), or detecting expression of the Notch target gene *Hes1* (van Es *et al.*, 2005), indicated uniform Notch1 activation in adenomas of APC (adenomatous polyposis coli)^{min} mice as well as in human colon cancer cell lines and primary human CRC tissue samples (Fernandez-Majada *et al.*, 2007) implying that Notch and Wnt signaling are simultaneously active in the proliferating adenoma cells. The great majority of human intestinal tumors display loss or mutations of the tumor suppressor APC. Mutations in the *APC* gene are responsible for the familial adenomatous polyposis syndrome, and are an early causative event in sporadic cancer development (reviewed by Nathke, 2004). The APC^{min} mice have been used extensively as a reasonable CRC model since these mice develop adenomatous polyps in the small intestine, which is however in contrast to human disease occurring predominantly in the colon. In addition, they do not progress into invasive or metastatic adenocarcinomas at significant frequency. But if additional mutations (e.g., K-ras^{G12D}) are introduced into these mice, the intestinal polyposis phenotypes are modified, and malignant adenocarcinomas develop (Sansom *et al.*, 2006).

Reedijk *et al.* (2008a) provide correlative evidence of active Notch signaling in human adenocarcinomas. A small cohort of colon tumors was analyzed for the expression of Notch receptors and ligands, modifiers, and downstream targets. Gene expression of both JAGGED ligands, NOTCH1, LFNG, and HES1 was detected by *in situ* hybridization and shown to be at comparable or greater levels than normally observed in cells of the crypt base. All of these findings emphasize the importance of understanding the functional and regulatory interactions amongst these two critical pathways under physiological, but more importantly in the tumor setting. In an elegant study performed by Fre *et al.* (2009) the interplay between the two signaling pathways was assessed *in vivo* by modulating Notch activity in mice carrying either a loss- or a gain-of-function mutation of Wnt signaling. The proliferative effect of active Notch signaling has on early intestinal progenitors requires Wnt signaling, whereas its influence on intestinal differentiation appears Wnt independent. This synergy was also observed in human intestinal adenomas. The analysis of HES1 expression in human colon cancer samples showed that 12 out of 15 polyps of both sporadic and hereditary low-grade adenomas present strong nuclear HES1 expression, whereas HES1 is either not detected or expressed at low levels in human adenocarcinomas. Similarly, HEY1, HEYL, and the Notch ligands

JAGGED1 and JAGGED2 were expressed at higher levels in human adenomas than carcinomas. These observations warrant the conclusion that elevated Notch signaling in benign adenomas may contribute to the initiation of CRC. On the other hand high Notch signals in adenomas could be interpreted to maintain a tumor suppressive function, whereas it seems to be dispensable at later stages of CRC development. Mechanistically it was shown that Jagged1 is transcriptionally controlled by the β-catenin/TCF4 complex, e.g., in hair follicles of the epidermis (Estrach *et al.*, 2006); it is possible that this mechanism is also conserved in the intestinal epithelium since promoter analysis of mammalian *Jagged1* orthologs revealed that this Notch ligand is an evolutionary conserved target of the WNT/β-catenin signaling pathway based on the conservation of two TCF/LEF-binding sites within the 5′-promoter region (Katoh, 2006). Guilmeau *et al.* (2010) provided correlative link between Wnt signaling driving colon tumor development and Jagged1 expression. The expression of JAGGED1 was restricted to enteroendocrine cells in normal human intestinal epithelium but the expression was elevated in about 50% of a human colon cancer tissue array of adenocarcinomas. Espinosa's group has published a recent study showing that β-catenin/TCF signaling is inducing Notch activation in CRC cells through direct regulation of Jagged1 expression (Rodilla *et al.*, 2009). They analyzed microarray data from Ls174T CRC cells and identified genes downstream to the β-catenin/TCF pathway. Genes directly regulated by the Jagged1/Notch signaling cascade included *Hes1*, *CD44, KLF5, NOX1, EpHB3*, and *SOX9*. siRNA knockdown conclusively established Jagged1 as a regulated Notch target downstream of β-catenin, and ectopic expression of Jagged1 was sufficient to block differentiation in transfected Ls174T/dnTCF4 CRC cells when Wnt/β-catenin signaling is switched off. In contrast to the effects of Jagged/Notch1 ICD, Hes1 expression alone was not sufficient to induce tumor growth *in vivo* in the absence of β-catenin/TCF activity suggesting that other target genes downstream of β-catenin participate in regulating tumorigenesis. The *in vivo* $APC^{min/+}$ mouse model confirmed that Jagged1 is strongly overexpressed in the tumor tissue of these mice compared with normal crypts concomitant with Notch1 activation. The complexity of the regulation of Notch ligand/receptor expression in CRC development is further highlighted by the following findings. β-catenin/TCF-LEF induces transcriptional activation of cell-cycle regulatory genes, like c-Myc (He *et al.*, 1998) and cyclin D1 (Tetsu and McCormick, 1999), and serves as the mechanistic basis for explaining the APC mutation-dependent initiation of colon adenomas with c-Myc being a mediator of the early stages of adenoma formation (Sansom *et al.*, 2007). APC mutant cells lacking c-Myc are not capable of initiating intestinal tumors in mice. Since *c-Myc* is a direct target gene of Notch1 activation in both human and murine T-ALL cell lines, where c-Myc is consistently overexpressed in a Notch1-dependent fashion

(Weng *et al.*, 2006), it would be interesting to assess whether c-Myc regulates Notch receptor expression in adenomas.

The recent findings summarized here using either human colon cancer cell lines or assaying primary human CRC samples elucidate that Notch and Wnt can signal cooperatively to control tumorigenesis in the intestine. Importantly, not all tumors with elevated Wnt signaling displayed elevated Jagged1 and there was no direct correspondence between nuclear β-catenin translocation in the tumors and Jagged1 expression. Thus, actual contribution of Notch signaling to Wnt-dependent intestinal tumorigenesis is still incompletely understood.

5. Notch Signaling in Pancreatic Cancer

Pancreatic ductal adenocarcinoma (PDAC) is a highly aggressive and devastating form of cancer. The overall 5-year survival rate is below 5% and the median survival after diagnosis is approximately 6 months. It is the fourth leading cause of cancer related death in the United States (Jemal *et al.*, 2009). Pancreatic cancer is one of the rare diseases where the progress in modern medicine is rather limited and consequently patient survival did not significantly improve over the last decades. Therefore it is important to develop a better understanding of the molecular mechanism that lead to the development and progression of the disease in order to identify novel potential therapeutic targets to interfere and/or prevent the disease.

Originally, Notch was only indirectly linked to pancreatic cancer. Most studies focused on the developmental role of Notch during embryonic pancreas development and its ability to influence differentiation processes of progenitor cells. The fact that early human pancreatic cancers changed their epithelial differentiation program fueled the notion that Notch might play a role during pancreatic cancer development and/or progression. The general principle is that a developing embryonic organ and a cancer share many similarities such as increasing the cell mass through proliferation, changing the differentiation status, inducing tissue remodeling through cell migration and angiogenesis. Therefore, it might not be surprising that developmental signaling pathways (including the Notch cascade), which are normally no longer expressed in adult terminally differentiated tissues get re-expressed during cancer development. Therefore, we will first describe the role of Notch during pancreas development followed by a discussion of recent work that more directly links Notch to pancreatic cancer.

In mice, the pancreas develops from Pdx1- and Ptf1a-expressing progenitors, which are found around E8.5 in distinct dorsal and ventral regions of the forgut endoderm (Edlund, 2002). Epithelial buds of these endodermal layers are visible at E10. Shortly after, both buds fuse to form one organ, in which cells proliferate and differentiate. By E13 most of the progenitor cells are fully committed and engaged into a differentiation program leading to endocrine, acinar, and ductal cells (Jorgensen *et al.*, 2007). The exocrine pancreas is composed of acinar and ductal cells, which produce and secrete digestive enzymes into the intestine, while the endocrine part of the pancreas is composed of several cell types forming the islets of Langerhans, which control glucose metabolism and secrete hormones into the blood stream (Edlund, 2002).

The role of Notch signaling during pancreas development was mainly established using genetic mouse models. Conventional gene targeted mice for *Dll1, RBP-J* (Apelqvist *et al.*, 1999), or the Notch target gene *Hes1* (Jensen *et al.*, 2000) exhibit accelerated differentiation of endocrine cells leading to pancreatic hypoplasia. Pancreas-specific inactivation of RBP-J led to a similar phenotype characterized by premature differentiation of pancreatic progenitors into endocrine cells and impaired development of the exocrine pancreas (Fujikura *et al.*, 2006; Nakhai *et al.*, 2008); however, this cannot be attributed to Notch since RBP-J cooperates with PTF1a (see below, chapter 7, and Beres *et al.*, 2006; Cras-Meneur *et al.*, 2009; Hori *et al.*, 2008; Masui *et al.*, 2007). Indeed, simultaneous inactivation of Notch1 and Notch2 did not significantly inhibit pancreatic development consistent with that seven has a Notch independent role in the development of the exocrine pancreas or that other Notch receptors might be involved in this process (Nakhai *et al.*, 2008). Interestingly, partial loss of γ-secretase or Notch2 in neurogenin3-positive progenitors allowed them to adopt the acinar fate, revealing an unexpected role for Notch in helping neurogenin3, a helix-loop helix transcription factor, commit progenitors to the endocrine fate by sequestering RBP-J away from PTF1a, freeing E2A for Ngn3 (Cras-Meneur *et al.*, 2009). Notch gain-of-function alleles expressed in PDX1 progenitors resulted in the inhibition of both exocrine and endocrine differentiation (Hald *et al.*, 2003; Murtaugh *et al.*, 2003). Taken together, these results were interpreted to mean that Notch signaling, and in particular, RBP-J, is an important regulator of pancreatic progenitor cells that have to choose between the endocrine and the acinar cell fate. How Notch regulates this cell fate decision is not fully understood. One possible mechanism suggested that Hes1 represses the expression of neurogenin3, which functions as a pro-endocrine factor (Lee *et al.*, 2001), and the cell-cycle regulator CDKI p57 (Georgia *et al.*, 2006), thereby preventing progenitor cells from exiting the cell cycle and from differentiating into the endocrine lineage.

In the exocrine pancreas of adult mice, Notch receptor expression is not or only barely detected (Fujikura *et al.*, 2007; Jensen *et al.*, 2000), whereas the expression of Hes1 is restricted to centro-acinar cells (Miyamoto *et al.*, 2003; Pasca di Magliano *et al.*, 2006), suggesting that under homeostatic conditions Notch signaling might not be very active. In contrast, in situations in which the pancreas is injured or has to regenerate such as in caerulein-induced acute pancreatitis, multiple components of the Notch signaling pathway (including Notch1, Notch2, and Hes1) are transiently upregulated (Gomez *et al.*, 2004; Jensen *et al.*, 2000). Moreover, mice carrying a tissue-specific inactivation of Notch1 in the pancreas exhibited impaired regeneration in a caerulein-induced pancreatitis model (Siveke *et al.*, 2008). Together, these results confirm that although Notch is not very active in homeostatic conditions, it seems to play important roles during tissue regeneration.

In more recent years, there is increasing evidence that link Notch more directly to the development and/or progression of pancreatic cancer. First, multiple Notch receptors, ligands, and down stream target genes have been shown to be expressed in early pancreatic lesions (known as pancreatic intraepithelial neoplasms, PanIN) as well as in PDAC tissue of mice and humans (Buchler *et al.*, 2005; Miyamoto *et al.*, 2003). A recent study analyzed more than 400 human cancer cell lines that were derived from different solid cancer types for their responsiveness to a selective pharmacological γ-secretase inhibitor (MRK-003). Remarkably, 50% of a cohort of 26 PDAC cell lines tested were sensitive to the inhibitor, which makes them the most responsive tumor type compared to any other type tested, including breast and non-small cell lung cancer (Plentz *et al.*, 2009). In addition, multiple genetically engineered mouse models for PDAC reveal increased expression of Notch1 and/or Hes1 in early metaplastic lesions (PanIN) as well as in fully developed PDACs (De La *et al.*, 2008; Habbe *et al.*, 2008; Hingorani *et al.*, 2003; Kimura *et al.*, 2007; Pasca di Magliano *et al.*, 2006; Sawey *et al.*, 2007), indicating that Notch expression might be an early event in the development of pancreatic cancer. Although the cell of origin leading to PanIN lesions is currently unknown, certain studies support a model in which acinar cells undergo acinar to ductal metaplasia, which subsequently will give rise to PanIN before progressing to PDACs (Habbe *et al.*, 2008; Hezel *et al.*, 2006; Zhu *et al.*, 2007). In this context it is interesting to note that forced N1ICD expression in explant cultures of adult mouse pancreas induced a similar metaplastic conversion from an acinar to a ductal cell-predominant epithelium (Miyamoto *et al.*, 2003). Curiously, acinar cells born from neurogenin3 progenitors lacking γ-secretase, or exposed to another selective γ-secretase inhibitor, DAPT, proliferate excessively and display elevated levels of apoptosis (Cras-Meneur *et al.*, 2009). This observation suggests that loss of γ-secretase (or Notch) in regenerating acinar tissue combined with additional loss of pro-apoptotic genes might also transform acinar cells.

Transforming growth factor-α (TGF-α)-induced EGF receptor signaling is frequently found in pancreatic cancers (Hruban *et al.*, 1999). Transgenic overexpression of *TGF-α* in mice also results in acinar to ductal metaplasia and correlated with increased Notch signaling. Importantly, TGF-α induced metaplasia was abolished by pharmacological inhibition of Notch signaling, suggesting that Notch mediates this TGF-α induced process (Miyamoto *et al.*, 2003). However, a recent study challenged this view by showing that pancreas-specific transgenic expression of *N1ICD* on its own is not sufficient to induce acinar to ductal metaplastic conversion or PanIN development (De La *et al.*, 2008). The reason for this discrepancy is currently unknown. However, the same group tested whether Notch could possibly cooperate with the *K-ras* proto-oncogene, which is mutated in most PDAC cases but also in early stage lesions indicating that activating *K-ras* mutations occur at an early stage during pancreatic carcinogenesis (Habbe *et al.*, 2008). Thus, simultaneous expression of NICD with an oncogenic form of K-ras in either pancreatic progenitors or mature acinar cells resulted in the development of PanIN lesions at time points where expression of NICD or K-ras alone did not lead to such lesions. These results strongly suggest that Notch and K-ras synergize and can cooperate to initiate pancreatic carcinogenesis in the mouse (De La *et al.*, 2008). Another interesting question is whether Notch signaling can also influence progression from PanIN lesion to PDAC. It may be possible to address this question by sequentially expressing NIC after K-ras induced PanIN lesions have developed. The group by Bardeesy and colleagues chose an alternative approach. They used a genetically engineered mouse model of PDAC (K-ras; p53Lox/+ mice) to investigate if γ-secretase mediated inhibition of Notch signaling influences the progression of PanINs to invasive PDAC. The genetically engineered mice were treated with a selective γ-secretase inhibitor or vehicle as control at a time point when the animals had developed isolated PanIN lesions but no PDACs. Analysis of these mice approximately 3 months after treatment revealed that roughly 30% of the vehicle-treated mice developed PDAC whereas none of the γ-secretase treated mice ($n = 25$) did (Plentz *et al.*, 2009). These results strongly suggest that Notch signaling in this experimental setting of murine PDAC can indeed promote progression from PanINs to PDAC and therefore could represent a therapeutic target. Although these studies are very encouraging, suggesting that Notch signaling is implicated in pancreatic carcinogenesis, many issues remain to be resolved. These include elucidation of the role Notch plays in a full-blown PDAC and whether interference with Notch influences disease outcome in PDACs, how do Notch receptors and ligands get re-expressed during the development of early PanIN lesions or during their progression to PDACs is also unclear, as is the question of which Notch target genes are activated during this process and what is their role. Trying to resolve these questions will certainly involve the use of further PDAC mouse models, but it will also be important to address these questions systematically in suitable human pancreatic cancer cell lines or primary human tumor samples to make the findings in the mouse relevant for the human disease.

6. Notch Signaling in Medulloblastoma

Medulloblastoma is the most common malignant brain tumor in children (Packer *et al.*, 1999). The 5-year survival rate for children with this aggressive neoplasm is high and approaches 90%. Nevertheless, the outcome for patients that are too young for radiation therapy is much worse. Moreover, the current form of therapy results often in severe long-term neurocognitive difficulties (Crawford *et al.*, 2007). The majority of medulloblastoma tumor cells appear to be undifferentiated cells which led to the suggestion that this neuroectodermal tumor arise from neuronal stem or precursor cells in the ventricular zone and the cerebellar external germinal layer (Oliver and Wechsler-Reya, 2004). Multiple signaling pathways, which are involved in regulating neural SC, are also aberrantly activated in medulloblastoma. One of these developmental pathways is the sonic hedgehog (shh) pathway. Aberrant shh signaling induces the expression of N-MYC, a protein that is frequently overexpressed in these tumors (Kenney *et al.*, 2003). Approximately 15% of the sporadic medulloblastoma patients have increased shh signaling due to inactivating mutations in the gene coding for the receptor Patched (Raffel *et al.*, 1997), activating mutations in the co-receptor smoothened or the downstream effector suppressor-of-Fused (Lee *et al.*, 2007b). Therefore, it is not surprising that most murine medulloblastoma models are based on the aberrant activation of shh signaling. Increased Wnt signaling due to mutations in the *APC, Axin*, or *β-catenin* genes have also been reported in some medulloblastomas (Dahmen *et al.*, 2001; Huang *et al.*, 2000; Zurawel *et al.*, 1998). Interestingly, nuclear β-catenin, an indicator for activation of the Wnt pathway, is found in approximately 25% of medulloblastoma and correlates with a better clinical outcome (Eberhart *et al.*, 2000; Ellison *et al.*, 2005). Wnt signaling is mostly seen in neoplasms that show no or low shh signaling, indicating that these pathways produced a distinct subtype of medulloblastoma (Clifford *et al.*, 2006). The Notch pathway was initially not associated with the development of medulloblastoma. However, the undifferentiated appearance of these tumor cells and the fact that Notch is involved in the maintenance of neural stem and progenitor cells motivated several groups to investigate the potential role of Notch in medulloblastoma (Gaiano and Fishell, 2002; Solecki *et al.*, 2001). Analysis of primary medulloblastoma tumor samples revealed increased mRNA expression of NOTCH2 but not NOTCH1. In 15% of the examined tumors increased NOTCH2 expression levels correlated with *NOTCH2* gene amplification suggesting that NOTCH2 may play a more important role for this neoplasm compared to the other Notch receptor family members. Moreover, increased Hes1 expression correlated with poor patient survival prognosis (Fan *et al.*, 2004). Interestingly, mice expressing a constitutively active form of the

Smoothened gene in cerebellar granule neurons develop a murine form of medulloblastoma, characterized by increased mRNA expression levels of Notch1 (75%), Notch2 (12.5%), Hes1 (46%), and Hes5 (71%) in the tumors (Hallahan *et al.*, 2004). These results suggest that in this experimental setting shh signaling can induce Notch signaling. How and why shh signaling induces expression of Notch is currently unknown. Numb has been shown to suppress shh signaling by targeting Gli1 for proteosomal degradation (Di Marcotullio *et al.*, 2006). Since Notch can negatively regulate Numb levels, it is conceivable that cells that activate shh and Notch will escape a Numb-mediated negative feedback loop that would counteract shh signaling. Whether shh signaling directly activates Notch, or vice versa, remains to be investigated.

Additional evidence that Notch signaling is involved in medulloblastoma is derived from experiments trying to interfere with the Notch cascade. Pharmacological inhibition of Notch activation or using soluble Delta ligands or siRNA approaches induced apoptosis and led to pronounced reduction of viable cells in medulloblastoma cell lines and/or primary explant cultures (Fan *et al.*, 2004; Hallahan *et al.*, 2004). Reciprocal gain-of-function studies overexpressing N2ICD promoted cell proliferation, soft agar colony formation, and tumor growth in xenotransplantation experiments. Somewhat surprisingly, similar experiments using N1ICD resulted in growth inhibition (Fan *et al.*, 2004). These results strongly suggest that both Notch receptor exhibit very distinct functions, in contradiction with the *in vivo* finding (Hallahan *et al.*, 2004). How the N2ICD growth promoting function differs mechanistically from a N1ICD-mediated growth inhibitory function is currently unknown and needs further investigation.

In recent years there is accumulating evidence supporting that medulloblastoma is a neoplasm that follows the CSC concept. Nodular/demoplastic medulloblastoma is mainly composed of two cell types, one that is undifferentiated and rapidly proliferating, while second cell type represents non-proliferative neurons suggesting that the undifferentiated cells have the capacity to differentiate and to enter growth arrest (Louis *et al.*, 2007). Moreover, medulloblastoma contain SC-like multi-potent cells that have the capacity to grow in neurosphere-cultures, an ability that is normally restricted to neural SC (Galli *et al.*, 2004; Hemmati *et al.*, 2003). The CSC marker CD133 is expressed in a subset of medulloblastoma cells in primary tumors as well as in established cell lines and correlates with increased ability to form neurospheres *in vitro* and tumors in xenografts (Eberhart, 2007; Singh *et al.*, 2003, 2004). Moreover, putative (CD133-positive) medulloblastoma CSC were reported to localize preferentially near endothelial cells and small vessels, a niche rich in Notch ligands (see the chapter 9). Although it is currently unclear whether CSC require the support of niche cells, it is tempting to speculate that blood vessel may in this situation serve as niche by providing a critical microenvironment supporting growth and survival of the CD133-positive cells. The concept that blood vessel may function as niche cells was previously suggested in the context of hematopoietic SC (Kiel and Morrison, 2008) and muscle satellite cells (see chapter 12). That Notch may play a role in this context was shown by pharmacological inhibition of γ-secretase, which resulted in the depletion of a $CD133^+$ and CSC-like population. More remarkable, loss of Notch signaling within this cell population inhibited medulloblastoma growth both *in vitro* and *in vivo* xenografts (Fan *et al.*, 2006). This concept remains to be proven, but if true it would open the possibility of an antiangiogenic therapy that could result in depleting cells with CSC-like properties. Although these results seem to be encouraging, additional systematic genetic, pharmacological, and antibody based loss-of-function experiments in established murine tumor models need to be performed to confirm that Notch signaling is indeed required or functionally involved in the maintenance of medulloblastoma CSC.

7. Notch and Its Tumor Suppressive Properties in the Skin

In all the previous sections of this book chapter, we exclusively described growth promoting and oncogenic roles of the Notch signaling pathway. However, Notch function can substantially differ and be dependent on cell type and tissue. Often the role of Notch signaling in a given tissue is unpredictable. In tissues in which Notch exhibits growth-promoting functions its physiological role is generally associated with regulating differentiation, proliferation, or survival of immature progenitors during development or tissue homeostasis. Although Notch is well known for its role in maintaining progenitor cells in an undifferentiated state (e.g., in the intestine and the brain) or to influence their cell fate choices (e.g., hematopoietic system, or the pancreas), Notch signaling can also induce terminal differentiation, which is accompanied with growth suppression. The skin is one of the best-studied examples of Notch exerting growth suppressive functions.

The skin and its appendages form a physical barrier that must be constantly renewed. The epidermis is composed of multiple layers of keratinocytes that are separated from the dermis by a basement membrane. Skin SC and transient amplifying cells are found within the basal cell layer of the epidermis, characterized by keratin 5 and 14 expression, and being the reservoir for epidermal SC. Keratinocytes that undergo cell-cycle arrest detach from the basement membrane and move upward to form a suprabasal spinous layer, characterized by a shift to keratin 1 and 10 expression.

Post-mitotic keratinocytes continue to migrate toward the outer surface of the skin to form the granular layer, characterized by cells that acquire lipid-containing granules that release their content in the intracellular space and thereby contribute to the barrier function of the skin. Granular layer cells synthesize Filagrin and Loricrin, which participate in the formation of the cornified envelope in the outermost layer before eliminating their nuclei and cytoplasmic organelles, a process known as cornification (reviewed in Lefort and Dotto (2004)) (Fig. 13.4A).

Human and mouse epidermis differ not only anatomically but also in their expression pattern for Notch receptors and ligands. NOTCH1, NOTCH2, and NOTCH3 mRNA are highly expressed in the basal cell layer and to a lesser extent in the suprabasal layer of the human epidermis (Thelu *et al.*, 2002). DLL1 expression was shown to be highest in regions where potential SC reside which led to the suggestion that DLL1-mediated Notch signaling induces SC to differentiate into transient amplifying cells (Lowell *et al.*, 2000). In murine skin, Notch1–3 and Jagged1 and 2 are preferentially expressed in the suprabasal layer (Rangarajan *et al.*, 2001). Despite these differences in Notch receptor and ligand expression between human and murine skin, cell culture experiments from both systems combined with genetic mouse studies suggest that Notch signaling induces differentiation in keratinocytes (Lowell *et al.*, 2000; Rangarajan *et al.*, 2001).

The *p63* gene is important for the self-renewing properties and stratification of keratinocytes in the skin. p63 is expressed in the proliferating compartment of the skin and is downregulated as soon as keratinocytes start to differentiate (Koster and Roop, 2004; McKeon, 2004). NOTCH1 and p63 negatively regulate each other and thereby regulate the balance between self-renewing and differentiation (Nguyen *et al.*, 2006). In the context of cancer, p63 is often upregulated in epithelial tumors, including squamous cell carcinomas (Westfall and Pietenpol, 2004), in which Notch receptor expression is often downregulated (Lefort and Dotto, 2004). Furthermore, reciprocal gain and loss of Notch1 function in primary keratinocytes showed that Notch1 regulates the cell regulator $p21^{WAF1/Cip1}$. N1ICD-RBP-J complexes bind directly to the promoter of the human $p21^{WAF1/Cip1}$ gene and thereby participate in its transcriptional regulation. In addition, Notch can indirectly activate $p21^{WAF1/Cip1}$ in a mechanism regulating the nuclear factors of activated T cells (NFAT; Mammucari *et al.*, 2005). The role of Notch1-mediated induction of $p21^{WAF1/Cip1}$, which could contribute to cell-cycle exit of differentiating keratinocytes, seems to be restricted to the murine skin and has so far not been reported or confirmed in human skin. Additional Notch-mediated mechanisms that help keratinocytes to differentiate are the induction of keratin1/10 and involucrin (Rangarajan *et al.*, 2001) as well as downregulation of integrin expression (Blanpain *et al.*, 2006; Rangarajan *et al.*, 2001). Taken together,

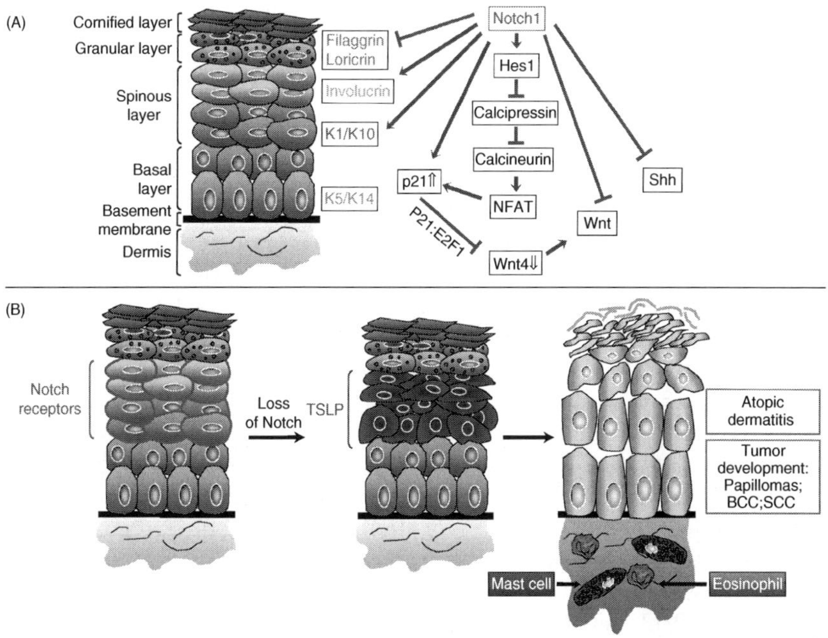

Figure 13.4 Notch signaling in the skin. (A) *Schematic representation of the murine skin depicting some proteins that are expressed in specific cellular layers.* The skin is a stratified squamous epithelium that is composed of multiple cell layers. The basal layer localizes to the basement membrane and consists mostly of transient amplifying cells with a few SC interspersed. The basal cell layer gives rise first to the spinous layer followed by the granular layer and then the cornified layer. Notch1 signaling induces the expression of early differentiation markers such as Keratin1 and Involucrin, and partially represses the expression of Loricrin and Filaggrin, two late differentiation markers. Moreover, Notch1 induces the expression of the cell-cycle regulator p21$^{CIP1/WAF}$ directly and through the activation of calcineurin/NFAT activity mediated by the downregulation of calcipressin via the Notch target gene Hes1. Both Wnt- and Shh-mediated signaling are normally suppressed in the murine epidermis via Notch1. The repression of the Wnt pathway is at least partially mediated by the downregulation of Wnt4 through a p21$^{CIP1/WAF}$:E2F1-dependent mechanism. (B) *Role of Notch as a tumor suppressor in the skin.* Skin specific loss of Notch signaling in the suprabasal layer leads to pronounced secretion of TSLP (thymic stromal lymphopoietin) by epithelial cells. The presence of TSLP in adult mice results in the recruitment of mast cells and eosinophiles within the dermis of Notch mutant mice, which increases massively in thickness. The infiltration of these cells contributes to massive inflammation and the development of an atopic dermatitis (AD)-like disease. The lack of Notch receptor expression in the skin leads also to spontaneous tumor development mostly being papillomas. A subset of these will progress to heavily vascularized basal cell carcinomas (BCC) and/or squamous cell carcinomas (SCC).

these studies strongly suggest that Notch indeed drives terminal differentiation processes in keratinocytes (Fig. 13.4A).

The data suggesting that Notch has tumor suppressive activities in the skin is mostly derived from genetic mouse studies and correlative expression studies of human skin lesions. Conditional inactivation of several signaling components of the Notch cascade including Notch1, Notch1 Notch2 Notch3 concomitantly, RBP-J, and Presinilin1 and 2 in mouse skin results in hyperproliferation of the skin, hair loss, and epidermal cyst formation within less than 4 weeks (Demehri *et al.*, 2008; Nicolas *et al.*, 2003; Pan *et al.*, 2004; Vauclair *et al.*, 2005; Yamamoto *et al.*, 2003). Skin tumors in mice lacking Notch1 develop only after a long latency period (approx. 12 months), but removal of additional Notch components accelerates time to tumor onset to as early as ~70 days in mice heterozygous for Notch2 and lacking Notch1 and Notch3 (Demehri *et al.*, 2009). The spontaneous tumors of these mice are papillomas with a subset progressing to heavily vascularized basal cell carcinoma-like tumors and a few squamous cell like tumors (Fig. 13.4B). This type of tumor in humans and mouse is frequently associated with aberrant shh signaling and indeed the hyperproliferative Notch1 deficient skin as well as the tumors showed increased Gli2 expression, which is a downstream component of the shh signaling cascade (Nicolas *et al.*, 2003). How loss of Notch1 resulted in upregulation of Gli2 is currently unknown. Hyperproliferative Notch1-deficient skin and tumors also exhibited increased β-catenin signaling activity suggesting that deregulated Wnt signaling may contribute to these skin lesions. This is consistent with a report showing that transgenic expression of a dominant negative form of MAML1, which inhibits Notch signaling mediated by all Notch receptors, develop squamous cell carcinoma that are characterized by the accumulation of nuclear β-catenin and cyclin D1 in tumor cells (Proweller *et al.*, 2006). Together, these results indicate that Notch might repress Wnt signaling in the normal skin (Nicolas *et al.*, 2003). Suppression of Wnt signaling by Notch was proposed to be mediated by an indirect mechanism in which $p21^{WAF1/Cip1}$ together with the E2F1 transcription factor bind to the promoter of Wnt4 and thereby downmodulate its expression (Devgan *et al.*, 2005), but this mechanism awaits confirmation (Fig. 13.4A).

The long latency of tumor onset in Notch1 deficient mice suggest that loss of Notch1 signaling on its own is not sufficient to develop skin tumors. During the latency period additional mutations are likely to accumulate thus sufficiently deregulating growth leading to tumor development. In this scenario Notch would cooperate with additional oncogenic mutations and thereby contribute to tumor development. This possibility was addressed using a classical chemical-induced carcinogenesis model. Treating skin with an initiating carcinogen 7,12-dimethylbenza anthracen (DMBA) generates cells carrying an initiating mutation in the *H-ras* gene. Subsequent continual exposure of the skin with a tumor-promoting agent (TPA) leads to the expansion of the mutated cells and eventually to tumor development. Application of this chemical carcinogen model to Notch1 deficient mice revealed that the frequency of mice that developed tumors as well as the tumor burden per mouse was tremendously increased in mice lacking Notch1 in the skin (Nicolas *et al.*, 2003). Repetitive administration of TPA alone on Notch1 deficient skin was not sufficient to induce tumors, but a single application of DMBA was sufficient to produce tumors in Notch1 deficient mice (Demehri *et al.*, 2009), suggesting that loss of Notch1 is not the primary oncogenic hit. Analysis of chimeric animals confirmed that Notch1 deficiency generates a "viscous cycle" of cytokines and growth factors acting as tumor promoting signals in the epidermis, affecting both Notch1 containing and Notch1 deleted cells harboring a Ras mutation (Demehri *et al.*, 2009). In addition, induced loss of Notch1 after papilloma formation showed that loss of Notch1 cooperates with oncogenic *H-ras* mutations to promote cancer (Demehri *et al.*, 2009), confirming a previous report (Nicolas *et al.*, 2003).

TPA induces hyperproliferation of the skin at least in part through inducing inflammation. It was demonstrated that skin specific loss of Notch1, or combined loss of Notch2 and Notch3, generated an inflammatory response in the skin, inducing also intense fibroplasia and angiogenesis (Demehri *et al.*, 2009). This might generate a chronic wound-like microenvironment which on a long-term basis may underlie tumor development following skin-specific inactivation of an increasing number of Notch alleles (Notch1, Notch 2, and Notch3) (Demehri *et al.*, 2009). One cytokine in particular correlates with accelerated tumor onset—thymic stromal lymphopoietin or TSLP. Its concentration in the serum, which causes an inflammatory response, is inversely correlated to the dose of Notch alleles (Demehri *et al.*, 2008). Mice lacking all three Notch receptors or RBP-J in the skin develop a sever form of atopic dermatitis-like disease characterized by acanthosis, spongiosis, and hyperkeratosis, as well as a massive dermal infiltration of eosinophils and mast cells (Demehri *et al.*, 2008; Dumortier *et al.*, 2010). Interestingly, a recent report indicated that Notch receptor expression is also downregulated in lesioned skin of human atopic dermatitis patients (Dumortier *et al.*, 2010). Moreover, mice with skin-specific compound loss of Notch develop lethal hematopoietic malignancies as a consequence of the high TSLP levels (Demehri *et al.*, 2008; Dumortier *et al.*, 2010). None of the mice lacking RBP-J or all Notch receptors or γ-secretase live long enough to develop skin tumors. How and whether TSLP-mediated inflammation influences tumor development of Notch deficient skin remains to be investigated (Fig. 13.4B).

The genetic mouse data seem to be consistent with observations in human skin cancer. Human basal cell carcinomas exhibit downregulated NOTCH1, NOTCH2, and JAGGED1 expression (Thelu *et al.*, 2002). Moreover, reduced expression of NOTCH1, NOTCH2, and HES1 was shown in a panel of human oral and skin squamous cell carcinoma cell lines, as well as in surgically excised squamous cell carcinomas from patients. In addition, suppression of Notch signaling in primary human keatinocytes that express an activated form of the *ras* gene is sufficient to cause aggressive squamous cell carcinomas in xenograft models (Lefort and Dotto, 2004), as seen with mouse keratinocytes (Nicolas *et al.*, 2003). Downregulation of NOTCH1 expression in human skin tumors may be linked to compromised p53 function, a regulator of NOTCH1 expression (Lefort *et al.*, 2007; Mandinova *et al.*, 2008; Yugawa *et al.*, 2007). Interestingly, a similar link between suppression of Notch and p53 activity was reported for Ewing's sarcoma (Ban *et al.*, 2008), possibly indicating that such a mechanism could be conserved in tissues where Notch has growth suppressive functions. Although the skin is clearly the best-studied organ system in which Notch exerts tumor suppressive properties, much more work needs to be done to map all the tissues where Notch loss may promote cancer. The prostate (Wang *et al.*, 2006), small cell lung cancer (Sriuranpong *et al.*, 2001, 2002) and hepatocellular carcinoma (Qi *et al.*, 2003) are all locations where loss of Notch signaling may promote dysplasia.

8. Therapeutically Targeting Notch in Cancer

Appreciation of the role Notch signaling plays in solid tumors is growing. Here we highlighted several important roles for this signaling pathway in cancer development. We reviewed evidence that Notch is constitutively active in several solid tumors and their CSC including colon, breast, and melanocytes cancers in which the association between Notch signaling and cancer development has been well documented. However, what has not been clearly defined so far is whether Notch activation is the cause or the effect in these solid tumors and whether it is just an essential factor for tumor growth utilized after the malignant changes were initiated. A true causative role for activated Notch in human carcinogenesis has so far only been demonstrated explicitly for human NOTCH1 in T-ALL cases.

Nevertheless, there is definitely a potential therapeutic benefit for targeting Notch in cancers including CSC depletion, reduced tumor angiogenesis, induced differentiation, and even cell death. The efficacy of targeting Notch in cancers will vary with the cancer types; even within the same cancer, targeting Notch may result in different effects on tumor subpopulations. Below we briefly review several possible strategies to target components of the Notch pathway in the context of cancer treatment (see Fig. 13.1B). Several strategies are currently tested in preclinical settings as well as in clinical trials. Inhibitory strategies include the following: (1) inhibitory antibodies (Abs) against individual Notch receptor and ligands with the aim to block specific receptor–ligand interactions (Hoey *et al.*, 2009; Noguera-Troise *et al.*, 2006; Ridgway *et al.*, 2006; Scehnet *et al.*, 2007); (2) receptor specific inhibitory Abs masking the S2 cleavage site thereby blocking ADAM-protease-mediated cleavage of the receptors (Li *et al.*, 2008); (3) various γ-secretase inhibitors (GSI) of different selectivity and efficacy, all preventing the S3 cleavage and thereby activation of all Notch receptors (Barten *et al.*, 2006; Real *et al.*, 2009; Staal and Langerak, 2008; Tolia and De Strooper, 2009; Wong *et al.*, 2003) and (4) novel stapled peptides blocking the formation of a functional NICD-MAML transcription complex (Moellering *et al.*, 2009) of unknown selectivity and low efficacy. Some of the more promising strategies are discussed in detail below.

The use of inhibitory Abs for Notch ligands has been shown to be effective in blocking *de novo* tumor blood vessel formation. DLL4, an important component of the Notch pathway, contributes to SC self-renewal and vascular development. DLL4 overexpression is found in tumor vasculature and in tumor cells to activate Notch signaling (Patel *et al.*, 2006; Yan *et al.*, 2001). Studies targeting blood vessel formation employing blocking Abs to DLL4 revealed substantial tumor growth reduction in cancer cell line-based xenograft models (Noguera-Troise *et al.*, 2006; Ridgway *et al.*, 2006; Scehnet *et al.*, 2007). The antitumor effect was shown to be the result of dysregulated angiogenesis characterized by increased sprouting in endothelial tip cells leading to non-functional vasculature in the tumor. Thus, inhibiting DLL4 disrupts productive angiogenesis in a manner distinct from traditional antiangiogenic therapies causing hyperproliferation of tumor vessels that leads to a reduction in tumor growth (Sainson and Harris, 2007; Thurston *et al.*, 2007). A land mark study by Hoey *et al.* (2009) demonstrate that blocking DLL4 signaling inhibits tumor growth through multiple mechanisms, including a reduction in CSC frequency. In addition to the previously described effect on deregulating angiogenesis by targeting DLL4 in the vasculature, they showed that selectively inhibiting DLL4 signaling in human tumor cells with a humanized anti-hDLL4 21M18 Ab leads to a decrease in colon tumor growth, a delay in tumor recurrence after chemotherapeutic treatment, and a decrease in the percentage of tumorigenic cells. Inhibition of DLL4-Notch signaling was envisioned as a tool for disrupting angiogenesis (Thurston *et al.*, 2007); both this utility as well as its efficacy in cancer treatment have yet to be proven effective in clinical trials. A report in *Nature* (Yan *et al.*, 2010) already raised some important safety concerns in the use of blocking DLL4 chronically. Using a rat model long term DLL4 blockade resulted in severe disruption of normal tissue homeostasis, caused pathological activation of endothelial cells and ultimately caused vascular neoplasms in skin, heart, and lung. In principle, specific inhibitors of individual Notch receptors may avoid or reduce the therapeutic complications caused by non-selective Notch inhibitors, and selective soluble Notch agonists would be useful experimental tools with possible therapeutic applications. Strategies used to inhibit Notch signaling in the laboratory include antisense RNA (Garces *et al.*, 1997),

RNA interference (Fan *et al.*, 2004; Purow *et al.*, 2005), soluble receptor decoys that act by sequestering Notch ligands (Garces *et al.*, 1997; Nickoloff *et al.*, 2002), and dominant negative peptides derived from MAML that decrease the transcriptional activation of target genes (Duncan *et al.*, 2005; Weng *et al.*, 2003). None of these research tools is close to being used in therapeutic applications. In a study recently published by Li and Zhou (Li *et al.*, 2008), a series of monoclonal Abs, which specifically inhibit or activate NOTCH3, was evaluated. The two potent blocking antibodies identified bind to overlapping epitopes on one face of the negative regulatory region (NRR) of NOTCH3 (see the chapter 2). These antibodies are highly specific for NOTCH3 and appear to mimic or inhibit certain effects of DSL ligands on cells. The small number of Abs assessed were potent and specific modulators of NOTCH3 S2 cleavage. The observation that the strongest inhibitory Abs all bind the NRR and not to the ligand binding domain will serve to focus further efforts to develop potent and selective Abs and small molecule modulators of human NOTCH receptors.

Inhibition of Notch signaling by pharmacological inhibition of the γ-secretase complex is a byproduct of efforts to treat or prevent Alzheimer's disease; several chemicals produced by different pharmaceutical companies have reached clinical trials for T-ALL. For example, the Notch inhibitor MK-0752 is currently under a phase I clinical trial for the treatment of T-ALL and advanced breast cancer (Clinical Trials ID: NCT00106145). The specificity, selectivity, and dosing strategies of GSI, which prevent the release of NICD, have been improving steadily over the last years. These agents target all Notch receptors since they all depend on γ-secretase (Doerfler *et al.*, 2001; Hadland *et al.*, 2001; Mizutani *et al.*, 2001; Saxena *et al.*, 2001). Even though there are at least six different γ-secretase complexes in humans, and subtype-specific inhibitors might be developed, Notch appears to be a substrate for each of these complexes. Most side effects observed with GSI can be attributed to Notch inhibition in other tissues, most frequently the intestine (Searfoss *et al.*, 2003). Recently however, Real *et al.* (2009) achieved a promising breakthrough using targeted combination therapy that may greatly improve the outcome in individuals with T-ALL. The therapy relies on the combination of a γ-secretase inhibitor with a glucocorticoid, dexamethasone. Dexamethasone counteracts lethal gut toxicity induced by the γ-secretase inhibitor. The authors outline how the combination therapy induces apoptosis in T-ALL cell lines, primary human T-ALL cells, and in xenografts of such T-ALL cell lines in mice to a much greater extent than either dexamethasone or the γ-secretase inhibitor alone. The approach, if it can be translated to human patients, might have real benefits over current treatment of T-ALL. The findings of Real *et al.* may have broader implications, as NOTCH1 signaling is involved in many cancers. By inference, the combination therapy might also be beneficial for solid tumors with aberrant NOTCH1 signaling.

A very recent publication converted the peptide MAML-based inhibitors into a drug-like molecule able to target the Notch/CSL transcription complex (Moellering *et al.*, 2009). As outlined in Chapter 2, the crystal structure of Notch/MAM/CSL identified a nearly continuous stretch of α-helices at the interface of the three proteins. Moellering *et al.* hypothesized that a helical peptide mimetic might be able to compete for binding to Notch ICD with full-length MAML1 and therefore inhibit transcriptional activation of Notch-targeted genes, which are cancer related. The researchers designed a series of six stapled α-helical peptides derived from MAML1, thus named for covalent backbone bonds stabilizing the helix. The stapled peptide is also more resistant to protease recognition and degradation. These stapled peptides were actively taken up by cells and entered the nucleus, where they can target the transcriptional process. *In vitro* cell culture studies confirmed that one peptide, SAHM1, prevented MAML1 from binding to the NICD-CSL complex, blocked expression of NOTCH1 target genes, and reduced proliferation of human T-ALL cell lines. The inhibitory effect was confirmed in a murine model of T-ALL reducing tumor burden significantly compared with vehicle. This strategy holds promise but its selectivity has not been tested, nor is it clear how its efficacy will improve to allow its adaptation to the clinic.

9. Concluding Remarks

In summary, it remains to be determined (a) which cancers are typified by active Notch signaling; (b) what specific roles are performed by components of the Notch signaling pathway in a given tumor (i.e., Notch1 versus Notch2); and (c) what genes or other pathways (i.e., Wnt or Hedgehog signaling) synergize with Notch where application of combinatorial therapy seems sensible. However, the road toward the development of safe and effective cell fate modifiers that target Notch signaling in cancer is open. Choosing the most promising agents for clinical development in specific indications, improving the available Notch-targeting biopharmaceuticals and designing new ones are the upcoming challenges. Progress in these areas will proceed hand in hand with advances in our understanding of the physiological roles of each Notch receptor and ligand and of the Notch signaling alterations associated with specific human diseases.

Acknowledgments

We would like to thank Luca Pellegrinet and Özden Yalçin for help in preparing the figures and Michela Marani for critically reading the manuscript.

REFERENCES

Al-Hajj, M., *et al.* (2003). Prospective identification of tumorigenic breast cancer cells. *Proc. Natl. Acad. Sci. U.S.A.* **100**, 3983–3988.

Apelqvist, A., *et al.* (1999). Notch signalling controls pancreatic cell differentiation. *Nature* **400**, 877–881.

Aubin-Houzelstein, G., *et al.* (2008). Melanoblasts' proper location and timed differentiation depend on Notch/RBP-J signaling in postnatal hair follicles. *J. Invest. Dermatol.* **128**, 2686–2695.

Balint, K., *et al.* (2005). Activation of Notch1 signaling is required for beta-catenin-mediated human primary melanoma progression. *J. Clin. Invest.* **115**, 3166–3176.

Ban, J., *et al.* (2008). EWS-FLI1 suppresses NOTCH-activated p53 in Ewing's sarcoma. *Cancer Res.* **68**, 7100–7109.

Barker, N., *et al.* (2007). Identification of stem cells in small intestine and colon by marker gene Lgr5. *Nature* **449**, 1003–1007.

Barten, D. M., *et al.* (2006). Gamma-secretase inhibitors for Alzheimer's disease: balancing efficacy and toxicity. *Drugs R. D.* **7**, 87–97.

Batlle, E., *et al.* (2002). Beta-catenin and TCF mediate cell positioning in the intestinal epithelium by controlling the expression of EphB/ephrinB. *Cell* **111**, 251–263.

Becker, D., *et al.* (1992). Inhibition of the fibroblast growth factor receptor 1 (FGFR-1) gene in human melanocytes and malignant melanomas leads to inhibition of proliferation and signs indicative of differentiation. *Oncogene* **7**, 2303–2313.

Bedogni, B., *et al.* (2008). Notch1 is an effector of Akt and hypoxia in melanoma development. *J. Clin. Invest.* **118**, 3660–3670.

Beres, T. M., *et al.* (2006). PTF1 is an organ-specific and Notch-independent basic helix-loop-helix complex containing the mammalian Suppressor of Hairless (RBP-J) or its paralogue, RBP-L. *Mol. Cell Biol.* **26**, 117–130.

Bittner, M., *et al.* (2000). Molecular classification of cutaneous malignant melanoma by gene expression profiling. *Nature* **406**, 536–540.

Blanpain, C., *et al.* (2006). Canonical notch signaling functions as a commitment switch in the epidermal lineage. *Genes Dev.* **20**, 3022–3035.

Borggrefe, T., Oswald, F. (2009). The Notch signaling pathway: transcriptional regulation at Notch target genes. *Cell Mol. Life Sci.* **66**, 1631–1646.

Bouras, T., *et al.* (2008). Notch signaling regulates mammary stem cell function and luminal cell-fate commitment. *Cell Stem Cell* **3**, 429–441.

Bray, S. J. (2006). Notch signalling: a simple pathway becomes complex. *Nat. Rev. Mol. Cell Biol.* **7**, 678–689.

Buchler, P., *et al.* (2005). The Notch signaling pathway is related to neurovascular progression of pancreatic cancer. *Ann. Surg.* **242**, 791–800, discussion 800–801.

Buono, K. D., *et al.* (2006). The canonical Notch/RBP-J signaling pathway controls the balance of cell lineages in mammary epithelium during pregnancy. *Dev. Biol.* **293**, 565–580.

Cannon-Albright, L. A., *et al.* (1996). A review of inherited predisposition to melanoma. *Semin. Oncol.* **23**, 667–672.

Clark, W. H. (1991). Tumour progression and the nature of cancer. *Br. J. Cancer.* **64**, 631–644.

Clifford, S. C., *et al.* (2006). Wnt/Wingless pathway activation and chromosome 6 loss characterize a distinct molecular sub-group of medulloblastomas associated with a favorable prognosis. *Cell Cycle* **5**, 2666–2670.

Collins, A. T., *et al.* (2005). Prospective identification of tumorigenic prostate cancer stem cells. *Cancer Res.* **65**, 10946–10951.

Cras-Meneur, C., *et al.* (2009). Presenilins, Notch dose control the fate of pancreatic endocrine progenitors during a narrow developmental window. *Genes Dev.* **23**, 2088–2101.

Crawford, J. R., *et al.* (2007). Medulloblastoma in childhood: new biological advances. *Lancet Neurol.* **6**, 1073–1085.

Dahmen, R. P., *et al.* (2001). Deletions of AXIN1, a component of the WNT/wingless pathway, in sporadic medulloblastomas. *Cancer Res.* **61**, 7039–7043.

Davies, H., *et al.* (2002). Mutations of the BRAF gene in human cancer. *Nature* **417**, 949–954.

De La, O. J., *et al.* (2008). Notch and Kras reprogram pancreatic acinar cells to ductal intraepithelial neoplasia. *Proc. Natl. Acad. Sci. U.S.A.* **105**, 18907–18912.

Demehri, S., *et al.* (2008). Notch-deficient skin induces a lethal systemic B-lymphoproliferative disorder by secreting TSLP, a sentinel for epidermal integrity. *PLoS Biol.* **6**, e123.

Demehri, S., *et al.* (2009). Epidermal Notch1 Loss Promotes Skin Tumorigenesis by Impacting the Stromal Microenvironment. *Cancer Cell* **16**, 55–66.

Demunter, A., *et al.* (2001). Analysis of N- and K-ras mutations in the distinctive tumor progression phases of melanoma. *J. Invest. Dermatol.* **117**, 1483–1489.

Devgan, V., *et al.* (2005). p21WAF1/Cip1 is a negative transcriptional regulator of Wnt4 expression downstream of Notch1 activation. *Genes Dev.* **19**, 1485–1495.

Di Marcotullio, L., *et al.* (2006). Numb is a suppressor of Hedgehog signalling and targets Gli1 for Itch-dependent ubiquitination. *Nat. Cell Biol.* **8**, 1415–1423.

Dievart, A., *et al.* (1999). Involvement of Notch1 in the development of mouse mammary tumors. *Oncogene* **18**, 5973–5981.

Doerfler, P., *et al.* (2001). Presenilin-dependent gamma -secretase activity modulates thymocyte development. *Proc. Natl. Acad. Sci. U.S.A.* **98**, 9312–9317.

Dumortier, A., *et al.* (2010). Atopic Dermatitis-Like Disease and Associated Lethal Myeloproliferative Disorder Arise from Loss of Notch Signaling in the Murine Skin. *PLoS ONE.* **5**, e9258–9271.

Duncan, A. W., *et al.* (2005). Integration of Notch and Wnt signaling in hematopoietic stem cell maintenance. *Nat. Immunol.* **6**, 314–322.

Eberhart, C. G. (2007). In search of the medulloblast: neural stem cells and embryonal brain tumors. *Neurosurg. Clin. N. Am.* **18**, 59–69, viii-ix.

Eberhart, C. G., *et al.* (2000). Nuclear localization and mutation of beta-catenin in medulloblastomas. *J. Neuropathol. Exp. Neurol.* **59**, 333–337.

Edlund, H. (2002). Pancreatic organogenesis–developmental mechanisms and implications for therapy. *Nat. Rev. Genet.* **3**, 524–532.

Ellisen, L. W., *et al.* (1991). TAN-1, the human homolog of the Drosophila notch gene, is broken by chromosomal translocations in T lymphoblastic neoplasms. *Cell* **66**, 649–661.

Ellison, D. W., *et al.* (2005). beta-Catenin status predicts a favorable outcome in childhood medulloblastoma: the United Kingdom Children's Cancer Study Group Brain Tumour Committee. *J. Clin. Oncol.* **23**, 7951–7957.

Estrach, S., *et al.* (2006). Jagged 1 is a beta-catenin target gene required for ectopic hair follicle formation in adult epidermis. *Development* **133**, 4427–4438.

Fan, X., *et al.* (2004). Notch1 and notch2 have opposite effects on embryonal brain tumor growth. *Cancer Res.* **64**, 7787–7793.

Fan, X., *et al.* (2006). Notch pathway inhibition depletes stem-like cells and blocks engraftment in embryonal brain tumors. *Cancer Res.* **66**, 7445–7452.

Farnie, G., *et al.* (2007). Novel cell culture technique for primary ductal carcinoma in situ: role of Notch and epidermal growth factor receptor signaling pathways. *J. Natl. Cancer. Inst.* **99**, 616–627.

Fernandez-Majada, V., *et al.* (2007). Nuclear IKK activity leads to dysregulated notch-dependent gene expression in colorectal cancer. *Proc. Natl. Acad. Sci. U.S.A.* **104**, 276–281.

Fre, S., *et al.* (2005). Notch signals control the fate of immature progenitor cells in the intestine. *Nature* **435**, 964–968.

Fre, S., *et al.* (2009). Notch and Wnt signals cooperatively control cell proliferation and tumorigenesis in the intestine. *Proc. Natl. Acad. Sci. U.S.A.* **106**, 6309–6314.

Fujikura, J., *et al.* (2006). Notch/Rbp-j signaling prevents premature endocrine and ductal cell differentiation in the pancreas. *Cell Metab.* **3**, 59–65.

Fujikura, J., *et al.* (2007). Rbp-j regulates expansion of pancreatic epithelial cells and their differentiation into exocrine cells during mouse development. *Dev. Dyn.* **236**, 2779–2791.

Gaiano, N., Fishell, G. (2002). The role of notch in promoting glial and neural stem cell fates. *Annu. Rev. Neurosci.* **25**, 471–490.

Gallahan, D., Callahan, R. (1997). The mouse mammary tumor associated gene INT3 is a unique member of the NOTCH gene family (NOTCH4). *Oncogene* **14**, 1883–1890.

Gallahan, D., *et al.* (1987). A new common integration region (int-3) for mouse mammary tumor virus on mouse chromosome 17. *J. Virol.* **61**, 218–220.

Gallahan, D., *et al.* (1996). Expression of a truncated Int3 gene in developing secretory mammary epithelium specifically retards lobular differentiation resulting in tumorigenesis. *Cancer Res.* **56**, 1775–1785.

Galli, R., *et al.* (2004). Isolation and characterization of tumorigenic, stem-like neural precursors from human glioblastoma. *Cancer Res.* **64**, 7011–7021.

Garces, C., *et al.* (1997). Notch-1 controls the expression of fatty acid-activated transcription factors and is required for adipogenesis. *J. Biol. Chem.* **272**, 29729–29734.

Georgia, S., *et al.* (2006). p57 and Hes1 coordinate cell cycle exit with self-renewal of pancreatic progenitors. *Dev. Biol.* **298**, 22–31.

Gomez, G., *et al.* (2004). Increased expression of hypoxia-inducible factor-1alpha, p48, and the Notch signaling cascade during acute pancreatitis in mice. *Pancreas* **28**, 58–64.

Gordon, W. R., *et al.* (2008). The molecular logic of Notch signaling–a structural and biochemical perspective. *J. Cell. Sci.* **121**, 3109–3119.

Guilmeau, S., *et al.* (2010). Heterogeneity of Jagged1 expression in human and mouse intestinal tumors: implications for targeting Notch signaling. *Oncogene* **29**, 992–1002.

Gutgemann, A., *et al.* (2001). Isolation of invasion-associated cDNAs in melanoma. *Arch. Dermatol. Res.* **293**, 283–290.

Habbe, N., *et al.* (2008). Spontaneous induction of murine pancreatic intraepithelial neoplasia (mPanIN) by acinar cell targeting of oncogenic Kras in adult mice. *Proc. Natl. Acad. Sci. U.S.A.* **105**, 18913–18918.

Hadland, B. K., *et al.* (2001). gamma-secretase inhibitors repress thymocyte development. *Proc. Natl. Acad. Sci. U.S.A.* **98**, 7487–7491.

Haines, N., Irvine, K. (2003). Glycosylation regulates Notch signalling. *Nat. Rev. Mol. Cell Biol.* **4**, 786–797.

Hald, J., *et al.* (2003). Activated Notch1 prevents differentiation of pancreatic acinar cells and attenuate endocrine development. *Dev. Biol.* **260**, 426–437.

Hallahan, A. R., *et al.* (2004). The SmoA1 mouse model reveals that notch signaling is critical for the growth and survival of sonic hedgehog-induced medulloblastomas. *Cancer Res.* **64**, 7794–7800.

Haltiwanger, R. S., Stanley, P. (2002). Modulation of receptor signaling by glycosylation: fringe is an O-fucose-beta1,3-N-acetylglucosaminyltransferase. *Biochim. Biophys. Acta* **1573**, 328–335.

Han, J., *et al.* (2009). The cytotoxicity of gamma-secretase inhibitor I to breast cancer cells is mediated by proteasome inhibition, not by gamma-secretase inhibition. *Breast Cancer Res.* **11**, R57.

Harrison, H., *et al.* (2010). Regulation of breast cancer stem cell activity by signaling through the Notch4 receptor. *Cancer Res.* **70**, 709–718.

He, T. C., *et al.* (1998). Identification of c-MYC as a target of the APC pathway. *Science* **281**, 1509–1512.

Hemmati, H. D., *et al.* (2003). Cancerous stem cells can arise from pediatric brain tumors. *Proc. Natl. Acad. Sci. U.S.A.* **100**, 15178–15183.
Hezel, A. F., *et al.* (2006). Genetics and biology of pancreatic ductal adenocarcinoma. *Genes Dev.* **20**, 1218–1249.
Hingorani, S. R., *et al.* (2003). Preinvasive and invasive ductal pancreatic cancer and its early detection in the mouse. *Cancer Cell* **4**, 437–450.
Hoek, K., *et al.* (2004). Expression profiling reveals novel pathways in the transformation of melanocytes to melanomas. *Cancer Res.* **64**, 5270–5282.
Hoey, T., *et al.* (2009). DLL4 blockade inhibits tumor growth and reduces tumor-initiating cell frequency. *Cell Stem Cell* **5**, 168–177.
Hori, K., *et al.* (2008). A nonclassical bHLH Rbpj transcription factor complex is required for specification of GABAergic neurons independent of Notch signaling. *Genes Dev.* **22**, 166–178.
Hruban, R. H., *et al.* (1999). Pathology of incipient pancreatic cancer. *Ann. Oncol.* **10 Suppl 4**, 9–11.
Hu, C., *et al.* (2006). Overexpression of activated murine Notch1 and Notch3 in transgenic mice blocks mammary gland development and induces mammary tumors. *Am. J. Pathol.* **168**, 973–990.
Huang, H., *et al.* (2000). APC mutations in sporadic medulloblastomas. *Am. J. Pathol.* **156**, 433–437.
Ignatova, T. N., *et al.* (2002). Human cortical glial tumors contain neural stem-like cells expressing astroglial and neuronal markers *in vitro*. *Glia* **39**, 193–206.
Imatani, A., Callahan, R. (2000). Identification of a novel NOTCH-4/INT-3 RNA species encoding an activated gene product in certain human tumor cell lines. *Oncogene* **19**, 223–231.
Jemal, A., *et al.* (2009). Cancer statistics, 2009. *CA Cancer J. Clin.* **59**, 225–249.
Jensen, J., *et al.* (2000). Control of endodermal endocrine development by Hes-1. *Nat. Genet.* **24**, 36–44.
Jhappan, C., *et al.* (1992). Expression of an activated Notch-related int-3 transgene interferes with cell differentiation and induces neoplastic transformation in mammary and salivary glands. *Genes Dev.* **6**, 345–355.
Jorgensen, M. C., *et al.* (2007). An illustrated review of early pancreas development in the mouse. *Endocr. Rev.* **28**, 685–705.
Katoh, M. (2006). Notch ligand, JAG1, is evolutionarily conserved target of canonical WNT signaling pathway in progenitor cells. *Int. J. Mol. Med.* **17**, 681–685.
Kenney, A. M., *et al.* (2003). Nmyc upregulation by sonic hedgehog signaling promotes proliferation in developing cerebellar granule neuron precursors. *Development* **130**, 15–28.
Kiaris, H., *et al.* (2004). Modulation of notch signaling elicits signature tumors and inhibits hras1-induced oncogenesis in the mouse mammary epithelium. *Am. J. Pathol.* **165**, 695–705.
Kidd, S., *et al.* (1986). Sequence of the notch locus of Drosophila melanogaster: relationship of the encoded protein to mammalian clotting and growth factors. *Mol. Cell Biol.* **6**, 3094–3108.
Kiel, M. J., Morrison, S. J. (2008). Uncertainty in the niches that maintain haematopoietic stem cells. *Nat. Rev. Immunol.* **8**, 290–301.
Kimura, K., *et al.* (2007). Activation of Notch signaling in tumorigenesis of experimental pancreatic cancer induced by dimethylbenzanthracene in mice. *Cancer Sci.* **98**, 155–162.
Klinakis, A., *et al.* (2006). Myc is a Notch1 transcriptional target and a requisite for Notch1-induced mammary tumorigenesis in mice. *Proc. Natl. Acad. Sci. U.S.A.* **103**, 9262–9267.
Kopan, R., Ilagan, M.X.G. (2009). The canonical Notch signaling pathway: unfolding the activation mechanism. *Cell* **137**, 216–233.

Kordon, E. C., *et al.* (1995). A novel non-mouse mammary tumor virus activation of the Int-3 gene in a spontaneous mouse mammary tumor. *J. Virol.* **69**, 8066–8069.

Korinek, V., *et al.* (1998). Depletion of epithelial stem-cell compartments in the small intestine of mice lacking Tcf-4. *Nat. Genet.* **19**, 379–383.

Koster, M. I., Roop, D. R. (2004). The role of p63 in development and differentiation of the epidermis. *J. Dermatol. Sci.* **34**, 3–9.

Krebs, L. T., *et al.* (2000). Notch signaling is essential for vascular morphogenesis in mice. *Genes Dev.* **14**, 1343–1352.

Krebs, L. T., *et al.* (2003). Notch signaling regulates left-right asymmetry determination by inducing Nodal expression. *Genes Dev.* **17**, 1207–1212.

Krejci, A., *et al.* (2009). Direct response to Notch activation: signaling crosstalk and incoherent logic. *Sci. Signal.* **2**, ra1.

Kumano, K., *et al.* (2008). Both Notch1 and Notch2 contribute to the regulation of melanocyte homeostasis. *Pigment Cell Melanoma Res.* **21**, 70–78.

Lawson, D. A., *et al.* (2007). Isolation and functional characterization of murine prostate stem cells. *Proc. Natl. Acad. Sci. U.S.A.* **104**, 181–186.

Lee, C. W., *et al.* (2008). Molecular dependence of estrogen receptor-negative breast cancer on a notch-survivin signaling axis. *Cancer Res.* **68**, 5273–5281.

Lee, J., *et al.* (2007). Bi-compartmental communication contributes to the opposite proliferative behavior of Notch1-deficient hair follicle and epidermal keratinocytes. *Development* **134**, 2795–2806.

Lee, J. C., *et al.* (2001). Regulation of the pancreatic pro-endocrine gene neurogenin3. *Diabetes* **50**, 928–936.

Lee, Y., *et al.* (2007). Loss of suppressor-of-fused function promotes tumorigenesis. *Oncogene* **26**, 6442–6447.

Lefort, K., Dotto, G. P. (2004). Notch signaling in the integrated control of keratinocyte growth/differentiation and tumor suppression. *Semin. Cancer Biol.* **14**, 374–386.

Lefort, K., *et al.* (2007). Notch1 is a p53 target gene involved in human keratinocyte tumor suppression through negative regulation of ROCK1/2 and MRCKalpha kinases. *Genes Dev.* **21**, 562–577.

Leong, K. G., *et al.* (2007). Jagged1-mediated Notch activation induces epithelial-to-mesenchymal transition through Slug-induced repression of E-cadherin. *J. Exp. Med.* **204**, 2935–2948.

Li, C., *et al.* (2007). Identification of pancreatic cancer stem cells. *Cancer Res.* **67**, 1030–1037.

Li, K., *et al.* (2008). Modulation of Notch signaling by antibodies specific for the extracellular negative regulatory region of NOTCH3. *J. Biol. Chem.* **283**, 8046–8054.

Liu, Z. J., *et al.* (2006). Notch1 signaling promotes primary melanoma progression by activating mitogen-activated protein kinase/phosphatidylinositol 3-kinase-Akt pathways and up-regulating N-cadherin expression. *Cancer Res.* **66**, 4182–4190.

Louis, D. N., *et al.* (2007). The 2007 WHO classification of tumours of the central nervous system. *Acta Neuropathol.* **114**, 97–109.

Lowell, S., *et al.* (2000). Stimulation of human epidermal differentiation by delta-notch signalling at the boundaries of stem-cell clusters. *Curr. Biol.* **10**, 491–500.

Mammucari, C., *et al.* (2005). Integration of Notch 1 and calcineurin/NFAT signaling pathways in keratinocyte growth and differentiation control. *Dev. Cell* **8**, 665–676.

Mandinova, A., *et al.* (2008). The FoxO3a gene is a key negative target of canonical Notch signalling in the keratinocyte UVB response. *EMBO J.* **27**, 1243–1254.

Massi, D., *et al.* (2006). Evidence for differential expression of Notch receptors and their ligands in melanocytic nevi and cutaneous malignant melanoma. *Mod. Pathol.* **19**, 246–254.

Masui, T., *et al.* (2007). Early pancreatic development requires the vertebrate Suppressor of Hairless (RBPJ) in the PTF1 bHLH complex. *Genes Dev.* **21**, 2629–2643.

McKeon, F. (2004). p63 and the epithelial stem cell: more than status quo? *Genes Dev.* **18**, 465–469.

Micchelli, C. A., Perrimon, N. (2006). Evidence that stem cells reside in the adult Drosophila midgut epithelium. *Nature* **439**, 475–479.

Miyamoto, Y., *et al.* (2003). Notch mediates TGF alpha-induced changes in epithelial differentiation during pancreatic tumorigenesis. *Cancer Cell* **3**, 565–576.

Mizutani, T., *et al.* (2001). Conservation of the biochemical mechanisms of signal transduction among mammalian Notch family members. *Proc. Natl. Acad. Sci. U.S.A.* **98**, 9026–9031.

Moellering, R. E., *et al.* (2009). Direct inhibition of the NOTCH transcription factor complex. *Nature* **462**, 182–188.

Monticone, M., *et al.* (2009). z-Leucinyl-leucinyl-norleucinal induces apoptosis of human glioblastoma tumor-initiating cells by proteasome inhibition and mitotic arrest response. *Mol. Cancer Res.* **7**, 1822–1834.

Morgan, T. H. (1917). The theory of the gene. *Am. Nat.* **51**, 513–544.

Moriyama, M., *et al.* (2006). Notch signaling via Hes1 transcription factor maintains survival of melanoblasts and melanocyte stem cells. *J. Cell Biol.* **173**, 333–339.

Murtaugh, L. C., *et al.* (2003). Notch signaling controls multiple steps of pancreatic differentiation. *Proc. Natl. Acad. Sci. U.S.A.* **100**, 14920–14925.

Nakamura, T., *et al.* (2007). Crosstalk between Wnt and Notch signaling in intestinal epithelial cell fate decision. *J. Gastroenterol.* **42**, 705–710.

Nakhai, H., *et al.* (2008). Conditional ablation of Notch signaling in pancreatic development. *Development* **135**, 2757–2765.

Nathke, I. S. (2004). The adenomatous polyposis coli protein: the Achilles heel of the gut epithelium. *Annu. Rev. Cell Dev. Biol.* **20**, 337–366.

Nguyen, B. C., *et al.* (2006). Cross-regulation between Notch and p63 in keratinocyte commitment to differentiation. *Genes Dev.* **20**, 1028–1042.

Nickoloff, B. J., *et al.* (2002). Jagged-1 mediated activation of notch signaling induces complete maturation of human keratinocytes through NF-kappaB and PPARgamma. *Cell Death Differ.* **9**, 842–855.

Nicolas, M., *et al.* (2003). Notch1 functions as a tumor suppressor in mouse skin. *Nat. Genet.* **33**, 416–421.

Noguera-Troise, I., *et al.* (2006). Blockade of Dll4 inhibits tumour growth by promoting non-productive angiogenesis. *Nature* **444**, 1032–1037.

O'Brien, C. A., *et al.* (2007). A human colon cancer cell capable of initiating tumour growth in immunodeficient mice. *Nature* **445**, 106–110.

Ohlstein, B., Spradling, A. (2006). The adult Drosophila posterior midgut is maintained by pluripotent stem cells. *Nature* **439**, 470–474.

Oliver, T. G., Wechsler-Reya, R. J. (2004). Getting at the root and stem of brain tumors. *Neuron* **42**, 885–888.

O'Neil, J., *et al.* (2007). FBW7 mutations in leukemic cells mediate NOTCH pathway activation and resistance to gamma-secretase inhibitors. *J. Exp. Med.* **204**, 1813–1824.

O'Neill, C. F., *et al.* (2007). Notch2 signaling induces apoptosis and inhibits human MDA-MB-231 xenograft growth. *Am. J. Pathol.* **171**, 1023–1036.

Packer, R. J., *et al.* (1999). Medulloblastoma: clinical and biologic aspects. *Neuro-oncology* **1**, 232–250.

Palomero, T., *et al.* (2006). NOTCH1 directly regulates c-MYC and activates a feed-forward-loop transcriptional network promoting leukemic cell growth. *Proc. Natl. Acad. Sci. U.S.A.* **103**, 18261–18266.

Pan, Y., *et al.* (2004). gamma-secretase functions through Notch signaling to maintain skin appendages but is not required for their patterning or initial morphogenesis. *Dev. Cell* **7**, 731–743.

Parr, C., *et al.* (2004). The possible correlation of Notch-1 and Notch-2 with clinical outcome and tumour clinicopathological parameters in human breast cancer. *Int. J. Mol. Med.* **14**, 779–786.

Pasca di Magliano, M., *et al.* (2006). Hedgehog/Ras interactions regulate early stages of pancreatic cancer. *Genes Dev.* **20**, 3161–3173.

Patel, N. S., *et al.* (2006). Up-regulation of endothelial delta-like 4 expression correlates with vessel maturation in bladder cancer. *Clin. Cancer Res.* **12**, 4836–4844.

Pear, W. S., *et al.* (1996). Exclusive development of T cell neoplasms in mice transplanted with bone marrow expressing activated Notch alleles. *J. Exp. Med.* **183**, 2283–2291.

Pece, S., *et al.* (2004). Loss of negative regulation by Numb over Notch is relevant to human breast carcinogenesis. *J. Cell. Biol.* **167**, 215–221.

Pinnix, C. C., Herlyn, M. (2007). The many faces of Notch signaling in skin-derived cells. *Pigment Cell Res.* **20**, 458–465.

Pinnix, C. C., *et al.* (2009). Active Notch1 confers a transformed phenotype to primary human melanocytes. *Cancer Res.* **69**, 5312–5320.

Pinto, D., *et al.* (2003). Canonical Wnt signals are essential for homeostasis of the intestinal epithelium. *Genes Dev.* **17**, 1709–1713.

Plentz, R., *et al.* (2009). Inhibition of gamma-secretase activity inhibits tumor progression in a mouse model of pancreatic ductalz adenocarcinoma. *Gastroenterology* **136**, 1741–1749e6

Postovit, L. M., *et al.* (2007). Targeting Nodal in malignant melanoma cells. *Expert Opin. Ther. Targets* **11**, 497–505.

Proweller, A., *et al.* (2006). Impaired notch signaling promotes de novo squamous cell carcinoma formation. *Cancer Res.* **66**, 7438–7444.

Purow, B. W., *et al.* (2005). Expression of Notch-1 and its ligands, Delta-like-1 and Jagged-1, is critical for glioma cell survival and proliferation. *Cancer Res.* **65**, 2353–2363.

Qi, R., *et al.* (2003). Notch1 signaling inhibits growth of human hepatocellular carcinoma through induction of cell cycle arrest and apoptosis. *Cancer Res.* **63**, 8323–8329.

Qin, J. Z., *et al.* (2004). p53-independent NOXA induction overcomes apoptotic resistance of malignant melanomas. *Mol. Cancer Ther.* **3**, 895–902.

Raafat, A., *et al.* (2004). Mammary development and tumorigenesis in mice expressing a truncated human Notch4/Int3 intracellular domain (h-Int3sh). *Oncogene* **23**, 9401–9407.

Raafat, A., *et al.* (2007). Kit and PDGFR-alpha activities are necessary for Notch4/ Int3-induced tumorigenesis. *Oncogene* **26**, 662–672.

Raafat, A., *et al.* (2009). Rbpj conditional knockout reveals distinct functions of Notch4/ Int3 in mammary gland development and tumorigenesis. *Oncogene* **28**, 219–230.

Radtke, F., Clevers, H. (2005). Self-renewal and cancer of the gut: two sides of a coin. *Science* **307**, 1904–1909.

Radtke, F., *et al.* (2006). From gut homeostasis to cancer. *Curr. Mol. Med.* **6**, 275–289.

Raffel, C., *et al.* (1997). Sporadic medulloblastomas contain PTCH mutations. *Cancer Res.* **57**, 842–845.

Rangarajan, A., *et al.* (2001). Notch signaling is a direct determinant of keratinocyte growth arrest and entry into differentiation. *EMBO J.* **20**, 3427–3436.

Raouf, A., *et al.* (2008). Transcriptome analysis of the normal human mammary cell commitment and differentiation process. *Cell Stem Cell* **3**, 109–118.

Raya, A., *et al.* (2003). Notch activity induces Nodal expression and mediates the establishment of left-right asymmetry in vertebrate embryos. *Genes Dev.* **17**, 1213–1218.

Real, P. J., *et al.* (2009). Gamma-secretase inhibitors reverse glucocorticoid resistance in T cell acute lymphoblastic leukemia. *Nat. Med.* **15**, 50–58.

Reedijk, M., *et al.* (2005). High-level coexpression of JAG1 and NOTCH1 is observed in human breast cancer and is associated with poor overall survival. *Cancer Res.* **65**, 8530–8537.

Reedijk, M., *et al.* (2008). Activation of Notch signaling in human colon adenocarcinoma. *Int. J. Oncol.* **33**, 1223–1229.

Reedijk, M., *et al.* (2008). JAG1 expression is associated with a basal phenotype and recurrence in lymph node-negative breast cancer. *Breast Cancer Res. Treat.* **111**, 439–448.

Reya, T., *et al.* (2001). Stem cells, cancer, and cancer stem cells. *Nature* **414**, 105–111.

Reynolds, T. C., *et al.* (1987). Analysis of DNA surrounding the breakpoints of chromosomal translocations involving the beta T cell receptor gene in human lymphoblastic neoplasms. *Cell* **50**, 107–117.

Ricci-Vitiani, L., *et al.* (2007). Identification and expansion of human colon-cancer-initiating cells. *Nature* **445**, 111–115.

Riccio, O., *et al.* (2008). Loss of intestinal crypt progenitor cells owing to inactivation of both Notch1 and Notch2 is accompanied by derepression of CDK inhibitors p27(Kip1) and p57(Kip2). *EMBO Rep.* **9**, 377–383.

Ridgway, J., *et al.* (2006). Inhibition of Dll4 signalling inhibits tumour growth by deregulating angiogenesis. *Nature* **444**, 1083–1087.

Rizzo, P., *et al.* (2008). Cross-talk between notch and the estrogen receptor in breast cancer suggests novel therapeutic approaches. *Cancer Res.* **68**, 5226–5235.

Robbins, J., *et al.* (1992). Mouse mammary tumor gene int-3: a member of the notch gene family transforms mammary epithelial cells. *J. Virol.* **66**, 2594–2599.

Rodilla, V., *et al.* (2009). Jagged1 is the pathological link between Wnt and Notch pathways in colorectal cancer. *Proc. Natl. Acad. Sci. U.S.A.* **106**, 6315–6320.

Rustighi, A., *et al.* (2009). The prolyl-isomerase Pin1 is a Notch1 target that enhances Notch1 activation in cancer. *Nat. Cell Biol.* **11**, 133–142.

Sainson, R. C., Harris, A. L. (2007). Anti-Dll4 therapy: can we block tumour growth by increasing angiogenesis? *Trends Mol. Med.* **13**, 389–395.

Sancho, E., *et al.* (2004). Signaling pathways in intestinal development and cancer. *Annu. Rev. Cell Dev. Biol.* **20**, 695–723.

Sansom, O. J., *et al.* (2006). Loss of Apc allows phenotypic manifestation of the transforming properties of an endogenous K-ras oncogene *in vivo*. *Proc. Natl. Acad. Sci. U.S.A.* **103**, 14122–14127.

Sansom, O. J., *et al.* (2007). Myc deletion rescues Apc deficiency in the small intestine. *Nature* **446**, 676–679.

Sawey, E. T., *et al.* (2007). Matrix metalloproteinase 7 controls pancreatic acinar cell transdifferentiation by activating the Notch signaling pathway. *Proc. Natl. Acad. Sci. U.S.A.* **104**, 19327–19332.

Saxena, M. T., *et al.* (2001). Murine Notch homologs (N 1-4) undergo Presenilin dependent proteolysis. *J. Biol. Chem.* **276**, 40268–40273.

Scehnet, J. S., *et al.* (2007). Inhibition of Dll4-mediated signaling induces proliferation of immature vessels and results in poor tissue perfusion. *Blood* **109**, 4753–4760.

Schier, A. F. (2003). Nodal signaling in vertebrate development. *Annu. Rev. Cell Dev. Biol.***19**, 589–621.

Schouwey, K., *et al.* (2007). Notch1 and Notch2 receptors influence progressive hair graying in a dose-dependent manner. *Dev. Dyn.* **236**, 282–289.

Searfoss, G.H., *et al.* (2003). Adipsin: a biomarker of gastrointestinal toxicity mediated by a functional gamma secretase inhibitor. J Biol Chem.

Singh, S. K., *et al.* (2003). Identification of a cancer stem cell in human brain tumors. *Cancer Res.* **63**, 5821–5828.

Singh, S. K., *et al.* (2004). Identification of human brain tumour initiating cells. *Nature* **432**, 396–401.

Siveke, J. T., *et al.* (2008). Notch signaling is required for exocrine regeneration after acute pancreatitis. *Gastroenterology* **134**, 544–555.

Solecki, D. J., *et al.* (2001). Activated Notch2 signaling inhibits differentiation of cerebellar granule neuron precursors by maintaining proliferation. *Neuron* **31**, 557–568.

Sriuranpong, V., *et al.* (2001). Notch signaling induces cell cycle arrest in small cell lung cancer cells. *Cancer Res.* **61**, 3200–3205.

Sriuranpong, V., *et al.* (2002). Notch signaling induces rapid degradation of achaete-scute homolog 1. *Mol. Cell Biol.* **22**, 3129–3139.

Staal, F. J., Langerak, A. W. (2008). Signaling pathways involved in the development of T-cell acute lymphoblastic leukemia. *Haematologica* **93**, 493–497.

Stahl, J. M., *et al.* (2004). Deregulated Akt3 activity promotes development of malignant melanoma. *Cancer Res.* **64**, 7002–7010.

Stylianou, S., *et al.* (2006). Aberrant activation of notch signaling in human breast cancer. *Cancer Res.* **66**, 1517–1525.

Tetsu, O., McCormick, F. (1999). Beta-catenin regulates expression of cyclin D1 in colon carcinoma cells. *Nature* **398**, 422–426.

Thelu, J., *et al.* (2002). Notch signalling is linked to epidermal cell differentiation level in basal cell carcinoma, psoriasis and wound healing. *BMC Dermatol.* **2**, 7.

Thompson, B. J., *et al.* (2007). The SCFFBW7 ubiquitin ligase complex as a tumor suppressor in T cell leukemia. *J. Exp. Med.* **204**, 1825–1835.

Thurston, G., *et al.* (2007). The Delta paradox: DLL4 blockade leads to more tumour vessels but less tumour growth. *Nat. Rev. Cancer* **7**, 327–331.

Tolia, A., De Strooper, B. (2009). Structure and function of gamma-secretase. *Semin. Cell Dev. Biol.***20**, 211–218.

Topczewska, J. M., *et al.* (2006). Embryonic and tumorigenic pathways converge via Nodal signaling: role in melanoma aggressiveness. *Nat. Med.* **12**, 925–932.

Uyttendaele, H., *et al.* (1996). Notch4/int-3, a mammary proto-oncogene, is an endothelial cell-specific mammalian Notch gene. *Development* **122**, 2251–2259.

van de Wetering, M., *et al.* (2002). The beta-catenin/TCF-4 complex imposes a crypt progenitor phenotype on colorectal cancer cells. *Cell* **111**, 241–250.

van Es, J. H., *et al.* (2005). Notch/gamma-secretase inhibition turns proliferative cells in intestinal crypts and adenomas into goblet cells. *Nature* **435**, 959–963.

Vauclair, S., *et al.* (2005). Notch1 is essential for postnatal hair follicle development and homeostasis. *Dev. Biol.* **284**, 184–193.

Vooijs, M., *et al.* (2007). Mapping the consequence of Notch1 proteolysis *in vivo* with NIP-CRE. *Development* **134**, 535–544.

Wang, X. D., *et al.* (2006). Notch signaling is required for normal prostatic epithelial cell proliferation and differentiation. *Dev. Biol.* **290**, 66–80.

Weeraratna, A. T., *et al.* (2002). Wnt5a signaling directly affects cell motility and invasion of metastatic melanoma. *Cancer Cell* **1**, 279–288.

Weijzen, S., *et al.* (2002). Activation of Notch-1 signaling maintains the neoplastic phenotype in human Ras-transformed cells. *Nat. Med.* **8**, 979–986.

Weng, A. P., *et al.* (2003). Growth suppression of pre-T acute lymphoblastic leukemia cells by inhibition of notch signaling. *Mol. Cell Biol.* **23**, 655–664.

Weng, A. P., *et al.* (2004). Activating mutations of NOTCH1 in human T cell acute lymphoblastic leukemia. *Science* **306**, 269–271.

Weng, A. P., *et al.* (2006). c-Myc is an important direct target of Notch1 in T-cell acute lymphoblastic leukemia/lymphoma. *Genes Dev.* **20**, 2096–2109.

Westfall, M. D., Pietenpol, J. A. (2004). p63: Molecular complexity in development and cancer. *Carcinogenesis* **25**, 857–864.

Wharton, K. A., *et al.* (1985). Nucleotide sequence from the neurogenic locus notch implies a gene product that shares homology with proteins containing EGF-like repeats. *Cell* **43**, 567–581.

Wong, K., *et al.* (2003). Notch ligation by Delta1 inhibits peripheral immune responses to transplantation antigens by a CD8+ cell-dependent mechanism. *J. Clin. Investig.* **112**, 1741–1750.

Yamaguchi, N., *et al.* (2008). NOTCH3 signaling pathway plays crucial roles in the proliferation of ErbB2-negative human breast cancer cells. *Cancer Res.* **68**, 1881–1888.

Yamamoto, N., *et al.* (2003). Notch/RBP-J signaling regulates epidermis/hair fate determination of hair follicular stem cells. *Curr. Biol.* **13**, 333–338.

Yan, M., *et al.* (2010). Chronic DLL4 blockade induces vascular neoplasms. *Nature* **463**, E6–E7.

Yan, X. Q., *et al.* (2001). A novel Notch ligand, Dll4, induces T-cell leukemia/lymphoma when overexpressed in mice by retroviral-mediated gene transfer. *Blood* **98**, 3793–3799.

Yang, Q., *et al.* (2001). Requirement of Math1 for secretory cell lineage commitment in the mouse intestine. *Science* **294**, 2155–2158.

Yugawa, T., *et al.* (2007). Regulation of Notch1 gene expression by p53 in epithelial cells. *Mol. Cell Biol.* **27**, 3732–3742.

Zecchini, V., *et al.* (2005). Notch signaling regulates the differentiation of post-mitotic intestinal epithelial cells. *Genes Dev.* **19**, 1686–1691.

Zhu, L., *et al.* (2007). Acinar cells contribute to the molecular heterogeneity of pancreatic intraepithelial neoplasia. *Am. J. Pathol.* **171**, 263–273.

Zou, G. M. (2007). Cancer stem cells in leukemia, recent advances. *J. Cell. Physiol.* **213**, 440–444.

Zurawel, R. H., *et al.* (1998). Sporadic medulloblastomas contain oncogenic beta-catenin mutations. *Cancer Res.* **58**, 896–899.

CHAPTER FOURTEEN

Biodiversity and Noncanonical Notch Signaling

Pascal Heitzler

Contents

Abstract

Early genetics in flies revealed that Notch is a complex pleiotropic locus. We now know that Notch is a receptor that plays prominent roles during development and functions locally in many tissues to instruct cell fate decisions. *Drosophila* has been an excellent model to identify genetically the elements that contribute to the canonical Notch signaling transduction machinery defined as DSL–Notch–CSL–MAML axis. This core pathway is required in many biological events in all animals. Though the canonical Notch pathway is relatively simple, and as the steps of the events are now more deeply understood, an increasing number of reports in the last decade show that many other molecules can influence Notch signaling, some by competing with a given element of the cascade. This may occur at any step bringing more diversity and plasticity to

Institut de Génétique et de Biologie Moléculaire et Cellulaire, Illkirch, France

Current Topics in Developmental Biology, Volume 92
ISSN 0070-2153, DOI 10.1016/S0070-2153(10)92014-0

the Notch pathway. Most of these regulatory molecules act in a context-specific manner and/or are themselves key regulators in other pathways, providing increasing examples of how connections among distinct pathway modulate each other ("cross talk"). The noncanonical signals discussed in this chapter are broadly defined and correspond to the following: DSL-independent activations, interactions with non-DSL ligands, CSL-independent signaling, signal transduction without cleavage, differential posttranslational modifications, competition/protection for a cofactor, and cross talk with other signaling pathways [Wnt, bone morphogenic protein (BMP), NF-κB, etc.]. Though some deemed controversial, these events may impact human diseases. Understanding the molecular nature of these events will allow avoidance of adverse effects during possible clinical treatments. In this review, we will focus on some noncanonical Notch activities and their *in vivo* significance during developmental and pathological processes.

1. A Wide Range of Notch-dependent Activities

Notch receptors trigger a wide range of cell fate choices through intercellular communications. Notch signaling is conserved across metazoan species and besides acquisition of specific cell fates and potentials enables also diverse cellular responses, like differentiation, proliferation, or apoptosis. Notch receptors function in multiple tissues and is used at various stages of development (Artavanis-Tsakonas *et al.*, 1999; Kopan and Ilagan, 2009; Lai, 2004). *Drosophila,* where only a single receptor exists, was the organism where Notch genetics was pioneered. The best-studied mechanism is lateral inhibition. In *Drosophila,* neural precursors segregate singly in a spaced pattern and are separated by intervening epidermal cells. Neural potential is given by the expression of proneural activators of the Achaete/Scute bHLH family, in a group of cells, called the proneural group. One neural cell is singled out via cell interactions that involve Notch and its ligand Delta. Both receptors and ligands are expressed in all equipotent cells of the proneural clusters, and mutual inhibition raises the threshold for the neural fate choice. When a cell begins to predominate as a possible neural precursor (in a process that is still not fully understood at the molecular level), its inhibitory signal persists preventing the others from realizing their neural potential. As a consequence, the other cells of the group become epidermal (Simpson, 1997). Flies mutant for Notch or Delta display an excess of neurons at the expense of epidermis, while constitutive Notch leads to a neural hypoplasia. As discussed throughout this book, activation leads to the cleavage and release of the intracellular domain of Notch (NICD) that associates with Su(H), the *Drosophila* CSL transcriptional mediator of lateral signaling, which in turn activates multiple genes of the E(spl) complex that encodes bHLH repressors of the HES family (Bray, 2006). These E(spl) factors repress Achaete/Scute, in most cells of the proneural group that adopt the epidermal cells, except the neural precursor(s). Thus, Notch signaling restricts the number of neurons in each proneural group through repression of Achaete/Scute. Lateral inhibition is a classical inhibitory function of canonical Notch signaling.

In worms (*Caenorhabditis elegans*), members of the Notch family includes Lin-12 and Glp-1 that share similar structural features with *Drosophila* Notch, save with less EGF (epidermal growth factor) repeats (see Chapter 2). Lin-12 is also required for several cell fate decisions using a mechanism of lateral inhibition. In the gonad, interactions between two initially equivalent cells, Z1.ppp and Z4.aaa, lead to one of them becoming an anchor cell (AC) and the ventral uterine (VU) precursor cell. Genetic studies on mosaic animals show that the signal from the AC is required for the VU fate, but if one cell is missing lin-12^{-}, it will always become an AC and the other cell always becomes a VU. When both cells are mutant for lin-12 they become AC; when lin-12 is activated, both cells become a VU (Greenwald and Seydoux, 1990). Therefore, lin-12 acts as a receptor for an "AC to VU" signal (Ambros, 1999; Chen and Greenwald, 2004; Greenwald, 1998; Komatsu *et al.*, 2008). Initially, lin-12 and lag-2, the ligand, are expressed in both Z1.ppp and Z4.aaa. During this lateral selection process, a feedback loop adapts the level of the receptor and the ligand. As one cell signals more effectively, it will produce more lag-2 and less Lin-12; the other cell will increase Lin-12 expression and reduce Lag-1, thus amplifying the signal to produce the AC/VU pair.

Notch is also required in an inductive signaling. Lag-2 is specifically expressed by a somatic distal tip cell that represses meiosis by activating the receptor Glp-1 in the germ line (Henderson *et al.*, 1994). During *Drosophila* wing development, directional activation of Notch occurs at the dorsoventral compartment boundary (Herranz and Milan, 2006; Irvine and Rauskolb, 2001; Koelzer and Klein, 2006). The LIM–homeodomain protein Apterous (Ap), the dorsal selector, induces the Notch effector Fringe (see Chapter 4) and the ligand Serrate within the whole dorsal territory. Fringe activity leads to a differential affinity of the ligands Delta and Serrate to Notch (Milan and Cohen, 2000), allowing a Serrate (dorsal) to Notch (ventral) signal and a reciprocal Delta (ventral) to Notch (dorsal) signal at the D/V boundary to turn on the expression of Wingless (Wg) into rows of cells. Wg in turn induces Serrate and Delta in nearby dorsal and ventral cells that adopt locally a wing margin identity. This Notch-dependent positive feedback loop maintains the induction of Notch ligands, maintains Wg and Cut expression, and promotes wing development (Milan and Cohen, 2000, 2003). Wg is required for cell survival, and both Wg and Notch synergistically promote tissue growth (Giraldez and Cohen, 2003). The timing of Notch activity is crucial for normal development. The vertebrate segmentation clock, described in

Chapter 10, provides a good example of how Notch signaling is involved at boundaries during somite formation.

Notch signaling misregulations are known to occur in human and to induce different pathologies and cancers (see Chapter 13). Another relevant Notch-dependent disease in human concerns cerebral autosomal dominant arteriopathy with subcortical infarcts and leukoencephalopathy. Mutations in some EGF repeats of Notch3 predispose individuals to dementia and strokes (Joutel *et al.*, 1996; Louvi *et al.*, 2006; Monet-Lepetre *et al.*, 2009). Other disorders were documented to result from deviant Notch signaling, including Alagille syndrome, B-cell chronic lymphocytic leukemia, basal cell carcinoma, cervical carcinomas, familial aortic valve disease, Hodgkin's lymphoma, neuroblastomas, prostate carcinomas, small-cell lung cancer, spondylocostal dysostosis, syndactyly, tetralogy of Fallot and others (Allenspach *et al.*, 2002; Demehri *et al.*, 2008; Kopan and Ilagan, 2009).

2. Noncanonical Notch Signaling and Bristle Patterning

In recent years, increasing examples of biological events have been reported that do not require the classical Delta-like DSL ligands nor the Su (H)-like CSL mediator. Some data reveal such a different pathway acting during *Drosophila* neurogenesis where multiple connections with the Wg signaling exist (Ramain *et al.*, 2001; Hayward *et al.*, 2008). The thorax of the flies exhibits two types of sensory organs, few large bristles, or macrochaetae, and numerous small bristles, or microchaetae. The microchaetae are arranged in rows, spaced roughly equally (Fig. 14.2A) due to the mechanism of lateral inhibition (Heitzler and Simpson, 1991). In the 1990s, we found new dominant alleles of Notch that constituted a nice allelic series. The mutants were named N^{Mcd}, where Mcd stands for microchaetae defective, since they lack between 60 and 99% of the microchaetae, whereas macrochaetae were not affected. Interestingly, the remaining bristles were organized in sparse but regular pattern (Fig. 14.2C), unlike the previously described gain-of-function Notch alleles (called Abruptex). Since the mutants have opposite phenotype to that of classical Notch loss-of-function mutants (Fig. 14.2B), the N^{Mcd} phenotype was first thought to represent simple over-activated lateral signaling. In such an hypothesis, removal of Su(H) or DSL ligands would be predicted to abolish the effects of the N^{Mcd} phenotype and adopt a neurogenic phenotype instead, lacking epidermal cells. Surprisingly, double mutant clones for N^{Mcd} and Su(H) or ligands display the N^{Mcd} phenotype with normal epidermis lacking microchaetae. Furthermore, the macrochaetae in such clones differentiate as a normal single bristle rather than as a neurogenic tuft. The N^{Mcd} phenotype is therefore independent of lateral signaling and does not require the ligands Delta/Serrate nor the effector Su(H), or the target E(spl)-C, in order to adopt the epidermal fate (Ramain *et al.*, 2001). This was interpreted as a confirmation that N^{Mcd} mutants affect a function of Notch distinct from lateral inhibition.

Deltex (Dx) was thought to participate in a signal transduction event downstream of Notch (Diederich *et al.*, 1994; Matsuno *et al.*, 1995, 1998) but is now known to function as a Ubiquitin ligase that regulates Notch endocytosis (Wilkin *et al.*, 2004, 2008). Loss of function of Dx displays an excess of microchaetae, whereas overexpression of Dx inhibits neurogenesis. Lack or reduced function of Dx restores nearly all microchaetae sensory organs in N^{Mcd} mitotic clones, suggesting that Dx is required for the N^{Mcd} phenotype (Ramain *et al.*, 2001). More recent studies on *Drosophila* wings (Hori *et al.*, 2004) revealed that Notch was activated by Dx via a Su(H)-independent but endocytic-dependent manner and that this activity required entry to a late-endosomal compartment, unlike the canonical Notch activity that can proceed as long as early endosomes can form. A balance between three E3 ubiquitin ligases, the ring domain Dx and two HECT domain E3 ubiquitin ligases, Su(dx) and DNedd4, can modulate Notch through altering its endosomal sorting (Wilkin *et al.*, 2004, 2008). Several lines of evidence suggest the existence of a Dx-dependent but CSL-independent Notch pathway in vertebrates (Hu *et al.*, 2003; Ordentlich *et al.*, 1998) and other reports provide a description of CSL-independent events that take place during *Drosophila* muscle development (Martinez Arias *et al.*, 2002; Rusconi and Corbin, 1998; Shawber *et al.*, 1996).

How is the signal transduction achieved in the N^{Mcd}-dependent antineural activity of Notch ? This important question remains unresolved. However, the dx-dependent antineural activity of Notch seems to be regulated by members of the Wg signaling cascade. Shaggy (Sgg), the *Drosophila* glycogen synthase kinase 3 (GSK3), plays a central regulatory role in Wnt signaling where it is repressed by the Wnt mediator phosphoprotein dishevelled (Dsh). Sgg loss was found to suppress the N^{Mcd} phenotype and might regulate the Deltex-dependent signal of Notch. GSK3β is able to bind the ANK repeat number 6 and phosphorylate the Notch2-ICD *in vivo* (Espinosa *et al.*, 2003), but this region can be substituted without any consequences to Notch2 activity (Kraman and McCright, 2005). The vast majority of the N^{Mcd} alleles encode truncated receptors that lack the C-terminal PEST (Fig. 14.1), a truncation that will stabilize NICD *in vivo* (Weng *et al.*, 2004). Furthermore, this missing region binds the phosphoprotein Dsh, itself a member of the Wnt signaling pathway known to affect the development of microchaetae (Axelrod *et al.*, 1996). When overexpressed, Dsh induces supernumerary microchaetae, but it cannot suppress the N^{Mcd} phenotype. In contrast, overexpression of Dsh partially or

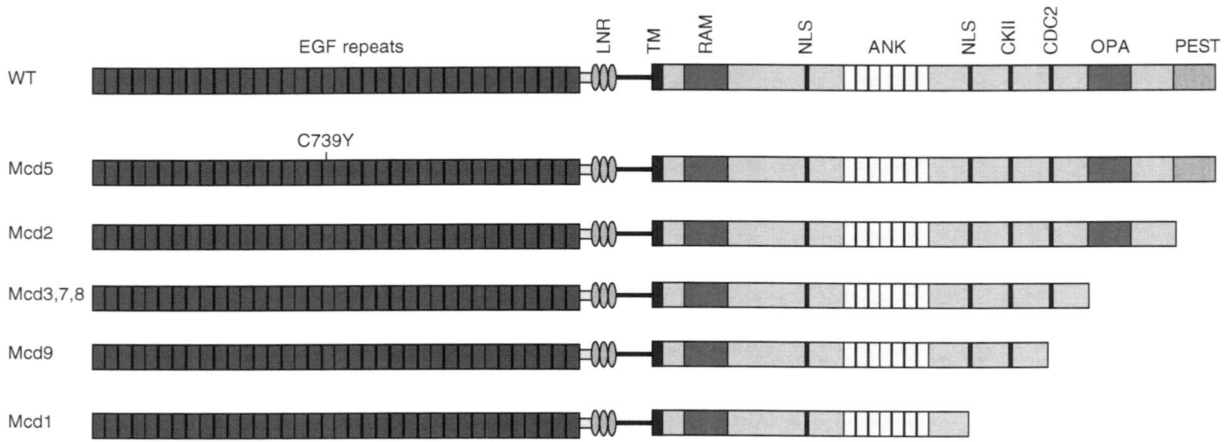

Figure 14.1 A schematic representation of the wild-type and N^{Mcd} mutant versions of the *Drosophila* Notch receptor, with EGF repeats, Lin12/Notch repeats (LNR), transmembrane domain (TM), RBPjκ association module (RAM), nuclear localization sites (NLS), ankyrin repeats (ANK), CKII and cdc2 phosphorylation sites, OPA and PEST sequences, indicated. All N^{Mcd} has C-terminal truncations except N^{Mcd} that has a CtoY substitution in the 18th EGF.

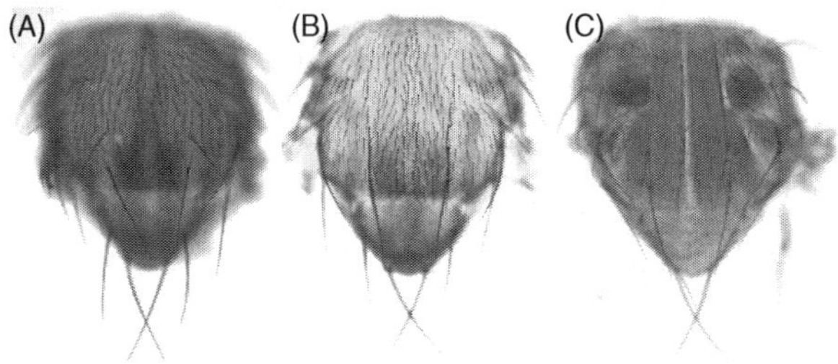

Figure 14.2 Dorsal view of *Drosophila* thoraces, showing (A) a wild-type density of the microchaetae or small sensory bristles. (B) The haploinsufficient phenotype of Notch (loss of function, $N^{55e11}/+$) shows a noticeable increase of the number of bristles. (C) The $N^{Mcd1}/+$ phenotype has opposite phenotype with loss of bristles.

completely rescues the Abruptex phenotypes, known to result from mutations in the 24–29th EGF repeats and that increase lateral signaling. Since the N^{Mcd5} mutant receptor carries a single C to Y substitution that disrupt the 18th EGF repeat (Fig. 14.1), this domain may also be involved in Dx-dependent Notch/Wg cross talk. Altogether, these results support the idea that the Wnt signaling controls the Dx-dependent pathway of Notch at different levels (Ramain *et al.*, 2001). A possible hypothesis presumes that in the absence of ligands or Su(H), a repressive, Dx-dependent function of Notch could maintain the cells in an uncommitted state. Dsh represses this hypothetical activity and cells become competent to consider alternative cell fate choices. Several recent papers continue to explore Notch/Wg interactions [reviewed in Hayward *et al.* (2008)]. Notch can regulate the activity, the localization, and the amount of both *Drosophila* Armadillo and its vertebrate counterpart, β-catenin, also in a manner independent of Su(H) (Fig. 14.3). Since Wnt and Notch signaling are so interconnected, the possibility that the two signaling systems act as one has been proposed (Hayward *et al.*, 2008). Wnt/Notch may function as a transistor in a noise-filtering device for cell fate assignments during development. Rather than defining the fate of a cell, Wnt/Notch determines the probability to convert an otherwise unreliable functional module into an effective one allowing stable inputs (Hayward *et al.*, 2008).

3. Noncanonical N Signals in Vertebrates

Increasing number of noncanonical Notch signals were observed and explored in mammals. In general, we define these as situations where signal transduction is possible independent of at least one component of the Notch core. These events include alternative ligands, alternative mediators, or proteolysis-independent Notch signals.

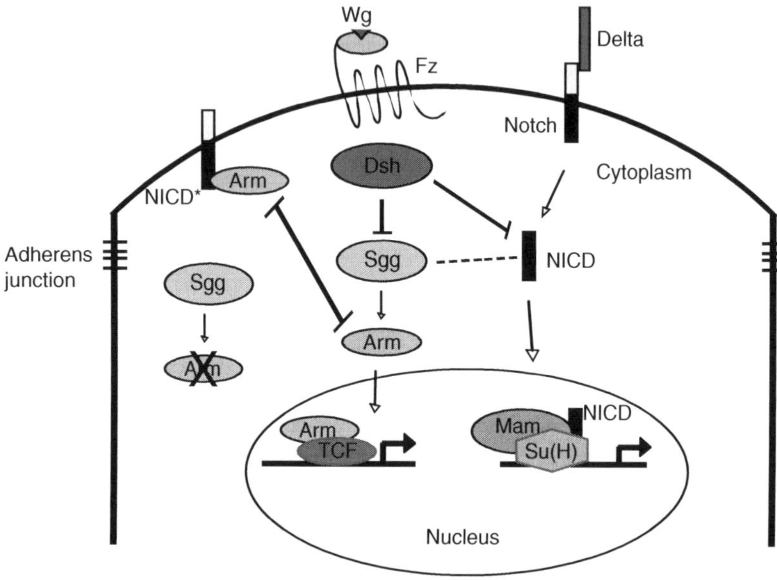

Figure 14.3 Schematic and simplified view on the Notch–Wg signaling pathways in *Drosophila*. In the absence of Wg, the GSK3 kinase shaggy (sgg) phosphorylates the cytosolic pool of Arm/β-catenin. In the absence of DSL ligands, Notch titrates Arm/β-catenin away from its "functional" pool in a complex that is degraded through the constitutive Notch endosomal trafficking circuitry near the adherens junctions.

3.1. Non-DSL Ligands

While DSL family members represent the classical Notch ligands, themselves type I transmembrane proteins, for activating Notch signaling through cell–cell *trans*-interactions (Bray, 2006; Heitzler and Simpson, 1991), they can also mediate *cis*-inhibition within the same cells (Fiuza *et al.*, 2010; Heitzler and Simpson, 1993; Micchelli *et al.*, 1997). Delta-like or Serrate-like (Jagged) ligands exhibit a number of conserved domains (Chapter 3). A growing repertoire of noncanonical ligands have been reported to activate Notch [see D'Souza *et al.* (2008) for review]. Some of them, such as Dlk-1 (Delta-like-1), DNER (Delta/Notch-like EGF-related receptor), or Jedi (Jagged and Delta protein), belong to membrane-tethered proteins that resemble Delta but lack the DSL motif. It is however not clear whether they compete with DSL ligands *in vivo*. By contrast, the F3/contactin, a structurally unrelated Notch ligand, is a glycosyl phosphatidylinositol-anchored neural cell adhesion molecule of the immunoglobulin superfamily. Like DSL molecules, F3/contactin binds the EGF repeats 11–12 of Notch and triggers γ-secretase-dependent nuclear translocation of the NICD. However, the F3/Notch signaling promotes oligodendrocyte precursor cell differentiation and myelination in the central nervous system, via Deltex1 rather than via the CSL (Hu *et al.*, 2003). Finally, a number of secreted proteins have been suggested to act as putative noncanonical ligands of Notch. In *Drosophila*, Scabrous was reported to bind Notch for patterning eye ommatidia and sensory bristles and to activate E(spl) bHLH targets (Powell *et al.*, 2001), but further analysis is required to ascertain whether Scabrous is truly a ligand or a modulator of Notch. Wg may bind Notch (Couso and Martinez-Arias, 1994; Wesley, 1999) but the functional significance of this remains controversial and none of the Wnt proteins has been shown to bind Notch in vertebrate (D'Souza *et al.*, 2008). CCN3 and MAGP-2 are the only secreted non-DSL ligands known in vertebrates. Interestingly, these two molecules are proposed to function as autocrine activators of Notch (Albig *et al.*, 2008; Gupta *et al.*, 2007; Minamizato *et al.*, 2007; Miyamoto *et al.*, 2006; Sakamoto *et al.*, 2002). At high concentrations, MAGP-2-induced Notch signaling can mediate nonenzymatic dissociation of Notch, raising the possibility that MAGP-2 or any protein that destabilizes the heterodimeric structure might activate signaling (D'Souza *et al.*, 2008).

3.2. Cross Talk with Other Nuclear Partners

Growing evidence points to interaction between Notch and other signaling pathways. In the next examples, we summarize recent findings.

3.2.1. NF-κB

NF-κB proteins constitute a family of dimeric transcriptional factors involved in regulating the immune functions. Members of the IκB family are negative regulators; IκB phosphorylation and degradation lead to nuclear translocation and DNA binding of NF-κB/Rel complexes and activation of NF-κB downstream genes. Interestingly, the NICD of Notch share structural features to IκB molecules, and CSL is similar to Rel proteins (see Chapter 2). When overexpressed, NICD can bind NF-κB through a poorly conserved region partially overlapping the RAM domain (Wang *et al.*, 2001). Like for the Wg pathway, NF-κB signaling meets Notch at multiple levels and bidirectional cross talk has been proposed. For example, overexpressed Notch enhances NF-κB activity by facilitating its nuclear retention (Shin *et al.*, 2006), but no evidence supports this as a function of endogenous NICD. Furthermore, the Notch-1 ICD potentiates direct transcriptional activation of Relb and NF-κb2, two NF-κB factors. Recent data show that Notch-1 ICD interacts with the IKK signalosome in mouse T-cell acute lymphoblastic leukemia (T-ALL) and in other cancers (Aifantis *et al.*, 2007; Vilimas *et al.*, 2007). However, the activated NF-κB is not sufficient to induce T-ALL. Interestingly, NF-κB/Rel proteins may directly repress the neural fate in *Drosophila* non-SOP cells while they participate to maintain high levels of sc expression in the SOP (Ayyar *et al.*, 2007). This activity is compatible with the known Notch functions during lateral signaling. Further studies will be required to sort out the precise biological scenario that occurs during normal development.

3.2.2. HIF-1α

Hypoxia and Notch pathways have also some connections and synergisms [see Poellinger and Lendahl (2008) for a review]. The transcription factor HIF-1α is a key component of the cellular hypoxic response. During prolonged hypoxia, HIF-1α/HIF-1β heterodimers binds to specific promoter elements, hypoxia response elements (HREs) that drive some physiological responses, including angiogenesis, erythropoiesis, and a switch to glycolytic metabolism. When associated with the NICD, HIF-1α can synergize activation of Notch targets such as Hey-1 (Chen *et al.*, 2007; Gustafsson *et al.*, 2005). Interestingly, some of these targets harbor both HRE- and CSL-binding sites, allowing a cooperative activation and hypoxia-induced Notch ligand Dll4 expression has been noticed (Diez *et al.*, 2007; Patel *et al.*, 2005). Recent reports indicate that both HIF-1α and Notch are substrates for the asparagine hydroxylase FIH-1 (factor inhibiting HIF-1α) (Coleman *et al.*, 2007; Zheng *et al.*, 2008). In the case of Notch, two asparagine residues from the NICD ankyrin repeats are concerned, and this post-transcriptional modification may decrease Notch signaling. Since FIH-1 binds Notch more efficiently than HIF-1α, a competition may occur when the NICD is present or overexpressed, leading to an under-hydroxylation on HIF-1α (Zheng *et al.*, 2008).

3.2.3. Smad

The transforming growth factor-beta (TGF-β)/BMP superfamily of signal molecules plays critical role during development and in pathogenesis of a number of diseases. TGF-β constitute secreted ligands that bind BMP transmembrane serine/threonine receptors. TGF-β associates to a BMP type II receptor dimer that then recruits a type I receptor dimer, forming a hetero–tetrameric complex with TGF-β. A conformational change occurs which allow type I receptor to phosphorylate the carboxy-terminus of R-Smads, including Smad2 or Smad3, the key intracellular mediators of TGF-β signals. Activated R-Smads associate to another partner, Smad4, and this complex travels to the nucleus where it binds a range of transcription factors to control expression of numerous target genes.

In the last few years, some reports described cross talk between Notch and TGF-β pathway [Klüppel and Wrana (2005) and Guo and Wang (2009) for reviews]. TGF-β can induce the transcription of Notch ligands

(Fig. 14.4). TGF-β-dependent activation has been found in several mammalian tissues. TGF-β induces epithelial-to-mesenchymal transition phenotypes in epithelial cells both *in vitro* and *in vivo*, a Smad3-dependent event that activates both Jagged and HES1 (Zavadil *et al.*, 2004). Another report correlates the TGF-β-induced expression of Jagged with the activation of the cell cycle inhibitor p21 and cytostasis in epithelial cells (Niimi *et al.*, 2007). Elevated TGF-β-dependent levels of Jagged and HES1 were detected in extracts from renal biopsies from diabetic nephropathy patients (Walsh *et al.*, 2008). Finally, in a nonmammalian model, Nodal induces Delta2 to establish a grid-like organization of distinct cell identities in the ascidian neural plate (Hudson *et al.*, 2007).

TGF-β is able to induced Hes-1 in C2C12 myogenic cells and as a consequence inhibit myogenic factors such as MyoD. This activity requires CSL, NICD, and Smad3. In C2C12 cells, Smad3 binds to the NICD and a Smad3–NICD–CSL complex is able to bind specifically to CSL-binding sites (Blokzijl *et al.*, 2003). Another report mentions that Smad1 may also bind the NICD to induce expression of Hey-1 (Dahlqvist *et al.*, 2003). Finally a similar observation was made in endothelial cells (EC). When EC cells are exposed to BMP ligands, Smad1 translocation leads to expression of ld1, an activator of EC migration. When Notch signaling is induced in EC

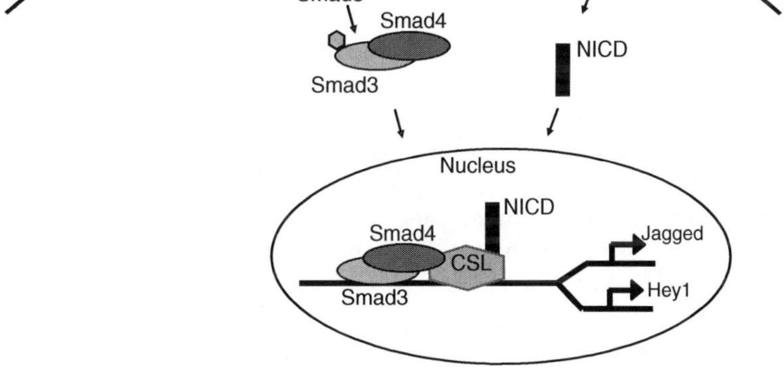

Figure 14.4 Schematic and simplified view on the Notch–TGF-β signaling pathways showing nuclear synergistic transcription activation between Smads and the CSL Notch mediator in vertebrates. This tissue-specific signature induces the transcription of the Notch ligand Jagged and the Notch target Hey1.

cells, a synergistic activation of Herp2, a Hes-related factor and negative regulator of EC migration, that promotes degradation of ld1 (Itoh *et al.*, 2004). These examples of TGF-β–Notch cross talk allows alternative or synergistic genetic programs controlling differentiation or migration of cells according to change in environment or physiology.

3.2.4. Ptf1a

Classical Notch signaling controls the endocrine/exocrine decision in early pancreatic progenitors (Jorgensen *et al.*, 2007). The pancreas primordia derive from the fusion of a dorsal and a ventral buds of Ptf1a-expressing progenitors that emerge from the foregut. The resulting organ then differentiates endocrine, acinar, and ductal cells. Notch is instrumental, controlling the initial endocrine versus exocrine decision. The Ngn3 bHLH transcription factor is required to commit the progenitors to the endocrine fate mainly through the activation of the paired box factor Pax6. Conversely, RBP-Jk interacts with Ptf1a and E12 to promote acinar differentiation (Jorgensen *et al.*, 2007). A recent study reveals that Ngn3 is not sufficient to commit cells to the endocrine fate and rather a noncanonical activity for Notch2 is required in order to allow $Ngn3^+$ progenitors to adopt a γ-secretase-dependent endocrine fate until Pax6 expression begins (Cras-Méneur *et al.*, 2009). During this transient developmental window, $Ngn3^+$ progenitors are bipotential. When presenilin amounts are limiting below a threshold, $Ngn3^+$ progenitors default to an acinar fate to form the bulk of the exocrine pancreas. In absence of presenilins (the catalytic component of the γ-secretase complex), all $Ngn3^+$ cells retain progenitor characteristics and finally contribute the exocrine mass instead of the endocrine fate. A speculative explanation how presenilins make a dosage-dependent contribution to endocrine determination in $Ngn3^+$ progenitors was proposed (Cras-Méneur *et al.*, 2009). Both Ngn3 and Ptf1a compete and court E12 to gain transcriptional activity, however Ngn3/E12 complexes form preferentially over E12/Ptf1a/RBP-Jk. Since the NICD and Ptf1a also compete for RBP-Jk in acinar progenitors (Beres *et al.*, 2006; Masui *et al.*, 2007) and RBP-Jk is dispensable for the development of Ngn3 progenitors, they speculate that NICD contributes to Ngn3/E12 complex formation by sequestering RBP-Jk, thus handicapping Ptf1a (Hori *et al.*, 2008). Presenilins modulate RBP-Jk availability. If NICD levels are sustained, the E12/Ptf1a/RBP-Jk complex cannot form, Ngn3/E12 complexes form, and no acinar differentiation occurs in $Ngn3^+$ cells. When presenilins appear to be limiting, cells would free RBP-Jk, allowing the formation of a complex with Ptf1a, lowering Ngn3/E12 complex formation and pushing the bipotent cells toward the acinar fate (Cras-Méneur *et al.*, 2009).

4. Misappropriation by Viruses

Some pathological virus infections can alienate CSL factors from their canonical function to serve the viruses' own advantages. Viruses like the Epstein–Barr virus (EBV), Kaposi's sarcoma-associated herpesvirus (KSHV), and adenovirus type 5 proteins are known to release multiple proteins necessary for cellular transformation that bind CSL and may modulate CSL-dependent transcription (Hsieh and Hayward, 1995). For example, the KSHV RTA factor and the EBV EBNA2 have antagonist consequence on B lymphocyte transformation, though they both recognize the same functional domain in CSL and, like the NICD, convert the CSL from a repressor to an activator.

5. New Insights on MAML

Recently, a noncanonical function of mam has been studied that enhances Hedgehog signaling specifically in stem cells of the *Drosophila* ovary (Vied and Kalderon, 2009). Rather than being a coactivator of Notch in these tissues, mam functions in follicle stem cells independently of Notch. There are increasing examples where mam-like transcriptional factors play unexpected roles in other signaling pathways including muscle differentiation and myopathies (MEF2C), tumor suppressor pathway (p53), and colon carcinoma survival (β-catenin). It is tempting to speculate that MAML proteins represent central coactivators in mediating signaling cross talks (McElhinny *et al.*, 2008). The MAML would be able to recruit appropriate Mediator and histone acetyltransferase complexes (Fryer *et al.*, 2002, 2004; Nam *et al.*, 2006, 2007). Similar to the Pitf1a story, NICD may compete MAML away from β-catenin, explaining some of the effects seen for the Notch/Wg integrative pathway (Hayward *et al.*, 2008).

6. Rheostat and Fine-tuning

Nemo-like kinases (NLK) represent a family of conserved protein kinases that function in various tissues and biological events. Though *Drosophila* nemo was suspected to be involved in Notch signaling (Verheyen *et al.*, 1996; Kankel *et al.*, 2007), a very recent study in zebra fish established the NICD of Notch1 as a target and its phosphorylation contributes to decrease the formation of the CSL–MAML–NICD transcriptional complex (Ishitani *et al.*, 2010). Knockdown Nlk leads to activation of HES genes and decrease neurogenesis, while ectopic expression of NLK decreases the HES genes. NLK represses Notch through the phosphorylation on multiple sites between the ANK repeats and the PEST domain of the NICD. The number rather the exact sites seems to be relevant for repression, providing a graded effect of a rheostat to fine-tune Notch activity (Ishitani *et al.*, 2010). NLK may be considered as an intrinsic regulator of the Notch pathway, perhaps acting as a rheostat.

7. Nonnuclear Mechanisms

Finally, few reports described Notch signal transduction events that are independent from the cleavage.

7.1. Abl

D-Abl, the *Drosophila* homolog of the vertebrate oncogene, is a cytoplasmic tyrosine kinase that may play a role in Notch signaling in axons. Notch and abl mutations interact synergistically to produce synthetic lethality with axonal defects rather than with neurogenic phenotypes. This genetic interaction reflects a function of Notch that is separable from its role in control of cell fate choices. Both Abl and Notch, as well as Dab (disabled) and Trio, two Abl accessory proteins are required in axon patterning and growth cones. Dab binds specifically the RAM domain of Notch and may constitute an adaptor protein for Notch to recruit Abl in a Su(H)-independent mechanism (Giniger, 1998; Le Gall *et al.*, 2008).

7.2. Akt

The PI3K–PTEN–Akt signaling pathway coordinates a variety of regulatory networks and intracellular signals. It is sensitive to external stimuli and regulate cell growth in promoting metabolic processes like glucose metabolism. When stimulated, the PI3K mediates phosphorylation of phosphatidylinositol to convert PIP2 into PIP3 that binds, with high affinity, the serine/threonine kinase Akt, a primary mediator of the PI3K cascade. Akt controls numerous cellular functions as metabolism, proliferation, growth, and survival. It has an antiapoptotic activity through the inhibition of many pro-apoptotic effectors. The PI3K-dependent Akt activation is antagonized by a phosphatase, where PIP3 represents a substrate. Since, the PI3K–Akt signaling pathway is activated in many tumors, PTEN is considered as an antitumoral protein.

Over the years, several authors attempted to link Notch and AKT. As we already developed in a previous section, Notch1 conveys the most prominent genetic T-ALL (Demarest *et al.*, 2008). However, many such oncogenic pathologies are not influenced by blocking the proteolytic activation of Notch1 with γ-secretase inhibitors. A recent work identifies the loss of PTEN as associated with resistance to Notch inhibition (Palomero *et al.*, 2007). In these cancers and also in normal tissues, Notch1 upregulates the activity of PI3K–Akt signaling pathway through HES1, which represses the expression of PTEN. In T-ALL, the loss of PTEN seems to confer resistance to γ-secretase inhibitors probably because the constitutive activity of the PI3K–Akt signal increases sufficient glucose metabolism to bypass the requirement of Notch1 signaling to support cell growth. This regulatory network is evolutionary conserved from *Drosophila* to human (Palomero *et al.*, 2007). In T-ALL cell lines, Notch1was reported to regulate the expression of cell cycle regulatory proteins cyclin D1, CDK2, and p21 (Guo *et al.*, 2009).

Similarly, Notch1 signaling is elevated in human melanoma and is a key effector of Akt in melanoma development (Bedogni *et al.*, 2008). In this model, however, hyperactivated Akt showed increases levels of Notch1-ICD, suggesting the existence of a reciprocal cross talk between Notch1 and PI3K–Akt pathways. This reciprocal activity requires NF-κB and seems to be cell type specific. In skin, where low-oxygen microenvironment occurs, Notch1 levels increase via stabilization of HIF-1α (Bedogni *et al.*, 2008). A similar Notch1 activity, influenced by tissue hypoxia, was recently identified in adenocarcinoma of the lung. In these cells, Akt-1 activation is a key mediator of Notch1 pro-survival effects through repression of PTEN and also via the unexpected induction of IGF-1R, the insulin-like growth factor 1 receptor (Eliasz *et al.*, 2010). The Notch-dependent Akt induction was reported to be mediated by an autocrine loop that sustains the transformation state in breast epithelial cells (Meurette *et al.*, 2009).

Another Notch-mediated signaling cascade has been recently proposed, promoting cell survival by inhibiting apoptosis triggered by neglect in mammalian cells. In this process, the NICD activates Akt that couples PI3K to the nutrient sensor kinase mTOR and its substrate-defining protein Rictor, depending on specific environmental cues. This activity does not required CSL-mediated transcription nor the nuclear localization of the NICD (Perumalsamy *et al.*, 2009) but rather the NICD antagonizes the pro-apoptotic Bcl-2 factor BAX that is required to convey multiprotein assemblies on mitochondria, committing cells to irrevocable damage. The mitochondrial remodeling proteins, mitofusins (Mfn-1 and Mfn-2), may act as key intermediates in the antiapoptotic cascade, connecting a noncanonical cytoplasmic Akt-dependent activity of the NICD to mitochondrial functions (Perumalsamy *et al.*, 2010). The Notch1-ICD seems to inhibit p53 via rapamycin (m-TOR) and confer protection against p53 cytotoxicity

(Mungamuri *et al.*, 2006). Since all these components are found in *Drosophila*, these antiapoptotic activities of Notch are likely to be conserved throughout animal evolution.

7.3. Titration of ß-catenin

As the team of A. Martinez-Arias further investigates the interaction between Notch and Wg, an unexpected level of regulation was discovered. They found that in the absence of Notch, an activated form of Arm/β-catenin induced tissue overgrowth in *Drosophila* epithelial cells (Sanders *et al.*, 2009). As distinct pools of Arm/β-catenin are involved in signaling and adhesion, they demonstrate that a novel function of Notch is to sequester a cell-surface-located pool of Arm and induce its degradation (Fig. 14.3). This cytoplasmic function of Notch does not require the CSL ligands but couples the constitutive turnover of Notch with Arm degradation in the endosomal compartment trafficking circuitry near the adherens junction. This NICD/β-catenin association may explain how the constitutive turnover of Notch may buffer the activation of Arm/β-catenin during development (Sanders *et al.*, 2009). In these conditions, Notch may act as a tumor suppressor function that modulates the oncogene activity of Arm/β-catenin. This tumor suppression function of Notch has also been observed to be linked with Arm/β-catenin in vertebrates (Nicolas *et al.*, 2003; Pan *et al.*, 2004). In this review, we also highlighted another cytoplasmic function of the NICD, eliciting PI3K–Akt-dependent tumorigenesis in several cancers and linking Notch to various aspects of cell survival versus apoptosis events. Further experiments are required to reconcile these two antagonistic cytoplasmic activities of Notch. Finally, another unexpected outcome from an integrated function of Notch emerged through the complex oscillating network where Notch, Wnt, and FGF signaling pathways are involved in the mechanism of the segmentation clock (see Chapter 10). Coupling the three oscillators through common intermediates may lead to synchronization and sustain robustness to the framework (Goldbeter and Pourquie, 2008).

8. Discussion

Notch mutants and most of the receptor transduction machinery were first identified in *Drosophila.* Due to partial redundancy among the Notch genes in vertebrates, the study of Notch receptors remains more sophisticated in Mouse or other mammalian models. *Drosophila* has only one receptor, and its powerful genetics continues to allow the identification of new modifiers that influence Notch signaling (Mummery-Widmer *et al.*,

2009). Though the functional aspects of the different Notch domains has given valuable information using some transgenics in flies (Rebay *et al.*, 1993; Langdon *et al.*, 2006; Le Gall *et al.*, 2008), a complete genetic analysis of a Notch receptor remains necessary, in order to map new or to dissect further extant functional domains of the receptor that will be involved specifically in the events described in this chapter. *Drosophila* is the most suitable model for such an analysis that will be of general interest.

Notch signaling is instrumental for normal development of many tissues, for cell fate choices that precede differentiation, for cell survival or apoptosis, for cell proliferation, or conversely to prevent or delay cell differentiation. To provide such a diversity of effects, Notch is implicated in a variety of mechanisms that require besides the classical core, an increasing number of noncanonical events. Depending on the context, Notch signaling is influenced by a growing number of new effectors, including even molecules that represent key components from other signaling pathways. The deviations occur through misappropriation of a given component of the Notch core, such as CSL, MAML, or NICD.

Non-DSL ligands are able to destabilize conformation of Notch to expose the metalloprotease cleavage site 2 and finally induce a CSL-dependent signal, like DSL ligands. Alternatively, noncanonical ligands can also induce another Notch signaling. The molecular logic that allows such discrimination of responses warrants further exciting investigations. The same question deserves attention for signaling events that may limit autocrine activation. How endocytosis is regulated in such cases was not systematically investigated, but all tethered noncanonical ligands may also facilitate transendocytosis since their intracellular domains contain, like for DSL ligands, the putative lysine-rich ubiquitination sites (D'Souza *et al.*, 2008). It will be of high interest to know whether vertebrate noncanonical ligands are antagonizing or promoting DSL functions when coexpression occurs. Intriguingly, some viruses, like EBV, encodes oncoproteins that bind and activate CSL in infected B immune cells in a Notch-independent fashion, for the profit of the viruses. Since MAML has been recently described as cofactors of other signal transduction pathways (McElhinny *et al.*, 2008), they represent putative targets for cross talk that may interfere with Notch.

For more general considerations, we have to admit that an increasing number of molecules behave as effectors through the activity of the NICD, including molecules that share identical or very close target sites on the NICD. As a consequence, one can expect that competition and mutual exclusion of effectors may also be relevant for alternative Notch activities. In *Drosophila* N^{Mcd} mutants, truncation of some phosphorylation sites and the PEST domain lead to the increasing stability of the NICD. This gain of stability may override the affinity of secondary partners of Notch that may control more subtle but important Notch function during normal development. A number of truncations of Notch1, similar to the *Drosophila* N^{Mcd} mutations, have been described in mouse and human that are associated with T-ALL leukemias (Feldman *et al.*, 2000; Hoemann *et al.*, 2000; Weng *et al.*, 2004; Chiang *et al.*, 2006). Context-specific availability or differential affinity of effectors will allow controlled fluctuations of the stability or turnover of the NICD. MAML has been reported to recruit the CycC:CDK8 dimer to phosphorylate the NICD. As a consequence, the MAML–CycC–CDK8 complex coordinates activation with the PEST-dependent turnover (Fryer *et al.*, 2004), by E3 ubiquitin ligases Sel10/Fbw7 (O'Neil *et al.*, 2007; Thompson *et al.*, 2007; Tsunematsu *et al.*, 2004). The NICD contains multiple phosphorylation sites, so that quantitative regulation of the stability versus degradation is possible. Extrinsic effectors may protect or elicit the NICD degradation. In this context, signaling cross talk could manipulate the Notch cascade like a rheostat to allow permissive or repressive threshold activities, required for alternative cell fate decisions (Ishitani *et al.*, 2010).

Cross talk between Notch and Wnt seems to be conserved throughout evolution. In an elegant work on mice genetics, a report described how presenilins modulate CSL availability in critical transcriptional complexes (Cras-Méneur *et al.*, 2009). In $Ngn3^+$ acinar progenitors, when the level of the NICD is sustained, Ngn3/E12 complexes form preferentially over E12/Ptf1a/CSL and no (acinar) differentiation occurs, a function that is then controlled by the Wnt/β-catenin signaling pathway (Murtaugh, 2008). Some parallels exist between the Ptf1a story and the N^{Mcd} ones that bring together a sequestration among transcriptional factors for a Wnt-dependent cell fate. The N^{Mcd} phenotype depends on both Dsh and Shaggy, two members of the Wg signaling cascade (Ramain *et al.*, 2001). We suggested that a constitutive function of Notch may repress neurogenesis independently of the classical core and that this function may prevent differentiation, a process that is override by the Wg signaling cascade.

Deviant Notch signaling is likely at the origin of a variety of human tumors that depend on context-specific situations. The accurate knowledge of levels of regulation of the Notch pathway, the molecular nature of synergism, or competition between Notch and other signaling pathways will be relevant to understand the pathological mechanisms. What are the specific conditions for "on/off" states for the receptor and also how Notch signaling is quantitatively regulated, indeed fine-tuned. A number of molecule inhibitors of the γ-secretase complex (Aster *et al.*, 2008; Grabher *et al.*, 2006) and the Notch transcription factor complex (Moellering *et al.*, 2009) have created the opportunity to develop molecular therapies. However, since cytoplasmic activities are relevant for controlling the tumor sensitivity, the knowledge of the relationships between the diverse Notch functions and their tissue specificities will secure from adverse effects of pharmaceutical therapeutics.

ACKNOWLEDGMENTS

I apologize to colleagues whose work was not cited. I am grateful to Inna Biryukova for artworks. This work was supported by the Association pour la Recherche sur le Cancer (ARC), the CNRS, the INSERM and the University of Strasbourg.

REFERENCES

Aifantis, I., Vilimas, T., and Buonamici, S. (2007). Notch, NF-kBs and the making of T cell leukemia. *Cell Cycle* **6**, 403–406.

Albig, A. R., Becenti, D. J., Roy, T. G., and Schiemann, W. P. (2008). Microfibril-associate glycoprotein-2 (MAGP-2) promotes angiogenic cell sprouting by blocking Notch signaling in endothelial cells. *Microvasc. Res.* **76**, 7–17.

Allenspach, E. J., Maillard, I., Aster, J. C., and Pear, W. S. (2002). Notch signaling in cancer. *Cancer Biol. Therapy* **1**, 466–476.

Ambros, V. (1999). Cell cycle-dependent sequencing of cell fate decisions in *Caenorhabditis elegans* vulva precursor cells. *Development* **126**, 1947–1956.

Artavanis-Tsakonas, S., Rand, M. D., and Lake, R. J. (1999). Notch signaling: cell fate control and signal integration in development. *Science* **284**, 770–776.

Aster, J. C., Pear, W. S., and Blacklow, S. C. (2008). Notch signaling in leukemia. *Annu. Rev. Pathol.* **3**, 587–613.

Axelrod, J. D., Matsuno, K., Artavanis-Tsakonas, S., and Perrimon, N. (1996). Interaction between Wingless and Notch signaling pathways mediated by dishevelled. *Science* **271**, 1826–1832.

Ayyar, S., Pistillo, D., Calleja, M., Brookfield, A., Gittins, K., Goldstone, C., and Simpson, P. (2007). NF-kB/Rel-mediated regulation of the neural fate in Drosophila. *PloS ONE* **11**, e1178.

Bedogni, B., Warneke, J. A., Nickoloff, B. J., Giaccia, A. J., and Broome Powell, M. (2008). Notch1 is an effector of Akt and hypoxia in melanoma development. *J. Clin. Invest.* **118**, 3660–3670.

Beres, T. M., Masui, T., Swift, G. H., Shi, L., Henke, R. M., and MacDonald, R. J. (2006). PTF1 is an organ-specific and Notch-independent basic helix-loop-helix complex containing the mammalian Suppressor of Hairless (RBP-J) or its paralogue, RBP-L. *Mol. Cell. Biol.* **26**, 117–130.

Blokzijl, A., Dahlqvist, C., Reissmann, E., Falk, A., Moliner, A., Lendhal, U., and Ibanez, C. F. (2003). Cross-talk between the Notch and the TGF-β signaling pathways mediated by interaction of the Notch intracellular domain with Smad3. *J. Cell Biol.* **163**, 723–728.

Bray, S. J. (2006). Notch signalling: a simple pathway becomes complex. *Nat. Rev. Mol. Cell Biol.* **7**, 678–689.

Chen, Y., De Marco, M. A., Graziani, I., Gazdar, A. F., Strack, P. R., Miele, L., and Bocchetta, M. (2007). Oxygen concentration determines the biological effects of Notch-1 signaling in adenocarcinoma of the lung. *Cancer Res.* **67**, 7954–7959.

Chen, N., and Greenwald, I. (2004). The lateral signal for LIN-12/Notch in *C. elegans* vulval development comprises redundant secreted and transmembrane DSL proteins. *Dev. Cell* **6**, 183–192.

Chiang, M. Y., Xu, M. L., Histen, G., Shestova, O., Roy, M., Nam, Y., Blacklow, S. C., Sacks, D. B., Pear, W. S., and Aster, J. C. (2006). Identification of a conserved negative regulatory sequence that influences the leukemogenic activity of NOTCH1. *Mol. Cell. Biol.* **26**, 6261–6271.

Coleman, M. L., McDonough, M. A., Hewitson, K. S., Coles, C., Mecinovic, J., Edelmann, M., Cook, K. M., Cockman, M. E., Lancaster, D. E., Kessler, B. M., et al. (2007). Asparaginyl hydroxylation of the Notch ankyrin repeat domain by factor inhibiting hypoxia-inducible factor. *J. Biol. Chem.* **282**, 24027–24038.

Couso, J. P., and Martinez-Arias, A. (1994). Notch is required for Wingless signaling in the epidermis of *Drosophila*. *Cell* **79**, 259–272.

Cras-Méneur, C., Li, L., Kopan, R., and Permutt, M. A. (2009). Presenilins, Notch dose control the fate of pancreatic endocrine progenitors during a narrow developmental window. *Genes Dev.* **23**, 2088–2101.

Dahlqvist, C., Blokzijl, A., Chapman, G., Falk, A., Dannaeus, K., Ibanez, C. F., and Lendhal, U. (2003). Functional Notch signaling is required for BMP4-induced inhibition of myogenic differentiation. *Development* **130**, 6089–6099.

Demarest, R. M., Ratti, F., and Capobianco, A. J. (2008). It's T-ALL about Notch. *Oncogene* **27**, 5082–5091.

Demehri, S., Liu, Z., Lee, J., Lin, M. H., Crosby, S. D., Roberts, C. J., Grigsby, P. W., Miner, J. H., Farr, A. G., and Kopan, R. (2008). Notch-deficient skin induces a lethal systemic B-lymphoproliferative disorder by secreting TSLP, a sentinel for epidermal integrity. *PLoS Biol.* **6**, e123.

Diederich, R. J., Matsuno, K., Hing, H., and Artavanis-Tsakonas, S. (1994). Cytosolic interaction between deltex and Notch ankyrin repeats implicated deltex in the Notch signalling pathway. *Development* **120**, 473–481.

Diez, H., Fischer, A., Winkler, A., Hu, C. J., Hatzopoulos, A. K., Breier, G., and Gessler, M. (2007). Hypoxia-mediated activation of Dll4-Notch-Hey2 signaling in endothelial progenitor cells and adoption of arterial cell fate. *Exp. Cell Res.* **313**, 1–9.

D'Souza, B., Miyamoto, A., and Weinmaster, G. (2008). The many facets of Notch ligands. *Oncogene* **27**, 5148–5167.

Eliasz, S., Liang, S., Chen, Y., De Marco, M. A., Machek, O., Skucha, S., Miele, L., and Bocchetta, M. (2010). Notch1 stimulates survival of lung adenocarcinoma cells during hypoxia by activating the IGF-1R pathway. *Oncogene* **29**, 2488–2498.

Espinosa, L., Ingles-Esteve, J., Aguilera, C., and Bigas, A. (2003). Phosphorylation by glycogen synthase kinase-3β down-regulates Notch activity, a link for Notch and Wnt pathways. *J. Biol. Chem.* **278**, 32227–32235.

Feldman, B. J., Hampton, T., and Clearly, M. L. (2000). A carboxy-terminal deletion mutant of Notch1 accelerates lymphoid oncogenesis in E2A-PBX1 transgenic mice. *Blood* **96**, 1906–1913.

Fiuza, U. M., Klein, T., Martinez-Arias, A., and Hayward, P. (2010). Mechanisms of ligand-mediated inhibition in Notch signaling activity in *Drosophila*. *Dev. Dyn.* **239**, 798–805.

Fryer, C. J., Lamar, E., Turbachova, I., Kintner, C., and Jones, K. A. (2002). Mastermind mediates chromatin-specific transcription and turnover enhancer complex. *Genes Dev.* **16**, 1397–1411.

Fryer, C. J., White, J. B., and Jones, K. A. (2004). Mastermind recruits CycC:CDK8 to phosphorylate the Notch ICD and coordinate activation with turnover. *Mol. Cell* **16**, 509–520.

Giniger, E. (1998). A role for Abl in Notch signaling. *Neuron* **20**, 667–681.

Giraldez, A. J., and Cohen, S. M. (2003). Wingless and Notch signaling provide cell survival cues and control cell proliferation during wing development. *Development* **130**, 6533–6543.

Goldbeter, A., and Pourquie, O. (2008). Modeling the segmentation clock as a network of coupled oscillations in the Notch, Wnt and FGF signaling pathways. *J. Theor. Biol.* **252**, 574–585.

Grabher, C., von Boehmer, H., and Look, A. T. (2006). Notch1 activation in the molecular pathogenesis of T-cell acute lymphoblastic leukaemia. *Nat. Rev. Cancer* **6**, 347–359.

Greenwald, I. (1998). LIN-12/Notch signaling: lessons from worms and flies. *Development* **125**, 3101–3109.
Greenwald, I., and Seydoux, G. (1990). Analysis of gain-of-function mutations in *lin-12* gene of *Caenorhabditis elegans*. *Nature* **346**, 197–199.
Guo, X., and Wang, X. F. (2009). Signaling cross-talk between TGF-β/BMP and other pathways. *Cell Res.* **19**, 71–88.
Guo, D., Ye, J., Dai, J., Li, L., Chen, F., Ma, D., and Ji, C. (2009). Notch1 regulates Akt signaling pathway and the expression of cell cycle regulatory proteins cyclin D1, CDK2 and p21 in T-ALL cell lines. *Leuk. Res.* **33**, 678–685.
Gupta, R., Hong, D., Iborra, F., Sarno, S., and Enver, T. (2007). NOV (CCN3) functions as a regulator of human hematopoietic stem or progenitor cells. *Science* **316**, 590–593.
Gustafsson, M. V., Zheng, X., Pereira, T., Gradin, K., Jin, S., Lundkvist, J., Ruas, J. L., Poellinger, L., Lendhal, U., and Bondesson, M. (2005). Hypoxia requires Notch signaling to maintain the undifferentiated cell state. *Dev. Cell* **9**, 617–628.
Hayward, P., Kalmar, T., and Martinez-Arias, A. (2008). Wnt/Notch signalling and information processing during development. *Development* **135**, 411–424.
Heitzler, P., and Simpson, P. (1991). The choice of cell fate in the epidermis of *Drosophila*. *Cell* **64**, 1083–1092.
Heitzler, P., and Simpson, P. (1993). Altered epidermal growth factor-like sequences provide evidence for a role of Notch as a receptor in cell fate decisions. *Development* **117**, 1113–1123.
Henderson, S. T., Gao, D., Lambie, E. J., and Kimble, J. (1994). lag-2 may encode a signaling ligand for the GLP-1 and LIN-12 receptors in *C. elegans*. *Development* **120**, 2913–2924.
Herranz, H., and Milan, M. (2006). Notch and the affinity boudaries in *Drosophila*. *BioEssays* **28**, 113–116.
Hoemann, C. D., Beaulieu, N., Girard, L., Rebai, N., and Jolicoeur, P. (2000). Two distinct Notch1 mutant alleles are involved in the induction of T-cell leukemia in c-myc transgenic mice. *Mol. Cell. Biol.* **20**, 3831–3842.
Hori, K., Cholewa-Waclaw, J., Nakada, Y., Glasgow, S. M., Masui, T., Henke, R. M., Wildner, H., Martarelli, B., Beres, T. M., and Epstein, J. A. (2008). A nonclassical bHLH Rbpj transcription factor complex is required for specification of GABAergic neurons independent of Notch signaling. *Genes Dev.* **22**, 166–178.
Hori, K., Fostier, M., Ito, M., Fuwa, T. J., Go, M. J., Okano, H., Baron, M., and Matsuno, K. (2004). *Drosophila* Deltex mediates Suppressor of Hairless-independent and late-endosomal activation of Notch signaling. *Development* **131**, 5527–5537.
Hsieh, J. J. D., and Hayward, S. D. (1995). Masking of the CBF1/RBPJk transcriptional repression domain by Epstein-Barr virus EBNA2. *Science* **268**, 560–563.
Hu, Q. D., Ang, B. T., Karsak, M., Hu, W. P., Cui, X. Y., Duka, T., Takeda, Y., Chia, W., Sankar, N., Ng, Y. K., Ling, E. A., Maciag, T., et al. (2003). F3/Contactin acts as a functional ligand for Notch during oligodendrocyte maturation. *Cell* **115**, 163–175.
Hudson, C., Lotito, S., and Yasuo, H. (2007). Sequential and combinatorial imputs from Nodal, Delta2/Notch and FGF/MEK/ERK signalling pathways establish a grid-like organisation of distinct cell identities in the ascidian neural plate. *Development* **134**, 3527–3537.
Irvine, K. D., and Rauskolb, C. (2001). Boundaries in development formation and function. *Annu. Rev. Cell Dev. Biol.* **17**, 189–214.
Ishitani, T., Hirao, T., Suzuki, M., Isoda, M., Ishitani, S., Harigaya, K., Kitagawa, M., Matsumoto, K., and Itoh, M. (2010). Nemo-like kinase suppresses Notch signalling by interfering with formation of the Notch active transcriptional complex. *Nat. Cell Biol.* **12**, 278–285.
Itoh, F., Itoh, S., Goumans, M. J., Valdimarsdottir, G., Iso, T., Dotto, G. P., Hamamori, Y., Kedes, L., Kato, M., and ten Dijke, Pt.P. (2004). Synergy and antagonism

between Notch and BMP receptor signaling pathways in endothelial cells. *EMBO J.* **23**, 541–551.
Jorgensen, M. C., Ahnfelt-Ronne, J., Hald, J., Madsen, O. D., Serup, P., and Hecksher-Sorensen, J. (2007). An illustrated review of early pancreas development in the mouse. *Endocr. Rev.* **28**, 685–705.
Joutel, A., Corpechot, C., Ducros, A., Vahedi, K., Chabriat, H., Mouton, P., et al. (1996). Notch3 mutations in CADASIL, a hereditary adult-onset condition causing stroke and dementia. *Nature* **383**, 707–710.
Kankel, M. W., et al. (2007). Investigating the genetic circuitry of mastermind in *Drosophila*, a Notch signal effector. *Genetics* **177**, 2493–2505.
Klüppel, M., and Wrana, J. L. (2005). Turning it up a Notch: cross-talk between TGFβ and Notch signaling. *BioEssays* **27**, 115–117.
Koelzer, S., and Klein, T. (2006). Regulation of expression of Vg and establishment of the dorsoventral compartment boundary in the wing imaginal disc by Suppressor of Hairless. *Dev. Biol.* **289**, 77–90.
Komatsu, H., Chao, M. Y., Larkins-Ford, J., Corkins, M. E., Somers, G. A., Tucey, T., Dionne, H. M., White, J. Q., Wani, K., Boxem, M., and Hart, A. C. (2008). OSM-11 facilitates LIN-12 Notch signaling during Caenorhabditis elegans vulval development. PLoS Biol. 6, e196.
Kopan, R., and Ilagan, M.X.G. (2009). The canonical Notch signaling pathway: unfolding the activation mechanism. *Cell* **137**, 216–233.
Kraman, M., and McCright, B. (2005). Functional conservation of Notch1 and Notch2 intracellular domains. *FASEB J.* **19**, 1311–1313.
Lai, E. C. (2004). Notch signaling: control of cell communication and cell fate. *Development* **131**, 965–973.
Langdon, T., Hayward, P., Brennan, K., Wirtz-Peitz, F., Sanders, P., Zecchini, V., Friday, A., Balayo, T., and Martinez Arias, A. (2006). Notch receptor encodes two structurally separable functions in *Drosophila*: a genetic analysis. *Dev. Dyn.* **235**, 998–1013.
Le Gall, M., De Mattei, C., and Giniger, E. (2008). Molecular separation of two signaling pathways for the receptor. Notch. *Dev. Biol.* **313**, 556–567.
Louvi, A., Arboleda-Velasquez, J. F., and Artavanis-Tsakonas, S. (2006). CADASIL: a critical look at a Notch disease. *Dev. Neurosci.* **28**, 5–12.
Martinez Arias, A., Zecchini, V., and Brennan, K. (2002). CSL-independent Notch signalling: a checkpoint in cell fate decisions during development ?. *Curr. Opin. Genet. Dev.* **12**, 524–533.
Masui, T., Long, Q., Beres, T. M., Magnuson, M. A., and Macdonald, R. J. (2007). Early pancreatic development requires the vertebrate Suppressor of Hairless (RBPJ) in the PTF1 bHLH complex. *Genes Dev.* **21**, 2629–2643.
Matsuno, K., Diederich, R. J., Go, M. J., Blaumueller, C. M., and Artavanis-Tsakonas, S. (1995). Deltex acts as a positive regulator of Notch signaling through interactions with the Notch ankyrin repeats. *Development* **121**, 2633–2644.
Matsuno, K., Eastman, D., Mitsiades, T., Quinn, A. M., Carcanciu, M. L., and Ordentlich, P. (1998). Human deltex is a conserved regulator of Notch signaling. *Nat. Genet.* **19**, 74–78.
McElhinny, A. S., Li, J. L., and Wu, L. (2008). Mastermind-like transcriptional co-activators: emerging roles in regulating cross talk among multiple signaling pathways. *Oncogene* **27**, 5138–5147.
Meurette, O., Stylianou, S., Rock, R., Collu, G. M., Gilmore, A. P., and Brennan, K. (2009). Notch activation induces Akt signaling via an autocrine loop to prevent apoptosis in breast epithelial cells. *Cancer Res.* **69**, 5015–5022.
Micchelli, C. A., Rulifson, E. J., and Blair, S. S. (1997). The function and regulation of cut expression on the wing margin of *Drosophila*: Notch, Wingless and a dominant negative role for Delta and Serrate. *Development* **124**, 1485–1495.

Milan, M., and Cohen, S. M. (2000). Temporal regulation of Apterous activity during the development of the *Drosophila* wing. *Development* **127**, 3069–3078.

Milan, M., and Cohen, S. M. (2003). A re-evaluation of the contributions of Apterous and Notch to the dorsoventral lineage restriction boundary in the *Drosophila* wing. *Development* **130**, 553–562.

Minamizato, T., Sakamoto, K., Liu, T., Kokubo, H., Katsube, K., Perbal, B., et al. (2007). CCN3/NOV inhibits BMP-2-induced osteoblast differentiation by interacting with BMP and Notch signaling pathways. *Biochem. Biophys. Res. Commun.* **354**, 567–573.

Miyamoto, A., Lau, R., Hein, P. W., Shipley, J. M., and Weinmaster, G. (2006). Microfibrillar proteins MAGP-1 and MAGP-2 induce Notch1 extracellular domain dissociation and receptor activation. *J. Biol. Chem.* **281**, 10089–10097.

Moellering, R. E., Cornejo, M., Davis, T. N., Del Bianco, C., Aster, J. C., Blacklow, S. C., Kung, A. L., Gilliland, D. G., Verdine, G. L., and Bradner, J. E. (2009). Direct inhibition of the Notch transcription factor complex. *Nature* **462**, 182–188.

Monet-Lepetre, M., Bardot, B., Lemaire, B., Domenga, V., Godin, O., Dichgans, M., Tournier-Lasserve, E., Cohen-Tannoudji, M., Chabriat, H., and Joutel, A. (2009). Distinct phenotypic and functional features of CADASIL mutations in the Notch3 ligand binding domain. *Brain* **132**, 1601–1612.

Mummery-Widmer, J. L., Yamazaki, M., Stoeger, T., Novatchkova, M., Bhalerao, S., Chen, D., Dietzl, G., Dickson, B. J., and Knoblich, J. A. (2009). Genome-wide analysis of Notch signalling in *Drosophila* by transgenic RNAi. *Nature* **458**, 987–992.

Mungamuri, S. K., Yang, X., Thor, A. D., and Somasundaram, K. (2006). Survival signaling by Notch1: mammalian target of rapamycin (mTOR)-dependent p53. *Cancer Res.* **66**, 4715–4724.

Murtaugh, L. C. (2008). The what, where, when and how Wnt/β-catenin signaling in pancreas development. *Organogenesis* **4**, 81–86.

Nam, Y., Sliz, P., Pear, W. S., Aster, J. C., and Blacklow, S. C. (2007). Cooperative assembly of higher-order Notch complexes functions as a switch to induce transcription. *Proc. Natl. Acad. Sci. U.S.A.* **104**, 2103–2108.

Nam, Y., Sliz, P., Song, L., Aster, J. C., and Blacklow, S. C. (2006). Structural basis for cooperativity in recruitment of MAML coactivators to Notch transcription complexes. *Cell* **124**, 973–983.

Nicolas, M., Wolfer, A., Raj, K., Kummer, J. A., Mill, P., van Noort, M., Hui, C. C., Clevers, H., Dotto, G. P., and Radtke, F. (2003). Notch1 functions as a tumor suppressor in mouse skin. *Nat. Genet.* **33**, 416–421.

Niimi, H., Pardali, K., Vanlandewijck, M., Heldin, C. H., and Moustakas, A. (2007). Notch signaling is necessary for epithelial growth arrest by TGF-β. *J. Cell Biol.* **176**, 695–707.

O'Neil, J., Grim, J., Strack, P., Rao, S., Tibbitts, D., Winter, C., Hardwick, J., Welcker, M., Meijerink, J. P., Pieters, R., et al. (2007). FBW7 mutations in leukemic cells mediate NOTCH pathway activation and resistance to γ-secretase inhibitors. *J. Exp. Med.* **204**, 1813–1824.

Ordentlich, P., Lin, A., Shen, C. P., Blaumueller, C., Matsuno, K., and Artavanis-Tsakonas, S. (1998). Notch inhibition of E47 supports the existence of a novel signaling pathway. *Mol. Cell. Biol.* **18**, 2230–2239.

Palomero, T., Sulis, M. L., Cortina, M., Real, P. J., Barnes, K., Ciofani, M., Caparros, E., Buteau, J., Brown, K., Perkins, S. L., Bhagat, G., Agarwal, A. M., et al. (2007). Mutational loss of PTEN induces resistance to Notch1 inhibition in T-cell leukemia. *Nat. Med.* **13**, 1203–1210.

Pan, Y., Lin, M. H., Tian, X., Cheng, H. T., Gridley, T., et al. (2004). γ-secretase functions through Notch signaling to maintain skin appendages but is not required for their patterning or initial morphogenesis.*Dev. Cell* **7**, 731–743.

Patel, N. S., Li, J. L., Generali, D., Poulsom, R., Cranston, D. W., and Harris, A. L. (2005). Up-regulation of Delta-like 4 ligand in human tumor vasculature and the role of basal expression in endothelial cell function. *Cancer Res.* **65**, 8690–8697.

Perumalsamy, L. R., Nagala, M., Banerjee, P., and Sarin, A. (2009). A hierarchical cascade activated by non-canonical Notch signaling and the mTOR-Rictor complex regulates neglect-induced death in mammalian cells. *Cell Death Diff.* **16**, 879–889.

Perumalsamy, L. R., Nagala, M., and Sarin, A. (2010). Notch-activated signaling cascade interacts with mitochondrial remodeling proteins to regulate cell survival. *Proc. Natl. Acad. Sci. U.S.A.* **107**, 6882–6887.

Poellinger, L., and Lendahl, U. (2008). Modulating Notch signaling by pathway-intrinsic and pathway-extrinsic mechanisms. *Curr. Opin. Genet. Dev.* **18**, 449–454.

Powell, P. A., Wesley, C., Spencer, S., and Cagan, R. L. (2001). Scabrous complexes with Notch to mediate boundary formation. *Nature* **409**, 626–630.

Ramain, P., Khechumian, K., Seugnet, L., Arbogast, N., Ackermann, C., and Heitzler, P. (2001). Novel Notch alleles reveal a Deltex-dependent pathway repressing neural fate. *Curr. Biol.* **11**, 1729–1738.

Rebay, I., Fehon, R. G., and Artavanis-Tsakonas, S. (1993). Specific truncations of Drosophila Notch define dominant activated and dominant negative forms of the receptor. *Cell* **74**, 319–329.

Rusconi, J. C., and Corbin, V. (1998). Evidence for a novel Notch pathway required for muscle precursor selection in *Drosophila. Mech. Dev.* **79**, 39–50.

Sakamoto, K., Yamaguchi, S., Ando, R., Miyawaki, A., Kabasawa, Y., Takagi, M., et al. (2002). The nephroblastoma overexpressed gene (NOV/ccn3) protein associates with Notch1 extracellular domain and inhibits myoblast differentiation via Notch signaling pathway. *J. Biol. Chem.* **277**, 29399–29405.

Sanders, P. G., Munoz-Descalzo, S., Balayo, T., Wirtz-Peitz, F., Hayward, P., and Martinez-Arias, A. (2009). Ligand-independent traffic of Notch buffers activated Armadillo in *Drosophila. PLoS Biol.* **7**, e1000169.

Shawber, C., Nofziger, D., Hsieh, J.J.D., Lindsell, C., Bogler, O., and Hayward, D. (1996). Notch signaling inhibits muscle cell differentiation through a CBF1-independent pathway. *Development* **122**, 3765–3773.

Shin, H. M., Minter, L. M., Cho, O. H., Gottipati, S., Fauq, A. H., Golde, T. E., Sonenshein, G. E., and Osborne, B. A. (2006). Notch1 augments NF-κB activity by facilitating its nuclear retention. *EMBO J.* **25**, 129–138.

Simpson, P. (1997). Notch signalling in development: on equivalence groups and asymmetric developmental potential. *Curr. Opin. Genet. Dev.* **7**, 537–542.

Thompson, B. J., Buonamici, S., Sulis, M. L., Palomero, T., Vilimas, T., Basso, G., Ferrando, A., and Aifantis, I. (2007). The SCFFBW7 ubiquitin ligase complex as a tumor suppressor in T cell leukemia. *J. Exp. Med.* **204**, 1825–1835.

Tsunematsu, R., Nakayama, K., Oike, Y., Nishiyama, M., Ishida, N., Hatakeyama, S., Bessho, Y., Kageyama, R., Suda, T., and Nakayama, K. I. (2004). Mouse Fw7/Sel10/Cdc4 is required for Notch degradation during vascular development. *J. Biol. Chem.* **279**, 9417–9423.

Verheyen, E. M., Purcell, K. J., Fortini, M. E., and Artavanis-Tsakonas, S. (1996). Analysis of dominant and suppressors of activated Notch in *Drosophila. Genetics* **144**, 1127–1141.

Vied, C., and Kalderon, D. (2009). Hedgehog-stimulated stem cells depend on non-canonical activity of the Notch co-activator Mastermind. *Development* **136**, 2177–2186.

Vilimas, T., Mascarenhas, J., Palomero, T., Mandal, M., Buonamici, S., Meng, F., Thompson, B., Spaulding, C., Macaroun, S., Alegre, M. L., Kee, B. L., Ferrando, A., et al. (2007). Targeting the NF-kB signaling pathway in Notch1-induced T-cell leukemia. *Nat. Med.* **13**, 70–77.

Walsh, D. W., Roxburgh, S. A., McGettigan, P., et al. (2008). Co-regulation of Gremlin and Notch signalling in diabetic nephropathy. *Biochim. Biophys. Acta* **1782**, 10–21.

Wang, J., Shelly, L., Miele, L., Boykins, R., Norcross, M. A., and Guan, E. (2001). Human Notch-1 inhibits NF-kB activity in the nucleus through a direct interaction involving a novel domain. *J. Immunol.* **167**, 289–295.

Weng, A. P., Ferrando, A. A., Lee, W., Morris, J. P., Silverman, L. B., Sanchez-Irizarry, C., Blacklow, S. C., Look, A. T., and Aster, J. C. (2004). Activating mutations of NOTCH1 in human T cell acute lymphoblastic leukemia. *Science* **306**, 269–271.

Wesley, C. (1999). Notch and Wingless regulate expression of cuticule patterning genes. *Mol. Cell. Biol.* **19**, 5743–5758.

Wilkin, M. B., Carbery, A. M., Gensch, N., Clemence, S., Motoki, M., Yamada, K., Hori, K., Taniguchi-Kanai, M., Franklin, E., Matsuno, K., and Baron, M. (2008). *Drosophila* HOPS and AP-3 complex genes are required for a Deltex-regulated activation of Notch in the endosomal trafficking pathway. *Dev. Cell* **15**, 762–772.

Wilkin, M., Tongngok, P., Fostier, M., Aslam, H., Mazaleyrat, S. L., Higgs, J., Myat, A., Evans, D.A.P., Cornell, M., and Baron, M. (2004). Regulation of Notch endosomal sorting and signaling by *Drosophila* Nedd4 family proteins. *Curr. Biol.* **14**, 2237–2244.

Zavadil, J., Cermak, L., Soto-Nieves, N., and Bottinger, E. P. (2004). Integration of TGF-β/Smad and Jagged1/Notch signalling in epithelial-to-mesenchymal transition. *EMBO J.* **23**, 1155–1165.

Zheng, X., Linke, S., Dias, J. M., Zheng, X., Gradin, K., Wallis, T. P., Hamilton, B. R., Gustafsson, M., Ruas, J. L., Wilkins, S., et al. (2008). Interaction with factor inhibiting HIF-1 defines an additional mode of cross-coupling between the Notch and hypoxia signaling pathways. *Proc. Natl. Acad. Sci. U.S.A.* **105**, 3368–3373.

Index

D

E

O

P

T

Contents of Previous Volumes

Volume 47

Volume 48

Volume 49

Volume 52

Volume 53

Volume 54

Volume 55

Volume 56

Volume 57

Volume 58

Volume 59

Volume 60

Volume 61

Volume 62

Volume 63

Volume 64

Volume 65

Volume 66

Volume 67

Volume 68

Volume 69

Volume 70

Volume 71

Volume 72

Volume 73

Volume 74

Volume 75

Volume 76

Volume 77

Volume 78

Volume 81

Volume 84

Volume 85

Volume 86

Volume 87

Volume 88

Volume 91

SnapShot: Notch Signaling Pathway

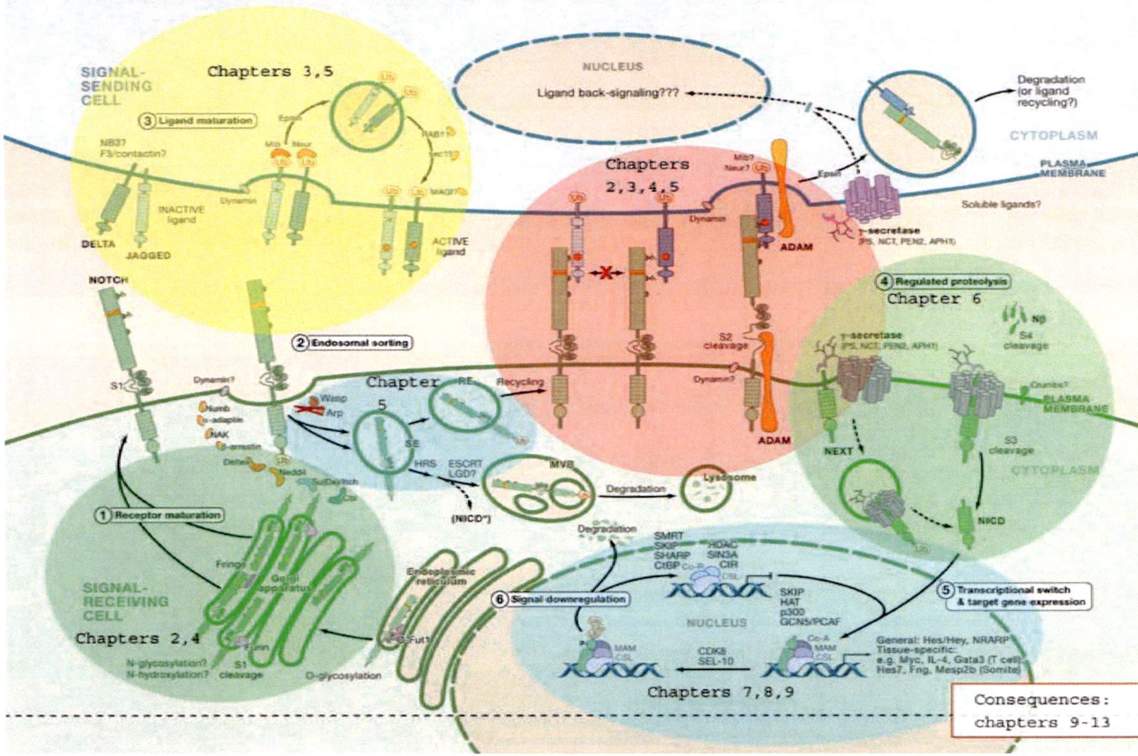

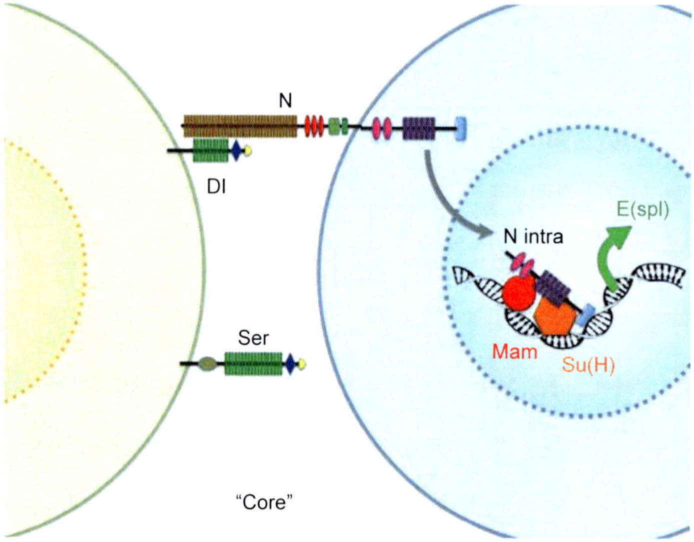

Spyros Artavanis-Tsakonas and Marc A. T. Muskavitch, Figure 1.5 Please refer to the legend in the text.

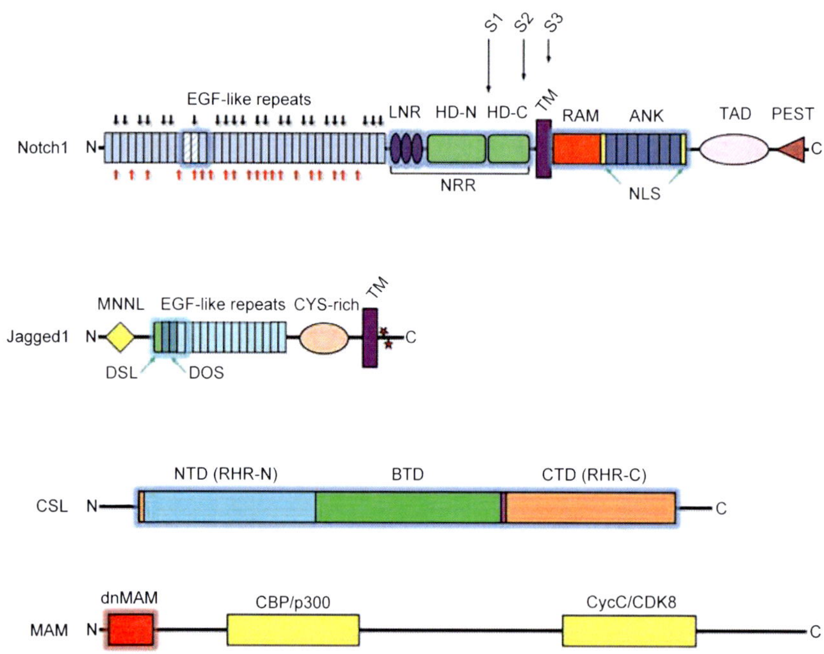

Rhett A. Kovall and Stephen C. Blacklow, Figure 2.1 Please refer to the legend in the text.

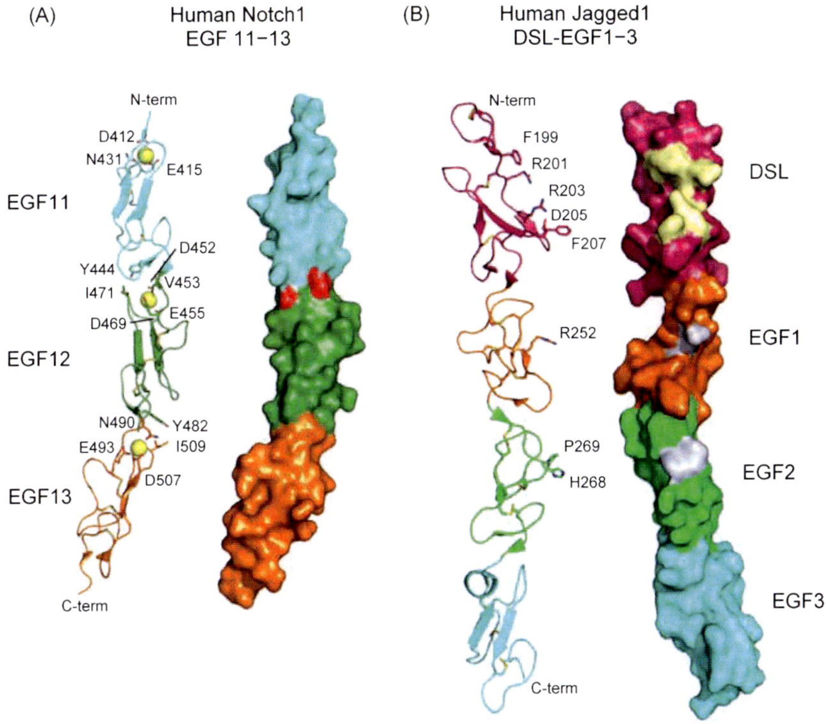

Rhett A. Kovall and Stephen C. Blacklow, Figure 2.2 Please refer to the legend in the text.

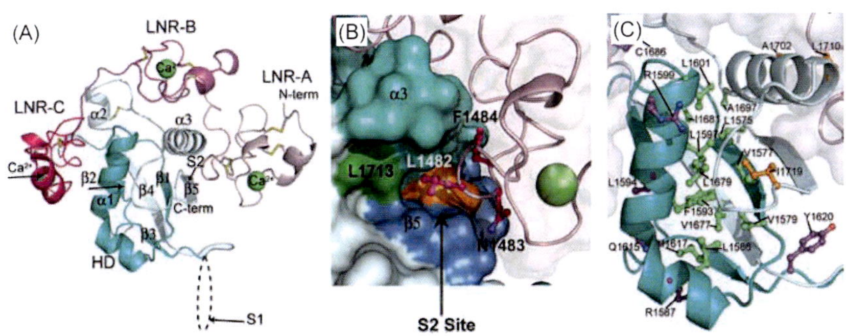

Rhett A. Kovall and Stephen C. Blacklow, Figure 2.3 Please refer to the legend in the text.

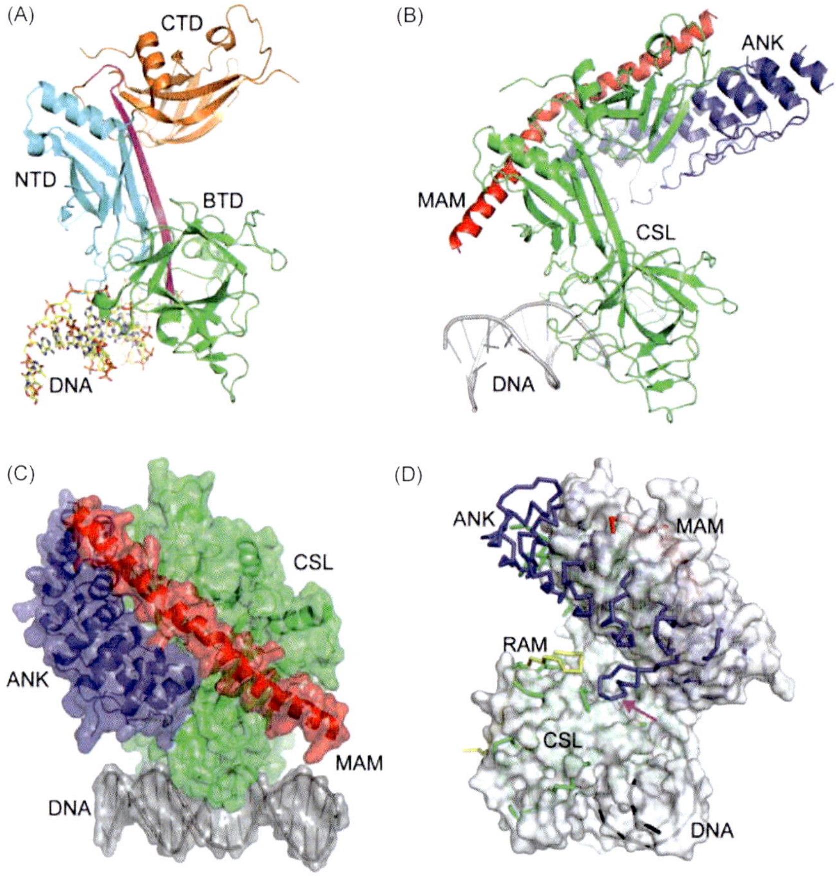

Rhett A. Kovall and Stephen C. Blacklow, Figure 2.4 Please refer to the legend in the text.

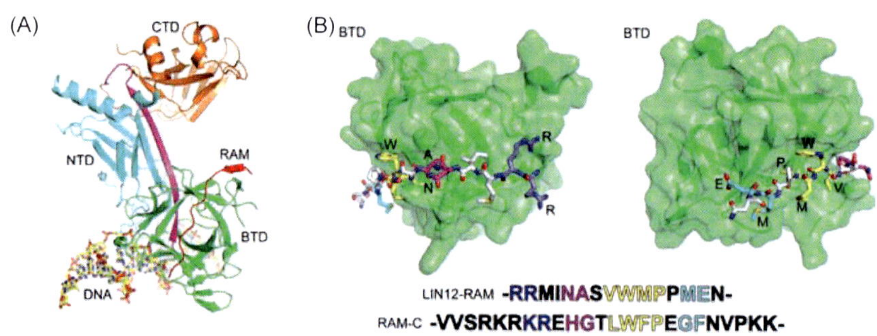

Rhett A. Kovall and Stephen C. Blacklow, Figure 2.5 Please refer to the legend in the text.

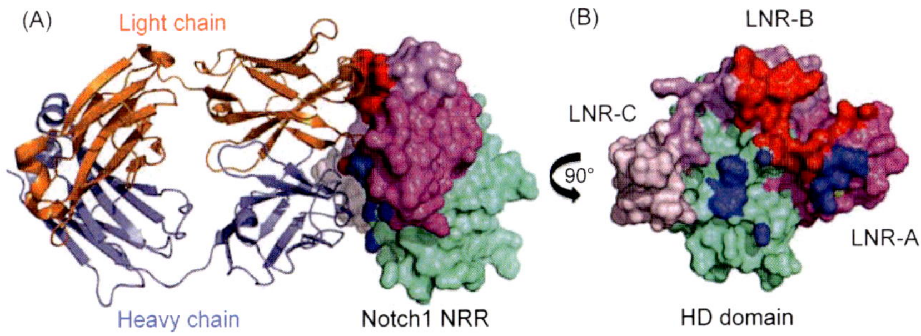

Rhett A. Kovall and Stephen C. Blacklow, Figure 2.7 Please refer to the legend in the text.

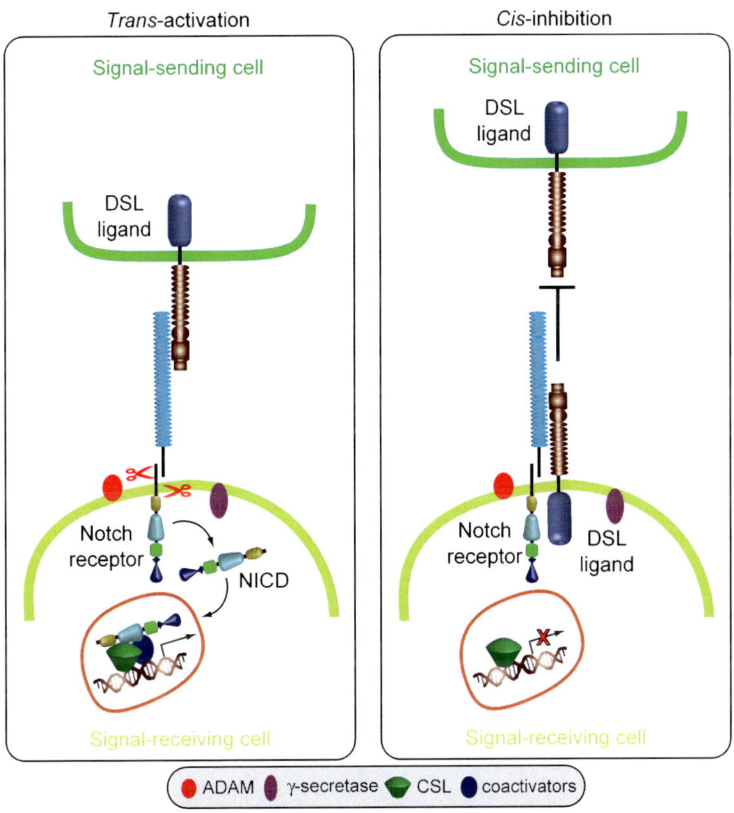

Brendan D'Souza *et al.*, Figure 3.1 Please refer to the legend in the text.

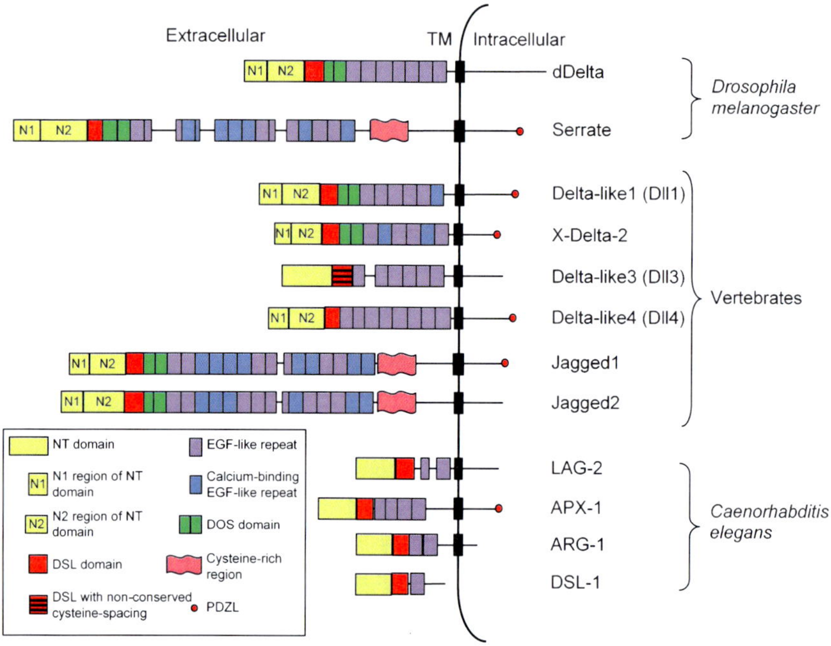

Brendan D'Souza *et al.*, Figure 3.2 Please refer to the legend in the text.

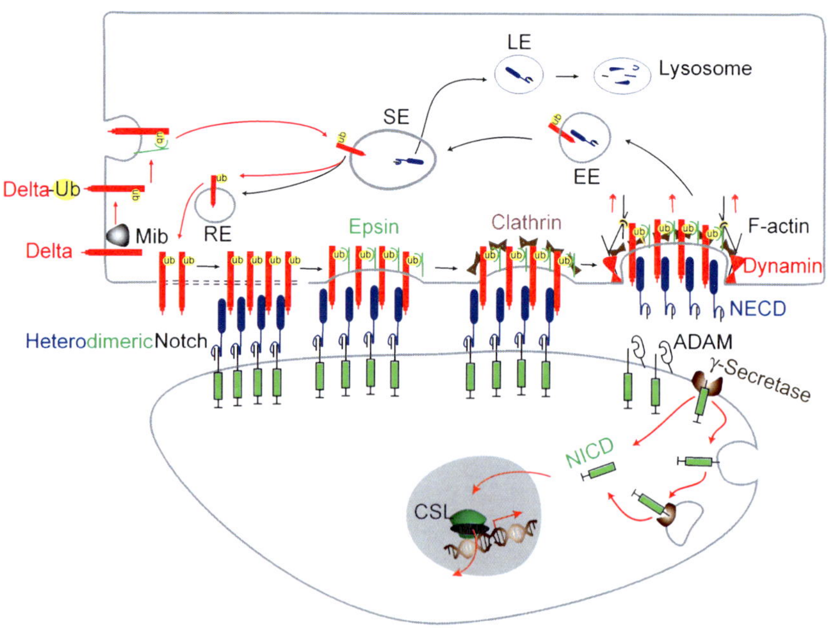

Brendan D'Souza *et al.*, Figure 3.3 Please refer to the legend in the text.

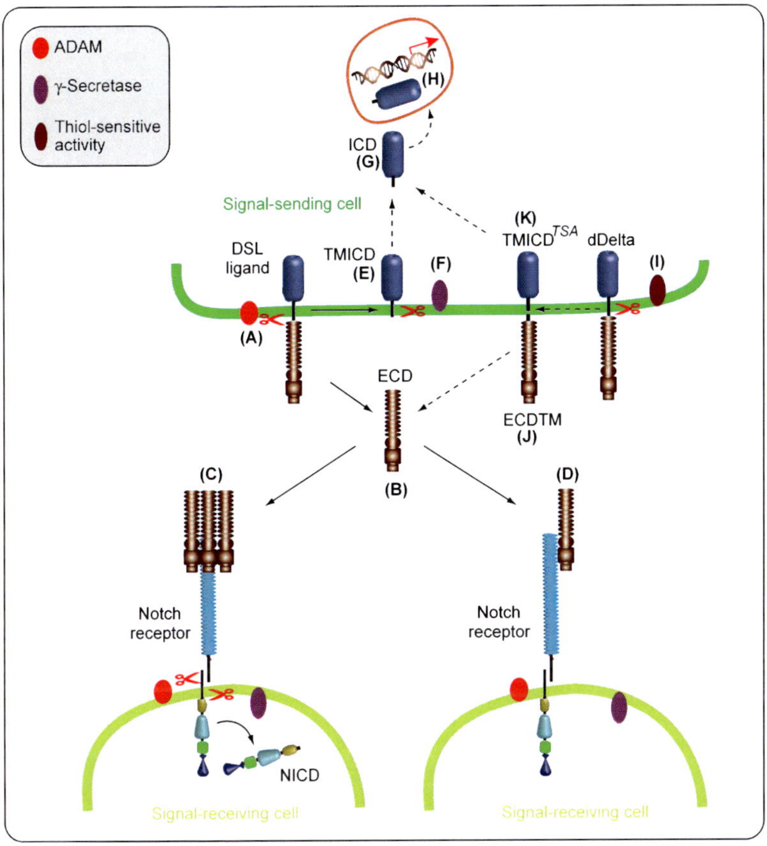

Brendan D'Souza *et al.*, Figure 3.4 Please refer to the legend in the text.

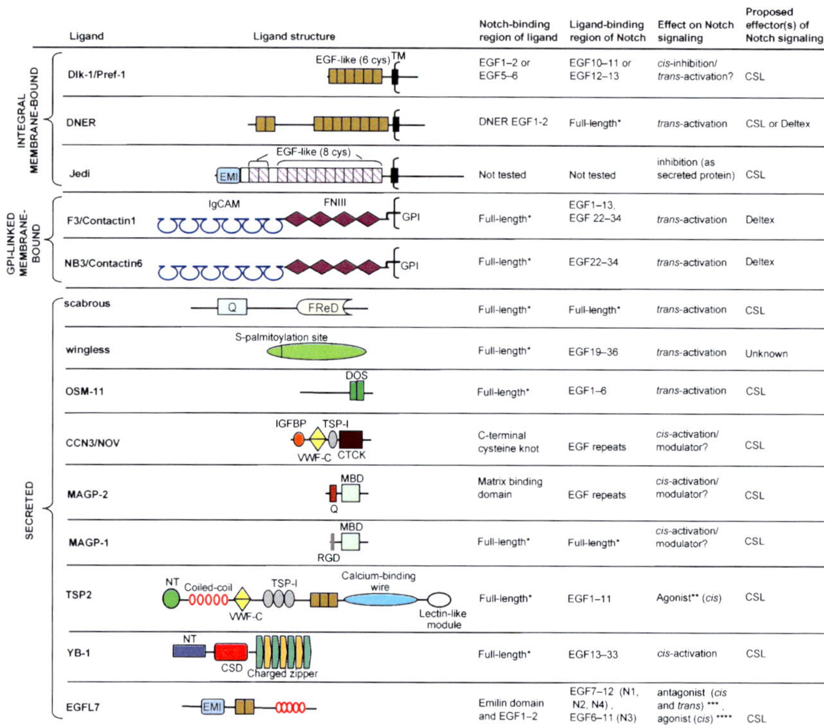

Brendan D'Souza *et al.*, Figure 3.5 Please refer to the legend in the text.

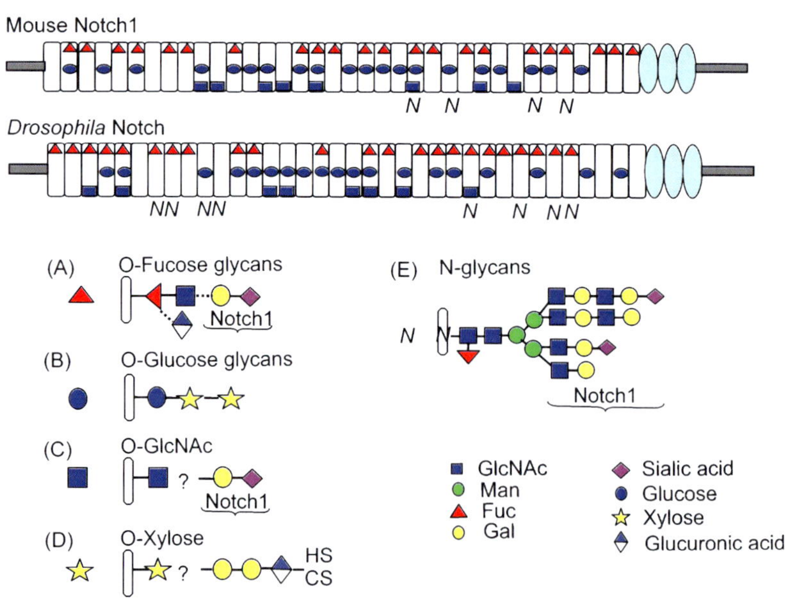

Pamela Stanley and Tetsuya Okajima, Figure 4.1 Please refer to the legend in the text.

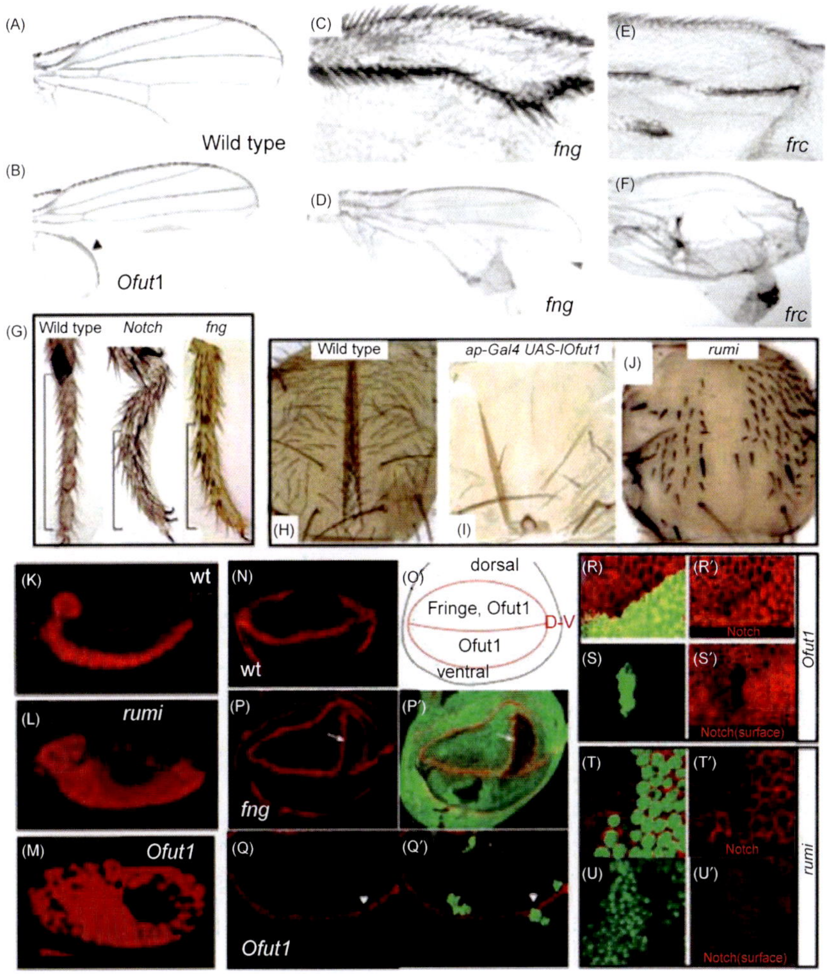

Pamela Stanley and Tetsuya Okajima, Figure 4.2 Please refer to the legend in the text.

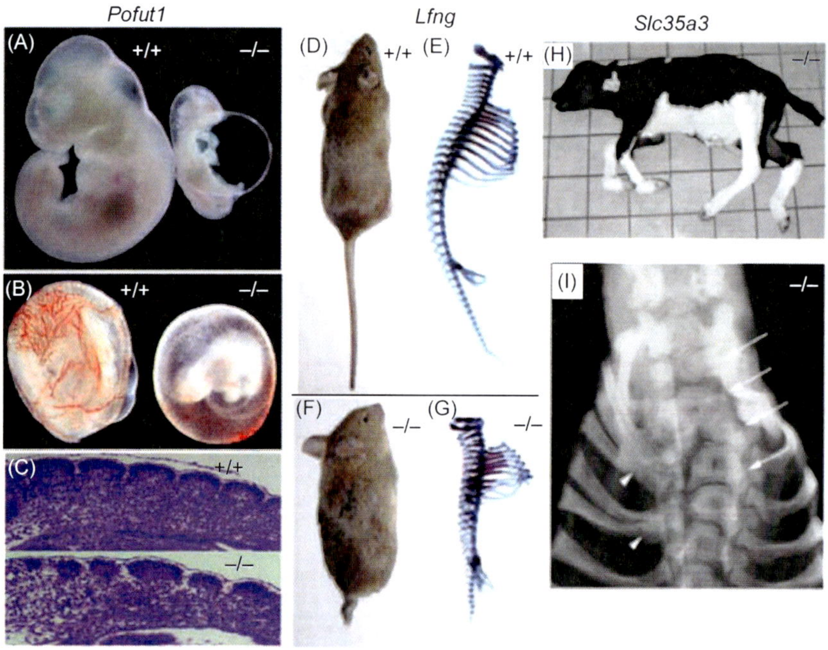

Pamela Stanley and Tetsuya Okajima, Figure 4.3 Please refer to the legend in the text.

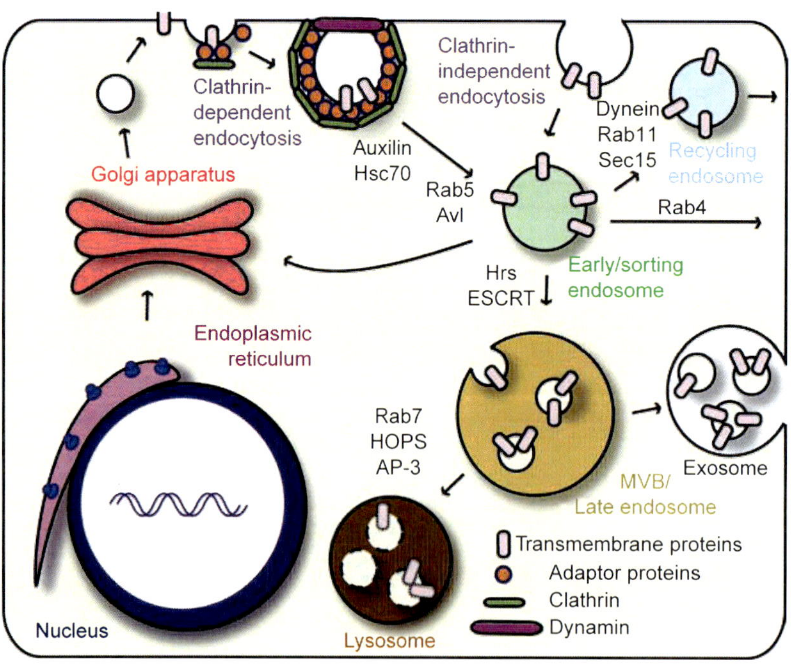

Shinya Yamamoto *et al.*, Figure 5.1 Please refer to the legend in the text.

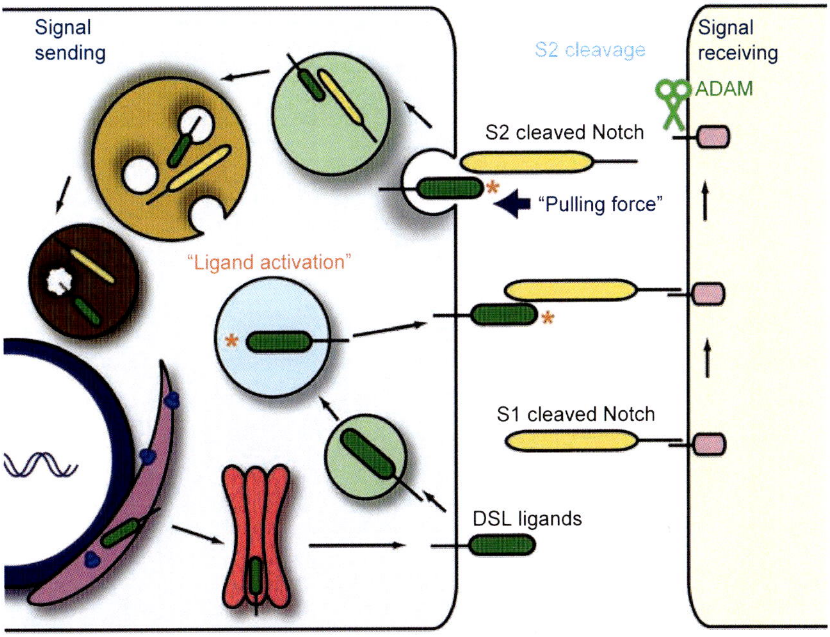

Shinya Yamamoto *et al.*, Figure 5.2 Please refer to the legend in the text.

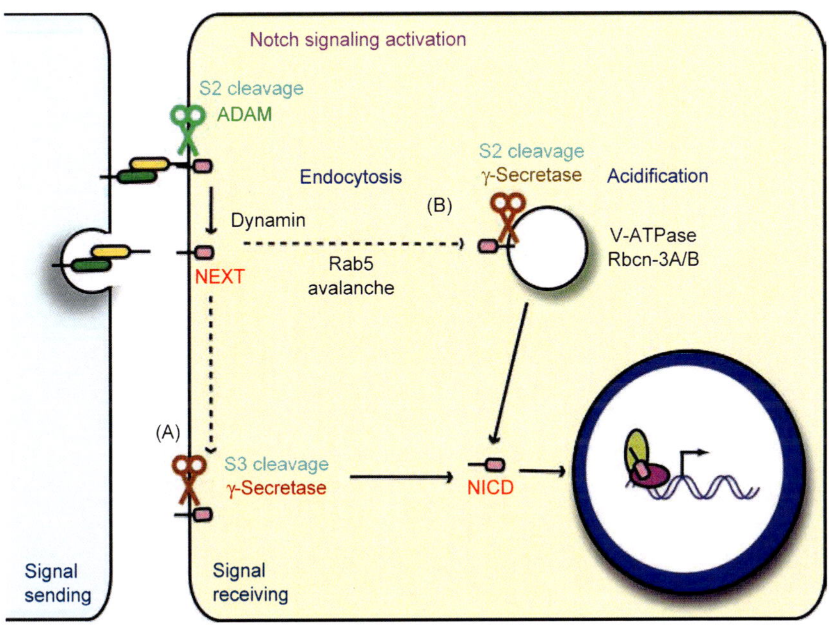

Shinya Yamamoto *et al.*, Figure 5.3 Please refer to the legend in the text.

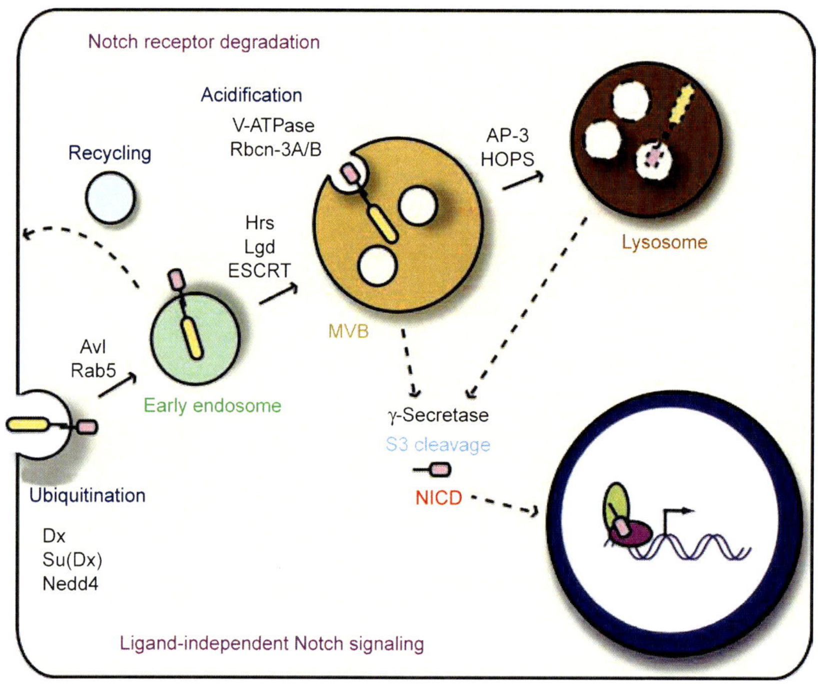

Shinya Yamamoto *et al.*, Figure 5.4 Please refer to the legend in the text.

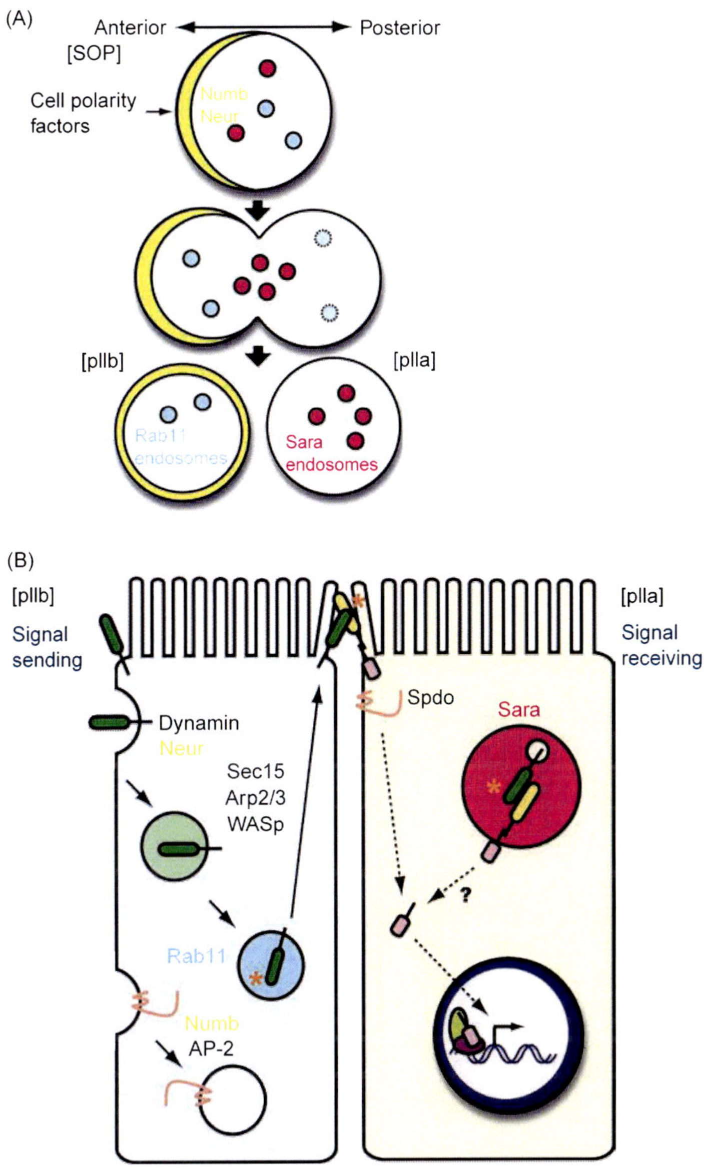

Shinya Yamamoto *et al.*, Figure 5.5 Please refer to the legend in the text.

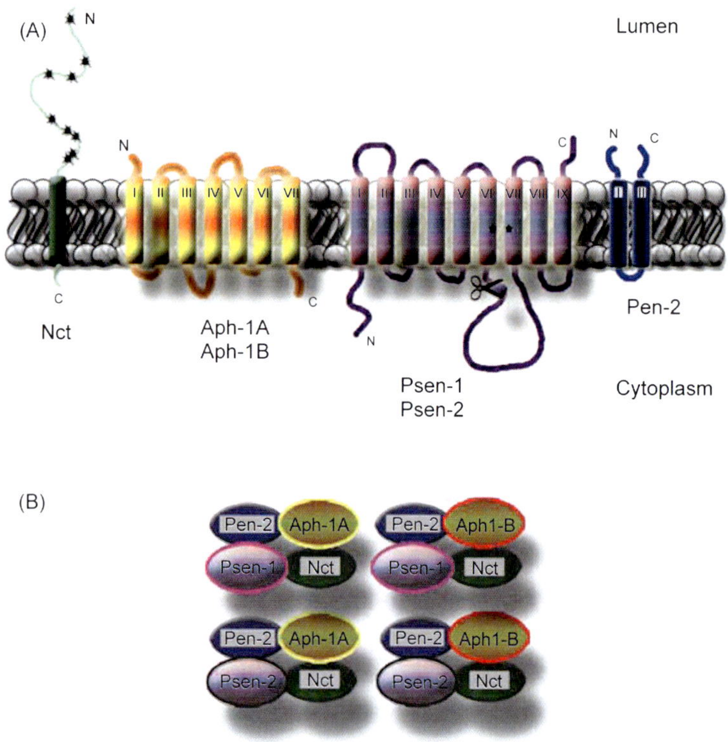

Ellen Jorissen and Bart De Strooper, Figure 6.1 Please refer to the legend in the text.

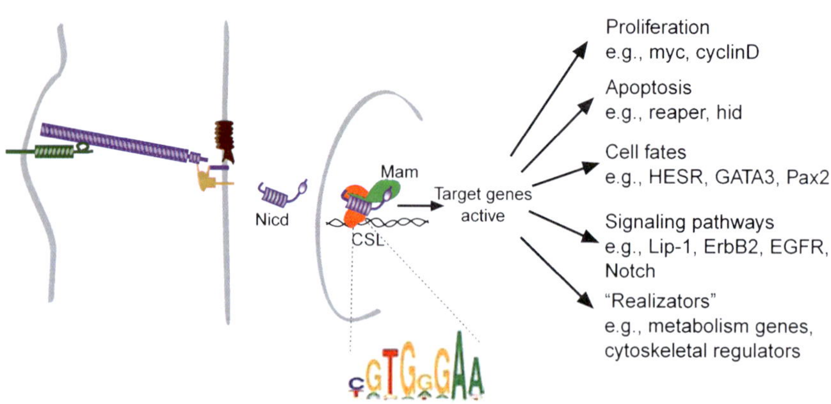

Sarah Bray and Fred Bernard, Figure 8.1 Please refer to the legend in the text.

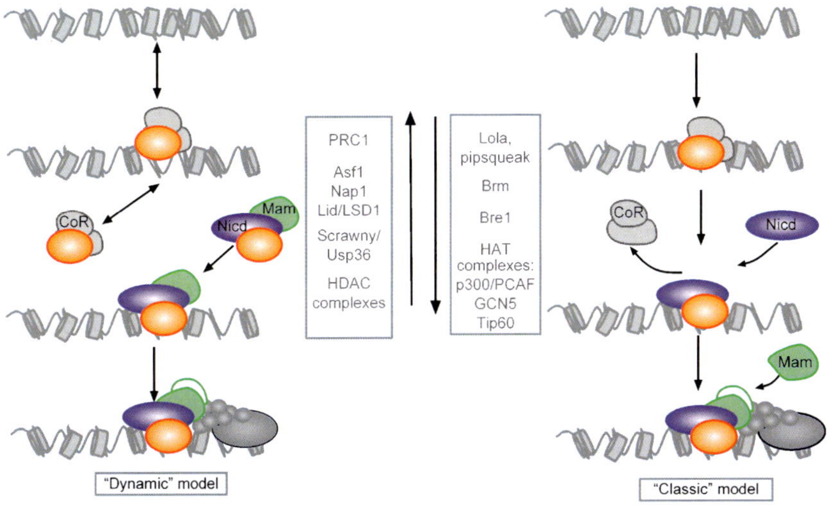

Sarah Bray and Fred Bernard, Figure 8.2 Please refer to the legend in the text.

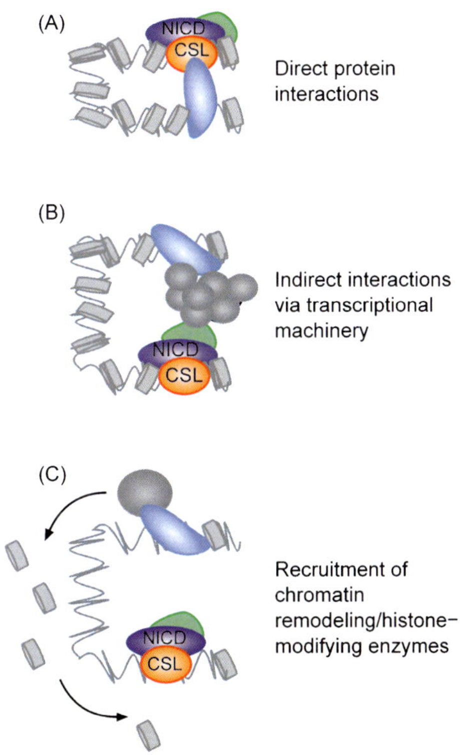

Sarah Bray and Fred Bernard, Figure 8.3 Please refer to the legend in the text.

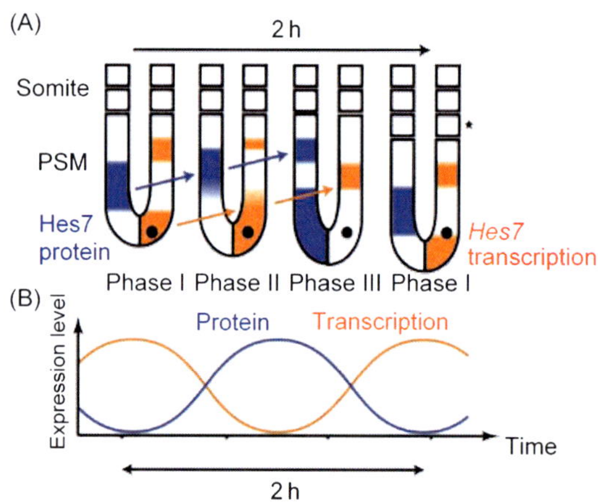

Ryoichiro Kageyama *et al.*, Figure 10.2 Please refer to the legend in the text.

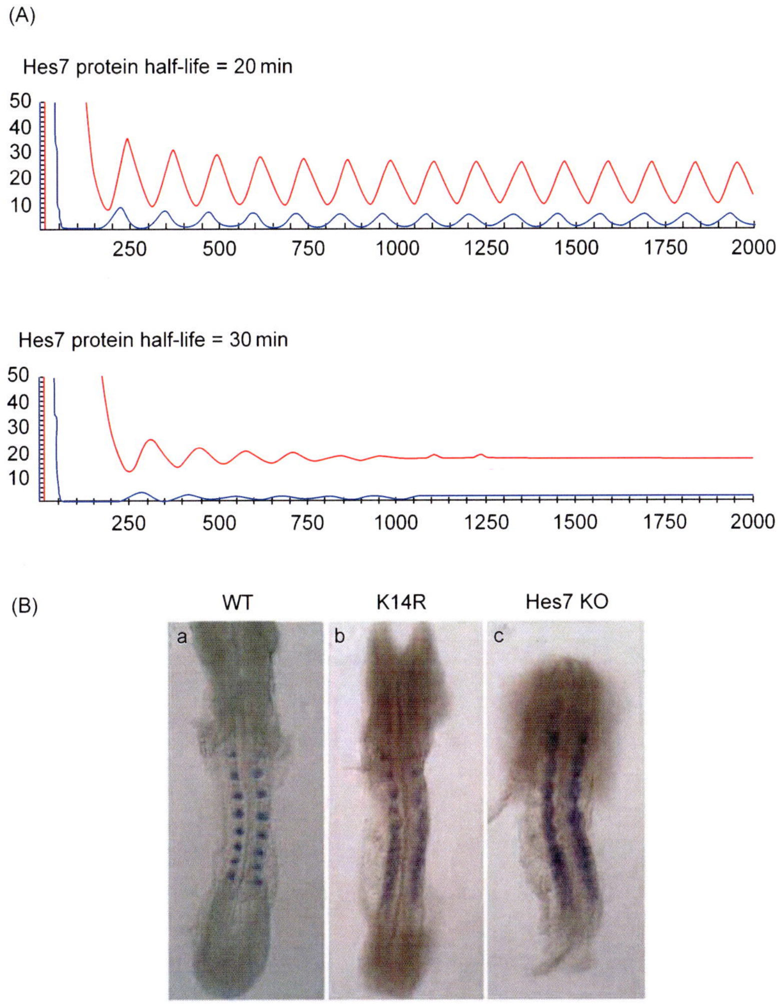

Ryoichiro Kageyama *et al.*, Figure 10.4 Please refer to the legend in the text.

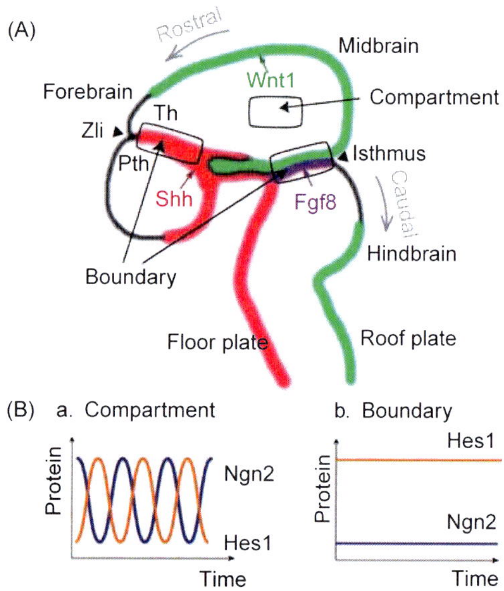

Ryoichiro Kageyama *et al.*, Figure 10.8 Please refer to the legend in the text.

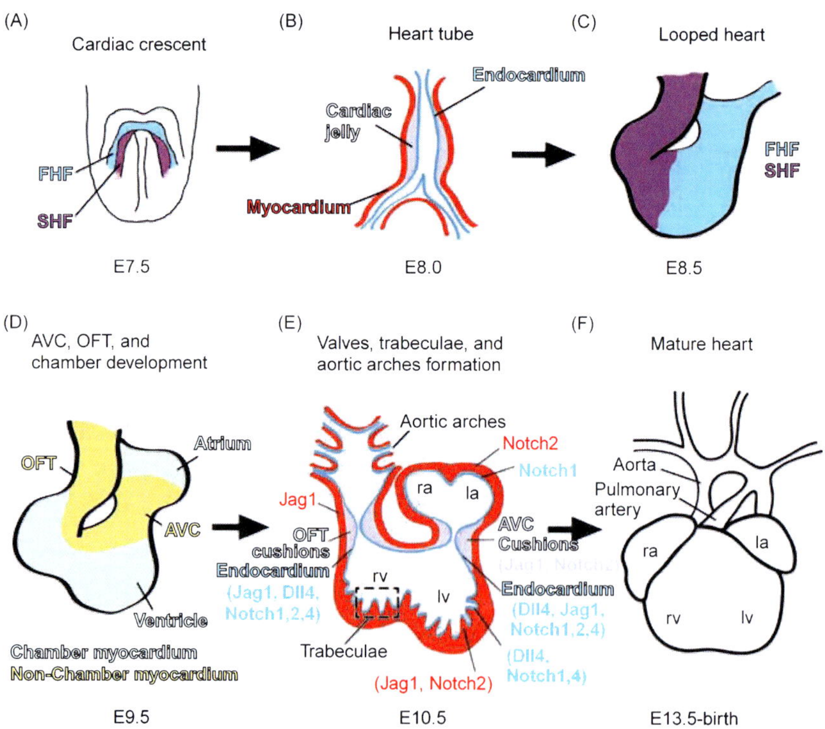

Donal MacGrogan *et al.*, Figure 11.1 Please refer to the legend in the text.

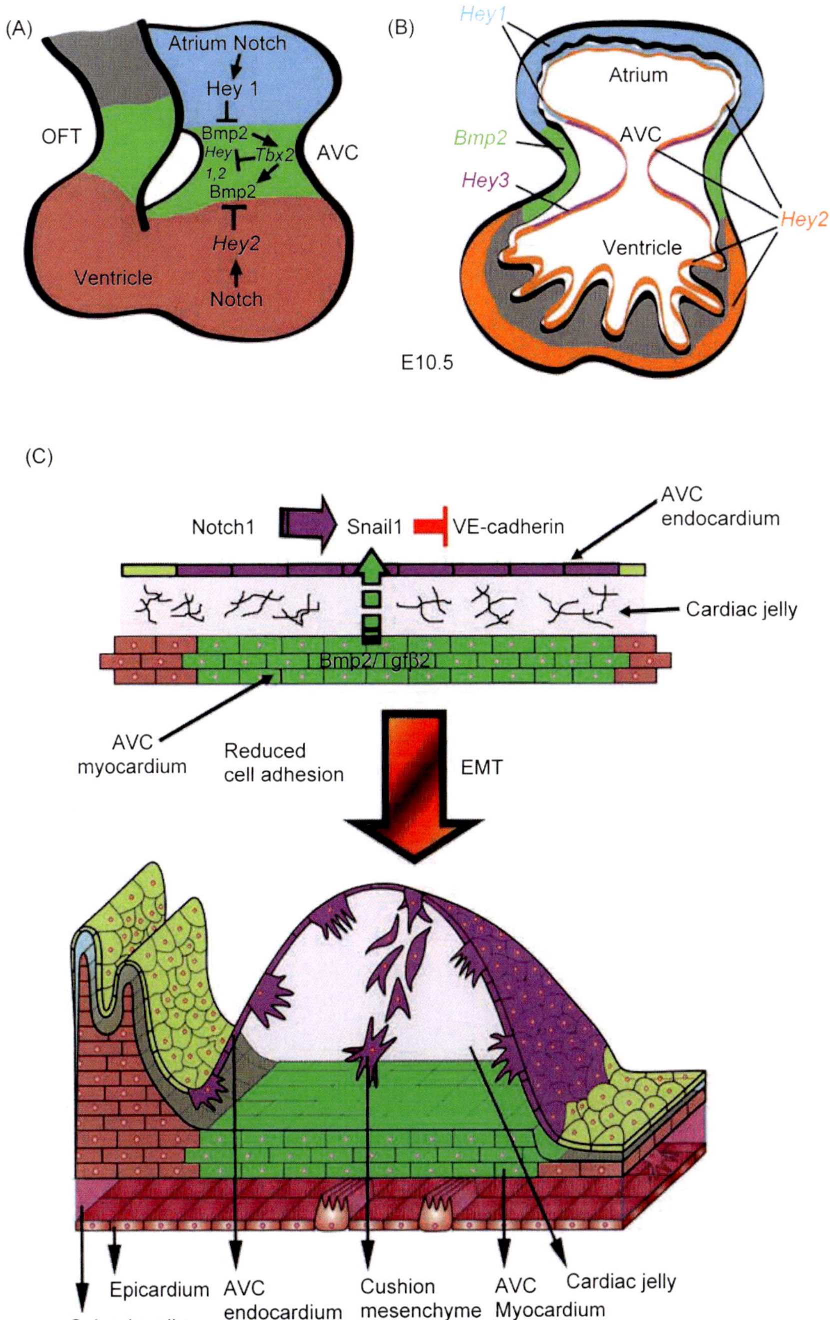

Donal MacGrogan *et al.*, Figure 11.2 Please refer to the legend in the text.

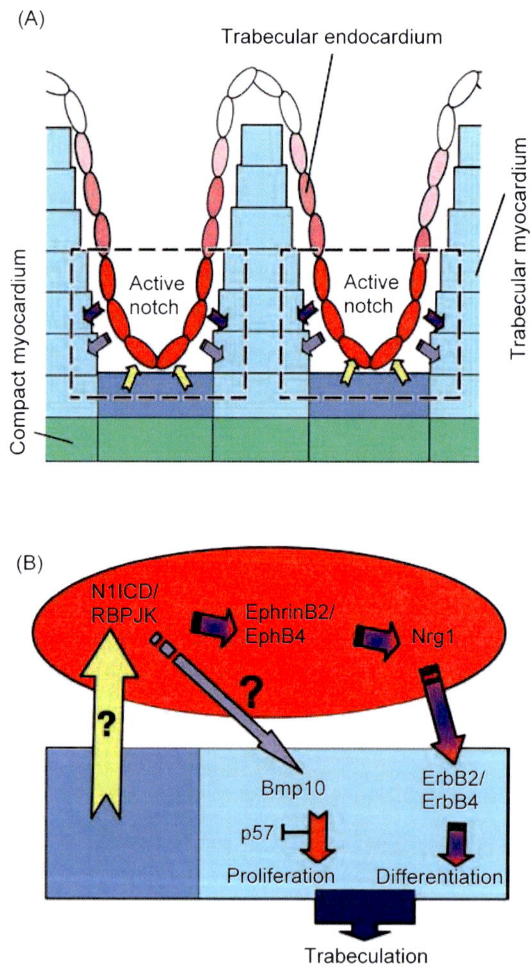

Donal MacGrogan *et al.*, Figure 11.3 Please refer to the legend in the text.

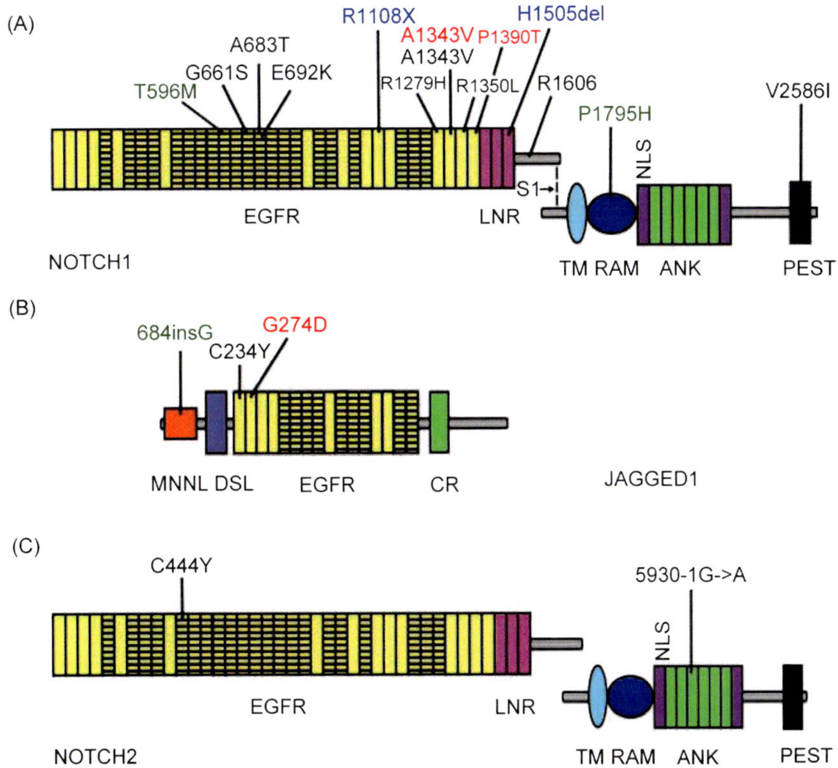

Donal MacGrogan *et al.*, Figure 11.4 Please refer to the legend in the text.

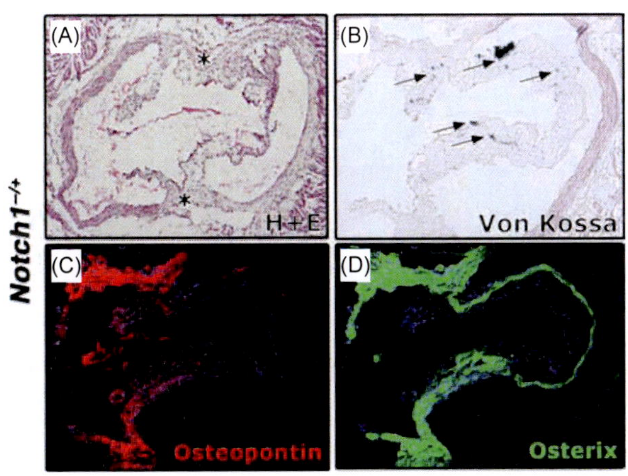

Donal MacGrogan *et al.*, Figure 11.5 Please refer to the legend in the text.

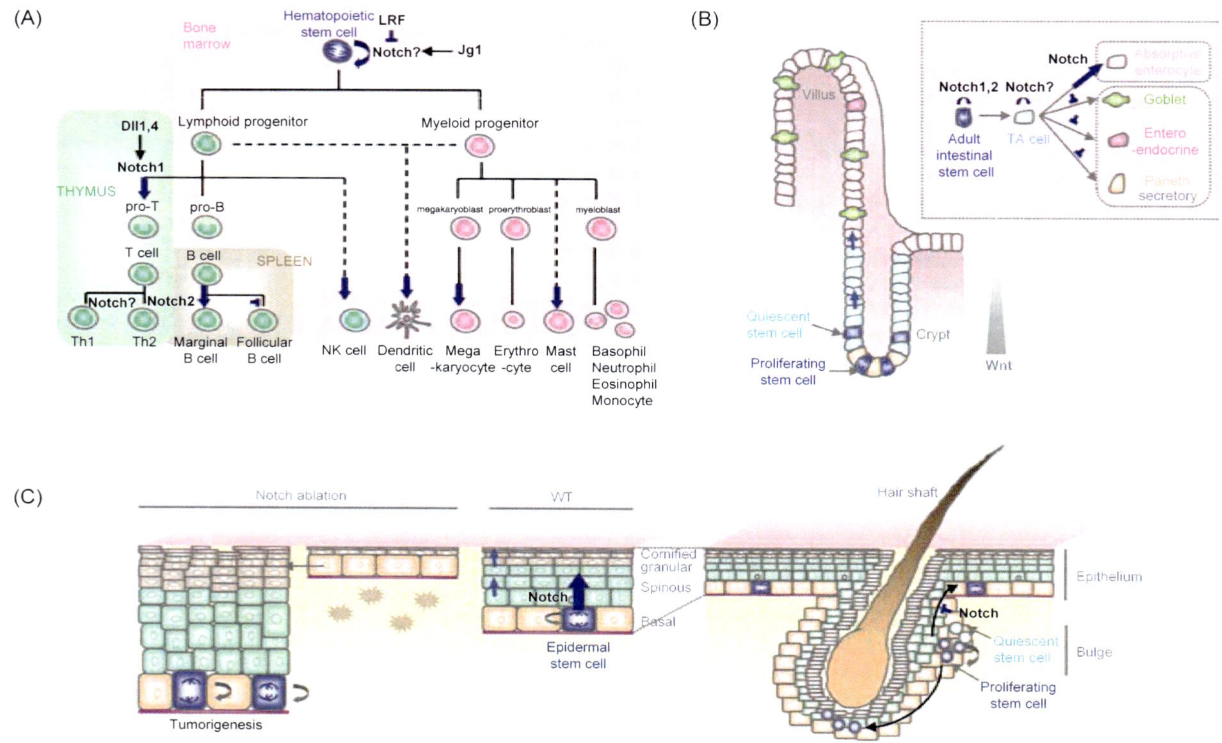

Jianing Liu *et al.*, Figure 12.1 (*Continued*)

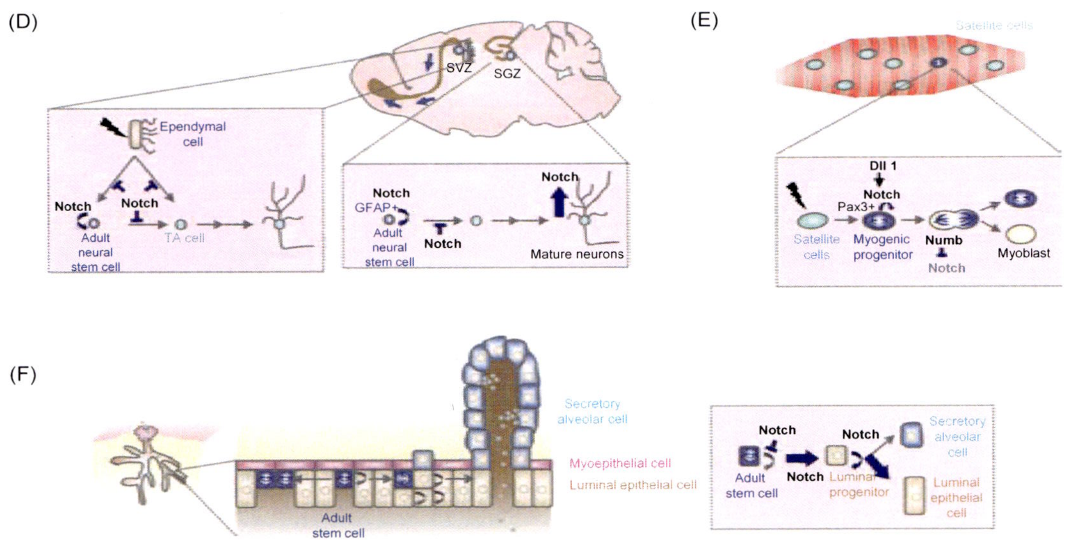

Jianing Liu *et al.*, Figure 12.1 Please refer to the legend in the text.

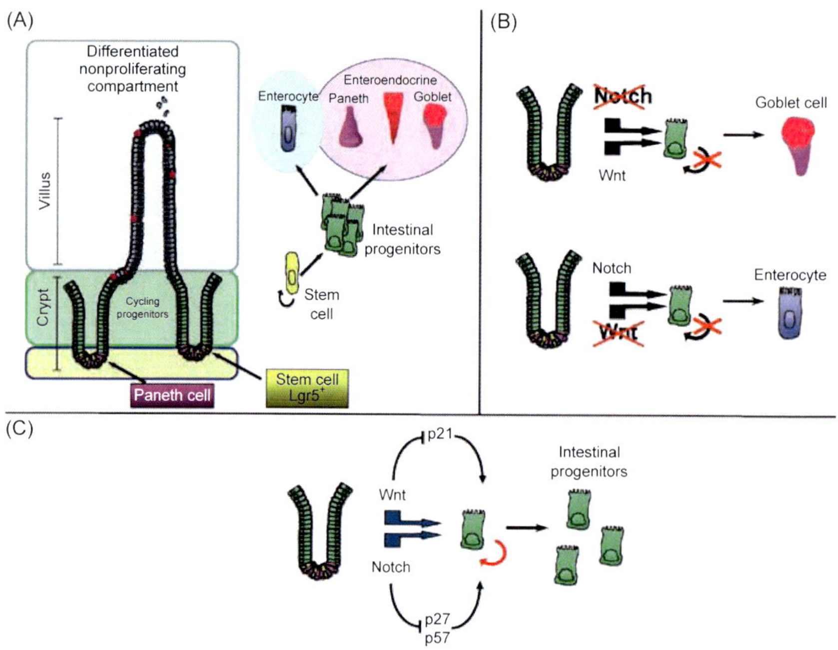

Ute Koch and Freddy Radtke, Figure 13.3 Please refer to the legend in the text.